Problem Books in Mathematics

Edited by P.R. Halmos

Springer
New York
Berlin
Heidelberg
Barcelona
Budapest
Hong Kong
London
Milan
Paris
Santa Clara
Singapore
Tokyo

Problem Books in Mathematics

Series Editor: P.R. Halmos

Polynomials
by *Edward J. Barbeau*

Problems in Geometry
by *Marcel Berger, Pierre Pansu, Jean-Pic Berry, and Xavier Saint-Raymond*

Problem Book for First Year Calculus
by *George W. Bluman*

Exercises in Probability
by *T. Cacoullos*

An Introduction to Hilbert Space and Quantum Logic
by *David W. Cohen*

Unsolved Problems in Geometry
by *Hallard T. Croft, Kenneth J. Falconer, and Richard K. Guy*

Problems in Analysis
by *Bernard R. Gelbaum*

Problems in Real and Complex Analysis
by *Bernard R. Gelbaum*

Theorems and Counterexamples in Mathematics
by *Bernard R. Gelbaum and John M.H. Olmsted*

Exercises in Integration
by *Claude George*

Algebraic Logic
by *S.G. Gindikin*

Unsolved Problems in Number Theory (2nd ed.)
by *Richard K. Guy*

An Outline of Set Theory
by *James M. Henle*

(continued after index)

Gábor J. Székely
Editor

Contests in Higher Mathematics

Miklós Schweitzer Competitions 1962–1991

With 39 Illustrations

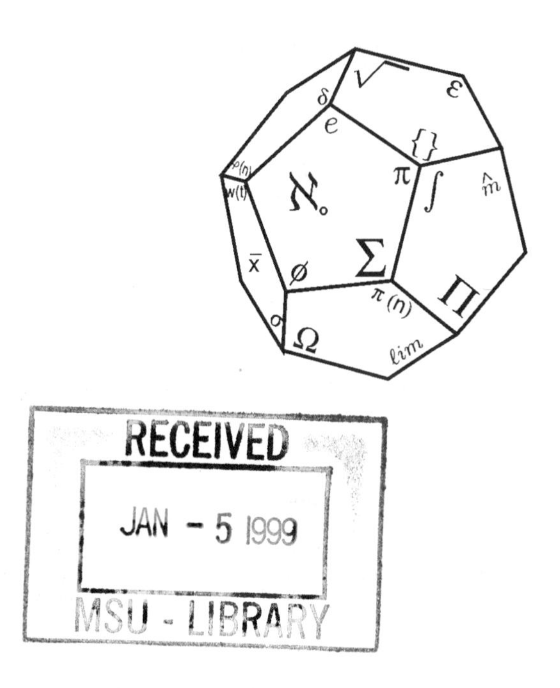

Springer

Gábor J. Székely
Department of Mathematics
Eötvös Loránd and Technical University
Múzeum krt. 6–8
1088 Budapest, Hungary

Series Editor:
Paul R. Halmos
Department of Mathematics
Santa Clara University
Santa Clara, CA 95053
USA

Mathematics Subject Classification (1991): 15-06, 05-06, 26-06, 51-06

Library of Congress Cataloging-in-Publication Data
Contests in higher mathematics:Miklos Schweitzer competitions,
1962–1991 / Gábor J. Székely (ed.).
p. cm. – (Problem books in mathematics)
Includes bibliographical references and index.
ISBN 0-387-94588-1
1. Mathematics – Competitions – Hungary. 2. Mathematics – Problems, exercises, etc. I. Schweitzer, Miklós, 1923–1945. II. Székely, Gábor J., 1947– . III. Series.
QA99.C62 1995
510′.76 – dc20 95-25361

Printed on acid-free paper.

Production managed by Hal Henglein; manufacturing supervised by Jeffrey Taub.
Camera-ready copy prepared from the editor's LaTeX files.
Printed and bound by R.R. Donnelley & Sons, Harrisonburg, VA.
Printed in the United States of America.

9 8 7 6 5 4 3 2 1

ISBN 0-387-94588-1 Springer-Verlag New York Berlin Heidelberg

Preface

"I had the opportunity to speak with Leo Szilárd about the contests of the Mathematical and Physical Society, and about the fact that the winners of these contests turned out later to be almost identical with the set of mathematicians and physicists who became outstanding ... "

(J. Neumann, in a letter to L. Fejér, Berlin, Dec. 7, 1929)

The solutions to deep scientific problems rarely come to us easily. Thus, it is important to motivate students to begin efforts on these kinds of problems. Scientific competition has proved to be an effective stimulant toward intellectual efforts. Successful examples include the "Concours" for admission to the "Grandes Écoles" in France, and the "Mathematical Tripos" in Cambridge, England. At the turn of the century, mathematical contests helped Hungary become one of the strongholds of the mathematical world.

With the revolution in 1848 and the Compromise in 1867, Hungary broke free from many centuries of rule by the Turks and then the Hapsburgs, and became a nation on equal footing with her neighbor, Austria. By the end of the 19th century, Hungary entered a period of cultural and economic progress. In 1891, Baron Loránd Eötvös, an outstanding Hungarian physicist, founded the Mathematical and Physical Society. In turn, the Society founded two journals: the *Mathematical and Physical Journal* in 1892 and the *Mathematical Journal for Secondary Schools* in 1893. This latter journal offered a rich variety of elementary problems for high school students. One of the first editors of the journal, László Rátz, later became the teacher of John Neumann and Eugene Wigner (a Nobel prize winner in physics). In 1894, the Society introduced a mathematical competition for high school students. Among the winners there were Lipót Fejér, Alfréd Haar, Tódor Kármán, Marcel Riesz, Gábor Szegő, Tibor Radó, Ede Teller, and many others who became world-famous scientists.

The success of high school competitions led the Mathematical Society to found a college-level contest. The first contest of this kind was organized in 1949 and named after Miklós Schweitzer, a young mathematician who died in the Second World War. Schweitzer placed second in the High School Contest in 1941, but the statutes of the fascist regime of that time prevented his admission to college. Schweitzer Contest problems are proposed and selected by the most prominent Hungarian mathematicians. Thus,

Schweitzer problems reflect the interest of these mathematicians and some aspects of the mainstream of Hungarian mathematics. The universities of Budapest, Debrecen, and Szeged have alternately been designated by the Society Presidium to conduct the Schweitzer Contests. The jury is chosen by the mathematics departments of the universities in question from among the mathematicians working in the host city. The jury sends out requests to leading Hungarian mathematicians to submit problems suitable for the contest. The list of problems selected by the jury is posted on the bulletin boards of mathematics departments and of local branches of the Mathematical Society (copies are available to anyone interested). Students may use any materials available in libraries or in their homes to solve the contest problems. In ten days the solutions are due, with the student's name, faculty, course, year, and university or high school recorded on the solution set.

The Schweitzer competition is one of the most unique in the world. Winners of the contests have gone on to become world-class scientists. Thus, the Schweitzer Contests are of interest to both math historians and mathematicians of all ages. They serve as reflections of Hungarian mathematical trends and as starting points for many interesting research problems in mathematics. The Schweitzer problems between 1949 and 1961 were previously published under the title *Contests in Higher Mathematics, 1949–1961* (Akadémiai Kiadó, Budapest, 1968; Chapter 4 of this book summarizes the mathematical work of M. Schweitzer). Our book is a continuation of that volume.

We hope that this collection of Schweitzer problems will serve as a guide for many young mathematicians and math majors. The large variety of research-level problems may spark the interest of seasoned mathematicians and historians of mathematics.

I wish to close by acknowledging the outstanding work of Dr. Marianna Bolla as Managing Editor. In addition, without the constant assistance of Dr. Dezső Miklós as Technical Editor, we could not have this book.

Bowling Green, OH
August 26, 1995

Gábor J. Székely

Contents

1. Problems of the Contests

The letter in parentheses after the text of a problem refers to the section in Chapter 3 containing its solution. The topics include these areas of mathematics:

A: Algebra
C: Combinatorics
F: Theory of Functions
G: Geometry
M: Measure Theory
N: Number Theory
O: Operators
P: Probability Theory
S: Sequences and Series
T: Topology
ℵ: Set Theory

Thus, for example, **P.3** refers to problem in "Probability Theory" section.

When available, the names of proposers are in brackets at the very end of each problem.

1962

1. Let f and g be polynomials with rational coefficients, and let F and G denote the sets of values of f and g at rational numbers. Prove that $F = G$ holds if and only if $f(x) = g(ax + b)$ for some suitable rational numbers $a \neq 0$ and b. (**N.1**) [E. Fried]
2. Determine the roots of unity in the field of p-adic numbers. (**A.1**) [L. Fuchs]
3. Let A and B be two Abelian groups, and define the sum of two homomorphisms η and χ from A to B by

$$a(\eta + \chi) = a\eta + a\chi \quad \text{for all } a \in A.$$

With this addition, the set of homomorphisms from A to B forms an Abelian group H. Suppose now that A is a p-group (p a prime number). Prove that in this case H becomes a topological group under the topology defined by taking the subgroups $p^k H$ ($k = 1, 2, \dots$) as a neighborhood base of 0. Prove that H is complete in this topology and that every connected component of H consists of a single element. When is H compact in this topology? (**A.2**) [L. Fuchs]
4. Show that

$$\prod_{1 \le x < y \le \frac{p-1}{2}} (x^2 + y^2) \equiv (-1)^{\left[\frac{p+1}{8}\right]} \pmod{p}$$

for every prime $p \equiv 3 \pmod 4$. ([.] is integer part.) (**N.2**) [J. Surányi]
5. Let f be a finite real function of one variable. Let $\overline{D}f$ and $\underline{D}f$ be its upper and lower derivatives, respectively, that is,

$$\overline{D}f(x) = \limsup_{\substack{h,k\to 0\\ h,k\ge 0\\ h+k>0}} \frac{f(x+h)-f(x-k)}{h+k}, \quad \underline{D}f(x) = \liminf_{\substack{h,k\to 0\\ h,k\ge 0\\ h+k>0}} \frac{f(x+h)-f(x-k)}{h+k}.$$

Show that $\overline{D}f$ and $\underline{D}f$ are Borel-measurable functions. (**M.1**) [Á. Császár]
6. Let E be a bounded subset of the real line, and let Ω be a system of (nondegenerate) closed intervals such that for each $x \in E$ there exists an $I \in \Omega$ with left endpoint x. Show that for every $\varepsilon > 0$ there exist a finite number of pairwise nonoverlapping intervals belonging to Ω that cover E with the exception of a subset of outer measure less than ε. (**M.2**) [J. Czipszer]
7. Prove that the function

$$f(\vartheta) = \int_1^{\frac{1}{\vartheta}} \frac{dx}{\sqrt{(x^2-1)(1-\vartheta^2 x^2)}}$$

(where the positive value of the square root is taken) is monotonically decreasing in the interval $0 < \vartheta < 1$. (**F.1**) [P. Turán]

8. Denote by $M(r, f)$ the maximum modulus on the circle $|z| = r$ of the transcendent entire function $f(z)$, and by $M_n(r, f)$ that of the nth partial sum of the power series of $f(z)$. Prove the existence of an entire function $f_0(z)$ and a corresponding sequence of positive numbers $r_1 < r_2 < \cdots \to +\infty$ such that

$$\limsup_{n\to\infty} \frac{M_n(r_n, f_0)}{M(r_n, f_0)} = +\infty.$$

(**F.2**) [P. Turán]

9. Find the minimum possible sum of lengths of edges of a prism all of whose edges are tangent to a unit sphere. (**G.1**) [Müller–Pfeiffer]

10. From a given triangle of unit area, we choose two points independently with uniform distribution. The straight line connecting these points divides the triangle, with probability one, into a triangle and a quadrilateral. Calculate the expected values of the areas of these two regions. (**P.1**) [A. Rényi]

1963

1. Show that the perimeter of an arbitrary planar section of a tetrahedron is less than the perimeter of one of the faces of the tetrahedron. (**G.2**) [Gy. Hajós]

2. Show that the center of gravity of a convex region in the plane halves at least three chords of the region. (**G.3**) [Gy. Hajós]

3. Let $R = R_1 \oplus R_2$ be the direct sum of the rings R_1 and R_2, and let N_2 be the annihilator ideal of R_2 (in R_2). Prove that R_1 will be an ideal in every ring $\bar{R}$ containing R as an ideal if and only if the only homomorphism from R_1 to N_2 is the zero homomorphism. (**A.3**) [Gy. Pollák]

4. Call a polynomial positive reducible if it can be written as a product of two nonconstant polynomials with positive real coefficients. Let $f(x)$ be a polynomial with $f(0) \neq 0$ such that $f(x^n)$ is positive reducible for some natural number n. Prove that $f(x)$ itself is positive reducible. (**A.4**) [L. Rédei]

5. Let H be a set of real numbers that does not consist of 0 alone and is closed under addition. Further, let $f(x)$ be a real-valued function defined on H and satisfying the following conditions:

$$f(x) \leq f(y) \text{ if } x \leq y \quad \text{and } f(x+y) = f(x) + f(y) \quad (x, y \in H).$$

Prove that $f(x) = cx$ on H, where c is a nonnegative number. (**F.3**) [M. Hosszú, R. Borges]

6. Show that if $f(x)$ is a real-valued, continuous function on the half-line $0 \leq x < \infty$, and

$$\int_0^\infty f^2(x)dx < \infty,$$

then the function

$$g(x) = f(x) - 2e^{-x}\int_0^x e^t f(t)dt$$

satisfies

$$\int_0^\infty g^2(x)dx = \int_0^\infty f^2(x)dx.$$

(**F.4**) [B. Szőkefalvi–Nagy]

7. Prove that for every convex function $f(x)$ defined on the interval $-1 \leq x \leq 1$ and having absolute value at most 1, there is a linear function $h(x)$ such that

$$\int_{-1}^1 |f(x) - h(x)|dx \leq 4 - \sqrt{8}.$$

(**F.5**) [L. Fejes–Tóth]

8. Let the Fourier series

$$\frac{a_0}{2} + \sum_{k\geq 1}(a_k \cos kx + b_k \sin kx)$$

of a function $f(x)$ be absolutely convergent, and let

$$a_k^2 + b_k^2 \geq a_{k+1}^2 + b_{k+1}^2 \quad (k = 1, 2, \dots).$$

Show that

$$\frac{1}{h}\int_0^{2\pi} (f(x+h) - f(x-h))^2\, dx \qquad (h > 0)$$

is uniformly bounded in h. (**S.1**) [K. Tandori]

9. Let $f(t)$ be a continuous function on the interval $0 \leq t \leq 1$, and define the two sets of points

$$A_t = \{(t, 0) : t \in [0, 1]\}, \quad B_t = \{(f(t), 1) : t \in [0, 1]\}.$$

Show that the union of all segments $\overline{A_t B_t}$ is Lebesgue-measurable, and find the minimum of its measure with respect to all functions f. (**M.3**) [Á. Császár]

10. Select n points on a circle independently with uniform distribution. Let P_n be the probability that the center of the circle is in the interior of the convex hull of these n points. Calculate the probabilities P_3 and P_4. (**P.2**) [A. Rényi]

1964

1. Among all possible representations of the positive integer n as $n = \sum_{i=1}^k a_i$ with positive integers k, $a_1 < a_2 < \dots < a_k$, when will the product $\prod_{i=1}^k a_i$ be maximum? (**C.1**)

2. Let p be a prime and let

$$l_k(x, y) = a_k x + b_k y \quad (k = 1, \dots, p^2),$$

be homogeneous linear polynomials with integral coefficients. Suppose that for every pair (ξ, η) of integers, not both divisible by p, the values $l_k(\xi, \eta)$, $1 \le k \le p^2$, represent every residue class mod p exactly p times. Prove that the set of pairs $\{(a_k, b_k) : 1 \le k \le p^2\}$ is identical mod p with the set $\{(m, n) : 0 \le m, n \le p-1\}$. (**N.3**)

3. Prove that the intersection of all maximal left ideals of a ring is a (two-sided) ideal. (**A.5**)
4. Let $A_1, A_2, \dots, A_n$ be the vertices of a closed convex n-gon K numbered consecutively. Show that at least $n-3$ vertices A_i have the property that the reflection of A_i with respect to the midpoint of $\overline{A_{i-1}A_{i+1}}$ is contained in K. (Indices are meant mod n.) (**G.4**)
5. Is it true that on any surface homeomorphic to an open disc there exist two congruent curves homeomorphic to a circle? (**G.5**)
6. Let $y_1(x)$ be an arbitrary, continuous, positive function on $[0, A]$, where A is an arbitrary positive number. Let

$$y_{n+1}(x) = 2\int_0^x \sqrt{y_n(t)}\,dt \quad (n = 1, 2, \dots).$$

Prove that the functions $y_n(x)$ converge to the function $y = x^2$ uniformly on $[0, A]$. (**S.2**)
7. Find all linear homogeneous differential equations with continuous coefficients (on the whole real line) such that for any solution $f(t)$ and any real number c, $f(t+c)$ is also a solution. (**F.6**)
8. Let F be a closed set in the n-dimensional Euclidean space. Construct a function that is 0 on F, positive outside F, and whose partial derivatives all exist. (**F.7**)
9. Let E be the set of all real functions on $I = [0, 1]$. Prove that one cannot define a topology on E in which $f_n \to f$ holds if and only if f_n converges to f almost everywhere. (**S.3**)
10. Let $\varepsilon_1, \varepsilon_2, \dots, \varepsilon_{2n}$ be independent random variables such that $P(\varepsilon_i = 1) = P(\varepsilon_i = -1) = 1/2$ for all i, and define $S_k = \sum_{i=1}^k \varepsilon_i$, $1 \le k \le 2n$. Let N_{2n} denote the number of integers $k \in [2, 2n]$ such that either $S_k > 0$, or $S_k = 0$ and $S_{k-1} > 0$. Compute the variance of N_{2n}. (**P.3**)

1965

1. Let p be a prime, n a natural number, and S a set of cardinality p^n. Let $\mathbf{P}$ be a family of partitions of S into nonempty parts of sizes divisible by p such that the intersection of any two parts that occur in any of the partitions has at most one element. How large can $|\mathbf{P}|$ be? (**N.4**)
2. Let R be a finite commutative ring. Prove that R has a multiplicative identity element (1) if and only if the annihilator of R is 0 (that is, $aR = 0$, $a \in R$ imply $a = 0$). (**A.6**)
3. Let $a, b_0, b_1, \dots, b_{n-1}$ be complex numbers, A a complex square matrix of order p, and E the unit matrix of order p. Assuming that the eigenvalues

of A are given, determine the eigenvalues of the matrix

$$B = \begin{pmatrix} b_0E & b_1A & b_2A^2 & \cdots & b_{n-1}A^{n-1} \\ ab_{n-1}A^{n-1} & b_0E & b_1A & \cdots & b_{n-2}A^{n-2} \\ ab_{n-2}A^{n-2} & ab_{n-1}A^{n-1} & b_0E & \cdots & b_{n-3}A^{n-3} \\ & & \ddots & & \\ ab_1A & ab_2A^2 & ab_3A^3 & \cdots & b_0E \end{pmatrix}.$$

(**O.1**)

4. The plane is divided into domains by n straight lines in general position, where $n \geq 3$. Determine the maximum and minimum possible number of angular domains among them. (We say that n lines are in general position if no two are parallel and no three are concurrent.) (**G.6**)

5. Let $A = A_1A_2A_3A_4$ be a tetrahedron, and suppose that for each $j \neq k$, $[A_j, A_{jk}]$ is a segment of length ρ extending from A_j in the direction of A_k. Let p_j be the intersection line of the planes $[A_{jk}A_{jl}A_{jm}]$ and $[A_kA_lA_m]$. Show that there are infinitely many straight lines that intersect the straight lines p_1, p_2, p_3, p_4 simultaneously. (**G.7**)

6. Consider the radii of normal curvature of a surface at one of its points P_0 in two conjugate directions (with respect to the Dupin indicatrix). Show that their sum does not depend on the choice of the conjugate directions. (We exclude the choice of asymptotic directions in the case of a hyperbolic point.) (**G.8**)

7. Prove that any uncountable subset of the Euclidean n-space contains an uncountable subset with the property that the distances between different pairs of points are different (that is, for any points $P_1 \neq P_2$ and $Q_1 \neq Q_2$ of this subset, $\overline{P_1P_2} = \overline{Q_1Q_2}$ implies either $P_1 = Q_1$ and $P_2 = Q_2$, or $P_1 = Q_2$ and $P_2 = Q_1$). Show that a similar statement is not valid if the Euclidean n-space is replaced with a (separable) Hilbert space. (**T.1**)

8. Let the continuous functions $f_n(x)$, $n = 1, 2, 3, \ldots$, be defined on the interval $[a, b]$ such that every point of $[a, b]$ is a root of $f_n(x) = f_m(x)$ for some $n \neq m$. Prove that there exists a subinterval of $[a, b]$ on which two of the functions are equal. (**S.4**)

9. Let f be a continuous, nonconstant, real function, and assume the existence of an F such that $f(x + y) = F[f(x), f(y)]$ for all real x and y. Prove that f is strictly monotone. (**F.8**)

10. A gambler plays the following coin-tossing game. He can bet an arbitrary positive amount of money. Then a fair coin is tossed, and the gambler wins or loses the amount he bet depending on the outcome. Our gambler, who starts playing with x forints, where $0 < x < 2C$, uses the following strategy: if at a given time his capital is $y < C$, he risks all of it; and if he has $y > C$, he only bets $2C - y$. If he has exactly $2C$ forints, he stops playing. Let $f(x)$ be the probability that he reaches $2C$ (before going bankrupt). Determine the value of $f(x)$. (**P.4**)

1966

1. Show that a segment of length h can go through or be tangent to at most $2[h/\sqrt{2}]+2$ nonoverlapping unit spheres. ([.] is integer part.) (**G.9**) [L. Fejes–Tóth, A. Heppes]
2. Characterize those configurations of n coplanar straight lines for which the sum of angles between all pairs of lines is maximum. (**G.10**) [L. Fejes–Tóth, A. Heppes]
3. Let $f(n)$ denote the maximum possible number of right triangles determined by n coplanar points. Show that

$$\lim_{n\to\infty}\frac{f(n)}{n^2}=\infty \quad\text{and}\quad \lim_{n\to\infty}\frac{f(n)}{n^3}=0.$$

(**G.11**) [P. Erdős]
4. Let I be an ideal of the ring of all polynomials with integer coefficients such that
   (a) the elements of I do not have a common divisor of degree greater than 0, and
   (b) I contains a polynomial with constant term 1.
   Prove that I contains the polynomial $1+x+x^2+\cdots+x^{r-1}$ for some natural number r. (**A.7**) [Gy. Szekeres]
5. A "letter T" erected at point A of the x-axis in the xy-plane is the union of a segment AB in the upper half-plane perpendicular to the x-axis and a segment CD containing B in its interior and parallel to the x-axis. Show that it is impossible to erect a letter T at every point of the x-axis so that the union of those erected at rational points is disjoint from the union of those erected at irrational points. (**M.4**) [Á. Császár]
6. A sentence of the following type is often heard in Hungarian weather reports: "Last night's minimum temperatures took all values between -3 degrees and $+5$ degrees." Show that it would suffice to say, "Both -3 degrees and $+5$ degrees occurred among last night's minimum temperatures." (Assume that temperature as a two-variable function of place and time is continuous.) (**T.2**) [Á. Császár]
7. Does there exist a function $f(x,y)$ of two real variables that takes natural numbers as its values and for which $f(x,y)=f(y,z)$ implies $x=y=z$? (**ℵ.1**) [A. Hajnal]
8. Prove that in a Euclidean ring R the quotient and remainder are always uniquely determined if and only if R is a polynomial ring over some field and the value of the norm is a strictly monotone function of the degree of the polynomial. (To be precise, there are two more trivial cases: R can also be a field or the null ring.) (**A.8**) [E. Fried]
9. If $\sum_{m=-\infty}^{+\infty}|a_m|<\infty$, then what can be said about the following expression?

$$\lim_{n\to\infty}\frac{1}{2n+1}\sum_{m=-\infty}^{+\infty}|a_{m-n}+a_{m-n+1}+\cdots+a_{m+n}|$$

(**S.5**) [P. Turán]

10. For a real number x in the interval $(0,1)$ with decimal representation

$$0.a_1(x)a_2(x)\dots a_n(x)\dots,$$

denote by $n(x)$ the smallest nonnegative integer such that

$$\overline{a_{n(x)+1}a_{n(x)+2}a_{n(x)+3}a_{n(x)+4}} = 1966.$$

Determine $\int_0^1 n(x)dx$. ($\overline{abcd}$ denotes the decimal number with digits a, b, c, d.) (**P.5**) [A. Rényi]

1967

1. Let

$$f(x) = a_0 + a_1x + a_2x^2 + a_{10}x^{10} + a_{11}x^{11} + a_{12}x^{12} + a_{13}x^{13} \qquad (a_{13} \neq 0)$$

and

$$g(x) = b_0 + b_1x + b_2x^2 + b_3x^3 + b_{11}x^{11} + b_{12}x^{12} + b_{13}x^{13} \qquad (b_3 \neq 0)$$

be polynomials over the same field. Prove that the degree of their greatest common divisor is at most 6. (**A.9**) [L. Rédei]

2. Let K be a subset of a group G that is not a union of left cosets of a proper subgroup. Prove that if G is a torsion group or if K is a finite set, then the subset

$$\bigcap_{k \in K} k^{-1}K$$

consists of the identity alone. (**A.10**) [L. Rédei]

3. Prove that if an infinite, noncommutative group G contains a proper normal subgroup with a commutative factor group, then G also contains an infinite proper normal subgroup. (**A.11**) [B. Csákány]

4. Let $a_1, a_2, \dots, a_N$ be positive real numbers whose sum equals 1. For a natural number i, let n_i denote the number of a_k for which $2^{1-i} \geq a_k > 2^{-i}$ holds. Prove that

$$\sum_{i=1}^{\infty} \sqrt{n_i 2^{-i}} \leq 4 + \sqrt{\log_2 N}.$$

(**A.12**) [L. Leindler]

5. Let f be a continuous function on the unit interval $[0,1]$. Show that

$$\lim_{n\to\infty} \int_0^1 \cdots \int_0^1 f\left(\frac{x_1 + \cdots + x_n}{n}\right) dx_1 \dots dx_n = f\left(\frac{1}{2}\right)$$

and

$$\lim_{n\to\infty} \int_0^1 \cdots \int_0^1 f(\sqrt[n]{x_1 \dots x_n}) dx_1 \dots dx_n = f\left(\frac{1}{e}\right).$$

(**P.6**)

6. Let A be a family of proper closed subspaces of the Hilbert space $H = l^2$ totally ordered with respect to inclusion (that is, if $L_1, L_2 \in A$, then either $L_1 \subset L_2$ or $L_2 \subset L_1$). Prove that there exists a vector $x \in H$ not contained in any of the subspaces L belonging to A. (**T.3**) [B. Szőkefalvi–Nagy]

7. Let U be an $n \times n$ orthogonal matrix. Prove that for any $n \times n$ matrix A, the matrices

$$A_m = \frac{1}{m+1} \sum_{j=0}^{m} U^{-j} A U^{j}$$

converge entrywise as $m \to \infty$. (**O.2**) [I. Kovács]

8. Suppose that a bounded subset S of the plane is a union of congruent, homothetic, closed triangles. Show that the boundary of S can be covered by a finite number of rectifiable arcs. (**G.12**) [L. Gehér]

9. Let F be a surface of nonzero curvature that can be represented around one of its points P by a power series and is symmetric around the normal planes parallel to the principal directions at P. Show that the derivative with respect to the arc length of the curvature of an arbitrary normal section at P vanishes at P. Is it possible to replace the above symmetry condition by a weaker one? (**G.13**) [A. Moór]

10. Let $\sigma(S_n, k)$ denote the sum of the kth powers of the lengths of the sides of the convex n-gon S_n inscribed in a unit circle. Show that for any natural number greater than 2 there exists a real number k_0 between 1 and 2 such that $\sigma(S_n, k_0)$ attains its maximum for the regular n-gon. (**G.14**) [L. Fejes–Tóth]

1968

1. Consider the endomorphism ring of an Abelian torsion-free (resp. torsion) group G. Prove that this ring is Neumann-regular if and only if G is a discrete direct sum of groups isomorphic to the additive group of the rationals (resp., a discrete direct sum of cyclic groups of prime order). (A ring R is called *Neumann-regular* if for every $\alpha \in R$ there exists a $\beta \in R$ such that $\alpha\beta\alpha = \alpha$.) (**A.13**) [E. Fried]

2. Let $a_1, a_2, \dots, a_n$ be nonnegative real numbers. Prove that

$$\left(\sum_{i=1}^{n} a_i\right)\left(\sum_{i=1}^{n} a_i^{n-1}\right) \le n \prod_{i=1}^{n} a_i + (n-1) \sum_{i=1}^{n} a_i^n .$$

(**S.6**) [J. Surányi]

3. Let K be a compact topological group, and let F be a set of continuous functions defined on K that has cardinality greater than continuum. Prove that there exist $x_0 \in K$ and $f \neq g \in F$ such that

$$f(x_0) = g(x_0) = \max_{x \in K} f(x) = \max_{x \in K} g(x).$$

(**T.4**) [I. Juhász]

4. Let f be a complex-valued, completely multiplicative, arithmetical function. Assume that there exists an infinite increasing sequence N_k of natural numbers such that

$$f(n) = A_k \neq 0 \quad \text{provided} \quad N_k \leq n \leq N_k + 4\sqrt{N_k}.$$

Prove that f is identically 1. (**N.5**) [I. Kátai]

5. Let k be a positive integer, z a complex number, and $\varepsilon < 1/2$ a positive number. Prove that the following inequality holds for infinitely many positive integers n:

$$\left| \sum_{0 \leq \ell \leq \frac{n}{k+1}} \binom{n-k\ell}{\ell} z^\ell \right| \geq \left(\frac{1}{2} - \varepsilon\right)^n .$$

(**F.9**) [P. Turán]

6. Let $\mathfrak{A} = \langle A; \dots \rangle$ be an arbitrary, countable algebraic structure (that is, $\mathfrak{A}$ can have an arbitrary number of finitary operations and relations). Prove that $\mathfrak{A}$ has as many as continuum automorphisms if and only if for any finite subset A' of A there is an automorphism $\pi_{A'}$ of $\mathfrak{A}$ different from the identity automorphism and such that

$$(x)\pi_{A'} = x$$

for every $x \in A'$. (**A.14**) [M. Makkai]

7. For every natural number r, the set of r-tuples of natural numbers is partitioned into finitely many classes. Show that if $f(r)$ is a function such that $f(r) \geq 1$ and $\lim_{r\to\infty} f(r) = +\infty$, then there exists an infinite set of natural numbers that, for all r, contains r-tuples from at most $f(r)$ classes. Show that if $f(r) \not\to +\infty$, then there is a family of partitions such that no such infinite set exists. (**C.2**) [P. Erdős, A. Hajnal]

8. Let n and k be given natural numbers, and let A be a set such that

$$|A| \leq \frac{n(n+1)}{k+1}.$$

For $i = 1, 2, \dots, n+1$, let A_i be sets of size n such that

$$|A_i \cap A_j| \leq k \qquad (i \neq j),$$

$$A = \bigcup_{i=1}^{n+1} A_i.$$

Determine the cardinality of A. (**C.3**) [K. Corrádi]

9. Let $f(x)$ be a real function such that

$$\lim_{x\to+\infty} \frac{f(x)}{e^x} = 1$$

and $|f''(x)| < c|f'(x)|$ for all sufficiently large x. Prove that

$$\lim_{x\to+\infty} \frac{f'(x)}{e^x} = 1.$$

(**F.10**) [P. Erdős]

10. Let h be a triangle of perimeter 1, and let H be a triangle of perimeter λ homothetic to h. Let $h_1, h_2, \ldots$ be translates of h such that, for all i, h_i is different from h_{i+2} and touches H and h_{i+1} (that is, intersects without overlapping). For which values of λ can these triangles be chosen so that the sequence $h_1, h_2, \ldots$ is periodic? If $\lambda \geq 1$ is such a value, then determine the number of different triangles in a periodic chain $h_1, h_2, \ldots$ and also the number of times such a chain goes around the triangle H. (**G.15**) [L. Fejes–Tóth]

11. Let $A_1, \ldots, A_n$ be arbitrary events in a probability field. Denote by C_k the event that at least k of $A_1, \ldots, A_n$ occur. Prove that

$$\prod_{k=1}^{n} P(C_k) \leq \prod_{k=1}^{n} P(A_k).$$

(**P.7**) [A. Rényi]

1969

1. Let G be an infinite group generated by nilpotent normal subgroups. Prove that every maximal Abelian normal subgroup of G is infinite. (We call an Abelian normal subgroup maximal if it is not contained in another Abelian normal subgroup.) (**A.15**) [J. Erdős]

2. Let $p \geq 7$ be a prime number, ζ a primitive pth root of unity, c a rational number. Prove that in the additive group generated by the numbers $1, \zeta, \zeta^2, \zeta^3 + \zeta^{-3}$ there are only finitely many elements whose norm is equal to c. (The norm is in the pth cyclotomic field.) (**A.16**) [K. Győry]

3. Let $f(x) \geq 0$ be a nonzero, bounded, real function on an Abelian group G, $g_1, \ldots, g_k$ are given elements of G and $\lambda_1, \ldots, \lambda_k$ are real numbers. Prove that if

$$\sum_{i=1}^{k} \lambda_i f(g_i x) \geq 0$$

holds for all $x \in G$, then

$$\sum_{i=1}^{k} \lambda_i \geq 0.$$

(**S.7**) [A. Máté]

4. Show that the following inequality holds for all $k \geq 1$, real numbers $a_1, a_2, \ldots, a_k$, and positive numbers $x_1, x_2, \ldots, x_k$.

$$\ln \frac{\sum_{i=1}^{k} x_i}{\sum_{i=1}^{k} x_i^{1-a_i}} \leq \frac{\sum_{i=1}^{k} a_i x_i \ln x_i}{\sum_{i=1}^{k} x_i}$$

(**S.8**) [L. Losonczi]

5. Find all continuous real functions f, g and h defined on the set of positive real numbers and satisfying the relation

$$f(x+y) + g(xy) = h(x) + h(y)$$

for all $x > 0$ and $y > 0$. (**F.11**) [Z. Daróczy]

6. Let x_0 be a fixed real number, and let f be a regular complex function in the half-plane Re $z > x_0$ for which there exists a nonnegative function $F \in L_1(-\infty, \infty)$ satisfying $|f(\alpha + i\beta)| \le F(\beta)$ whenever $\alpha > x_0$, $-\infty < \beta < +\infty$. Prove that

$$\int_{\alpha - i\infty}^{\alpha + i\infty} f(z)dz = 0.$$

(**F.12**) [L. Czách]

7. Prove that if a sequence of Mikusiński operators of the form $\mu e^{-\lambda s}$ (λ and μ nonnegative real numbers, s the differentiation operator) is convergent in the sense of Mikusiński, then its limit is also of this form. (**O.3**) [E. Gesztelyi]

8. Let f and g be continuous positive functions defined on the interval $[0, \infty)$, and let $E \subset [0, \infty)$ be a set of positive measure. Prove that the range of the function defined on $E \times E$ by the relation

$$F(x, y) = \int_0^x f(t)dt + \int_0^y g(t)dt$$

has a nonvoid interior. (**M.5**) [L. Losonczi]

9. In n-dimensional Euclidean space, the union of any set of closed balls (of positive radii) is measurable in the sense of Lebesgue. (**M.6**) [Á. Császár]

10. In n-dimensional Euclidean space, the square of the two-dimensional Lebesgue measure of a bounded, closed, (two-dimensional) planar set is equal to the sum of the squares of the measures of the orthogonal projections of the given set on the n-coordinate hyperplanes. (**M.7**) [L. Tamássy]

11. Let $A_1, A_2, \dots$ be a sequence of infinite sets such that $|A_i \cap A_j| \le 2$ for $i \ne j$. Show that the sequence of indices can be divided into two disjoint sequences $i_1 < i_2 < \dots$ and $j_1 < j_2 < \dots$ in such a way that, for some sets E and F, $|A_{i_n} \cap E| = 1$ and $|A_{j_n} \cap F| = 1$ for $n = 1, 2, \dots$. (**C.4**) [P. Erdős]

12. Let A and B be nonsingular matrices of order p, and let ξ and η be independent random vectors of dimension p. Show that if ξ, η and $\xi A + \eta B$ have the same distribution, if their first and second moments exist, and if their covariance matrix is the identity matrix, then these random vectors are normally distributed. (**P.8**) [B. Gyires]

1970

1. We have $2n+1$ elements in the commutative ring R:

$$\alpha, \alpha_1, \ldots, \alpha_n, \varrho_1, \ldots, \varrho_n .$$

Let us define the elements

$$\sigma_k = k\alpha + \sum_{i=1}^{n} \alpha_i \varrho_i^k .$$

Prove that the ideal $(\sigma_0, \sigma_1, \ldots, \sigma_k, \ldots)$ can be finitely generated. (**A.17**) [L. Rédei]

2. Let G and H be countable Abelian p-groups (p an arbitrary prime). Suppose that for every positive integer n,

$$p^n G \neq p^{n+1} G .$$

Prove that H is a homomorphic image of G. (**A.18**) [M. Makkai]

3. The traffic rules in a regular triangle allow one to move only along segments parallel to one of the altitudes of the triangle. We define the distance between two points of the triangle to be the length of the shortest such path between them. Put $\binom{n+1}{2}$ points into the triangle in such a way that the minimum distance between pairs of points is maximal. (**G.16**) [L. Fejes–Tóth]

4. If c is a positive integer and p is an odd prime, what is the smallest residue (in absolute value) of

$$\sum_{n=0}^{\frac{p-1}{2}} \binom{2n}{n} c^n \pmod{p}?$$

(**N.6**) [J. Surányi]

5. Prove that two points in a compact metric space can be joined with a rectifiable arc if and only if there exists a positive number K such that, for any $\varepsilon > 0$, these points can be connected with an ε-chain not longer than K. (**T.5**) [M. Bognár]

6. Let a neighborhood basis of a point x of the real line consist of all Lebesgue-measurable sets containing x whose density at x equals 1. Show that this requirement defines a topology that is regular but not normal. (**T.6**) [Á. Császár]

7. Let us use the word N-measure for nonnegative, finitely additive set functions defined on all subsets of the positive integers, equal to 0 on finite sets, and equal to 1 on the whole set. We say that the system $\mathfrak{A}$ of sets determines the N-measure μ if any N-measure coinciding with μ on all elements of $\mathfrak{A}$ is necessarily identical with μ. Prove the existence of an N-measure μ that cannot be determined by a system of cardinality less than continuum. (**M.8**) [I. Juhász]

8. Let $\pi_n(x)$ be a polynomial of degree not exceeding n with real coefficients such that

$$|\pi_n(x)| \leq \sqrt{1-x^2} \qquad \text{for} \qquad -1 \leq x \leq 1.$$

Then

$$|\pi_n'(x)| \leq 2(n-1).$$

(**F.13**) [P. Turán]

9. Construct a continuous function $f(x)$, periodic with period 2π, such that the Fourier series of $f(x)$ is divergent at $x = 0$, but the Fourier series of $f^2(x)$ is uniformly convergent on $[0, 2\pi]$. (**S.9**) [P. Turán]

10. Prove that for every ϑ, $0 < \vartheta < 1$, there exist a sequence λ_n of positive integers and a series $\sum_{n=1}^{\infty} a_n$ such that

(i) $\lambda_{n+1} - \lambda_n > (\lambda_n)^{\vartheta}$,

(ii) $\lim\limits_{r \to 1-0} \sum\limits_{n=1}^{\infty} a_n r^{\lambda_n}$ exists,

(iii) $\sum\limits_{n=1}^{\infty} a_n$ is divergent.

(**S.10**) [P. Turán]

11. Let $\xi_1, \xi_2, \ldots$ be independent random variables such that $E\xi_n = m > 0$ and $\mathrm{Var}(\xi_n) = \sigma^2 < \infty$ $(n = 1, 2, \ldots)$. Let $\{a_n\}$ be a sequence of positive numbers such that $a_n \to 0$ and $\sum_{n=1}^{\infty} a_n = \infty$. Prove that

$$P\left(\lim_{n\to\infty} \sum_{k=1}^{n} a_k \xi_k = \infty\right) = 1.$$

(**P.9**) [P. Révész]

12. Let $\vartheta_1, \ldots, \vartheta_n$ be independent, uniformly distributed, random variables in the unit interval $[0, 1]$. Define

$$h(x) = \frac{1}{n} \#\{k : \vartheta_k < x\}.$$

Prove that the probability that there is an $x_0 \in (0, 1)$ such that $h(x_0) = x_0$, is equal to $1 - \frac{1}{n}$. (**P.10**) [G. Tusnády]

1971

1. Let G be an infinite compact topological group with a Hausdorff topology. Prove that G contains an element $g \neq 1$ such that the set of all powers of g is either everywhere dense in G or nowhere dense in G. (**A.19**) [J. Erdős]

2. Prove that there exists an ordered set in which every uncountable subset contains an uncountable, well-ordered subset and that cannot be represented as a union of a countable family of well-ordered subsets. (**ℵ.2**) [A. Hajnal]

3. Let $0 < a_k < 1$ for $k = 1, 2, \ldots$. Give a necessary and sufficient condition for the existence, for every $0 < x < 1$, of a permutation π_x of the positive integers such that

$$x = \sum_{k=1}^{\infty} \frac{a_{\pi_x(k)}}{2^k}.$$

(**S.11**) [P. Erdős]

4. Suppose that V is a locally compact topological space that admits no countable covering with compact sets. Let **C** denote the set of all compact subsets of the space V and **U** the set of open subsets that are not contained in any compact set. Let f be a function from **U** to **C** such that $f(U) \subseteq U$ for all $U \in \mathbf{U}$. Prove that either

(i) there exists a nonempty compact set C such that $f(U)$ is not a proper subset of C whenever $C \subseteq U \in \mathbf{U}$,

(ii) or for some compact set C, the set

$$f^{-1}(C) = \bigcup \{U \in \mathbf{U} : f(U) \subseteq C\}$$

is an element of **U**, that is, $f^{-1}(C)$ is not contained in any compact set.

(**T.7**) [A. Máté]

5. Let $\lambda_1 \leq \lambda_2 \leq \ldots$ be a positive sequence and let K be a constant such that

$$\sum_{k=1}^{n-1} \lambda_k^2 < K\lambda_n^2 \quad (n = 1, 2, \ldots).$$

Prove that there exists a constant K' such that

$$\sum_{k=1}^{n-1} \lambda_k < K'\lambda_n \quad (n = 1, 2, \ldots).$$

(**S.12**) [L. Leindler]

6. Let $a(x)$ and $r(x)$ be positive continuous functions defined on the interval $[0, \infty)$, and let

$$\liminf_{x \to \infty} (x - r(x)) > 0.$$

Assume that $y(x)$ is a continuous function on the whole real line, that it is differentiable on $[0, \infty)$, and that it satisfies

$$y'(x) = a(x)y(x - r(x))$$

on $[0, \infty)$. Prove that the limit

$$\lim_{x \to \infty} y(x) \exp\left\{-\int_0^x a(u)du\right\}$$

exists and is finite. (**F.14**) [I. Győri]

7. Let $n \geq 2$ be an integer, let S be a set of n elements, and let A_i, $1 \leq i \leq m$, be distinct subsets of S of size at least 2 such that

$$A_i \cap A_j \neq \emptyset,\ A_i \cap A_k \neq \emptyset,\ A_j \cap A_k \neq \emptyset \quad \text{imply} \quad A_i \cap A_j \cap A_k \neq \emptyset.$$

Show that $m \leq 2^{n-1} - 1$. (**C.5**) [P. Erdős]

8. Show that the edges of a strongly connected bipolar graph can be oriented in such a way that for any edge e there is a simple directed path from pole p to pole q containing e. (A strongly connected bipolar graph is a finite connected graph with two special vertices p and q having the property that there are no points x, y, $x \neq y$, such that all paths from x to p as well as all paths from x to q contain y.) (**C.6**) [A. Ádám]

9. Given a positive, monotone function $F(x)$ on $(0, \infty)$ such that $F(x)/x$ is monotone nondecreasing and $F(x)/x^{1+d}$ is monotone nonincreasing for some positive d, let $\lambda_n > 0$ and $a_n \geq 0$, $n \geq 1$. Prove that if

$$\sum_{n=1}^{\infty} \lambda_n F\left(a_n \sum_{k=1}^{n} \frac{\lambda_k}{\lambda_n}\right) < \infty,$$

or

$$\sum_{n=1}^{\infty} \lambda_n F\left(\sum_{k=1}^{n} a_k \frac{\lambda_k}{\lambda_n}\right) < \infty,$$

then $\sum_{n=1}^{\infty} a_n$ is convergent. (**S.13**) [L. Leindler]

10. Let $\{\phi_n(x)\}$ be a sequence of functions belonging to $L^2(0,1)$ and having norm less than 1 such that for any subsequence $\{\phi_{n_k}(x)\}$ the measure of the set

$$\{x \in (0,1) : |\frac{1}{\sqrt{N}} \sum_{k=1}^{N} \phi_{n_k}(x)| \geq y\}$$

tends to 0 as y and N tend to infinity. Prove that ϕ_n tends to 0 weakly in the function space $L^2(0,1)$. (**M.9**) [F. Móricz]

11. Let C be a simple arc with monotone curvature such that C is congruent to its evolute. Show that under appropriate differentiability conditions, C is a part of a cycloid or a logarithmic spiral with polar equation $r = ae^{\vartheta}$. (**G.17**) [J. Szenthe]

1972

1. Let $\mathcal{F}$ be a nonempty family of sets with the following properties:

(a) If $X \in \mathcal{F}$, then there are some $Y \in \mathcal{F}$ and $Z \in \mathcal{F}$ such that $Y \cap Z = \emptyset$ and $Y \cup Z = X$.

(b) If $X \in \mathcal{F}$, and $Y \cup Z = X$, $Y \cap Z = \emptyset$, then either $Y \in \mathcal{F}$ or $Z \in \mathcal{F}$.

Show that there is a decreasing sequence $X_0 \supseteq X_1 \supseteq X_2 \supseteq \ldots$ of sets $X_n \in \mathcal{F}$ such that

$$\bigcap_{n=0}^{\infty} X_n = \emptyset.$$

(**C.7**) [F. Galvin]

2. Let $\leq$ be a reflexive, antisymmetric relation on a finite set A. Show that this relation can be extended to an appropriate finite superset B of A such that $\leq$ on B remains reflexive, antisymmetric, and any two elements of B have a least upper bound as well as a greatest lower bound. (The relation $\leq$ is extended to B if for $x, y \in A$, $x \leq y$ holds in A if and only if it holds in B.) (**ℵ.3**) [E. Fried]

3. Let λ_i $(i = 1, 2, \dots)$ be a sequence of distinct positive numbers tending to infinity. Consider the set of all numbers representable in the form

$$\mu = \sum_{i=1}^{\infty} n_i \lambda_i,$$

where $n_i \geq 0$ are integers and all but finitely many n_i are 0. Let

$$L(x) = \sum_{\lambda_i \leq x} 1 \qquad \text{and} \qquad M(x) = \sum_{\mu \leq x} 1.$$

(In the latter sum, each μ occurs as many times as its number of representations in the above form.)

Prove that if

$$\lim_{x \to \infty} \frac{L(x+1)}{L(x)} = 1,$$

then

$$\lim_{x \to \infty} \frac{M(x+1)}{M(x)} = 1.$$

(**F.15**) [G. Halász]

4. Let G be a solvable torsion group in which every Abelian subgroup is finitely generated. Prove that G is finite. (**A.20**) [J. Pelikán]

5. We say that the real-valued function $f(x)$ defined on the interval $(0, 1)$ is approximately continuous on $(0, 1)$ if for any $x_0 \in (0, 1)$ and $\varepsilon > 0$ the point x_0 is a point of interior density 1 of the set

$$H = \{x : \ |f(x) - f(x_0)| < \varepsilon\}.$$

Let $F \subset (0, 1)$ be a countable closed set, and $g(x)$ a real-valued function defined on F. Prove the existence of an approximately continuous function $f(x)$ defined on $(0, 1)$ such that

$$f(x) = g(x) \quad \text{for all} \quad x \in F.$$

(**M.10**) [M. Laczkovich, Gy. Petruska]

6. Let $P(z)$ be a polynomial of degree n with complex coefficients,

$$P(0) = 1, \quad \text{and} \quad |P(z)| \leq M \quad \text{for} \quad |z| \leq 1.$$

Prove that every root of $P(z)$ in the closed unit disc has multiplicity at most $c\sqrt{n}$, where $c = c(M) > 0$ is a constant depending only on M. (**F.16**) [G. Halász]

7. Let $f(x, y, z)$ be a nonnegative harmonic function in the unit ball of $\mathbb{R}^3$ for which the inequality $f(x_0, 0, 0) \le \varepsilon^2$ holds for some $0 \le x_0 < 1$ and $0 < \varepsilon < (1 - x_0)^2$. Prove that $f(x, y, z) \le \varepsilon$ in the ball with center at the origin and radius $(1 - 3\varepsilon^{1/4})$. (**F.17**) [P. Turán]

8. Given four points A_1, A_2, A_3, A_4 in the plane in such a way that A_4 is the centroid of the $\triangle A_1A_2A_3$, find a point A_5 in the plane that maximizes the ratio

$$\frac{\min_{1\le i<j<k\le 5} T(A_iA_jA_k)}{\max_{1\le i<j<k\le 5} T(A_iA_jA_k)}.$$

($T(ABC)$ denotes the area of the triangle $\triangle ABC$.) (**G.18**) [J. Surányi]

9. Let K be a compact convex body in the n-dimensional Euclidean space. Let $P_1, P_2, \ldots, P_{n+1}$ be the vertices of a simplex having maximal volume among all simplices inscribed in K. Define the points $P_{n+2}, P_{n+3}, \ldots$ successively so that P_k $(k > n+1)$ is a point of K for which the volume of the convex hull of $P_1, \ldots, P_k$ is maximal. Denote this volume by V_k. Decide, for different values of n, about the truth of the statement "the sequence $V_{n+1}, V_{n+2}, \ldots$ is concave." (**G.19**) [L. Fejes–Tóth, E. Makai]

10. Let $\mathcal{T}_1$ and $\mathcal{T}_2$ be second-countable topologies on the set E. We would like to find a real function σ defined on $E \times E$ such that

$$0 \le \sigma(x, y) < +\infty, \quad \sigma(x, x) = 0,$$
$$\sigma(x, z) \le \sigma(x, y) + \sigma(y, z) \quad (x, y, z \in E),$$

and, for any $p \in E$, the sets

$$V_1(p, \varepsilon) = \{x : \sigma(x, p) < \varepsilon\} \quad (\varepsilon > 0)$$

form a neighborhood base of p with respect to $\mathcal{T}_1$, and the sets

$$V_2(p, \varepsilon) = \{x : \sigma(p, x) < \varepsilon\} \quad (\varepsilon > 0)$$

form a neighborhood base of p with respect to $\mathcal{T}_2$. Prove that such a function σ exists if and only if, for any $p \in E$ and $\mathcal{T}_i$-open set $G \ni p$ $(i = 1, 2)$, there exist a $\mathcal{T}_i$-open set G' and a $\mathcal{T}_{3-i}$-closed set F with $p \in G' \subset F \subset G$. (**T.8**) [Á. Császár]

11. We throw N balls into n urns, one by one, independently and uniformly. Let $X_i = X_i(N, n)$ be the total number of balls in the ith urn. Consider the random variable

$$y(N, n) = \min_{1\le i\le n} |X_i - \frac{N}{n}|.$$

Verify the following three statements:

(a) If $n \to \infty$ and $N/n^3 \to \infty$, then

$$P\left(\frac{y(N,n)}{\frac{1}{n}\sqrt{\frac{N}{n}}} < x\right) \to 1 - e^{-x\sqrt{2/\pi}} \quad \text{for all } x > 0.$$

(b) If $n \to \infty$ and $N/n^3 \leq K$ (K constant), then for any $\varepsilon > 0$ there is an $A > 0$ such that

$$P(y(N,n) < A) > 1 - \varepsilon.$$

(c) If $n \to \infty$ and $N/n^3 \to 0$ then

$$P(y(N,n) < 1) \to 1.$$

(**P.11**) [P. Révész]

1973

1. We say that the rank of a group G is at most r if every subgroup of G can be generated by at most r elements. Prove that there exists an integer s such that for every finite group G of rank 2 the commutator series of G has length less than s. (**A.21**) [J. Erdős]
2. Let R be an Artinian ring with unity. Suppose that every idempotent element of R commutes with every element of R whose square is 0. Suppose R is the sum of the ideals A and B. Prove that $AB = BA$. (**A.22**) [A. Kertész]
3. Find a constant $c > 1$ with the property that, for arbitrary positive integers n and k such that $n > c^k$, the number of distinct prime factors of $\binom{n}{k}$ is at least k. (**N.7**) [P. Erdős]
4. Let $f(n)$ be the largest integer k such that n^k divides $n!$, and let $F(n) = \max_{2 \leq m \leq n} f(n)$. Show that

$$\lim_{n \to \infty} \frac{F(n) \log n}{n \log \log n} = 1.$$

(**N.8**) [P. Erdős]
5. Verify that for every $x > 0$,

$$\frac{\Gamma'(x+1)}{\Gamma(x+1)} > \log x.$$

(**F.18**) [P. Medgyessy]
6. If f is a nonnegative, continuous, concave function on the closed interval $[0,1]$ such that $f(0) = 1$, then

$$\int_0^1 x f(x) dx \leq \frac{2}{3}\left[\int_0^1 f(x) dx\right]^2.$$

(**F.19**) [Z. Daróczy]

7. Let us connect consecutive vertices of a regular heptagon inscribed in a unit circle by connected subsets (of the plane of the circle) of diameter less than 1. Show that every continuum (in the plane of the circle) of diameter greater than 4, containing the center of the circle, intersects one of these connected sets. (**G.20**) [M. Bognár]

8. What is the radius of the largest disc that can be covered by a finite number of closed discs of radius 1 in such a way that each disc intersects at most three others? (**G.21**) [L. Fejes–Tóth]

9. Determine the value of

$$\sup_{1\le\xi\le 2}\,[\log E\xi - E\log\xi],$$

where ξ is a random variable and E denotes expectation. (**P.12**) [Z. Daróczy]

10. Find the limit distribution of the sequence η_n of random variables with distribution

$$P\left(\eta_n = \arccos(\cos^2\frac{(2j-1)\pi}{2n})\right) = \frac{1}{n} \qquad (j=1,,\dots,n).$$

(arccos(.) denotes the main value.) (**P.13**) [B. Gyires]

1974

1. Let $\mathcal{F}$ be a family of subsets of a ground set X such that $\cup_{F\in\mathcal{F}}F = X$, and

(a) if $A, B \in \mathcal{F}$, then $A\cup B \subseteq C$ for some $C \in \mathcal{F}$;

(b) if $A_n \in \mathcal{F}$ $(n = 0, 1, \dots)$, $B \in \mathcal{F}$, and $A_0 \subset A_1 \subset \dots$, then, for some $k \ge 0$, $A_n \cap B = A_k \cap B$ for all $n \ge k$.

Show that there exist pairwise disjoint sets X_γ $(\gamma \in \Gamma)$, with $X = \cup\{X_\gamma : \gamma\in\Gamma\}$, such that every X_γ is contained in some member of $\mathcal{F}$, and every element of $\mathcal{F}$ is contained in the union of finitely many X_γ's. (**ℵ.4**) [A. Hajnal]

2. Let G be a 2-connected nonbipartite graph on $2n$ vertices. Show that the vertex set of G can be split into two classes of n elements each such that the edges joining the two classes form a connected, spanning subgraph. (**C.8**) [L. Lovász]

3. Prove that a necessary and sufficient condition for the existence of a set $S \subset \{1,\dots,n\}$ with the property that the integers $0, 1, \dots, n-1$ all have an odd number of representations in the form $x - y$, $x, y \in S$, is that $(2n - 1)$ has a multiple of the form $2\cdot 4^k - 1$. (**N.9**) [L. Lovász, J. Pelikán]

4. Let R be an infinite ring such that every subring of R different from $\{0\}$ has a finite index in R. (By the index of a subring, we mean the index of its additive group in the additive group of R.) Prove that the additive group of R is cyclic. (**A.23**) [L. Lovász, J. Pelikán]

5. Let $\{f_n\}_{n=0}^{\infty}$ be a uniformly bounded sequence of real-valued measurable functions defined on $[0,1]$ satisfying

$$\int_0^1 f_n^2 = 1.$$

Further, let $\{c_n\}$ be a sequence of real numbers with

$$\sum_{n=0}^{\infty} c_n^2 = +\infty.$$

Prove that some re-arrangement of the series $\sum_{n=0}^{\infty} c_n f_n$ is divergent on a set of positive measure. (**M.11**) [J. Komlós]

6. Let $f(x) = \sum_{n=1}^{\infty} a_n/(x+n^2)$, $(x \geq 0)$, where $\sum_{n=1}^{\infty} |a_n| n^{-\alpha} < \infty$ for some $\alpha > 2$. Let us assume that for some $\beta > 1/\alpha$, we have $f(x) = O(e^{-x^{\beta}})$ as $x \to \infty$. Prove that a_n is identically 0. (**S.14**) [G. Halász]

7. Given a positive integer m and $0 < \delta < \pi$, construct a trigonometric polynomial $f(x) = a_0 + \sum_{n=1}^{m}(a_n \cos nx + b_n \sin nx)$ of degree m such that $f(0) = 1$, $\int_{\delta \leq |x| \leq \pi} |f(x)|\, dx \leq c/m$, and $\max_{-\pi \leq x \leq \pi} |f'(x)| \leq c/\delta$, for some universal constant c. (**S.15**) [G. Halász]

8. Prove that there exists a topological space T containing the real line as a subset, such that the Lebesgue-measurable functions, and only those, extend continuously over T. Show that the real line cannot be an everywhere-dense subset of such a space T. (**T.9**) [Á. Császár]

9. Let A be a closed and bounded set in the plane, and let C denote the set of points at a unit distance from A. Let $p \in C$, and assume that the intersection of A with the unit circle K centered at p can be covered by an arc shorter than a semicircle of K. Prove that the intersection of C with a suitable neighborhood of p is a simple arc of which p is not an endpoint. (**T.10**) [M. Bognár]

10. Let μ and ν be two probability measures on the Borel sets of the plane. Prove that there are random variables $\xi_1, \xi_2, \eta_1, \eta_2$ such that

(a) the distribution of (ξ_1, ξ_2) is μ and the distribution of (η_1, η_2) is ν,

(b) $\xi_1 \leq \eta_1$, $\xi_2 \leq \eta_2$ almost everywhere, if and only if $\mu(G) \geq \nu(G)$ for all sets of the form $G = \cup_{i=1}^{k}(-\infty, x_i) \times (-\infty, y_i)$. (**P.14**) [P. Major]

1975

1. Show that there exists a tournament $(T, \to)$ of cardinality $\aleph_1$ containing no transitive subtournament of size $\aleph_1$. (A structure $(T, \to)$ is a *tournament* if $\to$ is a binary, irreflexive, asymmetric, and trichotomic relation. The tournament $(T, \to)$ is transitive if $\to$ is transitive, that is, if it orders T.) (**ℵ.5**) [A. Hajnal]

2. Let $\mathcal{A}_n$ denote the set of all mappings $f : \{1, 2, \ldots, n\} \to \{1, 2, \ldots, n\}$ such that $f^{-1}(i) := \{k : f(k) = i\} \neq \emptyset$ implies $f^{-1}(j) \neq \emptyset$, $j \in \{1, 2, \ldots, i\}$. Prove

$$|\mathcal{A}_n| = \sum_{k=0}^{\infty} \frac{k^n}{2^{k+1}}.$$

(**C.9**) [L. Lovász]

3. Let S be a semigroup without proper two-sided ideals, and suppose that for every $a, b \in S$ at least one of the products ab and ba is equal to one of the elements a, b. Prove that either $ab = a$ for all $a, b \in S$ or $ab = b$ for all $a, b \in S$. (**A.24**) [L. Megyesi]
4. Prove that the set of rational-valued, multiplicative arithmetical functions and the set of complex rational-valued, multiplicative arithmetical functions form isomorphic groups with the convolution operation $f \circ g$ defined by
$$(f \circ g)(n) = \sum_{d|n} f(d) g(\frac{n}{d}).$$
(We call a complex number *complex rational*, if its real and imaginary parts are both rational.) (**N.10**) [B. Csákány]
5. Let $\{f_n\}$ be a sequence of Lebesgue-integrable functions on $[0, 1]$ such that for any Lebesgue-measurable subset E of $[0, 1]$ the sequence $\int_E f_n$ is convergent. Assume also that $\lim_n f_n = f$ exists almost everywhere. Prove that f is integrable and $\int_E f = \lim_n \int_E f_n$. Is the assertion also true if E runs only over intervals but we also assume $f_n \geq 0$? What happens if $[0, 1]$ is replaced by $[0, \infty)$? (**M.12**) [J. Szűcs]
6. Let f be a differentiable real function, and let M be a positive real number. Prove that if
$$|f(x + t) - 2f(x) + f(x - t)| \leq M t^2 \quad \text{for all } x \text{ and } t,$$
then
$$|f'(x + t) - f'(x)| \leq M |t|.$$
(**F.20**) [J. Szabados]
7. Let $a < a' < b < b'$ be real numbers, and let the real function f be continuous on the interval $[a, b']$ and differentiable in its interior. Prove that there exist $c \in (a, b)$, $c' \in (a', b')$ such that
$$f(b) - f(a) = f'(c)(b - a),$$
$$f(b') - f(a') = f'(c')(b' - a'),$$
and $c < c'$. (**F.21**) [B. Szőkefalvi–Nagy]
8. Prove that if
$$\sum_{n=1}^{m} a_n \leq N a_m \quad (m = 1, 2, \dots)$$
holds for a sequence $\{a_n\}$ of nonnegative real numbers with some positive integer N, then $\alpha_{i+p} \geq p\alpha_i$ for $i, p = 1, 2, \dots$, where
$$\alpha_i = \sum_{n=(i-1)N+1}^{iN} a_n \quad (i = 1, 2, \dots).$$
(**S.16**) [L. Leindler]

9. Let l_0, c, α, g be positive constants, and let $x(t)$ be the solution of the differential equation

$$([l_0 + ct^\alpha]^2 x')' + g[l_0 + ct^\alpha]\sin x = 0, \quad t \geq 0, \quad -\frac{\pi}{2} < x < \frac{\pi}{2},$$

satisfying the initial conditions $x(t_0) = x_0$, $x'(t_0) = 0$. (This is the equation of the mathematical pendulum whose length changes according to the law $l = l_0 + ct^\alpha$.) Prove that $x(t)$ is defined on the interval $[t_0, \infty)$; furthermore, if $\alpha > 2$ then for every $x_0 \neq 0$ there exists a t_0 such that

$$\liminf_{t\to\infty} |x(t)| > 0.$$

(**F.22**) [L. Hatvani]

10. Prove that an idempotent linear operator of a Hilbert space is self-adjoint if and only if it has norm 0 or 1. (**O.4**) [J. Szűcs]

11. Let $X_1, X_2, \ldots, X_n$ be (not necessary independent) discrete random variables. Prove that there exist at least $n^2/2$ pairs (i,j) such that

$$H(X_i + X_j) \geq \frac{1}{3} \min_{1\leq k\leq n} \{H(X_k)\},$$

where $H(X)$ denotes the Shannon entropy of X. (**P.15**) [Gy. Katona]

12. Assume that a face of a convex polyhedron P has a common edge with every other face. Show that there exists a simple closed polygon that consists of edges of P and passes through all vertices. (**G.22**) [L. Lovász]

1976

1. Assume that R, a recursive, binary relation on $\mathbb{N}$ (the set of natural numbers), orders $\mathbb{N}$ into type ω. Show that if $f(n)$ is the nth element of this order, then f is not necessarily recursive. (**ℵ.6**) [L. Pósa]

2. Let G be an infinite graph such that for any countably infinite vertex set A there is a vertex p joined to infinitely many elements of A. Show that G has a countably infinite vertex set A such that G contains uncountably infinitely many vertices p joined to infinitely many elements of A. (**C.10**) [P. Erdős, A. Hajnal]

3. Let H denote the set of those natural numbers for which $\tau(n)$ divides n, where $\tau(n)$ is the number of divisors of n. Show that

(a) $n! \in H$ for all sufficiently large n,

(b) H has density 0.

(**N.11**) [P. Erdős]

4. Let $\mathbb{Z}$ be the ring of rational integers. Construct an integral domain I satisfying the following conditions:

(a) $\mathbb{Z} \subsetneqq I$;

(b) no element of $I \setminus \mathbb{Z}$ is algebraic over $\mathbb{Z}$ (that is, not a root of a polynomial with coefficients in $\mathbb{Z}$);

(c) I only has trivial endomorphisms.

(**A.25**) [E. Fried]

5. Let $S_\nu = \sum_{j=1}^{n} b_j z_j^\nu$ $(\nu = 0, \pm 1, \pm 2, \dots)$, where the b_j are arbitrary and the z_j are nonzero complex numbers. Prove that

$$|S_0| \le n \max_{0<|\nu|\le n} |S_\nu|.$$

(**S.17**) [G. Halász]

6. Let $0 \le c \le 1$, and let η denote the order type of the set of rational numbers. Assume that with every rational number r we associate a Lebesgue-measurable subset H_r of measure c of the interval $[0,1]$. Prove the existence of a Lebesgue-measurable set $H \subset [0,1]$ of measure c such that for every $x \in H$ the set

$$\{r : x \in H_r\}$$

contains a subset of type η. (**M.13**) [M. Laczkovich]

7. Let $f_1, f_2, \dots, f_n$ be regular functions on a domain of the complex plane, linearly independent over the complex field. Prove that the functions $f_i \overline{f}_k$, $1 \le i, k \le n$, are also linearly independent. (**F.23**) [L. Lempert]

8. Prove that the set of all linear combinations (with real coefficients) of the system of polynomials $\{x^n + x^{n^2}\}_{n=0}^{\infty}$ is dense in $C[0,1]$. (**F.24**) [J. Szabados]

9. Let D be a convex subset of the n-dimensional space, and suppose that D' is obtained from D by applying a positive central dilatation and then a translation. Suppose also that the sum of the volumes of D and D' is 1, and $D \cap D' \ne \emptyset$. Determine the supremum of the volume of the convex hull of $D \cup D'$ taken for all such pairs of sets D, D'. (**G.23**) [L. Fejes–Tóth, E. Makai]

10. Suppose that τ is a metrizable topology on a set X of cardinality less than or equal to continuum. Prove that there exists a separable and metrizable topology on X that is coarser than τ. (**T.11**) [I. Juhász]

11. Let $\xi_1, \xi_2, \dots$ be independent, identically distributed random variables with distribution

$$P(\xi_1 = -1) = P(\xi_1 = 1) = \frac{1}{2}.$$

Write $S_n = \xi_1 + \xi_2 + \dots + \xi_n$ $(n = 1, 2, \dots)$, $S_0 = 0$, and

$$T_n = \frac{1}{\sqrt{n}} \max_{0 \le k \le n} S_k .$$

Prove that $\liminf_{n\to\infty} (\log n)\, T_n = 0$ with probability one. (**P.16**) [P. Révész]

1977

1. Consider the intersection of an ellipsoid with a plane σ passing through its center O. On the line through the point O perpendicular to σ, mark the two points at a distance from O equal to the area of the intersection. Determine the loci of the marked points as σ runs through all such planes. (**G.24**) [L. Tamássy]

2. Construct on the real projective plane a continuous curve, consisting of simple points, which is not a straight line and is intersected in a single point by every tangent and every secant of a given conic. (**G.25**) [F. Kárteszi]

3. Prove that if a, x, y are p-adic integers different from 0 and $p|x$, $pa|xy$, then

$$\frac{1}{y}\,\frac{(1+x)^y-1}{x} \equiv \frac{\log(1+x)}{x} \pmod{a}.$$

(**N.12**) [L. Rédei]

4. Let $p > 5$ be a prime number. Prove that every algebraic integer of the pth cyclotomic field can be represented as a sum of (finitely many) distinct units of the ring of algebraic integers of the field. (**A.26**) [K. Győry]

5. Suppose that the automorphism group of the finite undirected graph $X = (P, E)$ is isomorphic to the quaternion group (of order 8). Prove that the adjacency matrix of X has an eigenvalue of multiplicity at least 4.

($P = \{1, 2, \dots, n\}$ is the set of vertices of the graph X. The set of edges E is a subset of the set of all unordered pairs of elements of P. The group of automorphisms of X consists of those permutations of P that map edges to edges. The adjacency matrix $M = [m_{ij}]$ is the $n \times n$ matrix defined by $m_{ij} = 1$ if $\{i, j\} \in E$ and $m_{ij} = 0$ otherwise.) (**A.27**) [L. Babai]

6. Let f be a real function defined on the positive half-axis for which $f(xy) = xf(y) + yf(x)$ and $f(x+1) \le f(x)$ hold for every positive x and y. Show that if $f(1/2) = 1/2$, then

$$f(x) + f(1-x) \ge -x\log_2 x - (1-x)\log_2(1-x)$$

for every $x \in (0, 1)$. (**F.25**) [Z. Daróczy, Gy. Maksa]

7. Let G be a locally compact solvable group, let $c_1, \dots, c_n$ be complex numbers, and assume that the complex-valued functions f and g on G satisfy

$$\sum_{k=1}^{n} c_k f(xy^k) = f(x)g(y) \quad \text{for all } x, y \in G.$$

Prove that if f is a bounded function and

$$\inf_{x\in G} \operatorname{Re} f(x)\chi(x) > 0$$

for some continuous (complex) character χ of G, then g is continuous. (**F.26**) [L. Székelyhidi]

8. Let $p \geq 1$ be a real number and $\mathbb{R}_+ = (0, \infty)$. For which continuous functions $g : \mathbb{R}_+ \to \mathbb{R}_+$ are the following functions all convex?

$$M_n(x) = \left[\frac{\sum_{i=1}^n g(\frac{x_i}{x_{i+1}}) x_{i+1}^p}{\sum_{i=1}^n g(\frac{x_i}{x_{i+1}})} \right]^{\frac{1}{p}},$$

$$x = (x_1, \dots, x_{n+1}) \in \mathbb{R}_+^{n+1}, \quad n = 1, 2, \dots$$

(**S.18**) [L. Losonczi]

9. Suppose that the components of the vector $\mathbf{u} = (u_0, \dots, u_n)$ are real functions defined on the closed interval $[a, b]$ with the property that every nontrivial linear combination of them has at most n zeros in $[a, b]$. Prove that if σ is an increasing function on $[a, b]$ and the rank of the operator

$$A(f) = \int_a^b \mathbf{u}(x) f(x) d\sigma(x), \quad f \in C[a, b],$$

is $r \leq n$, then σ has exactly r points of increase. (**F.27**) [E. Gesztelyi]

10. Let the sequence of random variables $\{X_m,\ m \geq 0\}$, $X_0 = 0$, be an infinite random walk on the set of nonnegative integers with transition probabilities

$$p_i = P(X_{m+1} = i + 1 \,|\, X_m = i) > 0, \quad i \geq 0,$$

$$q_i = P(X_{m+1} = i - 1 \,|\, X_m = i) > 0, \quad i > 0.$$

Prove that for arbitrary $k > 0$ there is an $\alpha_k > 1$ such that

$$P_n(k) = P\left(\max_{0 \leq j \leq n} X_j = k \right)$$

satisfies the limit relation

$$\lim_{L \to \infty} \frac{1}{L} \sum_{n=1}^{L} P_n(k) \alpha_k^n < \infty .$$

(**P.17**) [J. Tomkó]

1978

1. Let $\mathcal{H}$ be a family of finite subsets of an infinite set X such that every finite subset of X can be represented as the union of two disjoint sets from $\mathcal{H}$. Prove that for every positive integer k there is a subset of X that can be represented in at least k different ways as the union of two disjoint sets from $\mathcal{H}$. (**C.11**) [P. Erdős]

2. For a distributive lattice L, consider the following two statements:

(A) Every ideal of L is the kernel of at least two different homomorphisms.

(B) L contains no maximal ideal.

Which one of these statements implies the other?

(Every homomorphism φ of L induces an equivalence relation on L: $a \sim b$ if and only if $a\varphi = b\varphi$. We do not consider two homomorphisms different if they imply the same equivalence relation.) (**A.28**) [J. Varlet, E. Fried]

3. Let $1 < a_1 < a_2 < \cdots < a_n < x$ be positive integers such that $\sum_1^n 1/a_i \leq 1$. Let y denote the number of positive integers smaller than x not divisible by any of the a_i. Prove that

$$y > \frac{cx}{logx}$$

with a suitable positive constant c (independent of x and the numbers a_i). (**N.13**) [I. Z. Ruzsa]

4. Let $\mathbb{Q}$ and $\mathbb{R}$ be the set of rational numbers and the set of real numbers, respectively, and let $f : \mathbb{Q} \to \mathbb{R}$ be a function with the following property. For every $h \in \mathbb{Q}$, $x_0 \in \mathbb{R}$,

$$f(x+h) - f(x) \to 0$$

as $x \in \mathbb{Q}$ tends to x_0. Does it follow that f is bounded on some interval? (**F.28**) [M. Laczkovich]

5. Suppose that $R(z) = \sum_{n=-\infty}^{\infty} a_n z^n$ converges in a neighborhood of the unit circle $\{z : |z| = 1\}$ in the complex plane, and $R(z) = P(z)/Q(z)$ is a rational function in this neighborhood, where P and Q are polynomials of degree at most k. Prove that there is a constant c independent of k such that

$$\sum_{n=-\infty}^{\infty} |a_n| \leq ck^2 \max_{|z|=1} |R(z)|.$$

(**S.19**) [H. S. Shapiro, G. Somorjai]

6. Suppose that the function $g : (0,1) \to \mathbb{R}$ can be uniformly approximated by polynomials with nonnegative coefficients. Prove that g must be analytic. Is the statement also true for the interval $(-1,0)$ instead of $(0,1)$? (**F.29**) [J. Kalina, L. Lempert]

7. Let T be a surjective mapping of the hyperbolic plane onto itself which maps collinear points into collinear points. Prove that T must be an isometry. (**G.26**) [M. Bognár]

8. Let $X_1, \ldots, X_n$ be n points in the unit square ($n > 1$). Let r_i be the distance of X_i from the nearest point (other than X_i). Prove the inequality

$$r_1^2 + \cdots + r_n^2 \leq 4.$$

(**G.27**) [L. Fejes–Tóth, E. Szemerédi]

9. Suppose that all subspaces of cardinality at most $\aleph_1$ of a topological space are second-countable. Prove that the whole space is second-countable. (**T.12**) [A. Hajnal, I. Juhász]

10. Let Y_n be a binomial random variable with parameters n and p. Assume that a certain set H of positive integers has a density and that this density is equal to d. Prove the following statements:
(a) $\lim_{n\to\infty} P(Y_n \in H) = d$ if H is an arithmetic progression.
(b) The previous limit relation is not valid for arbitrary H.
(c) If H is such that $P(Y_n \in H)$ is convergent, then the limit must be equal to d.
(**P.18**) [L. Pósa]

1979

1. Let the operation f of k variables defined on the set $\{1, 2, \ldots, n\}$ be called *friendly* toward the binary relation ρ defined on the same set if

$$f(a_1, a_2, \ldots, a_k)\ \rho\ f(b_1, b_2, \ldots, b_k)$$

implies $a_i\ \rho\ b_i$ for at least one i, $1 \le i \le k$. Show that if the operation f is friendly toward the relations "equal to" and "less than," then it is friendly toward all binary relations. (**C.12**) [B. Csákány]

2. Let $\mathcal{V}$ be a variety of monoids such that not all monoids of $\mathcal{V}$ are groups. Prove that if $A \in \mathcal{V}$ and B is a submonoid of A, there exist monoids $S \in \mathcal{V}$ and C and epimorphisms $\varphi : S \to A$, $\varphi_1 : S \to C$ such that $((e)\varphi_1^{-1})\varphi = B$ (e is the identity element of C). (**A.29**) [L. Márki]

3. Let $g(n, k)$ denote the number of strongly connected, *simple* directed graphs with n vertices and k edges. (*Simple* means no loops or multiple edges.) Show that

$$\sum_{k=n}^{n^2-n} (-1)^k g(n, k) = (n-1)!.$$

(**C.13**) [A. A. Schrijver]

4. For what values of n does the group $SO(n)$ of all orthogonal transformations of determinant 1 of the n-dimensional Euclidean space possess a closed regular subgroup? ($G \le SO(n)$ is called *regular* if for any elements x, y of the unit sphere there exists a unique $\varphi \in G$ such that $\varphi(x) = y$.) (**A.30**) [Z. Szabó]

5. Give an example of ten different noncoplanar points $P_1, \ldots, P_5$, $Q_1, \ldots, Q_5$ in 3-space such that connecting each P_i to each Q_j by a rigid rod results in a rigid system. (**G.28**) [L. Lovász]

6. Let us define a pseudo-Riemannian metric on the set of points of the Euclidean space $\mathbb{E}^3$ not lying on the z-axis by the metric tensor

$$\begin{pmatrix} 1 & 0 & 0 \\ 0 & 1 & 0 \\ 0 & 0 & -\sqrt{x^2+y^2} \end{pmatrix},$$

where (x, y, z) is a Cartesian coordinate system in $\mathbb{E}^3$. Show that the orthogonal projections of the geodesic curves of this Riemannian space onto the (x, y)-plane are straight lines or conic sections with focus at the origin. (**G.29**) [P. Nagy]

7. Let T be a triangulation of an n-dimensional sphere, and to each vertex of T let us assign a nonzero vector of a linear space V. Show that if T has an n-dimensional simplex such that the vectors assigned to the vertices of this simplex are linearly independent, then another such simplex must also exist. (**C.14**) [L. Lovász]

8. Let $K_n (n = 1, 2, \dots)$ be periodical continuous functions of period 2π, and write

$$k_n(f; x) = \int_0^{2\pi} f(t)K_n(x - t)dt.$$

Prove that the following statements are equivalent:

(i) $\int_0^{2\pi} |k_n(f; x) - f(x)|\, dx \to 0 \ (n \to \infty)$ for all $f \in L_1[0, 2\pi]$.

(ii) $k_n(f; 0) \to f(0)$ for all continuous, 2π-periodic functions f.

(**S.20**) [V. Totik]

9. Let us assume that the series of holomorphic functions $\sum_{k=1}^{\infty} f_k(z)$ is absolutely convergent for all $z \in \mathbb{C}$. Let $H \subseteq \mathbb{C}$ be the set of those points where the above sum function is not regular. Prove that H is nowhere dense but not necessarily countable. (**S.21**) [L. Kérchy]

10. Prove that if a_i $(i = 1, 2, 3, 4)$ are positive constants, $a_2 - a_4 > 2$, and $a_1 a_3 - a_2 > 2$, then the solution $(x(t), y(t))$ of the system of differential equations

$$\dot{x} = a_1 - a_2 x + a_3 xy,$$
$$\dot{y} = a_4 x - y - a_3 xy \qquad (x, y \in \mathbb{R})$$

with the initial conditions $x(0) = 0$, $y(0) \geq a_1$ is such that the function $x(t)$ has exactly one strict local maximum on the interval $[0, \infty)$. (**F.30**) [L. Pintér, L. Hatvani]

11. Let $\{\xi_{kl}\}_{k,l=1}^{\infty}$ be a double sequence of random variables such that

$$E\xi_{ij}\xi_{kl} = O((\log(2|i-k|+2)\log(2|j-l|+2))^{-2}) \qquad (i, j, k, l = 1, 2, \dots).$$

Prove that with probability one,

$$\frac{1}{mn}\sum_{k=1}^{m}\sum_{l=1}^{n} \xi_{kl} \to 0 \quad \text{as } \max(m, n) \to \infty.$$

(**P.19**) [F. Móricz]

1980

1. For a real number x, let $\|x\|$ denote the distance between x and the closest integer. Let $0 \leq x_n < 1 \quad (n = 1, 2, \dots)$, and let $\varepsilon > 0$. Show that there exist infinitely many pairs (n, m) of indices such that $n \neq m$ and

$$\|x_n - x_m\| < \min\left(\varepsilon, \frac{1}{2|n-m|}\right).$$

(**C.15**) [V. T. Sós]

2. Let $\mathcal{H}$ be the class of all graphs with at most $2^{\aleph_0}$ vertices not containing a complete subgraph of size $\aleph_1$. Show that there is no graph $H \in \mathcal{H}$ such that every graph in $\mathcal{H}$ is a subgraph of H. (**ℵ.7**) [F. Galvin]

3. In a lattice, connect the elements $a \wedge b$ and $a \vee b$ by an edge whenever a and b are incomparable. Prove that in the obtained graph every connected component is a sublattice. (**A.31**) [M. Ajtai]

4. Let $T \in SL(n, \mathbb{Z})$, let G be a nonsingular $n \times n$ matrix with integer elements, and put $S = G^{-1}TG$. Prove that there is a natural number k such that $S^k \in SL(n, \mathbb{Z})$. (**N.14**) [Gy. Szekeres]

5. Let G be a transitive subgroup of the symmetric group S_{25} different from S_{25} and A_{25}. Prove that the order of G is not divisible by 23. (**A.32**) [J. Pelikán]

6. Let us call a continuous function $f : [a, b] \to \mathbb{R}^2$ *reducible* if it has a double arc (that is, if there are $a \leq \alpha < \beta \leq \gamma < \delta \leq b$ such that there exists a strictly monotone and continuous $h : [\alpha, \beta] \to [\gamma, \delta]$ for which $f(t) = f(h(t))$ is satisfied for every $\alpha \leq t \leq \beta$); otherwise f is irreducible. Construct irreducible $f : [a, b] \to \mathbb{R}^2$ and $g : [c, d] \to \mathbb{R}^2$ such that $f([a, b]) = g([c, d])$ and

(a) both f and g are rectifiable but their lengths are different;

(b) f is rectifiable but g is not.

(**F.31**) [Á. Császár]

7. Let $n \geq 2$ be a natural number and $p(x)$ a real polynomial of degree at most n for which

$$\max_{-1 \leq x \leq 1} |p(x)| \leq 1, \quad p(-1) = p(1) = 0.$$

Prove that then

$$|p'(x)| \leq \frac{n \cos \frac{\pi}{2n}}{\sqrt{1 - x^2 \cos^2 \frac{\pi}{2n}}} \qquad \left(-\frac{1}{\cos \frac{\pi}{2n}} < x < \frac{1}{\cos \frac{\pi}{2n}}\right).$$

(**F.32**) [J. Szabados]

8. Let $f(x)$ be a nonnegative, integrable function on $(0, 2\pi)$ whose Fourier series is $f(x) = a_0 + \sum_{k=1}^{\infty} a_k \cos(n_k x)$, where none of the positive integers n_k divides another. Prove that $|a_k| \leq a_0$. (**S.22**) [G. Halász]

9. Let us divide by straight lines a quadrangle of unit area into n subpolygons and draw a circle into each subpolygon. Show that the sum of the perimeters of the circles is at most $\pi\sqrt{n}$ (the lines are not allowed to cut the interior of a subpolygon). (**G.30**) [G. and L. Fejes–Tóth]

10. Suppose that the T_3-space X has no isolated points and that in X any family of pairwise disjoint, nonempty, open sets is countable. Prove that X can be covered by at most continuum many nowhere-dense sets. (**T.13**) [I. Juhász]

1981

1. We are given an infinite sequence of 1's and 2's with the following properties:

(1) The first element of the sequence is 1.

(2) There are no two consecutive 2's or three consecutive 1's.

(3) If we replace consecutive 1's by a single 2, leave the single 1's alone, and delete the original 2's, then we recover the original sequence.

How many 2's are there among the first n elements of the sequence? (**S.23**) [P. P. Pálfy]

2. Consider the lattice L of the contractions of a simple graph G (as sets of vertex pairs) with respect to inclusion. Let $n \geq 1$ be an arbitrary integer. Show that the identity

$$x \bigwedge \left(\bigvee_{i=0}^{n} y_i \right) = \bigvee_{j=0}^{n} \left(x \bigwedge \left(\bigvee_{\substack{0 \leq i \leq n \\ i \neq j}} y_i \right) \right)$$

holds if and only if G has no cycle of size at least $n+2$. (**C.16**) [A. Huhn]

3. Construct an uncountable Hausdorff space in which the complement of the closure of any nonempty, open set is countable. (**T.14**) [A. Hajnal, I. Juhász]

4. Let G be a finite group and $\mathcal{K}$ a conjugacy class of G that generates G. Prove that the following two statements are equivalent:

(1) There exists a positive integer m such that every element of G can be written as a product of m (not necessarily distinct) elements of $\mathcal{K}$.

(2) G is equal to its own commutator subgroup.

(**A.33**) [J. Dénes]

5. Let K be a convex cone in the n-dimensional real vector space $\mathbb{R}^n$, and consider the sets $A = K \cup (-K)$ and $B = (\mathbb{R}^n \setminus A) \cup \{0\}$ (0 is the origin). Show that one can find two subspaces in $\mathbb{R}^n$ such that together they span $\mathbb{R}^n$, and one of them lies in A and the other lies in B. (**G.31**) [J. Szűcs]

6. Let f be a strictly increasing, continuous function mapping $I = [0, 1]$ onto itself. Prove that the following inequality holds for all pairs $x, y \in I$:

$$1 - \cos(xy) \leq \int_0^x f(t) \sin(tf(t))dt + \int_0^y f^{-1}(t) \sin(tf^{-1}(t))dt\,.$$

(**F.33**) [Zs. Páles]

7. Let U be a real normed space such that, for any finite-dimensional, real normed space X, U contains a subspace isometrically isomorphic to

X. Prove that every (not necessarily closed) subspace V of U of finite codimension has the same property. (We call V of finite codimension if there exists a finite-dimensional subspace N of U such that $V+N=U$.) **(F.34)** [Á. Bosznay]

8. Let W be a dense, open subset of the real line $\mathbb{R}$. Show that the following two statements are equivalent:
(1) Every function $f:\mathbb{R}\to\mathbb{R}$ continuous at all points of $\mathbb{R}\setminus W$ and nondecreasing on every open interval contained in W is nondecreasing on the whole $\mathbb{R}$.
(2) $\mathbb{R}\setminus W$ is countable.
(T.15) [E. Gesztelyi]

9. Let $n\geq 2$ be an integer, and let X be a connected Hausdorff space such that every point of X has a neighborhood homeomorphic to the Euclidean space $\mathbb{R}^n$. Suppose that any discrete (not necessarily closed) subspace D of X can be covered by a family of pairwise disjoint, open sets of X so that each of these open sets contains precisely one element of D. Prove that X is a union of at most $\aleph_1$ compact subspaces. **(T.16)** [Z. Balogh]

10. Let P be a probability distribution defined on the Borel sets of the real line. Suppose that P is symmetric with respect to the origin, absolutely continous with respect to the Lebesgue measure, and its density function p is zero outside the interval $[-1,1]$ and inside this interval it is between the positive numbers c and d $(c<d)$. Prove that there is no distribution whose convolution square equals P. **(P.20)** [T. F. Móri, G. J. Székely]

1982

1. A map $F:P(X)\to P(X)$, where $P(X)$ denotes the set of all subsets of X, is called a *closure operation* on X if for arbitrary $A,B\subset X$, the following conditions hold:
(i) $A\subset F(A)$;
(ii) $A\subset B\Rightarrow F(A)\subset F(B)$;
(iii) $F(F(A))=F(A)$.
The cardinal number $\min\{|A|:A\subset X,\ F(A)=X\}$ is called the *density* of F and is denoted by $d(F)$. A set $H\subset X$ is called *discrete* with respect to F if $u\notin F(H-\{u\})$ holds for all $u\in H$. Prove that if the density of the closure operation F is a singular cardinal number, then for any nonnegative integer n, there exists a set of size n that is discrete with respect to F. Show that the statement is not true when the existence of an infinite discrete subset is required, even if F is the closure operation of a topological space satisfying the T_1 separation axiom. **(T.17)** [A. Hajnal]

2. Consider the lattice of all algebraically closed subfields of the complex field $\mathbb{C}$ whose transcendency degree (over $\mathbb{Q}$) is finite. Prove that this lattice is not modular. **(A.34)** [L. Babai]

3. Let $G(V,E)$ be a connected graph, and let $d_G(x,y)$ denote the length of the shortest path joining x and y in G. Let $r_G(x)=\max\{d_G(x,y):y\in V\}$ for $x\in V$, and let $r(G)=\min\{r_G(x):x\in V\}$. Show that if

$r(G) \geq 2$, then G contains a path of length $2r(G) - 2$ as an induced subgraph. (**C.17**) [V. T. Sós]

4. Let

$$f(n) = \sum_{\substack{p \mid n \\ p^\alpha \leq n < p^{\alpha+1}}} p^\alpha .$$

Prove that

$$\limsup_{n \to \infty} f(n) \frac{\log \log n}{n \log n} = 1.$$

(**N.15**) [P. Erdős]

5. Find a perfect set $H \subset [0,1]$ of positive measure and a continuous function f defined on $[0,1]$ such that for any twice differentiable function g defined on $[0,1]$, the set $\{x \in H : \ f(x) = g(x)\}$ is finite. (**M.14**) [M. Laczkovich]

6. For every positive α, natural number n, and at most αn points x_i, construct a trigonometric polynomial $P(x)$ of degree at most n for which

$$P(x_i) \leq 1, \quad \int_0^{2\pi} P(x) dx = 0, \quad \text{and} \quad \max P(x) > cn,$$

where the constant c depends only on α. (**F.35**) [G. Halász]

7. Let V be a bounded, closed, convex set in $\mathbb{R}^n$, and denote by r the radius of its circumscribed sphere (that is, the radius of the smallest sphere that contains V). Show that r is the only real number with the following property: for any finite number of points in V, there exists a point in V such that the arithmetic mean of its distances from the other points is equal to r. (**G.32**) [Gy. Szekeres]

8. Show that for any natural number n and any real number $d > 3^n/(3^n - 1)$, one can find a covering of the unit square with n homothetic triangles with area of the union less than d. (**G.33**)

9. Suppose that K is a compact Hausdorff space and $K = \cup_{n=0}^{\infty} A_n$, where A_n is metrizable and $A_n \subset A_m$ for $n < m$. Prove that K is metrizable. (**T.18**) [Z. Balogh]

10. Let $p_0, p_1, \ldots$ be a probability distribution on the set of nonnegative integers. Select a number according to this distribution and repeat the selection independently until either a zero or an already selected number is obtained. Write the selected numbers in a row in order of selection without the last one. Below this line, write the numbers again in increasing order. Let A_i denote the event that the number i has been selected and that it is in the same place in both lines. Prove that the events A_i $(i = 1, 2, \ldots)$ are mutually independent, and $P(A_i) = p_i$. (**P.21**) [T. F. Móri]

1983

1. Given n points in a line so that any distance occurs at most twice, show that the number of distances occurring exactly once is at least $[n/2]$. (**C.18**) [V. T. Sós, L. Székely]
2. Let I be an ideal of the ring R and f a nonidentity permutation of the set $\{1, 2, \ldots, k\}$ for some k. Suppose that for every $0 \neq a \in R$, $aI \neq 0$ and $Ia \neq 0$ hold; furthermore, for any elements $x_1, x_2, \ldots, x_k \in I$,

$$x_1 x_2 \cdots x_k = x_{1f} x_{2f} \cdots x_{kf}$$

holds. Prove that R is commutative. (**A.35**) [R. Wiegandt]
3. Let $f : \mathbb{R} \to \mathbb{R}$ be a twice differentiable, 2π-periodic even function. Prove that if

$$f''(x) + f(x) = \frac{1}{f(x + 3\pi/2)}$$

holds for every x, then f is $\pi/2$-periodic. (**F.36**) [Z. Szabó, J. Terjéki]
4. For which cardinalities κ do antimetric spaces of cardinality κ exist? (X, ϱ) is called an *antimetric space* if X is a nonempty set, $\varrho : X^2 \to [0, \infty)$ is a symmetric map, $\varrho(x, y) = 0$ holds iff $x = y$, and for any three-element subset $\{a_1, a_2, a_3\}$ of X

$$\varrho(a_{1f}, a_{2f}) + \varrho(a_{2f}, a_{3f}) < \varrho(a_{1f}, a_{3f})$$

holds for some permutation f of $\{1, 2, 3\}$. (**ℵ.8**) [V. Totik]
5. Let $g : \mathbb{R} \to \mathbb{R}$ be a continuous function such that $x + g(x)$ is strictly monotone (increasing or decreasing), and let $u : [0, \infty) \to \mathbb{R}$ be a bounded and continuous function such that

$$u(t) + \int_{t-1}^{t} g(u(s))ds$$

is constant on $[1, \infty)$. Prove that the limit $\lim_{t\to\infty} u(t)$ exists. (**F.37**) [T. Krisztin]
6. Let T be a bounded linear operator on a Hilbert space H, and assume that $\|T^n\| \leq 1$ for some natural number n. Prove the existence of an invertible linear operator A on H such that $\|ATA^{-1}\| \leq 1$. (**O.5**) [E. Druszt]
7. Prove that if the function $f : \mathbb{R}^2 \to [0, 1]$ is continuous and its average on every circle of radius 1 equals the function value at the center of the circle, then f is constant. (**F.38**) [V. Totik]
8. Prove that any identity that holds for every finite n-distributive lattice also holds for the lattice of all convex subsets of the $(n-1)$-dimensional Euclidean space. (For convex subsets, the lattice operations are the set-theoretic intersection and the convex hull of the set-theoretic union. We call a lattice *n-distributive* if

$$x \wedge \Big(\bigvee_{i=0}^{n} y_i\Big) = \bigvee_{j=0}^{n} \Big(x \wedge \Big(\bigvee_{\substack{0 \leq i \leq n \\ i \neq j}} y_i\Big)\Big)$$

holds for all elements of the lattice.) (**A.36**) [A. Huhn]

9. Prove that if $E \subset \mathbb{R}$ is a bounded set of positive Lebesgue measure, then for every $u < 1/2$, a point $x = x(u)$ can be found so that

$$|(x-h, x+h) \cap E| \geq uh$$

and

$$|(x-h, x+h) \cap (\mathbb{R} \setminus E)| \geq uh$$

for all sufficiently small positive values of h. (**M.15**) [K. I. Koljada]

10. Let R be a bounded domain of area t in the plane, and let C be its center of gravity. Denoting by T_{AB} the circle drawn with the diameter AB, let K be a circle that contains each of the circles T_{AB} $(A, B \in R)$. Is it true in general that K contains the circle of area $2t$ centered at C? (**G.34**) [J. Szűcs]

11. Let $M^n \subset \mathbb{R}^{n+1}$ be a complete, connected hypersurface embedded into the Euclidean space. Show that M^n as a Riemannian manifold decomposes to a nontrivial global metric direct product if and only if it is a real cylinder, that is, M^n can be decomposed to a direct product of the form $M^n = M^k \times \mathbb{R}^{n-k}$ $(k < n)$ as well, where M^k is a hypersurface in some $(k+1)$-dimensional subspace $E^{k+1} \subset \mathbb{R}^{n+1}$, $\mathbb{R}^{n-k}$ is the orthogonal complement of E^{k+1}. (**G.35**) [Z. Szabó]

12. Let $X_1, \dots, X_n$ be independent, identically distributed, nonnegative random variables with a common continuous distribution function F. Suppose in addition that the inverse of F, the quantile function Q, is also continuous and $Q(0) = 0$. Let $0 = X_{0:n} \leq X_{1:n} \leq \cdots \leq X_{n:n}$ be the ordered sample from the above random variables. Prove that if EX_1 is finite, then the random variable

$$\Delta = \sup_{0 \leq y \leq 1} \left| \frac{1}{n} \sum_{i=1}^{[ny]+1} (n+1-i)(X_{i:n} - X_{i-1:n}) - \int_0^y (1-u)\,dQ(u) \right|$$

tends to zero with probability one as $n \to \infty$. (**P.22**) [S. Csörgő, L. Horváth]

1984

1. Let κ be an arbitrary cardinality. Show that there exists a tournament $T_\kappa = (V_\kappa, E_\kappa)$ such that for any coloring $f : E_\kappa \to \kappa$ of the edge set E_κ, there are three different vertices $x_0, x_1, x_2 \in V_\kappa$ such that

$$x_0x_1, x_1x_2, x_2x_0 \in E_\kappa$$

and

$$|\{f(x_0x_1), f(x_1x_2), f(x_2x_0)\}| \leq 2.$$

(A *tournament* is a directed graph such that for any vertices $x, y \in V_\kappa$, $x \neq y$ exactly one of the relations $xy \in E_\kappa$, $yx \in E_\kappa$ holds.) (**C.19**) [A. Hajnal]

2. Show that there exist a compact set $K \subset \mathbb{R}$ and a set $A \subset \mathbb{R}$ of type F_σ such that the set

$$\{x \in \mathbb{R} : K + x \subset A\}$$

is not Borel-measurable (here $K + x = \{y + x : y \in K\}$). (**M.16**) [M. Laczkovich]

3. Let a and b be positive integers such that when dividing them by any prime p, the remainder of a is always less than or equal to the remainder of b. Prove that $a = b$. (**N.16**) [P. Erdős, P. P. Pálfy]

4. Let $x_1, x_2, y_1, y_2, z_1, z_2$ be transcendental numbers. Suppose that any 3 of them are algebraically independent, and among the 15 four-tuples only $\{x_1, x_2, y_1, y_2\}$, $\{x_1, x_2, z_1, z_2\}$, and $\{y_1, y_2, z_1, z_2\}$ are algebraically dependent. Prove that there exists a transcendental number t that depends algebraically on each of the pairs $\{x_1, x_2\}$, $\{y_1, y_2\}$, and $\{z_1, z_2\}$. (**A.37**) [L. Lovász]

5. Let $a_0, a_1, \ldots$ be nonnegative real numbers such that

$$\sum_{n=0}^{\infty} a_n = \infty .$$

For arbitrary $c > 0$, let

$$n_j(c) = \min\left\{k : c \cdot j \le \sum_{i=0}^{k} a_i\right\}, \quad j = 1, 2, \ldots .$$

Prove that if $\sum_{i=0}^{\infty} a_i^2 < \infty$, then there exists a $c > 0$ for which $\sum_{j=1}^{\infty} a_{n_j(c)} < \infty$, and if $\sum_{i=0}^{\infty} a_i^2 = \infty$, then there exists a $c > 0$ for which $\sum_{j=1}^{\infty} a_{n_j(c)} = \infty$. (**S.24**) [P. Erdős, I. Joó, L. Székely]

6. For which Lebesgue-measurable subsets E of the real line does a positive constant c exist for which

$$\sup_{-\infty < t < \infty} \left| \int_E e^{itx} f(x) dx \right| \le c \sup_{n = 0, \pm 1, \ldots} \left| \int_E e^{inx} f(x) dx \right|$$

for all integrable functions f on E? (**M.17**) [G. Halász]

7. Let V be a finite-dimensional subspace of $C[0,1]$ such that every nonzero $f \in V$ attains positive value at some point. Prove that there exists a polynomial P that is strictly positive on $[0,1]$ and orthogonal to V, that is, for every $f \in V$,

$$\int_0^1 f(x) P(x) dx = 0.$$

(**F.39**) [A. Pinkus, V. Totik]

8. Among all point lattices on the plane intersecting every closed convex region of unit width, which one's fundamental parallelogram has the largest area? (**G.36**) [L. Fejes–Tóth]

9. Let $X_0, X_1, \dots$ be independent, identically distributed, nondegenerate random variables, and let $0 < \alpha < 1$ be a real number. Assume that the series

$$\sum_{k=0}^{\infty} \alpha^k X_k$$

is convergent with probability one. Prove that the distribution function of the sum is continuous. (**P.23**) [T. F. Móri]

10. Let $X_1, X_2, \dots$ be independent random variables with the same distribution:

$$P(X_i = 1) = P(X_i = -1) = \frac{1}{2} \qquad (i = 1, 2, \dots).$$

Define

$$S_0 = 0, \qquad S_n = X_1 + X_2 + \dots + X_n \quad (n = 1, 2, \dots),$$
$$\xi(x, n) = |\{k : 0 \le k \le n,\ S_k = x\}| \qquad (x = 0, \pm 1, \pm 2, \dots),$$

and

$$\alpha(n) = |\{x : \xi(x, n) = 1\}| \qquad (n = 0, 1, \dots).$$

Prove that

$$P(\liminf \alpha(n) = 0) = 1$$

and that there is a number $0 < c < \infty$ such that $P(\limsup \alpha(n)/\log n = c) = 1$. (**P.24**) [P. Révész]

1985

1. Some proper partitions $P_1, \dots, P_n$ of a finite set S (that is, partitions containing at least two parts) are called *independent* if no matter how we choose one class from each partition, the intersection of the chosen classes is nonempty. Show that if the inequality

$$\frac{|S|}{2} < |P_1| \cdots |P_n| \tag{$*$}$$

holds for some independent partitions, then $P_1, \dots, P_n$ is maximal in the sense that there is no partition P such that $P, P_1, \dots, P_n$ are independent. On the other hand, show that inequality $(*)$ is not necessary for this maximality. (**C.20**) [E. Gesztelyi]

2. Let S be a given finite set of hyperplanes in $\mathbb{R}^n$, and let O be a point. Show that there exists a compact set $K \subseteq \mathbb{R}^n$ containing O such that the orthogonal projection of any point of K onto any hyperplane in S is also in K. (**G.37**) [Gy. Pap]

3. Let k and K be concentric circles on the plane, and let k be contained inside K. Assume that k is covered by a finite system of convex angular domains with vertices on K. Prove that the sum of the angles of the domains is not less than the angle under which k can be seen from a point of K. (**G.38**) [Zs. Páles]

4. Call a subset S of the set $\{1, \ldots, n\}$ *exceptional* if any pair of distinct elements of S are coprime. Consider an exceptional set with a maximal sum of elements (among all exceptional sets for a fixed n). Prove that if n is sufficiently large, then each element of S has at most two distinct prime divisors. (**N.17**) [P. Erdős]

5. Let $F(x, y)$ and $G(x, y)$ be relatively prime homogeneous polynomials of degree at least one having integer coefficients. Prove that there exists a number c depending only on the degrees and the maximum of the absolute values of the coefficients of F and G such that $F(x, y) \neq G(x, y)$ for any integers x and y that are relatively prime and satisfy $\max\{|x|, |y|\} > c$. (**A.38**) [K. Győry]

6. Determine all finite groups G that have an automorphism f such that $H \not\subseteq f(H)$ for all proper subgroups H of G. (**A.39**) [B. Kovács]

7. Let p_1 and p_2 be positive real numbers. Prove that there exist functions $f_i : \mathbb{R} \to \mathbb{R}$ such that the smallest positive period of f_i is p_i $(i = 1, 2)$, and $f_1 - f_2$ is also periodic. (**A.40**) [J. Rimán]

8. Let $2/(\sqrt{5}+1) \leq p < 1$, and let the real sequence $\{a_n\}$ have the following property: for every sequence $\{e_n\}$ of 0's and ±1's for which $\sum_{n=1}^{\infty} e_n p^n = 0$, we also have $\sum_{n=1}^{\infty} e_n a_n = 0$. Prove that there is a number c such that $a_n = cp^n$ for all n. (**S.25**) [Z. Daróczy, I. Kátai]

9. Let $D = \{z \in \mathbb{C}\colon |z| < 1\}$ and $D = \{w \in \mathbb{C}\colon |w| = 1\}$. Prove that if for a function $f : D \times B \to \mathbb{C}$ the equality

$$f\left(\frac{az+b}{\bar{b}z+\bar{a}}, \frac{aw+b}{\bar{b}w+\bar{a}}\right) = f(z, w) + f\left(\frac{b}{\bar{a}}, \frac{aw+b}{\bar{b}w+\bar{a}}\right)$$

holds for all $z \in D$, $w \in B$ and $a, b \in \mathbb{C}$, $|a|^2 = 1 + |b|^2$, then there is a function $L :]0, \infty[\to \mathbb{C}$ satisfying

$$L(pq) = L(p) + L(q) \quad \text{for all} \quad p, q > 0$$

such that f can be represented as

$$f(z, w) = L\Big(\frac{1-|z|^2}{|w-z|^2}\Big) \quad \text{for all} \quad z \in D, w \in B.$$

(**F.40**) [Gy. Maksa]

10. Show that any two intervals $A, B \subseteq \mathbb{R}$ of positive lengths can be countably disected into each other, that is, they can be written as countable unions $A = A_1 \cup A_2 \cup \ldots$ and $B = B_1 \cup B_2 \cup \ldots$ of pairwise disjoint sets, where A_i and B_i are congruent for every $i \in \mathbb{N}$. (**M.18**) [Gy. Szabó]

11. Let $\xi(E, \pi, B)$ $(\pi : E \to B)$ be a real vector bundle of finite rank, and let

$$\tau_E = V\xi \oplus H\xi \qquad (*)$$

be the tangent bundle of E, where $V\xi = \operatorname{Ker} d\pi$ is the vertical subbundle of τ_E. Let us denote the projection operators corresponding to the splitting (*) by v and h. Construct a linear connection ∇ on $V\xi$ such that

$$\nabla_X \vee Y - \nabla_Y \vee X = v[X,Y] - v[hX,hY].$$

(X and Y are vector fields on E, $[.,.]$ is the Lie bracket, and all data are of class $\mathcal{C}^\infty$). (**G.39**) [J. Szilasi]

12. Let $(\Omega, \mathcal{A}, P)$ be a probability space, and let $(X_n, \mathcal{F}_n)$ be an adapted sequence in $(\Omega, \mathcal{A}, P)$ (that is, for the σ-algebras $\mathcal{F}_n$, we have $\mathcal{F}_1 \subseteq \mathcal{F}_2 \subseteq \cdots \subseteq \mathcal{A}$, and for all n, X_n is an $\mathcal{F}_n$-measurable and integrable random variable). Assume that

$$E(X_{n+1}|\mathcal{F}_n) = \frac{1}{2}X_n + \frac{1}{2}X_{n-1} \qquad (n = 2, 3 \ldots).$$

Prove that $\sup_n E|X_n| < \infty$ implies that X_n converges with probability one as $n \to \infty$. (**P.25**) [I. Fazekas]

1986

1. If $(A, <)$ is a partially ordered set, its dimension, $\dim(A, <)$, is the least cardinal κ such that there exist κ total orderings $\{<_\alpha : \alpha < \kappa\}$ on A with $< = \bigcap_{\alpha<\kappa} <_\alpha$. Show that if $\dim(A, <) > \aleph_0$, then there exist disjoint $A_0, A_1 \subseteq A$ with $\dim(A_0, <), \dim(A_1, <) > \aleph_0$. (**ℵ.9**) [D. Kelly, A. Hajnal, B. Weiss]

2. Show that if $k \le n/2$ and $\mathcal{F}$ is a family of $k \times k$ submatrices of an $n \times n$ matrix such that any two intersect then

$$|\mathcal{F}| \le \binom{n-1}{k-1}^2.$$

(**C.21**) [Gy. Katona]

3. (a) Prove that for every natural number k, there are positive integers $a_1 < a_2 < \cdots < a_k$ such that $a_i - a_j$ divides a_i for all $1 \le i, j \le k$, $i \ne j$.

(b) Show that there is an absolute constant $C > 0$ such that $a_1 > k^{Ck}$ for every sequence $a_1, \ldots, a_k$ of numbers that satisfy the above divisibility condition. (**N.18**) [A. Balogh, I. Z. Ruzsa]

4. Determine all real numbers x for which the following statement is true: the field $\mathbb{C}$ of complex numbers contains a proper subfield F such that adjoining x to F we get $\mathbb{C}$. (**A.41**) [M. Laczkovich]

5. Prove the existence of a constant c with the following property: for every composite integer n, there exists a group whose order is divisible by n and is less than n^c, and that contains no element of order n. (**A.42**) [P. P. Pálfy]

6. Let U denote the set $\{f \in C[0,1] : |f(x)| \le 1 \text{ for all } x \in [0,1]\}$. Prove that there is no topology on $C[0,1]$ that, together with the linear structure of $C[0,1]$, makes $C[0,1]$ into a topological vector space in which the set U is compact. (**T.19**) [V. Totik]

7. Prove that the series $\sum_p c_p f(px)$, where the summation is over all primes, unconditionally converges in $L^2[0,1]$ for every 1-periodic function f whose restriction to $[0,1]$ is in $L^2[0,1]$ if and only if $\sum_p |c_p| < \infty$. (*Unconditional convergence* means convergence for all rearrangements.) (**F.41**) [G. Halász]

8. Let $a_0 = 0, a_1, \ldots, a_k$ and $b_0 = 0, b_1, \ldots, b_k$ be arbitrary real numbers.

(i) Show that for all sufficiently large n there exist polynomials p_n of degree at most n for which

$$p_n^{(i)}(-1) = a_i, \quad p_n^{(i)}(1) = b_i, \qquad i = 0, 1, \ldots, k, \tag{1}$$

and

$$\max_{|x|\le 1} |p_n(x)| \le \frac{c}{n^2}, \tag{2}$$

where the constant c depends only on the numbers a_i, b_i.

(ii) Prove that, in general, (2) cannot be replaced by the relation

$$\lim_{n\to\infty} n^2 \cdot \max_{|x|\le 1} |p_n(x)| = 0. \tag{3}$$

(**F.42**) [J. Szabados]

9. Consider a latticelike packing of translates of a convex region K. Let t be the area of the fundamental parallelogram of the lattice defining the packing, and let $t_{\min}(K)$ denote the minimal value of t taken for all latticelike packings. Is there a natural number N such that for any $n > N$ and for any K different from a parallelogram, $nt_{\min}(K)$ is smaller than the area of any convex domain in which n translates of K can be placed without overlapping? (By a *latticelike packing* of K we mean a set of nonoverlapping translates of K obtained from K by translations with all vectors of a lattice.) (**G.40**) [G. and L. Fejes–Tóth]

10. Let $X_1, X_2, \ldots$ be independent, identically distributed random variables such that $X_i \ge 0$ for all i. Let $EX_i = m$, $\mathrm{Var}(X_i) = \sigma^2 < \infty$. Show that, for all $0 < \alpha \le 1$,

$$\lim_{n\to\infty} n \,\mathrm{Var}\left(\left[\frac{X_1 + \cdots + X_n}{n}\right]^\alpha\right) = \frac{\alpha^2\sigma^2}{m^{2(1-\alpha)}}.$$

(**P.26**) [Gy. Michaletzki]

1987

1. Let us color the integers $1, 2, \ldots, N$ with three colors so that each color is given to more than $N/4$ integers. Show that the equation $x = y + z$ has a solution in which x, y, z are of distinct colors. (**C.22**) [Gy. Szekeres]

2. A binary relation $\prec$ is called a *quasi-order* if it is reflexive and transitive. The infimum of the quasi-order $(Q, \prec)$ is the greatest subset $J \subseteq Q$ such that

(i) for every $B \in Q$ there is an $A \in J$ with $A \prec B$, and

(ii) $A \prec B$, $A, B \in J$ imply $B \prec A$.

Let X be a finite, nonempty alphabet, let X^* be the set of all finite words from X, and let $\mathcal{P}$ be the set of infinite subsets of X^*. For $A, B \in \mathcal{P}$, let $A \prec B$ if every element of A is a (connected) subword of some element of B. Show that $(\mathcal{P}, \prec)$ has an infimum, and characterize its elements. (**ℵ.10**) [Gy. Pollák]

3. Let A be a finite simple groupoid such that every proper subgroupoid of A has cardinality one, the number of one-element subgroupoids is at least three, and the group of automorphisms of A has no fixed points. Prove that in the variety generated by A, every finitely generated free algebra is isomorphic to some direct power of A. (**A.43**) [Á. Szendrei]
4. Let the finite projective geometry P (that is, a finite, complemented, modular lattice) be a sublattice of the finite modular lattice L. Prove that P can be embedded in a projective geometry Q, which is a cover-preserving sublattice of L (that is, whenever an element of Q covers in Q another element of Q, then it also covers that element in L). (**A.44**) [E. Fried]
5. Let f and g be continuous real functions, and let $g \not\equiv 0$ be of compact support. Prove that there is a sequence of linear combinations of translates of g that converges to f uniformly on compact subsets of $\mathbb{R}$. (**F.43**) [Sz. Gy. Révész, V. Totik]
6. Is it true that if A and B are unitarily equivalent, self-adjoint operators in the complex Hilbert space $\mathcal{H}$, and $A \leq B$, then $A^+ \leq B^+$? (Here A^+ stands for the positive part of A.) (**O.6**) [L. Kérchy]
7. Let $x : [0, \infty) \to \mathbb{R}$ be a differentiable function satisfying the identity

$$x'(t) = -2x(t)\sin^2 t + (2 - |\cos t| + \cos t) \int_{t-1}^{t} x(s) \sin^2 s \, ds$$

on $[1, \infty)$. Prove that x is bounded on $[0, \infty)$ and that $\lim_{t\to\infty} x(t) = 0$. Does the conclusion remain true for functions satisfying the identity

$$x'(t) = -2x(t)t + (2 - |\cos t| + \cos t) \int_{t-1}^{t} x(s) s \, ds \, ?$$

(**F.44**) [L. Hatvani]

8. Let $c > 0$, $c \neq 1$ be a real number, and for $x \in (0, 1)$ let us define the function

$$f(x) = \prod_{k=0}^{\infty} (1 + cx^{2^k}).$$

Prove that the limit

$$\lim_{x\to 1-0} \frac{f(x^3)}{f(x)}$$

does not exist. (**F.45**) [V. Totik]

9. Show that there exists a constant c_k such that for any finite subset V of the k-dimensional unit sphere there is a connected graph G such that the set of vertices of G coincides with V, the edges of G are straight line

segments, and the sum of the kth powers of the lengths of the edges is less than c_k. (**G.41**) [V. Totik]

10. Let F be a probability distribution function symmetric with respect to the origin such that $F(x) = 1 - x^{-1}K(x)$ for $x \geq 5$, where

$$K(x) = \begin{cases} 1 & \text{if } x \in [5, \infty) \setminus \cup_{n=5}^{\infty}(n!, 4n!), \\ \frac{x}{n!} & \text{if } x \in (n!, 2n!], \quad n \geq 5, \\ 3 - \frac{x}{2n!} & \text{if } x \in (2n!, 4n!), \quad n \geq 5. \end{cases}$$

Construct a subsequence $\{n_k\}$ of natural numbers such that if $X_1, X_2, \ldots$ are independent, identically distributed random variables with distribution function F, then for all real numbers x

$$\lim_{x \to \infty} P\left\{ \frac{1}{n_k} \sum_{j=1}^{n_k} X_j < \pi x \right\} = \frac{1}{2} + \frac{1}{\pi} \arctan x.$$

(**P.27**) [S. Csörgő]

1988

1. Define a partial order on all functions $f : \mathbb{R} \to \mathbb{R}$ by the relation $f \prec g$ if $f(x) \leq g(x)$ for all $x \in \mathbb{R}$. Show that this partially ordered set contains a totally ordered subset of size greater than $2^{\aleph_0}$ but that the latter subset cannot be well-ordered. (**ℵ.11**) [P. Komjáth]

2. Suppose that a graph G is the union of three trees. Is it true that G can be covered by two planar graphs? (**C.23**) [L. Pyber]

3. Let G be a finite Abelian group and $x, y \in G$. Suppose that the factor group of G with respect to the subgroup generated by x and the factor group of G with respect to the subgroup generated by y are isomorphic. Prove that G has an automorphism that maps x to y. (**A.45**) [E. Lukács]

4. Let Φ be a family of real functions defined on a set X such that $k \circ h \in \Phi$ whenever $f_i \in \Phi$ $(i \in I)$ and $h : X \to \mathbb{R}^I$ is defined by the formula $h(x)_i = f_i(x)$, and

(1) k: $h(X) \to \mathbb{R}$ is continuous with respect to the topology inherited from the product topology of $\mathbb{R}^I$. Show that $f = \sup\{g_j : j \in J, g_j \in \Phi\} = \inf\{h_m : m \in M, h_m \in \Phi\}$ implies $f \in \Phi$. Does this statement remain true if (1) is replaced with the following condition?

(2) k: $\overline{h(X)} \to \mathbb{R}$ is continuous on the closure of $h(X)$ in the product topology.

(**T.20**) [Á. Császár]

5. Let us draw a circular disc of radius r around every integer point in the plane different from the origin. Let E_r be the union of these discs, and denote by d_r the length of the longest segment starting from the origin and not intersecting E_r. Show that

$$\lim_{r \to 0} (d_r - \frac{1}{r}) = 0.$$

(**G.42**) [M. Laczkovich]

6. Let $H \subset \mathbb{R}$ be a bounded, measurable set of positive Lebesgue measure. Prove that

$$\liminf_{t\to 0} \frac{\lambda((H+t)\setminus H)}{|t|} > 0,$$

where $H+t = \{x+t : x \in H\}$ and λ is the Lebesgue measure. (**M.19**) [M. Laczkovich]

7. Let S be the set of real numbers q such that there is exactly one 0–1 sequence $\{a_n\}$ satisfying

$$\sum_{n=1}^{\infty} a_n q^{-n} = 1.$$

Prove that the cardinality of S is $2^{\aleph_0}$. (**S.26**) [P. Erdős, I. Joó]

8. Let f and g be holomorphic functions on the open unit disc D, and suppose that $|f|^2 + |g|^2 \in \mathrm{Lip}1$. Prove that then $f, g \in \mathrm{Lip}\frac{1}{2}$.
A function $h : D \to \mathbb{C}$ is in the Lipα class if there is a constant K such that

$$|h(z) - h(w)| \le K|z-w|^{\alpha}$$

for every $z, w \in D$. (**F.46**) [L. Lempert]

9. We say that the point (a_1, a_2, a_3) is above (below) the point (b_1, b_2, b_3) if $a_1 = b_1$, $a_2 = b_2$ and $a_3 > b_3$ ($a_3 < b_3$). Let $e_1, e_2, \ldots, e_{2k}$ $(k \ge 2)$ be pairwise skew lines not parallel with the z-axis, and assume that among their orthogonal projections to the (x,y)-plane no two are parallel and no three are concurrent. Is it possible that going along any of the lines the points that are below or above a point of some other line e_i alternately follow one another? (**G.43**) [J. Pach]

10. Let $a \in \mathbb{C}$, $|a| \le 1$. Find all values of $b \in \mathbb{C}$ for which there exist probability measures with characteristic function ϕ satisfying $\phi(2) = a$ and $\phi(1) = b$. (**P.28**) [T. F. Móri]

1989

1. Let p be an arbitrary prime number. In the ring G of Gaussian integers, consider the subrings

$$A_n = \{pa + p^n bi : a, b \in \mathbb{Z}\}, \quad n = 1, 2, \ldots .$$

Let $R \subset G$ be a subring of G that contains A_{n+1} as an ideal for some n. Prove that this implies that one of the following statements must hold: $R = A_{n+1}$; $R = A_n$; or $1 \in R$. (**A.46**) [R. Wiegandt]

2. Let $n > 2$ be an integer, and let Ω_n denote the semigroup of all mappings $g : \{0,1\}^n \to \{0,1\}^n$. Consider the mappings $f \in \Omega_n$, which have the following property: there exist mappings $g_i : \{0,1\}^2 \to \{0,1\}$ $(i = 1, 2, \ldots, n)$ such that for all $(a_1, a_2, \ldots, a_n) \in \{0,1\}^n$,

$$f(a_1, a_2, \ldots, a_n) = (g_1(a_n, a_1), g_2(a_1, a_2), \ldots, g_n(a_{n-1}, a_n)).$$

Let Δ_n denote the subsemigroup of Ω_n generated by these f's. Prove that Δ_n contains a subsemigroup Γ_n such that the complete transformation semigroup of degree n is a homomorphic image of Γ_n. (**A.47**) [P. Dömösi]

3. Let $n_1 < n_2 < \dots$ be an infinite sequence of natural numbers such that $n_k^{1/2^k}$ tends to infinity monotone increasingly. Prove that $\sum_{k=1}^{\infty} 1/n_k$ is irrational. Show that this statement is best possible in a sense by giving, for every $c > 0$, an example of a sequence $n_1 < n_2 < \dots$ such that $n_k^{1/2^k} > c$ for all k but $\sum_{k=1}^{\infty} 1/n_k$ is rational. (**N.19**) [P. Erdős]

4. Cancelled.

5. Characterize the sets $A \subset \mathbb{R}$ for which

$$A + B = \{a + b : a \in A,\ b \in B\}$$

is nowhere-dense whenever $B \subset \mathbb{R}$ is a nowhere dense set. (**T.21**) [M. Laczkovich]

6. Find all functions $f : \mathbb{R}^3 \to \mathbb{R}$ that satisfy the parallelogram rule

$$f(x + y) + f(x - y) = 2f(x) + 2f(y), \qquad x, y \in \mathbb{R}^3,$$

and that are constant on the unit sphere of $\mathbb{R}^3$. (**F.47**) [Gy. Szabó]

7. Let K be a compact subset of the infinite-dimensional, real, normed linear space $(X, \|\cdot\|)$. Prove that K can be obtained as the set of all left limit points at 1 of a continuous function $g : [0,1[\to X$, that is, x belongs to K if and only if there exists a sequence $t_n \in [0,1[$ $(n = 1, 2, \dots)$ satisfying $\lim_{n\to\infty} t_n = 1$ and $\lim_{n\to\infty} \|g(t_n) - x\| = 0$. (**O.7**) [B. Garay]

8. For any fixed positive integer n, find all infinitely differentiable functions $f : \mathbb{R}^n \to \mathbb{R}$ satisfying the following system of partial differential equations:

$$\sum_{i=1}^{n} \partial_i^{2k} f = 0, \qquad k = 1, 2, \dots .$$

(**F.48**) [L. Székelyhidi]

9. Suppose that HTM is a direct complement to a vertical bundle VTM over the total space of the tangent bundle TM of the manifold M. Let v and h denote the projections corresponding to the decomposition $TTM = VTM \oplus HTM$. Construct a bundle involution $P : TTM \to TTM$ such that $P \circ h = v \circ P$ and prove that, for any pseudo-Riemannian metric given on the bundle VTM, there exists a unique metric connection ∇ such that

$$\nabla_X PY - \nabla_Y PX = P \circ h[X, Y]$$

if X and Y are sections of the bundle HTM, and

$$\nabla_X Y - \nabla_Y X = [X, Y]$$

if X and Y are sections of the bundle VTM. (**G.44**) [J. Szilasi]

10. Let $Y(k)$, $k = 1, 2, \ldots$ be an m-dimensional stationary Gauss–Markov process with zero expectation, that is, suppose that

$$Y(k+1) = A\,Y(k) + \varepsilon(k+1), \qquad k = 1, 2, \ldots$$

Let H_i denote the hypothesis $A = A_i$, and let $P_i(0)$ be the a priori probability of H_i, $i = 0, 1, 2$. The a posteriori probability $P_1(k) = P(H_1|Y(1), \ldots, Y(k))$ of hypothesis H_1 is calculated using the assumptions $P_1(0) > 0,\ P_2(0) > 0,\ P_1(0) + P_2(0) = 1$.
Characterize all matrices A_0 such that $P\{\lim_{k\to\infty} P_1(k) = 1\} = 1$ if H_0 holds. (**P.29**) [I. Fazekas]

1990

1. Let A be a finite set of points in the Euclidean space of dimension $d \geq 2$. For $j = 1, 2, \ldots, d$, let B_j denote the orthogonal projection of A onto the $(d-1)$-dimensional subspace given by the equation $x_j = 0$. Prove that

$$\prod_{j=1}^{d} |B_j| \geq |A|^{d-1}.$$

(**G.45**) [I. Z. Ruzsa]

2. Prove that for every positive number K, there are infinitely many positive integers m and N such that there are at least $KN/\log N$ primes among the integers $m+1, m+4, \ldots, m+N^2$. (**N.20**) [I. Z. Ruzsa]

3. Let $n = p^k$ (p a prime number, $k \geq 1$), and let G be a transitive subgroup of the symmetric group S_n. Prove that the order of the normalizer of G in S_n is at most $|G|^{k+1}$. (**A.48**) [L. Pyber]

4. Let P be a polynomial with all real roots that satisfies the condition $P(0) > 0$. Prove that if m is a positive odd integer, then

$$\sum_{k=0}^{m-1} \frac{f^{(k)}(0)}{k!} x^k > 0$$

for all real numbers x, where $f = P^{-m}$. (**F.49**) [J. Szabados]

5. We say that the real numbers x and y can be connected by a δ-chain of length k (where $\delta : \mathbb{R} \to (0, \infty)$ is a given function) if there exist real numbers $x_0, x_1, \ldots, x_k$ such that $x_0 = x$, $x_k = y$, and

$$|x_i - x_{i-1}| < \delta\left(\frac{x_{i-1} + x_i}{2}\right), \quad i = 1, \ldots, k.$$

Prove that for every function $\delta : \mathbb{R} \to (0, \infty)$ there is an interval in which any two elements can be connected by a δ-chain of length 4. Also, prove that we cannot always find an interval in which any two elements could be connected by a δ-chain of length 2. (**F.50**) [M. Laczkovich]

6. Find meromorphic functions ϕ and ψ in the unit disc such that, for any function f regular in the unit disc, at least one of the functions $f - \phi$ and $f - \psi$ has a root. (**F.51**) [G. Halász]

7. Denote by $B[0,1]$ and $C[0,1]$ the Banach space of all bounded functions and all continuous functions, respectively, on the interval $[0,1]$ with the supremum norm. Is there a bounded linear operator

$$T : B[0,1] \to C[0,1]$$

such that $Tf = f$ for all $f \in C[0,1]$? (**O.8**) [G. Halász]

8. Let $A_1^{(0)}, \dots, A_n^{(0)}$ be a sequence of $n \geq 3$ points in the Euclidean plane $\mathbb{R}^2$. Define the sequence $A_1^{(i)}, \dots, A_n^{(i)}$ ($i = 1, 2, \dots$) by induction as follows: let $A_j^{(i)}$ be the midpoint of the segment $A_j^{(i-1)}A_{j+1}^{(i-1)}$, where $A_{n+1}^{(i-1)} = A_1^{(i-1)}$. Show that, with the exception of a set of zero Lebesgue measure, for every initial sequence $(A_1^{(0)}, \dots, A_n^{(0)}) \in (\mathbb{R}^2)^n$, there exists a natural number N such that the points $A_1^{(N)}, \dots, A_n^{(N)}$ are consecutive vertices of a convex n-gon. (**G.46**) [B. Csikós]

9. Prove that if all subspaces of a Hausdorff space X are σ-compact, then X is countable. (**T.22**) [I. Juhász]

10. Let X and Y be independent identically distributed, real-valued random variables with finite expectation. Prove that

$$E|X + Y| \geq E|X - Y|.$$

(**P.30**) [T. F. Móri]

1991

1. To divide a heritage, n brothers turn to an impartial judge (that is, if not bribed, the judge decides correctly, so each brother receives $(1/n)$th of the heritage). However, in order to make the decision more favorable for himself, each brother wants to influence the judge by offering an amount of money. The heritage of an individual brother will then be described by a continuous function of n variables strictly monotone in the following sense: it is a monotone increasing function of the amount offered by him and a monotone decreasing function of the amount offered by any of the remaining brothers. Prove that if the eldest brother does not offer the judge too much, then the others can choose their bribes so that the decision will be correct. (**F.52**) [V. Totik]

2. Suppose that n points are given on the unit circle so that the product of the distances of any point of the circle from these points is not greater than two. Prove that the points are the vertices of a regular n-gon. (**G.47**) [L. I. Szabó]

3. Prove that if a finite group G is an extension of an Abelian group of exponent 3 with an Abelian group of exponent 2, then G can be embedded in some finite direct power of the symmetric group S_3. (**A.49**) [G. Czédli, B. Csákány]

4. Let $n \geq 2$ be an integer, and consider the groupoid $G = (\mathbb{Z}_n \cup \{\infty\}, \circ)$, where

$$x \circ y = \begin{cases} x+1 & \text{if } x = y \in \mathbb{Z}_n, \\ \infty & \text{otherwise.} \end{cases}$$

($\mathbb{Z}_n$ denotes the ring of the integers modulo n.) Prove that G is the only subdirectly irreducible algebra in the variety generated by G. (**A.50**) [Á. Szendrei]

5. Construct an infinite set $H \subseteq C[0,1]$ such that the linear hull of any infinite subset of H is dense in $C[0,1]$. (**F.53**) [V. Totik]

6. Let $\alpha > 0$ be irrational.

(a) Prove that there exist real numbers a_1, a_2, a_3, a_4 such that the function $f : \mathbb{R} \to \mathbb{R}$,

$$f(x) = e^x[a_1 + a_2 \sin x + a_3 \cos x + a_4 \cos(\alpha x)]$$

is positive for all sufficiently large x, and

$$\liminf_{x \to +\infty} f(x) = 0.$$

(b) Is the above statement true if $a_2 = 0$?

(**F.54**) [T. Krisztin]

7. Given $a_n \geq a_{n+1} > 0$ and a natural number μ, such that

$$\limsup_n \frac{a_n}{a_{\mu n}} < \mu,$$

prove that for all $\varepsilon > 0$ there exist natural numbers N and n_0 such that, for all $n > n_0$ the following inequality holds:

$$\sum_{k=1}^{n} a_k \leq \varepsilon \sum_{k=1}^{Nn} a_k .$$

(**S.27**) [L. Leindler]

8. Prove that if $\{a_k\}$ is a sequence of real numbers such that

$$\sum_{k=1}^{\infty} |a_k|/k = \infty \quad \text{and} \quad \sum_{n=1}^{\infty} \left(\sum_{k=2^{n-1}}^{2^n - 1} k(a_k - a_{k+1})^2 \right)^{1/2} < \infty,$$

then

$$\int_0^{\pi} \left| \sum_{k=1}^{\infty} a_k \sin(kx) \right| dx = \infty.$$

(**F.55**) [F. Móricz]

9. Let $h : [0, \infty) \to [0, \infty)$ be a measurable, locally integrable function, and write

$$H(t) := \int_0^t h(s) ds \qquad (t \geq 0).$$

Prove that if there is a constant B with $H(t) \leq Bt^2$ for all t, then

$$\int_0^\infty e^{-H(t)} \int_0^t e^{H(u)} du\, dt = \infty.$$

(**F.56**) [L. Hatvani, V. Totik]

10. Consider the equation $f'(x) = f(x+1)$. Prove that

(a) each solution $f : [0,\infty) \to (0,\infty)$ has an exponential order of growth, that is, there exist numbers $a > 0$, $b > 0$ satisfying $|f(x)| \leq a\, e^{bx}$, $x \geq 0$;

(b) there are solutions $f : [0,\infty) \to (-\infty,\infty)$ of nonexponential order of growth.

(**F.57**) [T. Krisztin]

11. Does there exist a bounded linear operator T on a Hilbert space H such that

$$\bigcap_{n=1}^{\infty} T^n(H) = \{0\} \quad \text{but} \quad \bigcap_{n=1}^{\infty} T^n(H)^{-} \neq \{0\},$$

where $^-$ denotes closure? (**O.9**) [L. Kérchy]

12. Let $X_1, X_2, \ldots$ be independent, identically distributed random variables such that, for some constant $0 < \alpha < 1$,

$$P\left\{X_1 = 2^{k/\alpha}\right\} = 2^{-k}, \quad k = 1, 2, \ldots$$

Determine, by giving their characteristic functions or any other way, a sequence of infinitely divisible, nondegenerate distribution functions G_n such that

$$\sup_{-\infty<x<\infty} \left| P\left\{\frac{X_1 + \cdots + X_n}{n^{1/\alpha}} \leq x\right\} - G_n(x)\right| \to 0 \quad \text{as} \quad n \to \infty.$$

(**P.31**) [S. Csörgő]

2. Results of the Contests

SUMMARY

Year of contest	Number of problems	Number of competitors	Number of correct solutions
1962	10	23	50
1963	10	56	402
1964	10	34	188
1965	10	52	321
1966	10	43	333
1967	10	34	258
1968	11	37	169
1969	12	33	191
1970	12	27	161
1971	11	26	159
1972	11	18	95
1973	10	35	184
1974	10	14	47
1975	12	63	441
1976	11	17	104
1977	10	30	138
1978	10	25	74
1979	11	34	84
1980	10	19	64
1981	10	42	137
1982	10	29	56
1983	12	28	138
1984	10	17	41
1985	12	34	183
1986	10	20	62
1987	10	21	84
1988	10	28	128
1989	10	40	224
1990	10	17	46
1991	12	17	74

List of Prize Winners and Honorably Mentioned Competitors

(Notations: *1. First prize, 2. Second prize, 3. Third prize, H. Honorably mentioned.*)

1962

1. Gábor Halász
2. –
3. –
H. Béla Bollobás, Árpád Elbert, István Juhász, György Petruska, Domokos Szász

1963

1. Árpád Elbert
2. Gyula Katona, Domokos Szász
3. –
H. Gábor Halász, János Komlós, Miklós Simonovits, József Szűcs

1964

1. Béla Bollobás, Péter Vámos
2. Gábor Halász, István Juhász, Gerzson Kéry, Miklós Simonovits, Domokos Szász
3. –

1965

1. Béla Bollobás, József Fritz, Miklós Simonovits
2. Gerzson Kéry, Attila Máté, József Pelikán
3. –
H. László Gerencsér, Előd Knuth, László Lovász, Lajos Pósa, György Vesztergombi

1966

1. Béla Bollobás, Endre Makai, Miklós Simonovits
2. László Lovász, Lajos Pósa
3. László Gerencsér, Miklós Laczkovich, József Pelikán, Ferenc Szigeti

1967

1. László Lovász, Attila Máté
2. László Gerencsér, Eörs Máté
3. –
H. László Babai, Róbert Freud, Miklós Laczkovich, Miklós Simonovits, Ferenc Szigeti

1968

1. Péter Gács, László Lovász, Endre Makai
2. Miklós Laczkovich, Lajos Pósa
3. László Babai

1969

1. Péter Gács, László Lovász
2. Miklós Laczkovich
3. László Babai, Endre Makai, József Pelikán, Lajos Pósa, Imre Z. Ruzsa

1970

1. László Lovász
2. László Babai, Miklós Laczkovich
3. Péter Gács, Endre Makai, József Pelikán, Lajos Pósa, Imre Z. Ruzsa

1971

1. Zsigmond Nagy, László Babai
2. Lajos Pósa
3. József Pelikán, Imre Z. Ruzsa, Jenő Deák

1972

1. Imre Z. Ruzsa
2. Zsigmond Nagy
3. László Babai, Péter Frankl, János Pintz

1973

1. László Babai, Péter Komjáth
2. János Pintz
3. Péter Frankl, Zsigmond Nagy, Imre Z. Ruzsa
H. Ervin Bajmóczi, Zoltán Balogh, József Beck, Ervin Győri, Emil Kiss, Tamás Móri, Zsolt Tuza, Endre Boros, László Lempert, János Revizcky

1974

1. László Lempert, Imre Z. Ruzsa
2. Péter Frankl
3. Ervin Győri

1975

1. Imre Z. Ruzsa
2. Péter Komjáth, László Lempert, Vilmos Totik
3. Zoltán Füredi, Gábor Somorjai
H. Ervin Bajmóczi, Ervin Győri, Tamás Móri, Péter Pál Pálfy, Mária Szendrei

1976

1. Imre Z. Ruzsa
2. Péter Komjáth
3. –
H. Vilmos Totik, Ervin Bajmóczi, Zoltán Füredi, Mihály Geréb, Ervin Győri, Emil Kiss, János Kollár

1977

1. Ferenc Göndőcs, Vilmos Totik
2. Zoltán Füredi
3. Gábor Czédli, Péter Pál Pálfy
H. Tamás Bara, Vilmos Komornik, András Sebő, Nándor Simányi, Sándor Veres

1978

1. Vilmos Totik
2. Zoltán Füredi, János Kollár, Nándor Simányi
3. Emil Kiss
H. Tibor Krisztin, Zoltán Magyar

1979

1. János Kollár, Péter Pál Pálfy
2. –
3. Mihály Geréb, Gábor Ivanyos, Ákos Seress
H. Emil Kiss, Tibor Krisztin, Zsolt Páles, Nándor Simányi

1980

1. János Kollár
2. Ákos Seress
3. Nándor Simányi
H. Gábor Ivanyos, Zoltán Magyar

1981

1. Gábor Ivanyos, Zoltán Magyar
2. Balázs Csikós, Ákos Seress
3. Péter Hajnal, Dezső Miklós, Márió Szegedy
H. Gábor Moussong, Zalán Bodó, András Zempléni

1982

1. Zoltán Magyar, Gábor Tardos
2. Ákos Seress
3. Balázs Csikós, András Szenes
H. Péter Hajnal, Dezső Miklós, Márió Szegedy, András Zempléni

1983

1. Zoltán Magyar
2. Balázs Csikós, Gábor Tardos
3. Márió Szegedy, József Varga
H. Péter Hajnal, Zoltán Buczolich, András Szenes, Ákos Seress

1984

1. Gábor Tardos
2. Márió Szegedy
3. Zoltán Buczolich
H. Balázs Csikós, András Szenes

1985

1. Gábor Tardos
2. András Szenes
3. Géza Bohus, Gábor Elek, Gyula Károlyi
H. Ferenc Beleznay, László Erdős, Miklós Mócsy, Tibor Ódor, Endre Szabó, László Szabó, Zoltán Szabó

1986

1. Gábor Tardos
2. Gábor Elek, Endre Szabó, Zoltán Szabó
3. László Erdős, Jenő Törőcsik
H. Gyula Károlyi, Ákos Magyar

1987

1. Ákos Magyar
2. István Sigray
3. Gyula Károlyi
H. László Erdős, Gábor Hetyei, Sándor Kovács, Jenő Törőcsik

1988

1. Géza Kós, Zoltán Szabó
2. László Erdős
3. István Sigray, Jenő Törőcsik
H. Sándor Kovács, Ákos Magyar, András Benczúr, Tamás Keleti, Miklós Mócsy

1989

1. László Erdős
2. Sándor Kovács
3. –
H. András Benczúr, György Birkás, András Bíró, Gábor Drasny, Géza Kós, Ákos Magyar, László Majoros, Miklós Mócsy, Tibor Szabó, Zoltán Szabó

1990

1. András Benczúr
2. Gábor Drasny
3. András Bíró
H. Tamás Hausel, Géza Makay

1991

1. András Bíró
2. –
3. Tamás Fleiner, Géza Kós
H. Mátyás Domokos, Gábor Hajdú, Gergely Harcos, Tamás Keleti, Vu Ha Van

3. Solutions to the Problems

3.1 ALGEBRA

Problem A.1. *Determine the roots of unity in the field of p-adic numbers.*

Solution. The p-adic number $a_{-m}p^{-m} + \cdots + a_{-1}p^{-1} + a_0 + a_1p + \cdots + a_np^n + \cdots \quad (0 \leqslant a_i < p)$ is a p-adic integer iff all the coefficients with negative index are equal to 0, and it is a unit in the ring of p-adic integers iff furthermore $a_0 \neq 0$. It is clear that every p-adic number α can be written in the form $\alpha = \beta p^r$, where β is a p-adic unit and r an integer. Since the product of p-adic units is again a p-adic unit, we deduce that a p-adic number can be a root of unity only in the case when it is a p-adic unit.

Every root of unity can be written in the form $\varepsilon = a + \alpha p^r$, where r is a positive integer, α is a p-adic unit, and $0 < a < p$. Now if $\varepsilon^n = 1$, then $a^n \equiv 1 \pmod p$ holds true. So the exponent of a is a divisor of n.

Consider the case $p \neq 2$. Let $\exp(a) = k$; we are going to show $k = n$. Let $\beta = \varepsilon^k$, and suppose $\beta \neq 1$. Then β is of the form $\beta = 1 + \gamma p^s$, where γ is a p-adic unit. If $k \neq n$ were the case, then some power of β would be 1. To show the impossibility of this, it is enough to show that a power with prime exponent of a number of the form $\beta = 1 + \gamma p^s$ (γ a p-adic unit) cannot be 1. Let q be a prime number; using the binomial theorem, we get

$$\beta^q = 1 + q\gamma p^s + \delta p^{2s} = 1 + \varphi p^s \qquad \text{if } q \neq p;$$

while in the case $q = p$, using $p > 2$ we get

$$\beta^p = 1 + \gamma p^{s+1} + \delta p^{2s+1} = 1 + \varphi p^{s+1},$$

where γ is a p-adic integer, whereas φ, as we can easily see, is a p-adic unit. This guarantees that β^q is different from 1. We have proved that in the field of p-adic numbers every root of unity is necessarily a $(p-1)$th root of unity. Now we show that such roots of unity indeed exist.

Let g_0 be a primitive root of unity mod p. We are going to prove that it is possible to determine numbers $0 \leqslant a_i < p$ in such a way that the p-adic number

$$\varepsilon = g_0 + a_1p + \cdots + a_np^n + \cdots$$

should be a $(p-1)$th primitive root of unity. It is clear that any power of ε with exponent less than $p-1$ is different from 1. So it is enough to show that for a suitable choice of the numbers a_j, the natural numbers $g_j = g_0 + a_1 p + \cdots + a_j p^j$ satisfy $g_j^{p-1} \equiv 1 \pmod{p^{j+1}}$. This will be proved by induction. For $j = 0$ the statement is true, as g_0 is a primitive root. Suppose that the statement holds for some j, that is,

$$g_j^{p-1} = 1 + c_j p^{j+1}.$$

Let a_{j+1} be the solution of the congruence $g_0 x \equiv c_j \pmod{p}$. Then

$$\begin{aligned} g_{j+1}^{p-1} &\equiv (g_j + a_{j+1}p^{j+1})^{p-1} \equiv 1 + c_j p^{j+1} - g_j a_{j+1} p^{j+1} \\ &\equiv 1 + (c_j - g_j a_{j+1}) p^{j+1} \equiv 1 \pmod{p^{j+2}}. \end{aligned}$$

Therefore ε and its powers are different $(p-1)$th roots of unity; their number is $p-1$. There can be no other roots of unity since the polynomial $x^{p-1}-1$ can have at most $p-1$ roots in a commutative field.

Let us now deal with the case $p = 2$. Similarly to the case of odd prime numbers, we can prove that a power with an odd exponent of a dyadic number of the form $1 + \cdots$ can be 1 only in case the number itself is 1. So if a dyadic number is a primitive n-th root of unity then n can have no odd prime divisors. There are two second roots of unity, 1 and the number $-1 = 1 + 2 + 2^2 + \cdots 2^r + \cdots$ as well. There can be no more second roots of unity, because the polynomial $x^2 - 1$ can have at most two roots in a commutative field.

We show that there are no other roots of unity in the field of dyadic numbers. Suppose there were another root of unity; then this would be necessarily a primitive root of unity belonging to some power of 2, so there would surely exist a primitive further root of unity, that is, a dyadic number η with $\eta^2 = -1$. Clearly, η must have the form $\eta = 1 + 2^r + \cdots$, where $r \geqslant 1$. This gives

$$-1 \equiv (1+2^r)^2 \equiv 1 + 2^{r+1} + 2^{2r} \pmod{2^{r+2}}.$$

Therefore,

$$-2 \equiv 2^{r+1} + 2^{2r} \pmod{4}, \quad \text{that is,} \quad 2^r + 2^{2r-1} \equiv 1 \pmod{2},$$

which is clearly impossible. This concludes the proof. □

Problem A.2. *Let A and B be two Abelian groups, and define the sum of two homomorphisms η and χ from A to B by*

$$a(\eta + \chi) = a\eta + a\chi \quad \text{for all } a \in A.$$

With this addition, the set of homomorphisms from A to B forms an Abelian group H. Suppose now that A is a p-group (p a prime number).

Prove that in this case H becomes a topological group under the topology defined by taking the subgroups $p^k H$ $(k = 1, 2, \dots)$ as a neighborhood base of 0. Prove that H is complete in this topology and that every connected component of H consists of a single element. When is H compact in this topology?

Solution. H is clearly a commutative group whose 0 element is the homomorphism mapping every element of A to $0 \in B$, and for $\eta \in H$, $-\eta$ is the mapping defined by $a(-\eta) = -a\eta$.

To prove that H is a topological group, we have to check the following:

(i) The intersection of the neighborhoods of 0 is 0 alone. Suppose namely that $\eta \in p^k H (k = 1, 2, \dots)$. As A is a p-group, the order of any $a \in A$ is of the form p^n for some n depending on a. Because $\eta \in p^n H$, we have $\eta = p^n \chi$ for some $\chi \in H$, and so

$$a\eta = a(p^n \chi) = (p^n a)\chi = 0\chi = 0,$$

proving $\eta = 0$.

(ii) To any neighborhood U of 0 there is a neighborhood V of 0 such that $V + (-V) \subseteq U$, a suitable choice being $V = U$.

(iii) If a is contained in some neighborhood U of 0, then there exists a neighborhood V of 0 such that $a + V \subseteq U$. Again the choice $V = U$ will do.

(iv) The intersection of any two neighborhoods of 0 contains a neighborhood of 0, because for any two neighborhoods, one of them contains the other.

We now come to the proof of completeness. We have to prove that if $\eta_1, \eta_2, \dots, \eta_n, \dots$ is a Cauchy sequence, then it converges to some element of H. By repeating members of the sequence, if necessary, we can assume that $\eta_{i+1} - \eta_i \in p^i H$, that is, $\eta_{i+1} = \eta_i + p^i \vartheta_i$. Furthermore, define η_0 as 0.

Now define mappings $\chi_i (i = 0, 1, 2, \dots)$ in the following way:
If the order of $a \in A$ is p^k then let

$$a\chi_i = a(\vartheta_i + p\vartheta_{i+1} + \cdots + p^{k-1}\vartheta_{i+k-1}) .$$

It is easy to check that χ_i is indeed a homomorphism from A to B and

$$\chi_i = \vartheta_i + p\chi_{i+1} .$$

We will show that the limit of the sequence $\eta_0, \eta_1, \dots, \eta_n, \dots$ is χ_0. For this it is enough to show $\chi_0 - \eta_i \in p^i H$, actually $\chi_0 - \eta_i = p^i \chi_i$, which we prove by induction. For $i = 0$, we have $\chi_0 - \eta_0 = \chi_0 - 0 = p^0 \chi_0$. Using our previous observations and the induction hypothesis, we have

$$\chi_0 - \eta_{i+1} = \chi_0 - \eta_i - p^i \vartheta_i = p^i \chi_i - p^i \vartheta_i = p^i (\chi_i - \vartheta_i) = p^{i+1} \chi_{i+1},$$

proving the completeness of H.

We now prove that the connected component of 0 consists of 0 alone; to prove this, it is enough to show that the intersection of all open-closed sets containing 0 is 0. As p^kH is an open subgroup, it is closed as well, and as the intersection of all of them (for $k = 1, 2, \dots$) is already 0 alone, we get the desired result.

For compactness we are going to prove the following result: H is compact if and only if the index of p^kH in H is finite for every k.

Necessity of this condition is easy to establish: the cosets $\chi + p^kH (\chi \in H)$ cover H. If H is compact, then already a finite number of them have to cover H, which is precisely what the condition says.

Suppose now that all the subgroups p^kH have finite index. We are going to show that if a family of closed subsets of H has the property that any finite number of them have a nonempty intersection, then the whole family has a nonempty intersection.

First, we show the following: let $U_\lambda (\lambda \in \Lambda)$ be subsets of the set H such that the intersection of any finite number of them is nonempty. Suppose that H is the disjoint union of two subsets S and T. Then at least one of the families $S \cap U_\lambda (\lambda \in \Lambda)$ and $T \cap U_\lambda (\lambda \in \Lambda)$ inherits the same intersection property. (In particular, either S or T meets all the U_λ.) Suppose that neither family inherited the property. Then we would have values $\lambda_1, \lambda_2, \dots, \lambda_r$ and $\lambda_{r+1}, \lambda_{r+2}, \dots, \lambda_{r+s}$ such that the intersection of the $S \cap U_{\lambda_i}$ and also of the $T \cap U_{\lambda_{r+j}}$ would be empty $(i = 1, 2, \dots, r,\ j = 1, 2, \dots, s)$. In other words, the sets

$$\left(\bigcap_{i=1}^{r} U_{\lambda_i}\right) \cap S \quad \text{and} \quad \left(\bigcap_{j=1}^{s} U_{\lambda_{r+j}}\right) \cap T$$

are empty, or equivalently

$$\left(\bigcap_{i=1}^{r} U_{\lambda_i}\right) \subseteq T \quad \text{and} \quad \left(\bigcap_{j=1}^{s} U_{\lambda_{r+j}}\right) \subseteq S.$$

Consequently,

$$\left(\bigcap_{i=1}^{r+s} U_{\lambda_i}\right) \subseteq T \cap S = \emptyset,$$

a contradiction. We then get the same type of statement for any decomposition of H into a finite number of pairwise disjoint subsets.

Let us now consider closed subsets $U_\lambda (\lambda \in \Lambda)$ of our topological group H satisfying the finite intersection condition. Repeatedly applying the above procedure, we see that there exists a sequence $\eta_1, \eta_2, \dots, \eta_k, \dots$ of elements of H for which

$$\eta_1 + pH \supseteq \eta_2 + p^2H \supseteq \cdots \supseteq \eta_k + p^kH \supseteq \dots \tag{1}$$

and such that for any k the intersection of the $U_\lambda (\lambda \in \Lambda)$ with $\eta_k + p^kH$ also satisfies the finite intersection condition; in particular, any U_λ and $\eta_k + p^kH$ have a common element $\chi_{k,\lambda}$.

In view of (1) and the completeness of H, the sequence $\eta_1, \eta_2, \ldots, \eta_k, \ldots$ has a limit η for which $\eta_k + p^k H = \eta + p^k H$. For any λ we have $\chi_{k,\lambda} \in \eta + p^k H$; thus the limit of the sequence $\chi_{1,\lambda}, \chi_{2,\lambda}, \ldots, \chi_{k,\lambda}, \ldots$ is η. Since U_λ is closed and $\chi_{k,\lambda} \in U_\lambda$, we get that η is contained in every U_λ, which completes the proof of the compactness of H. □

Problem A.3. *Let $R = R_1 \oplus R_2$ be the direct sum of the rings R_1 and R_2, and let N_2 be the annihilator ideal of R_2 (in R_2). Prove that R_1 will be an ideal in every ring $\bar{R}$ containing R as an ideal if and only if the only homomorphism from R_1 to N_2 is the zero homomorphism.*

Solution. First suppose there is a ring $\bar{R}$ containing R as an ideal and such that R_1 is not an ideal of $\bar{R}$. Since R is an ideal in $\bar{R}$, we have $R_1\bar{r} \subseteq R$ and $\bar{r}R_1 \subseteq R$ for every $\bar{r} \in \bar{R}$. Suppose for example, that $\bar{r}R_1 \subsetneq R_1$ for some $\bar{r} \in \bar{R}$. (The case $R_1\bar{r} \subsetneq R_1$ can be treated similarly.) Using the element $\bar{r}$ we will define a nonzero homomorphism from R_1 to N_2. The direct sum property implies that for every $r_1 \in R_1$ the element $\bar{r}r_1 \in R$ can be uniquely decomposed in the form $\bar{r}r_1 = g(r_1) + h(r_1)$, where $g(r_1) \in R_1$ and $h(r_1) \in R_2$. h will be the desired homomorphism. It is clearly defined for all elements of R_1. Furthermore,

(i) $h(r_1) \in N_2$ for every $r_1 \in R_1$.

Namely, for every $r_2 \in R_2$ we have

$$h(r_1)r_2 = g(r_1)r_2 + h(r_1)r_2 = (g(r_1) + h(r_1))r_2 = (\bar{r}r_1)r_2 = \bar{r}(r_1r_2) = 0$$

and

$$r_2h(r_1) = r_2(\bar{r}r_1) = (r_2\bar{r})r_1 \in RR_1 \subseteq R_1,$$

but on the other hand, $r_2h(r_1) \in R_2$, proving $r_2h(r_1) = 0$, which gives (i).

(ii) $h(r_1 + r_1') = h(r_1) + h(r_1')$ for every $r_1, r_1' \in R_1$.

Since

$$\begin{aligned}\bar{r}(r_1 + r_1') &= g(r_1 + r_1') + h(r_1 + r_1') = \bar{r}r_1 + \bar{r}r_1' \\ &= g(r_1) + h(r_1) + g(r_1') + h(r_1') = g(r_1) + g(r_1') + h(r_1) + h(r_1'),\end{aligned}$$

where $g(r_1) + g(r_1') \in R_1$ and $h(r_1) + h(r_1') \in R_2$, we get the validity of (ii).

(iii) $h(r_1r_1') = h(r_1)h(r_1')$ for every $r_1, r_1' \in R_1$.

We will show that both sides of the equality are equal to zero. For the right-hand side, this is clear because $h(r_1)$ is an annihilator of R_2 and $h(r_1') \in R_2$. For the left-hand side, we have to consider the product $\bar{r}(r_1r_1')$:

$$\bar{r}(r_1r_1') = (\bar{r}r_1)r_1' = (g(r_1) + h(r_1))r_1' = g(r_1)r_1' \in R_1,$$

so $h(r_1r_1') = 0$.

This establishes the first part of the statement.

For the reverse statement, we use the same idea. We assume that there is a nonzero homomorphism h from R_1 to N_2, and we construct the element $\bar{r}$ in such a way that it should produce h.

We denote by R_0 the zero ring over the additive group of the integers. The element of R_0 corresponding to $n \in \mathbb{Z}$ will be denoted by $\bar{n}$. The additive group of the ring $\bar{R}$ is defined by $\bar{R}^+ = R_0^+ \oplus R_1^+ \oplus R_2^+$, and for the elements of $\bar{R}$

$$r = \bar{n} + r_1 + r_2, \qquad r' = \bar{n'} + r_1' + r_2',$$

we define the multiplication by

$$rr' = r_1 r_1' + r_2 r_2' + nh(r_1') + n'h(r_1).$$

It is routine to check that $\bar{R}$ is a ring and R is a subring of it. In fact, R is even an ideal of $\bar{R}$ in view of $\bar{R}\bar{R} \subseteq R$.

To verify that R_1 is not an ideal of $\bar{R}$, take an $r_1 \in R_1$ with $h(r_1) \neq 0$. Then $\bar{1}r_1 = h(r_1) \notin R_1$ as $h(r_1) \in R_2$, and this shows that R_1 is not an ideal of $\bar{R}$. □

Problem A.4. *Call a polynomial positive reducible if it can be written as a product of two nonconstant polynomials with positive real coefficients. Let $f(x)$ be a polynomial with $f(0) \neq 0$ such that $f(x^n)$ is positive reducible for some natural number n. Prove that $f(x)$ itself is positive reducible.*

Solution. Consider the product representation of $f(x^n)$:

$$f(x^n) = \prod_{j=1}^{s} g_j(x), \tag{1}$$

where the $g_j(x)$ are polynomials with positive coefficients. Suppose that some of the polynomials $g_j(x)$ contain a term $c_{jk}x^k$ with nonvanishing c_{jk} such that $n \nmid k$. Then the product on the right-hand side of (1) contains the term $c_{jk} \prod_{i \neq j} g_i(0) x^k$. The coefficient of this term does not vanish, since $\prod_{i=1}^{s} g_i(0) = f(0) \neq 0$. Since the polynomials $g_j(x)$ all have positive coefficients, the right-hand side of (1) will contain the term x^k with a nonvanishing coefficient, which is a contradiction in view of $n \nmid k$.

This means that all the polynomials $g_j(x)$ have the form

$$g_j(x) = \sum_{i=0}^{l_j} b_{jl} x^{ln} \qquad (b_{jl} \geqslant 0, l = 0, 1, \ldots, l_j),$$

that is, $g_j(x)$ is actually a polynomial of x^n: $g_j(x) = \bar{g}_j(x^n)$. So upon substituting $x^n = y$, we get

$$f(y) = \prod_{j=1}^{s} \bar{g}_j(y),$$

proving the statement of the problem. □

Problem A.5. *Prove that the intersection of all maximal left ideals of a ring is a (two-sided) ideal.*

Solution. Let us denote by B the intersection of all maximal left ideals of the ring R (if there are no maximal left ideals, the statement is void). B is clearly a left ideal, so we only have to prove that $b \in B$ and $r \in R$ imply $br \in B$, or equivalently $br \notin B$ implies $b \notin B$. Since an element is not in B if and only if there is a maximal left ideal not containing it, we actually have to prove the following statement:

If for some elements $b \in B$, $r \in R$ there is a maximal left ideal M with $br \notin M$, then there exists a maximal left ideal N with $b \notin N$. Let

$$N = \{x \in R \mid xr \in M\}.$$

We are going to prove that N has the desired properties.

1. N is a left ideal: if $s \in R$ and $x \in N$, then $xr \in M$ implies $sxr \in M$ (M being a left ideal), which in turn implies $sx \in N$. Also $x \in N$, $y \in N$ clearly imply $x - y \in N$.
2. $b \notin N$ as $br \notin M$.
3. N is maximal. To prove this, we have to show that for any element a of R with $a \notin N$ the only left ideal of R containing both a and N is R itself. We have $ar \notin M$ since $a \notin N$. As M was maximal, every element of R is contained in the left ideal generated by ar and M; in particular, for every $c \in R$, cr can be written in the form

$$cr = yar + nar + m,$$

where $y \in R$, $m \in M$, and n is an integer. This implies

$$(c - ya - na)r = m \in M,$$

so

$$d = c - ya - na \in N,$$

but this shows that $c = ya+na+d$ is contained in the left ideal generated by a and N, which was to be proved. □

Remark. If R has an identity element, then B is of course the Jacobson radical.

Problem A.6. *Let R be a finite commutative ring. Prove that R has a multiplicative identity element (1) if and only if the annihilator of R is 0 (that is, $aR = 0$, $a \in R$ imply $a = 0$).*

Solution 1. If R has an identity element e, then $aR = 0$ $(a \in R)$ implies $0 = ae = a$, so the annihilator of R is indeed 0.

Conversely, suppose R is a finite commutative ring whose annihilator is 0. This implies that to any element a of R different from 0 there is an element b of R such that $ab \neq 0$. If $R = (0)$, then R surely has an identity element. So suppose $R \neq (0)$ and a_0 is an arbitrary element of R different from 0. Then the remark made above shows that for any natural number n we can find an element $a_n \in R$ such that the relations

$$a_0a_1 \neq 0,\ a_0a_1a_2 \neq 0, \dots,\ a_0a_1 \dots a_n \neq 0, \dots$$

hold true. Since R is finite, we have numbers m and n $(0 \leqslant m < n)$ such that

$$a_0a_1 \dots a_m = (a_0a_1 \dots a_m)(a_{m+1} \dots a_n)\,. \tag{1}$$

This means that the set E of elements $e \neq 0$ of the ring R for which there exists a $0 \neq d \in R$ such that

$$de = d \tag{2}$$

is nonempty. Choose an element e from E so that the number of elements $0 \neq d \in R$ corresponding to e that satisfy (2) should be maximal. (Such an e exists because R is finite.) We will show that e is an identity element of R.

Suppose, on the contrary, that for some $a \in R$ we have $ae - a \neq 0$. Then in view of (1) there are elements $r, s \in R$ such that

$$0 \neq (ae - a)r = (ae - a)rs \tag{3}$$

holds. (We allow r to be an empty product.)

Equation (3) implies

$$ar(e - es + s) = ar \neq 0 \tag{4}$$

consequently, $0 \neq e - es + s \in E$. If $de = d$, then

$$d(e - es + s) = de - des + ds = d - ds + ds = d; \tag{5}$$

furthermore,

$$are - ar = (ae - a)r \neq 0. \tag{6}$$

Equations (4), (5) and (6) together show that the element $e - es + s \in E$ satisfies condition (2) for more $d \in R$ than e. This contradiction shows the existence of an identity element in R. □

Solution 2. Suppose the commutative ring R with $n (\geqslant 2)$ elements has annihilator 0. Then R cannot be nilpotent. If R were nilpotent, there would exist an integer $k \geqslant 2$ such that $R^k = 0$, $R^{k-1} \neq 0$ would be satisfied, but this would imply that R^{k-1} is some nonzero annihilator of R, a contradiction.

Next, we show that there is an element a of R such that no power of a is equal to 0. Suppose that to every element a_i $(i = 1, 2, \dots, n)$ of R there exists a natural number l_i with $a_i^{l_i} = 0$. Let l be the maximum of the l_i $(i = 1, 2, \dots, n)$. If a product in R does not vanish, then every a_i can occur in it at most $(l-1)$ times. So every product with $n(l-1)+1$ factors vanishes; in other words, $R^{n(l-1)+1} = 0$, contrary to the nonnilpotency of R.

So suppose $a \in R$ is such that all the elements a^j $(j = 1, 2, \dots)$ are nonzero. Since R is finite, these powers cannot be all different, say $a^k = a^{k+l}$ $(l > 0)$. This implies $a^k = a^{k+l} = a^{k+2l} = a^{k+3l} = \dots$. Choosing m big enough to satisfy $ml > k$, we have

$$(a^{ml})^2 = a^{ml} \cdot a^{ml} = a^{k+ml} \cdot a^{ml-k} = a^k \cdot a^{ml-k} = a^{ml},$$

so a^{ml} is a nonzero idempotent of R.

Let us write $a^{ml} = e$ for simplicity. We can write R as a direct sum of the ideals A and B:

$$R = A \oplus B, \tag{7}$$

where A is the set of all elements of the form re $(r \in R)$ and B is the set of all elements of the form $r - re$ $(r \in R)$. Clearly, A and B are ideals of R. Any element r of R can be written in the form $r = re + (r - re)$, thus $R = A + B$. Furthermore, $ea = a$ for all a in A, whereas $eb = 0$ for all b in B. This implies $A \cap B = 0$, so (7) holds indeed. If $B = 0$ then $R = A$, so e is the desired identity element of R. If $B \neq 0$, then in view of $0 \neq e \in A$, R is a direct sum of two rings, each of which contains fewer than n elements.

Consider this case and suppose by induction that all commutative rings with fewer than n elements and with annihilator 0 have an identity element. (The one-element ring clearly has an identity element.) If $z \in A$ is such that $zA = 0$, then in view of $AB = 0$ we have $zR = z(A+B) = zA + zB = 0$, implying $z = 0$. Similarly, from $w \in B$, $wB = 0$ we get $w = 0$. This means that the annihilators of both A and B are 0, so by induction A has an identity element e_1 and B has an identity element e_2. But then $e_1 + e_2$ is an identity element of the ring $R = A \oplus B$. □

Remark. The statement of the problem is true for commutative Artinian rings, too (see *R. Baer, Inverses and zerodivisors, Bull. Amer. Math. Soc., 48 (1942), 630–638*).

Problem A.7. *Let I be an ideal of the ring of all polynomials with integer coefficients such that*

(a) the elements of I do not have a common divisor of degree greater than 0, and

(b) I contains a polynomial with constant term 1.

Prove that I contains the polynomial $1 + x + x^2 + \dots + x^{r-1}$ for some natural number r.

Solution. According to assumption (b), for a suitable $f \in \mathbb{Z}[x]$ we have

$$1 + xf \in I.$$

Because of

$$(1 + xf)x^r + x^{r+1}(-f) = x^r$$

for any $r \geqslant 0$, we have

$$x^r \in (1 + xf, x^{r+1}).$$

Repeatedly applying this observation, we get

$$(1 + xf, x^{r+1}) \supseteq (1 + xf, x^r) \supseteq \cdots \supseteq (1 + xf, x) \ni 1,$$

consequently,

$$1, x, \ldots, x^r \in (1 + xf, x^{r+1}),$$

which in turn means that we can find $g_r, h_r \in \mathbb{Z}[x]$ for which

$$1 + x + \cdots + x^r = (1 + xf)g_r + x^{r+1}h_r \tag{1}$$

holds true.

We claim that g_r and h_r can be chosen in such a way that

$$\deg h_r \leqslant \deg f \tag{2}$$

holds. To prove this, observe that the polynomial g_r figuring in (1) can be written as

$$g_r = x^{r+1}q + p$$

where $p, q \in \mathbb{Z}[x]$ and $\deg p \leqslant r$. From (1) we get

$$1 + x + \cdots + x^r = (1 + xf)p + x^{r+1}[(1 + xf)q + h_r],$$

and choosing p (resp., $(1 + xf)q + h_r$) as the new g_r (resp., h_r) we clearly get an h_r satisfying (2) because of $\deg p \leqslant r$.

Suppose $s > r$ and

$$1 + x + \cdots + x^s = (1 + xf)g_s + x^{s+1}h_s, \tag{3}$$

where $\deg h_s \leqslant \deg f$. Subtracting from (3) x^{s-r} times (1), we get

$$1 + x + \cdots + x^{s-r-1} = (1 + xf)(g_s - g_r x^{s-r}) + x^{s+1}(h_s - h_r).$$

This means that if there are indices $s > r$ such that

$$h_s - h_r \in I$$

holds, then

$$1 + x + \cdots + x^{s-r-1} \in I$$

is also true.

We prove that the ideal contains a nonzero constant polynomial. Take a $p \in I$ such that $p \neq 0$ and $\deg p$ is minimal. (Such a p exists in view of $1 + xf \neq 0$.) For any $q \in I$ let

$$q = pu + v,$$

where $u, v \in \mathbb{Q}[x]$ and $\deg v < \deg p$. Multiplying by a suitable nonzero integer α, we get

$$\alpha q = pu_1 + v_1,$$

where $u_1, v_1 \in \mathbb{Z}[x]$ and of course $\deg v_1 < \deg p$ still. This gives

$$v_1 = \alpha q - pu_1 \in I,$$

so by the choice of p and in view of $\deg v_1 < \deg p$, we have $v_1 = 0$. This means that to an arbitrary $q \in I$ we can find a $u \in \mathbb{Q}[x]$ for which $q = pu$. Let $p = \varphi p_1$, where p_1 is a primitive polynomial and $\varphi \in \mathbb{Z}$. Then $q = p_1(\varphi u)$, which in view of Gauss's lemma implies $\varphi u \in \mathbb{Z}[x]$. This in turn implies that for any $q \in I$ we have $p_1 \mid q$. In view of condition (a), this is only possible if $p_1 = 1$; consequently $\varphi \in I$, thus $(\varphi) \subseteq I$.

Finally, we show that there are values $s > r$ such that

$$h_s - h_r \in (\varphi)\,.$$

Indeed, any coefficient of a polynomial can take on at most φ values $\bmod \varphi$; furthermore, $\deg h_r \leqslant \deg f (r = 1, 2, \dots)$, so there exist values $r < s (\leqslant \varphi^{\deg f + 1} + 1)$ for which all coefficients of $h_s - h_r$ are divisible by φ. This proves the statement. □

Remark. The following, more general, statement can be proved:

Let R be a unique factorization domain whose proper factor rings are all finite. Let I be an ideal of the polynomial ring $R[x]$ satisfying the following two conditions:

(a) The elements of I do not have a common divisor of degree greater than 0.

(b) I contains a polynomial with constant term 1.

Then to every natural number N there exists a natural number $r \geqslant N$ for which $N \mid r + 1$ and

$$1 + x + \cdots + x^r \in I.$$

Problem A.8. *Prove that in a Euclidean ring R the quotient and remainder are always uniquely determined if and only if R is a polynomial ring over some field and the value of the norm is a strictly monotone function of the degree of the polynomial. (To be precise, there are two more trivial cases: R can also be a field or the null ring.)*

Solution. We shall work with the following definition of a Euclidean ring: let R be a ring and N the set of nonnegative integers. R is called a *Euclidean ring* if there exists a map $\varphi\colon R \to N$ having the following properties:

i) $\varphi(a) = 0 \iff a = 0$.

ii) For any $a, b \in R$ $(b \neq 0)$ there exist $q, r \in R$ such that

$$a = bq + r \quad \text{and} \quad \varphi(r) < \varphi(b)\,.$$

iii) The unicity of the quotient and remainder means that if $a = bq + r = bq_1 + r_1$ and $\varphi(r) < \varphi(b)$, $\varphi(r_1) < \varphi(b)$, then $r = r_1$ and $q = q_1$.

With these definitions, we will prove that if R is a (commutative) Euclidean ring whose quotient and remainder are unique, then R is the null ring (which consists of the single element 0), a (commutative) field, or a polynomial ring over a (commutative) field.

Note that although commutativity of R is usually included in the definition of a Euclidean ring, we did not assume it, and in the following 16-step proof the first 15 steps never use the commutativity of R. In the remarks, following the proof we shall devote some space to the noncommutative case.

Before embarking on the proof, let us make a simplification. If the values actually taken on by φ are $0 = n_0 < n_1 < \cdots < n_k < \ldots$, then instead of the value $\varphi(a) = n_k$ take the value $\varphi'(a) = k$. φ' is equivalent to φ in the sense that $\varphi(a) < \varphi(b)$ if and only if $\varphi'(a) < \varphi'(b)$ and φ' also satisfies conditions i), ii), and iii). So from now on we can assume $n_k = k$.

And now to the proof.

1. *R has no zerodivisors.*
If $ab = 0$, $a \neq 0$, $b \neq 0$ held, then the decomposition $0 = a{\cdot}0+0 = ab+0$ would contradict iii).

2. *If $c \neq 0$, then $\varphi(ac) \geqslant \varphi(a)$.*
Suppose $\varphi(ac) < \varphi(a)$, then $ac = ac + 0 = a \cdot 0 + ac$ would contradict iii).
Introduce the following notation:

$$T_i = \{a\colon a \in R,\ \varphi(a) \leqslant i\} \quad (i = 0, 1, 2, \ldots)\,.$$

Clearly,

$$T_0 = \{0\},\ T_0 \subseteq T_1 \subseteq \cdots \subseteq T_k \subseteq \ldots \quad \text{and} \quad R = \bigcup_{i=0}^{\infty} T_i\,.$$

3. *If $R = T_0$, then R is the null ring.*
This is clear.

In the following, we assume $R \neq T_0$.

4. *R has an identity 1 and $1 \in T_1$.*
Since $R \neq T_0$, there is an $a \neq 0$ with $a \in T_1$. If $a = aq + r$, we have $\varphi(r) < \varphi(a) = 1$, so $r = 0$, $a = aq$. Multiplying by some b and using step **1**, we get $ab = aqb$, $b = qb$, so q is a left-sided identity. Proceeding in the same way with an arbitrary c: $cb = cqb$, $c = cq$, so q is a two-sided identity; let us denote it by 1. If $\varphi(a) < \varphi(1)$, then $a = 1 \cdot a + 0 = 1 \cdot 0 + a$, so by unicity $a = 0$; consequently $\varphi(1) = 1$.

5. $\varphi(a-b) \leqslant max\{\varphi(a), \varphi(b)\}$.
Suppose $\max\{\varphi(a), \varphi(b)\} < \varphi(a-b)$. Then $a = (a-b)\cdot 1 + b = (a-b)\cdot 0 + a$, together with $1 \neq 0$, contradicts iii).

As a consequence, we have

6. *T_i is an additive group.*

7. $\varphi(-a) = \varphi(a)$.
$\varphi(-a) = \varphi(0-a) \leqslant \max\{\varphi(0), \varphi(a)\} = \varphi(a)$. In a similar way, we have $\varphi(a) \leqslant \varphi(-a)$, so $\varphi(a) = \varphi(-a)$.

8. *If $\varphi(b) < \varphi(a)$, then $\varphi(a-b) = \varphi(a)$.*
Step **5** gives $\varphi(a-b) \leqslant \varphi(a)$. On the other hand, $\varphi(a) = \varphi(b-(b-a)) \leqslant \max\{\varphi(b), \varphi(b-a)\} = \max\{\varphi(b), \varphi(a-b)\}$, so because $\varphi(b) < \varphi(a)$ we have $\varphi(a) \leqslant \varphi(a-b)$, and thus $\varphi(a-b) = \varphi(a)$.

9. *An element a of R is a (two-sided) unit if and only if $\varphi(a) = 1$ (equivalently, $a \in T_1$ and $a \neq 0$).*
Let $\varphi(a) = 1$, and let b be an arbitrary element of R. We have $b = aq + r$ with $\varphi(r) < \varphi(a) = 1$, so $r = 0$, $b = aq$. If, in particular, $b = 1$, then $1 = aa_1$, which also implies $a_1 = a_1aa_1$, $1 = a_1a$, and thus $a_1 = a^{-1}$. Conversely, if $ab = 1$, then step **1** gives $a \neq 0$, $b \neq 0$ and step **2** gives $\varphi(a) \leqslant \varphi(ab) = \varphi(1) = 1$, thus $\varphi(a) = 1$.

Steps **6** and **9** together imply

10. *T_1 is a (skew) field.*
If $R = T_1$, then we are again finished (and we see that actually every skew field — not only the commutative ones — satisfies i), ii), and iii) with $\varphi(a) = 1$, whenever $a \neq 0$). Suppose now that $R \supsetneqq T_1$. We will prove that R is the polynomial ring over T_1.

11. *If $\varphi(b) = 1$, then $\varphi(ab) = \varphi(a)$.*
Step **2** implies namely $\varphi(ab) \geqslant \varphi(a) = \varphi((ab)\cdot b^{-1}) \geqslant \varphi(ab)$, so $\varphi(ab) = \varphi(a)$.

The next statement is the converse of this.

12. *If for some $a \neq 0$ $\varphi(ab) = \varphi(a)$, then $\varphi(b) = 1$.*
Let $a = (ab)q + r$, where $\varphi(r) < \varphi(ab)$. Then $r = a(1-bq)$. If $\varphi(b) \neq 1$, then $1 - bq \neq 0$, so in view of step **2**, $\varphi(r) \geqslant \varphi(a)$, so $\varphi(a) < \varphi(ab)$, a contradiction.

13. *If $\varphi(x) = 2$, then $\varphi(x^k) = k+1$.*
We prove the statement by induction on k. For $k = 1$ the statement is clearly true. Suppose that $\varphi(x^{k-1}) = k$. Then, using steps **2** and **12**, we have $\varphi(x^k) = \varphi(x^{k-1}\cdot x) > \varphi(x^{k-1})$, so $\varphi(x^k) \geqslant k+1$. On the other hand, let $\varphi(a) = k+1$ and $a = x^{k-1}\cdot b + r$ where $\varphi(r) < \varphi(x^{k-1}) = k$. Then $\varphi(x^{k-1}b) = \varphi(a-r) = \varphi(a) = k+1 > k = \varphi(x^{k-1})$. Using step **12**, we get $b \notin T_1$. So if $b = xc + s$ with $\varphi(s) < \varphi(x) = 2$, then $\varphi(xc) = \varphi(b-s) = \varphi(b)$ because $\varphi(b) \geqslant 2 > \varphi(s)$. Thus $\varphi(xc) \geqslant 2$, and consequently $c \neq 0$. By substitution, we get $a = x^kc + x^{k-1}s + r$. Here $\varphi(a) = k+1$; in case $\varphi(s) = 1$, we have $\varphi(x^{k-1}s) = \varphi(x^{k-1}) = k$; in case $\varphi(s) = 0$, that is, $s = 0$, we have $\varphi(x^{k-1}\cdot s) = 0$, finally, we know $\varphi(r) < k$. All in all $\varphi(a) > \max\{\varphi(x^{k-1}s), \varphi(s)\}$, so using step **8**, $\varphi(x^k) \leqslant \varphi(x^kc) = \varphi(a - x^{k-1}s - r) = \varphi(a) = k+1$.

14. *If $\varphi(a) = k+1$, then a can be written in the form $a = \alpha_0 + x\alpha_1 +$*

$\cdots + x^k\alpha_k$, *where* $\alpha_i \in T_1$ $(i = 0, 1, \ldots, k)$ *and* $\alpha_k \neq 0$.

If $k = 0$, the statement is clear. Suppose it is true for $k - 1$. For $k > 0$, let $a = x^k b + r$, where $\varphi(r) < \varphi(x^k) = k + 1$, so $\varphi(x^k b) = \varphi(a - r) = \varphi(a) = k + 1 = \varphi(x^k)$, thus $\varphi(b) = 1$. This means $b \in T_1$, $b \neq 0$. Because $\varphi(r) \leqslant k$, we have $r = x^{k-1}\alpha_{k-1} + \cdots + x\alpha_1 + \alpha_0$ with $\alpha_i \in T_1$.

15. *x is transcendental over T_1.*

Suppose $x^k\alpha_k + \cdots + x\alpha_1 + \alpha_0 = 0$ $(\alpha_i \in T_1)$. Then in view of $\varphi(x^k\alpha_k) = \varphi(x^{k-1}\alpha_{k-1} + \cdots + x\alpha_1 + \alpha_0) \leqslant k$, $\alpha_k = 0$ follows. By repetition of the argument, we get $\alpha_{k-1} = 0, \ldots, \alpha_1 = 0, \alpha_0 = 0$.

16. *If R is commutative, then $R = T_1[x]$, and the Euclidean value of the norm is a strictly monotone function of the degree of the polynomial.*

This follows from steps **10**, **14**, and **15**.

Finally, we remark that in a polynomial ring over a commutative field quotient and remainder are indeed unique — the only possible quotient and remainder are the ones that we get using the usual division algorithm. □

Remarks.

1. The noncommutative case is tricky. The point is that although as a set R is the same as $T_1[x]$, R is not actually isomorphic to $T_1[x]$ equipped with the usual multiplication. If we define the product of two polynomials in the usual way, that is, the product of two polynomials

$$\sum_{i=0}^{n} x^i a_i \quad \text{and} \quad \sum_{i=0}^{m} x^i b_i$$

is defined as the polynomial

$$\sum_{i=0}^{m+n} x^i c_i,$$

where c_i is defined by the equation

$$c_i = a_0 b_i + a_1 b_{i-1} + \cdots + a_i b_0$$

(with $a_{-1} = a_{-2} = \cdots = b_{-1} = b_{-2} = \cdots = 0$), the statement of the problem does not remain true. But by suitably defining the product of two polynomials, the statement remains true. For arbitrary $a \in R$ there exist a^σ and a^τ which satisfy

$$ax = xa^\sigma + a^\tau \quad \text{where} \quad \varphi(a^\tau) < \varphi(x) = 2,$$

so $a^\tau \in T_1$. Clearly,

$$(a + b)^\sigma = a^\sigma + b^\sigma \text{ and } (a + b)^\tau = a^\tau + b^\tau.$$

Furthermore,

$$(ab)^\sigma = a^\sigma b^\sigma \text{ and } (ab)^\tau = a^\tau b^\tau.$$

So $a \to a^\sigma$ is an endomorphism and $\sigma : T_1 \to T_1$ is an automorphism, because for $a \in T_1$ $(a \neq 0)$ there exists an inverse of a, and since $1x = x1$ implies $1^\sigma = 1$ (so $T_1^\sigma \neq 0$), we get $1 = (aa^{-1})^\sigma = a^\sigma(a^{-1})^\sigma$, which shows $a^\sigma \in T_1$ and T_1 being a field, each endomorphism of it is actually an automorphism. If we now take a suitable automorphism $a \to a^\sigma$ of T_1 and a suitable derivation $a \to a^\tau$ (with respect to σ; by this we understand a map of T_1 into itself, satisfying $(a+b)^\tau = a^\tau + b^\tau$ and $(ab)^\tau = a^\tau b^\sigma + ab^\tau$) then, defining in the polynomial ring the product of two polynomials by $ax = xa^\sigma + a^\tau$, we can see that $T_1[x]_{\sigma,\tau}$ has the required properties.

2. In the formulation of the problem the words "strictly monotone" function of the degree are not superfluous. If, for instance, we have polynomials f and g with $\deg f = \deg g$, but $\varphi(f) > \varphi(g)$, say, then R is still Euclidean, but with a suitable constant α we have $\deg(\alpha f + g) = \deg f$ and $\alpha f + g = fq + r$, where $\deg r < \deg f$, but also $\alpha f + g = f\alpha + g$ where $\varphi(g) < \varphi(f)$, so in view of $r \neq g$ the remainder is unique. (We have in mind a φ with $\varphi(r) < \varphi(f)$, of course.)

Problem A.9. *Let*

$$f(x) = a_0 + a_1x + a_2x^2 + a_{10}x^{10} + a_{11}x^{11} + a_{12}x^{12} + a_{13}x^{13} \qquad (a_{13} \neq 0)$$

and

$$g(x) = b_0 + b_1x + b_2x^2 + b_3x^3 + b_{11}x^{11} + b_{12}x^{12} + b_{13}x^{13} \qquad (b_3 \neq 0)$$

be polynomials over the same field. Prove that the degree of their greatest common divisor is at most 6.

Solution. Let

$$f_1(x) = a_0 + a_1x + a_2x^2, \quad f_2(x) = a_{10} + a_{11}x + a_{12}x^2 + a_{13}x^3$$
$$g_1(x) = b_0 + b_1x + b_2x^2 + b_3x^3, \quad g_2(x) = b_{10} + b_{11}x + b_{12}x^2 + b_{13}x^3$$

(the problem says $b_{10} = 0$, but we will not use this). This implies

$$f(x) = f_1(x) + x^{10}f_2(x), \quad g(x) = g_1(x) + x^{10}g_2(x),$$

so

$$f(x)g_2(x) - g(x)f_2(x) = f_1(x)g_2(x) - f_2(x)g_1(x).$$

The right-hand side is divisible by the greatest common divisor of $f(x)$ and $g(x)$, since the left-hand side is divisible by it. But the polynomial on the right-hand side has degree 6, because $f_1(x)g_2(x)$ has degree at most 5 but $f_2(x)g_1(x)$ has degree precisely equal to 6 since $a_{13} \cdot b_3 \neq 0$. So the greatest common divisor of $f(x)$ and $g(x)$ divides a polynomial of degree 6; therefore its degree can be at most 6. □

Remarks.

1. The upper bound of 6 for the degree of the greatest common divisor cannot be improved as shown by the polynomials $x^{10}+x^{13}$ and x^3+x^{12}, whose greatest common divisor is x^3+x^6.
2. It is possible to prove in a similar manner the following generalization: if $f(x)$ and $g(x)$ are polynomials with coefficients in the same field, $\operatorname{grad} f = n$, $\operatorname{grad} g \leqslant n$, and $g(x)$ has a term of degree k, but the terms of degree $k+1, \dots, k+r-1$ are missing from both polynomials ($k+r \leqslant n$), then the greatest common divisor of f and g has degree at most $n-r$.

Problem A.10. *Let K be a subset of a group G that is not a union of left cosets of a proper subgroup. Prove that if G is a torsion group or if K is a finite set, then the subset*

$$\bigcap_{k \in K} k^{-1}K$$

consists of the identity alone.

Solution. Suppose there is an element $a \in G$, different from the identity element such that $a \in \bigcap_{k \in K} k^{-1}K$. Then for every $k \in K$ we have $ka \in K$, that is, $Ka \subseteq K$.

First we observe that if we have equality here, then K is the union of some left cosets of the subgroup H generated by a. Indeed, $Ka = K$ implies $Ka^i = K$ for all integers i, so $KH = K$, which in turn means $K = \cup_{k \in K} kH$.

Now we only have to prove that in the cases mentioned in the problem we indeed have $Ka = K$. This is clear if K is finite since Ka has the same cardinality as K and is a subset of it. On the other hand, if a has finite order n, then $K \supseteq Ka$ implies

$$K \supseteq Ka \supseteq Ka^2 \supseteq \cdots \supseteq Ka^n = K,$$

again proving $Ka = K$. □

Remark. If we assume that G is an Abelian group, then the assumption that G is a torsion group can be replaced by the weaker assumption that every element of K has finite order.

Problem A.11. *Prove that if an infinite, noncommutative group G contains a proper normal subgroup with a commutative factor group, then G also contains an infinite proper normal subgroup.*

Solution. First we prove a lemma:

Lemma. If the infinite group G has a proper normal subgroup N satisfying $N \nsubseteq Z(G)$ ($Z(G)$ denotes the centre of the group), then G has an infinite proper normal subgroup.

Proof. To prove the lemma, we can assume that N is finite. For every $g \in G$, the mapping

$$\varphi_g \; : n \mapsto g^{-1}ng \quad (n \in N)$$

is clearly an automorphism of N. Furthermore, the mapping $g \mapsto \varphi_g$ is a homomorphism of G to some subgroup Φ of the full group of automorphisms of N, and Φ is clearly finite in view of the finiteness of N. Because $N \nsubseteq Z(G)$, Φ contains nonidentity automorphisms so the kernel F of the homomorphism $g \mapsto \varphi_g$ is a proper normal subgroup of G that is infinite in view of $G/F \simeq \Phi$.

Now let H be a proper normal subgroup of the infinite, noncommutative group G such that G/H is commutative. If $H \nsubseteq Z(G)$, then we can apply the lemma. If, on the other hand, $H \subseteq Z(G)$ and h is any element of G not contained in $Z(G)$, then using the commutativity of G/H we see that $\langle H, h\rangle$ is a normal subgroup of G. Clearly, $\langle H, h\rangle \nsubseteq Z(G)$; on the other hand, $\langle H, h\rangle \neq G$ since G is noncommutative. Applying the lemma with $N = \langle H, h\rangle$ completes the proof. □

Remark. The condition of noncommutativity cannot be dropped from the statement of the problem, as shown by the group Z_{p^∞}. Conversely, the alternating group of countably infinite degree shows that the condition of the existence of a proper normal subgroup with commutative factor group cannot be dropped either.

Problem A.12. *Let $a_1, a_2, \dots, a_N$ be positive real numbers whose sum equals 1. For a natural number i, let n_i denote the number of a_k for which $2^{1-i} \ge a_k > 2^{-i}$ holds. Prove that*

$$\sum_{i=1}^{\infty} \sqrt{n_i 2^{-i}} \le 4 + \sqrt{\log_2 N}.$$

Solution. We know that

$$\sum_{i=1}^{\infty} \frac{n_i}{2^i} < \sum_{j=1}^{N} a_j = 1 \quad \text{and} \quad \sum_{i=1}^{\infty} n_i = N,$$

so, by applying Cauchy's inequality, we get the following estimate:

$$\begin{aligned}
\sum_{i=1}^{\infty}\sqrt{\frac{n_i}{2^i}} &= \sum_{i=1}^{[\log_2 N]}\sqrt{\frac{n_i}{2^i}} + \sum_{i=[\log_2 N]+1}^{\infty}\sqrt{\frac{n_i}{2^i}} \\
&\leqslant \left(\sum_{i=1}^{[\log_2 N]} 1\right)^{1/2}\left(\sum_{i=1}^{[\log_2 N]}\frac{n_i}{2^i}\right)^{1/2} + \left(\sum_{i=[\log_2 N]+1}^{\infty} n_i\right)^{1/2}\left(\sum_{i=[\log_2 N]+1}^{\infty}\frac{1}{2^i}\right)^{1/2} \\
&\leqslant \sqrt{\log_2 N} + \sqrt{N}\,\frac{1}{2^{\left(\frac{[\log_2 N]+1}{2}\right)}}\left(\sum_{i=1}^{\infty}\frac{1}{2^i}\right)^{1/2} \leqslant \sqrt{\log_2 N} + \sqrt{2}.
\end{aligned}$$

This proves a somewhat stronger form of the statement with $\sqrt{2}$ in place of 4. □

Remarks.

1. We can get the stronger estimate

$$\sum_{i=1}^{\infty}\sqrt{\frac{n_i}{2^i}} \leqslant \sqrt{\log_2 N} + O\left(\frac{\log_2\log_2 N}{\sqrt{\log_2 N}}\right)$$

in a similar way if we start with the decomposition

$$\sum_{i=1}^{\infty}\sqrt{\frac{n_i}{2^i}} = \sum_{i=1}^{K}\sqrt{\frac{n_i}{2^i}} + \sum_{i=K+1}^{\infty}\sqrt{\frac{n_i}{2^i}},$$

where $K = [\log_2 N + \log_2\log_2 N]$.

2. If we use Hölder's inequality instead of Cauchy's inequality, we get a similar estimate:

$$\sum_{i=1}^{\infty}\left(\frac{n_i}{2^i}\right)^{1/p} \leqslant \left(\log_2 N\right)^{1/q} + C(p),$$

where $p > 1$, $1/p + 1/q = 1$, and $C(p)$ is a positive constant depending only on p.

Problem A.13. *Consider the endomorphism ring of an Abelian torsion-free (resp. torsion) group G. Prove that this ring is Neumann-regular if and only if G is a discrete direct sum of groups isomorphic to the additive group of the rationals (resp., a discrete direct sum of cyclic groups of prime* *order). (A ring R is called Neumann-regular if for every $\alpha \in R$ there exists a $\beta \in R$ such that $\alpha\beta\alpha = \alpha$.)*

Solution. In the following, "group" will always mean "commutative group", where the operation is written as addition. Instead of *Neumann-regularity*, we shall just write *regularity*.

First, we prove the necessity of the condition. If the group G is torsion-free, call it G_0, if it is a torsion group, then it can be written as a (discrete) direct sum: $G = G_1 \oplus \cdots \oplus G_i \oplus \ldots$, where in the group G_i all elements have order a power of p_i (p_i denotes the ith prime number).

Let p be a prime number, and let Φ_p be the map of G defined by $\Phi_p \colon a \to pa$ ($a \in G$). Φ_p is clearly an endomorphism. Because of the regularity, there is an endomorphism Ψ_p with $\Phi_p\Psi_p\Phi_p = \Phi_p$. Then for any $a \in G$,

$$pa = p^2\Psi_p(a)\,. \tag{1}$$

It follows from (1) and the fact that Ψ_p is an endomorphism that if $p^2a = 0$, then $pa = 0$. Therefore every element ($\neq 0$) of G_i ($i > 0$) has order p_i.

In case $G = G_0$, we can cancel (1) by p, which shows that every element a of G_0 is divisible by every prime, thus G_0 is a divisible group. Furthermore — as G_0 is torsion-free — the quotient is uniquely determined. Let $p_0 = 0$, and denote by K_0 the field of rational numbers, whereas for $i = 1, 2, \ldots,$ denote by K_i the field with p_i elements. From our previous observations we see that G_i is a vector space over the field K_i ($i = 0, 1, 2, \ldots$). Invoking Zorn's lemma, we see that G_i has a basis that is just equivalent to the direct sum decomposition formulated in the problem.

Sufficiency will follow if we prove the following stronger statement: The endomorphism ring of the direct sum $G = G_0 \oplus G_1 \oplus \cdots \oplus G_i \oplus \ldots$ is regular if G_i is a vector space over K_i ($i = 0, 1, 2, \ldots$). Here "$\oplus$" can mean either discrete or complete direct sum. First, we prove that the endomorphism ring of each G_i is regular.

Let α be an endomorphism of G_i. Then α is a linear transformation of G_i as a vector space. Since in a vector space every subspace is a direct summand, there exist subspaces U_i, V_i such that

$$G_i = \text{Ker}\ \alpha \oplus U_i = V_i \oplus \text{Im}\ \alpha\,.$$

It is clear that α maps the elements of U_i onto Im α in a 1-to-1 way. Let us define $\beta \colon G_i \to G_i$ by the conditions $\beta(a) = 0$ for $a \in V_i$ and $\beta(a) = b$ for $a \in \text{Im}\ \alpha$, where b is the unique element of U_i with $\alpha(b) = a$. β is clearly an endomorphism of G_i with $\alpha\beta\alpha = \alpha$. This proves the regularity of the endomorphism rings of each of the G_i ($i = 0, 1, 2, \ldots$).

Now let α be an endomorphism of G. Since for every prime p, if $a \in G$ has order p, then $\alpha(a)$ again has order p or it is 0, α maps G_i to itself if $i \geqslant 1$. Now let $a \in G_0$, and suppose $\alpha(a) \in G_i$ ($i \geqslant 1$). We know $p_i\alpha(a) = 0$. As G_0 is a vector space over K_0, there is an $x \in G_0$ with $p_ix = a$. So $p_i^2\alpha(x) = 0$, thus $\alpha(x) \in G_i$, so $p_i\alpha(x) = \alpha(a) = 0$. *Applying* *this argument not only to α itself, but to α and subsequent projections to G_i $(i \geqslant 1)$, we see that each component of $\alpha(a)$ belonging to some G_i $(i \geqslant 1)$ is zero*, so α maps G_0 to itself.

These observations imply that every α uniquely determines a vector $(\alpha_0, \alpha_1, \alpha_2, \dots)$ where α_i is an endomorphism of G_i, and conversely every such vector uniquely determines an endomorphism α of G from which we get back the original vector. We clearly have $(\alpha\beta)_i = \alpha_i\beta_i$. Now let α be an endomorphism of G, $(\alpha_0, \alpha_1, \alpha_2, \dots)$ the corresponding vector. We have already proved the existence of endomorphism β_i of G_i such that $\alpha_i\beta_i\alpha_i = \alpha_i$ $(i = 0, 1, 2, \dots)$. The endomorphism β belonging to $(\beta_0, \beta_1, \beta_2, \dots)$ clearly satisfies $\alpha\beta\alpha = \alpha$, so the endomorphism ring of G is regular. □

Problem A.14. *Let $\mathfrak{A} = \langle A; \dots \rangle$ be an arbitrary, countable algebraic structure (that is, $\mathfrak{A}$ can have an arbitrary number of finitary operations and relations). Prove that $\mathfrak{A}$ has as many as continuum automorphisms if and only if for any finite subset A' of A there is an automorphism $\pi_{A'}$ of $\mathfrak{A}$ different from the identity automorphism and such that*

$$(x)\pi_{A'} = x$$

for every $x \in A'$.

Solution. Suppose first that the condition mentioned in the problem is not satisfied. Let A' be a finite subset of A with the property that every automorphism of $\mathfrak{A}$ fixing A' pointwise is necessarily the identity. Then for any two automorphisms π_1 and π_2 that satisfy $(x)\pi_1 = (x)\pi_2$ for all $x \in A'$, $\pi_1\pi_2^{-1}$ is the identity, so $\pi_1 = \pi_2$. The number of automorphisms of $\mathfrak{A}$ is therefore less than or equal to the number of mappings of A' to A, and so the number of automorphisms is countable.

Suppose now that the condition mentioned in the problem is satisfied. We can assume that $A = \{1, 2, \dots\}$. We define recursively an increasing sequence of finite subsets of A: $A_0 \subseteq A_1 \subseteq \dots \subseteq A_n \subseteq \dots$ and a sequence of automorphisms of $\mathfrak{A}$: $\pi_1, \dots, \pi_n, \dots$. Let $A_0 = \emptyset$. Now let $n \geqslant 0$, and suppose we have already defined A_n, which is finite, and π_i for all $i < n$. Let π_n be an automorphism of $\mathfrak{A}$ that fixes A_n pointwise and is not the identity, and let

$$A_{n+1} = \{1, \dots, n\} \cup \bigcup_{\substack{(\varepsilon_1, \dots, \varepsilon_n) \\ \varepsilon_i \in \{0,1\}}} (A_n)\pi_1^{\varepsilon_1} \dots \pi_1^{\varepsilon_n} \cup \{\min\{k \mid (k)\pi_n \neq k\}\}\,.$$

Then A_{n+1} is again finite, and by this procedure we have defined A_n and π_n for all n. The sequence so defined has the following properties:

α) π_n is the identity on A_n but is different from the identity on A_{n+1}.

β) If $\varepsilon_i \in \{0, 1\}$ $(i = 1, \dots, n)$, then

$$(A_n)\pi_1^{\varepsilon_1} \dots \pi_n^{\varepsilon_n} \subseteq A_{n+1}\,.$$

γ) $\cup_{n=1}^{\infty} A_n = A$ and all the A_n are finite.

Let $(\varepsilon_1, \dots, \varepsilon_n, \dots)$ be an arbitrary infinite sequence with $\varepsilon_n \in \{0, 1\}$ $(n = 1, 2, \dots)$. Let us define the product $\prod_{n=1}^{\infty} \pi_n = \pi$ in the following way. If $k \in A$, let

$$(k)\pi = (k)\pi_1^{\varepsilon_1} \dots \pi_{n_k}^{\varepsilon_{n_k}}, \tag{1}$$

where n_k denotes the smallest number for which we have $k \in A_{n_k}$. Because of γ), π is indeed a map of A into itself. For any $n > n_k$, we have

$$(k)\pi = (k)\pi_1^{\varepsilon_1} \dots \pi_n^{\varepsilon_n} = ((k)\pi_1^{\varepsilon_1} \dots \pi_{n_k}^{\varepsilon_{n_k}})\pi_{n_k+1}^{\varepsilon_{n_k+1}} \dots \pi_n^{\varepsilon_n} = (k)\pi_1^{\varepsilon_1} \dots \pi_{n_k}^{\varepsilon_{n_k}} \tag{2}$$

because in view of β) we have $(k)\pi_1^{\varepsilon_1} \dots \pi_{n_k}^{\varepsilon_{n_k}} \in A_{n_k+1}$ and in view of α) $\pi_{n_k+1}, \dots, \pi_n$ are all equal to the identity map on A_{n_k+1}. Taking into consideration that $\mathfrak{A}$ contains only finitary operations and relations, we see that in view of (2) π preserves operations and relations. For every $l \in A$ there exists an n with $l \in A_{n+1}$. Thus, there exists a k with $(k)\pi_1^{\varepsilon_1} \dots \pi_n^{\varepsilon_n} = l$. Because of α) we have for any $m > n$ $\quad (k)\pi_1^{\varepsilon_1} \dots \pi_m^{\varepsilon_m} = l$ so in view of (2) we have $(k)\pi = l$. Thus π is an automorphism. We are going to show that for $(\varepsilon_1, \dots, \varepsilon_n, \dots) \neq (\varepsilon_1', \dots, \varepsilon_n', \dots) \quad \pi = \prod_{n=1}^{\infty} \pi_n^{\varepsilon_n} \neq \pi' = \prod_{n=1}^{\infty} \pi_n^{\varepsilon_n'}$ is satisfied.

Let n be the smallest number for which $\varepsilon_n \neq \varepsilon_n'$. We may assume $\varepsilon_n = 0, \varepsilon_n' = 1$. Then $\pi_1^{\varepsilon_1} \dots \pi_{n-1}^{\varepsilon_{n-1}} = \pi_1^{\varepsilon_1'} \dots \pi_{n-1}^{\varepsilon_{n-1}'}$. Let $l \in A_{n+1}$ be an element with $(l)\pi_n \neq l$. Let k be the number for which $(k)\pi_1^{\varepsilon_1} \dots \pi_{n-1}^{\varepsilon_{n-1}} = l$. In view of α) and (2) we have $(k)\pi = (k)\pi_1^{\varepsilon_1} \dots \pi_n^{\varepsilon_n} = (k)\pi_1^{\varepsilon_1} \dots \pi_{n-1}^{\varepsilon_{n-1}} = l$, so $(k)\pi' = (l)\pi_n \neq (k)\pi = l$. This means that the number of different automorphisms of $\mathfrak{A}$ is at least as much as the number of infinite sequences of 0's and 1's. Taking into consideration the fact that A is a countable set, we see that the cardinality of the set of automorphisms of $\mathfrak{A}$ is indeed continuum. □

Remark. The following example shows that the statement of the problem is no longer true if we allow relations with infinitely many variables. Let $\mathfrak{A} = \langle A, R \rangle$, where A is the set of natural numbers and the relation $R(a_1, \dots, a_n, \dots)$ with countably many variables is true if and only if we have $a_n = n$ with finitely many exceptions. Then for any automorphism π of $\mathfrak{A}$, $(n)\pi = n$ holds with finitely many exceptions. Thus $\mathfrak{A}$ has only countably many automorphisms although the condition for finite subsets formulated in the problem is clearly satisfied by this $\mathfrak{A}$.

Problem A.15. *Let G be an infinite group generated by nilpotent normal subgroups. Prove that every maximal Abelian normal subgroup of G is infinite. (We call an Abelian normal subgroup maximal if it is not contained in another Abelian normal subgroup.)*

Solution. Let the group G be generated by nilpotent normal subgroups, and suppose it has a maximal Abelian subgroup A that is finite.

The centralizer C of A in G is a normal subgroup of G that has finite index in G. Indeed, if we map every element g of G to the mapping $x \mapsto g^{-1}xg$ of A, then this is a homomorphism from G to the automorphism group of A and the kernel of this homomorphism is C. Therefore G/C is finite.

Suppose B is an Abelian normal subgroup in G. Then we have $|B| \leqslant n$, where $n = |A| \cdot |G : C|$. Indeed, we have $C \cap B = A \cap B$, because otherwise the Abelian normal subgroup $A(C \cap B)$ of G would strictly contain A. Therefore $|B : A \cap B| \leqslant |G : C|$ because $B/A \cap B = B/C \cap B \simeq CB/C \leqslant G/C$, which gives $|B| = |A \cap B| \cdot |B : A \cap B| \leqslant |A| \cdot |G : C|$.

Let H be a nilpotent normal subgroup of G. We claim that the nilpotency class of H is at most $2n-2$. Consider namely the lower central series of H:

$$H = H_1 > H_2 > \cdots > H_k > H_{k+1} = 1.$$

It is well known that for the members of the lower central series, $[H_i, H_j] \leqslant H_{i+j}$ holds, so as soon as $r > k/2$ we have $[H_r, H_r] \leqslant H_{2r} = H_{k+1} = 1$, that is, the characteristic subgroup H_r of H is Abelian. By what we said previously, this implies $|H_r| \leqslant n$. This clearly implies that there can be at most n H_r such that r falls in the range $k+1 \geqslant r > k/2$, that is, $k+1 \leqslant 2n-1$ holds.

The group G is nilpotent. Indeed, for any elements $g_1, g_2, \ldots, g_{2n-1} \in G$,

$$[\ldots[[g_1, g_2], g_3], \ldots, g_{2n-1}] = 1$$

is true, because $g_1, g_2, \ldots, g_{2n-1}$ belong to some subgroup of G whose nilpotency class is at most $2n-2$.

It is true that $C = A$. If this were not true, then the nilpotent group G/A would contain the nonidentity normal subgroup C/A. By a well-known theorem, this implies that the center of G/A has a nonidentity intersection with C/A; in other words, there is an element $g \in C \setminus A$ for which Ag is contained in the center of G/A thus the subgroup $\langle A, g \rangle$ would be an Abelian normal subgroup properly containing A.

All this would imply that G is a finite group. Indeed, the subgroup C being identical with A is finite and we know that it has finite index in G, so $|G| = |C| \cdot |G : C|$ is finite. □

Problem A.16. *Let $p \geq 7$ be a prime number, ζ a primitive pth root of unity, c a rational number. Prove that in the additive group generated by the numbers $1, \zeta, \zeta^2, \zeta^3 + \zeta^{-3}$ there are only finitely many elements whose norm is equal to c. (The norm is in the pth cyclotomic field.)*

Solution. Let

$$L^{(k)}(x) = x_1 + \zeta^k x_2 + \zeta^{2k} x_3 + (\zeta^{3k} + \zeta^{-3k}) x_4 \quad (k = 1, \ldots, p-1).$$

It is enough to show that the inequality

$$|\, L^{(1)}(x) \ldots L^{(p-1)}(x) \,| \leqslant |c| \tag{1}$$

has only a finite number of rational integer solutions x_1, x_2, x_3, x_4.

For any such solutions

$$
\begin{aligned}
2^{p-1}|c| \geqslant 2^{p-1} \prod_{k=1}^{p-1} | L^{(k)}(x) | &\geqslant 2^{p-1} \prod_{k=1}^{p-1} | \operatorname{Im} L^{(k)}(x) | \\
&= \prod_{k=1}^{p-1} | (\zeta^k - \zeta^{-k})x_2 + (\zeta^{2k} - \zeta^{-2k})x_3 | \\
&= \prod_{k=1}^{p-1} (\zeta^k - \zeta^{-k}) \mid \prod_{k=1}^{p-1} (x_2 + (\zeta^k + \zeta^{-k})x_3) | \\
&= p\, |F(x_2, x_3)|.
\end{aligned}
$$

But for any k with $p \nmid k$ we know that $\zeta^k + \zeta^{-k}$ is an algebraic number of degree $(p-1)/2 \geqslant 3$, so by applying Thue's theorem we see that the homogeneous polynomial $F(x_2, x_3)$ with rational integer coefficients takes on rational integer values only at finitely many places, so there are only a finite number of possible values for x_2 and x_3.

Therefore, it is enough to show that for any fixed x_2 and x_3, the number of possible values of x_1 and x_4 is also finite. For $x_2 = x_3 = 0$, the proof of this statement is analogous to the previous proof. If one of x_2 and x_3 is different from 0, then the absolute value of each of the $L^{(k)}(x)$ is bounded from below in view of $|L^{(k)}(x)| \geqslant |\operatorname{Im} L^{(k)}(x)|$ by a positive bound independent of x_1 and x_4, and as the product of them is bounded from above in view of (1), their absolute value is also bounded from above by a bound independent of x_1 and x_4. But then the system of equations

$$
\begin{aligned}
x_1 + (\zeta^3 + \zeta^{-3})x_4 &= L^{(1)}(x) - (\zeta x_2 + \zeta^2 x_3), \\
x_1 + (\zeta^6 + \zeta^{-6})x_4 &= L^{(2)}(x) - (\zeta^2 x_2 + \zeta^4 x_3)
\end{aligned}
$$

guarantees that $|x_1|$ and $|x_4|$ are bounded from above, so the number of possible pairs x_1, x_4 is indeed finite. □

Problem A.17. *We have* $2n+1$ *elements in the commutative ring* R*:*

$$\alpha, \alpha_1, \ldots, \alpha_n, \varrho_1, \ldots, \varrho_n\,.$$

Let us define the elements

$$\sigma_k = k\alpha + \sum_{i=1}^{n} \alpha_i \varrho_i^k\,.$$

Prove that the ideal $(\sigma_0, \sigma_1, \ldots, \sigma_k, \ldots)$ *can be finitely generated.*

Solution. First, we are going to show that we can assume that R is a ring with identity. There is a standard way to embed an arbitrary ring R as an ideal into a ring with identity R^+: as a set R^+ consists of the ordered pairs (a, n) with $a \in R$ and n an integer, and the operations are defined in the following way:

$$(a,n) + (b,m) = (a+b, n+m), \qquad (a,n)(b,m) = (ab + ma + nb, nm).$$

R can be identified with the ideal of R^+ formed by the elements of the form $(a, 0)$, and a subset of R generates the same ideal in R and R^+. To verify this statement, take a subset H of R and denote the ideals generated by H in R and R^+ by I and J, respectively. Clearly, $I \subseteq R \cap J \subseteq J$. On the other hand, I consists of elements of the form $(a, 0)$, and multiplying these elements by elements of the form (b, m), we again get elements of I so I is an ideal of R^+, too. This proves $J \subseteq I$.

This shows that the ideal of R mentioned in the text of the problem is an ideal of R^+, too, and if it is finitely generated as an ideal of R^+, then the same elements clearly generate it as an ideal of R, too. Therefore, from now on we can safely assume that R is a ring with identity.

We are going to prove the following generalization of the problem:

Theorem. Suppose we have polynomials $f_i(x) \quad (i = 0, \ldots, n)$ with coefficients in the ring (with identity) R. Using these polynomials and fixed elements $\varrho_0, \ldots, \varrho_n$ of R, we form the elements

$$\sigma_k = \sum_{i=0}^{n} f_i(k)\varrho_i^k \qquad (k = 0, 1, 2, \ldots).$$

Then the ideal $(\sigma_0, \sigma_1, \ldots, \sigma_k, \ldots)$ is equal to the ideal $(\sigma_0, \ldots, \sigma_s)$ where $s = \sum(\deg f_i + 1)$.

We get the original problem as the special case where the polynomials are $f_0(x) = x$, $f_i(x) = \alpha_i$ $(i = 1, \ldots, n)$, and $\varrho_0 = 1$.

Proof. Let r_i be the degree of $f_i(x)$. If

$$F(x) = \prod_{i=0}^{n} (x - \varrho_i)^{r_i+1} = x^s - \sum_{j=0}^{s-1} \beta_j x^j$$

and D is the operator of taking the derivative with respect to x and then multiplying by x, then $D^m(x^k F(x))$ vanishes at the place ϱ_i if $m \leqslant r_i$. (We can see this if we take into consideration the fact that a root with multiplicity t of a polynomial $P(x)$ is also a root with multiplicity at least $(t-1)$ of the polynomial $D(P(x))$.) As $D^m(x^k) = k^m x^k$, we have

$$D^m(x^k F(x)) = (s+k)^m x^{s+k} - \sum_{j=0}^{s-1} \beta_j (j+k)^m x^{j+k},$$

and consequently

$$(s+k)^m \varrho_i^{s+k} = \sum_{j=0}^{s-1} \beta_j (j+k)^m \varrho_i^{j+k}.$$

Let

$$f_i(x) = \sum_{m=0}^{r_i} \gamma_{m,i} x^m .$$

Then

$$\begin{aligned}\sigma_{s+k} &= \sum_{i=0}^{n} f_i(s+k)\varrho_i^{s+k} = \sum_{i=0}^{n}\sum_{m=0}^{r_i} \gamma_{m,i}(s+k)^m \varrho_i^{s+k} \\ &= \sum_{i=0}^{n}\sum_{m=0}^{r_i}\sum_{j=0}^{s-1} \gamma_{m,i}\beta_j (j+k)^m \varrho_i^{j+k} = \sum_{j=0}^{s-1}\beta_j \sum_{i=0}^{n}\sum_{m=0}^{r_i} \gamma_{m,i}(j+k)^m \varrho_i^{j+k} \\ &= \sum_{j=0}^{s-1}\beta_j \sum_{i=0}^{n} f_i(j+k)\varrho_i^{j+k} = \sum_{j=0}^{s-1}\beta_j \sigma_{j+k} .\end{aligned}$$

This means that σ_{s+k} is contained in the ideal generated by σ_t with smaller indices, and this proves the theorem. □

Problem A.18. *Let G and H be countable Abelian p-groups (p an arbitrary prime). Suppose that for every positive integer n,*

$$p^n G \neq p^{n+1} G .$$

Prove that H is a homomorphic image of G.

Solution. We say that an element $g \in G$ ($g \neq 0$) is *of infinite height* if for every natural number n there is an $x \in G$ that satisfies $p^n x = g$. The elements of G that are of infinite height together with 0 form a subgroup A of G. Let $G^* = G/A$ be the factor group of G with respect to A. Clearly, G^* is a finite or countably infinite Abelian p-group. G^* contains no elements of infinite height. To see this, pick a $g^* \in G^*$ such that $p^n x^* = g^*$ has a solution $x^* \in G^*$ for every natural number n. Take an element $g \in G$ which is mapped to g^* by the natural homomorphism $G \to G^*$ and an $x \subset G$ which is mapped to x^* (for some fixed n). We see that $p^n x - g$ is mapped to 0, that is, $p^n - x \in A$, so $p^n x - g = p^n y$ for some $y \in G$, which gives $g = p^n(x - y)$, and this in turn implies $g \in A$, that is, $g^* = 0$.

Now a well-known theorem of Prüfer gives that G^* is a direct sum of cyclic groups (in the case where G^* is finite, this follows more easily from the fundamental theorem of finite Abelian groups).

Now we show that for every natural number n,

$$p^n G^* \neq p^{n+1} G^* .$$

We know that $p^n G \neq p^{n+1} G$, so for some $h \in G$ the equation $p^n h = p^{n+1} x$ has no solutions in G. We want to show that $p^n h^* = p^{n+1} y^*$ has no solutions $y^* \in G^*$ (h^* denotes the image of h in G^* with the natural homomorphism). Otherwise, for a suitable $y \in G$, $p^n h - p^{n+1} y$ would be mapped to $0 \in G^*$, so $p^n h - p^{n+1}y = p^{n+1}z$ would hold for some $z \in G$, giving the contradiction $p^n h = p^{n+1}(y+z)$. (This, incidentally, shows that G^* cannot be finite.)

Suppose the direct sum decomposition of G^* is

$$G^* = \bigoplus_{i=1}^{\infty} C_i,$$

where C_i is a cyclic group of order p^{k_i}. Then the set $\{k_1, k_2, \dots\}$ cannot be bounded. This follows from the fact that if, for some n, all C_i were of order at most p^n, then $p^n G^* = p^{n+1}G^* = 0$ would hold, contrary to our previous observation. Suppose c_i is an element generating C_i. Then every element $g^* \in G^*$ can be written uniquely in the form $g^* = \sum_{i=1}^{\infty} \alpha_i c_i$, where α_i is an integer taken modulo p^{k_i} and, with finitely many exceptions, all the α_i are equal to 0.

Now take a finite or countably infinite Abelian p-group H, the elements of which are $0, h_1, h_2, \dots$. Choose a series $c_{i_1}, c_{i_2}, \dots$ in such a way that the order of c_{i_j} is greater than the order of h_i. Then we have a homomorphism of C_{i_j} onto the cyclic subgroup generated by h_j. We map all the other C_l to 0. Since G^* is a direct sum of these cyclic subgroups, there is a homomorphism of G^* to H extending the above mentioned homomorphisms of the cyclic subgroups. As every h_j is the image of some element of C_{i_j}, this extended homomorphism is clearly *onto*. The composition of the natural homomorphism $G \to G^*$ and our homomorphism is the required homomorphism of G onto H. □

Problem A.19. *Let G be an infinite compact topological group with a Hausdorff topology. Prove that G contains an element $g \neq 1$ such that the set of all powers of g is either everywhere dense in G or nowhere dense in G.*

Solution. Suppose G is an infinite compact group with the property that the closure of every cyclic subgroup ($\neq 1$) has an inner point, that is, every closed subgroup ($\neq 1$) of G is open. We are going to prove that G has a dense cyclic subgroup.

First, let G be commutative. The closure of a cyclic subgroup $A \neq 1$ is open so it has a finite index n in G; therefore, the continuous isomorphism $\varphi : x \to x^n$ $(x \in G)$ maps G to this subgroup. $G^n \neq 1$ is a compact, thus closed, thus open subgroup of G, therefore $G^n \cap A$ is dense in G^n. So the cyclic subgroup $\varphi^{-1}(G^n \cap A)$ is dense in G.

Now we show that the index of every open subgroup of G is a power of the same prime number. If p is a prime divisor of n, then because $G^p \neq G$ the open subgroups G^{p^i} $(i = 0, 1, 2, \dots)$ form a strictly decreasing chain, so their intersection is a closed subgroup of infinite index, and so this intersection is the identity. The topology of G that we get by choosing the subgroups G^{p^i} as a base for neighborhoods of 1 is coarser than the original compact topology of G, so it is equal to it. Consequently, every open subgroup of G has index a power of p, since it contains a subgroup some G^{p^i}.

In the following, we do not suppose that G is commutative. The center Z of G is an open subgroup. Indeed, the closures of the cyclic subgroups $(\neq 1)$ of G are open subgroups and cover G, so finitely many of them cover G already, the intersection of these is a subset of Z, and therefore Z is an open subgroup.

The group G/Z is a p-group. Indeed, if every open subgroup of Z has index a power of p, and $H \leqslant G$ is a subgroup that contains Z as a subgroup of index q (q is prime, therefore H is commutative), then all elements $(\neq 1)$ of the factor group H^q/H^{q^2} $(\neq 1)$ formed using open subgroups of Z have order q; thus $p = q$ and consequently the finite group G/Z is a p-group.

Finally, we prove that G is actually commutative. Let V be a subgroup with $Z \leqslant V \leqslant G$ such that the center of G/Z is V/Z. Then for every $g \in G$, taking the endomorphism of V defined by $x \to [x, g]$ $(x \in V)$, the image of V will be finite (as the kernel of the endomorphism contains a subgroup Z that itself has finite index) so this image is the identity, thus $V = Z$ and thus $G = Z$, because G/Z is a finite p-group. □

Remarks.

1. Paragraphs 3, 5 and 6 of the elementary solution described above can be omitted if we use a theorem of Baer which states that if the factor group with respect to the center of a (discrete) group G is finite, then the commutator subgroup of G is also finite.
2. József Pelikán noticed the following. If we use the fact that every infinite compact group has a proper closed subgroup, then the statement of the problem can be reformulated in the following way:

Statement. If G is an infinite, compact, topological group, every closed subgroup $(\neq 1)$ of which is open, then G is topologically isomorphic to the topological group of the p-adic integers (p prime).

Problem A.20. *Let G be a solvable torsion group in which every Abelian subgroup is finitely generated. Prove that G is finite.*

Solution. A finitely generated Abelian torsion group is finite. Therefore, the problem is equivalent to the following one: If in a solvable torsion group every Abelian subgroup is finite, then G itself is finite. We are going to prove the following stronger statement: If in a solvable torsion group every Abelian normal subgroup is finite, then G itself is finite.

First we prove a lemma: If a group G has a commutative normal subgroup with a commutative factor group, and furthermore every Abelian normal subgroup of G is finite, then G itself is finite.

Let H be such a normal subgroup. Using Zorn's lemma, we see that the partially ordered set of all Abelian normal subgroups of G containing

H has a maximal element F. If an element g is permutable with every element of F, then F and g generate a commutative normal subgroup (it is normal because G/H is commutative), so by the maximality of F we have $g \in F$, that is, F is its own centralizer. Consequently, G/F is isomorphic to a subgroup of $\mathrm{Aut}\,(F)$ that is itself finite in view of the finiteness of F. Thus G is also finite.

Now we prove the statement using induction on the length of the derived series of the group G. If this length is 1, there is nothing to prove. If the length is 2, the statement follows from the previous lemma. In the general case, consider P, the last term of the derived series, different from the identity. P is a commutative normal subgroup (hence finite) and the length of the derived series of G/P is one less than the length of G. We only have to check that every commutative normal subgroup of G/P is finite. But if N/P is a commutative normal subgroup of G/P, then N satisfies the conditions of the lemma, so N is finite (hence N/P is also finite), completing the proof. □

Remark. In the problem, the condition on the solvability (or some weakened form of it) cannot be dropped: Novikov and Adian constructed examples of infinite groups where the order of the elements is finite (in fact bounded) but every Abelian subgroup is finite (actually, cyclic).

Problem A.21. *We say that the rank of a group G is at most r if every subgroup of G can be generated by at most r elements. Prove that there exists an integer s such that for every finite group G of rank 2 the commutator series of G has length less than s.*

Solution. We start by showing that if G is a finite group of rank 2, N a normal subgroup of it, and N^* the intersection of the normal subgroups of N that have prime index, then the following statement is true:

Statement. If in every rank-2 automorphism group of N/N^* the length of the commutator series is at most $n(>0)$, then $h(G)$, the length of the commutator series of G, is at most $n+2$.

Proof. The condition implies that the group obtained by restricting the inner automorphism of G/N^* to N/N^* has a commutator series whose length is at most n. In other words, taking the member $G^{(n)}$ of the commutator series $G \geqslant G' \geqslant G'' \geqslant \dots$ and forming the commutator group $[N, G^{(n)}]$, this will be a subgroup of N^*. Now in case $N \leqslant G^{(n)}$, applying this observation to N' in place of N we see that the group N'/N'^* is cyclic, for if a and b generate N, then the commutators $[a, b]$ and $[N', N]$ generate N', so N'/N'' is cyclic. Thus N'/N''^* is commutative, because — again by the first observation — N''/N''^* is a subgroup of the center of N'/N''^*, which gives $N''^* = N''$, and thus $N''' = N''$. Therefore, $h(G) \leqslant n+2$.

We now show that for every subgroup A of the group $L = GL_2(p)$ of 2×2 invertible matrices over the field of p elements (p prime), we have $h(A) \leqslant 5$ (this bound is not sharp).

In the group L of order $(p^2-1)p(p-1)$, the matrices of determinant 1 form a normal subgroup $S = SL_2(p)$ and the factor group L/S is commutative. If c is a generating element of the multiplicative group of the field with p elements, then the diagonal matrix whose diagonal entries are c and c^{-1} is an element of order $p-1$ in S. By multiplying with a generator element of the multiplicative group of the field with p^2 elements, we get an automorphism of order p^2-1 of the same field; this implies that S has an element of order $p+1$. As a consequence, we see that every Sylow subgroup of odd order of S is cyclic, since any odd prime power dividing $(p-1)p(p+1)$ divides precisely one of the three factors. Therefore, if A is a subgroup of rank 2 of L, then for every normal subgroup N of $A \cap S$, any automorphism group of N/N^* has a commutator series of length at most 2 (because N/N^* is a direct product of a cyclic group of odd order and at most two groups of order 2); therefore, by the statement first proved we have $h(A \cap S) \leqslant 4$, and consequently $h(A) \leqslant 5$.

Now for every normal subgroup N of a finite group G of rank 2, it is true that every Sylow subgroup of N/N^* is either of prime order or the direct product of two groups of prime order. Therefore every automorphism group B of rank 2 of N/N^* is isomorphic to a subgroup of the direct product of groups A of the abovementioned type, which implies $h(B) \leqslant 5$, and then, once more using the first proved statement, we get $h(G) \leqslant 7$ (which is again not the exact bound). □

Remarks.

1. Instead of the straightforward proof described above, we could have obtained the statement of the problem by applying the theorem of Blackburn on finite p-groups of rank 2 and the theorem of Zassenhaus on solvable matrix groups.
2. The statement of the problem does not remain true if we consider finite groups of rank 3 instead of finite groups of rank 2. A counterexample is provided by the p-Sylow subgroups of the automorphism group of the direct product of two cyclic groups of order p^n ($p > 2$ prime, $n = 1, 2, \dots$).

Problem A.22. *Let R be an Artinian ring with unity. Suppose that every idempotent element of R commutes with every element of R whose square is 0. Suppose R is the sum of the ideals A and B. Prove that $AB = BA$.*

Solution. We will use two lemmas.

Lemma 1. Every idempotent e of R lies in the center of R.

Proof. Let r be an arbitrary element of R. Then $(er - ere)^2 = 0$, and consequently $er - ere$ is permutable with e. Therefore,

$$er - ere = e(er - ere) = (er - ere)e = 0, \quad er = ere.$$

We get $re = ere$ in a similar fashion. Thus $er = re$, proving the lemma.

Lemma 2. If A is a nonnilpotent right ideal of a (right) Artinian ring R, then A can be written as a direct sum $A = eR \oplus N$, where e is an idempotent element of A and N is a suitable nilpotent right ideal of R.

Proof. The proof of this lemma can be found, for instance, in *A. Kertész, Vorlesungen über artinsche Ringe, Akadémiai Kiadó, Budapest, 1968; VEB, Deutsche Verlag, Berlin, 1975, Theorem 6.23, p.155.*

Turning to the proof of the statement, we see that it is enough to prove that $ab \in BA$ for any $a \in A$ and $b \in B$. Since R is the sum of the ideals A and B, we have elements $a_0 \in A$ and $b_0 \in B$ such that $1 = a_0 + b_0$. This gives

$$ab = 1 \cdot ab = (a_0 + b_0)^k ab = a_0^k ab + \Sigma b_i a_i \qquad (b_i \in B, a_i \in A).$$

If A is nilpotent, then for k sufficiently large we have $a_0^k ab = 0$, that is, $ab \in BA$. If A is not nilpotent, then using Lemma 2, we have $A = eR \oplus N$; in particular $a_0 = er + n \quad (r \in R, n \in N)$. Using Lemma 1, we get $a_0^k = ex_k + n^k$ for suitable elements x_k of R.

Since n is nilpotent, we have $n^k = 0$ for sufficiently large k, that is, $a_0^k = ex_k$, and consequently

$$ab = ex_k ab + \Sigma b_i a_i = b'e + \Sigma b_i a_i \in BA \quad (b' \in B)\,. \quad \square$$

Problem A.23. *Let R be an infinite ring such that every subring of R different from $\{0\}$ has a finite index in R. (By the index of a subring, we mean the index of its additive group in the additive group of R.) Prove that the additive group of R is cyclic.*

Solution. Choose an element $a \in R$ such that the order of a in the additive group R^+ is either infinite or a prime number p. Depending on these two cases, let M denote the ring of the integers or the finite field of p elements. Then the subring R_a generated by a consists of the elements of the form $f(a)$, where f is a polynomial with coefficients from M whose constant term is 0.

Suppose $f(a) \neq 0$ for every such $f \neq 0$. Then the elements of the form $f(a^2)$ form a subring that has infinite index in R_a, so also in R. Hence, there exists a polynomial $f \neq 0$ over M with constant term 0 such that $f(a) = 0$. Take such an f whose degree is the least possible. We have

$$f(x) = x(c + h(x)),$$

where $h \neq 0$ (because for $c \in M, c \neq 0$ we know $ca \neq 0$) and h is also a polynomial over M with zero constant term. Let $b = h(a)$. Since the degree of f was minimal, we have $b \neq 0$. Then

$$b^2 = h(a) \cdot h(a) = h(a)(c + h(a)) - c \cdot h(a) = -c \cdot h(a) = -c \cdot b.$$

This implies that the subring R_b generated by b consists of the elements of the form $c \cdot b \quad (c \in M)$ alone; therefore, the additive group R_b^+ is cyclic.

By the assumption of the problem R_b^+ has finite index in R^+, so R^+ is finitely generated and, as R^+ is infinite, R_b^+ is also infinite. So a could not have finite order in R^+, and consequently R^+ is torsion-free. By the fundamental theorem of finitely generated Abelian groups, R^+ can be written as a direct sum of infinite cyclic groups. But R^+ contains an infinite cyclic group of finite index, namely R_b^+, so the rank of R^+ is 1. Thus R^+ is an infinite cyclic group. □

Problem A.24. *Let S be a semigroup without proper two-sided ideals, and suppose that for every $a, b \in S$ at least one of the products ab and ba is equal to one of the elements a, b. Prove that either $ab = a$ for all $a, b \in S$ or $ab = b$ for all $a, b \in S$.*

Solution 1. S is clearly an idempotent semigroup. Applying a theorem of McLean (see *A. H. Clifford, and G. B. Preston, The Algebraic Theory of Semigroups, vol. I., AMS, Providence, R.I., 1961, p. 129*), we get that there is a congruence relation Θ on S such that S/Θ is a semilattice and each class of the congruence relation Θ is a rectangular band. It can be seen easily that S/Θ itself has no proper two-sided ideals, so it consists of one element; in other words, S is a rectangular band. Recalling the definition, this means that there are sets X and Y such that S is isomorphic to the semigroup $\langle X \times Y; \cdot \rangle$, where multiplication is defined by the rule

$$(x_1, y_1)(x_2, y_2) = (x_1, y_2) \qquad (x_i \in X, y_i \in Y).$$

But then the condition formulated in the problem can hold only in the case where either X or Y consists of a single element, and this proves the statement. □

Solution 2. From the second and first conditions of the problem, we see in turn that

(1) S is idempotent,

(2) for any elements $a, b \in S$, there exist elements $x, y \in S$ such that $b = xay$.

Let us define relations L, R on S in the following way:

$$aLb \overset{\text{def}}{\iff} ab = a,$$

$$aRb \overset{\text{def}}{\iff} ab = b.$$

L is clearly transitive, and in view of (1) it is also reflexive. Suppose for some $a, b \in S$ we have aLb. Then using (1) and (2), we see

$$a = ab = axay = axay^2 = (axay)y = ay,$$

$$b = xay = x(ay)(ay) = (xay)(ay) = ba,$$

therefore bLa. Thus, we have proved that L (and similarly R) is an equivalence relation. The definition of L and R and the second condition of the problem give in turn that

(3) aLb and aRb are both satisfied only in case $a = b$,

(4) for any elements $a, b \in S$, either aLb or aRb is satisfied.

But for equivalence relations L, R, (3) and (4) can be true simultaneously only in the case when one of L and R is the full relation $S \times S$. □

Problem A.25. *Let $\mathbb{Z}$ be the ring of rational integers. Construct an integral domain I satisfying the following conditions:*

(a) $\mathbb{Z} \subsetneq I$;

(b) no element of $I \setminus \mathbb{Z}$ is algebraic over $\mathbb{Z}$ (that is, not a root of a polynomial with coefficients in $\mathbb{Z}$);

(c) I only has trivial endomorphisms.

Solution 1. Choose a transcendental real number a, and let A be the set of numbers of the form $f(a)/g(a)$, where $f, g \in \mathbb{Z}[x]$ and the polynomial g is primitive, that is, the greatest common divisor of its coefficients is 1. A is an integral domain with identity that satisfies

a) $A \setminus \mathbb{Z}$ contains no algebraic number,

b) to every $\alpha \in A$ ($\alpha \neq 0$) there is a $\beta \in A$ ($\beta \neq 0$) such that $\alpha\beta \in \mathbb{Z}$.

Let $\mathcal{M}$ be the set of all integral domains I with $\mathbb{Z} < I < \mathbb{R}$ that satisfy a) and b). $\mathcal{M}$ satisfies the conditions of Zorn's lemma, so it contains a maximal element J. We are going to show that

c) for every $\alpha \in J$ ($\alpha > 0$) there exist elements $\beta_1, \beta_2, \ldots, \beta_k \in J$ and $\gamma_1, \gamma_2, \ldots, \gamma_k \in J$ such that

$$\alpha \sum_{i=1}^{k} \beta_i^2 = \sum_{j=1}^{n} \gamma_j^2 .$$

Indeed, let $\alpha \in J$, $\alpha > 0$ be arbitrary. If α is an integer or $\sqrt{\alpha} \in J$, then the statement holds trivially. Now let $\alpha \in J \setminus \mathbb{Z}$, $\alpha \neq \beta^2$ ($\beta \in J$), and look at the integral domain $J\sqrt{\alpha}$; this satisfies b). If $\beta + \gamma\sqrt{\gamma} \neq 0$, ($\beta, \gamma \in J$), then there is a $\delta \in J$ ($\delta \neq 0$) such that $\delta(\beta^2 - \gamma^2\alpha) \in \mathbb{Z}$ holds (we can assume $\beta^2 - \gamma^2\alpha \neq 0$) and so $(\beta + \gamma\sqrt{\alpha})(\delta\beta - \delta\sqrt{\alpha}) \in \mathbb{Z}$.

Since $J \lneqq J[\sqrt{\alpha}]$, the maximality of J implies that $J[\sqrt{\alpha}]$ cannot satisfy a). So there are $\beta, \gamma \in J$ with $\gamma \neq 0$ such that $\beta + \gamma\sqrt{\alpha}$ is algebraic, that is, there is a polynomial $f(x) = \sum_{i=0}^{N} a_i x^i \in \mathbb{Z}[x]$ with $f(\beta + \gamma\sqrt{\alpha}) = 0$. We have

$$f(\beta + \gamma\sqrt{\alpha}) = \sum_{i=0}^{N} a_i(\beta + \gamma\sqrt{\alpha})^i = C + D\sqrt{\alpha}\,,$$

where C and D, being polynomials with integer coefficients of β, γ and α, belong to J. If $D = 0$, then $C = 0$, so $f(\beta - \gamma\sqrt{\alpha}) = C - D\sqrt{\alpha} = 0$, that is, $\beta - \gamma\sqrt{\alpha}$ is also algebraic. This implies that $\gamma\sqrt{\alpha}$, consequently $\gamma^2\alpha$, is algebraic, so by a) $\gamma^2\alpha = n$ is an integer, and thus the statement of c) holds for α. If, on the other hand, $D \neq 0$, then $C + D\sqrt{\alpha} = 0$ implies $\alpha D^2 = C^2$, so c) is satisfied again.

Now we show that J satisfies the requirements of the problem. Let ϕ be a nontrivial endomorphism of J. Clearly, $\phi(1) = 1$, and so $\phi(n) = n$ for every $n \in \mathbb{Z}$. If $\alpha \in J$ ($\alpha \neq 0$), then b) implies $\phi(\alpha) \neq 0$. If $\alpha > 0$, then it is easy to deduce from c) that $\phi(\alpha) > 0$, which means that ϕ is order-preserving. But then ϕ can only be the identity that we easily verify if we extend in an operation-preserving way to the quotient field of J and observe that in this way ϕ becomes an order-preserving map, fixing all rational numbers. □

Solution 2. We prove that the integral domain

$$J = \mathbb{Z}\left[x, \frac{1}{x}, \frac{1}{x+3}, \frac{1}{x+10}\right]$$

satisfies the conditions of the problem. The elements of $J \setminus \mathbb{Z}$ are clearly transcendental, so we only have to show that J has no nontrivial endomorphism. J is evidently the set of rational functions of the form $f(x)/(x^k(x+3)^n(x+10)^m)$, where $f(x) \in \mathbb{Z}[x]$ and $k, n, m \geqslant 0$. This shows that the invertible elements of J are precisely the elements of the form $\varepsilon x^k(x+3)^n(x+10)^m$, where $\varepsilon = \pm 1$ and k, n, m are arbitrary integers. Let ϕ be an endomorphism of J. If ϕ is not identically zero, then $\phi(1) = 1$, so the image of an invertible element is invertible. This implies

$$\phi(x) = \varepsilon_1 x^{k_1}(x+3)^{n_1}(x+10)^{m_1}, \tag{1}$$

$$\phi(x+3) = \varepsilon_2 x^{k_2}(x+3)^{n_2}(x+10)^{m_2} = \phi(x) + 3, \quad \text{and} \tag{2}$$

$$\phi(x+10) = \varepsilon_3 x^{k_3}(x+3)^{n_3}(x+10)^{m_3} = \phi(x) + 10, \tag{3}$$

where $\varepsilon_i = \pm 1$ and k_i, n_i, m_i are suitable integers ($i = 1, 2, 3$). If $k_1 < 0$, then (1) and (2) imply that $k_2 = k_1$ (the right-hand side of (2) has a pole of order $-k_1$ in 0, thus the left-hand side too), which gives

$$\varepsilon_2 x^{k_2}(x+3)^{n_2}(x+10)^{m_2} = \varepsilon_1(x+3)^{n_1}(x+10)^{m_1} + 3x^{-k_1}. \tag{4}$$

Substituting $x = 0$, we get

$$\varepsilon_2 3^{n_2} 10^{m_2} = \varepsilon_1 3^{n_1} 10^{m_1},$$

which gives $\varepsilon_1 = \varepsilon_2$, $n_1 = n_2$, $m_1 = m_2$. But this is impossible in view of (4); therefore $k_1 \geqslant 0$. We get similarly $k_i \geq 0$, $n_i \geqslant 0$, and $m_i \geqslant 0$ ($i = 1, 2, 3$).

If $n_1 > 0$, then substituting in (3) $x = -3$ we get $n_3 = 0$ and $\varepsilon_3 \cdot (-3)^{k_3} \cdot 7^{m_3} = 0$, which is clearly impossible. Therefore, $n_1 = 0$ and similarly $m_1 = 0$. We can't have $k_1 = 0$, because then $\phi(x) = \varepsilon_1$ would give $\phi(x+3) = 2$ or 4, contradicting (2). Therefore $k_1 > 0$, and substituting $x = 0$ in (2), we get $k_2 = 0$ and $\varepsilon_2 \cdot 3^{n_2} \cdot 10^{m_2} = 3$. This gives $\varepsilon_2 = 1$, $n_2 = 1$, and $m_2 = 0$; therefore

$$\phi(x+3) = x + 3 = \phi(x) + 3, \qquad \text{that is,} \qquad \phi(x) = x.$$

Therefore, $\phi(a) = a$ holds for all $a \in J$, which was to be proved. □

Problem A.26. *Let $p > 5$ be a prime number. Prove that every algebraic integer of the pth cyclotomic field can be represented as a sum of (finitely many) distinct units of the ring of algebraic integers of the field.*

Solution. Put $\zeta = e^{2\pi i/p}$, and let us denote the pth cyclomatic field by $K_p = \mathbb{Q}(\zeta)$. It is well known that $1+\zeta, 1+\zeta^2, \ldots, 1+\zeta^{p-1}$ and ζ are units in K_p. So $\varepsilon_1 = (\zeta+\zeta^{-1})^2$ and $\varepsilon_2 = -(\zeta^2+\zeta^{-2})$ are also units. Furthermore,

$$\varepsilon_1 + \varepsilon_2 = 2\,. \tag{1}$$

We show that ε_1 and ε_2 are independent, that is, there are no rational integers a and b, at least one of them is nonzero, and such that

$$\varepsilon_1^a \varepsilon_2^b = 1\,. \tag{2}$$

Suppose the contrary. With the notation $4a = a'$, $-2b = b'$, we deduce from (2)

$$(\zeta+\zeta^{-1})^{a'} = -(\zeta^2+\zeta^{-2})^{b'}\,. \tag{3}$$

Let d denote the smallest positive integer such that $2^d \equiv 1 \pmod{p}$. Clearly, $3 \leqslant d \leqslant p-1$. It is well known that K_p is a normal extension of $\mathbb{Q}$ and the automorphisms of K_p are determined by $\zeta \to \zeta^i$ $(i = 1, 2, \ldots, p-1)$. Repeatedly applying the automorphism $\zeta \to \zeta^2$ to (3), we get

$$\begin{aligned}
(\zeta^2+\zeta^{-2})^{a'} &= (\zeta^{2^2}+\zeta^{-2^2})^{b'} \\
(\zeta^{2^2}+\zeta^{-2^2})^{a'} &= (\zeta^{2^3}+\zeta^{-2^3})^{b'} \\
&\ \vdots \\
(\zeta^{2^{d-1}}+\zeta^{-2^{d-1}})^{a'} &= (\zeta^{2^d}+\zeta^{-2^d})^{b'} = (\zeta+\zeta^{-1})^{b'}\,.
\end{aligned} \tag{4}$$

(3) and (4) impliy

$$(\zeta+\zeta^{-1})^{a'^d-b'^d} = 1\,.$$

If $a'^d \neq b'^d$, then $\zeta+\zeta^{-1}$ is a root of unity. But $\zeta+\zeta^{-1}$ is a real number, so the only way it can be a root of unity is for it to be equal to 1 or -1. But this would imply that ζ is a sixth or third root of unity, which is not possible. So $a'^d = b'^d$ and consequently $a' = \pm b' \neq 0$. But then (3) implies

$$\left[(\zeta+\zeta^{-1})(\zeta^2+\zeta^{-2})\right]^{a'} = 1 \tag{5}$$

or

$$\left(\frac{\zeta+\zeta^{-1}}{\zeta^2+\zeta^{-2}}\right)^{a'} = 1\,. \tag{6}$$

But the left-hand side of both (5) and (6) is a power with exponent $a' \neq 0$ of a real number that can be equal to 1 only if the numbers themselves are equal to 1 or -1. But this would imply that ζ is a root of a polynomial of degree at most 6 from $\mathbb{Z}[x]$, which is different from the seventh cyclomatic polynomial. Since this is impossible, ε_1 and ε_2 are indeed independent.

As is well known, $\zeta_1 = 1, \zeta_2 = \zeta, \ldots, \zeta_{p-1} = \zeta^{p-2}$ form an integer basis of K_p, which means that every algebraic integer of K_p can be represented as a linear combination of these units with integer coefficients. Therefore, every algebraic integer of K_p can be written as a sum of units of the form $\pm\zeta_k\varepsilon_1^j\varepsilon_2^i$ $(1 \leqslant k \leqslant p-1; i,j \in \mathbb{Z})$, and it follows from the previous observations that these units are pairwise different.

Now let α be an arbitrary algebraic integer in K_p, and suppose that

$$\alpha = \sum_{i=l}^{m}\sum_{j=n}^{q}\sum_{k=1}^{p-1} a_{ijk}\zeta_k\varepsilon_1^j\varepsilon_2^i\,, \tag{7}$$

where $a_{ijk}, l, m, n, q \in \mathbb{Z}$. We can assume that some $a_{ijk} \neq 0$ and that (7) is a representation for which the value of

$$M = \sum_{i,j,k} |a_{ijk}| \tag{8}$$

is minimal.

Using induction on M, we prove that α has a representation of the form

$$\alpha = \sum_{i=l}^{r}\sum_{j=n}^{s}\sum_{k=1}^{p-1} a'_{ijk}\zeta_k\varepsilon_1^j\varepsilon_2^i\,,$$

where $r, s \in \mathbb{Z}$ and $a'_{ijk} = -1$, 0, or 1. This is trivially true for $M = 1$. Suppose $M \geqslant 2$ and that the statement is true for all α such that

$$\sum_{i,j,k} |a_{ijk}| < M\,.$$

Let α be an algebraic integer in K_p that can be represented in the form (7) with property (8) (if such an element exists at all). In view of (1), clearly

$$2\zeta_k\varepsilon_1^j\varepsilon_2^i = \zeta_k\varepsilon_1^{j+1}\varepsilon_2^i + \zeta_k\varepsilon_1^j\varepsilon_2^{i+1}\,. \tag{9}$$

Applying (9) repeatedly to (7), any a_{ijk} with $|a_{ijk}| \geqslant 2$ can finally be reduced to -1, 0, or 1. Furthermore, by the minimality of M, $\sum_{i,j,k}|a_{ijk}|$ remains unchanged during the repeated applications of (9). So, after a finite number of steps, we arrive at

$$\alpha = \sum_{j=n}^{t}\sum_{k=1}^{p-1} a'_{ljk}\zeta_k\varepsilon_1^j\varepsilon_2^l + \sum_{i=l+1}^{u}\sum_{j=n}^{v}\sum_{k=1}^{p-1} b_{ijk}\zeta_k\varepsilon_1^j\varepsilon_2^i\,, \tag{10}$$

where $b_{ijk}, t, u, v \in \mathbb{Z}$ and $a'_{ijk} = -1$, 0, or 1. In addition,

$$\sum_{j,k}|a'_{ljk}| + \sum_{i,j,k}|b_{ijk}| = M$$

and

$$\sum_{j,k}|a'_{ljk}| \geqslant 1\,,$$

so we can apply the induction hypothesis to the second term of (10), and this concludes the proof. □

Problem A.27. *Suppose that the automorphism group of the finite undirected graph $X = (P, E)$ is isomorphic to the quaternion group (of order 8). Prove that the adjacency matrix of X has an eigenvalue of multiplicity at least 4. ($P = \{1, 2, \ldots, n\}$ is the set of vertices of the graph X. The set of edges E is a subset of the set of all unordered pairs of elements of P. The group of automorphisms of X consists of those permutations of P that map edges to edges. The adjacency matrix $M = [m_{ij}]$ is the $n \times n$ matrix defined by $m_{ij} = 1$ if $\{i, j\} \in E$ and $m_{ij} = 0$ otherwise.)*

Solution. Let π be a permutation of the set P, and let A_π be the corresponding permutation matrix, that is, the matrix that has 1 in the ith row and jth column if $i\pi = j$, and 0 otherwise. We see that A_π is an orthogonal matrix: $A_\pi^T = A_\pi^{-1}$. We also see easily that the element in the ith row and jth column of $A_\pi^{-1} M A_\pi$ is just $m_{i\pi, j\pi}$. This means that

(1) $\pi \in \mathrm{Aut}(X) \iff A_\pi^{-1} M A_\pi = M$.

We are going to use the following well-known (and trivial) fact:

(2) If $AB = BA$ (A and B are $n \times n$ matrices), then the subspace belonging to some eigenvalue of B ("*eigensubspace*") is an invariant subspace of A.

Let the eigensubspaces of M (in $\mathbb{R}^n$) be $V_1, \ldots, V_s$ and $\dim V_i = n_i$. The subspaces V_i are pairwise orthogonal (with respect to the usual scalar multiplication), and their direct sum is $\mathbb{R}^n$. (This follows from the fact that M is a real symmetric matrix.) So, applying (2) we see that the group of the orthogonal matrices that are permutable with M is a subgroup of the group $O(V_1) \times \cdots \times O(V_s)$ (here $\times$ denotes direct product, $O(V)$ the group of orthogonal transformations of the Euclidean space V; later $O(k)$ will denote the group of $k \times k$ real orthogonal matrices). In view of (1) $\mathrm{Aut}(X)$ is isomorphic to a subgroup of the group $O(n_1) \times \cdots \times O(n_s)$. Since $O(1) < O(2) < O(3)$, it will be enough to prove the following:

(3) The quaternion group is not isomorphic to any subgroup of $O(3) \times \cdots \times O(3)$. To prove this, suppose that the quaternion group $H = \{\pm 1, \pm i, \pm j, \pm k\}$ has an embedding

$$f : H \hookrightarrow O(3) \times \cdots \times O(3),$$

and let us denote by p_r the rth projection of the group $O(3) \times \cdots \times O(3)$ to $O(3)$. Now $f_r = f \circ p_r : H \to O(3)$ is a homomorphism. We claim the following:

(4) The kernel of any homomorphism $h : H \to O(3)$ contains $-1 \in H$. This implies (3) as it gives $-1 \in \ker f_r$ for all r, thus $-1 \in \ker f$, proving that f cannot be an injection. To prove (4), let us recall the full list of finite subgroups of $O(3)$:

(5) A finite subgroup of $O(3)$ is isomorphic to one of the following:

$$A_5 \times Z_2, S_4 \times Z_2, A_4 \times Z_2, Z_n \times Z_2, D_n \times Z_2, A_5, S_4, A_4, D_n, Z_n,$$

where A_t denotes the alternating group of degree t, S_t the symmetric group of degree t, Z_t the cyclic group of order t, and D_t the dihedral group of degree t (thus order $2t$) (see *H. S. M. Coxeter, Introduction to Geometry, Wiley, New York, 1969, Table III*).

(6) Cosequence: $O(3)$ has no subgroup isomorphic to H. So $h : H \to O(3)$ cannot be an embedding. Therefore $\ker h$ contains an $x \in H$ different from the identity. But then either x or x^2 is equal to -1, so necessarily $-1 \in \ker h$.

This concludes the proof. □

Problem A.28. *For a distributive lattice L, consider the following two statements:*

(A) Every ideal of L is the kernel of at least two different homomorphisms.

(B) L contains no maximal ideal.

Which one of these statements implies the other?

(Every homomorphism φ of L induces an equivalence relation on L: $a \sim b$ if and only if $a\varphi = b\varphi$. We do not consider two homomorphisms different if they imply the same equivalence relation.)

Solution.

(A) IMPLIES (B). Assume, by contradiction, that L contains a maximal ideal M. Suppose we have a homomorphism φ whose kernel is equal to M. We will show that the image of $L \setminus M$ by φ is a single point. Since we also know that under the equivalence relation induced by φ all elements of M are equivalent and no element of $L \setminus M$ is equivalent to an element of M, we get that there is at most one homomorphism whose kernel is M. Let a and b be two elem ents of $L \setminus M$. The following description of the ideal (M, a) generated by M and a is well known:

$$(M, a) = \{x \in L \mid \exists m \in M : x \leqslant a \vee m\}.$$

In view of the maximality of M, we have $(M, a) = L$. So for some element $m \in M$ we have $b \leqslant a \vee m$, therefore $b\varphi \leqslant a\varphi \vee m\varphi = a\varphi$. Similarly, $a\varphi \leqslant b\varphi$ consequently, $a\varphi = b\varphi$, as we stated.

(B) DOES NOT IMPLY (A). We construct a counterexample. Let

$$L = \{V \mid V \subset \mathbb{R} \text{ finite}\} \cup \{(-\infty, x] \cup V \mid x \in \mathbb{R}, V \subset \mathbb{R} \text{ finite}\}.$$

The operations on L should be the set-theoretic union and intersection. In this way, L is a sublattice of the lattice of subsets of $\mathbb{R}$ and therefore L is distributive. We claim that it has no maximal ideal, but on the other hand, the ideal $\{\emptyset\}$ is the kernel of a single homomorphism.

Let J be a proper ideal of L. Not all half-lines of the form $(-\infty, x]$ can occur in J because otherwise all finite subsets of $\mathbb{R}$ would be elements of J forcing $J = L$. So there is an $x \in \mathbb{R}$ such that $(-\infty, x] \notin J$ which of course implies $(-\infty, y] \notin J$ for all $y \geqslant x$. So the ideal

$$J^* = \{(-\infty, z] \cup V \mid z \leqslant x, V \subset \mathbb{R} \text{ finite}\} \cup \{V \mid V \subset \mathbb{R} \text{ finite}\}$$

is a proper ideal properly containing J. Consequently, J is not a maximal ideal.

To prove the other claim, let φ be a homomorphism whose kernel is $\{\emptyset\}$. We show that φ is a monomorphism, which means that the equivalence relation induced by φ is such that each element is equivalent only to itself, and this contradicts (A). Assume, on the contrary, that $a, b \in L$, $a \neq b$ but $a\varphi = b\varphi$. The symmetric difference of a and b is nonempty, so there is a $p \in \mathbb{R}$ with, say, $p \in a$ and $p \notin b$. This implies

$$\{p\}\varphi = (\{p\} \wedge a)\varphi = \{p\}\varphi \wedge a\varphi = \{p\}\varphi \wedge b\varphi = (\{p\} \wedge b)\varphi = \emptyset\varphi = 0,$$

in contradiction with $\ker\varphi = \{\emptyset\}$. Thus φ is indeed a monomorphism as claimed. □

Problem A.29. *Let $\mathcal{V}$ be a variety of monoids such that not all monoids of $\mathcal{V}$ are groups. Prove that if $A \in \mathcal{V}$ and B is a submonoid of A, there exist monoids $S \in \mathcal{V}$ and C and epimorphisms $\varphi : S \to A$, $\varphi_1 : S \to C$ such that $((e)\varphi_1^{-1})\varphi = B$ (e is the identity element of C).*

Solution. Take $D \in \mathcal{V}$, which is not a group. Then there exists an element $x \in D$ that is not invertible. The powers of x with positive integer exponents and the identity element form a submonoid of D, and the dichotomy underlying the previous description of this submonoid is a congruence relation of it. Taking the natural homomorphism, we see that the variety $\mathcal{V}$ contains the monoid $E = \{f, g\}$ $(f \neq g)$ with operation defined by $f \cdot f = f$, $f \cdot g = g \cdot f = g$, and $g \cdot g = g$.

Consider the set

$$S = \{(a, g) \mid a \in A\} \cup \{(b, f) \mid b \in B\}.$$

On the direct product $A \times E$ of the monoids A and E, we have

$$\begin{aligned} (a_1, g)(a_2, g) &= (a_1a_2, g), \quad (b_1, f)(b_2, f) = (b_1b_2, f), \\ (a_1, g)(b_1, f) &= (a_1b_1, g) \qquad (a_i \in A, b_i \in B), \end{aligned}$$

so S is actually a submonoid of $A \times E$ proving $S \in \mathcal{V}$. We see at the same time that the two sets defining S give rise to a congruence relation on S. Let φ_1 be the natural homomorphism with respect to this congruence relation and denote the image of S by C. Clearly, the set of the elements of S that are mapped to the identity element of C equals $\{(b, f) \mid b \in B\}$.

If we denote by φ the projection of an ordered pair to its first component, we see that $\varphi : S \to A$ is an epimorphism, and we have $\{(b, f) \mid b \in B\}\,\varphi = B$, which concludes the proof. □

Problem A.30. *For what values of n does the group $SO(n)$ of all orthogonal transformations of determinant 1 of the n-dimensional Euclidean space possess a closed regular subgroup? ($G \leq SO(n)$ is called regular if for any elements x, y of the unit sphere there exists a unique $\varphi \in G$ such that $\varphi(x) = y$.)*

Solution. The answer is that such a subgroup exists precisely for $n = 2$ and $n = 4$, namely for $n = 2$ the group SO(2) itself and for $n = 4$ the symplectic group denoted by Sp(1), which we get by taking the multiplications with quaternions whose absolute value is 1.

We show that there are no other solutions of the problem. For odd n, SO(n) certainly does not contain a regular subgroup since in this case every transformation $\varphi \in$ SO(n) has a fixed vector.

Suppose therefore that n is even, $n = 2m$. In this case, for every $\varphi \in G$ (G is subgroup in question) the space $\mathbb{R}^{2m}$ decomposes as the direct sum of m pairwise orthonal two-dimensional subspaces, invariant with respect to φ: $\mathbb{R}^{2m} = h_1 \oplus \cdots \oplus h_m$. Clearly, on every subspace h_i, φ induces a rotation φ_i with an angle α_i.

Now we prove two lemmas:

Lemma 1. For every $\varphi \in G$ there is a Jordan decomposition $\mathbb{R}^{2m} = h_1 \oplus \cdots \oplus h_m$ such that $\alpha_1 = \cdots = \alpha_m$.

Proof. There is a one-dimensional subgroup Φ of G such that $\varphi \in \Phi$ (this is well known). It is also well known that there exists an element $g \in \Phi$ whose powers are dense in Φ. Let $\mathbb{R}^{2m} = h_1 \oplus \cdots \oplus h_m$ be a Jordan decomposition with respect to g, and let the restriction of g to h_i be g_i. Then g_i is a rotation with angle β_i on h_i. At least one of the β_is is an *irrational multiple* of 2π, and consequently every other β_j is an irrational multiple of 2π; otherwise for some k, $g_j^k \neq \mathrm{id}$ would fix an element of h_j. This gives that the restriction Φ_i of Φ to h_i is the group SO(2) for all i. Let us consider the group homomorphism λ: $\Phi_i \to \Phi_j$, which maps every element $g_i^* \in \Phi_i$ (where $g^* \in \Phi$) to $g_j^* \in \Phi_j$. Thus λ is continuous and 1-to-1 in view of the regularity and compactness of G, so it produces an automorphism of the topological group SO(2). But there are just two such automorphisms: the identity and (identifying SO(2) with the unit circle of $\mathbb{C}$) the conjugation. These map a rotation with angle α to a rotation with angle α, so — using $\lambda(\varphi_i) = \varphi_j$ — we get $\alpha_i = \alpha_j$, which was to be proved.

This lemma immediately implies

Lemma 2. If some $\varphi \in G$ maps a unit vector to a vector orthogonal to it, then $\varphi^2 = -\mathrm{id}$ and φ maps any other vector y to a vector orthogonal to it; furthermore, the vectors y, $\varphi(y)$ span a plane that is invariant with respect to φ.

Proof. Let $\mathbb{R}^{2m} = h_1 \oplus \cdots \oplus h_m$ be the decomposition whose existence is assured by Lemma 1. Suppose that x is orthogonal to $\varphi(x)$ and $x = x_1 + \cdots + x_m$, $\varphi(x) = \varphi(x_1) + \cdots + \varphi(x_m)$ the Jordan decomposition with respect to ϕ and consider the inner product $(x, \varphi(x)) = \sum_{i=1}^{m}(x_i, \varphi(x_i)) = 0$. Lemma 1 guarantees that either $(x_i, \varphi(x_i)) \geqslant 0$ for all terms or $(x_i, \varphi(x_i))$ for all terms, therefore necessarily $(x_i, \varphi(x_i)) = 0$ for all terms. Thus, Lemma 1 implies that φ rotates with an angle $\pi/2$ in every subspace h_i, and from this the statements of the present lemma follow immediately.

Now the question formulated in the problem is answered by the following

Proposition. If the group $\mathrm{SO}(2m)$ ($2m \geqslant 4$) has a closed, regular subgroup G, then $2m = 4$ (furthermore, it is true that G is the universal covering group of $\mathrm{SO}(3)$, that is, $\mathrm{Sp}(1)$).

Proof. Let the unit vectors e_0, e_1 be arbitrary, but orthogonal, and let $\phi_1 \in G$ be the group element with $\varphi_1(e_0) = e_1$. Let e_2 be a unit vector that is orthogonal to the plane spanned by e_0, e_1, and let $e_3 = \varphi_1(e_2)$. By Lemma 2, the vectors e_0, e_1, e_2, e_3 are pairwise orthogonal (the subspace orthogonal to an invariant subspace is itself invariant) and denoting by $\varphi_i \in G$ the group element with $\varphi_i(e_0) = e_i$, then the following formulas are already true:

$$\varphi_1^2 = -\mathrm{id}, \quad \varphi_2^2 = -\mathrm{id}, \quad \varphi_3^2 = -\mathrm{id},$$

$$\varphi_1\varphi_2 = -\varphi_2\varphi_1 = \varphi_3,\ \varphi_1\varphi_3 = -\varphi_3\varphi_1 = -\varphi_2,\ \varphi_2\varphi_3 = -\varphi_3\varphi_2 = \varphi_1\,.$$

The formulas $\varphi_i\varphi_j = -\varphi_j\varphi_i$ in the last row follow immediately from the relation $(\varphi_i\varphi_j)^2 = -\mathrm{id}$. In the case $2m = 4$, the proof is finished.

It remains to show that the case $2m > 4$ cannot occur. Suppose $2m > 4$, and let K be the subspace spanned by e_0, e_1, e_2, e_3. If we choose a unit vector e_4 that is orthogonal to K and define $e_5 = \varphi_1(e_4)$, $e_6 = \varphi_2(e_4)$, $e_7 = \varphi_3(e_4)$, then the vectors e_4, e_5, e_6, e_7 are pairwise orthogonal and span a subspace K^*, orthogonal to K, consequently $e_0, e_1, \ldots, e_7$ is an orthonormal system of vectors. Denoting by $\varphi_i \in G$ the group element with $\varphi_i(e_0) = e_i$ ($i = 0, 1, \ldots, 7$), the transformations φ_i have the following multiplication table:

	φ_0	φ_1	φ_2	φ_3	φ_4	φ_5	φ_6	φ_7
φ_0	id	φ_1	φ_2	φ_3	φ_4	φ_5	φ_6	φ_7
φ_1	φ_1	$-$id	φ_3	$-\varphi_2$	φ_5	$-\varphi_4$	φ_7	$-\varphi_6$
φ_2	φ_2	$-\varphi_3$	$-$id	φ_1	φ_6	$-\varphi_7$	$-\varphi_4$	φ_5
φ_3	φ_3	φ_2	$-\varphi_1$	$-$id	φ_7	φ_6	$-\varphi_5$	$-\varphi_4$
φ_4	φ_4	$-\varphi_5$	$-\varphi_6$	$-\varphi_7$	$-$id	φ_1	φ_2	φ_3
φ_5	φ_5	φ_4	φ_7	$-\varphi_6$	$-\varphi_1$	$-$id	φ_3	$-\varphi_2$
φ_6	φ_6	$-\varphi_7$	φ_4	φ_5	$-\varphi_2$	$-\varphi_3$	$-$id	φ_1
φ_7	φ_7	φ_6	$-\varphi_5$	φ_4	$-\varphi_3$	φ_2	$-\varphi_1$	$-$id

From this multiplication table, we deduce the following identities:

$$\begin{aligned}(\varphi_4\varphi_5)\varphi_6 &= (\varphi_4\varphi_1\varphi_4)\varphi_6 = (-\varphi_4\varphi_4\varphi_1)\varphi_6 = \varphi_1\varphi_6 \\ &= \varphi_1(\varphi_2\varphi_4) = (\varphi_1\varphi_2)\varphi_4 = \varphi_3\varphi_4 = \varphi_7 \\ \varphi_4(\varphi_5\varphi_6) &= \varphi_4(\varphi_1\varphi_4\varphi_2\varphi_4) = \varphi_4(-\varphi_1\varphi_2\varphi_4\varphi_4) \\ &= \varphi_4(\varphi_1\varphi_2) = \varphi_4\varphi_3 = -\varphi_3\varphi_4 = -\varphi_7\,.\end{aligned}$$

This contradicts the associativity of the multiplication, so the case $2m > 4$ cannot occur. □

Problem A.31. *In a lattice, connect the elements $a \wedge b$ and $a \vee b$ by an edge whenever a and b are incomparable. Prove that in the obtained graph every connected component is a sublattice.*

Solution. Let us denote the fact that two elements a and b belong to the same connected component of the graph by $a \sim b$. The statement to be proved is that whenever $b_1, \dots, b_n$ is a path in the graph, we have $b_1 \sim b_1 \vee b_n$, which in turn reduces to $b_1 \vee b_{n-1} \sim b_1 \vee b_n$. We will denote the fact that two elements x and y are incomparable by $x \parallel y$.

Lemma. Let $x \parallel y, a = x \vee y, b = x \wedge y$, z arbitrary. Then one of the following two cases holds:

(a) $a \vee z \sim b \vee z$;

(b) one of a, b, z is $\sim a \vee z$ and one of a, b, z is $\sim b \vee z$.

This will indeed prove the statement of the problem. In the case $b_{n-1} = x \vee y$ and $b_n = x \wedge y$, we can apply the lemma with $a = bn - 1$, $b = b_n$, $z = b_1$. In case (a), the proof is finished immediately, and in case (b) we get that one of b_1, b_{n-1}, b_n is $\sim b_1 \vee b_n$, which suffices to conclude the proof. In the case $b_{n-1} = x \wedge y$ and $b_n = x \vee y$, we can proceed similarly.

We will prove the lemma in six steps. First, we deal with that part of the statement that refers to $a \vee z$.

1. If $a \gtrless z$, then $a \vee z = a, z$ so the first part of (b) holds.

2. Suppose $a \nleqslant z \vee y$. This gives $x \parallel z \vee y$ since $x \geqslant z \vee y$ would imply $x \geqslant y$, whereas in case $x \leqslant z \vee y$ we would have $a = x \vee y \leqslant z \vee y$. Therefore $a \vee z = x \vee y \vee z > x \wedge (y \vee z)$. In view of $x \wedge a = x$, we can further deduce $x \wedge (y \vee z) = x \wedge (a \wedge (y \vee z)) < x \vee (a \wedge (y \vee z))$ as soon as we establish $x \parallel a \wedge (z \vee y)$. To prove the latter statement, suppose first $x \leqslant a \wedge (z \vee y)$. Then in view of $y \leqslant a \wedge (z \vee y)$, we have $a = x \vee y \leqslant a \wedge (z \vee y)$, contrary to the initial assumption. On the other hand, if $x \geqslant a \wedge (z \vee y)$, then $y \leqslant a \wedge (z \vee y)$ implies $y \leqslant x$, which is not the case.

But $x \vee (a \wedge (y \vee z)) = a$ since $x, a \wedge (y \vee z) \leqslant a$ and $a \wedge (z \vee y) \geqslant y$, thus $x \vee (a \wedge (y \vee z)) \geqslant x \vee y = a$. We arrived at the conclusion $a \sim a \vee z$.

The case $a \nleqslant z \vee x$ can be treated similarly.

3. So we can assume $a \leqslant z \vee x$, $a \leqslant z \vee y$, and $a \parallel z$. Thus $x \leqslant z \vee y$, which implies $z \vee a = z \vee y$. Similarly $z \vee a = z \vee x$ consequently $z \vee x = z \vee y = z \vee a$. We can further assume $x \parallel z$ and $y \parallel z$. (Suppose namely $x \leqslant z$. This gives $a \vee z = y \vee z \sim a$. The case $y \leqslant z$ is similar, and in case $x \geqslant z$ or $y \geqslant z$ we get $a \geqslant z$.)

Suppose now $z \parallel (z \vee b) \wedge a$. This implies

$$\begin{aligned} z \vee a \sim z \wedge a &= ((z \vee b) \wedge z) \wedge a = ((z \vee b) \wedge a) \wedge z \\ &< ((z \vee b) \wedge a) \vee z = z \vee b . \end{aligned}$$

As to the last equality, $\leqslant$ is clear and we get the other direction from $b \leqslant (z \vee b) \wedge a$.

What if z and $(z \vee b) \wedge a$ are comparable? $z \leqslant (z \vee b) \wedge a \leqslant a$ is excluded, so this can happen only if $z > (z \vee b) \wedge a \geqslant b$, and consequently $z = z \vee b$.

4. Thus, we can assume $z = z \vee b$, $z \vee x = z \vee y = z \vee a$. If we have $z \wedge x = b$, we get $z \vee a = z \vee x > z \wedge x = b$, giving $z \vee a \sim b$, and similarly in the case $z \wedge y = b$.

5. In the remaining case let $t = ((x \wedge z) \vee (y \wedge z))$. Then $z \vee a = z \vee x > z \wedge x = x \wedge t$ since $z \geqslant t$ and $z \wedge x \leqslant t$. Certainly, $x \parallel t$ since $x \leqslant t \leqslant z$ would imply $x \leqslant z$ and $x \geqslant t$ implies $x \wedge y \geqslant y \wedge z > b$, which is impossible. So $x \wedge t < x \vee t = x \vee (y \wedge z)$. We have $x \parallel y \wedge z$ since $x \leqslant y \wedge z \leqslant y$ is impossible and $x \geqslant y \wedge z$ gives $x \wedge y \geqslant y \wedge z > b$, a contradiction.

So, finally, $x \vee (y \wedge z) > x \wedge (y \wedge z) = b$ since $b \leqslant z$. This gives $a \vee z \sim b$, proving the part of the lemma that refers to $a \vee z$.

6. Now let us deal with the statement concerning $b \vee z$. In case $b \gtrless z$, we have $b \vee z = b, z$. Otherwise, $b \vee z > b \wedge z$, and applying the part of the statement that has been already proved to the dual lattice, we get that $b \wedge z \sim$ to one of a, b, z or $b \wedge z \sim a \wedge z$. In case $a \gtrless z$, we have $a \wedge z = a, z$, whereas in case $a \parallel z$ using $a \vee z > a \wedge z$ we get $a \vee z \sim b \vee z$, and this finally concludes the proof of the lemma. □

Problem A.32. *Let G be a transitive subgroup of the symmetric group S_{25} different from S_{25} and A_{25}. Prove that the order of G is not divisible by 23.*

Solution. We assume that $23 \mid |G|$, and we try to reach a contradiction from that assumption. Denote the stabilizer of a point x by Stab (x). By the transitivity and our assumption, all the Stab (x) are isomorphic permutation groups containing a 23-cycle. Suppose that Stab (x) is not transitive. Then taking the fixed point y of Stab (x), we see that the fixed point of Stab (y) is x. That would give a pairing of the 25 points, which is impossible. So G is 2-transitive, and as Stab (x, y) contains a 23-cycle G is actually 3-transitive. We have seen that the 23-cycles of G act transitively so we can take G to be the group generated by them. We know then that $G \leqslant A_{25}$ and G is 3-transitive.

By a well-known theorem (see *H. Wielandt, Finite Permutation Groups, Academic Press, New York, 1964*, 13.10), the order of G cannot be divisible by 11. We know that Stab (x, y) has prime degree and is transitive. According to Burnside's theorem (Wielandt, 7.3) it is either 2-transitive or isomorphic to a subgroup of the group of all affine mappings: $z \mapsto az + b \quad (a, b \in \mathrm{GF}(23))$.

In the first case, G is 4-transitive, so its order is divisible by $25 \cdot 24 \cdot 23 \cdot 22$, therefore by 11, which is excluded. So Stab (x, y) is a subgroup of a group of order $2 \cdot 11 \cdot 23$. Again 11 is excluded. An element of order 2 would be of the form $z \mapsto -z + b$, which is a product of 11 transpositions and is thus not contained in A_{25}. Therefore, the order of Stab (x, y) is 23, G

is sharply 3-transitive, but then by the theorem of Zassenhaus *(Wielandt, 20.5)*, $25 - 1 = 24$ has to be a prime power, which is not the case. □

Problem A.33. *Let G be a finite group and $\mathcal{K}$ a conjugacy class of G that generates G. Prove that the following two statements are equivalent:*

(1) There exists a positive integer m such that every element of G can be written as a product of m (not necessarily distinct) elements of $\mathcal{K}$.

(2) G is equal to its own commutator subgroup.

Solution. Suppose first that (1) is satisfied, and consider the natural homomorphism $\varphi : G \to G/G'$. Since for any $h, k \in \mathcal{K}$ there exists a $g \in G$ such that $k = g^{-1}hg$ we have $\varphi(h^{-1}k) = \varphi(h^{-1}g^{-1}hg) = 1$. Thus $\mathcal{K}$ is contained in the coset $hG' = \varphi(h) \quad (h \in \mathcal{K})$.

Now let $g = h_1 \cdots h_m$ and $g' = h_1' \cdots h_m' \quad (h_1 \cdots h_m,\ h_1' \cdots h_m' \in \mathcal{K})$ be any two elements of G. By our previous remarks,

$$\varphi(g) = \varphi(h_1) \cdots \varphi(h_m) = \varphi(h)^m = \varphi(h_1') \cdots \varphi(h_m') = \varphi(g').$$

In other words G/G' has only one element, that is, $G = G'$.

Conversely, suppose (2) is satisfied and let us denote $|G|$ by n. Since $G = G'$, every element g of G can be written as a product of commutators:

$$g = a_1^{-1}b_1^{-1}a_1b_1 \cdots a_r^{-1}b_r^{-1}a_rb_r = a_1^{n-1}b_1^{n-1}a_1b_1 \cdots a_r^{n-1}b_r^{n-1}a_rb_r. \qquad (1)$$

Since $\mathcal{K}$ generates G and the inverse of an element of $\mathcal{K}$ can be replaced by its $(n-1)$th power, it is true that every $a_i,\ b_i\ (i = 1, \ldots, r)$ can be written as a product of elements of $\mathcal{K}$. Substituting these representations in (1), we can deduce the existence of a natural number k_g such that g can be written as a product of $k_g \cdot n$ elements of $\mathcal{K}$. But then for any $k \geqslant k_g$, we can also write g as a product of $k \cdot n$ elements of $\mathcal{K}$ by simply adding a suitable number of factors of the type $h^n = 1 \quad (h \in \mathcal{K})$. Let $l = \max\{k_g \mid g \in G\}$. Then every element of G can be written as a product of $m = n \cdot l$ elements of $\mathcal{K}$. □

Remark. In the proof of (2) $\Rightarrow$ (1) the only property of $\mathcal{K}$ that we used was the fact that $\mathcal{K}$ is a system of generators of G.

Problem A.34. *Consider the lattice of all algebraically closed subfields of the complex field $\mathbb{C}$ whose transcendency degree (over $\mathbb{Q}$) is finite. Prove that this lattice is not modular.*

Solution. For any set of numbers $H \subseteq \mathbb{C}$, let us denote by $A(H)$ the smallest algebraically closed subfield of $\mathbb{C}$ containing H. Let us consider a system of numbers $\alpha, \beta, \gamma_\xi \ (\xi < \omega_1)$ that are algebraically independent (over $\mathbb{Q}$). Since the algebraic closure of a countable set of numbers is countable, such a system does exist. We are going to show that for a

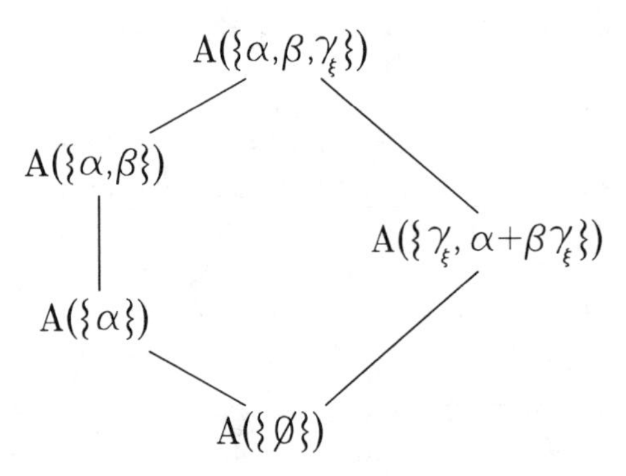

Figure A.1.

suitable $\xi < \omega_1$ the following diagram is actually the Hasse diagram of a sublattice of the lattice of algebraically closed subfields of $\mathbb{C}$, and this will prove the statement of the problem.

The containments indicated by the lines are evident. Furthermore,

$$A\,(A(\{\alpha\}) \cup A(\{\gamma_\xi, \alpha + \beta\gamma_\xi\})) = A(\{\alpha, \beta, \gamma_\xi\})$$

since the right-hand side clearly contains the left-hand side, while the left-hand side contains each of the numbers α, β, γ_ξ. We have $A(\{\alpha\}) \neq A(\{\alpha, \beta\})$ in view of the original assumption, so all that remains to be proved is to show that for some ξ we have

$$A(\{\alpha, \beta\}) \cap A(\{\gamma_\xi, \alpha + \beta\gamma_\xi\}) = A(\emptyset)\,.$$

For this, it is enough to prove

$$A(\{\alpha, \beta\}) \cap (A(\{\gamma_\xi, \alpha + \beta\gamma_\xi\}) \setminus A(\emptyset)) = \emptyset\,. \tag{1}$$

Suppose $\xi_1 \neq \xi_2$. Then the transcendency degree of

$$A\,(A(\{\gamma_{\xi_1}, \alpha + \beta\gamma_{\xi_1}\}) \cup A(\{\gamma_{\xi_2}, \alpha + \beta\gamma_{\xi_2}\})) = A(\{\gamma_{\xi_1}, \gamma_{\xi_2}, \alpha, \beta\})$$

is equal to 4; therefore, the transcendency degree of

$$A(\{\gamma_{\xi_1}, \alpha + \beta\gamma_{\xi_1}\}) \cap A(\{\gamma_{\xi_2}, \alpha + \beta\gamma_{\xi_2}\})$$

is equal to 0, which means

$$(A(\{\gamma_{\xi_1}, \alpha + \beta\gamma_{\xi_1}\}) \setminus A(\emptyset)) \cap (A(\{\gamma_{\xi_2}, \alpha + \beta\gamma_{\xi_2}\}) \setminus A(\emptyset)) = \emptyset\,.$$

Now the set $A(\{\alpha, \beta\})$ is countable and the sets $A(\{\gamma_\xi, \alpha + \beta\gamma_\xi\}) \setminus A(\emptyset)$ are pairwise disjoint, so there exists a ξ for which (1) holds, and this concludes the proof. $\square$

Problem A.35. *Let I be an ideal of the ring R and f a nonidentity permutation of the set $\{1,2,\ldots,k\}$ for some k. Suppose that for every $0 \neq a \in R$, $aI \neq 0$ and $Ia \neq 0$ hold; furthermore, for any elements $x_1, x_2, \ldots, x_k \in I$,*

$$x_1 x_2 \cdots x_k = x_{1f} x_{2f} \cdots x_{kf}$$

holds. Prove that R is commutative.

Solution. Let $m \in \{1,2,\ldots,k\}$ the smallest number that is not fixed by f. Clearly, $m < mf$. We then have $x_1 x_2 \cdots x_k = x_1 \cdots x_{m-1} x_{mf} \cdots x_{kf}$, that is, $x_1 \cdots x_{m-1}(x_m \cdots x_k - x_{mf} \cdots x_{kf}) = 0$ for any $x_1, \ldots, x_k \in I$. Fixing the elements $x_2, \ldots, x_k$ we see that $x_2 \cdots x_{m-1}(x_m \cdots x_k - x_{mf} \cdots x_{kf})$ annihilates I so has to be equal to 0. Repeating the argument, we get $x_m \cdots x_k = x_{mf} \cdots x_{kf}$ for any $x_m, \ldots, x_k \in I$.

Let $n = mf$. Then for every $a \in R$ and $x_m \cdots x_k \in I$, we get

$$a x_m x_{m+1} \cdots x_k = a x_{mf} x_{(m+1)f} \cdots x_{kf} = x_m x_{m+1} \cdots x_{n-1} a x_n x_{n+1} \cdots x_k.$$

(We get the second equality by using $a x_{mf} \in I$.) Rearranging,

$$(a x_m \cdots x_{n-1} - x_m \cdots x_{n-1} a) x_n \cdots x_k = 0$$

for any $a \in R$ and $x_m \cdots x_k \in I$. Again applying our previous argument, we get

$$a x_m \cdots x_{n-1} = x_m \cdots x_{n-1} a$$

for any $a \in R$ and $x_m \cdots x_{n-1} \in I$.

Now let $a, b \in R$ and $x_m \cdots x_{n-1} \in I$ be arbitrary elements. Then we have

$$\begin{aligned}(ab) x_m \cdots x_{n-1} &= x_m \cdots x_{n-1}(ab) = (x_m \cdots x_{n-1} a) b \\ &= (a x_m \cdots x_{n-1}) b = (a x_m) x_{m+1} \cdots x_{n-1} b \\ &= b (a x_m) x_{m+1} \cdots x_{n-1} = (ba) x_m \cdots x_{n-1}.\end{aligned}$$

Applying once more our previous argument to the resulting identity

$$(ab - ba) x_m \cdots x_{n-1} = 0 \qquad (a, b \in R,\ x_m \cdots x_{n-1} \in I),$$

we get $ab - ba = 0$ or $ab = ba$, which was to be proved. □

Remark. The proof did not use the fact that I was a two-sided ideal of the ring R, only the fact that I was a left ideal in the multiplicative semigroup of the ring R.

Problem A.36. *Prove that any identity that holds for every finite n-distributive lattice also holds for the lattice of all convex subsets of the $(n-1)$-dimensional Euclidean space. (For convex subsets, the lattice operations are the set-theoretic intersection and the convex hull of the set-theoretic union. We call a lattice n-distributive if*

$$x \wedge (\bigvee_{i=0}^{n} y_i) = \bigvee_{j=0}^{n} (x \wedge (\bigvee_{\substack{0 \leq i \leq n \\ i \neq j}} y_i))$$

holds for all elements of the lattice.)

Solution. Let us introduce some notation. Let E^{n-1} denote the Euclidean space of $n-1$ dimensions; for $X \subseteq E^{n-1}$, let $C(X)$ denote the convex hull of X; and let L be the lattice of convex subsets of E^{n-1}. For an arbitrary $H \subseteq E^{n-1}$, denote by $L(H)$ the partially ordered set $(\{X \cap H : X \in L\}, \subseteq)$. $L(H)$ is a complete lattice since it is closed with respect to forming arbitrary intersections and it has a greatest element. If we denote the lattice operations in $L(H)$ by $\vee_H$ and $\wedge_H$, respectively, then we see that for $X, Y \in L(H)$ we have $X \wedge_H Y = X \cap Y$ and $X \vee_H Y = C(X \cup Y) \cap H$. For a finite H, $L(H)$ is also finite since $|L(H)| \leqslant 2^{|H|}$.

First, we show that for a finite $H \subseteq E^{n-1}$, $L(H)$ is n-distributive. Let $X, Y_0, \ldots, Y_n \in L(H)$ be arbitrary elements of $L(H)$. By Carathéodory's theorem for an arbitrary subset U of E^{n-1},

$$C(U) = \bigcup_{\substack{V \subseteq U \\ |V| \leqslant n}} C(V).$$

Therefore

$$C(Y_0 \cup \cdots \cup Y_n) = \bigcup_{j=0}^{n} C\left(\bigcup_{\substack{i=0 \\ i \neq j}}^{n} Y_i\right)$$

also holds and so

$$X \wedge_H \bigvee_{i=0}^{n} Y_i = X \cap H \cap C\left(\bigcup_{i=0}^{n} Y_i\right) = H \cap X \cap H \cap \bigcup_{j=0}^{n} C\left(\bigcup_{\substack{i=0 \\ i \neq j}}^{n} Y_i\right)$$

$$= H \cap \bigcup_{j=0}^{n} \left(X \cap H \cap C(\bigcup_{\substack{i=0 \\ i \neq j}}^{n} Y_i)\right)$$

$$= \bigvee_{H\, j=0}^{n} \left(X \wedge_H \bigvee_{H\, \substack{i=0 \\ i \neq j}}^{n} Y_i\right).$$

That means that $L(H)$ is n-distributive.

Let p be a lattice-theoretic term in k variables, let $H \subseteq E^{n-1}$, and denote by p (resp., p_H) the term function induced by the term p on L

(resp., $L(H)$). For arbitrary $X_1, \ldots, X_k \in L$, $p_H(X_1 \cap H, \ldots, X_k \cap H)$ is a monotone function of H, that is, $H_1 \subseteq H_2$ implies

$$p_{H_1}(X_1 \cap H_1, \ldots, X_k \cap H_1) \subseteq p_{H_2}(X_1 \cap H_2, \ldots, X_k \cap H_2)\,.$$

This follows easily from the monotonicity of $\cup$, $\cap$, and C, using induction on the length of the term p.

We claim that for arbitrary $X_1, \ldots, X_k \in L$ and $h \in p(X_1, \ldots, X_k)$ there is a finite $H \subseteq E^{n-1}$ such that $h \in H$ and $h \in p_H(X_1 \cap H, \ldots, X_k \cap H)$. This statement will be proved by induction on the length of p. If p is just a variable, then we can choose $H = \{h\}$. Suppose that the statement is proved already for terms whose length is less than the length of p_n, and let $p = p' \wedge p''$. Then there are finite sets H_1 and H_2 such that $h \in p'_{H_1}(X_1 \cap H_1, \ldots, X_k \cap H_1)$, $h \in p''_{H_2}(X_1 \cap H_2, \ldots, X_k \cap H_2)$, and $h \in H_1$, $h \in H_2$. Let $H = H_1 \cup H_2$. Then we have

$$\begin{aligned} h &\in p'_{H_1}(X_1 \cap H_1, \ldots, X_k \cap H_1) \cap p''_{H_2}(X_1 \cap H_2, \ldots, X_k \cap H_2) \\ &\subseteq p'_H(X_1 \cap H, \ldots, X_k \cap H) \cap p''_H(X_1 \cap H, \ldots, X_k \cap H) \\ &= p_H(X_1 \cap H, \ldots, X_k \cap H)\,. \end{aligned}$$

Now let $p = p' \vee p''$. Then

$$h \in p(X_1, \ldots, X_k) \subseteq C\left(p'(X_1, \ldots, X_k) \cup p''(X_1, \ldots, X_k)\right).$$

By Carathéodory's theorem, there exist finitely many points $h'_1, \ldots, h'_u \in p'(X_1, \ldots, X_k)$ and $h''_1, \ldots, h''_v \in p''(X_1, \ldots, X_k)$ such that

$$h \in C\left(\{h'_1, \ldots, h'_u, h''_1, \ldots, h''_v\}\right) \quad \text{and} \quad u + v \leqslant n.$$

(Actually, we could obtain $u \leqslant 1$, $v \leqslant 1$, but we don't need this now.) By the induction hypothesis there exist finite sets $H_1, \ldots, H_u, G_1, \ldots, G_v \subseteq E^{n-1}$ such that $h'_i \in p'_{H_i}(X_1 \cap H_i, \ldots, X_k \cap H_i)$ and $h''_j \in p''_{G_j}(X_1 \cap G_j, \ldots, X_k \cap G_j)$ $(1 \leqslant i \leqslant n, 1 \leqslant j \leqslant v)$. Let $H = \{h\} \cup H_1 \cup \cdots \cup H_u \cup G_1 \cup \cdots \cup G_v$. Now

$$\begin{aligned} h &\in C\left(\{h'_1, \ldots, h'_u, h''_1, \ldots, h''_v\}\right) \cap H \\ &\subseteq C\left(p'_H(X_1 \cap H, \ldots, X_k \cap H) \cup p''_H(X_1 \cap H, \ldots, X_k \cap H)\right) \cap H \\ &= p_H(X_1 \cap H, \ldots, X_k \cap H)\,. \end{aligned}$$

Now let $p(x_1, \ldots, x_k) = q(x_1, \ldots, x_k)$ be an identity of lattices that is satisfied by every n-distributive lattice, and let $X_1, \ldots, X_k \in L$ be arbitrary. We have to prove $p(X_1, \ldots, X_k) = q(X_1, \ldots, X_k)$. To prove the inclusion $p(X_1, \ldots, X_k) \subseteq q(X_1, \ldots, X_k)$, take an arbitrary $h \in p(X_1, \ldots, X_k)$. Then for a suitable finite $H \subseteq E^{n-1}$ we have $h \in p_H(X_1 \cap H, \ldots, X_k \cap H)$. Since $L(H)$ is finite and n-distributive we have

$$\begin{aligned} h \in q_H(X_1 \cap H, \ldots, X_k \cap H) &\subseteq q_{E^{n-1}}(X_1 \cap E^{n-1}, \ldots, X_k \cap E^{n-1}) \\ &= q(X_1, \ldots, X_k)\,. \end{aligned}$$

This establishes the inclusion $p(X_1, \ldots, X_k) \subseteq q(X_1, \ldots, X_k)$, and the reverse inclusion can be proved in the same way. $\square$

Problem A.37. *Let $x_1, x_2, y_1, y_2, z_1, z_2$ be transcendental numbers. Suppose that any 3 of them are algebraically independent, and among the 15 four-tuples only $\{x_1, x_2, y_1, y_2\}$, $\{x_1, x_2, z_1, z_2\}$, and $\{y_1, y_2, z_1, z_2\}$ are algebraically dependent. Prove that there exists a transcendental number t that depends algebraically on each of the pairs $\{x_1, x_2\}$, $\{y_1, y_2\}$, and $\{z_1, z_2\}$.*

Solution. Take (nonzero) polynomials p, q, r with rational coefficients such that

$$\begin{aligned} p(x_1, x_2, y_1, y_2) &= 0\,, \\ q(x_1, x_2, z_1, z_2) &= 0\,, \\ r(y_1, y_2, z_1, z_2) &= 0\,. \end{aligned}$$

Since the polynomials in one variable,

$$p_1(X) = p(X, x_2, y_1, y_2)$$

and

$$q_1(X) = q(X, x_2, z_1, z_2),$$

have a common root, their resultant is zero:

$$R(p_1, q_1) = 0\,.$$

But it is well known that $R(p_1, q_1)$ is a polynomial with integer coefficients of the coefficients of p_1 and q_1. Therefore, it is a polynomial with rational coefficients of the numbers x_2, y_1, y_2, z_1, z_2:

$$R(p_1, q_1) = f(x_2, y_1, y_2, z_1, z_2)\,.$$

By the conditions, we know that x_2 does not depend algebraically on the numbers y_1, y_2, z_1 whereas z_2 does depend on them; consequently, x_2 does not depend algebraically on y_1, y_2, z_1, z_2. Therefore, $f(x_2, y_1, y_2, z_1, z_2) = 0$ implies that $f(X, y_1, y_2, z_1, z_2)$ is the (identically) zero polynomial of X.

Now take an arbitrary rational number u. Then we have

$$f(u, y_1, y_2, z_1, z_2) = 0,$$

and consequently the polynomials

$$p_2(X) = p(X, u, y_1, y_2) \quad \text{and} \quad q_2(X) = q(X, u, z_1, z_2)$$

have a common root t (since the resultant of the polynomials $p_2(X)$ and $q_2(X)$ is equal to $f(u, y_1, y_2, z_1, z_2)$).

Define now polynomials p^*, q^* in three variables as follows:

$$\begin{aligned} p^*(X, Y_1, Y_2) &= p(X, u, Y_1, Y_2)\,, \\ q^*(X, Y_1, Y_2) &= q(X, u, Y_1, Y_2)\,. \end{aligned}$$

It is clear that we can choose u in such a way that none of the polynomials is the zero polynomial (if, for every rational number u, at least one of them were the zero polynomial, then for every complex number u the same would hold, in particular for $u = x_2$ which is certainly not the case).

Then we have

$$p^*(t, y_1, y_2) = 0,$$

so t, y_1, y_2 are algebraically dependent. But y_1, y_2 are algebraically independent, so t is a transcendental number and depends algebraically on the pair $\{y_1, y_2\}$. Similarly, t depends algebraically on the pair $\{z_1, z_2\}$.

To prove that t depends algebraically on the pair $\{x_1, x_2\}$, suppose the contrary. Then $\{x_1, x_2, t\}$ would be a transcendency base of the algebraically closed field generated by $\{x_1, x_2, y_1, y_2\}$, and similarly it would be a transcendency base of the algebraically closed field generated by $\{x_1, x_2, z_1, z_2\}$. But these two fields cannot be isomorphic, since for example $\{x_1, x_2, y_1, z_1\}$ are algebraically independent. □

Problem A.38. *Let $F(x, y)$ and $G(x, y)$ be relatively prime homogeneous polynomials of degree at least one having integer coefficients. Prove that there exists a number c depending only on the degrees and the maximum of the absolute values of the coefficients of F and G such that $F(x, y) \neq G(x, y)$ for any integers x and y that are relatively prime and satisfy* $\max\{|x|, |y|\} > c$.

Solution. In the course of the proof, we will denote by c, c_1, c_2, c_3 numbers that depend only on the degrees and the maximum of the absolute values of the coefficients of F and G.

Let x and y be arbitrary relatively prime integers that are solutions of the equation

$$F(X, Y) = G(X, Y). \tag{1}$$

We will show that $\max\{|x|, |y|\} < c$ for a suitable constant c, and this is just the statement of the problem. We can suppose $xy \neq 0$ since otherwise in view of $(x, y) = 1$ we have $\max\{|x|, |y|\} \leqslant 1$. Let us denote the degrees of F and G by m and n, respectively, and let us assume $m \geqslant n$, say. Let $f(X) = F(X, 1)$ and $g(X) = G(X, 1)$. By assumption, F and G are relatively prime homogeneous polynomials so at least one of $f(X)$ and $g(X)$ is a nonconstant polynomial. Actually, we can assume that none of them is a constant. If $f(X)$ were a constant, then denoting the coefficient of X^n in G by a, we would get $a \neq 0$. Now (1) together with $(x, y) = 1$ implies $y \mid a$, and since $F(X, a) - G(X, a)$ is not the identically zero polynomial, we get $\max\{|x|, |y|\} < c_1$.

f and g are relatively prime because F and G are relatively prime. Thus, denoting by R the resultant of f and g, we have $R \neq 0$. Furthermore $|R| \leqslant c_2$. By a well-known theorem (see, for example, *L. Rédei, Algebra, Akadémiai Kiadó, Budapest, 1954, Thm. 196, p. 376*), there exist polynomials with integer coefficients $A(X)$ and $B(X)$ such that $\deg A < \deg g \leqslant n$,

$\deg B < \deg f \leqslant m$, and

$$A(X)f(X) + B(X)g(X) = R.$$

This implies that there exist homogeneous polynomials with integer coefficients $A_1(X,Y)$, $B_1(X,Y)$ and $A_2(X,Y)$, $B_2(X,Y)$ such that

$$A_1(X,Y)F(X,Y) + B_1(X,Y)G(X,Y) = R \cdot X^{m+n-1} \tag{2}$$

and

$$A_2(X,Y)F(X,Y) + B_2(X,Y)G(X,Y) = R \cdot Y^{m+n-1}. \tag{3}$$

But then, using (1), (2), (3), and the fact that $(x,y) = 1$, we see that $F(x,y)$ and $G(x,y)$ are different from 0 and both are divisors of R. So, in case $F(x,y) = G(x,y) = d$ we have $d \mid R$, and therefore $|d| \leqslant c_2$. This implies that x/y is a root of the polynomial with integer coefficients

$$h(t) = d^{m-n} f^n(t) - g^m(t),$$

which is not identically zero since f and g are relatively prime.

But then by Rolle's theorem we have $\max\{|x|, |y|\} < c_3$, and then with the choice $c = \max\{c_1, c_3\}$ we get the statement we wanted to prove. □

Remark. It is not hard to get explicit expressions for c_1, c_2, c_3, and c using the degrees of F and G and the maximum of the absolute values of their coefficients.

Problem A.39. *Determine all finite groups G that have an automorphism f such that $H \not\subseteq f(H)$ for all proper subgroups H of G.*

Solution. If φ has a fixed point $a \in G$, $a \neq 1$, then $\varphi(\langle a \rangle) = \langle a \rangle$ so $G = \langle a \rangle$, and as every subgroup of $\langle a \rangle$ is also fixed, $G = \langle a \rangle$ cannot have any proper subgroups, consequently it is a cyclic group of prime order.

Suppose now that φ is fixed-point-free. By a well-known result (see, for example, *D. Gorenstein, Finite Groups, Harper & Row, New York, 1968, Theorem 10.1.2*), for every prime p there exists a p-Sylow subgroup P of G with $\varphi(P) = P$, so G has to be a p-group. Furthermore, as the center is a characteristic subgroup and p-groups have nontrivial center, G has to be Abelian. Also, in an Abelian p-group, the elements of order p form a characteristic subgroup so every element of G ($\neq 1$) must have order p, that is, G is an elementary Abelian group:

$$G = \bigoplus_{i=1}^{n} Z_p. \tag{1}$$

We now prove that groups of the form (1) indeed possess an automorphism φ with the required property. G can be considered as a vector space of dimension n over the finite field GF(p), and automorphisms can be identified with the invertible linear transformations of this space. Furthermore, subgroups are the same as subspaces. Now for a subgroup H the condition $H \nleq \varphi(H)$ means that H is not an invariant subspace of φ. It is well known that a linear transformation with an irreducible characteristic polynomial can have no invariant subspace. It is also well known that for every n there exists a polynomial of degree n over GF(p) that is irreducible over GF(p). Finally, it is an elementary result that to every polynomial there is a linear transformation with the given polynomial as characteristic polynomial. Combining these observations, we see that a φ with the required properties exists indeed.

So the groups with the required property are the elementary Abelian groups (including the one-element group). □

Problem A.40. *Let p_1 and p_2 be positive real numbers. Prove that there exist functions $f_i : \mathbb{R} \to \mathbb{R}$ such that the smallest positive period of f_i is p_i $(i = 1, 2)$, and $f_1 - f_2$ is also periodic.*

Solution. If p_1/p_2 is rational, there exist positive integers m, n such that $np_1 = mp_2$. Let us take for $i = 1, 2$ the functions

$$f_i(x) = \begin{cases} 1, & \text{if } x = kp_i, \qquad k \in \mathbb{Z}, \\ 0, & \text{otherwise.} \end{cases}$$

The smallest positive period of f_i is p_i. But f_1, f_2, and consequently $f_1 - f_2$ are also periodic with period $np_1 = mp_2 = p$.

Suppose now that p_1/p_2 is irrational. Define for $i, j = 1, 2$

$$f_i(x) = \begin{cases} \alpha_j, & \text{if } x = \alpha_1 p_1 + \alpha_2 p_2, \quad \alpha_1, \alpha_2 \in \mathbb{Z}, \quad i \neq j, \\ 1, & \text{otherwise.} \end{cases}$$

f_i is well defined, because $x = \alpha_1 p_1 + \alpha_2 p_2 = \alpha_1' p_1 + \alpha_2' p_2$ $(\alpha_i, \alpha_i' \in \mathbb{Z},\ \alpha_i \neq \alpha_i')$ would imply that p_1/p_2 is rational.

Now f_i is periodic with period p_i, since

$$f_i(x + p_i) = \begin{cases} \alpha_j = f_i(x), & \text{if } x = \alpha_1 p_1 + \alpha_2 p_2, \\ 1 = f_i(x), & \text{otherwise.} \end{cases}$$

It is clear from the definition that $f_i(x) = 0 \iff x = \alpha_i p_i$, so the smallest positive period of f_i is p_i. On the other hand, $f_1 - f_2$ is periodic with period $p_1 + p_2$, since

$$(f_1 - f_2)(x + p_1 + p_2) = \begin{cases} (\alpha_2 + 1) - (\alpha_1 + 1) = (f_1 - f_2)(x), & \text{if } x = \alpha_1 p_1 + \alpha_2 p_2, \\ 1 - 1 = (f_1 - f_2)(x), & \text{otherwise.} \end{cases}$$ □

Problem A.41. *Determine all real numbers x for which the following statement is true: the field $\mathbb{C}$ of complex numbers contains a proper subfield F such that adjoining x to F we get $\mathbb{C}$.*

Solution. We will prove that a field F with the desired properties exists if and only if either x is transcendental or x is algebraic but some conjugate of x is nonreal. First, if x belongs to one of the classes described above, then there is an automorphism φ of $\mathbb{C}$ for which $x \notin \varphi(\mathbb{R})$. Then $F = \varphi(\mathbb{R})$ will be the suitable proper subfield: $\mathbb{C} = F(x)$.

Conversely, we show that if x is algebraic and all conjugates of x are real, then no subfield has the required property. Suppose on the contrary that $\mathbb{C} = L(x)$ for some proper subfield L of $\mathbb{C}$. Then x is also algebraic over L, so $\mathbb{C}$ is a finite extension of L. We use the following well-known theorem.

Theorem. If K is an algebraically closed field, $\operatorname{char} K = 0$, and K is a finite extension of some proper subfield L, then $| K : L |= 2$.

In view of this theorem, we have $| \mathbb{C} : L |= 2$. Next we prove that $i \notin L$. Suppose we have $x^2 + cx + d = 0$ with $c, d \in L$ and $i \in L$. This implies $x = c' + \sqrt{d'}$ for some $c', d' \in L$, and consequently $\mathbb{C} = L(d')$. But this in turn implies $\sqrt[4]{d} = a + b\sqrt{d}$, with suitable $a, b \in L$. Thus $\sqrt{d} = a^2 + b^2 d + 2ab\sqrt{d}$, so $a^2 + b^2 d = 0$ and $2ab = 1$, and therefore $4a^4 + 4a^2b^2 d = 0$ and $4a^2b^2 = 1$. This gives $4a^4 + d = 0$, $d = -4a^4$, so $\sqrt{d} = \pm i2a^2$, showing $\sqrt{d} \in L$. This leads to $L = L(\sqrt{d})$, and this contradiction proves $i \notin L$.

So $\mathbb{C} = L(i)$, which implies that L is real closed, thus L can be ordered. Let us consider $A = \{y \in L \mid y \text{ algebraic}\}$. Then A can be ordered, too. But a well-known theorem states that every ordering of any subfield of the field of all algebraic numbers is Archimedean, so A is a field with an Archimedean ordering, thus there exists an embedding $\varphi : A \to \mathbb{R}$. But then there exists an extension φ' of φ such that φ' is an embedding of all algebraic numbers to $\mathbb{C}$. Now $L(i) = \mathbb{C}$ implies that $A(i)$ is the set of all algebraic numbers. Clearly, $\varphi'(i) = \pm i$. It is also clear that φ' permutes the conjugates of x among themselves, so $\varphi'(x) \in \mathbb{R}$. Now we surely have $\varphi'(x) \in \varphi'(A(i))$, so $\varphi'(x) \in \varphi'(A(i)) \cap \mathbb{R} = \varphi'(A)$. Thus $\varphi'(x) \in \varphi'(A)$, that is, $x \in A$, which is a contradiction, proving our statement. □

Problem A.42. *Prove the existence of a constant c with the following property: for every composite integer n, there exists a group whose order is divisible by n and is less than n^c, and that contains no element of order n.*

Solution. If $n = p^k (k \geqslant 2)$, then $G = Z_p^k$ is suitable (Z_p denotes the cyclic group of order p). Suppose $n = p_1^{k_1} p_2^{k_2} \cdots p_r^{k_r}$ $(r \geqslant 2)$. We take

$$G = \mathrm{PSL}_2(q) \times Z_{\frac{n}{p_1 p_2}},$$

where q is a prime satisfying $q \equiv 1 \pmod{p_1}$ and $q \equiv -1 \pmod{p_2}$. Then we have

$$|G| = \frac{(q-1)q(q+1)}{2} \cdot \frac{n}{p_1 p_2} = \frac{q-1}{p_1} \cdot \frac{q+1}{p_2} \cdot \frac{1}{2} \cdot q \cdot n,$$

which is divisible by n and $|G| < (q^3 n / p_1 p_2)$. If there were an element of order n in G, then $\mathrm{PSL}_2(q)$ would contain an element of order $p_1 p_2$. But a well-known theorem of Dickson (see, for example *B. Huppert, Endliche Gruppen I, Springer, Berlin, 1967, Hauptsatz II.8.27., p. 213*) states that the order of any cyclic subgroup of $\mathrm{PSL}_2(q)$ divides one of q, $(q-1)/2$ and $(q+1)/2$. As these numbers are pairwise relatively prime and $p_1 | (q-1)/2$ and $p_2 | (q+1)/2$, we see that $\mathrm{PSL}_2(q)$ contains no element of order $p_1 p_2$; consequently, G contains no element of order n.

In our construction, q was an arbitrary prime number in a prescribed residue class mod $p_1 p_2$. A well-known theorem of Linnik (see, for example, *K. Prachar, Primzahlverteilung, Springer, Berlin, 1957, Satz X.4.1., p. 364*) states that the smallest such prime number satisfies $q < (p_1 p_2)^C$ with some constant C. Then we have

$$|G| < (p_1 p_2)^{3C} \frac{n}{p_1 p_2} = (p_1 p_2)^{3C-1} n \leqslant n^{3C}. \quad \square$$

Problem A.43. *Let A be a finite simple groupoid such that every proper subgroupoid of A has cardinality one, the number of one-element subgroupoids is at least three, and the group of automorphisms of A has no fixed points. Prove that in the variety generated by A, every finitely generated free algebra is isomorphic to some direct power of A.*

Solution. We will show that

Every subalgebra of a finite direct power of A is isomorphic to some direct power of A. $\quad (*)$

For an arbitrary natural number $\mathbf{k} \overset{\text{def}}{=} \{1, \ldots, k\}$ and for any $I \subseteq \mathbf{k}$, denote by pr_I the projection

$$A^k \to A^I : (a_1, \ldots, a_k) \to (a_i)_{i \in I}.$$

It is clear that if B is a subgroupoid of A^k then $\mathrm{pr}_I B$ is a subgroupoid of A^I.

Denote by U the set of those elements u of A for which $\{u\}$ is a subgroupoid of A. Furthermore, we shall use the following notation: if $B \subseteq A^k$, $1 \leqslant i < k$ and $a_{i+1}, \ldots, a_k \in A$, then

$$\begin{aligned} &B(x_1, \ldots x_i, a_{i+1}, \ldots, a_k) \\ &= \left\{(x_1, \ldots, x_i) \in A^i \mid (x_1, \ldots x_i, a_{i+1}, \ldots, a_k) \in B\right\}. \end{aligned}$$

It is clear that if B is a subgroupoid of A^k and $a_{i+1}, \dots, a_k \in U$, then $B(x_1, \dots x_i, a_{i+1}, \dots, a_k)$ is a subgroupoid of A^i. We shall denote the operation in the groupoid A by "$\circ$".

Our first observation is that $\circ$ is surjective. It cannot be constant because then A would have a single one-element subgroupoid. On the other hand, the set of the elements that we get as results of the operation $\circ$ is a subgroupoid, so it can only be A itself.

Now we prove two lemmas.

Lemma 1. For every natural number $n \geqslant 1$, if B is a subgroupoid of A^n with $U^n \subseteq B$, then $B = A^n$.

Proof. Suppose for some n there exists a subgroupoid B of A^n for which $U^n \subseteq B \subset A^n$ holds, and choose the minimal value of n, for which this can happen. Clearly, $n \geqslant 2$. Now for an arbitrary $u \in U$, $B(x_1, \dots, x_{n-1}, u)$ is a subgroupoid of A^{n-1}, containing U^{n-1}. So, by the minimality of n, we have $B(x_1, \dots, x_{n-1}, u) = A^{n-1}$. So U is a subset of the set

$$S = \left\{ a \in A \,|\, A^{n-1} x \{a\} \subseteq B \right\},$$

which implies $|S| \geqslant 3$. On the other hand, in view of $B \subset A^n$, we have $S \subset A$. Finally, using the surjectivity of $\circ$, we can easily check that S is a subgroupoid. The contradiction proves Lemma 1.

Lemma 2. A^2 has only the following subgroupoids:
(a) subgroupoids with one element;
(b) $\{u\} \times A$, respectively, $A \times \{u\}$ $(u \in U)$;
(c) automorphisms of A;
(d) A^2.

Proof. Suppose C is a subgroupoid of A^2 not in the list above. Clearly, $\mathrm{pr}_i C = A$ $(i = 1, 2)$; furthermore, for every element $u \in U$, $C(x, u)$ (and similarly $C(u, x)$) is one of the sets A, $\{v\}$ $(v \in U)$. By Lemma 1, $U^2 \not\subseteq C$, so by the previous observations only the following two cases are possible:

$$\begin{aligned} &C = (\{a\} \times A) \cup (A \times \{b\}) \cup C' \text{ for} \\ &\text{some } C' \subseteq (A \setminus U)^2 \text{ and } a, b \in U \end{aligned} \tag{1}$$

or

$$\begin{aligned} &C = \{(u, u\sigma) \,|\, u \in U\} \cup C' \text{ for some} \\ &C' \subseteq (A \setminus U)^2 \text{ and some permutation} \\ &\sigma \text{ of } U\,. \end{aligned} \tag{2}$$

In case (1), let $T = \{a \in A \,|\, (a, a) \in C\}$. Clearly, T is a subgroupoid of A. As $T \cap U = \{a, b\}$, T is a proper subgroupoid, so $|T| = 1$ and $a = b$. By our assumptions, there is an automorphism π of A for which $a \neq a\pi$. Consider the following set:

$$\tilde{C} = \left\{ x \in A \,|\, \exists\, y \in A \text{ such that } (x, y),\ (x\pi^{-1}, y\pi^{-1} \in C \right\}.$$

We can easily see that $\tilde{C}$ is a subgroupoid of A and $\tilde{C} \cap U = \{a, a\pi\}$. Thus $2 \leqslant |\tilde{C}| < |A|$, which is impossible.

In case (2), let

$$D = \{(x,y) \in A^2 \,|\, \exists\ z \in A \text{ such that } (x,z),\ (y,z) \in C\},$$

and denote by Δ the diagonal of A^2, that is, $\Delta = \{(a,a) \,|\, a \in A\}$. It is easy to verify that D is a subgroupoid of A^2 and $D \cap U^2 \subseteq \Delta$. Since, $\mathrm{pr}_1 C = \mathrm{pr}_2 C = A$ and C itself is not a permutation, we have $\Delta \subset D$. So the transitive closure of D is a nontrivial congruence of A, which is impossible.

Now we turn to the proof of $(*)$. We prove by induction on n that if B is a subgroupoid of A^n, then B is isomorphic to some direct power of A. The case $n = 1$ being trivial, suppose $n \geqslant 2$. If $B = A^n$, there is nothing to prove, and if for some index $i \in \mathbf{n}$ we have $|\mathrm{pr}_i B| = 1$, then $B \simeq \mathrm{pr}_{\mathbf{n}-\{i\}} B$ and by induction we are ready. So we can suppose $B \neq A^n$ and $\mathrm{pr}_i B = A$ for every $i \in \mathbf{n}$. Choose a minimal index set $I \subseteq \mathbf{n}$ such that $\mathrm{pr}_I B \neq A^I$, and let $k = |I|$, $I = \{i_1 < \cdots < i_k\}$, and $C = \mathrm{pr}_I B$. Clearly, $k > 1$ and

$$\mathrm{pr}_{\mathbf{k}-\{j\}} C = \mathrm{pr}_{I-\{i_j\}} B = A^{k-1} \text{ for every } j \in \mathbf{k}. \tag{3}$$

Consider the subgroupoid of A^2 of the form

$$C(u_1, \ldots, u_{k-2}, x, y) \quad (u_1, \ldots, u_{k-2} \in U).$$

Because of (3), $C(u_1, \ldots, u_{k-2}, x, y)$ has the property that both projections of it is A, so in view of Lemma 2 — it is either A^2 or an automorphism of A. Since Lemma 1 — and our assumption on C imply that $U^k \not\subseteq C$, there exists $(u'_1, \ldots, u'_{k-2}) \in U^{k-2}$ for which $C(u'_1, \ldots, u'_{k-2}, x, y)$ is an automorphism. Suppose that there exists also some $(u''_1, \ldots, u''_{k-2}) \in U^{k-2}$ for which $C(u''_1, \ldots, u''_{k-2}, x, y) = A^2$. Then in the sequence

$$C(u''_1, \ldots, u''_i, u'_{i+1}, \ldots u'_{k-2}, x, y)\ (i = 0, \ldots, k-2)$$

there are two consecutive terms such that the first is an automorphism of A and the second is A^2. So we can assume $u'_2 = u''_2, \ldots, u'_{k-2} = u''_{k-2}$. Look at the subgroupoid $C' = C(x, u'_2, \ldots, u'_{k-2}, y, u'_1)$ of A^2. Clearly $C'(u'_1, y) = 1$, $C'(u''_1, y) = A$. But by Lemma 2 we see that this cannot happen. So we know that

$$\begin{array}{l} C(u_1, \ldots, u_{k-2}, x, y) \text{ is an automorphism} \\ \text{of } A \text{ for every } u_1, \ldots, u_{k-2} \in U. \end{array} \tag{4}$$

Now let

$$\begin{aligned} D = \{(x,y) \in A^2 \,|\, \exists\ z_1, \ldots, z_{k-1} \in A, \text{ such that} \\ (z_1, \ldots, z_{k-1}, x),\ (z_1, \ldots, z_{k-1}, y) \in C\}. \end{aligned}$$

It is easy to check that D is a subgroupoid of A^2 and $\Delta \subseteq D$ (to prove this we use the following consequence of (3): $\mathrm{pr}_k C = A$). So by Lemma 2 either $D = \Delta$ or $D = A^2$.

Suppose that $D = A^2$, and fix arbitrary elements $u, v \in U$, $u \neq v$. Since $(u, v) \in D$, there exist elements $a_1, \ldots, a_{k-1} \in A$ such that

$$(a_1, \ldots, a_{k-1}, u),\ (a_1, \ldots, a_{k-1}, v) \in C\,.$$

So for the subgroupoid C^* of A^{k-1} defined by

$$C^* = C(x_1, \ldots, x_{k-1}, u) \cap (x_1, \ldots, x_{k-1}, v),$$

we have $(a_1, \ldots, a_{k-1}) \in C^*$. On the other hand, because of (4) we have

$$C^* \cap U^{k-1} = \emptyset, \tag{5}$$

which implies that not all of $a_1, \ldots, a_{k-1}$ are elements of U. For example, let $a_1 \in A \setminus U$. Since a_1 is contained in the subgroupoid $\mathrm{pr}_1 C^*$ of A, we have $\mathrm{pr}_1 C^* = A$. So for every $u_1 \in U$, $C^*(u_1, x_1, \ldots, x_{k-2}) \neq \emptyset$ holds. Let $l \leqslant k-2$ be the greatest index such that there exist elements $u_1, \ldots, u_l \in U$ with the property that the set

$$E = C^*(u_1, \ldots, u_l, x_1, \ldots, x_{k-l-1})$$

is nonempty. In the case $l < k-2$, the maximality of l, while in the case $l = k-2$ (5) shows that $(\mathrm{pr}_1 E) \cap U = \emptyset$. On the other hand $\mathrm{pr}_1 E \neq \emptyset$ since $E \neq \emptyset$. Thus $\mathrm{pr}_1 E$ is a proper subgroupoid of A which is different from the one-element subgroupoids $\{u\}$ $(u \in U)$. This contradiction proves $D = \Delta$.

The equality $D = \Delta$ shows that the projection $C \to \mathrm{pr}_{\mathbf{k}-\{k\}} C$ $(= A^{k-1})$ is injective, so it is an isomorphism. Consequently, the same is true for the projection $B \to \mathrm{pr}_{\mathbf{n}-\{i_k\}} B$. Using the induction hypothesis, we immediately get our claim for B. This concludes the proof of $(*)$.

Since in the variety generated by a finite algebra each finitely generated free algebra is isomorphic to a subalgebra of some finite direct power of the original algebra, $(*)$ proves the statement of the problem. □

Problem A.44. *Let the finite projective geometry P (that is, a finite, complemented, modular lattice) be a sublattice of the finite modular lattice L. Prove that P can be embedded in a projective geometry Q, which is a cover-preserving sublattice of L (that is, whenever an element of Q covers in Q another element of Q, then it also covers that element in L).*

Solution. Let us denote by 0 the smallest element of P. Let $L' = \{a \in L \mid 0 \leqslant a\}$. Then L' is a (clearly modular) cover-preserving sublattice of L. Evidently, it is enough to solve the problem for L' in place of L. Let Q be the sublattice of L' generated by those atoms of L' which are less than 1_P (1_P denotes the greatest element of P). Let δ be the dimension function on Q. Take a $q \in Q$. Then there is a finite set $\alpha_1, \ldots, \alpha_n \in Q$ of atoms of L' such that $1_Q = q \vee \alpha_1 \vee \cdots \vee \alpha_n$ and such that the value of n is minimal. Then

$$\delta(\alpha_1 \vee \cdots \vee \alpha_n) = n$$

(Q being modular) and $\delta(1_Q) = \delta(q) + n$. Using the well-known relation

$$\delta(x \vee y) + \delta(x \wedge y) = \delta(x) + \delta(y), \tag{1}$$

we get $\delta(q \wedge (\alpha_1 \vee \cdots \vee \alpha_n)) = 0$, that is, $\alpha_1 \vee \cdots \vee \alpha_n$ is a complement of q, and therefore Q is a projective geometry. Now let $\beta \in Q$ be an atom of Q. Then there exists an atom $\alpha \in L'$ of L' with $\alpha \leqslant \beta (\leqslant 1_P)$. So $\alpha \in Q$, and since β was an atom, $\beta = \alpha$, and thus β is an atom in L', too.

Now, if $q_1, q_2 \in Q$ and $q_1 \prec_Q q_2$ then there is an atom $\beta \in Q$ such that $q_2 = q_1 \vee \beta$ (using the fact that Q is a projective geometry), but then the modularity of L' implies that $q_1 \prec_{L'} q_2$ also holds. So Q is indeed a cover-preserving sublattice of L'.

We are going to embed P in Q. For an element $p \in P$ let $f(p) = \bigvee\{\alpha \in L' \mid \alpha \leqslant p;\ \alpha \text{ is an atom}\}$. The map f is injective. If namely for two elements $p_1, p_2 \in P$ we have $p_1 \nleqslant p_2$, then there is an element $r \in P$ such that $r \leqslant p_1$, and $r \wedge p_2 = 0$. Then there is an atom $\alpha \in L'$ with $\alpha \leqslant r$ and thus $\alpha \leqslant f(p_1)$ and $\alpha \wedge f(p_2) \leqslant r \wedge p_2 = 0$, proving $f(p_1) \neq f(p_2)$.

Next we prove $f(p_1 \vee p_2) = f(p_1) \vee f(p_2)$. The definition of f gives immediately $f(p_1 \vee p_2) \leqslant f(p_1) \vee f(p_2)$. To prove the inequality in the other direction, let $p_2' \leqslant p_2$ be an atom of P such that $p_1 \vee p_2' = p_1 \vee p_2$ and $p_1 \wedge p_2' = 0$. Clearly, $f(p_2') \leqslant f(p_2)$ and $f(p_1 \vee p_2') = f(p_1 \vee p_2)$, which means that for proving the inequality in the other direction we can assume $p_1 \wedge p_2 = 0$. Let $\alpha \leqslant p_1 \vee p_2$ be an atom of Q. We have to prove $\alpha \leqslant f(p_1) \vee f(p_2)$. In case $\alpha \leqslant p_1$ or $\alpha \leqslant p_2$, this is clear. Suppose $\alpha \wedge p_1 = \alpha \wedge p_2 = 0$. Let $\alpha_1 = p_1 \wedge (\alpha \vee p_2)$ and $\alpha_2 = p_2 \wedge (\alpha \vee p_1)$. Using the modularity of L', we see that α_1 and α_2 are atoms and $\alpha \vee p_2 = \alpha_1 \vee p_2$ and $\alpha \vee p_1 = \alpha_2 \vee p_1$ are true. Again using the relation (1) (where δ denotes the dimension function of L') for the elements $x = p_1 \vee \alpha$ and $y = p_2 \vee \alpha$, we get that the dimension in L' of the element $(p_1 \vee \alpha) \wedge (p_2 \vee \alpha)$ is two. Since $\alpha_1 \neq \alpha_2$ and $\alpha_1, \alpha_2 \leqslant (p_1 \vee \alpha) \wedge (p_2 \vee \alpha)$, we see that $\alpha_1 \vee \alpha_2 = (p_1 \vee \alpha) \wedge (p_2 \vee \alpha)$. But this implies $\alpha \leqslant \alpha_1 \vee \alpha_2$, thus $\alpha \leqslant f(p_1) \vee f(p_2)$.

It remains to be proved that $f(p_1) \wedge f(p_2) = f(p_1 \wedge p_2)$. It is evident from the definition of f that $f(p_1 \wedge p_2) \leqslant f(p_1) \wedge f(p_2)$. Since Q is a projective geometry, it is enough to prove that if $\beta \leqslant f(p_1) \wedge f(p_2)$ is an atom, then $\beta \leqslant f(p_1 \wedge p_2)$. But if for an atom β we have $\beta \leqslant f(p_1)$ and $\beta \leqslant f(p_2)$, then $\beta \leqslant p_1, \beta \leqslant p_2$, and consequently $\beta \leqslant p_1 \wedge p_2$, therefore $\beta \leqslant f(p_1 \wedge p_2)$, concluding the proof. $\square$

Problem A.45. *Let G be a finite Abelian group and $x, y \in G$. Suppose that the factor group of G with respect to the subgroup generated by x and the factor group of G with respect to the subgroup generated by y are isomorphic. Prove that G has an automorphism that maps x to y.*

Solution 1. It is enough to deal with the case where G is a p-group. Denote the subgroup generated by x (resp., y) by X (resp., Y). Suppose $a_1, \ldots, a_n$ is a base of G. Then every element of G can be written uniquely in the form $k_1 a_1 + \cdots + k_n a_n$, where $0 \leqslant k_i < \mathrm{o}(a_i)$. For the given element x, choose a base so that the number of nonzero coefficients k_i should be minimal. By multiplying with suitable integers relatively prime to p, we can achieve that all these nonzero coefficients are powers of p; let us arrange them in decreasing order:

$$x = p^{e_1} a_1 + \cdots + p^{e_k} a_k, \tag{1}$$

where $e_1 \geqslant e_2 \geqslant \cdots \geqslant e_k \geqslant 0$ $(k \leqslant n)$. Let us denote the order of a_i by p^{d_i} if $i > k$ and by $p^{e_i + f_i}$, where $f_i > 0$ if $i \leqslant k$.

We show that $e_1 > e_2 > \cdots > e_k$. Suppose, on the contrary that $e_i = e_{i+1}$, and let us say $f_i \geqslant f_{i+1}$ holds. Then, by replacing a_i by $a_i + a_{i+1}$, we would get a base giving fewer nonzero terms in representation (1). Next we show $f_1 > f_2 > \cdots > f_k$. Otherwise, assuming $f_i \leqslant f_{i+1}$, replacing a_{i+1} by $p^{e_i - e_{i+1}} a_i + a_{i+1}$ we would get a base giving less nonzero terms in the representation (1).

Now we determine the invariants of G/X (that is, the orders of the cyclic groups whose direct sum is isomorphic to G/X). Take

$$a_i' = p^{e_1 - e_i} a_1 + \cdots + p^{e_{i-1} - e_i} a_{i-1} + a_i \tag{2}$$

for $i \leqslant k$ and $a_i' = a_i$ for $i > k$. Then $a_1', a_2', \ldots, a_n'$ is a set of generators. Furthermore, $p^{e_i + f_{i+1}} a_i' = p^{f_{i+1}} x \in X$, so the order of the image of a_i' in the factor group is at most $p^{e_i + f_{i+1}}$ (we define f_{k+1} to be 0). Since $\mathrm{o}(x) = p^{f_1}$, we get that the images of the elements a_i' form a base of G/X and their respective orders are

$$p^{e_1 + f_2}, p^{e_2 + f_3}, \ldots, p^{e_{k-1} + f_k}, p^{e_k}, p^{d_{k+1}}, \ldots, p^{d_n}.$$

Therefore, if we know the isomorphism type of G and G/X, we can uniquely determine the exponents $f_1, e_1, \ldots, f_k, e_k$, namely, we omit the common invariants and we use the fact that

$$e_1 + f_1 > e_1 + f_2 > e_2 + f_2 > e_2 + f_3 > \cdots > e_{k-1} + f_k > e_k + f_k \geqslant 0.$$

Consequently, if G/X and G/Y are isomorphic, then in a suitable base with $\mathrm{o}(b_i) = \mathrm{o}(a_i)$ we have

$$y = p^{e_1} b_1 + \cdots + p^{e_k} b_k$$

with the same exponents $e_1, \ldots, e_k$ as in (1).

This means that the automorphism of G mapping a_i to b_i maps x to y. □

Solution 2. We assume again that G is a p-group and proceed by induction. We distinguish two cases according to whether X is contained in pG or not.

In the first case, we have $p(G/X) = pG/X$ and $(G/X)/p(G/X) \simeq G/pG$. In the second case, $p(G/X) = (pG + X)/X$ and $(G/X)/p(G/X) \simeq G/(pG + X)$ has an order smaller than G/pG; furthermore, $pG/pX = pG/(X \cap pG) \simeq (pG + X)/X = p(G/X)$. Since $G/X \simeq G/Y$, either the first case or the second case applies to both of them.

In the first case, $pG/X = p(G/X) \simeq p(G/Y) = pG/Y$, so by the induction hypothesis there is an automorphism of pG mapping x to y. It is easy to see that every automorphism of pG can be extended to an automorphism of G so in this case the proof is finished.

In the second case, $pG/pX \simeq pG/pY$ so by the previous statements we can assume $px = py$. It is easy to see that in the elementary Abelian group G/pG (which is actually a vector space over the field with p elements) there is a maximal subgroup M containing the image of neither x nor y. We have $|G : M| = p$, $M \cap X = pX = pY = M \cap Y$, and $G = M + X = M + Y$. Now it is clear that G has an automorphism that acts as the identity map on M and maps x to y. □

Remark. For p-groups a third possible way to solve the problem is to prove that the automorphism exists iff $h(p^k x) = h(p^k y)$ for each k, where $h(g) = h$ means that there is an element z of the group such that $p^h z = g$ and there is no z such that $p^{h+1} z = g$ (h is the "height" of g). An integer d is called a divisor of the element g of an Abelian group if there is an element z of the group such that $dz = g$. It is easy to see that for finite Abelian groups the above statement means that the automorphism exists iff the divisors of x and the divisors of y are the same. One of the contestants, Balázs Montágh, later proved the following generalization: If G is a finitely generated Abelian group and $x_1, x_2, \ldots, x_n$, $y_1, y_2, \ldots, y_n \in G$, then there exists an automorphism f of G such that $f(x_i) = y_i$ for each i iff the divisors of $a_1x_1 + \cdots + a_nx_n$ and the divisors of $a_1y_1 + \cdots + a_ny_n$ are the same.

Problem A.46. *Let p be an arbitrary prime number. In the ring G of Gaussian integers, consider the subrings*

$$A_n = \{pa + p^n bi : a, b \in \mathbb{Z}\}, \quad n = 1, 2, \ldots .$$

Let $R \subset G$ be a subring of G that contains A_{n+1} as an ideal for some n. Prove that this implies that one of the following statements must hold: $R = A_{n+1}$; $R = A_n$; or $1 \in R$.

Solution. Since $p \in A_{n+1}$ and A_{n+1} is an ideal in R, we have for any $x = a + bi \in R$, $pa + pbi = p(a + bi) \in A_{n+1}$. This implies $pb = p^{n+1}b'$, that is, $b = p^n b'$ for a suitable integer b'. Therefore, all elements of R are of the form $x = a + p^n bi$.

Now we distinguish two cases.

First, let us assume that for all elements $x = a + p^n bi$ of R we have $p \mid a$. Then $a = pa'$ for a suitable integer a', that is, $x = pa' + p^n bi \in A_n$. Since

x is arbitrary, this shows $R \leqslant A_n$. Clearly, A_{n+1} is an ideal in A_n, and the factor ring A_n/A_{n+1} has p elements. Thus $A_{n+1} \leqslant R \leqslant A_n$ implies either $R = A_{n+1}$ or $R = A_n$.

Now let us assume that there exists an element $x = a + p^n bi$ in R such that $p \nmid a$. Then for suitable integers u and v we have

$$pu + av = 1.$$

This implies

$$xv = av + p^n bvi = 1 - pu + p^n bvi.$$

In view of $p \in A_{n+1}$, this gives

$$1 + p^n bvi = xv + pu \in R + A_{n+1} = R\,.$$

With the notation $z = xv + pu$, we have

$$p^n bvi = z - 1,$$

and therefore

$$-p^{2n} b^2 v^2 = z^2 - 2z + 1.$$

Since $n \geqslant 1$ and $z \in R$, we finally get

$$1 = -p(p^{2n-1} b^2 v^2) - z^2 + 2z \in A_{n+1} + R + R = R\,,$$

that is, $1 \in R$. □

Problem A.47. *Let $n > 2$ be an integer, and let Ω_n denote the semigroup of all mappings $g : \{0,1\}^n \to \{0,1\}^n$. Consider the mappings $f \in \Omega_n$, which have the following property: there exist mappings $g_i : \{0,1\}^2 \to \{0,1\}$ $(i = 1, 2, \ldots, n)$ such that for all $(a_1, a_2, \ldots, a_n) \in \{0,1\}^n$,*

$$f(a_1, a_2, \ldots, a_n) = (g_1(a_n, a_1), g_2(a_1, a_2), \ldots, g_n(a_{n-1}, a_n))\,.$$

Let Δ_n denote the subsemigroup of Ω_n generated by these f's. Prove that Δ_n contains a subsemigroup Γ_n such that the complete transformation semigroup of degree n is a homomorphic image of Γ_n.

Solution. We prove a generalization of the problem, namely the one when $\{0,1\}$ is replaced by any group $(G,+)$ of order at least two. (The set $\{0,1\}$ with addition mod 2 is such a group.) This generalization causes no additional complication.

So let Ω_n be the semigroup of all mappings $G^n \to G^n$ with the operation

$$f, g \in \Omega_n,\ a_1, \ldots, a_n \in G\colon (fg)(a_1, \ldots, a_n) = g\,(f(a_1, \ldots, a_n))\,.$$

Let H denote the set of those mappings $f \in \Omega_n$ for which there exist mappings $g_1, \ldots, g_n : G^2 \to G$ such that for every $a_1, \ldots, a_n \in G$,

$$f(a_1, \ldots, a_n) = (g_1(a_n, a_1), g_2(a_1, a_2), \ldots, g_n(a_{n-1}, a_n)).$$

Let Δ_n denote the subsemigroup of Ω_n generated by H. Finally, let T_n denote the semigroup of transformations p: $\{1, \ldots, n\} \to \{1, \ldots, n\}$ with the operation

$$p, q \in T_n, \quad 1 \leqslant i \leqslant n: \quad (pq)(i) = q(p(i)).$$

Let us consider the mapping ψ: $T_n \to \Omega_n$ defined by

$$p \in T_n,\, a_1, \ldots, a_n \in G: \quad \psi(p)(a_1, \ldots, a_n) = (a_{p(1)}, \ldots, a_{p(n)}).$$

We shall prove that ψ has the following properties:
(1) ψ is a homomorphism;
(2) ψ is injective;
(3) $\psi(T_n) \subseteq \Delta_n$.

Then $\psi(T_n) = \Gamma_n$ will be a subsemigroup of Δ_n, and in view of the above-mentioned three properties the inverse of ψ maps $\psi(T_n)$ homomorphically *onto* T_n, which was the statement of the problem.

Let us now prove properties (1), (2), and (3).

(1) Let $p, q \in T_n$, $a_1, \ldots, a_n \in G$. Then

$$\begin{aligned}\psi(pq)(a_1, \ldots, a_n) &= (a_{(pq)(1)}, \ldots, a_{(pq)(n)}) \\ &= (a_{q(p(1))}, \ldots, a_{q(p(n))}) = \psi(q)(a_{(p(1)}, \ldots, a_{p(n)}) \\ &= \psi(q)[\psi(p)(a_1, \ldots, a_n)] = [\psi(p)\psi(q)](a_1, \ldots, a_n),\end{aligned}$$

which shows that ψ is a homomorphism.

(2) Let p and q be different elements of T_n. We show that $\psi(p) \neq \psi(q)$; in other words, ψ is injective. $p \neq q$ means that there is an i $(1 \leqslant i \leqslant n)$ for which $p(i) \neq q(i)$. Since G has at least two elements, it has elements $a_1, \ldots, a_n$ such that $a_{p(i)} \neq a_{q(i)}$. But then

$$\begin{aligned}\psi(p)(a_1, \ldots, a_n) &= (a(p(1), \ldots, a_{p(n)}) \\ &\neq (a(q(1), \ldots, a_{q(n)}) = \psi(q)(a_1, \ldots, a_n).\end{aligned}$$

(3) The proof that, for $p \in T_n$, $\psi(p) \in \Delta_n$ holds will be broken down into the following steps (a) toy (e).

(a) First, we prove the statement for the case when p is a transposition interchanging two (cyclically) neighboring element, that is, one of the transpositions $(1, 2)$, $(2, 3)$, ... ,$(n-1, n)$, $(n, 1)$. For instance, let $p = (1, 2)$. For the other cases, the proof is similar. We represent the mapping

$$\psi(p)(a_1, a_2, a_3, \ldots, a_n) = (a_2, a_1, a_3, \ldots, a_n)$$

as a product of elements of H. In what follows, '$\to$' always denotes a mapping in H.

$$\begin{aligned}
&(a_1, a_2, a_3, a_4, a_5, \dots, a_n) \to \\
&\to (-a_n, a_1, a_3 + a_2, a_4 - a_3, a_5 - a_4, \dots, a_n - a_{n-1}) \to \\
&\to (-a_n, a_1, a_3 + a_2, a_4 + a_2, a_5 - a_4, \dots, a_n - a_{n-1}) \to \cdots \to \\
&\to (-a_n, a_1, a_3 + a_2, a_4 + a_2, a_5 + a_2, \dots, a_n + a_2) \to \\
&\to (a_2, a_1, a_3 + a_2, a_4 + a_2, a_5 + a_2, \dots, a_n + a_2) \to \\
&\to (a_2, a_2 + a_1, a_3 + a_2 + a_1, a_4 - a_3, a_5 - a_4, \dots, a_n - a_{n-1}) \to \\
&\to (a_2, a_1, a_3, a_4 - a_3, a_5 - a_4, \dots, a_n - a_{n-1}) \to \\
&\to (a_2, a_1, a_3, a_4, a_5 - a_4, \dots, a_n - a_{n-1}) \to \cdots \to \\
&\to (a_2, a_1, a_3, a_4, a_5, \dots, a_n).
\end{aligned}$$

This procedure is correct for $n \geqslant 5$, and for $n < 5$ we can use a similar but simpler procedure.

(b) If p is an arbitrary permutation, then it can be represented as a product of transpositions interchanging neighboring elements. Since ψ is a homomorphism and the images of these transpositions are products of elements in H, the same is true for $\psi(p)$.

c) Suppose that $p \in T_n$ is such that $p(i) \neq i$ for some i, but $p(j) = j$ for every $j \neq i$. Without loss of generality, we can assume that $p(i) = 1$. Let us consider the following representation:

$$\begin{aligned}
&(a_1, a_2, a_3, \dots, a_{i-1}, a_i, a_{i+1}, \dots, a_n) \to \\
&\to (a_1, a_2 + a_1, a_3 - a_2, \dots, a_{i-1} - a_{i-2}, -a_{i-1}, a_{i+1}, \dots, a_n) \to \\
&\to (a_1, a_2 + a_1, a_3 + a_1, \dots, a_{i-1} - a_{i-2}, -a_{i-1}, a_{i+1}, \dots, a_n) \to \\
&\to \cdots \to \\
&\to (a_1, a_2 + a_1, a_3 + a_1, \dots, a_{i-1} + a_1, -a_{i-1}, a_{i+1}, \dots, a_n) \to \\
&\to (a_1, a_2, a_3 - a_2, \dots, a_{i-1} - a_{i-2}, a_1, a_{i+1}, \dots, a_n) \to \\
&\to (a_1, a_2, a_3, \dots, a_{i-1} - a_{i-2}, a_1, a_{i+1}, \dots, a_n) \to \cdots \to \\
&\to (a_1, a_2, a_3, \dots, a_{i-1}, a_1, a_{i+1}, \dots, a_n).
\end{aligned}$$

The case $i = 2, 3$ can again be handled in a similar but simpler way.

d) Now let p be a transformation that is identical on its image set: $p(p(i)) = p(i)$ $(i = 1, \dots, n)$. Let $p_{k,i}$ denote the following transformation of type (c):

$$p_{k,i}(i) = k, \quad p_{k,i}(j) = j, \text{ if } j \neq i.$$

Then this p can be represented in the following form:

$$p = p_{p(1),1} \cdots p_{p(n),n}.$$

To prove this, observe that on the right-hand side the first transformation from the left that changes i is $p_{p(i),i}$, and it maps i to $p(i)$.

If some later $p_{p(k),k}$ $(k > i)$ changes this element, then necessarily $k = p(i)$. But then $p_{p(k),k}$ maps $k = p(i)$ to $p(k) = p(p(i)) = p(i)$; thus, nevertheless, it actually fixes $p(i)$.

Now in view of (c), each of the $p_{p(k),k}$ can be represented as a product of elements of H, so p has such a representation as well.

e) To conclude the proof, let p be an arbitrary transformation from T_n. We shall show that there is a *permutation* q such that $r = qp$ is the identical map on its image set. This will establish the statement, since in this case $p = q^{-1}r$ (a permutation has an inverse), $\psi(p) = \psi(q^{-1})\psi(r)$, and (b) and (d) show that $\psi(q^{-1})$ and $\psi(r)$ belong to Δ_n, so the same is true for $\psi(p)$.

To construct the suitable q to the given p, we use induction. Let $p_0 = p$ and q_0 be the identical permutation. Suppose that we have already constructed p_i and q_i in such a way that q_i is a permutation and if $k \leqslant i$ belongs to the image of p, then $p_i(k) = k$. Now if $i + 1$ does not belong to the image of p, let $p_{i+1} = p_i$, $q_{i+1} = q_i$. However, if $i + 1$ does belong to the image of p then it also belongs to the image of p_i, so $p_i(j) = i + 1$ for some j and in this case let $p_{i+1} = (i + 1, j)p_i$, $q_{i+1} = (i + 1, j)q_i$, where $(i + 1, j)$ denotes the transposition of $i + 1$ and j. Then q_{i+1} is again a permutation, $p_{i+1}(i + 1) = i + 1$, and if for some element $k < i + 1$ of the image of p it was true that $p_i(k) = k$, then $p_{i+1}(k) = k$ also holds. The reason for this is that $k < i+1$ trivially implies $k \neq i+1$ and $k = j$ cannot hold either since $p_i(k) = k < i + 1 = p_i(j)$. Finally, we choose $q = q_n$. Then q is a permutation, and $p_n = q_n p = qp$ is the identical map on its image set. □

Problem A.48. *Let $n = p^k$ (p a prime number, $k \geq 1$), and let G be a transitive subgroup of the symmetric group S_n. Prove that the order of the normalizer of G in S_n is at most $|G|^{k+1}$.*

Solution. First, we show that if T is a minimal transitive subgroup of G, then T can be generated by k elements. Since T is transitive, we have $p^k \mid |T|$. Let P be a Sylow p-subgroup of T; we are going to show that P is transitive, thereby proving that T is necessarily a p-group.

Denoting the stabilizer of a point α in the group G by G_α, we have $|\, G : G_\alpha \,| = p^k$, so

$$p^k \mid |\, G : G_\alpha \,| \cdot |\, G_\alpha : G_\alpha \cap P \,| = |\, G : P \,| \cdot |\, P : G_\alpha \cap P \,| .$$

Since $(|\, G : P \,|, p) = 1$, we get $p^k \mid |\, P : G_\alpha \cap P \,|$, that is, $p^k \mid |\, P : P_\alpha \,|$, proving that P is indeed transitive.

Now let M be a maximal subgroup of T. Then M is not transitive, so MT_α cannot be transitive either. Therefore $MT_\alpha = M$, that is, $T_\alpha \leqslant M$. This shows that T_α is contained in the intersection of all maximal subgroups of T, the Frattini subgroup of T. Therefore $|\, T : \Phi(T) \,| \leqslant |\, T : T_\alpha \,| = p^k$, so by a well-known theorem of Burnside, T can indeed be generated by k elements: $T = \langle g_1, \ldots, g_k \rangle$.

We next observe that the elements of $N(G)$, the normalizer of G in S_n, act by way of conjugation on the elements of G, permuting the elements of G among themselves. The number of possible images of the k-tuple $(g_1, \ldots, g_k)$ is at most $|G|^k$. Conjugation by the elements $n, m \in N(G)$ has the same effect on $(g_1, \ldots, g_k)$ if and only if $n^{-1}m \in C(T)$, the centralizer of T in S_n. Thus $|N(G)| \leqslant |G|^k \, |C(T)|$. But the group $C(T)$ is easily shown to be semiregular. Suppose that $s \in C(T)$ fixes the point α and for any other point β choose a $t \in T$ with $\alpha t = \beta$. Then $\beta = \alpha t = (\alpha)sts^{-1} = (\beta)s^{-1}$, that is, $\beta s = \beta$, so s is the identity, proving semiregularity. This shows $|C(T))| \leqslant p^k$, consequently

$$|N(G)| \leqslant |G|^k \cdot p^k \leqslant |G|^{k+1} . \quad \square$$

Problem A.49. *Prove that if a finite group G is an extension of an Abelian group of exponent 3 with an Abelian group of exponent 2, then G can be embedded in some finite direct power of the symmetric group S_3.*

Solution. Applying the Schur–Zassenhaus theorem, we see that the extension is actually a split extension, that is, a semidirect product. Using the fundamental theorem of finite Abelian groups, we get that G actually has the form $G = \mathbb{Z}_3^m \overset{\xi}{\rtimes} \mathbb{Z}_2^n$, where ξ is a homomorphism $\xi : \mathbb{Z}_2^n \to \mathrm{Aut}(\mathbb{Z}_3^m) = \mathrm{GL}_m(3)$. ($\mathrm{GL}_m(3)$ is the group of automorphisms of $\mathbb{Z}_3^m$, considered as a vector space over the field of three elements.) Applying Maschke's theorem to the representation ξ, we see that there is a basis $b_1, b_2, \ldots, b_m$ of the vector space $\mathbb{Z}_3^m$, the elements of which are eigenvectors of all elements of $\xi(\mathbb{Z}_2^n)$. Let $a_1, a_2, \ldots, a_n$ be a basis in $\mathbb{Z}_2^n$, then b_j is an eigenvector of $\xi(a_i)$ with eigenvalue $c_{ij} \in \{1, 2\}$, say. Denote the elements $(1,2,3)$ and $(1,2)$ of S_3, by σ and τ, respectively. For an arbitrary $b = \sum_{i=1}^m \lambda_i b_i \in \mathbb{Z}_3^m$, let

$$\bar{b} = (\sigma^{\lambda_1}, \ldots, \sigma^{\lambda_m}, 1, \ldots, 1) \in S_3^{m+n} ,$$

and let

$$\hat{a}_i = (\tau^{c_{i1}-1}, \tau^{c_{i2}-1}, \ldots, \tau^{c_{im}-1}, 1, \ldots, \overset{m+i}{\breve{\tau}}, 1, \ldots, 1) \in S_3^{m+n} .$$

Then $B = \{\bar{b} \mid b \in \mathbb{Z}_3^m\}$ is a normal subgroup of S_3^{m+n}, isomorphic to $\mathbb{Z}_3^m$. The mapping $a_i \mapsto \hat{a}_i$ can be extended to a homomorphism $\mathbb{Z}_2^n \to S_3^{m+n}$, and then $A = \{\hat{a} \mid a \in \mathbb{Z}_2^n\}$ is a subgroup of S_3^{m+n}, isomorphic to $\mathbb{Z}_2^n$. Let us consider the homomorphism $\eta : A \to \mathrm{Aut}(B)$ mapping each $x \in A$ to the conjugation by x. Then for the subgroup BA of S_3^{m+n} we have $BA \simeq B \overset{\eta}{\rtimes} A$. On the other hand, for any $a \in \mathbb{Z}_2^n$ and $b \in \mathbb{Z}_3^m$ we have

$$\overline{b^{\xi(a)}} = \bar{b}^{\eta(\hat{a})}$$

(it is enough to check this for basis elements b_i and a_j). This shows that indeed $G \simeq BA$. □

Remark. Instead of referring to the Schur–Zassenhaus theorem, we can simply consider a Sylow 2-subgroup.

Problem A.50. *Let $n \geq 2$ be an integer, and consider the groupoid $G = (\mathbb{Z}_n \cup \{\infty\}, \circ)$, where*

$$x \circ y = \begin{cases} x+1 & \text{if } x = y \in \mathbb{Z}_n, \\ \infty & \text{otherwise.} \end{cases}$$

($\mathbb{Z}_n$ denotes the ring of the integers modulo n.) Prove that G is the only subdirectly irreducible algebra in the variety generated by G.

Solution. ($\mathcal{G}$ will denote the groupoid G considered as a member of the variety of the problem.)

The unary operation taking the constant value ∞ is a polynomial expression in $\mathcal{G}$, for example,

$$\infty = (x \circ x) \circ x. \tag{0}$$

(This unary operation will also be denoted by ∞.)

Furthermore, it is easy to check that for the polynomial expressions

$$f(x) = x \circ x \qquad \text{and} \qquad x \bigvee y = f^{n-1}(x \circ y) \tag{1}$$

of $\mathcal{G}$ the following hold:

$$\vee \text{ is a semilattice operation and } \infty \text{ is the greatest element with respect to it,} \tag{2}$$

$$f \text{ is an automorphism with respect to } \bigvee, \tag{3}$$

and the following identities hold:

$$f^n(x) = x, \quad f^k(x) \bigvee f^l(x) = \infty \quad \text{if} \quad k \neq l \quad (0 \leqslant k, l < n). \tag{4}$$

It is also true that the original operation $\circ$ can be expressed using $\vee$ and f since the identity

$$x \circ y = f(x \vee y) \tag{5}$$

holds true in $\mathcal{G}$.

Now take any algebra $\mathcal{C} = (C; \circ)$ in the variety generated by $\mathcal{G}$, and consider the algebra $\mathcal{C}' = (C; \vee, f, \infty)$ whose basic operations are the polynomial expressions of $\mathcal{C}$ defined according to (0) and (1). As all the properties discussed above are described by identities, they hold simultaneously for $\mathcal{C}$ (resp. $\mathcal{C}'$). Identity (5) (together with the definitions included in (0) and

(1)) ensures that a map $\varphi : C \to \mathbb{Z}_n \cup \{\infty\}$ produces a homomorphism $\varphi : \mathcal{C} \to \mathcal{G}$ if and only if $\varphi : \mathcal{C}' \to \mathcal{G}'$ is a homomorphism. So instead of $\mathcal{G}$ and $\mathcal{C}$, we can work with the algebras $\mathcal{G}'$ and $\mathcal{C}'$, which can be handled more comfortably. Also denote by $\leqslant$ the natural partial order of the semilattice $(C; \vee)$.

Let a be an arbitrary element of the set C, different from ∞, and consider the following subsets of C:

$$C_i = \{c \in C \mid c \leqslant f^i(a)\} \quad (i \in \mathbb{Z}_n), \quad C_\infty = C \setminus \bigcup_{i \in \mathbb{Z}_n} C_i.$$

We will show that $\{C_0, \ldots, C_{n-1}, C_\infty\}$ is a partition of C. Clearly, the union of these sets is C, and none of them is empty, since $f^i(a) \in C_i$ ($i \in \mathbb{Z}_n$) and $\infty \in C_\infty$. If $c \in C_i \cap C_j$ for some $0 \leqslant i \leqslant j < n$, that is, $c \leqslant f^i(a)$ and $c \leqslant f^j(a)$, then using (2) and (3) we get

$$f^{j-i}(c) \bigvee c \leqslant f^{j-i}(f^i(a)) \bigvee f^j(a) = f^j(a) \bigvee f^j(a) = f^j(a) < \infty,$$

which, in view of (4), implies $j - i = 0$. So $C_0, \ldots, C_{n-1}$ are pairwise disjoint, and C_∞ is clearly disjoint from each of them.

So the following map $\varphi : \mathcal{C}' \to \mathcal{G}'$ is well defined:

$$c\varphi_a = i \qquad \text{if} \qquad c \in C_i.$$

Clearly, φ_a is surjective, it is permutable with the operation f, and $\infty\varphi_a = \infty$. For arbitrary $c, d \in C$ and $l \in \mathbb{Z}_n$, we have

$$(c \bigvee d)\varphi_a = l \;\Leftrightarrow\; c \bigvee d \leqslant f^l(a) \;\Leftrightarrow\; c, d \leqslant f^l(a) \;\Leftrightarrow\; c\varphi_a = l, d\varphi_a = l.$$

This implies that φ_a is permutable with $\vee$ too, so $\varphi_a : \mathcal{C}' \to \mathcal{G}'$ is a surjective homomorphism.

As $a\varphi_a = 0$, we have for every element $b \in C$ the following: $a\varphi_a = b\varphi_a \;\Leftrightarrow\; b \leqslant a$. So if a, b are different elements of the algebra $\mathcal{C}'$ (say $a \neq \infty$), then in case $b \nleqslant a$ we have $a\varphi_a \neq b\varphi_a$, whereas in case $b < a$ (this implies $b \neq \infty$) we have $a\varphi_b \neq b\varphi_b$. Consequently, the homomorphisms $\varphi_a (a \in C \setminus \{\infty\})$ give a representation of $\mathcal{C}'$ as a subdirect power of $\mathcal{G}'$. This shows that $\mathcal{C}'$ can be subdirectly irreducible only in case it is isomorphic to $\mathcal{G}'$.

It is easy to check that the algebra $\mathcal{G}'$ is simple, so necessarily subdirectly irreducible. Namely, if for some congruence relation σ of $\mathcal{G}'$ we have $a \;\sigma\; b, a \neq b$ (say $a \neq \infty$), then $a = a \vee a \;\sigma\; a \vee b = \infty$ so for any $0 \leqslant k < n$ we have $f^k(a) \;\sigma\; f^k(\infty) = \infty$. □

3.2 COMBINATORICS

Problem C.1. *Among all possible representations of the positive integer n as $n = \sum_{i=1}^{k} a_i$ with positive integers k, $a_1 < a_2 < \cdots < a_k$, when will the product $\prod_{i=1}^{k} a_i$ be maximum?*

Solution. First, we are going to investigate the properties of the extremal integer sets. Suppose that $A_n = \{a_1, a_2, \ldots, a_k\}$ is extremal.

Claim 1. There are no integers i and j such that $a_1 < i < j < a_k$ and $i \notin A_n$, $j \notin A_n$.

If it is not the case, then there are some elements $a_r, a_s \in A_n$ such that $a_r + 1 \notin A_n$, $a_s - 1 \notin A_n$, and $a_r + 3 \le a_s$. Then

$$(a_r + 1)(a_s - 1) = a_r a_s + (a_s - a_r) - 1 \ge a_r a_s + 2 > a_r a_s,$$

so replacing the elements a_r, a_s of A_n by $a_r + 1$ and $a_s - 1$, the sum of the elements of A_n does not change, but their product increases, a contradiction to the extremality of A_n.

Claim 2. $2 \le a_1$ if $k > 1$.

If $a_1 = 1$, then replacing the elements a_1, a_k of A_n by $a_k + 1$ gives us a "better" system, a contradiction to the choice of A_n.

Claim 3. $a_1 = 2$ or $a_1 = 3$ if $n \ge 5$.

If $a_1 > 4$, then replacing the element a_1 of A_n by 2 and $a_1 - 2$, the sum of the elements of A_n does not change, but their product increases since

$$2(a_1 - 2) = 2a_1 - 4 = a_1 + (a_1 - 4) > a_1,$$

a contradiction to the extremality of A_n.

If $a_1 = 4$, then $k > 1$ by $n \ge 5$, and replacing the elements a_1, a_2 of A_n by 2, $a_1 - 1$, and $a_2 - 1$, the sum of the elements of A_n does not change, but their product increases, since

$$\begin{aligned} 2(a_1 - 1)(a_2 - 1) &= 2a_1a_2 - 2(a_1 + a_2) + 2 = a_1a_2 + (a_1 - 2)(a_2 - 2) - 2 \\ &\ge a_1a_2 + 6 - 2 = a_1a_2 + 4 > a_1a_2, \end{aligned}$$

a contradiction to the extremality of A_n.

Claim 4. If $a_1 = 3$ and i is an integer such that $a_1 < i < a_k$, $i \notin A_n$ then $i = a_k - 1$.

If not, then $i + 2 \in A_n$ by claim 1, and replacing $i + 2$ by 2 and i, the sum of the elements of A_n does not change, but their product increases, since

$$2i \ge i + a_1 + 1 = i + 4 > i + 2,$$

a contradiction to the extremality of A_n.

Now, we are ready to determine the extremal sets A_n. Let $A(i,j,l)$ denote the set $\{i, i+1, \dots, l-1, l+1, \dots, i+j-1, i+j\}$, and let $s(i,j,l)$ denote the sum of the elements of $A(i,j,l)$. For $k = 2, 3, \dots$, let $s(2, k+2, k+2) = 2+3+\dots+k+(k+1) = \binom{k+2}{2} - 1 = L_k$. By claims 1–4, A_n has at least two elements if $n \geq 5$, and if $k \geq 2$ then the possible extremal sets are as follows: $A(2, k+2, k+2), A(2, k+2, k+1), \dots, A(2, k+2, 3), A(2, k+2, 2), A(3, k+3, k+1)$. Obviously, $s(2, k+2, k+2) = L_k, s(2, k+2, k+1) = L_k + 1, \dots, s(2, k+2, 3) = L_k + (k-1), s(2, k+2, 2) = L_k + k$ and $s(3, k+3, k+1) = L_k + (k+1) = L_{k+1} - 1$, so the sums of the elements of the above sets are the integers in the interval $[L_k, L_{k+1})$, and each integer appears exactly once. Thus, for any $n \geq 5$, we have exactly one set of type above and this is the extremal set A_n for this n. If $1 \leq n \leq 4$, then obviously $A_n = \{n\}$ is the only extremal set. □

Remark. Consider the following generalization of the problem: Let $f(x)$ be an arbitrary function that is strictly concave from below in the interval $(0, +\infty)$ ($f(x) = \log x$ in the problem above). Let $n = \sum_{i=1}^{k} a_i$, where $a_1 < a_2 < \dots < a_k$ are positive integers, k is not fixed. When will the sum $\sum_{i=1}^{k} f(a_k)$ be maximum for a given n? It can be proved that also in this case every extremal set can be obtained from a sequence of consecutive natural numbers deleting at most one element.

Problem C.2. *For every natural number r, the set of r-tuples of natural numbers is partitioned into finitely many classes. Show that if $f(r)$ is a function such that $f(r) \geq 1$ and $\lim_{r\to\infty} f(r) = +\infty$, then there exists an infinite set of natural numbers that, for all r, contains r-tuples from at most $f(r)$ classes. Show that if $f(r) \not\to +\infty$, then there is a family of partitions such that no such infinite set exists.*

Solution. Let $\mathbb{N}$ denote the set of natural numbers. We will use the following well-known theorem of Ramsey. If the r-tuples of natural numbers are divided into finitely many classes, then there is an infinite subset $\mathbb{N}'$ of $\mathbb{N}$ such that every r-tuple in $\mathbb{N}'$ is contained in the very same class.

To prove the first statement, we define a sequence $\mathbb{N}_0 \supseteq \dots \supseteq \mathbb{N}_r \supseteq \dots$ of subsets and a sequence $x_1, \dots, x_r, \dots$ of natural numbers by induction on r. Let $\mathbb{N}_0 = \mathbb{N}$. Suppose that $r \geq 0$ and that $\mathbb{N}_r$ and x_i $(i < r)$ are defined, $\mathbb{N}_r$ is infinite. Let x_r be an arbitrary element of $\mathbb{N}_r$ and let $\mathbb{N}'_r = \mathbb{N}_r - \{x_1, \dots, x_r\}$. For any set $A' \subseteq \{x_1, \dots, x_r\}$, divide the $(r-s)$-tuples of $\mathbb{N}'_r$ into finitely many classes where $s = |A'|$. Put two $(r-s)$-tuples into the same class if and only if adding the elements of A' to them the resulting r-tuples are in the same class of the original partition. Applying Ramsey's theorem 2^r times, we get that there is a subset $\mathbb{N}_{r+1}$ of the set $\mathbb{N}'_r$ such that

$$|A| = |B| = r, \tag{1}$$

$$A, B \subseteq \{x_1, \dots, x_r\} \cup \mathbb{N}_{r+1}, \quad A \cap \{x_1, \dots, x_r\} = B \cap \{x_1, \dots, x_r\} \tag{2}$$

imply that A and B are in the very same class.

So, we defined the sequences $\mathbb{N}_0 \supseteq \cdots \supseteq \mathbb{N}_r \supseteq \ldots; x_1, \ldots, x_r, \ldots$. Let $X = \{x_1, \ldots, x_r, \ldots\}$. Now, (1) and (2) imply that X has the following properties.

(*) Suppose that $|A| = |B| = r, \quad A, B \subseteq X, \quad A \cap \{x_1, \ldots, x_r\} = B \cap \{x_1, \ldots, x_r\}$ for any r. Then A and B are in the very same class. Since $f(r) \geq 1$ and $f(r) \to +\infty$ as $r \to +\infty$, thus there exists a monotone subsequence x_{r_k} such that

$$|\{x_{r_k} : r_k \leq r\}| \leq \log_2 f(r).$$

Let $X' = \{x_{r_1}, x_{r_2}, \ldots\}$. This set X' is obviously infinite and meets the requirements of the problem by (*).

Now, we are to prove a statement that is stronger than the second part of the problem. We show that a set of cardinality continuum has the desired partition, as well. Let $S = [0, 1]$. We divide the set of r-element subsets of S into r classes for any r. Let $X = \{x_0, \ldots, x_{r-1}\}$ be an r element subset of S with $x_0 < \cdots < x_{r-1}$. We put X into the ith class if exactly i of the intervals $(x_j, x_{j+1}) \quad (j = 0, 1, \ldots, r-2)$ are longer than $1/r$. If S' is an arbitrary infinite subset of S, then using the fact that S' has an accumulation point, it is easy to see that for any i there is a number r_0 such that for $r > r_0$ the set S' contains a set out of the ith class of the r-tuples, and so our proof is complete. □

Problem C.3. *Let n and k be given natural numbers, and let A be a set such that*

$$|A| \leq \frac{n(n+1)}{k+1}.$$

For $i = 1, 2, \ldots, n+1$, let A_i be sets of size n such that

$$|A_i \cap A_j| \leq k \qquad (i \neq j),$$

$$A = \bigcup_{i=1}^{n+1} A_i.$$

Determine the cardinality of A.

Solution 1. Let ϕ_x denote the number of sets A_i containing the point x. Obviously,

$$\sum_{x \in A_j} \phi_x = \sum_{i=1}^{n+1} |A_i \cap A_j| = n + \sum_{i \neq j} |A_i \cap A_j|,$$

and so

$$\sum_{x \in A_j} \phi_x \leq n + nk = n(k+1). \tag{1}$$

Add these up for all j. On the left-hand side, we get

$$\sum_{j=1}^{n+1} \sum_{x \in A_j} \phi_x = \sum_{x \in A} \sum_{x \in A_j} 1 = \sum_{x \in A} \phi_x^2.$$

Estimating it by the inequality between arithmetic and quadratic means, we get

$$\sum_{x\in A} \phi_x^2 \geq |A| \left(\frac{\sum_{x\in A} \phi_x}{|A|} \right)^2 = \frac{1}{|A|} \left(\sum_{i=1}^{n+1} |A_j| \right)^2 = \frac{n^2(n+1)^2}{|A|}.$$

Adding up the right-hand sides of (1) for all j, we get $n(n+1)(k+1)$. Thus,

$$\frac{n^2(n+1)^2}{|A|} \leq n(n+1)(k+1),$$

that is,

$$|A| \geq \frac{n(n+1)}{k+1}.$$

Since the opposite inequality was supposed, it implies the equality above. Naturally, $k+1|n(n+1)$ is needed for the existence of such a family of sets. □

Solution 2. We will use the following theorem: Let $F_1, \dots, F_N$ be arbitrary polynomials of some events $A_1, \dots, A_n$, and let $c_1, \dots, c_N$ be arbitrary real numbers. Then, the inequality

$$\sum_{k=1}^{N} c_k p(F_k) \geq 0$$

holds in any probability space and for any events $A_1, \dots, A_n$ provided that it holds in the trivial probability space.

Thus, it is sufficient to verify that the inequality

$$p(A_1 \cup \dots \cup A_{n+1}) \geq \frac{2k+1}{(k+1)^2} \sum_{i=1}^{n+1} p(A_i) - \frac{2}{(k+1)^2} \sum_{1\leq i<j\leq n+1} p(A_i \cap A_j)$$

holds when $A_i = \emptyset$ or Ω. If $A_i = \emptyset$ for all i, then we get $0 \geq 0$, which holds. If exactly l of the A_i's are Ω ($l \geq 1$), then we have to prove that

$$1 \geq \frac{2k+1}{(k+1)^2} l - \frac{2}{(k+1)^2} \binom{l}{2},$$

that is,

$$\frac{l^2-l}{(k+1)^2} - \frac{2k+1}{(k+1)^2} l + 1 = \left(\frac{l}{k+1} - 1 \right)^2 \geq 0,$$

which holds as well. Now, let $A_i \quad (i = 1, \dots, n+1)$ be the sets given in the problem, and consider the probability space such that $\Omega = A = \cup_{i=1}^{n+1} A_i$

and the probability of each point is $1/|A|$. So, even in this space,

$$\begin{aligned} p(A) = 1 &\geq \frac{2k+1}{(k+1)^2} \sum_{i=1}^{n+1} \frac{|A_i|}{|A|} - \frac{2}{(k+1)^2} \sum_{1 \leq i < j \leq n+1} \frac{|A_i \cap A_j|}{|A|} \\ &\geq \frac{2k+1}{(k+1)^2} \frac{(n+1)n}{|A|} - \frac{2}{(k+1)^2} \binom{n+1}{2} \frac{k}{|A|}. \end{aligned}$$

Reducing this, we obtain that $|A| \geq n(n+1)/(k+1)$. We supposed that $|A| \leq n(n+1)/(k+1)$, so $|A| = n(n+1)/(k+1)$. □

Problem C.4. *Let $A_1, A_2, \ldots$ be a sequence of infinite sets such that $|A_i \cap A_j| \leq 2$ for $i \neq j$. Show that the sequence of indices can be divided into two disjoint sequences $i_1 < i_2 < \ldots$ and $j_1 < j_2 < \ldots$ in such a way that, for some sets E and F, $|A_{i_n} \cap E| = 1$ and $|A_{j_n} \cap F| = 1$ for $n = 1, 2, \ldots$.*

Solution. Suppose that for $k \geq 3$, there are some finite disjoint sets E_k and F_k such that

$$\text{either } |A_i \cap E_k| = 1 \text{ or } |A_i \cap F_k| = 1, \tag{1}$$

$$\text{if } i > k \text{ then } |A_i \cap E_k| \leq 1 \text{ or } |A_i \cap F_k| \leq 1. \tag{2}$$

For $k = 3$, it is easy to construct such sets E_3 and F_3. If we show that there are some finite disjoint sets $E_{k+1} \supseteq E_k$ and $F_{k+1} \supseteq F_k$ satisfying (1) and (2) for A_{k+1}, then the sets $E = \cup_{k=1}^{\infty} E_k$ $F = \cup_{k=1}^{\infty} F_k$ have the desired properties.

First, suppose that A_{k+1} meets both E_k and F_k. According to (2), it is possible only if, say, $|A_{k+1} \cap E_k| = 1$, and then $E_{k+1} = E_k$ and $F_{k+1} = F_k$ satisfy the conditions (1) and (2).

Now, suppose that A_{k+1} does not meet at least one of the sets E_k and F_k, say, E_k. Consider all the sets A_i that meet $E_k \cup F_k$ in at least three elements. Since the given three elements can be contained in finitely many sets A_i only, the number of these sets is finite. Let us denote them by $A_{i_1}, \ldots, A_{i_m}$. Let e_{k+1} denote an arbitrary element of the infinite set

$$A_{k+1} - (\bigcup_{r=1}^{m} A_{i_r}) \bigcup (\bigcup_{i=1}^{k} A_i) \bigcup F_k,$$

and let $E_{k+1} = E_k \cup \{e_{k+1}\}$, $F_{k+1} = F_k$. These sets satisfy (1) obviously. Furthermore, if $i > k+1$ and $e_{k+1} \notin A_i$, then (2) holds as well, and if $e_{k+1} \in A_i$ then A_i differs from the sets $A_{i_1}, \ldots, A_{i_m}$ by the choice of e_{k+1}, and so $|A_i \cap (E_k \cup F_k)| \leq 2$, which yields (2). □

Problem C.5. *Let $n \geq 2$ be an integer, let S be a set of n elements, and let A_i, $1 \leq i \leq m$, be distinct subsets of S of size at least 2 such that*

$$A_i \cap A_j \neq \emptyset,\ A_i \cap A_k \neq \emptyset,\ A_j \cap A_k \neq \emptyset \quad \textit{imply} \quad A_i \cap A_j \cap A_k \neq \emptyset.$$

Show that $m \leq 2^{n-1} - 1$.

Solution 1. We will prove the statement by induction on n. It obviously holds for $n = 2$. Assume that $n > 2$, and let $A_1 \neq S$ be a maximal element of the set

$$K = \{A_i : 1 \leq i \leq k\},$$

that is, if $A_1 \subseteq A_i$, then either $i = 1$ or $A_i = S$. Choose an arbitrary element x of the set $S - A_1$, and let

$$\begin{aligned}
K_1 &= \{A_i \in K : x \notin A_i\}, \\
K_2 &= \{A_i \in K : x \in A_i, A_1 \cap A_i = \emptyset\}, \\
K_3 &= \{A_i \in K : x \in A_i, A_1 \subseteq A_i\}, \\
K_4 &= \{A_i \in K : x \in A_i, A_1 \cap A_i \neq \emptyset, A_1 \not\subseteq A_i\}.
\end{aligned}$$

Obviously,

$$K = K_1 \cup K_2 \cup K_3 \cup K_4. \tag{1}$$

We will estimate the cardinality of the set K. The inductional hypothesis implies that

$$|K_1| \leq 2^{n-2} - 1. \tag{2}$$

Let $l = |A_1|$. Then

$$|K_2| \leq 2^{n-l-1} - 1 \tag{3}$$

since every element of K is a set of at least two elements. By the maximality of A_1, the only element that may be contained in K_2 is S, that is,

$$|K_3| \leq 1. \tag{4}$$

Finally, the elements of the set K_4 can meet the sets $S - A_1$ and A_1 in at most 2^{n-l-1} and $2^l - 2$ distinct sets, respectively (they cannot meet A_1 in A_1 or $\emptyset$). Pairing these intersections in all possible ways, we get

$$|K_4| \leq 2^{n-l-1}(2^l - 2) = 2^{n-1} - 2^{n-l}.$$

Actually, this estimate can be sharpened. If $X \in K_4$, then

$$Y = (X - A_1) \cup (A_1 - X) \notin K_4$$

since the sets A_1, X, Y do not satisfy the intersection conditions for the sets A_i (the sets $A_1 \cap X, A_1 \cap Y, X \cap Y$ are not empty, but the set $A_1 \cap X \cap Y$ is empty). Since the mapping

$$X \to Y = (X - A_1) \cup (A_1 - X)$$

is one-to-one for a fixed set A_1 — and if $x \in X$, $A_1 \cap X \neq \emptyset$, and $A_1 \subseteq X$, then the same holds for the set Y so the cardinality of K_4 is at most half of the estimate above, that is,

$$|K_4| \leq 2^{n-2} - 2^{n-l-1}. \tag{5}$$

Summarizing (1) to (5), we get the desired inequality, $|K| \leq 2^{n-1} - 1$. □

Solution 2. We will prove the statement by induction on n. It obviously holds for $n = 2$. Assume that $n > 2$. We distinguish two cases.

Case 1. There are no i and j such that $A_i \cup A_j = S$ and $A_i \cap A_j$ is of one element.

In this case, it is not difficult to show the statement. Considering an arbitrary element of the set S, the number of sets A_i not containing x is at most $2^{n-2} - 1$ by the inductional hypothesis. The number of sets $X \subseteq S$ containing x is 2^{n-1}. At most half of them, that is, at most 2^{n-2}, appear as a set A_i, since if $X = A_i$, then according to the assumption above, there is no j such that $A_j = (S - X) \cup \{x\}$. Thus, the number of sets A_i is at most $2^{n-1} - 1$.

Case 2. There is an element $x \in S$ such that $A_1 \cup A_2 = S$ and $A_1 \cap A_2 = \{x\}$.

Let r and s denote the cardinality of the sets A_1 and A_2, respectively. Clearly, $r + s = n + 1$. The number of the sets A_i such that $A_i \subseteq A_1$ is at most $2^{r-1} - 1$ by the inductional hypothesis. Similarly, the number of the sets A_i such that $A_i \subseteq A_2$ is at most $2^{s-1} - 1$. If A_i is not a subset of A_1 or A_2, then $A_1 \cap A_i \neq \emptyset$, $A_2 \cap A_i \neq \emptyset$, and, of course, $A_1 \cap A_2 \neq \emptyset$. Thus $A_1 \cap A_2 \cap A_i \neq \emptyset$ by the conditions of the problem, that is, $x \in A_i$. Now, $A_i = \{x\} \cup (A_i - A_1) \cup (A_i - A_2)$, and since the nonempty sets $A_i - A_1$ and $A_i - A_2$ can be chosen in $2^{s-1} - 1$ and $2^{r-1} - 1$ ways, respectively, the number of these sets A_i is at most $(2^{s-1} - 1)(2^{r-1} - 1)$. Adding up the partial results obtained, we get that the number of all sets A_i is at most $2^{n-1} - 1$. □

Remark. It can be shown that if $k = 2^{n-1} - 1$, then, necessarily, the sets A_i are exactly the sets of at least two elements containing a given element $x \in S$. It can be shown by induction, say, like we proceeded in the first solution.

Problem C.6. *Show that the edges of a strongly connected bipolar graph can be oriented in such a way that for any edge e there is a simple directed path from pole p to pole q containing e. (A strongly connected bipolar graph is a finite connected graph with two special vertices p and q having the property that there are no points x, y, $x \neq y$, such that all paths from x to p as well as all paths from x to q contain y.)*

Solution. First, we prove that there is a real-valued function f defined on the vertices such that

(i) $f(p) = 1, f(q) = 0$,
(ii) if $f(x) = f(y)$, then $x = y$,
(iii) if $x \neq p$, then there is an edge xy such that $f(y) > f(x)$,
(iv) if $x \neq q$, then there is an edge xy such that $f(y) < f(x)$.

The graph G is connected so there is a path $p = x_0, \ldots, x_m = q$ from p to q. In this subset, $f(x_k) = 1 - k/m$ is a desired function.

Now, suppose that f is a desired function on a proper subset $V_1 \subset V(G)$. We show that f can be extended for some bigger set. Let $z \in V_2 = V(G) - V_1$. The graph G is connected so there is a path from z to p. Let x_1 be the last point of this path belonging to V_2, and let x_0 be the next point of this path. The graph G is strongly connected so there is an $x_1 - p$ or $x_1 - q$ path not containing x_0. Let $x_1, x_2, \ldots$ be such a path, and let $n > 1$ be the smallest index i such that $x_i \in V_1$ (there is such a point since p, q are in V_1). The points x_n, x_0 are distinct elements of V_1 so the open interval $f(x_0), f(x_n)$ is nonempty by (ii) and it contains infinitely many elements even if we delete the values of f in V_1. Thus, we can assign a strictly monotone subsequence of $n-1$ members to the points $x_1, \ldots, x_{n-1}$ so that the sequence $f(x_0), f(x_1), \ldots, f(x_n)$ is strictly monotone. Now, (i)–(iv) hold in the extended domain as well: (i) and (ii) hold obviously, as well, as (iii) and (iv) for $x \in V_1$. Finally, if $x = x_i$ $(1 \leq i \leq n-1)$, then x_{i-1} and x_{i+1} is an appropriate choice for y, respectively.

Since G is finite, it implies that there is a function f satisfying (i)–(iv).

Now, let us orient the edges of G as follows. An edge joining x and y should be oriented from x to y if and only if $f(x) > f(y)$. We show that this orientation has the desired properties. Let e be an edge from some x_1 to some y_1. For $n \geq 1$, let x_{n+1} be a neighbor of x_n such that $f(x_{n+1}) > f(x_n)$ if $x_n \neq p$. (Such a point exists by (iii).) Similarly, let y_{n+1} be a neighbor of y_n such that $f(y_{n+1}) < f(y_n)$ if $y_n \neq q$. (Such a point exists by (iv).) Since G is finite, after a while we get that $x_r = p$, $y_s = q$ for some $r, s \geq 1$. Then $x_r, x_{r-1}, \ldots, x_2, x_1, y_1, y_2, \ldots, y_{s-1}, y_s$ is a desired path through e. □

Problem C.7. *Let $\mathcal{F}$ be a nonempty family of sets with the following properties:*

(a) If $X \in \mathcal{F}$, then there are some $Y \in \mathcal{F}$ and $Z \in \mathcal{F}$ such that $Y \cap Z = \emptyset$ and $Y \cup Z = X$.

(b) If $X \in \mathcal{F}$, and $Y \cup Z = X$, $Y \cap Z = \emptyset$, then either $Y \in \mathcal{F}$ or $Z \in \mathcal{F}$.

Show that there is a decreasing sequence $X_0 \supseteq X_1 \supseteq X_2 \supseteq \ldots$ of sets $X_n \in \mathcal{F}$ such that

$$\bigcap_{n=0}^{\infty} X_n = \emptyset.$$

Solution. We will show that the statement of the problem holds even if the condition (a) is replaced by the following weaker condition:

(a') If $X \in \mathcal{F}$, then there are some $Y \in \mathcal{F}$ and $Z \in \mathcal{F}$ such that $Y \cap Z = \emptyset$, $Y \cup Z \subseteq X$.

We will prove it by contradiction. Suppose that the statement does not hold, that is, if $X_0 \supseteq X_1 \supseteq X_2 \supseteq \dots$ for some sets $X_n \in \mathcal{F}$, then $\cap_{n=0}^{\infty} X_n \neq \emptyset$.

1. For any set $A \in \mathcal{F}$, there is a set $A' \in \mathcal{F}$ such that $A' \subseteq A$ and $A' = \cup_{i=1}^{\infty} A_i$ where the sets A_i are pairwise disjoint elements of $\mathcal{F}$. By condition (a'), there exist sets $A_1, A_1' \in \mathcal{F}$ such that $A_1 \cup A_1' \subseteq A$, $A_1 \cap A_1' = \emptyset$. Similarly, there exist sets $A_2, A_2' \in \mathcal{F}$ such that $A_2 \cup A_2' \subseteq A_1$, $A_2 \cap A_2' = \emptyset$, and so on. We are done if $\cup_{i=1}^{\infty} A_i \in \mathcal{F}$. If it is not the case then by condition b), the sets $A_0 = A - \cup_{i=1}^{\infty} A_i \in \mathcal{F}$ and $A' = A = \cup_{i=0}^{\infty} A_i$ have the desired properties.
2. For any set $A \in \mathcal{F}$ there are some sets $B, C \in \mathcal{F}$ such that $B \cap C = \emptyset$, $B \cup C \subseteq A$ and if $B \subseteq P \subseteq B \cup C$ then $P \in \mathcal{F}$.
   Consider the sets A_i existing by 1. We try to define a sequence of sets N_i as follows. Let $N_1 = A'$, and if N_{i-1} is defined, then let N_i be an arbitrary set satisfying the following four conditions:

$$N_i \in \mathcal{F},$$
$$N_i \subseteq N_{i-1},$$
$$N_i \subseteq \bigcup_{j=i}^{\infty} A_j,$$
$$N_i \cap A_j \in \mathcal{F} \quad \text{for infinitely many} \quad j.$$

According to the third condition, $\cap_{i=1}^{\infty} N_i = \emptyset$ if the sequence is infinite, so there is an n such that N_n is defined but N_{n+1} cannot be defined. Let $B = N_n \cap A_n$ and $C = N_n \cap A_k$ for some index $k > n$ such that $N_n \cap A_k \in \mathcal{F}$. Suppose that there is a set P such that $B \subseteq P \subseteq B \cup C$ but $P \notin \mathcal{F}$. Then $N_n - P \in \mathcal{F}$, and it is easy to see that $N_{n+1} = N_n - P$ would be an appropriate choice. ($B \in \mathcal{F}$ by the choice $P = B$.)

Thus, for $A \in \mathcal{F}$, there are some sets $B_1, C_1 \in \mathcal{F}$ such that $B_1 \cup C_1 \subseteq A$, $B_1 \cap C_1 = \emptyset$. Similarly, for $C_1 \in \mathcal{F}$, there are some sets $B_2, C_2 \in \mathcal{F}$ such that $B_2 \cup C_2 \subseteq C_1$, $B_2 \cap C_2 = \emptyset$, and so forth. Let $P_i = \cup_{j=i}^{\infty} B_j$ for $i = 1, 2, \dots$. Then $P_i \supseteq P_{i+1}$, $\cap_{i=1}^{\infty} P_i = \emptyset$, and $P_i \in \mathcal{F}$ because $B_i \subseteq P_i \subseteq B_i \cup C_i$, a contradiction. □

Remark. The problem was motivated by the following set theoretical game. There are two players: Black and White. Let X_1 be a given set. White partitions it into two sets, Y_1 and Z_1. Then Black chooses one of the sets Y_1, Z_1, let us denote this set by X_2, and partitions it into some sets Y_2 and Z_2. Then White chooses one of these sets Y_2 and Z_2 (let us denote it by X_3) and partitions it into some sets Y_3 and Z_3, and so on. Thus, the players construct a countably infinite sequence $X_1, X_2, X_3, \dots$ of sets. White wins if $\cap_{i=1}^{\infty} X_i \neq \emptyset$ and Black wins if $\cap_{i=1}^{\infty} X_i = \emptyset$. If the statement of the problem were false, that is, if there were a family $\mathcal{F}$ of sets satisfying (a) and (b) such that $\cap_{i=1}^{\infty} X_i \neq \emptyset$ for any decreasing sequence of sets X_i, then White would have a winning strategy for any set $X_1 \in \mathcal{F}$: White would partition X_{2k+1} so that $Y_{2k+1}, Z_{2k+1} \in \mathcal{F}$ and, from among

Y_{2k} and Z_{2k}, would choose a set contained in $\mathcal{F}$. Finding this kind of set families is motivated by the fact that we do not know who has a winning strategy in sets X_1 of different cardinalities.

Problem C.8. *Let G be a 2-connected nonbipartite graph on $2n$ vertices. Show that the vertex set of G can be split into two classes of n elements each such that the edges joining the two classes form a connected, spanning subgraph.*

Solution. Let F be a spanning tree of the graph G. The vertices of F can be colored with red and blue so that if two vertices are of the same color then they are not adjacent. Actually, F has exactly two colorings like this, which can be obtained from each other by interchanging the colors red and blue. Notice that the statement of the problem is equivalent to the following one: *There exists a spanning tree F in G such that the numbers of the red and blue vertices are the same.*

Let F_1 and F_2 be two spanning trees in G, and let x be a vertex that is a leaf (a vertex of degree one) in both spanning trees. We say that the spanning trees F_1 and F_2 are *neighboring at x* if $F_1 - x = F_2 - x$. We will prove the following lemma, which implies the statement above, as we will see.

Lemma. Let F and F' be two spanning trees of G, and let T be a common subtree of them. Then there is a sequence $F_0 = F, F_1, \ldots, F_k = F'$ of spanning trees such that $T \subseteq F_i$ and F_i and F_{i+1} are neighboring at some vertex not in T.

Proof. We prove the lemma by "backward" induction on the number of vertices in T. If T has $2n-1$ vertices, then F and F' are neighboring. Suppose now that T has $2n-l$ vertices ($l \geq 2$). Let xy and uv be an edge of F and F' leaving T, respectively, such that $x, u \in T$. If $xy = uv$ then $T' = T+xy$ is a common subtree and there is a desired sequence of spanning trees by the inductional hypothesis. If $y \neq v$, then let F'' be a spanning tree containing the tree $T+xy+uv$ as a subgraph. Then $T+xy \subseteq F \cap F''$, and so by the inductional hypothesis, there is a sequence $F = F_0, F_1, \ldots, F_m = F''$ of spanning trees such that F_i and F_{i+1} are neighboring ($i = 0, 1, \ldots, m-1$) and $T \subseteq T+xy \subseteq F_i$. Similarly, $T+uv \subseteq F' \cap F''$ and so by the inductional hypothesis, there is a sequence $F'' = F_m, F_{m+1}, \ldots, F_k = F'$ of spanning trees such that F_i and F_{i+1} are neighboring ($i = m, m+1, \ldots, k-1$) and $T \subseteq T + uv \subseteq F_i$, and $F = F_0, F_1, \ldots, F_k = F'$ is the desired sequence of spanning trees.

Finally, suppose that $xy \neq uv$ but $y = v$. Since $G - y$ is connected, it has a vertex $z \notin T$ joined to a vertex $w \in T$. Now, let F'' and F''' each be a spanning tree containing $T + xy + wz$ and $T + uv + wz$, respectively. Then, as above, there is a desired sequence of spanning trees from F to F'', from F'' to F''', and from F''' to F', and the union of these sequences is a desired sequence from F to F'.

Since G is nonbipartite, it contains a cycle C of odd length. Let x be a vertex of C, and let y and z be its neighbors in C. Let T_0 be a spanning tree of $G - x$ containing $C - x$. Let $F = T_0 + xy$ and $F' = T_0 + xz$. Color F so that x is red.

According to the lemma, there is a sequence $F = F_0, F_1, \ldots, F_k = F'$ of spanning trees such that F_i and F_{i+1} are neighboring at some vertex $x_i \neq x$. Color F_i with colors red and blue so that the colorings of F_i and F_{i+1} differ from each other at most in the color of x_i. It defines some colorings of the spanning trees $F_1, \ldots, F_k = F'$ starting with the coloring of $F_0 = F$ such that x is red in every coloring. Clearly, every vertex got different colors in the colorings of F and F', except x. Let a_i denote the number of red vertices in the coloring of F_i. Then

$$|a_{i+1} - a_i| \leq 1,$$
$$a_0 + a_k = 2n + 1.$$

Thus,

$$\text{either} \qquad a_0 \leq n < a_k \qquad \text{or} \qquad a_k \leq n < a_0,$$

so there is an index $0 \leq i \leq k$ such that $a_i = n$, and this is what we wanted to prove. □

Problem C.9. *Let $\mathcal{A}_n$ denote the set of all mappings $f : \{1, 2, \ldots, n\} \to \{1, 2, \ldots, n\}$ such that $f^{-1}(i) := \{k : f(k) = i\} \neq \emptyset$ implies $f^{-1}(j) \neq \emptyset$, $j \in \{1, 2, \ldots, i\}$. Prove*

$$|\mathcal{A}_n| = \sum_{k=0}^{\infty} \frac{k^n}{2^{k+1}}.$$

Solution 1. Let a_n denote the cardinality of the set $\mathcal{A}_n$. Then the number of mappings $f \in \mathcal{A}_n$ such that $f(i) = 1$ holds for exactly l integers i is equal to $\binom{n}{l} a_{n-l}$. Since l is positive for any function f, we have

$$a_n = \sum_{l=1}^{n} \binom{n}{l} a_{n-l} = \sum_{l=0}^{n-1} a_l$$

for $n \geq 1$.

Using the notation $b_n = a_n/n!$, we have $b_0 = 1$ and

$$b_n = \sum_{l=0}^{n-1} \frac{1}{(n-l)!} b_l.$$

Clearly, $b_n < 2^n$, so if $|z| < 1/2$, then the series $\sum_{n=0}^{\infty} b_n z^n$ is absolutely convergent. For the sum $f(z)$ of this series, we have

$$\begin{aligned} f(z) = \sum_{n=0}^{\infty} b_n z^n &= 1 + \sum_{n=1}^{\infty}\sum_{l=0}^{n-1} \frac{1}{(n-l)!} b_l z^n \\ &= 1 + \sum_{l=0}^{\infty}\left(\sum_{n=l+1}^{\infty} \frac{1}{(n-l)!} z^n\right) b_l \\ &= 1 + \sum_{l=0}^{\infty}(e^z - 1) b_l z^l = 1 + (e^z - 1) f(z). \end{aligned}$$

It implies that

$$\begin{aligned} f(z) &= \frac{1}{2}\cdot\frac{1}{1 - e^z/2} = \sum_{k=0}^{\infty} \frac{e^{kz}}{2^{k+1}} = \sum_{k=0}^{\infty}\sum_{n=0}^{\infty} \frac{1}{n!}\frac{k^n}{2^{k+1}} z^n \\ &= \sum_{n=0}^{\infty}\left(\frac{1}{n!}\sum_{k=0}^{\infty}\frac{k^n}{2^{k+1}}\right) z^n. \end{aligned}$$

Thus,

$$b_n = \frac{1}{n!}\sum_{k=0}^{\infty} \frac{k^n}{2^{k+1}},$$

which we wanted to prove. □

Solution 2. Let $s_{n,i}$ denote the number of surjective mappings of the set $\{1, 2, \ldots, n\}$ onto the set $\{1, 2, \ldots, i\}$. Then the cardinality of the set $\mathcal{A}_n$ is

$$\sum_{i=1}^{n} s_{n,i}.$$

Counting the mappings of the set $\{1, 2, \ldots, n\}$ onto the set $\{1, 2, \ldots, k\}$, we get the equality

$$k^n = \sum_{i=1}^{k} \binom{k}{i} s_{n,i}.$$

Now, we have to prove that

$$\sum_{i=1}^{n} s_{n,i} = \sum_{k=1}^{\infty}\sum_{i=1}^{k} \frac{\binom{k}{i}}{2^{k+1}} s_{n,i},$$

that is, using $s_{n,n+1} = s_{n,n+2} = \cdots = 0$ and supposing that the rearrangement does not change the sum, we have to prove that

$$\sum_{i=1}^{n} s_{n,i} = \sum_{i=1}^{n}\left(\sum_{k=i}^{\infty} \frac{\binom{k}{i}}{2^{k+1}}\right) s_{n,i}.$$

In the parentheses, we have the sum of the terms of a negative binomial distribution ($\binom{k}{i}/2^{k+1}$ is the probability of the event that flipping a coin, we get the $(i+1)$st head in the $(k+1)$st flip), and so the coefficient of every term $s_{n,i}$ on the right is 1. Since the series in the parentheses are absolutely convergent, the rearrangement did not change the sum, and the proof is complete. □

Problem C.10. *Let G be an infinite graph such that for any countably infinite vertex set A there is a vertex p joined to infinitely many elements of A. Show that G has a countably infinite vertex set A such that G contains uncountably infinitely many vertices p joined to infinitely many elements of A.*

Solution. We prove the statement by contradiction. Assume that G is a graph with vertex set V such that

(1) for any countably infinite vertex set $A \subseteq V$, the set H_A of vertices $p \in V - A$ joined to infinitely many elements of A is a nonempty, countable set.

The set H_A is obviously infinite since if not, then for the set $B = A \cup H_A$, there is no vertex $p \in V - B$ joined to infinitely many elements of B. It implies, that deleting finitely many vertices from G, the resulting graph has property (1).

Notice that the vertex set V is clearly uncountable since (1) holds for $A = V$ as well.

Now, we show that we can delete finitely many vertices from G so that the resulting graph G_1 has no finite vertex set X covering V with the exception of a countable set. (A vertex set X *covers* the union of the sets of neighbors of the elements of X.) Suppose it is not the case. Then there is a finite set X_0 covering all elements of V except a countable set. Deleting X_0, the resulting graph still must have a finite set X_1 with the same property, and so on. So if we do not get a desired graph after finitely many steps then we can define the pairwise disjoint finite sets X_n covering V with the exception of some countable sets. Let $A = \cup_{n=0}^{\infty} X_n$. Then every element of V (except the elements of a countable set) is joined to infinitely many elements of A, that is, $V - H_A$ is countable. Since H_A is countable by (1), it implies that V is countable, a contradiction.

As we have seen, G_1 has property (1) as well, so we may assume that $G_1 = G$.

Let $A \subset V$ be an arbitrary countably infinite set. We show that there is a countably infinite set $F(A) \subset V - A$ such that $H_A \subset F(A)$ and for any finite subset $\{a_1, \dots, a_n\}$ of A, the set $F(A)$ has infinitely many elements not joined to any vertex a_i $(i = 1, \dots, n)$. Let $A = \{a_1, a_2, \dots\}$ and for each n, put infinitely many elements of $V - A$ not joined to any element of $\{a_1, \dots, a_n\}$ into $F(A)$. It can be done, as we have seen. Then take the union of H_A and the set of the elements obtained.

Now, let $A_0 \subset V$ be an arbitrary countably infinite set, and let

$$A_i = F(A_0 \cup A_1 \cup \cdots \cup A_{i-1}) \quad (i = 1, 2, \dots),$$
$$A_\omega = F\left(\bigcup_{i=0}^{\infty} A_i\right).$$

For any element $p \in A_\omega$ and for any index i, p is joined to finitely many elements of A_i. Really, $p \in A_\omega \subset V - \cup_{i=0}^{\infty} A_i \subset V - A_{i+1}$, and so $p \notin F(A_0 \cup A_1 \cup \cdots \cup A_{i-1}) \supset H_{A_i}$.

Let $\cup_{i=0}^{\infty} A_i = \{a_1, a_2, \dots\}$, $A_\omega = \{b_1, b_2, \dots\}$. Let us define an infinite set $C = \{c_1, c_2, \dots\}$ such that if $C \subset \cup_{i=0}^{\infty} A_i$ and $n < m$, then c_m is joined to neither a_n nor b_n. It yields the desired contradiction since $p \notin \cup_{i=0}^{\infty} A_i$ for $p \in H_C$ (if $p = a_n$, then it is joined to at most n elements c_m), and so, $p \in H_C \subset H_{\cup_{i=0}^{\infty} A_i} \subset F\left(\cup_{i=0}^{\infty} A_i\right) = A_\omega$. However, it is a contradiction again, since if $p = b_n$, then it is joined to at most the elements $c_1, c_2, \dots, c_n$ of C.

Thus, let c_1 be an arbitrary element of A_0, and suppose that $c_1, c_2, \dots, c_n$ are given so that $\{a_1, a_2, \dots, a_n, c_1, c_2, \dots, c_n\} \subset A_0 \cup A_1 \cup \cdots \cup A_N$ for some N. Then, A_{N+1} has an infinite subset T such that the elements of T are not joined to any of the elements $a_1, a_2, \dots, a_n$. As we have seen above, each of the elements $b_1, b_2, \dots, b_n$ is joined to finitely many elements of T, so by choosing c_{n+1} out of the rest of T, we meet the requirements. □

Remark. Assuming the continuum hypothesis, it is possible to construct a graph G satisfying the conditions of the problem in which every countably infinite set A has a countably infinite subset B such that H_B is countable.

Problem C.11. *Let $\mathcal{H}$ be a family of finite subsets of an infinite set X such that every finite subset of X can be represented as the union of two disjoint sets from $\mathcal{H}$. Prove that for every positive integer k there is a subset of X that can be represented in at least k different ways as the union of two disjoint sets from $\mathcal{H}$.*

Solution. The solution is based on the following theorem of *Ramsey*.

Theorem. Let Y be an arbitrary infinite set and let n be an arbitrary natural number. Subdivide the family of all subsets of n elements of Y into two classes. Then, there exists an infinite set $Y' \subset Y$ such that every subset of n elements of the set Y' belongs to the very same class.

For a set A, let $[A]^n$ denote the set of subsets of n elements of the set A. For a given natural number k, we show that there is a natural number $m \geq k$ and an infinite set $Z \subset Y$ such that $[Z]^m \subset \mathcal{H}$. It implies the statement of the problem, since then any subset of $2m$ elements of the set Z can be obtained as the union of two disjoint members of $\mathcal{H}$ in at least $\binom{2m}{m} \geq k$ different ways.

We find a desired set Z as follows. Subdivide the set $[X]^2$ into two classes depending on whether or not the elements are in $\mathcal{H}$. By Ramsey's theorem above, there exists an infinite set $X_2 \subset X$ such that

$$\text{either} \qquad [X_2]^2 \subset \mathcal{H} \qquad \text{or} \qquad [X_2]^2 \cap \mathcal{H} = \emptyset.$$

Then classify the subsets of three elements of X_2 based on whether or not they belong to $\mathcal{H}$. Again by Ramsey's theorem above, there exists an infinite set $X_3 \subset X_2$ such that

$$\text{either} \qquad [X_3]^3 \subset \mathcal{H} \qquad \text{or} \qquad [X_3]^3 \cap \mathcal{H} = \emptyset.$$

Proceeding further, we obtain a sequence $X \supset X_2 \supset \cdots \supset X_{2k}$ of infinite subsets of X with the following property:

$$\text{either} \qquad [X_m]^m \subset \mathcal{H} \qquad \text{or} \qquad [X_m]^m \cap \mathcal{H} = \emptyset$$

for $2 \le m \le 2k$. For such an m, we naturally have

$$\text{either} \qquad [X_{2k}]^m \subset \mathcal{H} \qquad \text{or} \qquad [X_{2k}]^m \cap \mathcal{H} = \emptyset$$

as well. We are done if we can prove that the second possibility cannot be the case. Actually, choose a subset A of $2k$ elements of the set X_{2k}. By the conditions of the problem (which are used only at this point), $A = B \cup C$ for some sets $B, C \in \mathcal{H}$. One of these sets, say B, has at least k elements. Then for $m = |B|$, we have $B \in [X_{2k}]^m \cap \mathcal{H}$, that is, $[X_{2k}]^m \cap \mathcal{H} \ne \emptyset$, and so $[X_{2k}]^m \subset \mathcal{H}$, which we wanted to prove. □

Remarks.

1. We did not use the fact that the finite subsets of X can be obtained as the union of two *disjoint* sets in $\mathcal{H}$; this part of the condition can be omitted. Furthermore, it is sufficient to assume that for a fixed integer c, the finite subsets of X can be obtained as the union of at most c members of $\mathcal{H}$.
2. Considering natural numbers instead of sets and sum instead of union, we get the following analogous problem: let H be an arbitrary set of natural numbers. Suppose that every natural number is the sum of two elements of H. Does it imply that for any natural number k, there is a number that can be obtained as the sum of two elements of H in at least k different ways? This problem of S. Sidon is open. However, the multiplicative version of the problem (where we say "product" instead of "sum" in all cases) has been answered affirmatively. It follows from the statement of the original problem, as follows: let X be the set of all prime numbers, and assign the set of prime divisors to any natural number in H. Thus, we obtain a family $\mathcal{H}$ meeting the conditions of the original problem. Thus, there exists a set $A \subset X$ that can be obtained as the union of two disjoint members of $\mathcal{H}$ in at least k different ways. Then, the product of the prime divisors in A can be obtained as the product of two elements of H in at least k different ways.

Problem C.12. *Let the operation f of k variables defined on the set $\{1, 2, \ldots, n\}$ be called friendly toward the binary relation ρ defined on the same set if*

$$f(a_1, a_2, \ldots, a_k) \, \rho \, f(b_1, b_2, \ldots, b_k)$$

implies $a_i \, \rho \, b_i$ for at least one i, $1 \leq i \leq k$. Show that if the operation f is friendly toward the relations "equal to" and "less than," then it is friendly toward all binary relations.

Solution. Since f is friendly with the relations "equal to" and "less than," we have

$$f(1, \ldots, 1) < f(2, \ldots, 2) < \cdots < f(n, \ldots, n)$$

and thus $f(i, \ldots, i) = i$ for all i. Hence, the following property holds:

(*) If $f(a_1, \ldots, a_k) = a$, then there is an i such that $a_i = a$ (since $f(a_1, \ldots, a_k) = a = f(a, \ldots, a)$ and f is friendly with the relation "equal to").

It is sufficient to prove that if $f(a_1, \ldots, a_k) = a$ and $f(b_1, \ldots, b_k) = b$ then there is a subscript i such that $a_i = a$ and $b_i = b$. Let $I = \{i \mid a_i = a\}$ and $J = \{j \mid b_j = b\}$. Property (*) shows that $I, J \neq \emptyset$; however, what we need is that $I \cap J \neq \emptyset$. This is obviously true for $n = 1$. Now, let $n \geq 2$, and assume that $I \cap J = \emptyset$. We distinguish two cases:

Case 1. $a \neq b$. Let $c_i = b$ for $i \in I$ and $c_i = a$ otherwise. Furthermore, let $d_i = a$ if $i \in J$ and $d_i = b$ otherwise. Then $f(c_1, \ldots, c_k) \neq f(a_1, \ldots, a_k) = a$ (since $c_i = b \neq a = a_i$ if $i \in I$ and $c_i = a \neq a_i$ if $i \notin I$), and similarly $f(d_1, \ldots, d_k) \neq f(b_1, \ldots, b_k) = b$. Property (*) implies that $f(c_1, \ldots, c_k)$ and $f(d_1, \ldots, d_k)$ could only be a or b and thus $f(c_1 \ldots, c_k) = b$ and $f(d_1 \ldots, d_k) = a$. Without loss of generality, we may assume that $a < b$, that is, $f(d_1, \ldots, d_k) < f(c_1, \ldots, c_k)$. Since f is friendly with the relation "less than" we must have an i such that $d_i < c_i$. On the other hand, however, if $i \in I$ or $i \in J$ then $c_i = d_i$; otherwise, $c_i = a < b = d_i$, a contradiction.

Case 2. $a = b$. Let $c_i = c(\neq a)$ if $i \in I$ and $c_i = a$ otherwise; similarly $d_i = c$ if $i \in J$ and $d_i = a$ otherwise; and finally $e_i = a$ if $i \in J$ and $e_i = c$ otherwise. Similarly to the previous case, we have $f(c_1, \ldots, c_k) \neq f(a_1, \ldots, a_k) = a$, $f(d_1, \ldots, d_k) \neq f(b_1, \ldots, b_k) = b = a$ and $f(e_1, \ldots, e_k) \neq f(d_1, \ldots, d_k)$. However, all the values $f(c_1, \ldots, c_k)$, $f(d_1, \ldots, d_k)$ and $f(e_1, \ldots, e_k)$ must be either a or c, and thus $f(c_1, \ldots, c_k) = f(d_1, \ldots, d_k) = c$, and so $f(e_1, \ldots, e_k) = a$. Again, it is enough to see the case when $a < b$, that is, $f(e_1, \ldots, e_k) < f(c_1, \ldots, c_k)$, which implies the existence of an index i such that $e_i < c_i$. However, we have $c_i = e_i = c$ if $i \in I$, $c_i = e_i = a$ if $i \in J$, and $c_i = a < c = e_i$ otherwise. This is a contradiction again, which makes the proof complete. □

Problem C.13. *Let $g(n,k)$ denote the number of strongly connected, simple directed graphs with n vertices and k edges. (Simple means no loops or multiple edges.) Show that*

$$\sum_{k=n}^{n^2-n}(-1)^k g(n,k)=(n-1)!.$$

Solution 1. We prove the statement by induction on n. For $n=2$, the statement is obviously true. Let us write the left-hand side in the form

$$\sum_G(-1)^{|E(G)|}, \tag{1}$$

where G runs over the graphs described above. For every G in the summation (1), consider the edges of G having one end vertex at n. We distinguish two cases:

Case 1. G has exactly two edges incident to n and the other endpoint of both of them is the same vertex p.

In this case, $G\backslash\{n\}$ is strictly connected as well, and so

$$(-1)^{|E(G\backslash\{n\})|}=(-1)^{|E(G)|}.$$

Thus, for a fixed p, the sum of these terms is $(n-2)!$ according to the induction assumption and summing them up for $p=1,\dots,(n-1)$, we get exactly $(n-1)!$.

Case 2. There are more than two edges of G incident to n, or there are two such edges of it having different other endvertices.

There must be two vertices $p\neq q$, $1\le p,q\le n-1$, such that $pn\in E(G)$ and $nq\in E(G)$. In this case, the graph having the same edge set as G, except that it has the edge pq if G does not have it, or it does not have this edge if G does have it, is strongly connected if and only if G is strongly connected; thus we can pair these graphs such that every such graph G not containing the edge pq has its pair $G+pq$. For every graph, we have a pair, and the sum in (1) is 0 for such a pair; thus, the sum is 0 for the graphs in this case.

Summing up $(-1)^{|E(G)|}$ for the graphs in Cases 1 and 2, we get that (1) is $(n-1)!$. □

Solution 2. If $\mathcal{G}$ is a set of graphs satisfying the properties of the problem, let us define

$$\mu(\mathcal{G})=\sum_{G\in\mathcal{G}}(-1)^{|E(G)|}$$

For an arbitrary graph G, assign the permutation $\{a_1=1,a_2,\dots,a_n\}$ of the vertex set such that if a_k is already chosen we pick a_{k+1} as the smallest number having an edge from the set $\{a_1, a_2, \ldots, a_k\}$ to it. (We always have such an edge and vertex since the graph G is strongly connected.)

For any given permutation π, let $\mathcal{G}_\pi$ denote the set of graphs having π as the assigned permutation. Since we have

$$\sum_G (-1)^{|E(G)|} = \sum_{\pi \in S_{n-1}} \mu(\mathcal{G}_\pi),$$

and since the number of permutations of the set $\{2, \ldots, n\}$ is $(n-1)!$, it suffices to prove that $\mu(\mathcal{G}_\pi) = 1$ for all π.

Let us color the edges of these graphs with blue and red as follows: let the edges going along the permutation π be colored blue and the other ones colored red (that is, the blue edges are of form $a_i a_j$ for $i < j$). Clearly,

($*$) for every vertex $v \neq 1$, there is a blue edge pointing to v, but there is no blue edge $a_k a_l$ if $a_l < a_j$ for some $k < j < l$.

If $\mathcal{G}_\alpha$ is the set of graphs corresponding to a fixed set α of blue edges, that is, having exactly those edges blue that belong to α (provided we have the fixed permutation π), then we have

$$\mu(\mathcal{G}_\pi) = \sum_{\mathcal{G}_\alpha \subseteq \mathcal{G}_\pi} \mu(\mathcal{G}_\alpha),$$

Let now us consider these sums $\mu(\mathcal{G}_\alpha)$ now. If in any graph in $\mathcal{G}_\alpha$ there is a red edge $a_i a_j$ such that $i - j > 1$, then pick the edge having the smallest j and among them that having the biggest i. Let $\mathcal{G}_\alpha^{i,j}$ be the subset of $\mathcal{G}_\alpha$ in which the graphs have $a_i a_j$ as the red edge picked, and let $G \in \mathcal{G}_\alpha^{i,j}$. The way we chose j ensures that the edges $a_i a_j$, $a_j a_{j-1}, \ldots, a_2 a_1$ are edges in G and furthermore — because of the way we constructed π — we can reach a_{i-1} from the vertex $a_1 = 1$ through blue edges only. The graph, having the same edge set as G with the exception of the edge $a_i a_{i-1}$, is strongly connected if and only if G is strongly connected. Thus, we can pair the graphs of $\mathcal{G}_\alpha^{i,j}$ so that the pairs differ only in the edge $a_i a_{i-1}$. Such a pair contributes 0 to the sum; thus, $\mu(\mathcal{G}_\alpha^{i,j}) = 0$ and $\sum_{i,j} \mu(\mathcal{G}_\alpha^{i,j}) = 0$.

There is exactly one set of red edges not containing any red edge to pick up, namely $\{a_i a_{i-1} \mid i = 2, \ldots, n\}$, and so

$$\mu(\mathcal{G}_\alpha) = (-1)^{|\alpha|}(-1)^{n-1} = (-1)^{|\alpha|-(n-1)}.$$

It is easy to see that any set of edges satisfying property ($*$) and having all edges going along the permutation can be chosen as the set α of blue edges. Thus, we have

$$\begin{aligned}
\mu(\mathcal{G}_\pi) &= \sum_{\mathcal{G}_\alpha \subseteq \mathcal{G}_\pi} (-1)^{|\alpha|-(n-1)} \\
&= \sum_{l_2=1}^{p_2} \sum_{l_3=1}^{p_3} \cdots \sum_{l_n=1}^{p_n} \binom{p_j}{l_j} (-1)^{l_j-1} = \prod_{j=2}^{n} \sum_{l_j=1}^{p_j} (-1)^{l_j-1} \binom{p_j}{l_j},
\end{aligned}$$

where p_k denotes the maximal in-degree of vertex k. Since we have

$$\sum_{l_j=1}^{p_j}(-1)^{l_j-1}\binom{p_j}{l_j} = (-1)\left(\sum_{l_j=0}^{p_j}(-1)^{l_j}\binom{p_j}{l_j} - \binom{p_j}{0}\right) = (-1)(0-1) = 1,$$

we get $\mu(\mathcal{G}_\pi) = 1$, and topthe proof is complete. □

Problem C.14. *Let T be a triangulation of an n-dimensional sphere, and to each vertex of T let us assign a nonzero vector of a linear space V. Show that if T has an n-dimensional simplex such that the vectors assigned to the vertices of this simplex are linearly independent, then another such simplex must also exist.*

Solution 1. Let $A = (a_0, \ldots, a_n)$ be a simplex in T such that the vectors $v(a_0), \ldots, v(a_n)$ assigned to the vertices of A are linearly independent. We will call an $(n-1)$-dimensional simplex in T *red* if for any index $0 \le i \le n-1$, only one vector in the set

$$X_i = \langle v(a_0), \ldots, v(a_i)\rangle - \langle v(a_0), \ldots, v(a_{i-1})\rangle$$

is assigned to the vertices of it. (Here, $\langle v_1, \ldots, v_k\rangle$ denotes the subspace generated by the vectors $v_1, \ldots, v_k$.) Obviously, $(a_0, \ldots, a_n)$ is red.

Lemma. If the vectors assigned to the vertices of an n-dimensional simplex in T are not linearly independent, then the number of its red faces is 0 or 2.

Proof. If the set $v(S)$ of vectors assigned to the vertices of S does not contain any vector in one of the sets $X_0, \ldots, X_{n-1}$, then obviously S has no red face. If $v(S)$ contains an element of each of the sets $X_0, \ldots, X_{n-1}$ and a vector not in $X_0 \cup \cdots \cup X_{n-1}$ then the vectors assigned to the vertices of S are linearly independent, a contradiction. So, we may assume that $v(S)$ contains an element of each set X_j $(j = 0, 1, \ldots, n-1)$ and that it contains exactly two elements of, say, X_i and exactly one element of the other X_j's. Then S has two red faces by which can be obtained from S, deleting one of the vertices with vectors in X_i assigned to it. The proof of the lemma is complete.

Now, consider the graph G whose vertices are the n-dimensional simplices of T and where two vertices are joined by an edge if they share a common red face. In this graph, the degree of the simplex A is 1. Thus, G contains one more vertex of odd degree. By the above lemma, the vectors assigned to the vertices of the corresponding simplex are linearly independent which we wanted to prove. □

Solution 2. We prove the following, slightly more general, statement: If K is an arbitrary n-dimensional complex with boundary 0 (that is, K is a cycle mod 2), and if we assign nonzero vectors of the linear space V to the 0-dimensional simplices (that is, the vertices) of K, then the number of n-dimensional simplices with linearly independent assigned vectors cannot be 1.

We prove the statement by induction on n. It obviously holds for $n = 1$. Suppose that $n \geq 2$. Suppose that $\dim v(A) = n + 1$ for $A \in K$. We have to show that K contains one more such simplex. Let B be an $(n-1)$-dimensional face of A. If $C \neq A$ is a simplex such that B is a face of it, then $v(C) \not\subseteq v(B)$ implies that $\dim v(C) > \dim v(B) = n$, that is, C has the desired properties. So, we may assume that if $C \neq A$ and B is a face of A, then $v(C) \subseteq v(B)$.

Let $K_0 = \{S \in K | v(S) \subseteq v(B)\}$ and let $K_1 = \delta(K_0)$, where δ denotes the boundary. Obviously, $\delta(K_1) = \delta(\delta(K_0)) = 0$, and since $A \notin K_0$ and $\delta(K) = 0$, B is thus the face of an odd number of simplices in K_0, that is, $B \in K_1$. By the induction hypothesis, there exists a simplex $B_1 \neq B$ in K_1 such that $\dim v(B_1) = \dim v(B)$, and so $v(B_1) = v(B)$. Since $B_1 \in \delta(K_0)$ but $B_1 \notin \delta(K)$, there is a thus simplex A_1 in K such that B_1 is a face of A_1 but $v(A_1) \not\subseteq v(B)$, and so $\dim v(A_1) > \dim v(B_1) = n$. So, the vectors assigned to the vertices of A_1 are linearly independent. To make the proof complete, notice that it is impossible that $A = A_1$ because if it is the case then A has two faces B and B_1 such that $v(B) = v(B_1)$ and so $\dim v(A) = \dim v(B) = n$, a contradiction. □

Problem C.15. *For a real number x, let $\|x\|$ denote the distance between x and the closest integer. Let $0 \leq x_n < 1 \quad (n = 1, 2, \dots)$, and let $\varepsilon > 0$. Show that there exist infinitely many pairs (n, m) of indices such that $n \neq m$ and*

$$\|x_n - x_m\| < \min\left(\varepsilon, \frac{1}{2|n-m|}\right).$$

Solution. Let us fix $\varepsilon > 0$. We show that if n is large enough, $0 \leq x_i < 1 \quad (i = 1, \dots, n)$, then there exists a pair (i, j) of indices such that $\|x_i - x_j\| < \min(\varepsilon, 1/over2|i-j|)$. It suffices since it implies that by subdividing the infinite sequence $x_1, x_2, \dots$ into blocks of n terms, we can find a desired pair of indices in each block and the difference $|i - j|$ does not change if the indices in the block are replaced by the indices in the whole sequence.

Let $f : \{1, \dots, n\} \to \{1, \dots, n\}$ be a permutation such that

$$0 \leq x_{f(1)} \leq \dots \leq x_{f(n)} < 1.$$

We may assume that

$$\|x_{f(i)} - x_{f(i+1)}\| < \min\left(\varepsilon, \frac{1}{2|f(i) - f(i+1)|}\right)$$

for $i = 1, \dots, n$ since if it is not the case then we are done. Let A denote the set of the indices $1 \leq i \leq n$ such that $\|x_{f(i)} - x_{f(i+1)}\| \geq \varepsilon$. Since the cyclically taken intervals $[x_{f(1)}, x_{f(2)}), [x_{f(2)}, x_{f(3)}), \ldots, [x_{f(n)}, x_{f(1)})$ cover the cyclic interval $[0,1)$

$$1 \geq \sum_{i=1}^{n} \|x_{f(i)} - x_{f(i+1)}\| \geq \sum_{i \in A} \varepsilon + \sum_{i \notin A} \frac{1}{2|f(i) - f(i+1)|},$$

and $|A| \leq 1/\varepsilon$. That is,

$$1 - |A|\varepsilon \geq \frac{1}{2} \sum_{i \notin A} \frac{1}{|f(i) - f(i+1)|}.$$

Applying the inequality between the arithmetic and harmonic means, we get the inequality

$$\sum_{i \notin A} |f(i) - f(i+1)| \geq \frac{(n - |A|)^2}{2(1 - |A|\varepsilon)}.$$

In the sum $\sum_{i=1}^{n} |f(i) - f(i+1)|$, every integer $j \in \{1, \ldots, n\}$ appears exactly twice with positive or negative signs, so that the total sum of the signs is 0. Obviously, the sum is maximum when the large numbers have coefficient $+2$ and the small numbers have coefficient -2. Thus,

$$\sum_{i=1}^{n} |f(i) - f(i+1)| \leq \frac{n^2}{2}$$

when n is even, and

$$\sum_{i=1}^{n} |f(i) - f(i+1)| \leq \frac{n^2 - 1}{2}$$

when n is odd. Combining this fact with the preceding inequality, we obtain $1 - |A|\varepsilon \geq (1 - |A|/n)^2$, which is a contradiction if $|A| \geq 1$ and n is sufficiently large. If $A = \emptyset$, then we get $1 = 1$, which means that n is even and the numbers $|f(i) - f(i+1)|$ are equal (mean inequality). However, if $f(i) > n/2$, then $f(i)$ appears twice with positive sign, that is, $f(i) > f(i-1)$, $f(i) > f(i+1)$, implying that $f(i-1) = f(i+1)$, which is possible only if $n \leq 2$. But it is easy to see that equality cannot hold in this case either, and we got the desired contradiction in all cases. □

Remark. The existence of a pair (i,j) of indices such that $\|x_i - x_j\| < \min(\varepsilon, 1/c|i-j|)$ can be proved if $c \leq \sqrt{5}$ (the proof is much more complicated). The statement is false if $c > \sqrt{5}$ since it would imply that for an irrational number α, there are infinitely many rational numbers p/q such that $|\alpha - p/q| < 1/cq^2$.

Problem C.16. *Consider the lattice L of the contractions of a simple graph G (as sets of vertex pairs) with respect to inclusion. Let $n \geq 1$ be an arbitrary integer. Show that the identity*

$$x \bigwedge \left(\bigvee_{i=0}^{n} y_i \right) = \bigvee_{j=0}^{n} \left(x \bigwedge \left(\bigvee_{\substack{0 \leq i \leq n \\ i \neq j}} y_i \right) \right)$$

holds if and only if G has no cycle of size at least $n+2$.

Solution. It can be seen that $a, b \in \vee_{i=0}^{k} z_i$ holds if and only if there is a natural number l and a path $a = u_o, u_1, \ldots, u_l = b$ such that for every $0 \leq j < l$, $(u_i, u_{j+1}) \in z_i$ for some $0 \leq i \leq k$. Furthermore, $(a, b) \in z_1 \wedge z_2$ holds if and only if there is a natural number l and a path $a = u_o, u_1, \ldots, u_l = b$ such that $(u_i, u_{i+1}) \in z_1$ and $(u_i, u_{i+1}) \in z_2$ for every $0 \leq i < l$. The inequality $\geq$ always holds in equality (1), so what we have to prove is that the inequality $\leq$ holds if and only if G has no cycle of at least $n+2$ vertices. Suppose that $v_0, \ldots, v_k$ is a cycle with $k \geq n+1$. If x contracts (v_0, v_k), y_i contracts (v_i, v_{i+1}) $(0 \leq i < n)$, and y_n contracts the vertices $v_n, v_{n+1}, \ldots, v_k$, then the left-hand side of (1) is x and the right-hand side of it is not since (v_0, v_k) is not contracted in the right-hand side.

Conversely, suppose that G has no cycle of at least $n+2$ vertices, and let $(a, b) \in x \wedge \vee_{i=0}^{n} y_i$. We show that (a, b) is an element of the right-hand side of (1) as well. The assumption $(a, b) \in x \wedge \vee_{i=0}^{n} y_i$ implies that there is a path $a = v_0, v_1, \ldots, v_k = b$ in G such that $(v_t, v_{t+1}) \in x$ and $(v_t, v_{t+1}) \in \wedge_{i=0}^{n} y_i$ for $0 \leq t < k$. Let us fix t. There is a path $v_t = u_0, u_1, \ldots, u_l = v_{t+1}$ such that for every $0 \leq s < l$, we have $(u_s, u_{s+1}) \in y_m$ for some $0 \leq\leq n$. Since the vertices $u_0, u_1, \ldots, u_l$ constitute a cycle, $l+1 < n+2$, that is, $l \leq n$ and some of the y's do not appear among these y_m. Thus,

$$(v_t, v_{t+1}) \in \bigvee_{i=0,\ldots,j-1,j+1,\ldots n} y_i$$

for some j, and hence (a, b) is contained in the right-hand side of (1). □

Problem C.17. *Let $G(V, E)$ be a connected graph, and let $d_G(x, y)$ denote the length of the shortest path joining x and y in G. Let $r_G(x) = \max\{d_G(x, y) : y \in V\}$ for $x \in V$, and let $r(G) = \min\{r_G(x) : x \in V\}$. Show that if $r(G) \geq 2$, then G contains a path of length $2r(G) - 2$ as an induced subgraph.*

Solution 1. Let $n = r(G)$, and let x be a vertex of G such that $r_G(x) = n$ and the cardinality of the set $\{y : d(x, y) = n\}$ is minimum. Let x_n be a vertex such that $d(x, x_n) = n$, and let $x = x_0, x_1, y = x_2, \ldots, x_n$ be a path joining x and x_n. We show that if G does not contain a path of length $2n-2$ as an induced subgraph then $r_G(y) = n$ and that $d(x,z) < n (z \in V)$ implies that $d(y,z) < n$, which contradicts the choice of x, and $d(y, x_n) = n-2$.

$r_G(y) = n$ follows from the other statement, so it is sufficient to prove that one.

Since $d(y,z) \leq d(x,z) + 2$, it is sufficient to consider the cases $d(x,z) = n-2$ and $d(x,z) = n-1$. Let $x = z_0, z_1, \ldots, z_k = z$ $(k = n-2$ or $n-1)$ be a path of length $d(x,z)$ between x and z. By the minimality of the paths, if $x_i = y_j$ $\quad (i > 0, j > 0)$ then $i = j$, but then the length of the path $y = x_2, \ldots, x_i = z_i, \ldots, z_k$ is at most k and we are done. Similarly, if $x_i z_j$ is an edge of the graph, then $i - 1 \leq j$. If this is an edge of G and $i \geq 2$, then the length of the path $y = x_2, \ldots, x_i, z_j, \ldots, z_k$ is at most k $(< n)$ and if $i = 1$ then the length of the path $y, x_1, z_j, \ldots, z_k$ is $k - j + 2$. We are done if either $j \geq 2$ or $k = n-2$, so we may assume that $j = 1$ and $k = n-1$ and then $x_n, \ldots, x_1, z_1, \ldots, z_k$ is an induced path of length $2n-2$. If G does not contain any edge $x_i z_j$, then $x_n, \ldots, x_1, x_0, z_1, \ldots, z_k$ is an induced path of length $2n-2$ or $2n-1$, which we wanted to prove. □

Solution 2. (sketch) Every graph G with $r(G) = n$ contains a connected, induced subgraph G' such that $r(G') = n$ and $r(G'') < n$ for every connected, induced, proper subgraph G'' of G', and it is sufficient to take such a graph G'.

1. If G' is a path, then its length is $n-1$.
2. If $G' - \{x\}$ is not connected, then it has two components and one of them is a path.
3. Let C be the set of vertices x such that there is no vertex y such that $G - \{y\}$ is not connected and the component of $G - \{y\}$ containing x is a path. Then C induces a 2-connected subgraph.
4. For every vertex $x \in C$, we get a path of length k if we delete the vertices of the component of $G - \{x\}$ containing $C - \{x\}$.
5. For every vertex $x \in C$, there is exactly one vertex $y \in C$ such that $d(x,y) = n-k$.
6. For such a pair of vertices $\{x,y\}$, the graph induced by $C - \{x,y\}$ is not connected, that is, joining x and y in two components, we get a cycle of length at least $2(n-k)$.
7. For some $0 \leq k \leq n-2$, the graph G' is a cycle of length $2(n-k)$ with a pending path of length k at each vertex of it. □

Remarks.

1. The second solution shows that the statement is sharp; it is false for path of length $2n-1$.

2. There is an infinite graph G such that $r(G) = 3$, and every induced path in G is of length at most 3.

Problem C.18. *Given n points in a line so that any distance occurs at most twice, show that the number of distances occurring exactly once is at least $[n/2]$.*

Solution. Suppose that the points are on the real line at the numbers $p_1 < p_2 < \cdots < p_n$. For any index $1 \le i \le n$, let us take the set A_i of lengths of segments from p_i to the right, that is, $A_i = \{p_j - p_i : i < j\}$. Obviously, $|A_i| = n - i$.

We prove (by contradiction) that $|A_i \cap A_j| \le 1$ for $i < j$. Suppose that $u, v \in A_i \cap A_j$, $u \ne v$, and, say, $u = p_{k_1} - p_i = p_{k_2} - p_j$ and $v = p_{m_1} - p_i = p_{m_2} - p_j$. But then, the distance $p_j - p_i = p_{k_2} - p_{k_1} = p_{m_2} - p_{m_1}$ occurs three times, a contradiction.

Now we estimate the number of distances occurring from below. As we have seen,

$$\begin{aligned}|A_i - (A_1 \cup \cdots \cup A_{i-1})| &= |A_i - ((A_i \cap A_1) \cup \cdots \cup (A_i \cap A_{i-1}))| \\ &= |A_i| - |(A_i \cap A_1) \cup \cdots \cup (A_i \cap A_{i-1})| \\ &\ge n - i - (i-1) = n - 2i + 1.\end{aligned}$$

Hence,

$$\begin{aligned}&|A_1 \cup \cdots \cup A_n| \\ &= |A_1 \cup (A_2 - A_1) \cup (A_3 - (A_1 \cup A_2)) \cup \cdots \cup (A_n - (A_1 \cup \cdots \cup A_{n-1}))| \\ &\ge n - 1 + n - 3 + n - 5 + \cdots + n - 2[n/2] + 1,\end{aligned}$$

that is,

$$|A_1 \cup \cdots \cup A_n| \ge \begin{cases} n^2/4 & \text{if } n \text{ is even,} \\ (n^2-1)/4 & \text{if } n \text{ is odd.} \end{cases}$$

Now let d_1 and d_2 denote the number of distances occurring once and twice, respectively. Obviously,

$$d_1 + 2d_2 = \binom{n}{2}$$

and

$$d_1 + d_2 \ge \begin{cases} n^2/4 & \text{if } n \text{ is even,} \\ (n^2-1)/4 & \text{if } n \text{ is odd.} \end{cases}$$

Subtracting the equality from the double of the inequality, we obtain the desired inequality, $d_1 \ge [n/2]$. □

Problem C.19. *Let κ be an arbitrary cardinality. Show that there exists a tournament $T_\kappa = (V_\kappa, E_\kappa)$ such that for any coloring $f : E_\kappa \to \kappa$ of the edge set E_κ, there are three different vertices $x_0, x_1, x_2 \in V_\kappa$ such that*

$$x_0x_1, x_1x_2, x_2x_0 \in E_\kappa$$

and

$$|\{f(x_0x_1), f(x_1x_2), f(x_2x_0)\}| \le 2.$$

(A tournament is a directed graph such that for any vertices $x, y \in V_\kappa$, $x \neq y$ exactly one of the relations $xy \in E_\kappa$, $yx \in E_\kappa$ holds.)

Solution. By a famous theorem of Erdős and Rado, there exists a triangle-free graph $G = (V, F)$ with chromatic number 2^κ. Order the vertex set V of this graph G in an arbitrary way, and let $<$ denote this ordering. Now we define a tournament on V as follows. For $x, y \in V$, $x < y$, let $xy \in E$ if $xy \in F$ and let $yx \in E$ if $xy \notin F$.

We show that the resulting tournament (V, E) meets the requirements of the problem. Let $f : E \to \kappa$ be an arbitrary coloring of E. For every $x \in V$, consider the set $A_x = \{f(yx) : y < x, yx \in E\} \subset \kappa$. Let us fix a vertex $x \in V$. The cardinality of the set $\{y \in V : y < x, yx \in F \text{ or } x < y, xy \in F\}$ is greater than 2^κ, so there is a vertex x' such that $A_x = A_{x'}$ and $x'x \in E$ if $x' < x$, $xx' \in E$ if $x < x'$. Let x_1 and x_2 denote the smaller and the bigger element from among x and x', respectively. Then $f(x_1x_2) \in A_{x_1} = A_{x_2}$, that is, there exists a vertex $x_0 \in V$ such that

$$x_0 < x_1, \quad x_0x_1 \in E, \quad \text{and} \quad f(x_0x_1) = f(x_1x_2).$$

However, by the definition of E,

$$x_0x_1, x_1x_2 \in F,$$

and so

$$x_2x_0 \notin F$$

and

$$x_2x_0 \in E,$$

that is, the vertices x_0, x_1, x_2 have the desired properties. □

Remark. It is easy to see that we cannot replace 2 by 1 in the statement. For any tournament $T = (V, E)$ and for any ordering $<$ of V, define a coloring f of E as follows: if $x < y$, then let $f(xy) = 0$ if $xy \in E$ and let $f(yx) = 1$ if $yx \in E$. Obviously, there are no three vertices $x, y, z \in V$ such that $xy, yz, zx \in E$ and f is constant on these three edges.

Problem C.20. *Some proper partitions $P_1, \dots, P_n$ of a finite set S (that is, partitions containing at least two parts) are called independent if no matter how we choose one class from each partition, the intersection of the chosen classes is nonempty. Show that if the inequality*

$$\frac{|S|}{2} < |P_1| \cdots |P_n| \qquad (*)$$

holds for some independent partitions, then $P_1, \dots, P_n$ is maximal in the sense that there is no partition P such that $P, P_1, \dots, P_n$ are independent.

On the other hand, show that inequality $(*)$ is not necessary for this maximality.

Solution. We prove the statement by contradiction. Suppose that (1) holds and that there is a proper partition P_0 such that $P_0, P_1, \ldots, P_n$ are independent. Let us choose one of the classes of each partition and take the intersection of them. For the sake of brevity, let us call these intersections class intersections. Obviously, the total number of class intersections is $|P_0||P_1|\ldots|P_n|$ and any two class intersections are disjoint since among the classes constituting any two class intersections there are two classes belonging to the same partition, and so they are disjoint. On the other hand, all the class intersections are nonempty since $P_0, P_1, \ldots, P_n$ are independent. Thus, we have

$$|P_0||P_1|\ldots|P_n| \leq |S|. \tag{2}$$

Combining (1) and (2) and using the fact that $|S| \neq 0$, we obtain that $|P_0| < 2$, which is a contradiction to the assumption that P_0 is a proper partition. So, we proved that (1) implies the maximality.

On the other hand, the next example shows that (1) is not necessary for the maximality. Let $S = \{a_1, \ldots, a_8\}$. Let

$$P_1 = \{S_{11}, S_{12}\}, \quad P_2 = \{S_{21}, S_{22}\}$$

be the partitions where

$$\begin{aligned} S_{11} &= \{a_1, a_2, a_3, a_4\}, \\ S_{12} &= \{a_5, a_6, a_7, a_8\}, \\ S_{21} &= \{a_1, a_5\}, \\ S_{22} &= \{a_2, a_3, a_4, a_6, a_7, a_8\}. \end{aligned}$$

Then

$$S_{11} \cap S_{21} = \{a_1\}, \tag{3}$$

and the other class intersections are not empty either, that is, P_1, P_2 are independent. Equation (1) does not hold since $|P_1||P_2| = 4 = |S|/2$, and we show that $\{P_1, P_2\}$ is maximal. Suppose that P_0, P_1, P_2 are independent for some proper partition $P_0 = \{S_{01}, S_{02}, \ldots\}$. Then the intersection of S_{01} with the set (3) is nonempty and so $a_1 \in S_{01}$. We similarly obtain $a_1 \in S_{02}$. But then $S_{01} \cap S_{02} \neq \emptyset$, a contradiction to the assumption that P_0 is a partition. ☐

Problem C.21. *Show that if $k \leq n/2$ and $\mathcal{F}$ is a family of $k \times k$ submatrices of an $n \times n$ matrix such that any two intersect then*

$$|\mathcal{F}| \leq \binom{n-1}{k-1}^2.$$

Solution. For any $k \times k$ submatrix $M \in \mathcal{F}$, let R_M and C_M denote the k-tuple of its rows and columns, respectively. Obviously, R_M and C_M determine M in a unique way. The condition of the problem says that

$$R_{M_1} \cap R_{M_2} \neq \emptyset \qquad \text{and} \qquad C_{M_1} \cap C_{M_2} \neq \emptyset$$

for any two matrices $M_1, M_2 \in \mathcal{F}$. Let us take the families

$$\mathcal{R} = \{R_M : M \in \mathcal{F}\} \qquad \text{and} \qquad \mathcal{C} = \{C_M : M \in \mathcal{F}\}.$$

Then $\mathcal{R}$ and $\mathcal{C}$ are families of subsets of k elements of a set of n elements such that any two members of $\mathcal{R}$ and $\mathcal{C}$ have nonempty intersection, respectively. Thus,

$$|\mathcal{R}|, |\mathcal{C}| \leq \binom{n-1}{k-1}$$

by the famous Erdős–Ko–Rado theorem, and so

$$|\mathcal{F}| \leq \binom{n-1}{k-1}^2.$$

Obviously, the bound $\binom{n-1}{k-1}^2$ is sharp since this is exactly the number of $k \times k$ submatrices containing a given element. □

Problem C.22. *Let us color the integers $1, 2, \ldots, N$ with three colors so that each color is given to more than $N/4$ integers. Show that the equation $x = y + z$ has a solution in which x, y, z are of distinct colors.*

Solution. We prove the statement by contradiction. Suppose that we can color the integers $1, 2, \ldots, N$ with red, green, and blue so that each color is given to more than $N/4$ integers and there are no x, y, z of distinct colors, where $x = y + z$. We may assume that 1 is red. Then by our hypothesis, there are no green and blue integers such that their difference is 1. We will call a nonempty set $S \subseteq \{1, \ldots, N\}$ an *interval* if it consists of consecutive integers, that is, if there are some integers $1 \leq a \leq b \leq N$ such that $S = \{s : a \leq s \leq b\}$. If we delete the red integers, then the remaining integers can be partitioned into intervals. Furthermore, each interval is monochromatic by the observation above, since if an interval contains some integers $x < y$ of distinct colors, then there are two consecutive integers of distinct colors among $x, x + 1, \ldots, y$.

Suppose that there exist intervals of at least two integers in both colors green and blue. Let A and B denote the longest green and blue interval, respectively. If $a \in A$ and $b \in B$, then $|a - b| (= a - b$ or $b - a)$ is not red. Hence, the set $C = \{|a - b| : a \in A, b \in B\}$ does not contain any red integer. If $A = [a_1, a_2], B = [b_1, b_2]$, then

$$C = \begin{cases} [b_1 - a_2, b_2 - a_1] & \text{if} \quad a_2 < b_1, \\ [a_1 - b_2, a_2 - b_1] & \text{if} \quad b_2 < a_1, \end{cases}$$

so it is an interval of $|A| + |B| - 1 > |A|, |B|$ integers. The set C does not contain any red integers, so it is monochromatic, as we have seen above. But it is a contradiction to the maximal choice of the intervals A and B.

On the other hand, at least one of the colors green and blue contains two consecutive integers, since otherwise the number of red integers would be greater than or equal to the number of green or blue integers. In that case, the number of blue or green integers would be at most $N/2$, a contradiction to the assumption that each color is given to more than $N/4$ integers.

Suppose now that there are no two consecutive green integers but there are two consecutive blue integers, that is, $|B| \geq 2$ for the longest blue interval B. If $1 \leq s < N$ is green, then $s \geq 2$ and $s-1$ and $s+1$ are red. Suppose that the distance between any two green integers is at least 3. Then the intervals $[s-1, s+1]$ are pairwise disjoint for the green integers s and each of them contains two red integers. If N is not green, then it implies that the number of red integers is at least double the number of the green integers, which is impossible if each color is given to more than $N/4$ integers. If N is green, then it implies that $n_r \geq 2n_g - 1$, where n_r and n_g are the number of red and green integers, respectively. If 2 is not green, then 1 is not counted in the intervals above, so $n_r \geq 2n_g$, a contradiction again. If 2 is green, then take a blue interval $B = [b_1, b_2]$ such that $|B| \geq 2, b_2 < N$. Now, $b_2 - 1$ is blue, $b_2 + 1$ is red, 2 is green, and they constitute a solution to the equation $x = y + z$ such that x, y, z are of distinct colors, a contradiction.

Thus, we may assume that there is a green integer s such that $s + 2$ is green as well. Either $b_1 > s + 2$ or $b_2 < s$, since $|B| \geq 2$ and $B \cap \{s, s+2\} = \emptyset$. We may assume that $b_1 > s + 2$. Consider the set

$$C = \{b - s : b \in B\} \cup \{b - s - 2 : b \in B\}.$$

Since $b_1 < b_2$ so C is an interval such that $|C| = |B| + 2$. The interval C does not contain any red integers, so it is monochromatic. But C is not green because $|C| > 2$ and not blue because of the maximal choice of B, a contradiction. □

Problem C.23. *Suppose that a graph G is the union of three trees. Is it true that G can be covered by two planar graphs?*

Solution. The answer is no. We construct a graph that is the union of three acyclic graphs but that cannot be covered by two planar graphs. Any acyclic graph can be extended to a tree, so it implies the statement.

Let A and B be a set of n and $\binom{n}{3}$ elements, respectively, where the value of n will be determined later. Let $A \cup B$ denote the vertex set of the graph. Let us choose three elements of A in all possible ways and join each of these triples to an element of B so that distinct triples are joined to distinct vertices in B. Let G denote the resulting bipartite graph. It can be obtained as the union of three acyclic graphs in the following way: for any vertex $b \in B$, the three edges incident to b are put into three distinct subgraphs. Any vertex $b \in B$ is of degree one in each of the three resulting subgraphs, so these subgraphs do not contain any cycle.

Suppose that G is the union of two planar graphs G_1 and G_2. We may assume that G_1 and G_2 have no common edge.

For every vertex $b \in B$, let us proceed as follows. Delete one of the edges of G incident to b so that the remaining two edges belong to the same graph G_i. (It can be done since two of the edges incident to B always belong to the very same subgraph.) Now, replace the vertex b and the remaining two edges incident to b by one edge joining the end vertices of these two edges in A. If G_1 and G_2 are planar, then the resulting graphs are as well. The resulting graphs are defined on A. The number of replacements is $\binom{n}{3}$, and an edge is obtained in at most $n-2$ different ways since a couple can be extended to a triple in at most this many ways. So, the resulting graph H has at least

$$\frac{\binom{n}{3}}{n-2} = \frac{n(n-1)}{6}$$

edges. It is known that a planar graph of n vertices has at most $3n-6$ edges (if $n \geq 3$). Thus, H has at most $6n-12$ vertices. From these estimates, we obtain

$$\frac{n(n-1)}{6} \leq 6n - 12,$$

that is,

$$n^2 - 37n + 72 \leq 0.$$

But this inequality does not hold if $n \geq 35$. Thus, G cannot be covered by two planar graphs if $n \geq 35$. □

3.3 THEORY OF FUNCTIONS

Problem F.1. *Prove that the function*

$$f(\vartheta) = \int_1^{\frac{1}{\vartheta}} \frac{dx}{\sqrt{(x^2-1)(1-\vartheta^2 x^2)}}$$

(where the positive value of the square root is taken) is monotonically decreasing in the interval $0 < \vartheta < 1$.

Solution 1. Substitute

$$t = \sqrt{\frac{x^2-1}{1-\vartheta^2 x^2}}\,.$$

While x increases in the interval $\left(1, \dfrac{1}{\vartheta}\right)$, the value t increases in $(0, +\infty)$. Differentiating the relation

$$\vartheta^2 t^2 x^2 - t^2 + x^2 - 1 = 0,$$

we obtain

$$\frac{dx}{dt} = \frac{t(1-\vartheta^2 x^2)}{x(1+\vartheta^2 t^2)}$$

and, consequently,

$$\begin{aligned} f(\vartheta) &= \int_1^{\frac{1}{\vartheta}} \frac{dx}{\sqrt{(x^2-1)(1-\vartheta^2 x^2)}} \\ &= \int_1^{\frac{1}{\vartheta}} \frac{dx}{t(1-\vartheta^2 x^2)} = \int_0^{+\infty} \frac{dx}{dt} \cdot \frac{dt}{t(1-\vartheta^2 x^2)} \\ &= \int_0^{+\infty} \frac{dt}{x(1+\vartheta^2 t^2)} = \int_0^{+\infty} \frac{dt}{\sqrt{(1+t^2)(1+\vartheta^2 t^2)}}\,. \end{aligned}$$

Now, for increasing ϑ the integrand decreases. Since the limits of integration are independent of ϑ, the integral is also monotonically decreasing. □

Solution 2. Map the first quadrant of the z-plane (excluding the points $z = 1$ and $z = 1/\vartheta$ by semicircles open from below) to the w-plane with the help of the function

$$w = \int_0^z \frac{d\zeta}{\sqrt{(1-\zeta^2)(1-\vartheta^2\zeta^2)}}$$

(taking the value of the square root that is positive on the positive half-axis). If, starting from 0, $z = x + iy$ runs through the segment $0 \le z \le 1$, then starting from 0, $w = u + iv$ obviously runs through the segment

$$0 \le w \le \int_0^1 \frac{dx}{\sqrt{(1-x^2)(1-\vartheta^2 x^2)}} = A.$$

If $1 \le z \le \dfrac{1}{\vartheta}$, then it is clear that w runs through the segment

$$u = A, \quad 0 \le v \le \int_0^{\frac{1}{\vartheta}} \frac{dx}{\sqrt{(x^2-1)(1-\vartheta^2 x^2)}} = B.$$

If z runs through the segment $1/\vartheta \le z < \infty$, then w runs from the point $A + Bi$ along the horizontal line $v = B$ to the point $(A - C) + Bi$, where

$$C = \int_{\frac{1}{\vartheta}}^{+\infty} \frac{dx}{\sqrt{(x^2-1)(\vartheta^2 x^2 - 1)}}.$$

On the other hand, if, starting from 0, z runs through the positive part of the imaginary axis, then, starting from 0, w runs through the segment

$$u = 0, \qquad 0 \le v \le D,$$

where

$$D = \int_0^{+\infty} \frac{dy}{\sqrt{(1+y^2)(1+\vartheta^2 y^2)}}.$$

Since the mapping is domain preserving, we have the relations

$$A = C, \qquad B = D,$$

the latter of which implies the statement. □

Problem F.2. *Denote by $M(r, f)$ the maximum modulus on the circle $|z| = r$ of the transcendent entire function $f(z)$, and by $M_n(r, f)$ that of the nth partial sum of the power series of $f(z)$. Prove the existence of an entire function $f_0(z)$ and a corresponding sequence of positive numbers $r_1 < r_2 < \cdots \to +\infty$ such that*

$$\limsup_{n\to\infty} \frac{M_n(r_n, f_0)}{M(r_n, f_0)} = +\infty.$$

Solution. By a theorem of Fejér, there exists a power series

$$f(z) = \sum_{n=0}^{\infty} a_n z^n,$$

which defines a regular function in the disc $|z| < 1$ such that $|f(z)| \le 1$ and such that the sequence of the partial sums

$$s_r(z) = \sum_{n=0}^{r} a_n z^n$$

is unbounded at the point $z = 1$.

We shall use this function $f(z)$ for constructing the function $f_0(z)$ that meets the requirements of the problem.

We define sequences of numbers n_k, m_k, and c_k as follows. Let $n_0 = m_0 = c_0 = 0$. Then suppose that n_{k-1}, m_{k-1}, and c_{k-1} have already been defined.

We begin by defining m_k. Since $\limsup_{n\to+\infty} s_n(1) = \infty$, there is an m_k with

$$|s_{m_k}(1)| > k \sum_{l=0}^{k-1} \max_{|z|=k} |s_{n_l}(z)|;$$

in addition, we may assume that $m_k > n_{k-1}$.

We now define c_k. Since $s_{m_k}(z)$ is continuous, for c_k sufficiently close to 1 we also have

$$|s_{m_k}(c_k)| > k \sum_{l=0}^{k-1} \max_{|z|=k} |s_{n_l}(z)|.$$

We additionally assume that c_k is real, further

$$c_k > c_{k-1} \quad \text{and} \quad c_k \geq 1 - \frac{1}{2^k}.$$

Finally, we define n_k in the following way. Since $s_n(z)$ tends to $f(z)$ in the disc $|z| < 1$, for sufficiently large n_k we have

$$\max_{|z|=c_k} |s_{n_k}(z)| < 1.$$

On the other hand, the absolute convergence of the power series of $f(z)$ implies the convergence of $\sum_{n=0}^{\infty} |a_n| c_k^n$; thus, for sufficiently large n_k,

$$\sum_{n>n_k} |a_n| c_k^n < 1.$$

In addition to these two inequalities, we require that n_k satisfy $n_k > m_k$.

Now consider the power series

$$\sum_{n=1}^{n_1} a_n z^n + \sum_{n=n_1+1}^{n_2} a_n \left(\frac{z}{2}\right)^n + \dots$$
$$= \sum_{k=1}^{\infty} \sum_{n=n_{k-1}+1}^{n_k} a_n \left(\frac{z}{k}\right)^n = \sum_{k=1}^{\infty} \left\{ s_{n_k}\left(\frac{z}{k}\right) - s_{n_{k-1}}\left(\frac{z}{k}\right) \right\}.$$

We show that the function $f_0(z)$ defined by this power series has the desired property.

Prescribing an arbitrarily large positive R, for $k > 2R$ and $|z| \leq R$ we have $|a_n (z/k)^n| \leq |a_n|(1/2^n)$. Since $\sum_{n=0}^{\infty} |a_n|(1/2^n)$ is convergent, the power series of $f_0(z)$ is absolutely convergent in the disc $|z| < R$. Thus $f_0(z)$ is a transcendental entire function.

We give an upper estimate for $f_0(z)$ on the circle $|z| = kc_k$. In view of our previous remarks,

$$
\begin{aligned}
|f_0(z)| \leq & \left| \sum_{l=1}^{k-1} \left\{ s_{n_l}\left(\frac{z}{k}\right) - s_{n_{l-1}}\left(\frac{z}{k}\right) \right\} - s_{n_{k-1}}\left(\frac{z}{k}\right) \right| \\
& + \left| s_{n_k}\left(\frac{z}{k}\right) \right| + \left| \sum_{l=k+1}^{\infty} \sum_{n=n_{l-1}+1}^{n_k} a_n \left(\frac{z}{k}\right)^n \right| \\
\leq & \, 2\sum_{l=0}^{k-1} \max_{|z|=kc_k} |s_{n_l}(z)| + \max_{|z|=c_k} |s_{n_k}(z)| + \sum_{l=k+1}^{\infty} \sum_{n=n_{l-1}+1}^{n_k} |a_n| \left(\frac{kc_k}{l}\right)^n \\
\leq & \, 2\sum_{l=0}^{k-1} \max_{|z|=k} |s_{n_l}(z)| + \max_{|z|=c_k} |s_{n_k}(z)| + \sum_{n>n_k} |a_n| c_k^n \\
\leq & \, 2\sum_{l=0}^{k-1} \max_{|z|=k} |s_{n_l}(z)| + 1 + 1.
\end{aligned}
$$

Setting $\sum_{l=0}^{k-1} \max_{|z|=k} |s_{n_l}(z)| = T_k$, we obtain

$$M(kc_k, f_0) \leq 2T_k + 2.$$

Next we give a lower estimate for the corresponding partial sum on the circle $|z| = kc_k$; since we are concerned with the maximum, it is sufficient to do this at the point $z = kc_k$.

The modulus of the m_kth partial sum =

$$
\begin{aligned}
&= \left| s_{m_k}\left(\frac{z}{k}\right) - s_{n_{k-1}}\left(\frac{z}{k}\right) + \sum_{l=1}^{k-1} \left\{ s_{n_l}\left(\frac{z}{k}\right) - s_{n_{l-1}}\left(\frac{z}{k}\right) \right\} \right| \\
&\geq |s_{m_k}(c_k)| - 2\sum_{l=0}^{k-1} \max_{|z|=k} |s_{n_l}(z)| \geq (k-2)T_k;
\end{aligned}
$$

that is,

$$M_{m_k}(kc_k, f_0) \geq (k-2)T_k.$$

Now, since $\limsup_{l\to+\infty} |s_{n_l}(1)| = +\infty$, the numbers

$$T_k = \sum_{l=0}^{k-1} \max_{|z|=k} |s_{n_l}(z)|$$

tend to $+\infty$. Thus, for sufficiently large k we have $T_k > 1$ and, consequently,

$$\frac{M_{m_k}(kc_k, f_0)}{M(kc_k, f_0)} \geq \frac{(k-2)T_k}{2(T_k+1)} > \frac{k-2}{4}.$$

Therefore, taking $r_{m_k} = kc_k$, and for n different from the m_k choosing r_n with the consideration of monotonicity but otherwise arbitrarily, we obtain

$$\limsup_{n\to+\infty} \frac{M_n(r_n, f_0)}{M(r_n, f_0)} \geq \limsup_{k\to+\infty} \frac{M_{m_k}(r_{m_k}, f_0)}{M(r_{m_k}, f_0)} \geq \limsup_{k\to+\infty} \frac{k-2}{4} = +\infty . \quad \square$$

Problem F.3. *Let H be a set of real numbers that does not consist of 0 alone and is closed under addition. Further, let $f(x)$ be a real-valued function defined on H and satisfying the following conditions:*

$$f(x) \leq f(y) \ \text{ if } x \leq y \quad \text{and } f(x+y) = f(x) + f(y) \quad (x, y \in H).$$

Prove that $f(x) = cx$ on H, where c is a nonnegative number.

Solution. Let x_0 be an element of H other than 0, and let $c = f(x_0)/x_0$. It can be seen by induction that $f(nx) = nf(x)$ $(n = 1, 2, \dots)$ for every x in H. Therefore

$$f(nx_0) = cnx_0. \tag{1}$$

Let y be an arbitrary element of H. Then there is a positive integer n_0 such that $(y+n_0x_0)x_0 > 0$. If n is a sufficiently large positive integer, then there exist two positive integers m_n and μ_n such that

$$m_n x_0 \leq n(y + n_0 x_0) \leq \mu_n x_0 \quad \text{and} \quad |m_n - \mu_n| = 1. \tag{2}$$

In view of the monotonicity of the function $f(x)$, we have

$$f(m_n x_0) \leq f\left(n(y + n_0 x_0)\right) \leq f(\mu_n x_0),$$

so by (1),

$$cm_n x_0 \leq nf(y + n_0 x_0) \leq c\mu_n x_0.$$

It follows that

$$c\frac{m_n}{n} x_0 \leq f(y + n_0 x_0) \leq c\frac{\mu_n}{n} x_0. \tag{3}$$

According to (2),

$$\frac{m_n}{n} x_0 \to y + n_0 x_0, \quad \frac{\mu_n}{n} x_0 \to y + n_0 x_0.$$

Thus, from (3) we obtain that $f(y + n_0 x_0) = c(y + n_0 x_0)$. Consequently, $f(y) = f(y+n_0x_0) - f(n_0x_0) = c(y+n_0x_0) - cn_0x_0 = cy$. The monotonicity of the function $f(x)$ yields $c \geq 0$. $\square$

Problem F.4. *Show that if $f(x)$ is a real-valued, continuous function on the half-line $0 \leq x < \infty$, and*

$$\int_0^\infty f^2(x)dx < \infty,$$

then the function

$$g(x) = f(x) - 2e^{-x} \int_0^x e^t f(t)dt$$

satisfies

$$\int_0^\infty g^2(x)dx = \int_0^\infty f^2(x)dx.$$

Solution. We assume that the function $f(x)$ is square integrable in the Lebesgue sense on the half-line $(0, \infty)$. The relation

$$f(x) - g(x) = 2e^{-x} \int_0^x e^t f(t)\, dt \tag{1}$$

implies that

$$(f(x) - g(x))' = f(x) + g(x) \quad \text{almost everywhere.} \tag{2}$$

Using the Schwarz inequality, we obtain

$$\begin{aligned}
e^{-\omega}\left|\int_0^\omega e^t f(t)\, dt\right| \leq{}& e^{-\omega}\left|\int_0^{\omega/2} e^t f(t)\, dt\right| + e^{-\omega}\left|\int_{\omega/2}^{\omega} e^t f(t)\, dt\right| \\
\leq{}& e^{-\omega}\left(\int_0^{\omega/2} e^{2t}\, dt\right)^{\frac{1}{2}}\left(\int_0^{\omega/2} f^2(t)\, dt\right)^{\frac{1}{2}} \\
&+ e^{-\omega}\left(\int_{\omega/2}^{\omega} e^{2t}\, dt\right)^{\frac{1}{2}}\left(\int_{\omega/2}^{\omega} f^2(t)\, dt\right)^{\frac{1}{2}} \\
\leq{}& e^{-\omega/2}\int_0^\infty f^2(x)\, dx + \int_{\omega/2}^\infty f^2(t)\, dt,
\end{aligned}$$

whence it follows that

$$\lim_{\omega\to\infty} e^{-\omega} \int_0^\omega e^t f(t)\, dt = 0. \tag{3}$$

Relation (1) assures that $f(x) - g(x)$, and therefore $(1/2)\,(f(x) - g(x))^2$, is absolutely continuous on each bounded subinterval of $(0, \infty)$. By (1) and (2),

$$\begin{aligned}
\int_0^\omega \left(f^2(x) - g^2(x)\right)\, dx &= \int_0^\omega \left(\frac{(f(x) - g(x))^2}{2}\right)' dx \\
&= \left[\frac{(f(x) - g(x))^2}{2}\right]_0^\omega = 2e^{-2\omega}\left(\int_0^\omega e^t f(t)\, dt\right)^2.
\end{aligned}$$

Making use of (3), the statement follows. □

Problem F.5. *Prove that for every convex function $f(x)$ defined on the interval $-1 \le x \le 1$ and having absolute value at most 1, there is a linear function $h(x)$ such that*

$$\int_{-1}^{1} |f(x) - h(x)| dx \le 4 - \sqrt{8}.$$

Solution. We prove more than stated. We establish the existence of a constant k such that

$$\int_{-1}^{1} |f(x) - k| \, dx \le 4 - \sqrt{8}. \tag{1}$$

Without loss of generality, we may assume that $f(x)$ is continuous even at the endpoints of the interval $[-1, 1]$ and that $f(-1) \ge f(1)$. Since $f(x)$ is continuous and convex on $[-1, 1]$, there is a largest interval $[c_1, c_2]$ $(-1 \le c_1 \le c_2 \le 1)$ on which $f(x)$ is minimal. Introduce the notation

$$\min_{x \in [-1,1]} f(x) = p, \qquad \max_{x \in [-1,1]} f(x) = q.$$

Let $\phi_1(y)$ be the inverse of the restriction to $[-1, c_1]$ of the function $f(x)$, and let

$$\phi_2(y) = \begin{cases} f^{-1}(y) & \text{if } \ p \le y \le f(1), \\ 1 & \text{if } \ f(1) \le y \le q, \end{cases}$$

where $f^{-1}(y)$ denotes the inverse of the restriction to $[c_2, 1]$ of the function $f(x)$. Obviously, the function $\phi(y) = \phi_2(y) - \phi_1(y)$ is continuous and strictly increasing on the interval $[p, q]$; further $\phi(p) = \phi_2(p) - \phi_1(p) = c_2 - c_1$ and $\phi(q) = \phi_2(q) - \phi_1(q) = 2$. We distinguish between two cases.

a. If $c_2 - c_1 \le 1$, then by the above properties of the function $\phi(y)$ there is one and only one number k $(\in [p, q])$ with $\phi(k) = 1$. It can be shown that k satisfies (1). Put $\phi_1(k) = d$, $\phi_2(k) = e$, $D = (d, k)$, $E = (e, k)$, $G = (-1, k)$, $H = (1, k)$, $A = (-1, 1)$, and $B = (1, 1)$. The lines AD and BE intersect at a point F. By the convexity of $f(x)$ and the relation $|f(x)| \le 1$, it is obvious that the graphs of the restrictions of $f(x)$ to the segments $[-1, d]$, $[d, e]$, and $[e, 1]$ lie in the triangles AGD, DFE, and EHB, respectively. Therefore,

$$\begin{aligned}\int_{-1}^{1} |f(x) - k| \, dx &= \int_{-1}^{d} |f(x) - k| \, dx + \int_{d}^{e} |f(x) - k| \, dx + \int_{e}^{1} |f(x) - k| \, dx \\ &\le t(AGD_\triangle) + t(DFE_\triangle) + t(EHB_\triangle),\end{aligned}$$

where t stands for area. If F lies above the line $y = -1$, then — DE being the mid-parallel of the triangle ABF — the altitudes perpendicular to GH of the triangles in question are equal, so the sum of the areas of the triangles is

$$\frac{1}{2}m \cdot GD + \frac{1}{2}m \cdot DE + \frac{1}{2}m \cdot EH = \frac{1}{2}m(GD + DE + EH) = m \leq 1, \quad (2)$$

where $m = BH$. If F lies below the line $y = -1$, then (denoting by J the point of intersection of the line $y = -1$ and the segment AF, and further denoting by K the point of intersection of the line $y = -1$ and the segment BF) it is obvious that the graph of the restriction to $[d, e]$ of $f(x)$ remains in the trapezoid T determined by the vertices D, J, K, E and, consequently,

$$\int_{-1}^{1} |f(x) - k|\, dx \leq t(AGD_\triangle) + t(T) + t(EHB_\triangle),$$

while $m = BH \geq 1$. Then

$$t(AGD_\triangle) + t(EHB_\triangle) = \frac{1}{2}m \cdot GD + \frac{1}{2}m \cdot EH = \frac{1}{2}m.$$

Since $JK = (2m - 2)m^{-1}$, we have

$$t(T) = \frac{1}{2}\left(\frac{2m-2}{m} + 1\right)(2 - m).$$

It follows that

$$\begin{aligned}\int_{-1}^{1} |f(x) - k|\, dx &\leq \frac{4m - 2 - m^2}{m} = 4 - \left(\frac{2}{m} + m\right) \\ &\leq 4 - 2\left(\frac{2}{m}m\right)^{1/2} = 4 - \sqrt{8}. \qquad (3)\end{aligned}$$

Relations (2) and (3) imply (1).

b. If $c_2 - c_1 > 1$, then let $k = p$. Setting $A = (-1, 1)$, $B = (1, 1)$, $D = (c_1, p)$, $E = (c_2, p)$, $G = (-1, p)$, and $H = (1, p)$, it follows as before that

$$\begin{aligned}\int_{-1}^{1} |f(x) - p|\, dx &\leq t(AGD_\triangle) + t(EHB_\triangle) \\ &\leq \frac{1}{2}m \cdot GD + \frac{1}{2}m \cdot EH = \frac{1}{2}m\,(1 - (c_2 - c_1)) < 1,\end{aligned}$$

which completes the proof. □

Remarks.

1. Several participants have noted that the estimate cannot be improved in general. For instance, in the case of the function

$$f(x) = \begin{cases} -1 & \text{if } 0 \leq x \leq 1 - \sqrt{2}/2, \\ 1 + \sqrt{8}(x - 1) & \text{if } 1 - \sqrt{2}/2 < x \leq 1, \end{cases}$$

the estimate is the best possible.

2. Paul Turán called the attention of the organizing committee to the fact that S. Bernstein *(Doklady Akad. Nauk SSSR, 1927, 405–407)* had proved the following theorem:

Theorem. Let $f(x)$ be an $n+1$ times differentiable function on the interval $[-1,1]$ satisfying the condition $f^{(n+1)}(x)>0$ on $[-1,1]$. The expression

$$\int_{-1}^{1} |f(x)-R_n(x)|\,dx,$$

where $R_n(x)$ denotes a polynomial of degree n, is minimal when $R_n(x)$ is the Lagrange interpolation polynomial of degree n that coincides with $f(x)$ at the points $\cos(h\pi/(n+2))$ $(h=1,2,\dots n+1)$.

Problem F.6. *Find all linear homogeneous differential equations with continuous coefficients (on the whole real line) such that for any solution $f(t)$ and any real number c, $f(t+c)$ is also a solution.*

Solution. As usual in the literature, we restrict attention to differential equations with the coefficient of the term of highest order identical to 1.

Let the differential equation

$$y^{(n)}(x)+f_1(x)y^{(n-1)}(x)+\dots+f_n(x)y(x)=0 \tag{1}$$

have the desired properties, and let $\phi(x)$ be a solution. Let c be an arbitrary real number. Obviously,

$$\frac{d^i\phi(x+c)}{dx^i}=\left(\frac{d^i\phi(t)}{dt^i}\right)_{t=x+c} \qquad (i=1,2,\dots,n); \tag{2}$$

since together with $\phi(x)$ $\phi(x+c)$ also satisfies (1), it follows that

$$\left(\frac{d^n\phi(t)}{dt^n}\right)_{t=x+c}+f_1(x)\left(\frac{d^{n-1}\phi(t)}{dt^{n-1}}\right)_{t=x+c}+\dots+f_n(x)\left(\phi(t)\right)_{t=x+c}=0,$$

that is,

$$\phi^{(n)}(t)+f_1(t-c)\phi^{(n-1)}(t)+\dots+f_n(t-c)\phi(t)=0,$$

and this is true for any real constant c and all real values of t. This means that all solutions of (1), so for example, n linearly independent solutions of (1), satisfy the differential equation

$$y^{(n)}(x)+f_1(x-c)y^{(n-1)}(x)+\dots+f_n(x-c)y(x)=0; \tag{3}$$

hence — as is well known — it follows that the coefficient functions of the differential equations (1) and (3) coincide:

$$f_i(x)=f_i(x-c) \qquad (i=1,2,\dots,n). \tag{4}$$

Choosing $x = 0$, we obtain $f_i(0) = f_i(-c)$ for every c $(i = 1, 2, \ldots, n)$. Thus $f_i(x) \equiv f_i(0)$, and so the differential equation (1) has constant coefficients.

Conversely, if (1) has constant coefficients, then from (2) it follows easily that together with any solution $\phi(x)$ $\phi(x + c)$ is also a solution. □

Remarks.

1. Assuming the conditions for a single c only, from (4) we obtain that the coefficient functions are periodic with period c.
2. If, for any solution $\phi(x)$, $\phi(x + c_1)$ and $\phi(x + c_2)$ are also solutions, then together with $\phi(x)$ obviously $\phi((x + c_1) + c_2) = \phi(x + c_1 + c_2)$ is also a solution. Consequently, assuming the conditions for a (finite or infinite) set $\{c_\alpha\}$, the condition will also be fulfilled by the elements of the smallest additive semigroup that contains the numbers c_α. In view of the previous remark, this implies that it is sufficient to assume the conditions only for values of c that generate an additive semigroup containing a sequence that tends to zero.

Problem F.7. *Let F be a closed set in the n-dimensional Euclidean space. Construct a function that is 0 on F, positive outside F, and whose partial derivatives all exist.*

Solution. Define a function $\phi_r(y)$ in the following way:

$$\phi_r(y) = \begin{cases} e^{\frac{1}{y-r^2}} & \text{if} \quad |y| < r^2, \\ 0 & \text{if} \quad |y| \geq r^2. \end{cases}$$

Obviously, $\phi_r(y)$ is positive if $|y| < r^2$, and infinitely differentiable in the intervals $(-\infty, r^2)$ and $(r^2, +\infty)$. Further, it is clear that $(d^k\phi_r(y)/dy^k)$ has the form $R_k(y)\phi_r(y)$, where $R_k(y)$ is a rational fractional function; so

$$\lim_{x \to r^2+0} \frac{d^k\phi_r(y)}{dy^k} = \lim_{x \to r^2-0} \frac{d^k\phi_r(y)}{dy^k} = 0.$$

Therefore $(d^k\phi_r(y)/dy^k)$ is continuous in $(-\infty, r^2)$ and $(r^2, +\infty)$ and has right-hand and left-hand limits at the point $y = r^2$, whence it follows by induction that $\phi_r(y)$ has continuous derivatives of any order at $y = r^2$ and, consequently, at all points y.

Denote by $\mathbb{E}_n$ the n-dimensional Euclidean space. For $a \in \mathbb{E}_n$, $x \in \mathbb{E}_n$, $a = (a_1, a_2, \ldots, a_n)$, $x = (x_1, x_2, \ldots, x_n)$ put

$$y_a(x) = (x_1 - a_1)^2 + (x_2 - a_2)^2 + \cdots + (x_n - a_n)^2.$$

Obviously, for every x, all partial derivatives of $y_a(x)$ exist.

Finally, let $G \subset \mathbb{E}_n$ be an open ball of center a and radius r chosen arbitrarily, and let

$$f_G(x) = \phi_r(y_a(x)).$$

Since $\phi_r(y)$ is infinitely differentiable for any y and all partial derivatives of $y_a(x)$ exist for any x, all partial derivatives of $f_G(x)$ exist for any x. Moreover, it is evident that if $x \in G$ then $0 \le y_a(x) < r^2$, so $f_G(x) > 0$, while if $x \notin G$ then $f_G(x) = 0$. Finally, each partial derivative of $f_G(x)$ is bounded since it is continuous everywhere and zero outside G.

The complement $\bar{F}$ of F relative to $\mathbb{E}_n$ is an open set, so it can be represented in the form $\bar{F} = \cup_{k=1}^{\infty} G_k$, where $G_1, G_2, \dots$ are open balls.

Since each partial derivative of the functions $f_{G_1}(x)$, $f_{G_2}(x)$, ... is bounded, there exist positive constants c_k $(k = 1, 2, \dots)$ such that all partial derivatives of order not higher than k of the function $f_{G_k}(x)$ (including $f_{G_k}(x)$ itself) have absolute value less than $(1/c_k)(1/2^k)$ for every x. Then, obviously, the function series

$$\sum_{k=1}^{\infty} c_k f_{G_k}(x)$$

is absolutely convergent; we set

$$F(x) = \sum_{k=1}^{\infty} c_k f_{G_k}(x).$$

Forming a partial derivative of order i of the terms of the series, the definition of c_k ensures that, beginning with the ith term, the absolute value of each term can be majorized by $1/2^k$, so the sum of the partial derivatives is absolutely convergent. It follows that all partial derivatives of $F(x)$ exist. Finally, it is clear that $F(x) = 0$ for $x \in F$ and $F(x) > 0$ for $x \notin F$, thus the function $F(x)$ has the required properties. □

Remark. Let F_1 and F_2 be disjoint closed sets in the n-dimensional Euclidean space. By a similar method, one can construct an infinitely differentiable function that is equal to 0 on F_1, equal to 1 on F_2, and positive and less than 1 outside F_1 and F_2.

Problem F.8. *Let f be a continuous, nonconstant, real function, and assume the existence of an F such that $f(x+y) = F[f(x), f(y)]$ for all real x and y. Prove that f is strictly monotone.*

Solution. Suppose that f is not strictly monotone. Then by the continuity of f there exist real numbers $s_1 < s_2$ with $f(s_1) = f(s_2)$. If $\varepsilon > 0$ is arbitrary, then by the continuity of f there are values $t_1 < t_2$ in the closed interval $[s_1, s_2]$ such that $t_2 - t_1 < \varepsilon$ and $f(t_1) = f(t_2)$. Then, however,

$$\begin{aligned} f[t + (t_2 - t_1)] &= f[(t - t_1) + t_2] = F[f(t - t_1), f(t_2)] \\ &= F[f(t - t_1), f(t_1)] = f[(t - t_1) + t_1] = f(t) \end{aligned}$$

for all real values t. Thus $\tau = t_2 - t_1$ is a period of f. Since $\tau < \varepsilon$ where $\varepsilon > 0$ is arbitrary, it follows that the continuous function f has arbitrarily small periods. Hence f is constant, contrary to the assumption. □

Problem F.9. *Let k be a positive integer, z a complex number, and $\varepsilon < 1/2$ a positive number. Prove that the following inequality holds for infinitely many positive integers n:*

$$\left| \sum_{0 \le \ell \le \frac{n}{k+1}} \binom{n-k\ell}{\ell} z^\ell \right| \ge \left(\frac{1}{2} - \varepsilon\right)^n .$$

Solution. Put

$$a_n(z) = a_n = \sum_{0 \le \ell \le \frac{n}{k+1}} \binom{n-k\ell}{\ell} z^\ell, \quad n = 1, 2, \ldots .$$

We have to prove that $\limsup \sqrt[n]{|a_n|} \ge 1/2$.

If $|w + w^{k+1}z| < 1$, then

$$\frac{1}{1-(w+w^{k+1}z)} = 1 + \sum_{m=1}^{\infty} \sum_{\ell=0}^{m} \binom{m}{\ell} z^\ell w^{m+\ell k} = 1 + \sum_{n=1}^{\infty} a_n w^n. \qquad (1)$$

Therefore, the series $1 + \sum_{n=1}^{\infty} a_n w^n$ is the power series of the function $1/(1-(w+w^{k+1}z))$ at the point $w = 0$. According to the Cauchy–Hadamard criterion, it is sufficient to establish that the radius of convergence of the power series is not greater than 2.

For this purpose, it is sufficient to show that some zero of the polynomial $1 - w - w^{k+1}z$ $(k \ge 1)$ has absolute value not greater than 2. If $|1/z| \le 2^{k+1}$, then this follows from the fact that the product of the zeros of the polynomial equals $1/z$. If $|1/z| > 2^{k+1}$, then $|w^{k+1}z| < 1 \le |1-w|$ on the circle $|w| = 2$, so by Rouché's theorem the polynomial considered and the polynomial $1 - w$ have the same number of zeros inside this circle.

In the case $z = -1/4$, $k = 1$, using relation (1), it is easy to see that $a_n(-1/4) = (n+1)/2^n$. This shows that the assertion concerning the lim sup cannot be improved. □

Remarks.

1. One participant proved the following, stronger, statement: $\limsup |a_n|^{1/n} \ge k/(k+1)$, and this is sharp for every k; $\limsup |a_n(z)|^{1/n}$ assumes its minimum for $z = (-k^k/(k+1)^{k+1})$. (We omit the proof, which is rather long.)
2. Several participants proved that the a_n satisfy the following recursive definition:

$$a_0 = \cdots = a_k = 1, \quad a_{n+1} = a_n + a_{n-k}z \quad \text{if } n \ge k.$$

Problem F.10. *Let $f(x)$ be a real function such that*

$$\lim_{x\to+\infty} \frac{f(x)}{e^x} = 1$$

and $|f''(x)| < c|f'(x)|$ for all sufficiently large x. Prove that

$$\lim_{x\to+\infty} \frac{f'(x)}{e^x} = 1.$$

Solution 1. Suppose that $|f''(x)| < c|f'(x)|$ for $x \geq x_0$. We first show that $x \geq x_0$ and $t < 1/c$ imply

$$|f'(x+t) \leq \frac{1}{1-ct}|f'(x)|. \tag{1}$$

We may assume that $|f'(x+t)| > |f'(x)|$, since otherwise (1) is trivially fulfilled. Put

$$t_0 = \min\{t' : t' > 0, |f(x+t')| = |f(x+t)|\}.$$

Since the function $|f'(\xi)|$ is continuous and does not intersect the horizontal line of ordinate $|f'(x+t)|$ for $\xi \in [x, x+t_0)$ while it remains under this line at the point x, therefore $|f'(\xi)| \leq |f'(x+t)|$ on the whole interval $[x, x+t_0]$. From the Lagrange mean value theorem,

$$\begin{aligned}|f'(x+t)| - |f'(x)| \leq |f'(x+t_0)| - |f'(x)| &= t_0|f''(\xi)| \\ \leq t_0c|f'(\xi)| &\leq t_0c|f'(x+t)|,\end{aligned}$$

which gives (1).

From (1) we see that if $f'(x) = 0$ for some $x \geq x_0$, then $|f'(x')| = 0$ for all $x' \geq x$; but this is impossible since then $e^{-x}f(x)$ would tend to 0. Consequently, $f'(x)$ is either positive for all $x \geq x_0$ or negative for all $x \geq x_0$. However, because $e^{-x}f(x) \to 1$, the function $f(x)$ cannot be monotone decreasing. Thus

$$f'(x) > 0 \tag{2}$$

for all $x \geq x_0$.

From (1) and (2), we obtain that in case $x \geq x_0 + 1/c$ and $t < 1/c$,

$$(1-ct)f'(x+u) \leq f'(x) \leq \frac{1}{1-ct}f'(x-u)$$

for $0 \leq u \leq t$. Hence, by integration,

$$(1-ct)[f(x+t) - f(x)] \leq tf'(x) \leq \frac{1}{1-ct}[f(x) - f(x-t)],$$

that is,

$$\frac{1-ct}{t}\left[\frac{f(x+t)}{e^{x+t}}e^{t}-\frac{f(x)}{e^{x}}\right]\le\frac{f'(x)}{e^{x}}$$
$$\le\frac{1}{t(1-ct)}\left[\frac{f(x)}{e^{x}}-\frac{f(x-t)}{e^{x-t}}e^{-t}\right]. \tag{3}$$

We first consider the left-hand side of (3). For fixed t,

$$\lim_{x\to\infty}\left(\frac{f(x+t)}{e^{x+t}}e^{t}-\frac{f(x)}{e^{x}}\right)=e^{t}-1.$$

Hence

$$\liminf_{x\to\infty}\frac{f'(x)}{e^{x}}\ge(1-ct)\frac{e^{t}-1}{t},$$

and therefore

$$\liminf_{x\to\infty}\frac{f'(x)}{e^{x}}\ge\lim_{t\to+0}\left[(1-ct)\frac{e^{t}-1}{t}\right]=1.$$

From the right-hand side of (3), it similarly follows that

$$\limsup_{x\to\infty}\frac{f'(x)}{e^{x}}\le\lim_{t\to+0}\left[\frac{1}{1-ct}\frac{1-e^{-t}}{t}\right]=1,$$

which proves the statement. □

Solution 2. We shall make use of the following theorem, which is well known and easy to prove.

Theorem. Let g be a two times differentiable function such that $\lim_{x\to+\infty}g(x)$ exists and is finite, whereas $|g''(x)|\le C$ if $x\ge x_0$ for suitable real numbers C and x_0. Then $\lim_{x\to+\infty}g'(x)=0$.

Now let $g(x)=(f(x)/e^{x})$. Then

$$g'(x)=\frac{f'(x)-f(x)}{e^{x}},\quad g''(x)=\frac{f''(x)-2f'(x)+f(x)}{e^{x}}.$$

By assumption, $\lim_{x\to+\infty}g(x)=1$. From the other assumption, it follows that for sufficiently large x the derivative $f'(x)$ is of constant sign. Since $\lim_{x\to+\infty}f(x)=+\infty$, it must be positive. Thus, if x_1 is sufficiently large and $x\ge x_1$, then

$$|f'(x)-f'(x_1)|\le\int_{x_1}^{x}|f''(t)|\,dt\le c\int_{x_1}^{x}f'(t)\,dt=f(x)-f(x_1).$$

Therefore $|f''(x)|\le c'f(x)$ and $|f'(x)|\le c'f(x)$ with a suitable constant c' and, consequently, $|g''(x)|\le(3c'+1)g(x)$ if $x\ge x_1$. It follows that, for some C and x_0, $|g''(x)|\le C$ if $x\ge x_0$.

By the theorem stated at the beginning,

$$\lim_{x\to+\infty} g'(x) = \lim_{x\to+\infty} \frac{f'(x)-f(x)}{e^x} = 0,$$

whence $\lim_{x\to+\infty}(f'(x)e^x) = 1$. □

Remark. In some sense, it is necessary to assume that $|f''/f|$ and $|g''|$ are bounded. To show this, let

$$g(x) = 1 + \frac{\sin x^2}{x}.$$

Then

$$g'(x) = \frac{-\sin x^2}{x^2} + 2\cos x^2$$

and

$$\liminf_{x\to+\infty} g'(x) = -2 < 2 = \limsup_{x\to+\infty} g'(x).$$

Problem F.11. *Find all continuous real functions f, g and h defined on the set of positive real numbers and satisfying the relation*

$$f(x+y) + g(xy) = h(x) + h(y)$$

for all $x > 0$ and $y > 0$.

Solution. Choose $y = 1$ in the equation, and then

$$g(x) = h(x) - f(x+1) + h(1) \tag{1}$$

for $x > 0$. Substituting (1) into the original equation, we obtain

$$h(x) + h(y) - h(xy) = f(x+y) - f(xy+1) + h(1). \tag{2}$$

Put

$$H(x,y) = h(x) + h(y) - h(xy).$$

Then

$$H(xy,z) + H(x,y) = H(x,yz) + H(y,z) \tag{3}$$

for any triple of positive numbers x, y, z. By relation (2),

$$H(x,y) = f(x+y) - f(xy+1) + h(1);$$

putting this into (3), we find that

$$f(xy+z) - f(xy+1) + f(yz+1) = f(x+yz) + f(y+z) - f(x+y) \quad (x,y,z>0). \tag{4}$$

Since f is continuous on the set of positive numbers, passing to the limit $z \to 0\,(z > 0)$, from (4) it follows that

$$f(xy) - f(xy+1) + f(1) = f(x) + f(y) - f(x+y). \tag{5}$$

Introduce the notations

$$f^*(t) = f(t) - f(t+1) + f(1) \qquad (t > 0)$$

and

$$F(x,y) = f(x) + f(y) - f(x+y).$$

Then

$$F(x+y,z) + F(x,y) = F(x,y+z) + F(y,z) \quad (x,y,z > 0) \tag{6}$$

and, by (5),

$$F(x,y) = f^*(xy). \tag{7}$$

Putting (7) into (6), we obtain

$$f^*(xz+yz) + f^*(xy) = f^*(xy+xz) + f^*(yz) \quad (x,y,z > 0). \tag{8}$$

Hence, choosing $z = 1/y$ and writing

$$u = \frac{x}{y}, \qquad v = xy, \tag{9}$$

it follows that

$$f^*(u+1) + f^*(v) = f^*(u+v) + f^*(1) \tag{10}$$

for all positive values of u and v, since for positive u and v the system of equations (9) can be satisfied by suitable positive values x and y. From (10), interchanging u and v,

$$f^*(u+1) + f^*(v) = f^*(v+1) + f^*(u),$$

whence, taking $v = 1$,

$$f^*(u+1) = f^*(u) + f^*(2) - f^*(1).$$

Substituting this into (10),

$$f^*(u+v) + f^*(1) = f^*(u) + f^*(v) + f^*(2) - f^*(1),$$

whence, by the continuity of f^*,

$$f^*(t) = \alpha t + \beta,$$

where α and β are constants. Then in view of (5),

$$\alpha xy + \beta = f(x) + f(y) - f(x+y),$$

and therefore, with the notation $\tilde{f}(x) = f(x) = (\alpha/2)x^2 - \beta$, we find that $\tilde{f}(x+y) = \tilde{f}(x) + \tilde{f}(y)$. Thus $\tilde{f}(x) = \gamma x$ and

$$f(x) = -\frac{\alpha}{2}x^2 + \gamma x + \beta. \tag{11}$$

Putting (11) into (2), we see that the function

$$\tilde{h}(x) = h(x) + \frac{\alpha}{2}x^2 - \gamma x - \delta \quad (\delta = \frac{\alpha}{2} - \gamma - \beta + h(1))$$

satisfies the equation

$$\tilde{h}(x) + \tilde{h}(y) = \tilde{h}(xy) \qquad (x, y > 0),$$

which yields $\tilde{h}(x) = \kappa \ln x$. Consequently,

$$h(x) = -\frac{\alpha}{2}x^2 + \gamma x + \kappa \ln x - \delta. \tag{12}$$

Finally, from (1), using (11) and (12), it follows that

$$g(x) = \kappa \ln x + \alpha x - 2\delta - \beta. \tag{13}$$

We have shown that the solutions of the equation can only be the functions of the forms (11), (12), and (13). On the other hand, it is easy to see that the functions (11), (12), and (13) are solutions of the equation for any choice of constants α, β, γ, δ, and κ. □

Remark. Most participants first show that f, g, and h are two times continuously differentiable functions and then reduce the problem to a differential equation. Several of them note that when using this method it is sufficient to assume the integrability of f, g, and h on every bounded, closed subinterval of the set of positive numbers.

Problem F.12. *Let x_0 be a fixed real number, and let f be a regular complex function in the half-plane Re $z > x_0$ for which there exists a non-negative function $F \in L_1(-\infty, \infty)$ satisfying $|f(\alpha + i\beta)| \leq F(\beta)$ whenever $\alpha > x_0$, $-\infty < \beta < +\infty$. Prove that*

$$\int_{\alpha - i\infty}^{\alpha + i\infty} f(z) dz = 0.$$

Solution. Let $x_0 < \alpha_1 < \alpha_2$. Let $\{\beta_n\}$ and $\{\gamma_n\}$ be sequences of real numbers tending to $+\infty$ such that $F(\beta_n) \to 0$ and $F(-\gamma_n) \to 0$. By the Cauchy integral theorem,

$$\int_{\alpha_1 - i\gamma_n}^{\alpha_1 + i\beta_n} f(z)\,dz + \int_{\alpha_1 + i\beta_n}^{\alpha_2 + i\beta_n} f(z)\,dz + \int_{\alpha_2 + i\beta_n}^{\alpha_2 - i\gamma_n} f(z)\,dz + \int_{\alpha_2 - i\gamma_n}^{\alpha_1 - i\gamma_n} f(z)\,dz = 0$$

(the path of integration is always the connecting segment). Since

$$\left|\int_{\alpha_1+i\beta_n}^{\alpha_2+i\beta_n} f(z)\,dz\right| = \left|\int_{\alpha_1}^{\alpha_2} f(\alpha+i\beta_n)\,d\alpha\right| \le (\alpha_2-\alpha_1)F(\beta_n) \to 0,$$

and similarly

$$\left|\int_{\alpha_2-i\gamma_n}^{\alpha_1-i\gamma_n} f(z)\,dz\right| \le (\alpha_2-\alpha_1)F(-\gamma_n) \to 0,$$

therefore

$$\int_{\alpha_1-i\infty}^{\alpha_1+i\infty} f(z)\,dz = \int_{\alpha_2-i\infty}^{\alpha_2+i\infty} f(z)\,dz, \tag{1}$$

which means that the integral in question is independent of α. (Here the improper integrals exist since the integrand admits an integrable majorant.) Denote by A the common value of the integrals appearing in (1).

Apply our result to the function $f(z)/z$ (which is analytic, for instance, in the half-plane $\operatorname{Re} z \ge 1$). Then

$$B = \int_{\alpha-i\infty}^{\alpha+i\infty} \frac{f(z)}{z}\,dz = i\int_{-\infty}^{\infty} \frac{f(\alpha+i\beta)}{\alpha+i\beta}\,d\beta; \quad \alpha > \max\{1, x_0\}$$

is independent of α. Let $\alpha \to \infty$. Since

$$\left|\frac{f(\alpha+i\beta)}{\alpha+i\beta}\right| \le F(\beta) \qquad (\alpha > 1),$$

integration and transition to the limit can be interchanged by the theorem of Lebesgue, and we obtain $B = 0$. Then

$$A = A - \alpha B = \int_{\alpha-i\infty}^{\alpha+i\infty} f(z)\left(1-\frac{\alpha}{z}\right)dz = -\int_{-\infty}^{\infty} f(\alpha+i\beta)\frac{\beta}{\alpha+i\beta}\,d\beta.$$

Here, again, $F(\beta)$ is a common majorant of the integrands, and for each fixed value of β the integrand tends to 0 as $\alpha \to \infty$. Consequently, by the theorem of Lebesgue, the integral tends to 0, whence $A = 0$. □

Remark. All participants first show that the integral under consideration is independent of α. Then some of them choose the function $f(z)/z$ and the way described above, while others work with the function $e^{-tz}f(z)$, where $t > 0$ is a real number. They establish similarly (using the fact that this function also satisfies the conditions of the problem and tends to 0 as $\operatorname{Re} z \to \infty$) that

$$\int_{\alpha-i\infty}^{\alpha+i\infty} e^{-tz}f(z)\,dz = 0;$$

hence, letting $t \to 0$, the desired equation follows.

Problem F.13. *Let $\pi_n(x)$ be a polynomial of degree not exceeding n with real coefficients such that*

$$|\pi_n(x)| \le \sqrt{1-x^2} \qquad \textit{for} \qquad -1 \le x \le 1.$$

Then

$$|\pi_n'(x)| \le 2(n-1).$$

Solution. The background of the problem is provided by the Markov inequality: If the polynomial $P(x)$ of degree n has absolute value less than 1 through the interval $(-1,1)$, then its derivative has absolute value less than n^2 on the same interval. Of similar type is Bernstein's theorem: If a trigonometric polynomial of order n has absolute value not greater than 1, then its derivative has absolute value not greater than n.

We have strengthened the hypothesis of Markov's theorem and wish to prove an estimate much sharper than that of the Markov theorem. To this end, we shall need Bernstein's theorem cited above as well as the following theorem:

Theorem. If a polynomial $Q(x)$ of degree k satisfies

$$|Q(x)| < \frac{1}{\sqrt{1-x^2}} \qquad \text{for} \quad -1 < x < 1,$$

then

$$|Q(x)| < k+1 \qquad \text{for the same } x.$$

For both theorems, see for example, *I. P. Natanson, Constructive Function Theory 1–3, 1964–65 (translated from the Russian), sections V.1 and VI.6.*

Now we prove the statement of the problem. We may assume that $\pi_n(x)$ is nonconstant. From the relations $\pi_n(\pm 1) = 0$, it follows that $\pi_n(x) = (1-x^2)f(x)$, where $f(x)$ is a polynomial of degree not exceeding $n-2$ and

$$|f(x)| < \frac{1}{\sqrt{1-x^2}} \qquad \text{for} \quad -1 < x < 1,$$

whence $|f(x)| \le n-1$. Let $x = \cos\vartheta$. It is well known that in this case $f(x) = f(\cos\vartheta) = F(\vartheta)$ is a trigonometric polynomial of order not exceeding $n-2$. Write $G(\vartheta) = F(\vartheta)\sin\vartheta$. Since $G(\vartheta)$ arises from $f(x)\sqrt{1-x^2}$ by the substitution $x = \cos\vartheta$, therefore $|G(\vartheta)| \le 1$. The trigonometric polynomial $F(\vartheta)$ has order not greater than $n-2$, so $G(\vartheta)$ has order not greater than $n-1$, that is,

$$|G'(\vartheta)| \le n-1. \tag{1}$$

We calculate the derivative of $\pi_n(\cos\vartheta)$ with respect to ϑ in two ways. On the one hand, we have

$$\frac{d}{d\vartheta}\pi_n(\cos\vartheta) = \pi_n'(\cos\vartheta)(-\sin\vartheta). \tag{2}$$

On the other hand,

$$\begin{aligned}\frac{d}{d\vartheta}\pi_n(\cos\vartheta) &= \frac{d}{d\vartheta}(G(\vartheta)\sin\vartheta) = G'(\vartheta)\sin\vartheta + G(\vartheta)\cos\vartheta \\ &= G'(\vartheta)\sin\vartheta + F(\vartheta)\sin\vartheta\cos\vartheta. \qquad (3)\end{aligned}$$

Comparing (2) and (3),

$$-\pi_n'(\cos\vartheta) = G'(\vartheta) + F(\vartheta)\cos\vartheta, \qquad (4)$$

that is,

$$|\pi_n'(x)| \le (n-1) + (n-1)|x| = (1+|x|)(n-1) \le 2(n-1). \qquad (5)$$

The proof is complete. □

Remark. It should be noted that for $\sin\vartheta = 0$, we cannot divide by $\sin\vartheta$, but the continuity of $\pi_n'(x)$ ensures that the assertion of the problem is true also in this case. Equality can only hold if $\pi_n(x) = (1-x^2)Q_{n-2}(x)$ or the negative of this, where $Q_{n-2}(x)$ stands for the Chebyshev polynomial of the second kind of degree $n-2$. This case can also be characterized by the relation $G(\vartheta) = \pm\sin(n-1)\vartheta$. Then in the Bernstein inequality, for the values ϑ corresponding to $|x| = 1$, we have equality, and in the triangle inequality applied in (5) the terms have equal sign, so equality really holds for these π_n.

Problem F.14. *Let $a(x)$ and $r(x)$ be positive continuous functions defined on the interval $[0,\infty)$, and let*

$$\liminf_{x\to\infty}(x - r(x)) > 0.$$

Assume that $y(x)$ is a continuous function on the whole real line, that it is differentiable on $[0,\infty)$, and that it satisfies

$$y'(x) = a(x)y(x - r(x))$$

on $[0,\infty)$. Prove that the limit

$$\lim_{x\to\infty} y(x)\exp\left\{-\int_0^x a(u)du\right\}$$

exists and is finite.

Solution. Integrating (2), we obtain

$$y(x) = y(u) + \int_u^x a(t)y(t - r(t))\,dt \quad (x \ge u \ge 0). \qquad (3)$$

If $y(x)$ is any continuous function on the interval $(-\infty, 0]$, then it is easy to see that $y(x)$ can be uniquely extended to the whole real line so that (3) is valid. In fact, suppose that x_0 is the supremum of those values up to which unique extension is possible. Then in the neighborhood of x_0 of some radius δ, the function $r(t)$ is greater than some positive ε. Denote by η the smaller of the numbers ε and δ. Then, by (3), the values of $y(x)$ taken for $x \leq x_0 - \eta/2$ uniquely determine the values of $y(x)$ in the interval $(x_0 - \eta/2, x_0 + \eta/2)$. This contradicts the choice of x_0.

A similar reasoning shows that if $y(x)$ is positive on $(-\infty, 0]$ then it is positive on the whole real line. In this case, with the help of (2) we obtain that $y(x)$ is monotonically increasing on $[0, \infty)$. Put

$$z(x) = y(x)e^{-\int_0^x a(t)\,dt} \qquad (x \geq 0). \tag{4}$$

Differentiating and using (2) we obtain

$$z'(x) = a(x)e^{-\int_0^x a(t)\,dt}\left[y(x - r(x)) - y(x)\right]. \tag{5}$$

To prove the statement of the problem, first suppose that $y(x)$ is positive on $(-\infty, 0]$. Then, as we have seen, $y(x)$ is positive everywhere and increasing for $x \geq 0$. Thus, $z(x)$ is positive for all $x \geq 0$; further, $z(x)$ is monotone decreasing for sufficiently large x. Equation (1) implies that $x - r(x) > 0$ if x is large; so from (5) by the monotonicity of $y(x)$ it follows that $z'(x)$ is negative. Since $z(x)$ is positive and decreasing, $\lim_{x\to\infty} z(x)$ exists.

To prove the general case, represent the function $y(x)$ on $(-\infty, 0]$ in the form

$$y(x) = y_1(x) - y_2(x), \tag{6}$$

where $y_1(x)$ and $y_2(x)$ are positive, continuous functions. As we have seen, $y_1(x)$ and $y_2(x)$ can be extended to the whole real line so that they satisfy the differential equation (2) for $x \geq 0$. By the uniqueness of the solution of (2) mentioned above, it is also clear that (6) remains valid on the whole real line. Defining the functions $z_1(x)$ and $z_2(x)$ in analogy with (4), it follows as above that the limits $\lim_{x\to\infty} z_1(x)$ and $\lim_{x\to\infty} z_2(x)$ exist. This implies the existence of the limit

$$\lim_{x\to\infty} z_(x) = \lim_{x\to\infty} (z_1(x) - z_2(x)).$$

The solution of the problem is complete. □

Problem F.15. *Let λ_i $(i = 1, 2, \dots)$ be a sequence of distinct positive numbers tending to infinity. Consider the set of all numbers representable in the form*

$$\mu = \sum_{i=1}^{\infty} n_i \lambda_i,$$

where $n_i \geq 0$ are integers and all but finitely many n_i are 0. Let

$$L(x) = \sum_{\lambda_i \leq x} 1 \qquad \text{and} \qquad M(x) = \sum_{\mu \leq x} 1.$$

(In the latter sum, each μ occurs as many times as its number of representations in the above form.) Prove that if

$$\lim_{x \to \infty} \frac{L(x+1)}{L(x)} = 1,$$

then

$$\lim_{x \to \infty} \frac{M(x+1)}{M(x)} = 1.$$

Solution. (When in the solution we speak of a μ, we mean not only its value but also its representation by a fixed sequence $\{n_i\}$, keeping in mind that a number can possibly be represented in several ways.)

If $\mu_1 = \sum n_i^{(1)} \lambda_i$ and $\mu_2 = \sum n_i^{(2)} \lambda_i$, then let $\mu_1 | \mu_2$ mean that $n_i^{(1)} \leq n_i^{(2)}$ $(i = 1, 2, \dots)$. Then

$$\mu = \sum_{\substack{n \geq 1 \text{ integer }, \lambda_i \\ n\lambda_i | \mu}} \lambda_i$$

and

$$\sum_{\mu \leq x} \mu = \sum_{\mu \leq x} \sum_{\substack{n \geq 1, \lambda_i \\ n\lambda_i | \mu}} \lambda_i.$$

In the inner sum, $\mu' = \mu - n\lambda_i$ is also a μ-number; let us sum with respect to this. Then

$$\sum_{\mu \leq x} \mu = \sum_{\mu' \leq x} \sum_{\substack{n \geq 1, \lambda_i \\ n\lambda_i \leq x - \mu'}} \lambda_i.$$

With the notation

$$\sum_{\substack{n \geq 1, \lambda_i \\ n\lambda_i \leq y}} \lambda_i = \mathcal{L}(y)$$

(replacing μ' by μ) we have

$$\sum_{\mu \leq x} \mu = \sum_{\mu \leq x} \mathcal{L}(x - \mu). \tag{1}$$

Apply this to $x+1$ and subtract the two relations from each other:

$$\sum_{x < \mu \leq x+1} \mu = \sum_{\mu \leq x} (\mathcal{L}(x+1-\mu) - \mathcal{L}(x-\mu)) + \sum_{x < \mu \leq x+1} \mathcal{L}(x+1-\mu).$$

The left-hand side is at least $x\,(M(x+1)-M(x))$, and the second sum on the right-hand side is at most $\mathcal{L}(1)\,(M(x+1)-M(x))$ (since $\mathcal{L}(y)$ is increasing). Consequently,

$$(x-\mathcal{L}(1))\,(M(x+1)-M(x)) \le \sum_{\mu\le x} (\mathcal{L}(x+1-\mu)-\mathcal{L}(x-\mu))\,.$$

Here $\mathcal{L}(y)$ is expressed through the sequence λ_i alone, and from the hypothesis relating to the latter we shall deduce that

$$\frac{\mathcal{L}(y+1)}{\mathcal{L}(y)} \to 1 \qquad (y\to\infty), \tag{2}$$

whence for sufficiently large K, $\mathcal{L}(y+1)-\mathcal{L}(y)\le\varepsilon\mathcal{L}(y)$ if $y\ge K$. Therefore, decomposing the sum based on whether $\mu\le x-K$ or $\mu>x-K$, and using (1) again,

$$\begin{aligned}\sum_{\mu\le x-K} (\mathcal{L}(x+1-\mu)-\mathcal{L}(x-\mu)) &\le \varepsilon \sum_{\mu\le x-K} \mathcal{L}(x-\mu)\\ &\le \varepsilon\sum_{\mu\le x}\mathcal{L}(x-\mu)=\varepsilon\sum_{\mu\le x}\mu\le\varepsilon x M(x)\end{aligned}$$

and

$$\sum_{x-K<\mu\le x} (\mathcal{L}(x+1-\mu)-\mathcal{L}(x-\mu)) \le \mathcal{L}(K+1)M(x).$$

We have thus obtained

$$\begin{gathered}(x-\mathcal{L}(1))\,(M(x+1)-M(x)) \le \varepsilon x M(x)+\mathcal{L}(K+1)M(x),\\ \frac{M(x+1)-M(x)}{M(x)} \le \frac{\varepsilon x+\mathcal{L}(K+1)}{x-\mathcal{L}(1)},\\ \limsup_{x\to\infty}\frac{M(x+1)-M(x)}{M(x)}\le\varepsilon,\end{gathered}$$

and since $\varepsilon>0$ is arbitrary,

$$\limsup_{x\to\infty}\frac{M(x+1)-M(x)}{M(x)}=0.$$

It remains to prove (2). We may write

$$\begin{aligned}\mathcal{L}(x)=\sum_{\lambda_i\le x}\lambda_i\sum_{1\le n\le\frac{x}{\lambda_i}}1 &\ge \sum_{\lambda_i\le\frac{x}{2}}\lambda_i\frac{x}{2\lambda_i}+\sum_{\frac{x}{2}<\lambda_i\le x}\lambda_i\\ &\ge\frac{x}{2}L(\tfrac{x}{2})+\frac{x}{2}\left(L(x)-L(\tfrac{x}{2})\right)=\frac{1}{2}xL(x).\end{aligned} \tag{3}$$

Now

$$\mathcal{L}(x+1)-\mathcal{L}(x)=\sum_{\substack{n\ge1,\lambda_i\\ x<n\lambda_i\le x+1}}\lambda_i.$$

For a fixed $\varepsilon > 0$, decompose the sum according to the cases $\lambda_i \le \varepsilon(x+1)$ and $\lambda_i > \varepsilon(x+1)$:

$$\sum_{\lambda_i \le \varepsilon(x+1)} \lambda_i \sum_{\frac{x}{\lambda_i} < n \le \frac{x+1}{\lambda_i}} 1 \le \frac{1}{\min \lambda_i} \sum_{\lambda_i \le \varepsilon(x+1)} \lambda_i$$

(since the interval corresponding to n has length not greater than $1/\min \lambda_i =$ constant). Further, using (3),

$$\frac{1}{\min \lambda_i} \sum_{\lambda_i \le \varepsilon(x+1)} \lambda_i \le \frac{\varepsilon(x+1)}{\min \lambda_i} L(\varepsilon(x+1))$$
$$\le \frac{2\varepsilon x}{\min \lambda_i} L(x) \le \frac{4\varepsilon}{\min \lambda_i} \mathcal{L}(x).$$

In the other part, $n \le (x+1)/\varepsilon(x+1) = 1/\varepsilon$, so

$$\sum_{1 \le n \le \frac{1}{\varepsilon}} \sum_{\frac{x}{n} < \lambda_i \le \frac{x+1}{n}} \lambda_i \le \sum_{1 \le n \le \frac{1}{\varepsilon}} \frac{x+1}{n} \left(L\left(\frac{x+1}{n}\right) - L\left(\frac{x}{n}\right) \right).$$

The inner sum consists of finitely many terms (for fixed $\varepsilon > 0$) and, by the assumption, each of them satisfies the relation

$$L\left(\frac{x+1}{n}\right) - L\left(\frac{x}{n}\right) \le L\left(\frac{x}{n} + 1\right) - L\left(\frac{x}{n}\right) = o\left(L\left(\frac{x}{n}\right)\right) = o(L(x))$$

as $x \to \infty$. Therefore, again by (3), the entire sum is $o((x+1)L(x)) = o(\mathcal{L}(x))$. As a result,

$$\mathcal{L}(x+1) - \mathcal{L}(x) \le \frac{4\varepsilon}{\min \lambda_i} \mathcal{L}(x) + o(\mathcal{L}(x)),$$

whence

$$\frac{\mathcal{L}(x+1) - \mathcal{L}(x)}{\mathcal{L}(x)} \to 0.$$

The proof is complete. □

Remark. The solution above is based on relation (1). Application of this relation is motivated by the following argument.

If $\lambda_i = \log p_i$, where p_i is the ith prime, then the numbers μ are the $\log n$ where $n \ge 1$ are integers, each appearing once. Although the condition on $L(x)$ does not hold in this case, namely $L(x+1)/L(x) \to e$, it is natural to start from the formula applied in prime number theory (for example, in the proof of Chebysev's theorem), the formula corresponding to the prime factorization of $n!$.

Problem F.16. *Let $P(z)$ be a polynomial of degree n with complex coefficients,*

$$P(0) = 1, \quad \text{and} \quad |P(z)| \leq M \quad \text{for} \quad |z| \leq 1.$$

Prove that every root of $P(z)$ in the closed unit disc has multiplicity at most $c\sqrt{n}$, where $c = c(M) > 0$ is a constant depending only on M.

Solution 1. It is sufficient to examine the multiplicity of number 1. In fact, if we prove something for 1 then we may apply the result to the polynomial $p(z) = P(\alpha z)$ with $|\alpha| \leq 1$, and in this way we obtain the same estimate for all roots lying in the unit disc.

The idea of the solution is the following. We consider the integral

$$F(P) = \int_0^{2\pi} \log |P(e^{i\phi})| \, d\phi$$

and show that it exists and is nonnegative. Then we estimate it from above, once in the neighborhood of 1 with the aid of the multiplicity of 1 and the degree of P, and once at other points using the condition $|P(z)| \leq M$.

It is sufficient to prove the existence of the integral for polynomials of the form $z - z_0$. If

$$P(z) = c \prod_{i=1}^{n} (z - z_i),$$

then

$$\log |P(z)| = \log |c| + \sum_{i=1}^{n} \log |z - z_i|.$$

The existence of

$$\int_0^{2\pi} \log |e^{i\phi} - z_0| \, d\phi$$

is evident if $|z_0| \neq 1$. Next let $|z_0| = 1$. Without loss of generality, we may assume that $z_0 = 1$ (a substitution $\phi = \eta + \phi_0$ takes them into each other). Then

$$\log |e^{i\phi} - 1| = 2 \sin \frac{\phi}{2},$$

and

$$\int_0^{2\pi} \log \left(2 \sin \frac{\phi}{2} \right) d\phi$$

really exists and is equal to 0.

Next, compute the integral

$$f(\alpha) = \int_0^{2\pi} \log \left| 1 - \frac{e^{i\phi}}{\alpha} \right| d\phi \qquad (\alpha \neq 0)$$

for $\alpha \neq 1$. Obviously, its value depends on the absolute value of α only (again, a substitution as above may be applied), so it is the same for the numbers $\alpha\varepsilon_1, \alpha\varepsilon_2, \ldots, \alpha\varepsilon_n$, where the ε_j are the nth unit roots. Therefore,

$$\begin{aligned} \eta f(\alpha) = \sum_{j=1}^{n} f(\alpha\varepsilon_j) &= \int_0^{2\pi} \log\left|\prod_{j=1}^{n}\left(1 - \frac{e^{i\phi}}{\alpha\varepsilon_j}\right)\right| d\phi \\ &= \int_0^{2\pi} \log\left|1 - \frac{e^{in\phi}}{\alpha^n}\right| d\phi. \end{aligned}$$

Now, if $|\alpha| > 1$, then for $n \to \infty$ the integral on the right-hand side tends to zero since $1 - e^{in\phi}/\alpha^n \to 1$ uniformly; thus $f(\alpha) = 0$. On the other hand, if $|\alpha| < 1$ then

$$n\left(f(\alpha) + 2\pi \log|\alpha|\right) = \int_0^{2\pi} \log\left|\alpha^n - e^{in\phi}\right| d\phi \to 0$$

since $1 - |\alpha|^n < |\alpha^n - e^{in\phi}| < 1 + |\alpha|^n$; so in this case $f(\alpha) = -2\pi \log|\alpha|$ is only possible. In each of the three cases, we have $f(\alpha) \geq 0$.

In our case, the relation $P(0) = 1$ implies

$$P(z) = \prod_{j=1}^{n}\left(1 - \frac{z}{z_j}\right),$$

whence

$$F(P) = \sum_{j=1}^{n} f(z_j) \leq 0. \tag{1}$$

Now let $P(z) = (z-1)^k Q(z) = a_0 + a_1 z + \cdots + a_n z^n$, where $Q(1) \neq 0$. We estimate $F(P)$ with the help of k, n, and M. Let

$$F(P) = \int_0^{2\pi} = \int_{-\varepsilon}^{\varepsilon} + \int_{\varepsilon}^{2\pi-\varepsilon} = F_1 + F_2.$$

Then

$$F_2 \leq \int_0^{2\pi} \log M \, d\phi = 2\pi \log M. \tag{2}$$

We split F_1 again into two parts:

$$F_1 = \int_{-\varepsilon}^{\varepsilon} \log|(z-1)^k| \, d\phi + \int_{-\varepsilon}^{\varepsilon} \log|Q(e^{i\phi})| \, d\phi = F_3 + F_4. \tag{3}$$

Clearly,

$$\begin{aligned} F_3 = k\int_{-\varepsilon}^{\varepsilon} \log\left(2\sin\frac{|\phi|}{2}\right) d\phi &< 2k\int_0^{\varepsilon} \log\phi \, d\phi \\ &= 2k\varepsilon(\log\varepsilon - 1). \end{aligned} \tag{4}$$

For estimating F_4, we need an estimate of Q, which we obtain from the coefficients of the expansion of Q about 1. Let $Q(1+z) = R(z)$. We calculate the coefficients of R from those of P using the formula $R(z) = P(z+1)/z^k$:

$$\begin{aligned} P(z+1) &= \sum_{j=0}^{n} a_j (z+1)^j = \sum_{j=0}^{n} \sum_{m=0}^{j} a_j \binom{j}{m} z^m \\ &= \sum_{m=k}^{n} z^m \sum_{j=m}^{n} a_j \binom{j}{m}, \end{aligned}$$

since for $m < k$ the coefficient of z^m is 0 by our assumption. Thus

$$R(z) = \sum_{m=0}^{n-k} b_m z^m, \quad b_m = \sum_{j=m+k}^{n} a_j \binom{j}{m+k}. \tag{5}$$

Further, by the Cauchy inequalities, $|a_j| = |P^{(j)}(0)|/j! \leq M$.

Putting this into (5), we find

$$|b_m| \leq M \sum_{j=m+k}^{n} \binom{j}{m+k} = M \binom{n+1}{m+k+1}. \tag{6}$$

If $|z| = \delta$ and $\delta(n-k)/(k+2) < 1$, then in view of (6),

$$\begin{aligned} \frac{|R(z)|}{M} &\leq \sum_{m=0}^{\infty} \delta^m \binom{n+1}{m+k+1} \\ &= \binom{n+1}{k+1} \left(1 + \delta \frac{n-k}{k+2} + \delta^2 \frac{n-k}{k+2} \frac{n-k-1}{k+3} + \dots \right) \\ &\leq \binom{n+1}{k+1} \sum_{j=0}^{\infty} \left(\delta \frac{n-k}{k+2} \right)^j = \binom{n+1}{k+1} \frac{1}{1 - \delta \frac{n-k}{k+2}}. \end{aligned}$$

Since $|e^{i\phi} - 1| = 2\,|\sin(\phi/2)| \leq |\phi|$, therefore if $\varepsilon < ((k+2)/2(n-k))$, then for $|\phi| \leq \varepsilon$ we have

$$\left|Q(e^{i\phi})\right| = \left|R(e^{i\phi} - 1)\right| \leq M \binom{n+1}{k+1} \frac{1}{1 - \varepsilon \frac{n-k}{k+2}} < 2M \binom{n+1}{k+1}.$$

If $k \geq 2$, then using the relation $t! > t^t e^{-t}$ we obtain

$$\begin{aligned} \binom{n+1}{k+1} &= \frac{(n+1)n(n-1)\dots(n-k+1)}{(k+1)!} \\ &= \frac{(n^2-1)n(n-2)\dots(n-k+1)}{(k+1)!} < \frac{n^{k+1}}{(k+1)!} < \left(\frac{en}{k+1} \right)^{k+1}. \end{aligned}$$

Thus

$$\log\left|Q(e^{i\phi})\right| < \log(2M) + (k+1)\log\frac{en}{k+1}$$

and, consequently,

$$F_4 < 2\varepsilon\left[\log(2M) + (k+1)\log\frac{en}{k+1}\right].$$

Collecting everything, by (1), (2), (3), and (4)

$$(\pi+\varepsilon)\log M + \varepsilon\log\frac{2en}{k+1} + \varepsilon k\log\frac{\varepsilon n}{k+1} \geq 0. \tag{7}$$

Now if $n = k^2/2c$, let $\varepsilon = c/k$ (this fulfills the condition $\varepsilon < ((k+2)/2(n-k)))$. Then (7) becomes

$$\left(\pi + \frac{c}{k}\right)\log M + \frac{c}{k}\log\frac{ek^2}{c(k+1)} + c\log\frac{k}{2(k+1)} \geq 0.$$

We only make things worse if we also replace $k+1$ by k. Moreover, $c/k = k/2n \leq 1/2$ gives $\pi + c/k < 4$, $c/k\log(ek/c) < (c/k)\cdot(ek/c) = e < 3$, so finally

$$\begin{aligned} 4\log M + 3 &\geq c\log 2, \\ c &< 8\log M + 6. \end{aligned} \tag{8}$$

Since $k = \sqrt{2cn}$, relation (8) means that we have proved the assertion of the problem with

$$c(M) = \sqrt{16\log M + 12}. \quad \square$$

Solution 2. It is sufficient to study the multiplicity of 1 (see the previous solution). We establish the following lemma:

Lemma. Assume that the polynomial $\omega_n(z)$ of degree n satisfies $|\omega_n(z)| \leq M$ if $|z| \leq 1$. Let $z = 1$ be a root of multiplicity ℓ for $\omega_n(z)$. Then there exists an absolute constant c_1 ($c_1 = (4e/\pi) + \varepsilon$) such that for $z = e^{i\psi}$, $|\psi| < \ell\pi/2n$ we have

$$\frac{\omega_n(z)}{(z-1)^\ell} < M\left(\frac{c_1 n}{\ell}\right)^\ell.$$

Proof. Setting $\omega_n^2(z) = \omega_{2n}(z)$, the assertion takes the form

$$\left|\frac{\omega_{2n}(z)}{(z-1)^{2\ell}}\right| < M^2\left(\frac{c_1 n}{\ell}\right)^{2\ell},$$

in which form we shall prove it for all polynomials of degree $2n$. Let the roots of the equation $z^{2n} + 1 = 0$ be

$$z_k = e^{i\frac{2k-1}{2n}\pi} \quad (k = 1, 2, \ldots, 2n).$$

Proposition. There exist complex numbers a_k $(k = \ell, \ell+1, \dots, 2n-\ell+1)$ such that

$$\frac{\omega_{2n}(z)}{(z-1)^{2\ell}} = \sum_{k=\ell}^{2n-\ell+1} \left(\frac{z - \frac{1}{z}}{z_k - \frac{1}{z_k}} + 1 \right) \frac{a_k}{2} \omega_{2n}(z_k) \prod_{\substack{j=\ell \\ j \neq k, 2n+1-k}}^{2n-\ell+1} (z - z_j).$$

Suppose that $\omega_{2n}(z_\nu) \neq 0$ $(\ell \leq \nu \leq 2n-\ell+1)$. (Such ν obviously exists since ω_{2n} has at most $2n-2\ell$ roots different from 1.) For $\ell \leq i \leq 2n-\ell+1$, $i \neq \nu$, put

$$a_i = \frac{1}{(z_i - 1)^{2\ell} \prod_{\substack{j=\ell \\ j \neq i, 2n-i+1}}^{2n-\ell+1} (z_i - z_j)}.$$

Then for $z = z_i$ $(i \neq \nu)$ the left-hand side is equal to the right-hand side. Indeed, the terms of the sum appearing on the right-hand side are zero except for the term with $k = i$; if $k \neq 2n - i + 1$, then this is true because of the factor

$$\prod_{\substack{j=\ell \\ j \neq k, 2n+1-k}}^{2n-\ell+1} (z_i - z_j) = 0,$$

while if $k = 2n - i + 1$, then because of the factor

$$\frac{z_i - \frac{1}{z_i}}{z_k - \frac{1}{z_k}} + 1 = 0.$$

For $k = i$, the two sides are equal because of the definition of a_i. Now choose the remaining a_ν so that the coefficients of $z^{2n-2\ell+1}$ and $1/z$ are 0. This can be done because the two coefficients are the negative of each other since

$$\prod_{\substack{j=\ell \\ j \neq k, 2n+1-k}}^{2n-\ell+1} (-z_j) = 1$$

for all k; together with each $-z_j$ appearing in the product, its reciprocal $-z_{2n+1-j}$ also appears.

Thus, the condition

$$\frac{1}{2} \sum_{k=\ell}^{2n-\ell+1} a_k \omega_{2n}(z_k) = 0$$

is to be fulfilled and, as $\omega_{2n}(z_\nu) \neq 0$, this is possible if we choose a_ν properly. But then both the left-hand and the right-hand members are polynomials of degree $2n - 2\ell$ which coincide at $2n - 2\ell + 1$ points (the points $z = z_i$, $\ell \leq i \leq 2n - 2\ell + 1$, $i \neq \nu$), so they are equal identically. (Of course, hence it follows that they are equal for $z = z_\nu$ and therefore, similar to the case $i \neq \nu$,

$$a_\nu = \frac{1}{(z_\nu - 1)^{2\ell} \prod\limits_{\substack{j=\ell \\ j\neq \nu, 2n-\nu+1}}^{2n-\ell+1} (z_\nu - z_j)}$$

also holds.)

Since $|z_i^2 - 1| = |z_i - 1/z_i| = |z_i - z_{2n+1-i}|$ and

$$\prod_{\substack{j=1 \\ j\neq i}}^{2n} (z_i - z_j) = \left|(z^{2n}+1)'_{z=z_i}\right| = 2n,$$

we have

$$\left|(z_i^2 - 1) \prod_{\substack{j=1 \\ j\neq i, 2n+1-i}}^{2n} (z_i - z_j)\right| = 2n.$$

Relying on this, and setting $\phi = \pi/n$, we obtain the following estimate of $|a_i|$ (for $\ell \leq i \leq n$, but also for $i > n$ because of the symmetry):

$$|a_i| = \left|\frac{1}{(z_i - 1)^{2\ell} \prod\limits_{\substack{j=\ell \\ j\neq i, 2n+1-i}}^{2n-\ell+1} (z_i - z_j)}\right|$$

$$= \left|\frac{\left[\prod\limits_{j=1}^{\ell-1}(z_i - z_j) \prod\limits_{j=2n-\ell+2}^{2n} (z_i - z_j)\right](z_i^2 - 1)}{(z_i - 1)^{2\ell}}\right| \cdot \frac{1}{2n}$$

$$\leq \frac{\prod\limits_{j=1}^{\ell-1}[(i-j)\phi] \prod\limits_{j=0}^{\ell-2}[(i+j)\phi] \cdot 2}{\left[\left(i-\frac{1}{2}\right)\frac{\phi}{2}\right]^{2\ell-1}} \cdot \frac{1}{2n}$$

$$\leq \frac{\left(i-\frac{1}{2}\right)^{2\ell-2} \cdot 2^{2\ell-1} \cdot \phi^{2\ell-2}}{\left(i-\frac{1}{2}\right)^{2\ell-1} \cdot \phi^{2\ell-1} \cdot n} \leq 2^{2\ell}.$$

(We have made use of the relation $\tau/2 \leq \sin\tau \leq \tau,\ \tau \leq \pi/2$.)

Now we have to give an upper estimate of the expression

$$\left|\prod_{\substack{j=\ell\\ j\neq i,2n-i+1}}^{2n-\ell+1}(z-z_j)\right|\leq 4\left|\prod_{j=\ell}^{2n-\ell+1}(z-z_j)\right|$$

for each $z=e^{i\psi}$, $|\psi|\leq(\ell\pi/2n)$. Considering pairs and setting $z=e^{i\psi}$, $z_j=e^{i\vartheta}$, we find that

$$\begin{aligned}|(z-z_j)(z-z_{2n-j+1})| &= 4\sin\frac{\vartheta-\psi}{2}\sin\frac{\vartheta+\psi}{2}\\ &= 2(\cos\psi-\cos\vartheta)\leq 2(1-\cos\vartheta);\end{aligned}$$

that is, in the interval $|\psi|\leq(\ell\pi/2n)$ the expression attains its maximum at the point $z=1$. On the other hand,

$$\begin{aligned}4\left|\prod_{j=\ell}^{2n-\ell+1}(1-z_j)\right| &= 4\frac{\left|\prod_{j=1}^{2n}(1-z_j)\right|}{\left|\prod_{j=1}^{\ell-1}(1-z_j)\right|^2}\leq\frac{4(1^{2n}+1)}{\left[\prod_{j=1}^{\ell-1}(j-\frac{1}{2})\frac{\phi}{2}\right]^2}\\ &\leq\frac{8}{\left[\frac{1}{2}(\ell-2)!\left(\frac{\phi}{2}\right)^{\ell-1}\right]^2}\leq\left(\frac{c_0 n}{\ell\pi}\right)^{2\ell-2}.\end{aligned}$$

Since $z=e^{i\psi}$, where $|\psi|\leq(\ell/2)\cdot(\pi/n)$, we have

$$\frac{z-\frac{1}{z}}{z_i-\frac{1}{z_i}}=\left|\frac{2\,\mathrm{Im}z}{2\,\mathrm{Im}z_i}\right|\leq 1$$

for all $i\geq\ell$, and therefore

$$\left|\frac{\omega_{2n}(z)}{(z-1)^{2\ell}}\right|\leq n\cdot 2\cdot\frac{2^{2\ell}}{2}\cdot M^2\cdot\left(\frac{c_0 n}{\ell\pi}\right)^{2\ell-2}<M^2\left(\frac{c_1 n}{\ell}\right)^{2\ell},$$

which proves the lemma.

To prove the assertion of the problem, let

$$\begin{aligned}g(\vartheta) &= |\omega_n(e^{i\vartheta})|^2,\\ \omega_n(z) &= c\prod_{|z_\nu|<1}(z-z_\nu)\prod_{|z_\mu|\geq 1}(z-z_\mu),\\ \omega_n^*(z) &= c\prod_{|z_\nu|<1}(1-z\bar{z}_\nu)\prod_{|z_\mu|\geq 1}(z-z_\mu).\end{aligned}$$

Then by *Example 43* in *Gy. Pólya, and G. Szegő, Problems and Theorems in Analysis, Springer, Berlin, 1976, vol. 2, p. 82*, $|\omega_n(z)|=|\omega_n^*(z)|$ for $|z|=1$ (this is evident), and

$$\frac{\omega_n(0)}{\omega_n^*(0)}=\prod_{|z_\nu|<1}|z_\nu|\leq 1,$$

whence

$$|\omega_n^*(0)| \geq |\omega_n(0)| = 1,$$

and by *Example 53* in the same *vol. 2, p. 84*,

$$\frac{1}{2\pi}\int_0^{2\pi} \log g(\vartheta)\, d\vartheta = \log |\omega_n^*(0)|^2 \geq \log 1 = 0.$$

On the other hand, $g(\vartheta) \leq M^2$ for all ϑ, and for $|\vartheta| \leq (\ell\pi/2n)$

$$g(\vartheta) = |\omega_{2n}(e^{i\vartheta})| \leq M^2 \left(\frac{c_1 n}{\ell}|z-1|\right)^{2\ell} \leq M^2 \left(\frac{c_1 n}{\ell}\vartheta\right)^{2\ell},$$

so in this interval

$$\log g(\vartheta) \leq 2\log M + 2\ell \log \frac{c_1 n}{\ell}\vartheta.$$

We use this relation only for $|\vartheta| \leq (\ell/nc_1)\,(< (\ell/n)\cdot(\pi/2))$. At the remaining points, we apply $\log g(\vartheta) \leq 2\log M$ to obtain

$$\begin{aligned} 0 \leq \int_0^{2\pi} \log g(\vartheta)\, d\vartheta &\leq 2\pi \cdot 2\log M + 2\cdot 2\ell \int_0^{\frac{\ell}{nc_1}} \log\left(\frac{c_1 n}{\ell}\vartheta\right) d\vartheta \\ &= 2\pi\cdot 2\log M + 2\cdot 2\ell \cdot \frac{\ell}{c_1 n}\int_0^1 \log\tau\, d\tau = 4\pi\log M + 4\frac{\ell^2}{c_1 n}(-1), \end{aligned}$$

that is,

$$\begin{aligned} 4\frac{\ell^2}{c_1 n} &\leq 4\pi \log M, \\ \ell^2 &\leq \pi c_1 \log M \cdot n, \\ \ell &\leq \sqrt{\pi c_1}\sqrt{\log M}\sqrt{n} = c(M)\sqrt{n}. \end{aligned}$$

The proof is complete. □

Remark. If n is large, then we may choose $c_1 = 4e/\pi + \varepsilon$.

Problem F.17. *Let $f(x, y, z)$ be a nonnegative harmonic function in the unit ball of $\mathbb{R}^3$ for which the inequality $f(x_0, 0, 0) \leq \varepsilon^2$ holds for some $0 \leq x_0 < 1$ and $0 < \varepsilon < (1-x_0)^2$. Prove that $f(x, y, z) \leq \varepsilon$ in the ball with center at the origin and radius $(1 - 3\varepsilon^{1/4})$.*

Solution. We regard f only in the interior of the unit ball. We write x_0 instead of $(x_0, 0, 0)$ and x instead of (x, y, z), and let $|x|$ denote the length of the vector (x, y, z) in $\mathbb{R}^3$. Let $0 \leq A < 1$, $0 \leq B < 1$, $\max(A, B) < R < 1$, and suppose that $|x| \leq A$ and $|x_0| \leq B$. By Poisson's formula (see, for example, the reference in Remark 1 later) the values of f inside the sphere

$S(0,R)$ with center at the origin and of radius R are given by the values on $S(0,R)$:

$$f(x) = \frac{R^2 - |x|^2}{4\pi R} \int_{S(0,R)} \frac{f(\xi)}{|x-\xi|^3} dS_\xi$$
$$= \frac{R^2 - |x|^2}{4\pi R} \int_{S(0,R)} \left(\frac{|x_0 - \xi|}{|x-\xi|}\right)^3 \cdot \frac{f(\xi)}{|x_0-\xi|^3} dS_\xi,$$

where S_ξ denotes the surface element on $S(0,R)$.

Since $|x_0 - \zeta| \le R + |x_0|$, $|x - \zeta| \ge R - |x|$, we obtain from the nonnegativity of f that

$$f(x) \le \left(\frac{R+|x_0|}{R-|x|}\right)^3 \cdot \frac{R^2-|x|^2}{4\pi R} \int_{S(0,R)} \frac{f(\xi)}{|x_0-\xi|^3} dS_\xi$$
$$= \left(\frac{R+|x_0|}{R-|x|}\right)^3 \cdot \frac{R^2-|x|^2}{R^2-|x_0|^2} f(x_0)$$
$$= \frac{(R+|x_0|)^2}{R-|x_0|} \cdot \frac{R+|x|}{(R-|x|)^2} f(x_0) \le f(x_0) \frac{(R+B)^2}{R-B} \frac{R+A}{(R-A)^2}.$$

This holds for all $R < 1$, hence with the notations $1-A = \alpha$ and $1-B = \beta$ we get by letting R tend to $1-0$,

$$f(x) \le f(x_0) \frac{(1+B)^2}{1-B} \frac{1+A}{(1-A)^2} = f(x_0) \frac{(2-\beta)^2(2-\alpha)}{\alpha^2\beta} \le f(x_0) \frac{8}{\alpha^2\beta}.$$

Thus, for $\alpha^2\beta \ge 8\varepsilon$ we have $f(x) \le (1/\varepsilon) \cdot f(x_0)$. In the problem $\varepsilon < (1-x_0)^2$, hence $|x_0| < 1 - \varepsilon^{1/2}$ and $|x| < 1 - 3\varepsilon^{1/4}$, so we can choose $\alpha = 3\varepsilon^{1/4}$ and $\beta = \varepsilon^{1/2}$, because then $\alpha^2\beta = 9\varepsilon > 8\varepsilon$. Then from the preceding estimate

$$f(x) \le \frac{1}{\varepsilon} \cdot \varepsilon^2 = \varepsilon. \quad \square$$

Remarks.

1. The above solution is virtually the same as the usual proof for Harnack's inequality (see, for example, Theorem 1.18 in the book *W. K. Hayman, and P. B. Kennedy, Subharmonic Functions, Math. Soc. Monographs, 9, London, Academic Press, 1976*) according to which if f is a nonnegative harmonic function in the unit ball of $\mathbb{R}^m$, then for $|\xi| < \rho < 1$

$$\frac{1-\rho}{(1+\rho)^{m-1}} f(0) \le f(\xi) \le \frac{1+\rho}{(1-\rho)^{m-1}} f(0).$$

2. The statement of the theorem is a certain strengthening of the maximum principle for harmonic functions. In fact, the maximum principle asserts that if f is nonnegative and harmonic in the unit ball, then $f(x_0) = 0$ implies that f vanishes identically. The problem yields that if f is nonnegative and harmonic in the unit ball and $f(x_0)$ is "small," then in a disk of radius "almost 1" around the origin f is also "small." In this form, the statement is valid on an arbitrary, bounded, connected domain (this version is also often called Harnack's theorem), which can be seen by a standard argument using chains of overlapping disks.

Problem F.18. *Verify that for every $x > 0$,*

$$\frac{\Gamma'(x+1)}{\Gamma(x+1)} > \log x.$$

Solution. It is known (see, for example, *1.7.(3)* in *Erdélyi et al., Bateman Manuscript Project, McGraw-Hill, New York, 1953*) that if $x > 0$, then $\Gamma(x) > 0$ and

$$\frac{\Gamma'(x)}{\Gamma(x)} = -C - \frac{1}{x} + \sum_{v=1}^{\infty}\left(\frac{1}{v} - \frac{1}{x+v}\right),$$

where C is Euler's constant. From this, we get

$$\left(\frac{\Gamma'(x)}{\Gamma(x)}\right)' = \sum_{v=0}^{\infty}\frac{1}{(x+v)^2} > 0;$$

that is,

$$\frac{\Gamma'(x)}{\Gamma(x)}$$

is strictly increasing. On the other hand,

$$\int_x^{x+1}\frac{\Gamma'(t)}{\Gamma(t)}dt = \log\Gamma(x+1) - \log\Gamma(x) = \log x,$$

where we used $\Gamma(x+1) = x\Gamma(x)$ for all x. Hence, by the mean value theorem,

$$\log x = \int_x^{x+1}\frac{\Gamma'(t)}{\Gamma(t)}dt = \frac{\Gamma'(\xi)}{\Gamma(\xi)} \qquad \text{for some } x < \xi < x+1,$$

and so

$$\frac{\Gamma'(x+1)}{\Gamma(x+1)} > \frac{\Gamma'(\xi)}{\Gamma(\xi)} = \log x\,. \quad \square$$

Problem F.19. *If f is a nonnegative, continuous, concave function on the closed interval $[0,1]$ such that $f(0) = 1$, then*

$$\int_0^1 xf(x)dx \le \frac{2}{3}\left[\int_0^1 f(x)dx\right]^2.$$

Solution. Let $A = \int_0^1 f(x)dx$ and $B = \int_0^1 xf(x)dx$. Integrating by parts, we obtain

$$B = A - \int_0^1\left(\int_0^x f(t)dt\right)dx. \tag{1}$$

Since f is concave, its curve lies above the chord joining the points $(0, f(0))$ and $(x, f(x))$, that is, for $0 \le t \le x$,

$$f(t) \ge \frac{f(x)-1}{x}t + 1, \tag{2}$$

where we have also taken into account that $f(0) = 1$. If we integrate (2), we obtain

$$\int_0^x f(t)dt \ge \frac{f(x)-1}{x} \cdot \frac{x^2}{2} + x = \frac{1}{2}xf(x) + \frac{1}{2}x\,. \tag{3}$$

Using (1) and (2), we arrive at

$$A - B = \int_0^1 \left(\int_0^x f(t)dt \right) \ge \frac{1}{2}B + \frac{1}{4},$$

that is,

$$B \le \frac{2}{3}\left(A - \frac{1}{4}\right). \tag{4}$$

From the inequality $0 \le (2A-1)^2 = 4A^2 - 4A + 1$, it follows that $A - 1/4 \le A^2$. Substituting this into (4), we finally arrive at

$$B \le \frac{2}{3}\left(A - \frac{1}{4}\right) \le \frac{2}{3}A^2, \tag{5}$$

and this is what we had to prove. □

Remarks.

1. The proof does not use the positivity of f.

2. The equality

$$B = \frac{2}{3}A^2 \tag{6}$$

occurs if and only if

$$f(x) = 1 - x. \tag{7}$$

Clearly, (7) implies (6). Conversely, if (6) holds, then by (5)

$$A = \frac{1}{2}. \tag{8}$$

Furthermore, because of the continuity of f, we must have equality in (2) for all $0 < t < x$. But then

$$\frac{f(t)-1}{t} = \frac{f(x)-1}{x} = c = \text{constant},$$

and so f is of the form $f(x) = cx + 1$. Using (8), we can conclude

$$\frac{1}{2} = A = \int_0^1 f(x)dx = \frac{c}{2} + 1,$$

from which $c = -1$ follows. Thus, $f(x) = 1 - x$, as we have claimed.

3. The statement is a special case of the following theorem that can be found in the book *I. M. Jaglom, and V. I. Boltjainskii, Convex Figures (in Russian), Gostehizdat, Moscow, 1951*: If the boundary of a convex domain contains a segment of length 1, then the distance of the weight point of the domain from that segment is at most 2/3 times the area.

Problem F.20. *Let f be a differentiable real function, and let M be a positive real number. Prove that if*

$$|f(x+t) - 2f(x) + f(x-t)| \leq M\,t^2 \quad \textit{for all } x \textit{ and } t,$$

then

$$|f'(x+t) - f'(x)| \leq M\,|t|.$$

Solution. We shall prove more; namely, we shall not assume the differentiability of f in advance. It will be enough to assume its continuity.

The inequality

$$|f(x+t) - 2f(x) + f(x-t)| \leq M \cdot t^2$$

means that for all x and t we have

$$f(x+t) - 2f(x) + f(x-t) \leq M \cdot t^2 \tag{1}$$

and

$$f(x+t) - 2f(x) + f(x-t) \geq -M \cdot t^2. \tag{2}$$

Let $h_1(x) = f(x) - (M/2)x^2$. For this function, we obtain from (1)

$$h_1(x+t) - 2h_1(x) + h_1(x-t) = f(x+t) - 2f(x) + f(x-t) - Mt^2 \leq 0.$$

This means that h_1 is a continuous function with nonnegative second-order symmetric differences. We know that then h_1 is concave. The same argument based on (2) yields the convexity of $h_1(x) = f(x) + (M/2)x^2$. It is well known that h_1 and h_2 then have one-sided derivatives for which

$$h_1^{(-)}(x) \geq h_1^{(+)}(x), \tag{3}$$

$$h_2^{(-)}(x) \leq h_2^{(+)}(x). \tag{4}$$

It follows that $f^{(-)}$ and $f^{(+)}$ also exist at every point and satisfy

$$f^{(-)}(x) \equiv h_1^{(-)}(x) + Mx, \tag{5}$$

$$f^{(+)}(x) \equiv h_1^{(+)}(x) + Mx, \tag{6}$$

$$f^{(-)}(x) \equiv h_2^{(-)}(x) - Mx, \tag{7}$$

$$f^{(+)}(x) \equiv h_2^{(+)}(x) - Mx. \tag{8}$$

By (3), (5), and (6) we have $f^{(-)} \geq f^{(+)}$, while (4), (7), and (8) yield $f^{(-)} \leq f^{(+)}$. Thus, $f^{(-)} = f^{(+)}$, which means that f is differentiable, and so both h_1 and h_2 are also differentiable.

Since the derivative of a differentiable concave (convex) function is decreasing (increasing), we can conclude for $x \le y$ that

$$h_1'(y) - h_1'(x) = f'(y) - f'(x) - M \cdot (y - x) \le 0 \tag{9}$$

and

$$h_2'(y) - h_2'(x) = f'(y) - f'(x) + M \cdot (y - x) \ge 0. \tag{10}$$

Equations (9) and (10) can be summarized as

$$|f'(y) - f'(x)| \le M|y - x|,$$

and this is equivalent to the statement. □

Remark. The statement can be generalized as follows. Let

$$\Delta_t^r f(x) = \sum_{k=0}^{r} (-1)^k \binom{r}{k} f\left(x + (\tfrac{r}{2} - k)t\right)$$

be the rth symmetric difference of f. If

$$|\Delta_t^r(x)| \le Mt^r$$

for all x and t, then

$$|\Delta_t^{r-j} f^{(j)}(x)| \le Mt^{r-j}, \qquad 1 \le j < r.$$

This can be derived from the formula

$$\Delta_t^k f(x) = \int_0^t \cdots \int_0^t f^{(k)}\left(x - \frac{k}{2}t + u_1 + \cdots + u_k\right) du_1 \cdots du_k\,,$$

provided the $(k-1)$th derivative of f is absolutely continuous.

Problem F.21. *Let $a < a' < b < b'$ be real numbers, and let the real function f be continuous on the interval $[a, b']$ and differentiable in its interior. Prove that there exist $c \in (a, b)$, $c' \in (a', b')$ such that*

$$f(b) - f(a) = f'(c)(b - a),$$

$$f(b') - f(a') = f'(c')(b' - a'),$$

and $c < c'$.

Solution 1. First, we verify the following statement.

Statement. Let p, q, q' be real numbers such that $p < q < q'$. Suppose that the real function f is continuous on the interval $[p, q']$ and differentiable in its interior. Then for every $r \in (p, q)$ for which

$$f(q) - f(p) = f'(r)(q - p)$$

holds, there exists an $r' \in (p, q')$ with

$$f(q') - f(p) = f'(r')(q' - p)$$

and $r' > r$.

Proof. To prove this, we may assume $f(p) = f(q') = 0$, since otherwise we can work with the function

$$f(x) - f(p) - (x - p)\frac{f(q') - f(p)}{q' - p}$$

If $f(q) = 0$, then the statement is obvious, so we can also assume that $f(q) > 0$. Let us choose the point $r \in (p, q)$ so that

$$f(q) = f(q) - f(p) = f'(r)(q - p).$$

If now $f(r) \leq 0$, then in the interval $[r, q)$ there is a point p' such that $f(p') = 0$. Therefore in the interval (p', q') there is an r' with $f(r') = 0$, and this r' satisfies the requirements.

We have to examine the case when $f(r) > 0$. Since

$$f'(r) = \frac{f(q) - f(p)}{q - p} > 0,$$

there is a $p' \in (r, q')$ with the property that the ratio

$$\frac{f(p') - f(r)}{p' - r}$$

is bigger than 0, that is, $f(p') > f(r)$. But then there is a point q'' in the interval (p', q') for which $f(q'') = f(r)$, and so for a suitable point r' of the interval (r, q'') we have $f(r') = 0$, and with this the statement is verified.

Now we turn to the solution of the problem. Let us choose the point $d \in (a', b)$ so that

$$f(b) - f(a') = f'(d)(b - a').$$

On applying the above statement, we get a $c' \in (a', b')$ with the property

$$f(b') - f(a') = f'(c')(b' - a')$$

and $c' > d$, and it similarly follows that there is a $c \in (a, b)$ for which

$$f(b) - f(a) = f'(c)(b - a)$$

and $d > c$. This proves the assertion of the problem. □

Remark. It can be seen from this proof that the assertion is valid in the cases $a \leq a' < b < b'$ or $a < a' < b \leq b'$ as well.

Solution 2. Let

$$D = \frac{f(b) - f(a)}{b - a}, \quad D' = \frac{f(b') - f(a')}{b' - a'},$$

$$r = \inf\{c \in (a,b) : f'(c) = D\}, \qquad r' = \sup\{c' \in (a',b') : f'(c') = D'\}.$$

Our aim is to prove that $r < r'$. Suppose, on the contrary, that $r \geq r'$. Then

$$a < a' < r' \leq r < b < b'.$$

Because of the Darboux property of derivative functions, $f'(x)$ lies on one side of D for all $x \in (a, r)$, and we may suppose that for all such x we have $f'(x) > D$. Likewise, in $(r, b')(\subseteq (r', b'))$ the derivative $f'(x)$ lies on one side of D'. We distinguish two cases.

Case 1. $f'(x) > D'$ for all $x \in (r, b')$. By the mean value theorem, we have

$$f(r) - f(a) > D(r - a), \quad f(b') - f(r) > D'(b' - r).$$

Since by definition

$$f(b) - f(a) = D(b - a), \quad f(b') - f(a') = D'(b' - a'),$$

we get by subtraction

$$f(b) - f(r) < D(b - r), \quad f(r) - f(a') < D'(r - a'). \tag{1}$$

On the other hand, again using the mean value theorem

$$f(b) - f(r) > D'(b - r), \quad f(r) - f(a') > D(r - a'). \tag{2}$$

Since $b - r > 0$ and $r - a' > 0$, we can conclude $D > D'$ from the left sides of (1) and (2), while from their right sides it follows that $D' > D$. This contradiction proves our claim in Case 1.

Case 2. $f'(x) < D'$ for all $x \in (r, b')$. Similarly as before, we get

$$D'(b - a') < f(b) - f(a') < D(b - a'),$$

and from here $f'(b) < D' < D < f'(a')$. Let $T \in (D', D)$ such that $T \neq f'(r)$. Because of the Darboux property, there must be a place $t \in (a', b)$ where $f'(t) = T$. This, however, can be neither in (a', r) (because there $f'(x) > D > T$) nor in (r, b) (because there $f'(x) < D' < T$). Neither can t be equal to r (because of the choice of T), and we have arrived again at a contradiction. This proves the claim in Case 2. $\square$

Problem F.22. *Let l_0, c, α, g be positive constants, and let $x(t)$ be the solution of the differential equation*

$$([l_0 + ct^\alpha]^2 x')' + g[l_0 + ct^\alpha] \sin x = 0, \quad t \geq 0, \quad -\frac{\pi}{2} < x < \frac{\pi}{2},$$

satisfying the initial conditions $x(t_0) = x_0$, $x'(t_0) = 0$. (This is the equation of the mathematical pendulum whose length changes according to the law $l = l_0 + ct^\alpha$.) Prove that $x(t)$ is defined on the interval $[t_0, \infty)$; furthermore, if $\alpha > 2$ then for every $x_0 \neq 0$ there exists a t_0 such that

$$\liminf_{t\to\infty} |x(t)| > 0.$$

Solution. With the notation $y = [l_0 + ct^\alpha]^2 x'$, equation (1) transforms to

$$\begin{aligned} x' &= \frac{y}{[l_0 + ct^\alpha]^2} \\ y' &= -g[l_0 + ct^\alpha] \sin x \qquad \left(t \geq 0,\ |x| < \frac{\pi}{2},\ y \in \mathbb{R}\right). \end{aligned}$$

Applying the Picard–Lindelöf theorem to the latter system, we get that the solution satisfying the given initial conditions exists in a right neighborhood of t_0. From the theorem of "continuation up to the boundary" (see the book *L. Sz. Pontriagin, Ordinary Differential Equations, Addison–Wesley, London, 1962*), we can conclude that if $x(t)$ exists on the interval $[t_0, T)$ but it cannot be continued beyond T, then we must have $|x(t)| \to \pi/2$ as $t \to T - 0$, since it follows from equation (1) that $x'(t)$ is bounded on bounded intervals.

Let us introduce the notation $l(t) = l_0 + ct^\alpha$ and consider the function

$$V(t, x, x') = \frac{l(t)}{g}(x')^2 + 2(1 - \cos x),$$

which is nothing else than the mechanical energy of the pendulum divided by $l(t)/2$. From equation (1), we get by differentiation that $v(t) = V(t, x(t), x'(t))$ is a nonincreasing function. Therefore,

$$2 > v(t_0) = 2(1 - \cos x_0) \geq v(t) \geq 2(1 - \cos x(t)),$$

which makes $|x(t)| \to \pi/2$, as $t \to T - 0$ is impossible. Thus, $x(t)$ is defined on the whole interval $[t_0, \infty)$.

To prove the second statement, let us start from the equation

$$x(t) = x_0 - g \int_{t_0}^{t} \frac{1}{[l_0 + cs^\alpha]^2} \int_{t_0}^{s} [l_0 + c\tau^\alpha] \sin x(\tau) d\tau ds,$$

which follows from (1) by integrating twice and taking into account the initial conditions. If $\alpha > 2$, then

$$\int_{t_0}^{\infty} \frac{1}{[l_0 + ct^\alpha]^2} \int_{t_0}^{t} [l_0 + cs^\alpha] ds dt < \infty,$$

and so for large t_0

$$|x(t)| \geq |x_0| - g \int_{t_0}^{t} \frac{1}{[l_0 + cs^\alpha]^2} \int_{t_0}^{s} [l_0 + c\tau^\alpha] d\tau ds > \frac{|x_0|}{2}$$

for all t on the interval $[t_0, \infty)$, and this is what we needed to prove. □

Remark. It can be proved that if $0 < \alpha \leq 2$, then for any solution of (1) that is defined on the whole interval $[t_0, \infty)$, we have

$$\lim_{t \to \infty} x(t) = 0$$

(see *L. Hatvani, On absence of asymptotic stability with respect to a part of the variables, J. Anal. Math. Mech., 40 (1976), 223–225*).

Problem F.23. *Let $f_1, f_2, \ldots, f_n$ be regular functions on a domain of the complex plane, linearly independent over the complex field. Prove that the functions $f_i \overline{f}_k$, $1 \leq i, k \leq n$, are also linearly independent.*

Solution. First, we prove the following lemma:

Lemma. Let f_i, g_i, $1 \leq i \leq n$ be regular functions in the domain D, and suppose that not every g_i vanishes identically. If $\sum_{i=1}^n f_i \overline{g}_i = 0$, then the f_i's are linearly dependent.

Proof. Indeed, for every $h \neq 0$ and $z \in D$

$$0 = \sum_{i=1}^{n} \frac{f_i(z+h)\overline{g_i(z+h)} - f_i(z)\overline{g_i(z)}}{h}$$

$$= \sum_{i=1}^{n} \overline{g_i(z+h)} \frac{f_i(z+h) - f_i(z)}{h} + \sum_{i=1}^{n} f_i(z) \left(\overline{\frac{g_i(z+h) - g_i(z)}{h}} \right) \cdot \frac{\overline{h}}{h}.$$

Letting h tend to zero first on the real and then on the imaginary axis, we obtain

$$\sum_{i=1}^{n} f_i'(z)\overline{g_i(z)} + \sum_{i=1}^{n} f_i(z)\overline{g_i'(z)} = 0$$

and

$$\sum_{i=1}^{n} f_i'(z)\overline{g_i(z)} - \sum_{i=1}^{n} f_i(z)\overline{g_i'(z)} = 0$$

for all $z \in D$, from which $\sum_{i=1}^n f_i' \overline{g}_i = 0$ follows. Repeating this argument we get $\sum_{i=1}^n f_i^{(m)} \overline{g}_i = 0$ for every m.

Without loss of generality, we can assume that $0 \in D$. Let

$$f_i(z) = \sum_{j=0}^{\infty} a_j^{(i)} z^j \quad \text{and} \quad g_i(z) = \sum_{j=0}^{\infty} b_j^{(i)} z^j.$$

These series converge in some disk $\{z : |z| < \delta\}$. We may also assume that $b_0^{(i)} \neq 0$ for at least one i (in the opposite case, we may factor out an appropriate power of z from every g_i). For every m, the function $\sum_{i=1}^{n} f_i^{(m)} \overline{g}_i$ takes the value

$$m! \sum_{i=1}^{n} a_m^{(i)} \overline{b}_0^{(i)} = 0$$

at $z = 0$, and so $\sum_{i=1}^{n} \overline{b}_0^{(i)} f_i = 0$ because every coefficient in its power series vanishes. This proves our lemma.

Now we prove that if, besides the assumption of the theorem, the functions g_j, $1 \leq j \leq m$ are also regular and linearly independent on D, then the same is true of the system $f_i \overline{g}_j$, $1 \leq i \leq n$, $1 \leq j \leq m$. This is obviously stronger than what the problem asked for.

In fact, suppose that $\sum_{i=1}^{n} \sum_{j=1}^{m} c_{ij} f_i \overline{g}_j = 0$. Then $\sum_{i=1}^{n} f_i \overline{(\sum_{j=1}^{m} \overline{c_{ij}} g_j)} = 0$, and so it follows from our lemma that $\sum_{j=1}^{m} \overline{c_{ij}} g_j = 0$ for all i. From this, we get $c_{ij} = 0$ for every i and j because of the assumed linear independence of the functions g_j, and this is exactly what we wanted to prove. □

Problem F.24. *Prove that the set of all linear combinations (with real coefficients) of the system of polynomials $\{x^n + x^{n^2}\}_{n=0}^{\infty}$ is dense in $C[0,1]$.*

Solution. Let P be the set of the linear combinations. Since the set of polynomials is dense in $C[0,1]$ (Weierstrass's theorem), it is enough to prove that every power x^n, $n = 0, 1, 2, \ldots$, can be uniformly approximated by polynomials from P.

For $n = 0, 1$, we have $x^n \in P$. If $n > 1$, then let

$$\begin{aligned} f_m^n(x) &= \sum_{i=0}^{m-1} (-1)^i \frac{m-i}{m} \left(x^{n^{2^i}} + x^{n^{2^{i+1}}}\right) \\ &= x^n + \frac{1}{m} \sum_{i=1}^{m} (-1)^{i+1} x^{n^{2^i}} := x^n + \frac{1}{m} \cdot A(x). \end{aligned}$$

Obviously, $f_m^n \in P$. The sum defining $A(x)$ consists of terms that alternate in sign and decrease in absolute value, and the first term lies in between 0 and 1. Hence, $0 \leq A(x) \leq 1$. As a consequence,

$$|x^n - f_m^n(x)| \leq \frac{1}{m}$$

for all $x \in [0,1]$, which proves our claim. □

Remark. The problem can be generalized as follows. Let $g(n)$ be a strictly increasing function defined on and taking values from the set of nonnegative integers. Then the linear combinations of the system $\{x^n + x^{g(n)}\}$ form a dense set in $C[0,1]$. The proof goes along the lines presented above.

Problem F.25. *Let f be a real function defined on the positive half-axis for which $f(xy) = xf(y) + yf(x)$ and $f(x+1) \le f(x)$ hold for every positive x and y. Show that if $f(1/2) = 1/2$, then*

$$f(x) + f(1-x) \ge -x\log_2 x - (1-x)\log_2(1-x)$$

for every $x \in (0,1)$.

Solution. Let $\varphi(n) = f(n)/n$. Then $\varphi(nm) = \varphi(n) + \varphi(m)$ and $(n+1)\varphi(n+1) \le n\varphi(n)$ hold for all $n, m \in \mathbb{N}$. Hence, there exists a $c \in \mathbb{R}$, for which $\varphi(n) = c\log_2 n$ for every $n \in \mathbb{N}$ (*see p. 19* in *J. Aczél, and Z. Daróczy, On Measures of Information and Their Characterizations, Academic Press, New York, 1975*). From here, because $f(1/2) = 1/2$ and

$$0 = f(1) = \frac{1}{2}f(2) + 2f\left(\frac{1}{2}\right) = \frac{1}{2}(f(2)+2),$$

we get

$$f(n) = -n\cdot\log_2 n \quad (n \in \mathbb{N}).$$

Since for every positive x and y

$$f(x+y) - f(x) - f(y) = y\left[f\left(\frac{x}{y}+1\right) - f\left(\frac{x}{y}\right)\right] \le 0,$$

we have

$$f(x+y) \le f(x) + f(y).$$

Let $n \in \mathbb{N}$ and $0 < x < 1$. Then using what we have obtained so far, we can write

$$\begin{aligned}
0 &= f(1) = f[(x+1-x)^n] \\
&= f\left[\sum_{k=0}^{n}\binom{n}{k}x^k(1-x)^{n-k}\right] \le \sum_{k=0}^{n} f\left[\binom{n}{k}x^k(1-x)^{n-k}\right] \\
&= -\sum_{k=0}^{n}\binom{n}{k}\log_2\binom{n}{k}\cdot x^k(1-x)^{n-k} + \sum_{k=0}^{n}\binom{n}{k}f(x^k(1-x)^{n-k}) \\
&= I + II.
\end{aligned}$$

From the functional equation on f, we obtain

$$f(x^k) = kx^{k-1}f(x) \qquad \text{and} \qquad f((1-x)^k) = k(1-x)^{k-1}f(1-x),$$

hence the sum representing II can be seen to be equal to $nf(x)+nf(1-x)$, which yields

$$f(x) + f(1-x) \geq \frac{1}{n} \sum_{k=0}^{n} \binom{n}{k} x^k (1-x)^{n-k} \log_2 \binom{n}{k} =: S_n(x). \quad (1)$$

Now let

$$p_k^{(n)}(x) = \binom{n}{k} x^k (1-x)^{n-k}, \qquad k = 0, 1, \ldots, n.$$

Since these (for a fixed x) take their maximum for either $k = [nx]$ or $k = [nx] + 1$, we easily obtain from Stirling's formula for the factorials appearing in the binomial coefficients that $\lim_{n\to\infty} p_k^{(n)}(x) = 0$ uniformly in k. Using this and $\lim_{t\to 0+0} t \log_2 t = 0$, we get

$$\lim_{n\to\infty} \frac{1}{n} \sum_{k=0}^{n} p_k^{(n)}(x) \log_2 p_k^{(n)}(x) = 0. \quad (2)$$

On the other hand,

$$\frac{1}{n} \sum_{k=0}^{n} p_k^{(n)}(x) \log_2 p_k^{(n)}(x)$$

$$= \frac{1}{n} \sum_{k=0}^{n} \binom{n}{k} x^k (1-x)^{n-k} \cdot \left[\log_2 \binom{n}{k} + k \log_2 x + (n-k) \log_2 (1-x)\right]$$
$$= S_n(x) + x \log_2 x + (1-x) \log_2 (x-1),$$

from where, with the aid of (2), we can conclude that

$$\lim_{n\to\infty} S_n(x) = -x \log_2 x - (1-x) \log_2 (1-x).$$

Taking (1) into account, we get the statement. □

Problem F.26. *Let G be a locally compact solvable group, let $c_1, \ldots, c_n$ be complex numbers, and assume that the complex-valued functions f and g on G satisfy*

$$\sum_{k=1}^{n} c_k f(xy^k) = f(x) g(y) \quad \text{for all } x, y \in G.$$

Prove that if f is a bounded function and

$$\inf_{x\in G} Re\, f(x)\chi(x) > 0$$

for some continuous (complex) character χ of G, then g is continuous.

We will give two solutions. Neither of them will use the local compactness, and the second solution will not use the solvability of G either. Hence, the claim is true on any topological group.

Solution 1. It is known (see for, example, *F. R. Greenleaf, Invariant Means on Topological Groups, Van Nostrand, Princeton, 1969*) that because of the solvability of G there exists a right-invariant mean m on the space of complex-valued, bounded functions on G. That is, m has the following properties: m is (complex) linear; $m(1) = 1$; $m(\overline{f}) = \overline{m(f)}$; if $f \geq 0$, then $m(f) \geq 0$; furthermore, $m(f_x) = m(f)$ for every bounded function f defined on G, where $f_x(t) = f(tx)$, $x, t \in G$.

Let f, g, and χ be as in the problem. Since $\operatorname{Re} f\chi \geq \inf_{x\in G} \operatorname{Re} f(x)\chi(x) > 0$, we have

$$m(\operatorname{Re} f\chi) \geq m(\inf_{x\in G} \operatorname{Re} f(x)\chi(x)) = \inf_{x\in G} \operatorname{Re} f(x)\chi(x) > 0,$$

and from this

$$m(f\chi) = m(\operatorname{Re} f\chi) + im(\operatorname{Im} f\chi) \neq 0.$$

Now let $x, y \in G$ be arbitrary, and let us multiply the equality

$$\sum_{k=1}^{n} c_k f(xy^k) = f(x)g(y)$$

by $\chi(x)$:

$$\sum_{k=1}^{n} c_k f(xy^k)\chi(x) = f(x)\chi(x)g(y).$$

On the left-hand side, we make use of the identity

$$\chi(x) = \chi(xy^k)\chi(y^{-k}) = \chi(xy^k)\overline{\chi}^k(y),$$

by which we obtain

$$\sum_{k=1}^{n} c_k f(xy^k)\chi(xy^k)\overline{\chi}^k(y) = f(x)\chi(x)g(y). \tag{1}$$

This holds for all $x, y \in G$. Apply now to both sides as functions of x the mean m, and make use of

$$m_x(f(xy^k)\chi(xy^k)) = m(f\chi) \neq 0$$

(here m_x denotes that the argument has to be considered as a function of x). It follows that

$$\sum_{k=1}^{n} c_k\overline{\chi}^k(y) = g(y), \tag{2}$$

from which the continuity of g is obvious. $\square$

Remark. One can prove with the same method the generalization that one obtains by allowing different functions f_k on the left of the assumed equality.

Solution 2. In the second solution, we will not use the existence of the mean m; otherwise, the argument that follows is similar to the one above. We start from the identity

$$\sum_{k=1}^{n} c_k f(xy^k)\chi(xy^k)\overline{\chi}^k(y) = f(x)\chi(x)g(y)$$

that we obtained in the first solution in (1). Applying this with $x = y^{-1}, \ldots, x = y^{-m}$ and adding the resulting equalities together, we obtain with

$$S_m(y) = \sum_{j=1}^{m} f(xy^{-j})\chi(xy^{-j})$$

the identity

$$\sum_{k=1}^{n} c_k \left(S_m(y) + \left(\sum_{j=0}^{k-1} - \sum_{j=m-k+1}^{m} \right) f(xy^j)\chi(xy^j) \right) \overline{\chi}^k(y) = S_m(y)g(y). \tag{3}$$

By the assumption,

$$\operatorname{Re} S_m(y) \geq mc \to \infty \qquad \text{as} \quad m \to \infty,$$

where c denotes a positive lower bound on the real parts of the products $f(x)\chi(x)$. Hence, the numbers $|S_m(y)|$ tend to infinity as $m \to \infty$. Now if we divide (3) by $S_m(y)$ and let l tend to infinity, then we again arrive at (2) in view of the boundedness of f. This completes the proof. □

Problem F.27. *Suppose that the components of the vector* $\mathbf{u} = (u_0, \ldots, u_n)$ *are real functions defined on the closed interval* $[a, b]$ *with the property that every nontrivial linear combination of them has at most* n *zeros in* $[a, b]$. *Prove that if* σ *is an increasing function on* $[a, b]$ *and the rank of the operator*

$$A(f) = \int_a^b \mathbf{u}(x) f(x) d\sigma(x), \quad f \in C[a, b],$$

is $r \leq n$, *then* σ *has exactly* r *points of increase.*

Solution. By the assumption that there is a $0 \neq c \in \mathbb{R}^{n+1}$ orthogonal onto the range of A, that is, $(c, A(f)) = 0$ for every $f \in C[a, b]$. This means that

$$\int_a^b (c, u(t)) f(t) d\sigma(t) = 0$$

holds for every f continuous on $[a, b]$. In particular,

$$\int_a^b (c, u(t))^2 d\sigma(t) = 0,$$

from which it follows that every point of increase of σ is a zero of $(c, u(t))$. Since $c \neq 0$, the assumption in the problem implies that σ has at most n points of increase. Let these be $x_1 < \cdots < x_m$, $m \leq n$, and let the corresponding jumps be $\Delta\sigma_1, \ldots, \Delta\sigma_m$. Then

$$A(f) = \sum_{i=1}^{m} u(x_i) f(x_i) \Delta\sigma_i.$$

Because of the assumption made on u, the vectors $u(x_1), \ldots, u(x_m)$ are easily seen to be linearly independent — furthermore $\Delta\sigma_i > 0$ — for every $i = 1, \ldots, m$. Hence the range of A is the subspace spanned by the vectors $u(x_1), \ldots, u(x_m)$. Then its rank is m, and this was to be proved. □

Problem F.28. *Let $\mathbb{Q}$ and $\mathbb{R}$ be the set of rational numbers and the set of real numbers, respectively, and let $f : \mathbb{Q} \to \mathbb{R}$ be a function with the following property. For every $h \in \mathbb{Q}$, $x_0 \in \mathbb{R}$,*

$$f(x + h) - f(x) \to 0$$

as $x \in \mathbb{Q}$ tends to x_0. Does it follow that f is bounded on some interval?

Solution. The answer is no. Consider the following function:

$$f\left(\frac{p}{q}\right) = \log\log 2q,$$

where p/q is a rational number written in the form where the denominator is positive and the fraction cannot be simplified. This is not bounded on any interval. On the other hand, if $x = p/q$ and $h = k/m$, then $x + h = (pm + qk)/qm$, where the last fraction may be simplified. In any case, $f(x + h) \leq \log\log 2qm$, and so

$$f(x + h) - f(x) \leq \log \frac{\log 2q + \log m}{\log 2q}.$$

Now if $x \to x_0$, then $q \to \infty$, and so

$$\limsup_{x \to x_0} f(x + h) - f(x) \leq 0.$$

On applying this with h and x_0 replaced by $-h$ and $x_0 + h$, respectively, we obtain the opposite inequality,

$$\liminf_{x \to x_0} f(x + h) - f(x) \geq 0.$$

Thus, this f is a counterexample. □

Problem F.29. *Suppose that the function $g : (0,1) \to \mathbb{R}$ can be uniformly approximated by polynomials with nonnegative coefficients. Prove that g must be analytic. Is the statement also true for the interval $(-1,0)$ instead of $(0,1)$?*

Solution. Let $p_1, p_2, \ldots$ be polynomials with nonnegative coefficients, and assume that $\lim_{n\to\infty} p_n(x) = g(x)$ pointwise on $(0,1)$. We shall prove that g is analytic. This is stronger than what the problem asked for, since we will not use the assumption that the limit is uniform.

We claim that the polynomials p_n are uniformly bounded on every disc $\{z : |z| \le \rho\}$ for every $\rho < 1$. In fact, let $p_n(z) = \sum a_k z^k$. Then for $|z| \le \rho$,

$$|p_n(z)| \le \sum a_k |z|^k \le \sum a_k \rho^k = p_n(\rho) < K = K(\rho),$$

since the sequence $\{p_n(\rho)\}$ is convergent, and hence it is bounded.

We also know that the sequence $\{p_n\}$ is convergent on a segment of the unit disk, hence we can invoke Vitaly's theorem to conclude that $\{p_n\}$ is convergent inside the unit disc and that the convergence is uniform on every compact subset of the open unit disk. This implies that $G(z) = \lim_{n\to\infty} p_n(z) = g(z)$ is analytic in the unit disk, and so its restriction to $(0,1)$ also has this property. This proves the first part of the problem.

The answer for the second part is negative. In fact, we shall prove much more, namely that the following result holds:

Result. A continuous function $g : [-1,0] \to \mathbb{R}$ can be uniformly approximated by polynomials with nonnegative coefficients if and only if $g(0) \ge 0$.

The necessity of the condition is obvious, hence we only have to deal with its sufficiency. Let G be the set of all continuous functions on $[-1,0]$ that are the uniform limit of some sequence of polynomials with nonnegative coefficients. Obviously, G is closed for addition, multiplication and forming uniform limits; furthermore, G contains every polynomial with nonnegative coefficients.

First, we prove that the function $-x$ is in G. This immediately follows from the fact that the polynomials $x(1+x)^n - x$ have nonnegative coefficients and $|x(1+x)^n| < (1/n)$ for every $-1 \le x \le 0$ (this can be easily verified by differentiation). However, then $-x^n = (-x)x^{n-1}$ also belongs to G for every $n \ge 1$, and so G contains every polynomial whose constant term is nonnegative.

Now let g be an arbitrary continuous function on $[-1,0]$ with $g(0) \ge 0$. By Weierstrass's theorem, there is a sequence of polynomials p_n converging uniformly to f on $[-1,0]$. But then the same is true of the sequence with terms

$$P_n(x) = p_n(x) - p_n(0) + g(0),$$

and these polynomials are from G. Hence, g also belongs to G, which proves the sufficiency part in our claim. $\square$

Problem F.30. *Prove that if a_i $(i = 1, 2, 3, 4)$ are positive constants, $a_2 - a_4 > 2$, and $a_1 a_3 - a_2 > 2$, then the solution $(x(t), y(t))$ of the system of differential equations*

$$\dot{x} = a_1 - a_2 x + a_3 xy,$$
$$\dot{y} = a_4 x - y - a_3 xy \qquad (x, y \in \mathbb{R})$$

with the initial conditions $x(0) = 0$, $y(0) \geq a_1$ is such that the function $x(t)$ has exactly one strict local maximum on the interval $[0, \infty)$.

Solution. x strictly increases in a neighborhood of $t = 0$, hence it is enough to prove that $\dot{x}$ changes sign exactly once.

Let us draw the pieces of the hyperbolas $\dot{x} = 0$ and $\dot{y} = 0$ lying in the first quadrant of the phase plane (x, y) (see Figure F.1):

$$\dot{x} = 0 : a_3 x \left(y - \frac{a_2}{a_3} \right) + a_1 = 0$$
$$\dot{y} = 0 : -a_3 \left(x + \frac{1}{a_3} \right) \left(y - \frac{a_4}{a_3} \right) - \frac{a_4}{a_3} = 0.$$

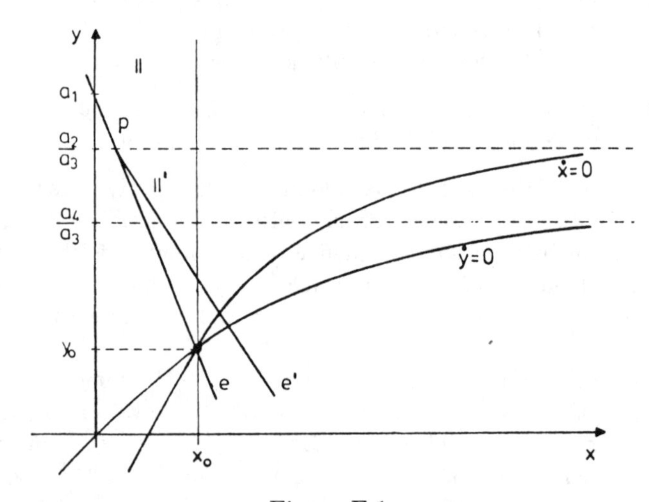

Figure F.1.

Denote (x_0, y_0) the intersection of these. The trajectory of the solution cannot leave the region I bounded by the curves $\dot{x} = 0$, $\dot{y} = 0$, $y \geq y_0$. Therefore, it is enough to prove that $y(t) > y_0$, $t \geq 0$, and the trajectory intersects the curve $\dot{x} = 0$. If the trajectory intersecting the half-line $x = x_0$, $y > y_0$ leaves the region II bounded by the lines $x = 0$, $x = x_0$, $e = (a_4 - a_2)x - y + a_1 = 0$, then it must intersect the curve $\dot{x} = 0$, namely in the opposite case

$$y(t) \downarrow z_1, \ x(t) + y(t) \downarrow \text{const} > 0, \quad \text{and so} \quad x(t) \nearrow x_1 \quad (t \to \infty).$$

From this, it follows that $x_1 = x_0$ and $y_1 = y_0$ because the point (x_1, y_1) must lie on the curves $\dot{x} = 0$, $\dot{y} = 0$. But this is a contradiction for $x(t) > x_0$ for large t.

Next, we show that the trajectory must indeed intersect the half-line $x = x_0$, $y > y_0$. Write the function $\dot{y}/\dot{x}$ in the form

$$\frac{\dot{y}}{\dot{x}} = \frac{-a_3xy + a_4x - y}{a_3xy - a_2x + a_1} = -1 + \frac{(a_4 - a_2)x - y + a_1}{a_3xy - a_2x + a_1}.$$

From this, it follows that the values of $\dot{y}/\dot{x}$ on the line e would be equal to -1, which is bigger than $a_4 - a_2$. Hence the trajectory does not intersect the line e. We can also see that the function $\dot{y}/\dot{x}$ is a decreasing function of x in the region $II' = \{(x, y) \in II : y < a_2/a_3\}$ along the lines $(a_4 - a_2)x - y + C = 0$ $(a_1 < C = \text{const})$. Since

$$\begin{aligned}\left.\frac{\dot{y}}{\dot{x}}\right|_{x = x_0} &= \frac{-a_3x_0y + a_4x_0 - y}{a_3x_0y - a_2x_0 + a_1} \\ &= \frac{-a_3x_0(y - y_0) - (y - y_0)}{a_3x_0(y - y_0)} = -\left(1 + \frac{1}{a_3x_0}\right),\end{aligned}$$

it follows that

$$\left.\frac{\dot{y}}{\dot{x}}\right|_{(x,y) \in II'} \geq -\left(1 + \frac{1}{a_3x_0}\right).$$

Now we show that

$$1 + \frac{1}{a_2x_0} < a_2 - a_4.$$

Simple calculation gives the formula

$$x_0 = \frac{(a_1a_3 - a_2) + \sqrt{(a_1a_3 - a_2)^2 + 4a_1a_3(a_2 - a_4)}}{2a_3(a_2 - a_4)},$$

from which $a_3x_0 \geq (a_1a_2 - a_2)/(a_2 - a_4)$ follows. Thus, to conclude the required inequality, it is enough to verify

$$a_1a_3 - a_2 > \frac{1}{1 - \frac{1}{a_2 - a_4}},$$

which is obvious from our assumptions.

Now let e' be a line through the point P, the tangent of which lies in between $-(a_2 - a_4)$ and $-(1 + 1/a_3x_0)$. Since

$$\left.\frac{\dot{y}}{\dot{x}}\right|_{(x,y) \in e' \cap II'} \geq -\left(1 + \frac{1}{a_3x_0}\right),$$

the trajectory coming from the region II cannot intersect e' from above. If it does not intersect the line $x = x_0$ either, then $(x(t), y(t)) \to (x_1, y_1) \neq (x_0, y_0)$, $(t \to \infty)$ would follow, which is a contradiction.

With this, the proof is done. $\square$

Problem F.31. *Let us call a continuous function $f : [a,b] \to \mathbb{R}^2$ reducible if it has a double arc (that is, if there are $a \le \alpha < \beta \le \gamma < \delta \le b$ such that there exists a strictly monotone and continuous $h : [\alpha, \beta] \to [\gamma, \delta]$ for which $f(t) = f(h(t))$ is satisfied for every $\alpha \le t \le \beta$); otherwise f is irreducible. Construct irreducible $f : [a,b] \to \mathbb{R}^2$ and $g : [c,d] \to \mathbb{R}^2$ such that $f([a,b]) = g([c,d])$ and*

(a) both f and g are rectifiable but their lengths are different;

(b) f is rectifiable but g is not.

Solution. In both a) and b) the curve f will be the following: Let $A \subseteq [0,1]$ be Cantor's ternary set, and let E be the unit interval $[0,1]$ considered as a curve on the plane. It is known that there is a continuous and increasing surjection $\varphi : A \to E$ (write $x \in A$ in the ternary form $x = 0.\varepsilon_1\varepsilon_2\dots$ where each ε_j is 0 or 2, and then let $\varphi(x)$ be defined by the binary form $0.(\varepsilon_1/2)(\varepsilon_2/2)\dots$). Let φ be the restriction of f to A. In the complementary intervals of A of lengths 3^{-n}, let f run through circles of radius 3^{-n} in a continuous fashion. It can be easily seen that f is continuous, irreducible, and rectifiable.

In case (a), let $g : [-1,1] \to \mathbb{R}^2$ be defined on $[-1,0]$ by $g(t) = E(-t)$, and on $[0,1]$ let g coincide with f. Then g is irreducible, but its length is longer than that of f by 1.

In case (b), let us divide the set of dyadic rational numbers of $[-1,1]$, from which there exists a monotone correspondence with the complementary intervals of A and hence with the circles that f runs through, into the sets $S_0, S_1, \dots$, each of which is dense in $[-1,1]$. Let us also divide the parameter interval of $g : [-1,1] \to \mathbb{R}^2$ into the consecutive intervals I_n of length 2^{-n}, $n = 0, 1, \dots$. On I_1, let g run from $E(0)$ to $E(1/2)$ running also through the circles corresponding to the points of $S_1 \cap I_1$. On I_2, let g run from $E(1/2)$ to $E(0)$ running also through the circles corresponding to the points of $S_2 \cap I_2$. On I_3, let g run from $E(0)$ to $E(1/3)$ and so on. Finally, on I_0 let us define g so that it goes from $E(1)$ to $E(0)$, and besides the circles corresponding to the points of S_0, it should also run through every other circle through which it has not yet run. Obviously, the continuity of g has to be checked only at the point 1, which can be easily done. Since each set S_i is dense, g is irreducible. But it is not rectifiable, because for every n we can write into it a polygon (with vertices $E(0)$, $E(1/2)$, $E(0)$, $E(1/3)$, ..., $E(1/n)$, $E(0)$) of length $2\,(1/2 + 1/3 + \cdots + 1/n)$. □

Remark. If we like, we can also assume that $[a,b] = [c,d]$.

Problem F.32. *Let $n \ge 2$ be a natural number and $p(x)$ a real polynomial of degree at most n for which*

$$\max_{-1 \le x \le 1} |p(x)| \le 1, \quad p(-1) = p(1) = 0.$$

Prove that then

$$|p'(x)| \le \frac{n \cos\frac{\pi}{2n}}{\sqrt{1 - x^2 \cos^2\frac{\pi}{2n}}} \qquad \left(-\frac{1}{\cos\frac{\pi}{2n}} < x < \frac{1}{\cos\frac{\pi}{2n}}\right).$$

Solution. Let $c = \cos(\pi/2n)$. It is enough to prove that for $|x| \le 1/c$ we have $|p(x)| \le 1$, because then by applying the classical Bernstein inequality on the interval $[-1/c, 1/c]$, it follows that

$$|p'(x)| \le \frac{n}{\sqrt{\left(\frac{1}{c} - x\right)\left(\frac{1}{c} + x\right)}} = \frac{nc}{\sqrt{1 - c^2x^2}} \qquad \left(|x| < \frac{1}{c}\right).$$

Let τ, $|\tau| > 1$ be the number outside $[-1, 1]$ with the smallest absolute value for which $|p(\tau)| = 1$. We have to show that $\tau \ge 1/c$. If we apply a linear transformation, this amounts to the same as proving that if $|p(x)| \le 1$ for $-1 \le x \le 1$ and $|p(1)| = 1$, then for $c < x \le 1$ the polynomial $p(x)$ cannot vanish, that is, $p(x) \ne 0$.

Suppose that this is not true, and let $T_n(x) = \cos(n \arccos x)$ be the Chebyshev polynomial of degree n. We know that for some sequence $1 = x_0 > x_1 > \cdots > x_n = -1$, we have $T_n(x_i) = (-1)^i$. Furthermore, $T_n(c) = 0$, and T_n is positive to the right of c. We shall arrive at a contradiction by proving that the polynomial $p - T_n$ has $n + 1$ zeros. One zero is $x = 1$. If $i > 0$, then there is a zero in the interval $[x_i, x_{i+1}]$. If this happens to be one of the endpoints of this interval, then it is a double zero, and hence there are at least $n - 1$ zeros in $[-1, x_1]$ (counting the possible root x_1 only once). By our assumption, there is an $x_1 < x < 1$ for which $0 = p(x) < T_n(x)$, so $p - T_n$ must vanish in at least one point of $[x_1, x]$ (recall that $p(x_1) - T_n(x_1) \ge 0$). Thus, $p - T_n$ has at least $n + 1$ zeros. But since its degree is at most n, this can only happen if $p \equiv T_n$, which is not possible, because $p(x) < T_n(x)$. The obtained contradiction proves our claim. □

Remark. It can be proven that equality occurs if and only if $p(x) = \pm T_n(x/c)$.

Problem F.33. *Let f be a strictly increasing, continuous function mapping $I = [0, 1]$ onto itself. Prove that the following inequality holds for all pairs $x, y \in I$:*

$$1 - \cos(xy) \le \int_0^x f(t)\sin(tf(t))dt + \int_0^y f^{-1}(t)\sin(tf^{-1}(t))dt\,.$$

Solution. Let

$$\begin{aligned} T_1 &= \{(u, v) : 0 \le u \le x, \quad 0 \le v \le y\}, \\ T_2 &= \{(u, v) : 0 \le u \le x, \quad 0 \le v \le f(u)\}, \\ T_3 &= \{(u, v) : 0 \le v \le y, \quad 0 \le u < f^{-1}(v)\}. \end{aligned}$$

Then $T_1 \subset T_2 \cup T_3$, $T_2 \cap T_3 = \emptyset$, and each of these sets is a Borel set, hence for every nonnegative and Lebesgue-integrable function g on $[0, 1] \times [0, 1]$, we have

$$\int_0^y \int_0^x g(u, v)dudv \le \int_0^x \int_0^{f(u)} g(u, v)dvdu + \int_0^y \int_0^{f^{-1}(v)} g(u, v)dudv.$$

Thus, with the choice

$$g(x,y) = \frac{\partial^2(1-\cos(xy))}{\partial y \partial x} = \sin xy + xy\cos xy \geq 0,$$

we get

$$1-\cos(xy) \leq \int_0^x f(u)\sin(u\cdot f(u))du + \int_0^y f^{-1}(v)\sin(v\cdot f^{-1}(v))dv\,. \quad \square$$

Problem F.34. *Let U be a real normed space such that, for any finite-dimensional, real normed space X, U contains a subspace isometrically isomorphic to X. Prove that every (not necessarily closed) subspace V of U of finite codimension has the same property. (We call V of finite codimension if there exists a finite-dimensional subspace N of U such that $V+N=U$.)*

Solution. By applying induction, we may assume that V has codimension 1. Let W be a finite-dimensional normed space, and consider $W\oplus W$ with the norm $\|(w_1,w_2)\| = \max\{\|w_1\|,\|w_2\|\}$. Let $W_1\oplus W_2$ be the isometric isomorphic image of $W\oplus W$ in U. It is enough to show that $V\cap(W_1\oplus W_2)$ contains a subspace that is isometrically isomorphic to W.

If $V\cap(W_1\oplus W_2) = W_1\oplus W_2$, then we are done. If $V\cap(W_1\oplus W_2)$ has codimension 1 in $W_1\oplus W_2$, then there is an $0\neq f: W_1\oplus W_2\to\mathbb{R}$ linear functional with kernel $V\cap(W_1\oplus W_2)$. With the notations $f_1(w_1) = f(w_1\oplus 0)$, $f_2(w_2) = f(w_2\oplus 0)$ we get two linear functionals on W_1 and W_2, respectively, and

$$f(w_1\oplus w_2) = f_1(w_1) + f_2(w_2).$$

Without loss of generality, we can assume that $\|f_1\|\geq\|f_2\|$ and $0 < \|f_1\|$. Consider the linear functional

$$\lambda f_1 \to -\lambda\|f_2\|.$$

By the Hahn–Banach theorem, there is a $\varphi\in W_1^{**}$ for which $\|\varphi\| = \|f_2\|/\|f_1\|$ and $\varphi(f_1) = -\|f_2\|$. If $\|f_2\|\neq 0$, then let $B(g) = \varphi(g)\cdot f_2/\|f_2\|$ $(g\in W_1^*)$, while in the case $\|f_2\| = 0$ let $B\equiv 0$. This $B: W_1^*\to W_2^*$ is a linear mapping of norm at most 1. Since our spaces are of finite dimensions, we have $B = A^*$ for some linear operator $A: W_2\to W_1$.

Since $\|A\| = \|A^*\| = \|B\|\leq 1$, the mapping $w_2\to Aw_2\oplus w_2$ is an isometry of W_2 into $W_1\oplus W_2$. Furthermore, the range of this mapping is part of $V\cap(W_1\oplus W_2)$ because

$$f_1(Aw_2) + f_2(w_2) = (A^*f_1)(w_2) + f_2(w_2) = -f_2(w_2) + f_2(w_2) = 0.$$

Thus, we have found an isometry from W_2 into $V\cap(W_1\oplus W_2)$, by which we verified that $V\cap(W_1\oplus W_2)$ contains a subspace that is isometrically isomorphic to W. $\square$

Problem F.35. *For every positive α, natural number n, and at most αn points x_i, construct a trigonometric polynomial $P(x)$ of degree at most n for which*

$$P(x_i) \le 1, \quad \int_0^{2\pi} P(x)dx = 0, \quad \text{and} \quad \max P(x) > cn,$$

where the constant c depends only on α.

Solution. We shall construct a P of degree $2n$ instead of n, but this is the same as if we solved the problem with 2α instead of α.

Without loss of generality, we can assume that the midpoint of the largest contiguous interval determined by the points x_i is 0 (if this was not the case, then we would have to translate the trigonometric polynomial to be constructed below). This means that there is no x_i in the interval $(-\pi/\alpha n, \pi/\alpha n)$. Let

$$P_1(x) = \frac{1}{2} + \sum_{v=1}^{n} \cos vx = \frac{\sin\left(n+\frac{1}{2}\right)x}{2\sin\frac{x}{2}} = D_n(x)$$

be the nth Dirichlet kernel. For this $P_1(0) = n + (1/2)$. Furthermore, P_1 monotonically decreases on $(0, \pi/n)$, for $\pi \ge |x| > \pi/n$, its absolute value is smaller than $1/2\sin(\pi/2n) \sim n/\pi$, and at the point $\pi/\alpha n$ its value is

$$\frac{\sin\left(n+\frac{1}{2}\right)\frac{\pi}{\alpha n}}{2\sin\frac{\pi}{2\alpha n}} \sim n\frac{\sin\frac{\pi}{\alpha}}{\frac{\pi}{\alpha}},$$

where the coefficient of n is smaller than 1 and where $A \sim B$ means that the ratio A/B tends to 1 as $n \to \infty$. Hence, there exists a $c < 1$ such that $P_1(x) - cn$ is negative at every x_i but $P_1(0) - cn > n(1-c)$ is positive.

Let

$$P_2(x) = \frac{1}{n+1}\sum_{v=0}^{n} D_v(x) = \frac{2}{n+1}\left\{\frac{\sin\frac{1}{2}(n+1)x}{2\sin\frac{x}{2}}\right\}^2$$

be the Fejér kernel. We have

$$P_2(0) = \frac{n+1}{2} \quad \text{and} \quad \frac{1}{\pi}\int_0^{2\pi} P_2(x)dx = 1.$$

Therefore, if $P_3(x) = P_2(x)(P_1(x) - cn)$, then

$$P_3(0) > c'n^2, \qquad P_3(x_i) < 0,$$

and

$$\left|\frac{1}{2\pi}\int_0^{2\pi} P_3(x)dx\right| \le c''n\frac{1}{\pi}\int_0^{2\pi} P_2(x)dx = c''n.$$

Finally, let

$$P(x) = \left(P_3(x) - \frac{1}{2\pi}\int_0^{2\pi} P_3(x)dx\right) \cdot \frac{1}{c''n}.$$

Then

$$\int_0^{2\pi} P(x)dx = 0, \qquad P(x_i) \leq 1,$$

because of the previous estimate, and

$$P(0) > \frac{c'}{c''}n - 1. \quad \square$$

Problem F.36. *Let $f : \mathbb{R} \to \mathbb{R}$ be a twice differentiable, 2π-periodic even function. Prove that if*

$$f''(x) + f(x) = \frac{1}{f(x + 3\pi/2)}$$

holds for every x, then f is $\pi/2$-periodic.

Solution. Since $1/f(x + 3\pi/2)$ is defined and f is continuous for all x, f must have the same sign on $\mathbb{R}$. From the parity of f, we get $f''(-x) = f''(x)$. On applying the identity of the problem at $-x$, we get

$$f''(-x) + f(x) = \frac{1}{f(-x + 3\pi/2)},$$

therefore

$$f(x + \frac{3\pi}{2}) = f(-x + \frac{3\pi}{2}) = f(x - \frac{3\pi}{2})$$

holds for every x. Thus, f is 3π-periodic, and since it is also 2π-periodic, it follows that π is a period of f. Hence, the condition in the problem can be written in the form

$$f''(x) + f(x) = \frac{1}{f(x + \pi/2)}.$$

Let us consider the function $g(x) = f(x + \pi/2)$. This is also even:

$$g(-x) = f(-x + \frac{\pi}{2}) = f(x - \frac{\pi}{2}) = f(x + \frac{\pi}{2}) = g(x).$$

In view of $g'(x) = f'(x + \pi/2)$ and $g''(x) = f''(x + \pi/2)$, we can write

$$f''(x) + f(x) = \frac{1}{g(x)}, \tag{1}$$

$$g''(x) + g(x) = \frac{1}{f(x)}. \tag{2}$$

Multiplying here by g and f and subtracting the second equality from the first, we get

$$0 = f''g - fg'' = (f'g - fg')',$$

and so the function $c = f'g - fg'$ is constant. Furthermore, since the derivative of an even function is odd, we find that c must be odd, hence $c = 0$. Since g does not have a zero, we can conclude that $(f/g)' = c/g^2 = 0$, and therefore f/g is constant.

f is continuous and periodic, so it takes its maximum and minimum at some points x_1 and x_0. Hence $g(x_0) = f(x_0 + \pi/2) \geq f(x_0)$ and $g(x_1) = f(x_1 + \pi/2) \leq f(x_1)$. These inequalities and the fact that the ratio f/g is constant show that this constant is equal to 1, which means that $f(x) = g(x)$ for every x, and this is exactly what we had to prove. □

Problem F.37. *Let $g : \mathbb{R} \to \mathbb{R}$ be a continuous function such that $x+g(x)$ is strictly monotone (increasing or decreasing), and let $u : [0,\infty) \to \mathbb{R}$ be a bounded and continuous function such that*

$$u(t) + \int_{t-1}^{t} g(u(s))ds$$

is constant on $[1,\infty)$. Prove that the limit $\lim_{t\to\infty} u(t)$ exists.

Solution. It is enough to prove the existence of $\lim_{t\to\infty}(u(t) + g(u(t)))$ because $x + g(x)$ is continuous and strictly increasing.

Consider the function $v(t) = \int_{t-1}^{t} g(u(s))ds$ on the interval $[1,\infty)$. For this,

$$v'(t) = g(u(t)) - g(u(t-1))\,.$$

Since u is bounded, say $u : [0,\infty) \to [a,b]$, we get that v' is also bounded. Hence, v is uniformly continuous on $[0,\infty)$. But we have assumed that $u(t) + v(t)$ is constant, hence the uniform continuity of u also follows. On the other hand, g is uniformly continuous on $[a,b]$, which yields the uniform continuity of v' via the preceding formula.

Next we show that v' has limit zero at infinity. To this end, it is enough to show that the integral $\int_1^\infty (v'(s))^2 ds$ is finite, because the integrand is nonnegative and uniformly continuous. We can write

$$\begin{aligned}
\int_1^t (v'(s))^2 ds &= \int_1^t \Big(g(u(s)) - g(u(s-1))\Big)^2 ds \\
&= 2\int_1^t g(u(s))(g(u(s)) - g(u(s-1)))ds \\
&\quad - \int_1^t g(u(s))^2 ds + \int_1^t g(u(s-1))^2 ds \\
&= 2\int_1^t g(u(s))v'(s)ds - \int_{t-1}^t g(u(s))^2 ds + \int_0^1 g(u(s))^2 ds\,.
\end{aligned}$$

Since g is bounded, the second and third terms are bounded with a bound independent of t. The same is true of the first term, namely, using $v' = -u'$ (which follows from the assumption that $u(t) + v(t)$ is constant), we can get

$$2\int_1^t g(u(s))v'(s)ds = -2\int_1^t g(u(s))u'(s)ds = -2\int_{u(1)}^{u(t)} g(x)dx,$$

and g is bounded in $[a, b]$. With this, we have verified that the above integral is finite, furthermore

$$\lim_{t\to\infty} u'(t) = -\lim_{t\to\infty} v'(t) = 0.$$

Let us now notice that

$$u(t) + g(u(t)) = (u(t) + v(t)) + (g(u(t)) - \int_{t-1}^t g(u(s))ds)$$

$$= (u(t) + v(t)) + \int_{t-1}^t (g(u(t)) - g(u(s)))ds.$$

By our assumption, the first term is constant. For the second term, we get from the mean value theorem that for every $s \in [t-1, t)$ there is a $\chi \in (s, t)$ with

$$|u(t) - u(s)| = |(t-s)u'(\chi)| \le |u'(\chi)|.$$

Therefore,

$$\lim_{t\to\infty} \sup_{s\in[t-1,t]} |u(t) - u(s)| = 0.$$

Hence, the integrand in the preceding formula, and together with it also the second term, tends to zero as $t \to \infty$. Thus, the limit $\lim_{t\to\infty}(u(t)+g(u(t)))$ exists, and the proof is complete. □

Problem F.38. *Prove that if the function $f : \mathbb{R}^2 \to [0, 1]$ is continuous and its average on every circle of radius 1 equals the function value at the center of the circle, then f is constant.*

Solution. It is enough to prove that for every $1 > a > 0$ the function

$$g_a(z) = g(z) = \frac{1}{4a^2}\int_{-a}^a\int_{-a}^a f(z + x + iy)dxdy$$

is constant, since then we get the constancy of f for $a \to 0$. Note that g also satisfies the assumptions of the problem, namely by Fubini's theorem,

$$\begin{aligned}\frac{1}{2\pi}\int_0^{2\pi} g(z + e^{it})dt &= \frac{1}{2\pi}\int_0^{2\pi}\frac{1}{4a^2}\int_{-a}^a\int_{-a}^a f(z + e^{it} + x + iy)dxdydt \\ &= \frac{1}{4a^2}\int_{-a}^a\int_{-a}^a\frac{1}{2\pi}\int_0^{2x} f(z + x + iy + e^{it})dtdxdy \\ &= \frac{1}{4a^2}\int_{-a}^a\int_{-a}^a f(z + x + iy)dxdy = g(z).\end{aligned} \tag{1}$$

Obviously, g is uniformly continuous. Hence there is a positive function δ such that

$$\lim_{\varepsilon \to 0} \delta(\varepsilon) = 0,$$

and for every $z, z' \in \mathbb{R}^2 = \mathbb{C}$

$$|g(z) - g(z')| \leq \delta(|z - z'|). \tag{2}$$

Now let $\mathcal{G}$ be the set of all $g : \mathbb{R}^2 \to [0, 1]$ that satisfy the mean value property (1) and the smoothness property (2) (with the above δ, which we consider to be fixed). $\mathcal{G}$ consists of uniformly bounded and uniformly equicontinuous functions, hence $\mathcal{G}$ is compact with respect to the uniform norm. Thus, the functional $g(1) - g(0)$ attains its supremum α on $\mathcal{G}$ for some g_0. Since $\mathcal{G}$ is clearly translation and rotation invariant, we can conclude that for every g in $\mathcal{G}$ and for every z, z' with $|z - z'| = 1$ the inequality

$$|g(z) - g(z')| \leq \alpha \tag{3}$$

holds. But then, on applying the mean value property (1) for g_0, we get that

$$\alpha = g_0(1) - g_0(0) = \frac{1}{2\pi} \int_0^{2\pi} (g_0(1 + e^{it}) - g_0(e^{it}))dt \leq \frac{1}{2\pi} \int_0^{2\pi} \alpha dt = \alpha,$$

which, in view of the continuity of the functions involved, is only possible if we have equality everywhere under the integral sign, that is,

$$g_0(1 + e^{it}) - g_0(e^{it}) \equiv \alpha.$$

In particular, $g_0(2) - g_0(1) = \alpha$.

Applying the same procedure to this equality, we can conclude $g_0(3) - g_0(2) = \alpha$, and in general $g_0(k + 1) - g_0(k) = \alpha$ for every $k = 1, 2, \ldots$. Adding these together, we arrive at $g_0(n) - g_0(0) = n\alpha$, which, taking into account the boundedness of g_0, is only possible if $\alpha = 0$.

Return now to (3). This says that for every g in $\mathcal{G}$ we have $g(z) = g(z')$ provided $|z - z'| = 1$. But every two points on the plane can be joined by a chain of points with consecutive distance 1; hence, we can conclude that every function in $\mathcal{G}$ is constant. Thus, the function g_a from the beginning of the proof is also constant because it is a member of $\mathcal{G}$, and this proves our claim. □

Remarks.

1. The conclusion holds if we assume only the nonnegativity of f. This is a result of H. Shockey, and J. Deny.
2. The statement is a certain strenghtening of the fact that a bounded and harmonic function on the plane is constant; namely, the harmonic functions are exactly the functions that have the mean value property on every circle (not just on those with radius one).
3. The statement is the continuous variant of the following well-known problem, which can be solved along the lines discussed above: If we write numbers from $[0, 1]$ into the lattice points of the plane in such a way that every number equals the average of the four neighboring ones, then all the numbers are the same.

Problem F.39. *Let V be a finite-dimensional subspace of $C[0,1]$ such that every nonzero $f \in V$ attains positive value at some point. Prove that there exists a polynomial P that is strictly positive on $[0,1]$ and orthogonal to V, that is, for every $f \in V$,*

$$\int_0^1 f(x)P(x)dx = 0.$$

Solution. Let $f_1, \ldots, f_k$ be a basis in V, and let us define a mapping $f : [0,1] \to \mathbb{R}^k$ by $f(x) = (f_1(x), \ldots, f_k(x))$. If $\lambda = (\lambda_1, \ldots, \lambda_k)$, then

$$\lambda^T f(x) = \sum \lambda_i f_i(x) \in V,$$

and hence there is an $x \in [0,1]$ for which

$$\lambda^T f(x) > 0.$$

That is, if L is any half-space containing the origin, then $f(x)$ is an inner point of L for some x. Hence, the origin is contained in the convex hull of

$$\{f(x) | x \in [0,1]\}.$$

But then there exists an $\varepsilon > 0$ such that this set contains the ball with center at the origin and of radius ε. Let us consider the points of the form

$$(\varepsilon e_1, \ldots, \varepsilon e_k),$$

where $e_i = \pm 1$. For any choice of the values $e_1, \ldots, e_k$, there are numbers $c_1 \geq 0, \ldots, c_{k+1} \geq 0$, $\sum c_i = 1$, $x_1, \ldots, x_{k+1} \in [0,1]$ for which

$$\sum c_j f_i(x_j) = \varepsilon e_i.$$

Approximating first the distribution $\sum_j c_j \delta_{x_j}$ by positive functions, and then these functions by polynomials, we get the existence of a strictly positive polynomial p for which

$$\left| \sum c_j f_i(x_j) - \int_0^1 f_i(x)p(x)dx \right| < \frac{\varepsilon}{2}.$$

Now let $J \subset \{1, \ldots, k\}$. Let us choose the numbers e_j as follows:

$$e_i = 1 \quad \text{if} \quad i \in J; \qquad e_i = -1 \quad \text{if} \quad i \notin J,$$

and let p_J denote the polynomial p corresponding to this choice. Then

$$\int p_J f_i > \frac{\varepsilon}{2}, \quad i \in J,$$

and

$$\int p_J f_i < -\frac{\varepsilon}{2}, \quad i \notin J.$$

Hence, in every quadrant of the space $\mathbb{R}^k$, there is a point of the form

$$\left(\int p_J f_1, \ldots, \int p_J f_k\right).$$

Thus, the origin belongs to the convex hull of these points, that is, with some $d_J > 0$,

$$\int \left(\sum_J d_J p_J\right) f_i = 0, \quad i = 1, \ldots, k,$$

and at the same time the polynomial $\sum_J d_J p_J$ is strictly positive on $[0,1]$. □

Remark. For more information see *A. Pinkus, and V. Totik, One-sided L^1-approximation, Can. Bull. Math., 25 (1986), 84–90.*

Problem F.40. *Let $D = \{z \in \mathbb{C}\colon |z| < 1\}$ and $D = \{w \in \mathbb{C}\colon |w| = 1\}$. Prove that if, for a function $f : D \times B \to \mathbb{C}$, the equality*

$$f\left(\frac{az+b}{\bar{b}z+\bar{a}}, \frac{aw+b}{\bar{b}w+\bar{a}}\right) = f(z,w) + f\left(\frac{b}{\bar{a}}, \frac{aw+b}{\bar{b}w+\bar{a}}\right) \tag{1}$$

holds for all $z \in D$, $w \in B$, and $a, b \in \mathbb{C}$, $|a|^2 = 1 + |b|^2$, then there is a function $L :]0, \infty[\to \mathbb{C}$ satisfying

$$L(pq) = L(p) + L(q), \quad \text{for all} \quad p, q > 0,$$

such that f can be represented as

$$f(z,w) = L\Big(\frac{1-|z|^2}{|w-z|^2}\Big), \quad \text{for all} \quad z \in D, w \in B.$$

Solution. First, we prove that if f satisfies equality (1), then there exists a $\varphi : D \to \mathbb{C}$ such that

$$f(z,w) = \varphi(z\overline{w}), \quad (z,w) \in D \times B, \tag{2}$$

$$\varphi\left(\frac{t(1-s)+s(1-\overline{s})}{t\overline{s}(1-s)+1-\overline{s}}\right) = \varphi(t) + \varphi(s), \quad t, s \in D. \tag{3}$$

Let $a^2 = \overline{w}$ hold for $(z,w) \in D \times B$, $b = 0$, and $a \in \mathbb{C}$. Then, from equality (1) it follows that

$$f(z\overline{w}, 1) = f(z,w) + f(0,1). \tag{4}$$

By substituting $z = 0$, $w = 1$, we get $f(0,1) = 0$, and so, by (4), (2) holds for the following function $\varphi : D \to \mathbb{C}$:

$$\phi(z) = f(z,1), \qquad z \in D.$$

From (1) by (2), we get

$$\varphi\left(\frac{az+b}{\bar{b}z+\bar{a}} \cdot \frac{\overline{aw}+\bar{b}}{b\overline{w}+a}\right) = \varphi(z\overline{w}) + \varphi\left(\frac{b}{\bar{a}} \cdot \frac{\overline{aw}+\bar{b}}{b\overline{w}+a}\right),$$
$$(z,w) \in D \times B, \quad |a|^2 = 1 + |b|^2. \tag{5}$$

Let $t, s \in D$. If we substitute

$$a = \frac{1}{\sqrt{1-|s|^2}}, \quad b = as, \quad w = \frac{1-s}{1-\bar{s}}, \quad z = tw$$

for (5), then (3) follows from the fact $(\bar{a} \cdot \overline{w} + \bar{b})/(b\overline{w} + a) = 1$.

Let $A = \{u \in \mathbb{C} : \ \text{Re}\, u > 0\}$ and

$$\psi(u) = \varphi(\frac{u-1}{u+1}) \qquad u \in A. \tag{6}$$

Substituting for $u, v \in A$: $t = (u-1)/(u+1)$, $s = (v-1)/(v+1)$, from (3), we obtain

$$\psi(u\text{Re}\, v + i \ \text{Im}\, v) = \psi(u) + \psi(v), \qquad u, v \in A. \tag{7}$$

We are going to show that

$$\psi(u) = \psi(\text{Re}\ u), \quad u \in A. \tag{8}$$

Let $x \in (0, +\infty)$, $y \in \mathbb{R}$. Then from (7) it follows that

$$\psi(1+2iy) = \psi((1+iy)1+iy) = \psi(1+iy) + \psi(1+iy) = 2\psi(1+iy)$$

and

$$\begin{aligned}\psi(1+2iy) &= \psi(2) + \psi(1+2iy) - \psi(2) = \psi(2+2iy) - \psi(2)\\ &= \psi((1+iy)2) - \psi(2) = \psi(1+iy) + \psi(2) - \psi(2) = \psi(1+iy).\end{aligned}$$

Hence, $\psi(1+iy) = 0$, and from (7) we get

$$\psi(x+iy) = \psi(x \cdot 1 + iy) = \psi(x) + \psi(1+iy) = \psi(x).$$

Finally, let $L(p) = \psi(p)$, $p \in (0, +\infty)$. Then from (6), it obviously follows that

$$L(pq) = L(p) + L(q), \quad p, q \in (0, \infty).$$

On the other hand, from (2), (6), and (8), we obtain

$$\begin{aligned}f(z,w) = \varphi(z\overline{w}) &= \psi\Big(\frac{1+z\overline{w}}{1-z\overline{w}}\Big) = \psi\Big(\text{Re}\frac{1+z\overline{w}}{1-z\overline{w}}\Big)\\ &= \psi\Big(\frac{1-|z|^2}{|w-z|^2}\Big) = L\Big(\frac{1-|z|^2}{|w-z|^2}\Big),\end{aligned}$$

for all $(z,w) \in D \times B$. $\square$

Problem F.41. *Prove that the series $\sum_p c_p f(px)$, where the summation is over all primes, unconditionally converges in $L^2[0,1]$ for every 1-periodic function f whose restriction to $[0,1]$ is in $L^2[0,1]$ if and only if $\sum_p |c_p| < \infty$. (Unconditional convergence means convergence for all rearrangements.)*

Solution. Let $f_p(x) = f(px)$. Suppose first that the sum in question is finite. Since we have $\|f_p\| = \|f\|$, where $\|\cdot\|$ denotes the $L^2[0,1]$-norm, we get

$$\sum \|c_p f_p\| \leq \sum |c_p| \|f_p\| = \|f\| \sum |c_p| < \infty.$$

Hence the completeness of $L^2[0,1]$ implies the convergence of $\sum c_p f_p$ in any rearrangement.

Conversely, suppose that $\sum |c_p|$ is divergent. Let A_p denote the norm-preserving operation $f \to f_p$. If δ is an arbitrary number, there is a finite subset p_i, $i \in I$, of the primes such that $|\sum_{i \in I} c_{p_i}| > \delta$. Let $S_I = \sum_{i \in I} c_{p_i} A_{p_i}$ and $H = \{\prod_{i \in J} p_i : J \subset I\}$. If $n \in H$, then let $J_n = \{i \in I : p_i | n\}$, $\overline{n} = \prod \{p_i : i \in I \setminus J_n\}$, $v_n = \sum \{c_p : p \in J_n\}$.

Consider the functions $e^{2\pi int} \in L^2[0,1]$, $n \in H$. These form an orthonormal system, and $A_p(e^{2\pi int}) = e^{2\pi inpt}$. The function $\sum_{n \in H} e^{2\pi int}$ has norm $2^{|H|/2}$. Apply the operator S_I to this function. Then for $n \in H$, the coefficient of $e^{2\pi int}$ in the resulting function is

$$\sum_{p_i | n} c_{p_i}$$

if $lp_i = n$ for some $i \in I$, and 0 otherwise. Hence, the coefficient is exactly v_n. Thus,

$$\|S_I(\sum_{n \in H} e^{2\pi int})\|^2 \geq \sum_{n \in H} v_n^2 = \frac{1}{2} \sum_{n \in H} (v_n^2 + v_{\overline{n}}^2)$$

$$\geq \frac{1}{4} \sum_{n \in H} (v_n + v_{\overline{n}})^2 = 2^{|H|-2} (\sum_{n \in I} c_{p_i})^2,$$

that is,

$$\|S_I\| \geq \frac{1}{2} \delta.$$

Since here δ is arbitrary, we can form a sequence of operators S_{I_k} with some increasing sequence $I_1 \subset I_2 \subset \cdots$ of subsets of the natural numbers, the union of which contains every natural number such that the norms $\|S_{I_k}\|$ tend to infinity as $k \to \infty$. Hence by the Banach–Steinhaus theorem, there is an f for which the sequence $S_{I_k}(f)$, $k = 1, 2, \ldots$ is not convergent in $L^2[0,1]$, and this proves the necessity of the condition. $\square$

Problem F.42. *Let $a_0 = 0, a_1, \ldots, a_k$ and $b_0 = 0, b_1, \ldots, b_k$ be arbitrary real numbers.*

(i) Show that for all sufficiently large n there exist polynomials p_n of degree at most n for which

$$p_n^{(i)}(-1) = a_i, \quad p_n^{(i)}(1) = b_i, \qquad i = 0, 1, \ldots, k, \tag{1}$$

and

$$\max_{|x| \le 1} |p_n(x)| \le \frac{c}{n^2}, \tag{2}$$

where the constant c depends only on the numbers a_i, b_i.

(ii) Prove that, in general, (2) cannot be replaced by the relation

$$\lim_{n \to \infty} n^2 \cdot \max_{|x| \le 1} |p_n(x)| = 0. \tag{3}$$

Solution. Let us search p_n in the form

$$p_n(x) = \frac{1}{n^2} \sum_{s=0}^{2k+1} c_{s,n} T_{(2m+1)s}(x), \tag{4}$$

where

$$m = \left[\frac{n}{4k+2} - \frac{1}{2} \right] \quad (n > 4k - 1), \tag{5}$$

and $T_j(x) = \cos(j \arccos x)$ is the Chebyshev polynomial of degree j. Then the degree of p_n is at most n, and conditions (1) lead to the system of equations

$$\sum_{s=0}^{2k+1} c_{s,n} T^{(i)}_{(2m+1)s}(-1) = a_i n^2, \ \sum_{s=0}^{2k+1} c_{s,n} T^{(i)}_{(2m+1)s}(1) = b_i n^2, \ i = 0, \ldots, k. \tag{6}$$

We are going to prove that for large enough n this system can be uniquely solved.

First of all, because $T_j^{(i)}(-1) = (-1)^{i+j} T_j^{(i)}(1)$, we get from (6) by addition that

$$\sum_{s=0}^{k} c_{2s,n} T^{(i)}_{2(2m+1)s}(1) = \frac{(-1)^i a_i + b_i}{2} n^2, \quad i = 0, \ldots, k. \tag{7}$$

If we differentiate the well-known differential equation

$$(1 - x^2) T_j''(x) - x T_j'(x) + j^2 T_j(x) = 0$$

for the Chebyshev polynomials $(i-1)$ times and substitute $x = 1$, we arrive at

$$\begin{aligned} T_j^{(i)}(1) &= \frac{j^2 - (i-1)^2}{2i - 1} T_j^{(i-1)}(1) = \cdots = \frac{j^2(j^2 - 1) \cdots (j^2 - (i-1)^2)}{(2i-1)!!} \\ &= \frac{j^{2i}}{(2i-1)!!} + O(j^{2i-2}), \qquad i = 1, \ldots, k; j \to \infty, \end{aligned}$$

whence from the above equations and from (5) the equations in (7) take the form

$$\sum_{s=0}^{k} c_{2s,n}\{s^{2i}+O(n^{-2})\} = \frac{(-1)^i a_i + b_i}{2^{2i}(2m+1)^{2i}}(2i-1)!!n^2, \qquad i=0,\dots,k. \quad (8)$$

Now, if n tends to infinity, then in the limit this takes the form (recall that $a_0 = b_0 = 0$)

$$\sum_{s=0}^{k} c_{2s}s^{2i} = \frac{b_1 - a_1}{2}(2k+1)\delta_{i1}, \qquad i=0,\dots,k,$$

which has a unique solution, for its determinant is the Vandermonde determinant formed of the elements $1^2, 2^2, \dots, k^2$, and so it is different from zero. But then (8) has a unique solution for sufficiently large n, and $c_{2s,n} = O(1)$, $s = 0,\dots,k$. Similarly, it follows from (6) that $c_{2s+1,n} = O(1)$, $s=0,\dots,k$. Hence, (4) satisfies (2) whenever we choose c so large that $c \geq \sum_{s=0}^{2k+1} |c_{s,n}|$ is satisfied.

The fact that in general (3) cannot hold follows from Markov's inequality:

$$\max_{x\in[-1,1]} |p_n'| \leq n^2 \max_{x\in[-1,1]} |p_n|.$$

Indeed, using this we can see that (3) would imply

$$\lim_{n\to\infty} \max_{|x|\leq 1} |p_n'(x)| = 0,$$

which contradicts (1) unless $a_1 = 0$ and $b_1 = 0$. □

Problem F.43. *Let f and g be continuous real functions, and let $g \not\equiv 0$ be of compact support. Prove that there is a sequence of linear combinations of translates of g that converges to f uniformly on compact subsets of $\mathbb{R}$.*

Solution. If $\{g_n\}$ is a sequence of linear combinations of translates of g for which $|g_n(x) - f(x)| \leq 1/n$ for every x in the interval $[-n, n]$, then this sequence uniformly converges to f on every compact subset of the real line. Hence, it is enough to verify that any f can be arbitrarily well approximated on any finite closed interval I by linear combinations of translates of g.

Let us suppose for an indirect proof that this is not the case, and I is such an interval, that the linear hull of translates of g is not dense in $C(I)$. Then, by the Hahn–Banach theorem, there is a linear functional $L \in C^*(I)$ that vanishes on every translate of g. By the Riesz representation theorem, L can be identified by integration with respect to a signed measure μ with support in I. Let $h(x) = g(-x)$. Then

$$(h*d\mu)(t) = \int_{-\infty}^{\infty} h(t-x)d\mu(x) = \int_{-\infty}^{\infty} g(x-t)d\mu(x) = \int_{-n}^{n} g(x-t)d\mu(x) = 0.$$

Taking the Fourier transform here (which is possible because h and μ have compact support), we find $\mathcal{F}(h)\cdot\mathcal{F}(\mu)\equiv 0$. However, the functions $\mathcal{F}(h)$ and $\mathcal{F}(\mu)$ are analytic, hence one of them must be identically 0. But this is a contradiction since neither h nor μ is identically zero. This proves the statement of the problem. □

Problem F.44. *Let $x:[0,\infty)\to\mathbb{R}$ be a differentiable function satisfying the identity*

$$x'(t) = -2x(t)\sin^2 t + (2-|\cos t|+\cos t)\int_{t-1}^{t} x(s)\sin^2 s\,ds$$

on $[1,\infty)$. Prove that x is bounded on $[0,\infty)$ and that $\lim_{t\to\infty} x(t)=0$. Does the conclusion remain true for functions satisfying the identity

$$x'(t) = -2x(t)t + (2-|\cos t|+\cos t)\int_{t-1}^{t} x(s)s\,ds\,?$$

Solution. Let us consider the equation

$$x'(t) = -2a(t)x(t) + (2-|\cos t|+\cos t)\int_{t-1}^{t} a(s)x(s)ds,$$

where $a(t)\geq 0$ is a continuous function. If $x(t)$ is a solution, then let us estimate the upper right derivative $D^+|x(t)|$ of $|x(t)|$:

$$\begin{aligned}
D^+|x(t)| &\leq -2a(t)|x(t)| + (2-|\cos t|+\cos t)\int_{t-1}^{t} a(s)|x(s)|ds \\
&= -2a(t)|x(t)| + 2\int_{t-1}^{t} a(s)|x(s)|ds \\
&\quad - (|\cos t|-\cos t)\int_{t-1}^{t} a(s)|x(s)|ds \\
&= -2\left(\int_{-1}^{0} a(t)|x(t)|ds - \int_{-1}^{0} a(t+s)|x(t+s)|ds\right) \\
&\quad - (|\cos t|-\cos t)\int_{t-1}^{t} a(s)|x(s)|ds \\
&= -2\frac{d}{dt}\int_{-1}^{0}\int_{t+s}^{t} a(u)|x(u)|duds \\
&\quad -(|\cos t|-\cos t)\int_{t-1}^{t} a(s)|x(s)|ds.
\end{aligned}$$

Hence,

$$D^+\left(|x(t)|+2\int_{-1}^{0}\int_{t+s}^{t} a(u)|x(u)|du\,ds\right) \leq -(|\cos t|-\cos t)\int_{t-1}^{t} a(s)|x(s)|ds. \tag{1}$$

Since the right-hand side of (1) is nonpositive, we get that the function

$$|x(t)| + 2\int_{-1}^{0}\int_{t+s}^{t} a(u)|x(u)|duds \tag{2}$$

is decreasing, and so $|x(t)|$ is bounded.

For $k = 0, 1, \ldots$, let $H_k = [(2k+1)\pi - 1, (2k+1)\pi + 1]$. We show that

$$\max_{t\in H_k}\int_{t-1}^{t} a(s)|x(s)|ds \to 0 \quad (k \to \infty). \tag{3}$$

If we suppose that on the contrary, (3) does not hold, then there is an $\alpha > 0$ and a sequence $\{t_n\}$ such that $t_n \in \cup_{k=0}^{\infty} H_k$, $t_n \to \infty$, and

$$\int_{t_n-1}^{t_n} a(s)|x(s)|ds \geq \alpha.$$

Then at least one of the equalities

$$\int_{t_n-1}^{t_n-1/2} a(s)|x(s)|ds \geq \alpha/2, \quad \int_{t_n-1/2}^{t_n} a(s)|x(s)|ds \geq \alpha/2$$

holds. Since $|\cos t| - \cos t > 0$ on the interval $H_k = [(2k+1)\pi - 3/2, (2k+1)\pi + 3/2]$, there exists a $\beta > 0$ such that at every point of an interval of length $1/2$ and containing t_n, we have

$$(|\cos t| - \cos t)\int_{t-1}^{t} a(s)|x(s)|ds > \beta.$$

However, this and (1) imply

$$|x(t)| + 2\int_{-1}^{0}\int_{t+s}^{t} a(u)|x(u)|duds \to -\infty \qquad (t \to \infty),$$

which is impossible. Hence, (3) holds.

From (3) it follows that

$$\max_{t\in H_k}\int_{-1}^{0}\int_{t+s}^{t} a(u)|x(u)|duds \to 0 \qquad (k \to \infty). \tag{4}$$

Using (4) and the monotonicity and nonnegativity of the function in (2), we can see that to prove the existence of the limit of $x(t)$ at infinity it is enough to show that there is a sequence $\{t_n\}$ of points of the set $\cup_{k=0}^{\infty} H_k$ for which $t_n \to \infty$, $x(t_n) \to 0$ as $n \to \infty$. If there was not such a sequence, then there would be a $k_0 \in \mathbb{N}$ and a $\gamma > 0$ with the property that $|x(t)| > \gamma$ for every $t \in H_k$ and $k \geq k_0$. This, however, contradicts (3), whether $a(t) = \sin^2 t$ or $a(t) = t$.

This proves the claim of the problem, and we have also obtained that the same conclusion holds if we use the equation in the second half of the problem. □

Problem F.45. *Let $c > 0$, $c \neq 1$ be a real number, and for $x \in (0,1)$ let us define the function*

$$f(x) = \prod_{k=0}^{\infty} (1 + cx^{2^k}).$$

Prove that the limit

$$\lim_{x \to 1-0} \frac{f(x^3)}{f(x)}$$

does not exist.

Solution. We argue indirectly, hence let us assume that the above limit exists. In general, let S be the set of all positive real numbers q for which the limit

$$\lim_{x \to 1-0} \frac{f(x^q)}{f(x)} =: g(q)$$

exists. Thus, the indirect assumption is that $3 \in S$. First, we are going to show that this implies $S = \mathbb{R}_+$, and we also get an explicit representation on the function g.

Note that $2 \in S$ because

$$\lim_{x \to 1-0} \frac{f(x^2)}{f(x)} = \lim_{x \to 1-0} \frac{1}{1 + cx} = \frac{1}{1+c}. \tag{1}$$

Furthermore, by substitution it is easy to verify that S is a multiplicative subgroup of the real field, and

$$g(q_1)g(q_2) = g(q_1 q_2) \qquad \text{for} \quad q_1, q_2 \in S. \tag{2}$$

The numbers $2^m 3^l$, $m, l = 0, \pm 1, \ldots$ form a subgroup S_1 of S that is dense on the positive real line (use the fact that by the prime factorization theorem the number $\log 2 / \log 3$ is irrational, hence numbers of the form $m \log 2 + l \log 3$ form a dense set on $\mathbb{R}$).

Using the monotonicity of f, we can see that $g(q) \leq 1$ if $q \geq 1$ and $g(q) \geq 1$ if $q \leq 1$. This implies $g(q) = q^\gamma$ for $q \in S_1$ for some $\gamma < 0$. In fact, let $g(2) = a$, $g(3) = b$, and let us choose γ to satisfy $a = 2^\gamma$. If $b \neq 3^\gamma$, say $b < 3^\gamma$, then by choosing a number $q = 2^m 3^l$, $l < 0$, in the interval $[1, 2]$ such that $(b/3^\gamma)^l > 1$, we get a $q \in S_1$ with $q \geq 1$ and $g(q) > 1$, which is not possible. Hence, $g(q) = q^\gamma$ for $q = 2, 3$, from which the same follows for all $q \in S_1$ by the group property (2).

Now it is easy to show that $S = \mathbb{R}_+$ and that for every $q \in \mathbb{R}_+$ we have

$$g(q) = q^\gamma. \tag{3}$$

In fact, if $q \in \mathbb{R}_+$ is arbitrary, then for every $\varepsilon > 0$ there are $q_1, q_2 \in S_1$ such that

$$q_1 < q < q_2, \qquad 1 \leq \frac{g(q_1)}{g(q_2)} < 1 + \varepsilon.$$

But the monotonicity of f implies that

$$g(q_1) \geq \limsup_{x\to 1-0} \frac{f(x^q)}{f(x)} \geq \liminf_{x\to 1-0} \frac{f(x^q)}{f(x)} \geq g(q_2),$$

and here the left- and right-hand sides can be arbitrarily close to q^γ if ε is chosen sufficiently small. This verifies (3) for all q. We will not use it, but it is clear from (1) that $\gamma = -\log_2(1+c)$.

Since

$$\lim_{x\to 1-0} \frac{(1-x^q)^\gamma}{(1-x)^\gamma} = q^\gamma,$$

we can conclude from (3) that f is of the form

$$f(x) = (1-x)^{-\rho} L(x) \tag{4}$$

(with $\rho = -\gamma$), where L is slowly varying inthe sense that

$$\lim_{x\to 1-0} \frac{L(x^q)}{L(x)} = 1$$

for every $q > 0$.

Let

$$f(x) = \sum_{k=0}^{\infty} a_k x^k, \qquad 0 < x < 1,$$

and $s_n = \sum_{k=0}^{n} a_k$. By a well-known Tauberian theorem (see *G. H. Hardy, Divergent Series, Clarendon Press, Oxford, 1989, Ch. VII, Theorem 108*), we can derive from (4) with any fixed $q \in (0,1)$ and some $p > 0$ that

$$s_n \sim p f(q^{1/n}),$$

where $a_n \sim b_n$ means that the ratio a_n/b_n tends to 1. However,

$$s_{2^m} = \prod_{0}^{m-1} (1+c) + c = (1+c)^m + c$$

and

$$s_{2^m+2^{m-1}} = (1+c)^m + c(1+c)^{m-1} + c^2,$$

from which

$$1 + \frac{c}{1+c} \sim \frac{s_{2^m+2^{m-1}}}{s_{2^m}} \sim \frac{f(q^{1/(2^{m-1}\cdot 3)})}{f(q^{1/(2^{m-1}\cdot 2)})}$$

$$\sim \frac{(1-q^{1/(2^{m-1}\cdot 3)})^{-\rho}}{(1-q^{1/(2^{m-1}\cdot 2)})^{-\rho}} \sim \left(\frac{2\cdot 2^{m-1}}{3\cdot 2^{m-1}}\right)^{-\rho} \sim \left(\frac{3}{2}\right)^{\rho}$$

follows if $m \to \infty$. Hence,

$$1 + \frac{c}{1+c} = \left(\frac{3}{2}\right)^{\rho},$$

and by (1) and (4),

$$1 + c = 2^\rho.$$

Thus, $1 + 2c = 3^\rho$, and ρ must satisfy the equation

$$3^\rho - 2 \cdot 2^\rho + 1 = 0,$$

which certainly holds for $\rho = 0$ and $\rho = 1$, but does not hold for any other ρ, because the function $3^t - 2 \cdot 2^t + 1$ is convex on $[0, \infty)$. If $\rho = 0$, then $c = 0$; while if $\rho = 1$, then $c = 1$, and these values for c were not allowed. The obtained contradiction proves the claim. □

Remark. Note that if $c = 1$, then $f(x) = 2/(1 - x)$, and the limit

$$\lim_{x \to 1-0} \frac{f(x^q)}{f(x)}$$

exists for every $q > 0$.

Problem F.46. *Let f and g be holomorphic functions on the open unit disc D, and suppose that $|f|^2 + |g|^2 \in$ Lip1. Prove that then $f, g \in$ Lip$\frac{1}{2}$. A function $h : D \to \mathbb{C}$ is in the Lipα class if there is a constant K such that*

$$|h(z) - h(w)| \leq K|z - w|^\alpha$$

for every $z, w \in D$.

Solution. We shall use the following Hardy–Littlewood theorem: if a function $h : D \to \mathbb{C}$ satisfies

$$|h'(z)| < \frac{M}{\sqrt{1 - |z|}} \tag{1}$$

with some constant M, then $h \in$ Lip$\frac{1}{2}$. In fact, in order to verify the Hardy-Littlewood theorem, it is enough to show that if $|z - z'| \leq \delta < 1/2$, then

$$|f(z) - f(z')| \leq C\sqrt{\delta}. \tag{2}$$

But if $w = (1 - \delta)z$, $w' = (1 - \delta)z'$, then (1) easily implies that

$$\begin{aligned} |f(z) - f(w)| &\leq \int_{(1-\delta)|z|}^{|z|} \frac{M}{\sqrt{1-t}} dt \\ &= 2M(\sqrt{|z|} - \sqrt{|(1 - \delta)|z|}) \leq \frac{M\delta}{\sqrt{1 - (1 - \delta)|z|}} \leq M\sqrt{\delta}. \end{aligned}$$

A similar argument shows that

$$|f(z') - f(w')| \leq C\sqrt{\delta} \qquad \text{and} \qquad |f(w) - f(w')| \leq C\sqrt{\delta},$$

from which (2) follows.

We shall prove that

$$|f'(z)|^2 + |g'(z)|^2 < \frac{K}{1-|z|},$$

where K is the Lipschitz constant corresponding to $|f|^2 + |g|^2$. From here our statement follows by the Hardy–Littlewood theorem.

Let $z \in D$ and $r > 0$ be such that the circle with radius r and center at z is contained in D. Let us apply Parseval's identity on this circle:

$$\frac{1}{2\pi}\int_0^{2\pi} |f(z+e^{i\theta}r)|^2 d\theta = \sum_{n=0}^{\infty} \left|\frac{f^{(n)}(z)}{n!}\right|^2 r^{2n} \geq |f(z)|^2 + |f'(z)|^2 r^2,$$

or in rearranged form,

$$\frac{1}{2\pi}\int_0^{2\pi} (|f(z+e^{i\theta}r)|^2 - |f(z)|^2) d\theta \geq |f'(z)|^2 r^2.$$

Applying the same inequality with f replaced by g, and adding the two together we, get

$$\frac{1}{2\pi}\int_0^{2\pi} \left((|f(z+e^{i\theta}r)|^2 + |g(z+e^{i\theta}r)|^2) - (|f(z)|^2 + |g(z)|^2)\right) d\theta$$

$$\geq (|f'(z)|^2 + |g'(z)|^2) r^2.$$

But because $|f|^2 + |g|^2 \in \mathrm{Lip}1$, the modulus of the integrand on the left-hand side is at most $K|e^{i\theta}r| = Kr$, hence the integral is at most $2\pi Kr$. Thus,

$$\frac{1}{2\pi} 2\pi Kr = Kr \geq (|f'(z)|^2 + |g'(z)|^2) r^2,$$

that is,

$$|f'(z)|^2 + |g'(z)|^2 \leq \frac{K}{r}.$$

Since r can be any number smaller than $1-|z|$, by letting r tend to $1-|z|$ we obtain

$$|f'(z)|^2 + g'(z)|^2 \leq \frac{K}{1-|z|},$$

and this is what we had to prove. □

Remark. The example $f \equiv 0$, $g(z) = (1-z)^{1/2}$ shows that the conclusion is sharp.

Problem F.47. *Find all functions $f : \mathbb{R}^3 \to \mathbb{R}$ that satisfy the parallelogram rule*

$$f(x+y) + f(x-y) = 2f(x) + 2f(y), \qquad x, y \in \mathbb{R}^3,$$

and that are constant on the unit sphere of $\mathbb{R}^3$.

Solution. Suppose that the function $f : \mathbb{R}^3 \to \mathbb{R}$ satisfies the functional equation

$$f(x+y) + f(x-y) = 2f(x) + 2f(y), \qquad x, y \in \mathbb{R}^3, \tag{1}$$

and is constant on the unit sphere

$$f(u) = c, \qquad u \in \mathbb{R}^3, \quad \|u\| = 1. \tag{2}$$

We shall prove that, with the help of an additive function $a : \mathbb{R} \to \mathbb{R}$, f can be written in the following form:

$$f(x) = a\left(\|x\|^2\right), \qquad x \in \mathbb{R}^3. \tag{3}$$

We begin by showing that f takes equal values on vectors of equal norms. So first let $x, y \in \mathbb{R}^3$, $\|x\| = \|y\| < 1$. Then there is a vector $z \in \mathbb{R}^3$ such that $\|x\|^2 + \|z\|^2 = 1 = \|y\|^2 + \|z\|^2$ and $z \perp x$, $z \perp y$. By the Pythagorean theorem, $\|x \pm z\| = 1 = \|y \pm z\|$, and therefore by (1)–(2),

$$\begin{aligned} 2f(x) &= f(x+z) + f(x-z) - 2f(z) = c + c - 2f(z) \\ &= f(y+z) + f(y-z) - 2f(z) = 2f(y), \end{aligned}$$

that is, $f(x) = f(y)$.

On the other hand, setting $y = 0$ in (1),

$$2f(x) = f(x+0) + f(x-0) = 2f(x) + 2f(0),$$

whence $f(0) = 0$. Thus the substitution $y = x$ leads to the relation

$$f(2x) = f(x+x) + f(x-x) = 2f(x) + 2f(x) = 4f(x), \quad x \in \mathbb{R}^3.$$

Consequently, if for some positive number r and all $x, y \in \mathbb{R}^3$ satisfying $\|x\| = \|y\| < r$ we have $f(x) = f(y)$, then for any $x', y' \in \mathbb{R}^3$ satisfying $\|x'\| = \|y'\| < 2r$, with the notation $x = x'/2$, $y = y'/2$, we obtain $\|x\| = \|y\| < r$ and therefore also

$$f(x') = f(2x) = 4f(x) = 4f(y) = f(2y) = f(y').$$

So, by induction, $f(x) = f(y)$ whenever $x, y \in \mathbb{R}^3$ and $\|x\| = \|y\|$.

By what we have proved, $f(x)$ depends only on $\|x\|$ or equivalently, same, on $\|x\|^2$; in other words, there exists a function $a : \mathbb{R} \to \mathbb{R}$ such that

$$f(x) = a\left(\|x\|^2\right), \qquad x \in \mathbb{R}^3.$$

It remains to verify the additiveness of the function a. To this end, fix $\lambda, \mu \in \mathbb{R}_+$ arbitrarily, and choose vectors $x, y \in \mathbb{R}^3$ so that $\lambda = \|x\|^2$, $\mu = \|y\|^2$, and $x \perp y$. Then, in view of (1) and the Pythagorean theorem,

$$\begin{aligned} 2a(\lambda) + 2a(\mu) &= 2a\left(\|x\|^2\right) + 2a\left(\|y\|^2\right) = 2f(x) + 2f(y) \\ &= f(x+y) + f(x-y) = a\left(\|x+y\|^2\right) + a\left(\|x-y\|^2\right) \\ &= a\left(\|x\|^2 + \|y\|^2\right) + a\left(\|x\|^2 + \|y\|^2\right) = 2a(\lambda+\mu). \end{aligned}$$

Thus a is additive on $\mathbb{R}_+$ and, as we may choose its value on $\mathbb{R}_-$ arbitrarily, setting $a(-\lambda) = -a(\lambda)$, $\lambda \in \mathbb{R}_+$, we get the additive function desired.

Conversely, simple substitution shows that the functions of the form (3) are solutions of the problem (1)–(2). □

Remarks.

1. We obtain the same solution if we assume (1) only for y satisfying $\|y\| = 1$ and, in a more general way, consider vectors of a real inner product space of dimension at least three. Of course, in this case we encounter further, essential difficulties.
2. The two-dimensional case has proved to be still harder. It is an open question whether in this case the problem (1)–(2) admits solutions different from (3).

Problem F.48. *For any fixed positive integer n, find all infinitely differentiable functions $f : \mathbb{R}^n \to \mathbb{R}$ satisfying the following system of partial differential equations:*

$$\sum_{i=1}^{n} \partial_i^{2k} f = 0, \qquad k = 1, 2, \ldots .$$

Solution. Let $n \in \mathbb{N}$, and consider the system of partial differential equations

$$\sum_{i=1}^{n} \partial_i^{2k} f = 0, \qquad k = 1, 2, \ldots \tag{1/n}$$

for the infinitely differentiable unknown function $f : \mathbb{R}^n \to \mathbb{R}$. Obviously, linear combinations of any partial derivatives of solutions are solutions again. We show that there is a universal solution in the sense that all solutions are linear combinations of partial derivatives of this universal solution. We begin with an important observation.

Lemma 1. If f is a solution of system $(1/n)$, then $\partial_i^{2n} f = 0$ $(i = 1, 2, \ldots, n)$, so f is a polynomial of degree not greater than $2n - 1$ in each variable.

Proof. For the commuting partial differential operators ∂_1^2, ∂_2^2, ∂_n^2, consider the power sums

$$P_k = \partial_1^{2k} + \partial_2^{2k} + \cdots + \partial_n^{2k}, \quad k = 1, 2, \ldots,$$

and the elementary symmetric polynomials

$$S_1 = \partial_1^2 + \partial_2^2 + \cdots + \partial_n^2$$
$$S_2 = \partial_1^2\partial_2^2 + \partial_1^2\partial_3^2 + \cdots + \partial_{n-1}^2\partial_n^2$$
$$S_3 = \partial_1^2\partial_2^2\partial_3^2 + \partial_1^2\partial_2^2\partial_4^2 + \cdots + \partial_{n-2}^2\partial_{n-1}^2\partial_n^2$$
$$\cdots\cdots\cdots\cdots$$
$$S_n = \partial_1^2\partial_2^2 \ldots \partial_n^2.$$

Then the differential equations can be written in the form

$$P_k f = 0, \qquad k = 1, 2, \ldots .$$

Further, by the Newton formulas,

$$P_1 - S_1 = 0,$$
$$P_2 - S_1P_1 + 2S_2 = 0,$$
$$\cdots\cdots\cdots\cdots\cdots\cdots\cdots\cdots$$
$$P_n - S_1P_{n-1} + \cdots + (-1)^{n-1}S_{n-1}P_1 + (-1)^n nS_n = 0.$$

Hence, it follows that

$$S_1 f = P_1 f = 0,$$
$$S_2 f = \frac{1}{2}(S_1P_1f - P_2f) = 0,$$
$$\cdots\cdots\cdots\cdots\cdots\cdots\cdots\cdots$$
$$S_n f = \frac{1}{n}(S_{n-1}P_1f - S_{n-2}P_2f + \cdots + (-1)^{n-1}P_nf) = 0.$$

Now, for fixed $i \in \{1, 2, \ldots, n\}$, consider the differential operator

$$P(\partial_i^2) = (\partial_i^2 - \partial_1^2)(\partial_i^2 - \partial_2^2)\ldots(\partial_i^2 - \partial_n^2)$$
$$= \partial_i^{2n} - S_1\partial_i^{2n-2} + S_2\partial_i^{2n-4} - \cdots + (-1)^n S_n .$$

We see that $P(\partial_i^2) \equiv 0$ and, therefore,

$$\partial_i^{2n} f = S_1\partial_i^{2n-2} f - S_2\partial_i^{2n-4} f + \cdots + (-1)^{n+1}S_n f = 0.$$

Lemma 2. Let $n \in \mathbb{N}$, and let the function $Q_n : \mathbb{R}^n \to \mathbb{R}$ be defined by the following determinant:

$$Q_n(x_1, x_2, \ldots, x_n) = \det \begin{pmatrix} x_1^{2n-1} & x_1^{2n-3} & \cdots & x_1 \\ x_2^{2n-1} & x_2^{2n-3} & \cdots & x_2 \\ \cdots & \cdots & \ddots & \cdots \\ x_n^{2n-1} & x_n^{2n-3} & \cdots & x_n \end{pmatrix}.$$

Then $f = Q_n$ is a solution of the system $(1/n)$.

Proof. We proceed by induction for n.
If $n = 1$, then the assertion is obvious:

$$Q_1(x_1) = x_1, \quad Q_1''(x_1) = 0.$$

Suppose the assertion is true for $n-1$, that is,

$$\sum_{i=1}^{n-1} \partial_i^{2k} Q_{n-1}(x_1, x_2, \ldots, x_{n-1}) = 0, \quad k = 1, 2, \ldots .$$

Expand the determinant Q_n according to the first column:

$$Q_n(x_1, x_2, \ldots, x_n) = \sum_{j=1}^{n} (-1)^{j-1} x_j^{2n-1} Q_{n-1}(x_1, \ldots, x_{j-1}, x_{j+1}, \ldots, x_n).$$

It follows that

$$\begin{aligned}
&\sum_{i=1}^{n} \partial_i^{2k} Q_n(x_1, x_2, \ldots, x_n) \\
&= \sum_{i=1}^{n} \sum_{j=1}^{n} (-1)^{j-1} \partial_i^{2k} \left[x_j^{2n-1} Q_{n-1}(x_1, \ldots, x_{j-1}, x_{j+1}, \ldots, x_n) \right] \\
&= \sum_{j=1}^{n} \left[(-1)^{j-1} \left(\frac{d^{2k}}{dx_j^{2k}} x_j^{2n-1} \right) Q_{n-1}(x_1, \ldots, x_{j-1}, x_{j+1}, \ldots, x_n) \right] \\
&+ \sum_{j=1}^{n} \left[(-1)^{j-1} x_j^{2n-1} \sum_{i \neq j} \partial_i^{2k} Q_{n-1}(x_1, \ldots, x_{j-1}, x_{j+1}, \ldots, x_n) \right].
\end{aligned}$$

The second sum vanishes by the induction hypothesis. Obviously, the first sum also vanishes if $k \geq n$, whereas for $k < n$ it is the expansion according to the first column of the following determinant:

$$(2n-1)(2n-2)\ldots(2n-2k) \cdot \det \begin{pmatrix} x_1^{2n-2k-1} & x_1^{2n-3} & \cdots & x_1 \\ x_2^{2n-2k-1} & x_2^{2n-3} & \cdots & x_2 \\ \cdots & \cdots & \ddots & \cdots \\ x_n^{2n-2k-1} & x_n^{2n-3} & \cdots & x_n \end{pmatrix}.$$

This, too, is zero since the first and $(k+1)$th columns of the determinant coincide.

Theorem. The function $f : \mathbb{R}^n \to \mathbb{R}$ satisfies the system of partial differential equations $(1/n)$ if and only if it is a linear combination of partial derivatives of Q_n.

Proof. Lemma 2 shows that Q_n satisfies the system $(1/n)$. Therefore, each of its partial derivatives and all linear combinations of them will also be solutions.

The converse can be proved by induction for n.

For $n = 1$, it is obvious, since $f'' = 0$ implies

$$f(x_1) = ax_1 + b = a\,Q_1(x_1) + b\,\partial_1 Q_1(x_1)\,.$$

Suppose that the assertion is true for all natural numbers not greater than n. Let f be a solution of the system $(1/n+1)$. By Lemma 1, $\partial_{n+1}^{2n+2} f = 0$, and so $\partial_{n+1}^{2n+1} f$ is independent of x_{n+1}:

$$\partial_{n+1}^{2n+1} f(x_1, x_2, \ldots, x_n, x_{n+1}) = \phi_0(x_1, x_2, \ldots, x_n).$$

Since together with any solution its partial derivatives are also solutions, ϕ_0 satisfies $(1/n+1)$ and, since it does not depend on x_{n+1}, it satisfies $(1/n)$ as well. Thus, by the induction hypothesis, we have the representation

$$\phi_0(x_1, x_2, \ldots, x_n) = \sum_{\alpha} c_0^{\alpha} \partial_1^{\alpha_1} \partial_2^{\alpha_2} \ldots \partial_n^{\alpha_n} Q_n(x_1, x_2, \ldots, x_n)$$

with suitable constants $c_0^{\alpha} \in \mathbb{R}$ $(\alpha \in \mathbb{N}_0^n)$ among which only finitely many are different from 0. On the other hand, we note that

$$Q_n(x_1, x_2, \ldots, x_n) = \frac{(-1)^n}{(2n+1)!} \partial_{n+1}^{2n+1} Q_{n+1}(x_1, x_2, \ldots, x_{n+1}),$$

which can be seen from the expansion of Q_{n+1} according to its last row:

$$Q_{n+1}(x_1, x_2, \ldots, x_{n+1}) = \sum_{k=0}^{n} x_{n+1}^{2k+1} (-1)^k D_k(n).$$

Here $D_k(n)$ is the respective minor and, obviously, $D_n(n) = Q_n(x_1, x_2, \ldots, x_n)$.

Comparing the above results it follows that

$$\phi_0 = \sum_{\alpha} c_0^{\alpha} \frac{(-1)^n}{(2n+1)!} \partial_1^{\alpha_1} \partial_2^{\alpha_2} \ldots \partial_n^{\alpha_n} \partial_{n+1}^{2n+1} Q_{n+1}\,.$$

Next, define the function $g_0 : \mathbb{R}^{n+1} \to \mathbb{R}$ by the relation

$$g_0 = \sum_{\alpha} c_0^{\alpha} \frac{(-1)^n}{(2n+1)!} \partial_1^{\alpha_1} \partial_2^{\alpha_2} \ldots \partial_n^{\alpha_n} \partial_{n+1}^{0} Q_{n+1}\,.$$

Clearly, g_0 is a solution of $(1/n+1)$, and $\partial_{n+1}^{2n+1}(f - g_0) = \phi_0 - \phi_0 = 0$. Thus $f_1 = f - g_0$ is a solution of $(1/n+1)$ and $\partial_{n+1}^{2n+1} f_1 = 0$, that is, $\partial_{n+1}^{2n} f_1$ does not depend on x_{n+1}:

$$\partial_{n+1}^{2n} f_1(x_1, x_2, \ldots, x_{n+1}) = \phi_1(x_1, x_2, \ldots, x_n).$$

Pursuing the process, we obtain the functions $\phi_0, \phi_1, \ldots, \phi_{2n}$, $g_0, g_1, \ldots$, g_{2n}, and $f_1, f_2, \ldots, f_{2n+1}$. Here each of the functions $g_0, g_1, \ldots, g_{2n}$ is a linear combination of partial derivatives of Q_{n+1}. So with the help of Lemma 2, we see by induction that for $k = 1, 2, \ldots, 2n$ the function $f_{k+1} = f_k - g_k$ is a solution of the system $(1/n+1)$, and $\partial_{n+1}^{2n-k+1} f_{k+1} = \partial_{n+1}^{2n-k+1}(f_k - g_k) = \phi_k - \phi_k = 0$. Consequently, f_{2n+1} does not already depend on x_{n+1}, that is,

$$f_{2n+1}(x_1, x_2, \ldots, x_n, x_{n+1}) = \phi_{2n+1}(x_1, x_2, \ldots, x_n)$$

where, similar to the arguments above, it can be proved that ϕ_{2n+1} is a linear combination of partial derivatives of Q_{n+1}. Finally, the representation

$$f = g_0 + g_1 + \cdots + g_{2n} + f_{2n+1}$$

proves the theorem. □

Problem F.49. *Let P be a polynomial with all real roots that satisfies the condition $P(0) > 0$. Prove that if m is a positive odd integer, then*

$$\sum_{k=0}^{m-1} \frac{f^{(k)}(0)}{k!} x^k > 0$$

for all real numbers x, where $f = P^{-m}$.

Solution. Let

$$F(x) = \sum_{k=0}^{m-1} \frac{f^{(k)}(0)}{k!} x^k,$$

and consider the polynomial $Q(x) = P^m(x) \cdot F(x)$. Denoting by n the degree of P, from the factor P^m (counting multiplicities) we obtain $n \cdot m$ real roots of Q. Assume, contrary to the assertion of the problem, that F also has real roots. We distinguish between two cases:

1. F has degree $m - 1$; then F has at least two real roots since its degree is even.
2. F has degree not greater than $m - 2$.

According to Rolle's theorem, we can count at least $mn + 1$ roots (with multiplicities) of Q' in the first case, and at least mn roots in the second case. We note that because of the relations $P(0) \neq 0$ and $F(0) \neq 0$, the root 0 could be counted at most once. Zero is actually a root of Q', of multiplicity not less than $m - 1$ at that, since for $j = 1, 2, \ldots, m - 1$ we have

$$\begin{aligned} Q^{(j)}(0) &= \sum_{l=0}^{j} \binom{j}{l} \left[(P^m)^{(l)}(0)\right] \cdot \left[F^{(j-l)}(0)\right] \\ &= \sum_{l=0}^{j} \binom{j}{l} \left[(P^m)^{(l)}(0)\right] \cdot \left[f^{(j-l)}(0)\right] = (P^m \cdot f)^{(j)}(0) = 0, \end{aligned}$$

the function $P^m \cdot f$ being equal to the constant 1.

Consequently, at least $m-2$ roots can be joined to those counted so far. Thus Q' has at least $nm+m-1$ roots in case 1, and at least $nm+m-2$ roots in case 2. Since, however, the degree of Q' is equal to $nm+m-2$ in case 1 and is not greater than $nm+m-3$ in case 2, we obtain that Q' is identically 0. Therefore, Q is a constant polynomial, and hence it is identically 0, contrary to the relation $Q(0) \neq 0$.

Thus, F has no real roots and, since $F(0) = P^{-m}(0) > 0$, we have $F(x) > 0$ for all real values of x. □

Problem F.50. *We say that the real numbers x and y can be connected by a δ-chain of length k (where $\delta : \mathbb{R} \to (0, \infty)$ is a given function) if there exist real numbers $x_0, x_1, \ldots, x_k$ such that $x_0 = x$, $x_k = y$, and*

$$|x_i - x_{i-1}| < \delta\left(\frac{x_{i-1} + x_i}{2}\right), \quad i = 1, \ldots, k.$$

Prove that for every function $\delta : \mathbb{R} \to (0, \infty)$ there is an interval in which any two elements can be connected by a δ-chain of length 4. Also, prove that we cannot always find an interval in which any two elements could be connected by a δ-chain of length 2.

Solution. To prove the first assertion, using the Baire category theorem, choose a positive integer n and an interval I such that the set $H_n = \{x\colon \delta(x) > 1/n\}$ is dense in I; we may also assume that the length $|I| < 1/10n$. Let K be the middle third of I; we show that in it, any two elements can be connected by a δ-chain of length 4. So let $x.y \in K$. If $c \in K \cap H_n$ and $b = c + (y-x)/2$, then reflection in b followed by reflection in c give translation by $x-y$. If we insert this between two reflections in a, we obtain a translation by $y-x$, which carries x into y. In order that in this way we obtain a δ-chain connecting x and y, we have to care only about the following two things: first, $a \in H_n$ should be satisfied and, second, $2a - x$ should fall in the $\delta(b)/2$-neighborhood of b. Since, however, H_n is dense in I, this can easily be achieved.

Proving the second half of the problem requires a bit longer argument. Let B be the set of all real numbers that can be written using a finite number of binary digits. We specify certain pairs of points of B so that for suitable δ they cannot be connected by a δ-chain of length at most 2, and every interval contain such a pair of points. To this end, let $P = \cup_{i=1}^{\infty} P_i$, where

$$P_i = \left\{\left(\frac{-2^{2i}}{2^{2i}}, \frac{-2^{2i}+2}{2^{2i}}\right), \ldots, \left(\frac{2^{2i}-2}{2^{2i}}, \frac{2^{2i}}{2^{2i}}\right)\right\}.$$

We first define δ on elements of B in the following obvious manner: if $x \in B$ ends in $1/2^i$, then let $\delta(x) = 1/2^{i+1}$ (if x is an integer, we take $i = 0$). Then, evidently, pairs of points of P cannot be connected by δ-chains of length 1, nor by those of length 2 and passing through an element of B.

We extend δ to $\mathbb{R}$ in the following way. B is an additive subgroup of $\mathbb{R}$, so $\mathbb{R}$ is a disjoint union of residue classes of the form $a+B$. If $a \notin B$, we define δ on $a+B$ so that for $b \in a+B_i$ the value $\delta(b)$ be smaller than $\min\{2|x-b|, 2|y-b|\}$ for all $(x,y) \in \cup_{j=1}^{i} P_j$, where $B = \cup_{i=0}^{\infty} B_i$, and B_i consists of those numbers whose fractional part can be written using exactly i digits. It is easy to see that δ may be defined in this way. To finish the proof, we verify that for any $(x,y) \in P$ and $z \in a+B$ the numbers $x, 2z, y$ cannot form a δ-chain. Suppose that $(x,y) \in P_i$. Then $(y-x)/2 = 1/2^i$. If $x, 2z, y$ form a δ-chain, then

$$\delta\left(\frac{x}{2}+z\right) > |x-2z| = 2\left|x-\left(\frac{x}{2}+z\right)\right|,$$

whence by the definition of δ for $j \geq i$ we obtain $x/2+z \notin a+B_j$. We similarly obtain $y/2+z \notin a+B_j$. Then, however, the fractional part of $(y-x)/2$ can be written using at most $i-1$ digits, which contradicts the relation $(y-x)/2 = 1/2^i$. □

Remark. It is not known what happens if we require the existence of δ-chains of length not greater than 3.

Problem F.51. *Find meromorphic functions ϕ and ψ in the unit disc such that, for any function f regular in the unit disc, at least one of the functions $f-\phi$ and $f-\psi$ has a root.*

Solution. Introduce the notation

$$\begin{aligned}\kappa(z) &= \psi(z)-\phi(z) \not\equiv 0,\\ g(z) &= f(z)-\phi(z),\\ h(z) &= \frac{g(z)}{\kappa(z)}.\end{aligned}$$

Examine the functions that satisfy the following conditions:
1. $f = h\kappa + \phi$ is regular;
2. $h\kappa$ is nowhere 0;
3. $(h-1)\kappa$ is nowhere 0.

We claim that there exist functions ϕ and κ, meromorphic in the unit disc, with $\kappa(z) \neq 0$ if $|z| < 1$ and such that the above set of conditions cannot be satisfied by any h. This immediately gives the assertion of the problem.

If we choose ϕ and κ so that the locations and orders of their poles coincide, then the fulfillment of condition 1 implies the regularity of h. If all three conditions are satisfied, then the values of $h(z)$ can be 1 only at the poles of κ, and even this can be executed by choosing κ and ϕ properly. Assuming first-order poles, we only have to take care that, denoting by $A(z_0)$ and $B(z_0)$ the residues of κ and ϕ at the pole z_0, respectively, the value $B(z_0)/A(z_0)$ is different from -1. Actually, then $h(z_0) = -B(z_0)/A(z_0) \neq 1$ by condition 1.

Thus, the regular function h does not take on values 0 and 1 in the unit disc. By Schottky's theorem, it follows that $|h(z)|$ remains below a bound depending on $|h(0)|$ and $|z|$ only. We show that this is impossible. Consider the following sequences of functions:

$$\phi_n = \phi = \frac{1}{z} + \frac{1}{z - \frac{1}{2}},$$
$$\kappa_n = \phi = \frac{(1-z)^n}{z(z-\frac{1}{2})} \quad .$$

Calculate the residues of the two poles:

$$A_n(0) = -2, \quad B_n(0) = 1, \quad A_n\left(\frac{1}{2}\right) = 2^{-n+1}, \quad B_n\left(\frac{1}{2}\right) = 1.$$

Hence, it is easy to see that the functions meet the requirements stated so far. If, however, the three conditions above are fulfilled, then $h_n(0) = 1/2$ and $h_n(1/2) = -2^{n-1}$, contrary to the conclusion of the Schottky theorem. Therefore, if n is sufficiently large, the statement of the problem holds for κ_n, ϕ_n and the function $\psi_n = \kappa_n + \phi_n$ obtained from them. □

Problem F.52. *To divide a heritage, n brothers turn to an impartial judge (that is, if not bribed, the judge decides correctly, so each brother receives $(1/n)$th of the heritage). However, in order to make the decision more favorable for himself, each brother wants to influence the judge by offering an amount of money. The heritage of an individual brother will then be described by a continuous function of n variables strictly monotone in the following sense: it is a monotone increasing function of the amount offered by him and a monotone decreasing function of the amount offered by any of the remaining brothers. Prove that if the eldest brother does not offer the judge too much, then the others can choose their bribes so that the decision will be correct.*

Solution 1. In terms of functions, the problem can be expressed in the following way. We are given the continuous functions

$$g_1(x_1, \ldots, x_n), \ldots, g_2(x_1, \ldots, x_n), \ldots, g_n(x_1, \ldots, x_n)$$

defined on $[0, \infty)^n$ (here $g_j(x_1, \ldots, x_n)$ is the deviation of the heritage of the jth brother from $1/n$ times the whole heritage, provided that the judge is offered x_1 units of currency by the first brother, x_2 units by the second, etc.), the sum of the functions being 0, and the function $g_j(x_1, \ldots, x_n)$ being strictly increasing in the variable x_j but strictly decreasing in all other variables x_k, $k \neq j$. Further, we know that the judge is originally impartial, that is, $g_j(0, \ldots, 0) = 0$ for all j. We have to show that there exists $a_n > 0$ such that for any $0 \leq x_n \leq a_n$ there are values $x_1 = x_1(x_n)$, $\ldots$,$x_{n-1} = x_{n-1}(x_n)$ with $g_j(x_1, \ldots, x_n) = 0$ for all j. We prove more (to be exact, we must prove more in order that the proof below remain valid), namely, that there even exist $x_1 = x_1(x_n), \ldots, x_{n-1} = x_{n-1}(x_n)$, which, besides satisfying the relations just mentioned, tend to 0 as $x_n \to 0$.

We apply induction for n. If $n = 1$, there is nothing to prove. Assume that the assertion holds for $n-1$ functions. We claim that there is a number $b > 0$ such that if $0 \leq x_2, \ldots, x_n \leq b$ are arbitrary, then there exists one and only one $y = y(x_2, \ldots, x_n)$ satisfying $g_1(y, x_2, \ldots, x_n) = 0$, whereas y is a continuous, strictly increasing function of the variables $x_2, \ldots, x_n$. Really, the uniqueness of y follows from g_1 being strictly increasing in its first variable. If $x_2 = x_3 = \cdots = x_n = 0$, then we may choose $y = 0$, while for other values the existence of y can be seen as follows. Since $g_1(1, 0, \ldots, 0) > 0$, therefore the continuity assumed ensures the existence of $b > 0$ such that for $0 \leq x_2, \ldots, x_n \leq b$ we have

$$g_1(1, x_2, \ldots, x_n) > 0.$$

On the other hand,

$$g_1(0, x_2, \ldots, x_n) < 0.$$

So, again by continuity, the y above must exist. If $x_j' > x_j$, $j \neq 1$, then

$$\begin{aligned}
&g_1(y(x_2, \ldots, x_j', \ldots, x_n), x_2, \ldots, x_j', \ldots, x_n) \\
&= 0 = g_1(y(x_2, \ldots, x_n), x_2, \ldots, x_n) \\
&> g_1(y(x_2, \ldots, x_n), x_2, \ldots, x_j', \ldots, x_n),
\end{aligned}$$

and this shows that $y(x_2, \ldots, x_n)$ is an increasing function of x_j. We next verify the continuity of y. Since

$$g_1(y(x_2, \ldots, x_n) - \varepsilon, x_2, \ldots, x_n) < 0 < g_1(y(x_2, \ldots, x_n) + \varepsilon, x_2, \ldots, x_n),$$

there is a $\delta > 0$ such that for $|x_j - x_j'| \leq \delta$, $j = 2, 3, \ldots, n$, we have

$$g_1(y(x_2, \ldots, x_n) - \varepsilon, x_2', \ldots, x_n') < 0 < g_1(y(x_2, \ldots, x_n) + \varepsilon, x_2', \ldots, x_n').$$

By the foregoing, this proves the inequalities

$$y(x_2, \ldots, x_n) - \varepsilon < y(x_2', \ldots, x_n') < y(x_2, \ldots, x_n) + \varepsilon.$$

After these preparations, consider the $n - 1$ functions

$$h_j(x_2, \ldots, x_n) = g_j(y(x_2, \ldots, x_n), x_2, \ldots, x_n), \quad j = 2, \ldots, n.$$

By the properties of y, they are continuous, have sum 0, and satisfy the relations $h_j(0, \ldots, 0) = 0$ for all j. We claim that they are strictly monotone in the required sense. Actually, if $k \neq j$ and x_k increases, then $y(x_2, \ldots, x_n)$ also increases and therefore h_j decreases. Thus, if x_j increases, then all h_k, $k \neq j$, will decrease, so h_j must increase since the sum of the h_l is zero. Consequently, the $n - 1$ functions h_j satisfy the hypotheses of the problem. So, by the induction hypothesis, there exists $a'_n > 0$ such that for any $0 \le x_n \le a'_n$ there are $x_2 = x_2(x_n), \dots ,x_{n-1} = x_{n-1}(x_n)$ satisfying $h_j(x_2, \dots, x_n) = 0$ for all $j \ge 2$, and here each $x_j(x_n)$ tends to 0 as $x_n \to 0$. Hence, there exists $\min\{a'_n, b\} > a_n > 0$, where b is the constant used during the preparation, such that if $0 \le x_n \le a_n$, then $0 \le x_j(x_n) \le b$ for $j = 2, \dots, n-1$. Then, however, with the values

$$x_1 = x_1(x_n) := y(x_2(x_n), \dots, x_{n-1}(x_n), x_n),$$
$$x_2 = x_2(x_n), \dots, x_{n-1} = x_{n-1}(x_n),$$

the relation $g_j(x_1, \dots, x_n) = 0$ is valid for all j (for $j = 1$ this follows from the definition of y), and here each $x_j(x_n)$ tends to 0 as $x_n \to 0$ (we again make use of the continuity of y). Thus, we have proved the statement also for n functions. □

Solution 2. Let g_i $(i = 1, 2, \dots, n)$ be the functions defined in Solution 1.

Let $e_i = (0, \dots, 0, 1, 0, \dots, 0)$ be the ith unit vector in $\mathbb{R}^n$ (the ith coordinate is 1, the others are 0). By the assumptions, $g_1(e_i) < 0$ if $i > 1$. In view of the continuity of g_1, there is a number $\varepsilon_i > 0$ such that $0 \le x_1 < \varepsilon_i$ implies $g_1(x_1e_1 + e_i) < 0$. Set $\varepsilon = \min\{\varepsilon_i : 2 \le i \le n\}$. We show that if $0 \le x_1 < \varepsilon$, then there exist nonnegative numbers $x_2, x_3, \dots, x_n$ such that $g_i(x_1, x_2, \dots, x_n) = 0$ for $1 \le i \le n$. To this end, it is sufficient to prove that $g_i(x_1, x_2, \dots, x_n) = 0$ for $i \ge 2$ (recall that the sum of all functions g_i is zero).

Let $0 \le x_1 < \varepsilon$ be an arbitrary number, which will be fixed in what follows. Set

$$G(x_2, \dots, x_n) = \max_{2 \le i \le n} g_i(x_1, x_2, \dots, x_n)$$

and

$$H = \{(x_2, \dots, x_n) \in \mathbb{R}^{n-1} \ge 0 : G(x_1, x_2, \dots, x_n) \le 0\}.$$

Then

a. $H \subset [0,1]^{n-1}$, since if $x_i \ge 1$ for some $i \ge 2$ then $0 > g_1(x_1e_1 + e_i) \ge g_1(x_1, x_2, \dots, x_n)$. Therefore $\sum_i g_i \equiv 0$ gives $G(x_2, \dots, x_n) > 0$.
b. H is closed, being a level set of a continuous function.
c. H is nonempty, since $(0, 0, \dots, 0) \in H$. It follows that $g_1(x_1, x_2, \dots, x_n)$ takes its minimum on H at some point $(x_2^0, x_3^0, \dots, x_n^0)$.
We prove that

$$g_i(x_1, x_2^0, x_3^0, \dots, x_n^0) = 0 \quad \text{for all} \quad i \ge 2.$$

By the definition of H, it is clear that

$$g_i(x_1, x_2^0, \dots, x_n^0) \le 0 \quad (i = 2, \dots, n).$$

Suppose that for some $i \ge 2$ we have

$$g_i(x_1, x_2^0, \dots, x_n^0) < 0.$$

Then making x_i^0 a little larger, we remain in H, but the value of g_1 becomes smaller, which contradicts the choice of $(x_2^0, \dots, x_n^0)$. This contradiction proves the statement. □

Remarks.

1. It is easy to show by examples that the assertion becomes false if we replace the condition of strict monotonicity by monotonicity in the wide sense.
2. Here is a simple example that if $a > 0$ is arbitrary then a correct decision cannot already be achieved in case the eldest (say, the nth) brother offers the judge at least a units of currency: let

$$g_j(x_1, \dots, x_n) = (n-2)x_j + \frac{ax_j}{x_j+1} - x_1 - \dots - x_{j-1} - x_{j+1} - \dots - x_n$$

if $j = 1, 2, \dots, n-1$, and let

$$g_n(x_1, \dots, x_n) = (n-1)x_n - \frac{ax_1}{x_1+1} - -\frac{ax_2}{x_2+1} - \dots - \frac{ax_{n-1}}{x_{n-1}+1}.$$

Problem F.53. *Construct an infinite set $H \subseteq C[0,1]$ such that the linear hull of any infinite subset of H is dense in $C[0,1]$.*

Solution. Let $\{g_k\}_{k=1}^{\infty}$ be dense in $C[0,1]$ $(g_k \not\equiv 0)$. We show that the set $H = \{h_n\}_{n=2}^{\infty}$, where

$$h_n = \sum_{k=1}^{\infty} \left(\frac{g_k}{\|g_k\|_\infty} \right) \frac{1}{n^k} \quad ,$$

meets the requirements.

If L is any bounded linear functional on $C[0,1]$, set

$$h_L(z) = \sum_{k=1}^{\infty} \left(\frac{Lg_k}{\|g_k\|_\infty} \right) z^k.$$

Then h_L is analytic in the unit disc, and $Lh_n = h_L(1/n)$. Let

$$H' = \{h_{n_1}, h_{n_2}, h_{n_3}, \dots\}$$

be an infinite subset of H, and let $L \in C[0,1]^*$ be any functional that annihilates H'. Then $h_L(1/n_m) = Lh_{n_m} = 0$ for all m, so the analyticity of h_L yields $h_L \equiv 0$. Hence $Lg_k \equiv 0$ $(k = 1, 2, \dots)$, that is, $L \equiv 0$. This, however, means exactly that the linear hull of H' is dense in $C[0,1]$. □

Remark. Of course, the proof works in any separable Banach space.

Problem F.54. *Let $\alpha > 0$ be irrational.*

(a) *Prove that there exist real numbers a_1, a_2, a_3, a_4 such that the function $f : \mathbb{R} \to \mathbb{R}$,*

$$f(x) = e^x[a_1 + a_2 \sin x + a_3 \cos x + a_4 \cos(\alpha x)]$$

is positive for all sufficiently large x, and

$$\liminf_{x \to +\infty} f(x) = 0.$$

(b) *Is the above statement true if $a_2 = 0$?*

Solution.

(a) Put

$$f(x) = e^x(2 - \cos(x - 2\pi a) - \cos \alpha x),$$

where a is to be defined later. Then $f(x) \geq 0$, while $f(x) = 0$ if $x = 2k\pi$ and $x - 2\pi a = 2n\pi$ for some integers k and n. Hence $\alpha(n + a) = k$. Assuming that also $f(x') = 0$ for some $x' \neq x$, we obtain $\alpha(n' + a) = k'$ for some integers $n' \neq n$ and $k' \neq k$. The relations $\alpha(n+a) = k$, $\alpha(n'+a) = k'$ yield $\alpha = (k - k')/(n - n')$, a contradiction. Thus $f(x) > 0$ for all sufficiently large values of x.

Choose $a \in [0, 1)$ so that, for some sequence $\{n_k\}$ of natural numbers,

$$\left|\frac{n_k}{\alpha} - a\right| \leq \frac{e^{-n_k\pi/\alpha}}{n_k} \quad (\text{mod } 1).$$

The existence of an a of this kind can be established as follows. Let K be the circle of perimeter 1, and $p_0 \in K$. Starting from p_0, lay $1/\alpha$ on K (in a given direction) n times to obtain p_n. Denote by I_n the interval

$$\left[p_n - \frac{e^{-n\pi/\alpha}}{n}, p_n + \frac{e^{-n\pi/\alpha}}{n}\right]$$

on K. Let $n_0 = 1$, and suppose that $\{n_k\}_{k=0}^m$ are given. Define n_{m+1} so that $n_{m+1} > n_m$ and $I_{n_{m+1}} \subset I_{n_m}$. There is an n_{m+1} of this kind since $\{p_n\}_{n=k}^\infty$ is dense in K and $e^{-n\pi/\alpha}/n \to 0$ $(n \to \infty)$. We have $I_{n_0} \supset I_{n_1} \supset \ldots$, and $|I_n| \to 0$ $(n \to \infty)$. Let $a = \cap_{k=0}^\infty I_{n_k}$. Then

$$\begin{aligned} f\left(\frac{2n_k\pi}{\alpha}\right) &= e^{2n_k\pi/\alpha}\left[2 - \cos\left(\frac{2n_k\pi}{\alpha} - 2\pi a\right) - \cos 2n_k\pi\right] \\ &= e^{2n_k\pi/\alpha}\left[1 - \cos 2\pi\left(\frac{n_k}{\alpha} - a - m_k\right)\right], \end{aligned}$$

where m_k is an integer satisfying the relation

$$\left|\frac{n_k}{\alpha} - a - m_k\right| \leq \frac{e^{-n_k\pi/\alpha}}{n_k}.$$

Therefore, using the inequality $\cos u \geq 1 - u^2/2$, we obtain

$$f\left(\frac{2n_k\pi}{\alpha}\right) \leq e^{2n_k\pi/\alpha}\frac{1}{2}\left[2\pi\left(\frac{n_k}{\alpha} - a - m_k\right)\right]^2$$
$$\leq \frac{2\pi^2}{n_k^2} \to 0 \qquad (k \to \infty).$$

Hence, $\liminf_{x\to+\infty} f(x) = 0$.

(b) We show that in the linear hull of the functions e^x, $e^x \cos x$, $e^x \cos \alpha x$ there is a function with the required properties if and only if for any $\varepsilon > 0$ the inequality

$$\left|\alpha - \frac{m}{n}\right| < \varepsilon \frac{e^{-\pi n/2}}{n} \qquad (*)$$

holds for infinitely many rational numbers $\dfrac{m}{n}$ $(m, n \in \mathbb{N})$. Let

$$g(x) = e^x(a_1 + a_3 \cos x + a_4 \cos \alpha x)$$

be such a function. Then, necessarily, $a_1 = |a_3| + |a_4|$, and $a_i \neq 0$ for $i = 1, 3, 4$. Let $\{x_k\}$ be a sequence that tends to ∞ and satisfies $g(x_k) \to 0$ as $k \to \infty$. Since $a_1 = |a_3| + |a_4|$, we have

$$x_k = \pi n_k + \delta_k, \qquad \alpha x_k = \pi m_k + \Delta_k,$$

where n_k and m_k are natural numbers, $n_k \to \infty$, $m_k \to \infty$, $\delta_k \to 0$ and $\Delta_k \to 0$ as $k \to \infty$. Let $\beta \in (0,1)$ and $b = \min\{|a_3|, |a_4|\}$. If k is sufficiently large, then

$$\begin{aligned}
\beta \geq g(x_k) &= e^{\pi n_k + \delta_k}\left(a_1 + a_3\cos(\pi n_k + \delta_k) + a_4 \cos(\pi m_k + \Delta_k)\right)\\
&= e^{\pi n_k + \delta_k}\left(a_1 - |a_3|\cos\delta_k - |a_4|\cos\Delta_k\right)\\
&= e^{\pi n_k + \delta_k}\left(a_1 - |a_3| + |a_3|\frac{\delta_k^2}{2} + o(\delta_k^2) - |a_4| + |a_4|\frac{\Delta_k^2}{2} + o(\Delta_k^2)\right)\\
&\geq e^{\pi n_k - 1}\frac{1}{3}\left(|a_3|\delta_k^2 + |a_4|\Delta_k^2\right) > e^{\pi n_k}\frac{b}{3e}\left(\delta_k^2 + \Delta_k^2\right),
\end{aligned}$$

since $\cos u = 1 - u^2/2 + o(u^2)$ for $u \to 0$. Hence, for sufficiently large k,

$$|\delta_k|, |\Delta_k| \leq \sqrt{\frac{3e\beta}{b}} e^{-\pi n_k/2}$$

and, therefore,

$$\begin{aligned}
\left|\alpha - \frac{m_k}{n_k}\right| &= \left|\frac{\pi m_k + \Delta_k}{\pi n_k + \delta_k} - \frac{m_k}{n_k}\right| = \left|\frac{\Delta_k - \frac{m_k}{n_k}\delta_k}{\left(\pi + \frac{\delta_k}{n_k}\right)n_k}\right|\\
&\leq \frac{|\Delta_k| + 2\alpha|\delta_k|}{(\pi - 1)n_k} \leq \frac{2\alpha + 1}{\pi - 1}\sqrt{\frac{3e\beta}{b}}\frac{e^{-\pi n_k/2}}{n_k}.
\end{aligned}$$

Since $\beta \in (0,1)$ is arbitrary, we obtain that for any $\varepsilon > 0$ the relation $(*)$ is valid for infinitely many rational numbers $\frac{m}{n}$.

Next let $\varepsilon > 0$ be fixed, and suppose that $(*)$ has infinitely many rational solutions m_k/n_k, $m_k \in \mathbb{N}$, $n_k \in \mathbb{N}$, $k = 1, 2, \dots$. Passing to subsequences if necessary, we may assume that all the n_k have the same parity, all the m_k have the same parity, and

$$\left|\alpha - \frac{m_k}{n_k}\right| < \varepsilon \frac{e^{-\pi n_k/2}}{n_k} \qquad (k = 1, 2, \dots).$$

Let $a_1 = 2$, and furthermore,

$$a_3 = \begin{cases} 1 & \text{if the members of } \{n_k\} \text{ are odd}, \\ -1 & \text{if the members of } \{n_k\} \text{ are even} \end{cases}$$

and

$$a_4 = \begin{cases} 1 & \text{if the members of } \{m_k\} \text{ are odd}, \\ -1 & \text{if the members of } \{m_k\} \text{ are even}. \end{cases}$$

Then $g(x) = e^x(a_1 + a_3 \cos x + a_4 \cos \alpha x) > 0$ for $x > 0$, and

$$\begin{aligned} g(n_k\pi) &= e^{n_k\pi}\left[2 + a_3 \cos n_k\pi + a_4 \cos \alpha n_k \pi\right] \\ &= e^{n_k\pi}\left[1 - \cos \pi(\alpha n_k - m_k)\right] \\ &= e^{n_k\pi}\left[\frac{1}{2}\pi^2(\alpha n_k - m_k)^2 + o\left((\alpha n_k - m_k)^2\right)\right] \\ &\le e^{n_k\pi}\pi^2\varepsilon^2 e^{-\pi n_k} = \varepsilon^2\pi^2 \end{aligned}$$

for sufficiently large k. Consequently,

$$\liminf_{x \to +\infty} g(x) \le \varepsilon^2\pi^2.$$

Thus, if for any $\varepsilon > 0$ the relation $(*)$ has infinitely many solutions, then

$$\liminf_{x \to +\infty} g(x) = 0.$$

If α is an algebraic number then, by the Thue–Siegel–Roth theorem, for any $\delta > 0$ there are only finitely many m/n such that

$$\left|\alpha - \frac{m}{n}\right| < \frac{1}{n^{2+\delta}}.$$

So, if $(*)$ for any $\varepsilon > 0$ has infinitely many rational solutions m/n, then α must be transcendent. There is an α of this kind. For instance, let $N > e^{\pi/2}$ be an integer, $n_1 = N$, and $n_{k+1} = N^{n_k}$, $k = 1, 2, \dots$, and further let $\alpha = \sum_{i=1}^{\infty} 1/n_i$. Then with the notation

$$\alpha_k = \sum_{i=1}^{k} \frac{1}{n_i} = \frac{m_k}{n_k},$$

we have

$$\alpha - \alpha_k < \frac{2}{n_{k+1}} = \frac{2}{N^{n_k}}$$

and, therefore,

$$\left|\alpha - \frac{m_k}{n_k}\right| < \frac{2}{N^{n_k}} = \frac{2n_k}{(Ne^{-\pi/2})^{n_k}} = \frac{e^{-\pi n_k/2}}{n_k}.$$

Since

$$\frac{2n_k}{(Ne^{-\pi/2})^{n_k}} \to 0, \qquad k \to \infty,$$

it follows for any $\varepsilon > 0$ that

$$\left|\alpha - \frac{m_k}{n_k}\right| < \varepsilon \frac{e^{-\pi n_k/2}}{n_k}$$

for all sufficiently large indices k. □

Problem F.55. *Prove that if $\{a_k\}$ is a sequence of real numbers such that*

$$\sum_{k=1}^{\infty} |a_k|/k = \infty \quad \textit{and} \quad \sum_{n=1}^{\infty} \left(\sum_{k=2^{n-1}}^{2^n-1} k(a_k - a_{k+1})^2 \right)^{1/2} < \infty,$$

then

$$\int_0^{\pi} \left| \sum_{k=1}^{\infty} a_k \sin(kx) \right| dx = \infty.$$

Solution. Although the condition

$$\lim_{k\to\infty} a_k = 0 \tag{1}$$

does not appear in the statement of the problem, by the well-known Cantor–Lebesgue theorem, this follows from the fact that the series

$$\sum_{k=1}^{\infty} a_k \sin kx \tag{2}$$

is convergent almost everywhere. We note that for the application of this theorem it would be sufficient if the series (2) were convergent on a set of positive measure. To make reference easier, we list the remaining conditions:

$$\sum_{k=1}^{\infty} \frac{|a_k|}{k} = \infty, \tag{3}$$

$$\sum_{n=1}^{\infty}\left(\sum_{k=2^{n-1}}^{2^n-1} k|\Delta a_k|^2\right)^{1/2} < \infty, \tag{4}$$

where

$$\Delta a_k := a_k - a_{k+1} \qquad (k = 1, 2, \dots).$$

From (4) it follows that the sequence $\{a_k\}$ has bounded variation, that is,

$$\sum_{k=1}^{\infty} |\Delta a_k| < \infty. \tag{5}$$

Really, by the Cauchy inequality,

$$\begin{aligned}
\sum_{k=1}^{\infty} |\Delta a_k| &= \sum_{n=1}^{\infty} \sum_{k=2^{n-1}}^{2^n-1} |\Delta a_k| \\
&\le \sum_{n=1}^{\infty}\left(2^{n-1} \sum_{k=2^{n-1}}^{2^n-1} |\Delta a_k|^2\right)^{1/2} \\
&\le \sum_{n=1}^{\infty}\left(\sum_{k=2^{n-1}}^{2^n-1} k|\Delta a_k|^2\right)^{1/2}.
\end{aligned}$$

Consider the nth partial sum of (2). By Abel's rearrangement, we obtain

$$\sum_{k=1}^{n} a_k \sin kx = \sum_{k=1}^{n} \tilde{D}_k(x)\Delta a_k + a_{n+1}\tilde{D}_n(x), \tag{6}$$

where $\tilde{D}_n(x)$ is the conjugate Dirichlet kernel:

$$\tilde{D}_n(x) := \sum_{k=1}^{n} \sin kx = \frac{\cos\frac{x}{2} - \cos\left(n+\frac{1}{2}\right)x}{2\sin\frac{x}{2}} \qquad (n = 1, 2, \dots).$$

Introduce the notation

$$\bar{D}_n(x) := -\frac{\cos\left(n+\frac{1}{2}\right)x}{2\sin\frac{x}{2}} \qquad (n = 0, 1, \dots).$$

Then

$$\tilde{D}_n(x) = \bar{D}_n(x) - \bar{D}_0(x) \quad (n = 0, 1, \dots;\ \tilde{D}_0(x) = 0),$$

and from (6) we derive

$$\begin{aligned}
\sum_{k=1}^{n} a_k \sin kx &= \sum_{k=1}^{n} \bar{D}_k(x)\Delta a_k - \bar{D}_0(x)\sum_{k=1}^{n}\Delta a_k + a_{n+1}\bar{D}_n(x) - a_{n+1}\bar{D}_0(x) \\
&= \sum_{k=1}^{n} \bar{D}_k(x)\Delta a_k - a_1\bar{D}_0(x) + a_{n+1}\bar{D}_n(x) \\
&= \sum_{k=0}^{n} \bar{D}_k(x)\Delta a_k + a_{n+1}\bar{D}_n(x),
\end{aligned}$$

where the convention $a_0 := 0$ and its consequence $\Delta a_0 = -a_1$ are used. It follows that the series (2) is convergent:

$$\sum_{k=1}^{\infty} a_k \sin kx = \sum_{k=0}^{\infty} \bar{D}_k(x) \Delta a_k =: f(x) \tag{7}$$

for every x with the possible exception of the case $x = 0 \pmod{2\pi}$.

In the following, we make use of an inequality of the Sidon type: for any integer $n \geq 2$ and numerical sequence $\{b_k\}$,

$$\int_{\pi/n}^{\pi} \left| \sum_{k=n}^{2n-1} b_k \bar{D}_k(x) \right| dx \leq C \left(\sum_{k=n}^{2n-1} k b_k^2 \right)^{1/2}, \tag{8}$$

where C is a positive constant. To see this, we first apply the Cauchy–Schwarz inequality and then exploit the orthogonality of the system $\{\cos\left(k+\frac{1}{2}\right)x\}$:

$$\int_{\pi/n}^{\pi} \left| \sum_{k=n}^{2n-1} b_k \bar{D}_k(x) \right| dx = \int_{\pi/n}^{\pi} \left| \sum_{k=n}^{2n-1} b_k \frac{\cos\left(k+\frac{1}{2}\right)x}{2\sin\frac{x}{2}} \right| dx$$

$$\leq \left(\int_{\pi/n}^{\pi} \frac{dx}{(2\sin\frac{x}{2})^2} \right)^{1/2} \left(\int_0^{\pi} \left(\sum_{k=n}^{2n-1} b_k \cos\left(k+\frac{1}{2}\right)x \right)^2 dx \right)^{1/2}$$

$$\leq C n^{\frac{1}{2}} \left(\sum_{k=n}^{2n-1} b_k^2 \right)^{1/2} \leq C \left(\sum_{k=n}^{2n-1} k b_k^2 \right)^{1/2}.$$

Let $s \geq 1$ be an integer. According to (7),

$$\int_{\pi 2^{-s}}^{\pi} |f(x)|\, dx \geq \sum_{j=1}^{2^s-1} \int_{\pi/(j+1)}^{\pi/j} \left| \sum_{k=0}^{j-1} \bar{D}_k(x)\Delta a_k \right| dx$$

$$\sum_{j=1}^{2^s-1} \int_{\frac{\pi}{j+1}}^{\frac{\pi}{j}} \left| \sum_{k=j}^{\infty} \bar{D}_k(x)\Delta a_k \right| dx := I_1 - I_2\,. \tag{9}$$

Using the inequality

$$\left| \bar{D}_k(x) + \frac{1}{x} \right| \leq k+1 \quad (0 < x \leq \pi\,;\, k = 0, 1, \ldots),$$

we obtain that

$$I_1 \geq \sum_{j=1}^{2^s-1} \int_{\pi/(j+1)}^{\pi/j} \left| \sum_{k=0}^{j-1} \Delta a_k \right| \frac{dx}{x} - \sum_{j=1}^{2^s-1} \int_{\pi/(j+1)}^{\pi/j} \sum_{k=0}^{j-1} (k+1)\, |\Delta a_k|\, dx$$

$$:= I_{11} - I_{12}\,.$$

Since

$$\ln\left(1+\frac{1}{j}\right) \geq \frac{1}{j} - \frac{1}{j(j+1)},$$

by the foregoing

$$I_{11} \geq \sum_{j=1}^{2^s-1} \left(\frac{|a_j|}{j} - \frac{|a_j|}{j(j+1)}\right).$$

It is easy to see that

$$\sum_{j=1}^{2^s-1} \frac{|a_j|}{j(j+1)} \leq \max_{j\geq 1} |a_j| \sum_{j=1}^{\infty} \frac{1}{j(j+1)} \leq \sum_{k=1}^{\infty} |\Delta a_k|,$$

whence

$$I_{11} \geq \sum_{j=1}^{2^s-1} \frac{|a_j|}{j} - \sum_{k=1}^{\infty} |\Delta a_k|.$$

Similarly,

$$I_{12} = \pi \sum_{j=1}^{2^s-1} \sum_{k=0}^{j-1} \frac{k+1}{j(j+1)} |\Delta a_k| \leq \pi \sum_{k=0}^{\infty} |\Delta a_k| \leq 2\pi \sum_{k=1}^{\infty} |\Delta a_k|.$$

Therefore

$$I_1 \geq \sum_{j=1}^{2^s-1} \frac{|a_j|}{j} - (1+2\pi) \sum_{k=1}^{\infty} |\Delta a_k|. \tag{10}$$

We turn to estimating I_2:

$$\begin{aligned}
I_2 &= \sum_{l=1}^{s} \sum_{j=2^{l-1}}^{2^l-1} \int_{\pi/(j+1)}^{\pi/j} \left| \left(\sum_{k=j}^{2^l-1} + \sum_{n=l+1}^{\infty} \sum_{k=2^{n-1}}^{2^n-1} \right) \bar{D}_k(x) \Delta a_k \right| dx \\
&\leq \sum_{l=1}^{s} \sum_{j=2^{l-1}}^{2^l-1} \int_{\pi/(j+1)}^{\pi/j} \left| \sum_{k=j}^{2^l-1} \bar{D}_k(x) \Delta a_k \right| dx \\
&\quad + \sum_{l=1}^{s} \sum_{j=2^{l-1}}^{2^l-1} \sum_{n=l+1}^{\infty} \int_{\pi/(j+1)}^{\pi/j} \left| \sum_{k=2^{n-1}}^{2^n-1} \bar{D}_k(x) \Delta a_k \right| dx =: I_{21} + I_{22}.
\end{aligned}$$

Making use of the elementary inequality

$$\sin x \geq \frac{2}{\pi} x \qquad \left(0 \leq x \leq \frac{\pi}{2}\right),$$

we obtain

$$\begin{aligned}
I_{21} &\leq \sum_{l=1}^{s} \sum_{j=2^{l-1}}^{2^l-1} \int_{\pi/(j+1)}^{\pi/j} \frac{1}{2\sin\frac{x}{2}} \sum_{k=2^{l-1}}^{2^l-1} |\Delta a_k| \, dx \\
&\leq \sum_{l=1}^{s} \int_{\pi 2^{-l}}^{\pi 2^{-l+1}} \frac{\pi}{2x} \sum_{k=2^{l-1}}^{2^l-1} |\Delta a_k| \, dx = \frac{\pi \ln 2}{2} \sum_{k=1}^{2^s-1} |\Delta a_k|.
\end{aligned}$$

We now apply (8):

$$\begin{aligned} I_{22} &= \sum_{l=1}^{s} \sum_{n=l+1}^{\infty} \int_{\pi 2^{-l}}^{\pi 2^{-l+1}} \left| \sum_{k=2^{n-1}}^{2^n-1} \bar{D}_k(x) \Delta a_k \right| dx \\ &\le \sum_{n=2}^{\infty} \sum_{l=1}^{n-1} \int_{\pi 2^{-l}}^{\pi 2^{-l+1}} \left| \sum_{k=2^{n-1}}^{2^n-1} \bar{D}_k(x) \Delta a_k \right| dx \\ &= \sum_{n=2}^{\infty} \int_{\pi 2^{-n+1}}^{\pi} \left| \sum_{k=2^{n-1}}^{2^n-1} \bar{D}_k(x) \Delta a_k \right| dx \\ &\le C \sum_{n=2}^{\infty} \left(\sum_{k=2^{n-1}}^{2^n-1} k |\Delta a_k|^2 \right)^{1/2}. \end{aligned}$$

Consequently,

$$I_2 \le \frac{\pi \ln 2}{2} \sum_{k=1}^{2^s-1} |\Delta a_k| + C \sum_{n=2}^{\infty} \left(\sum_{k=2^{n-1}}^{2^n-1} k |\Delta a_k|^2 \right)^{1/2}. \tag{11}$$

Relying on (9), (10), and (11), we find that

$$\begin{aligned} &\int_{\pi 2^{-s}}^{\pi} \left| \sum_{k=1}^{\infty} a_k \sin kx \right| dx \ge \sum_{k=1}^{2^s-1} \frac{|a_k|}{k} \\ &- \left(1 + 2\pi + \frac{\pi \ln 2}{2} \right) \sum_{k=1}^{\infty} |\Delta a_k| - C \sum_{n=2}^{\infty} \left(\sum_{k=2^{n-1}}^{2^n-1} k |\Delta a_k|^2 \right)^{1/2}. \end{aligned}$$

In view of conditions (3), (4), and (5), this already yields the desired conclusion. □

Problem F.56. *Let $h : [0, \infty) \to [0, \infty)$ be a measurable, locally integrable function, and write*

$$H(t) := \int_0^t h(s) ds \qquad (t \ge 0).$$

Prove that if there is a constant B with $H(t) \le Bt^2$ for all t, then

$$\int_0^{\infty} e^{-H(t)} \int_0^t e^{H(u)} du \, dt = \infty.$$

Solution 1. Interchanging the order of integration, we obtain

$$\begin{aligned} \int_0^{\infty} e^{-H(t)} \int_0^t e^{H(u)} \, du \, dt &= \int_0^{\infty} \int_0^t e^{[-H(t)-H(u)]} \, du \, dt \\ &= \int_0^{\infty} \int_u^{\infty} e^{[-H(t)-H(u)]} \, dt \, du \\ &= \int_0^{\infty} \int_u^{\infty} \exp\left[- \int_u^t h(s) \, ds \right] dt \, du. \end{aligned}$$

It is sufficient to prove the existence of a constant $c > 0$ such that

$$I(T) := \int_T^{2T} \int_u^{u+1/T} \exp\left[-\int_u^t h(s)\,ds\right] dt\,du \geq c$$

for large T. We have

$$\begin{aligned}I(T) &\geq \int_T^{2T} \int_u^{u+1/T} \exp\left[-\int_u^{u+1/T} h(s)\,ds\right] dt\,du \\ &= \frac{1}{T}\int_T^{2T} \exp\left[-\int_u^{u+1/T} h(s)\,ds\right] du.\end{aligned}$$

Introduce the notation

$$Q_t := \left\{u \in [T, 2T]\colon \int_u^{u+1/T} h(s)\,ds \geq 10B\right\},$$

and find an upper estimate for the Lebesgue measure $\mu(Q_T)$ of the set Q_T:

$$\begin{aligned}10B\mu(T) &\leq \int_T^{2T}\int_u^{u+1/T} h(s)\,ds\,du \leq \int_T^{2T+1/T}\int_{s-1/T}^{s} h(s)\,du\,ds \\ &= \frac{1}{T}\int_T^{2T+1/T} h(s)\,ds \leq \frac{1}{T}B\left(2T + \frac{1}{T}\right)^2,\end{aligned}$$

whence

$$\mu(Q_T) \leq \frac{1}{10}\left(2T + \frac{4}{T} + \frac{1}{T^3}\right).$$

Therefore,

$$\begin{aligned}I(T) &\geq \frac{1}{T}\int_{[T,2T]\setminus Q_T} e^{-10B}\,du \geq \frac{e^{-10B}}{T}(T - \mu(Q_T)) \\ &\geq e^{-10B}\left(1 - \frac{9}{10}\right) =: c > 0,\end{aligned}$$

provided that $T \geq 1$. □

Solution 2. We first show that if $g\colon [1,\infty) \to (0,\infty)$ is measurable and locally integrable, further

$$\int_1^t g(s)\,ds \leq B_1 t^2 \qquad (B_1 = \text{constant})$$

for every t, then

$$\int_1^\infty \frac{1}{g(s)}\,ds = \infty.$$

Indeed, let $T > 1$ be any real number. By the Bunyakovski–Schwarz inequality

$$T^2 = \left(\int_T^{2T} 1\,dt\right)^2 = \left(\int_T^{2T} \sqrt{g(t)}\frac{1}{\sqrt{g(t)}}\,dt\right)^2$$
$$\leq \left(\int_T^{2T} g(t)\,dt\right)\left(\int_T^{2T} \frac{1}{g(t)}\,dt\right) \leq B_1 \cdot 4T^2 \int_T^{2T} \frac{1}{g(t)}\,dt\,.$$

So

$$\int_T^{2T} \frac{1}{g(t)}\,dt \geq \frac{1}{4B_1} \qquad (T > 0),$$

which yields the assertion.

Let

$$g(t) := \frac{e^{H(t)}}{\int_0^t e^{H(u)}\,du} \qquad (t \geq 1)\,.$$

According to the problem, we have to prove that

$$\int_1^\infty \frac{1}{g(t)}\,dt = \infty.$$

To this end, by the above remark, it is sufficient to verify the inequality

$$\int_1^t g(s)\,ds \leq B_1 t^2 \qquad (t \geq 1)$$

with a suitable B_1. This is simple:

$$\int_1^t g(s) = \int_1^t \frac{e^{H(s)}}{\int_0^t e^{H(u)}\,du}\,ds$$
$$= \ln \int_0^t e^{H(u)}\,du - \ln \int_0^1 e^{H(u)}\,du.$$

Since H is monotone increasing,

$$\int_1^t g(s) \leq \ln\left[te^{H(t)}\right] = \ln t + H(t)$$
$$\leq Bt^2 + \ln t \leq B_1 t^2$$

with a suitable B_1. $\square$

Problem F.57. *Consider the equation $f'(x) = f(x+1)$. Prove that*

(a) *each solution $f : [0,\infty) \to (0,\infty)$ has an exponential order of growth, that is, there exist numbers $a > 0$, $b > 0$ satisfying $|f(x)| \le a\, e^{bx}$, $x \ge 0$;*

(b) *there are solutions $f : [0,\infty) \to (-\infty,\infty)$ of nonexponential order of growth.*

Solution. In the statement of the problem, a constant c has been omitted. The statement is correctly, $f'(x) = cf(x+1)$, where $0 < c \le 1/e$.

Then there exists a $\lambda > 0$ such that $\lambda = ce^{\lambda}$ and so $e^{\lambda x}$ is a positive solution. On the other hand, the equation $f'(x) = f(x+1)$ has no positive solution f. If it had one, then f would be strictly increasing, and the Lagrange mean value theorem would give $f(1) > f(1) - f(0) = f'(\xi) = f(\xi+1) > f(1)$, a contradiction. The equation $f'(x) = cf(x+1)$ has a positive solution on $[0,\infty)$ if and only if $c \le 1/e$ (see *T. Krisztin, Exponential bound for positive solutions of functional differential equations, unpublished manuscript*).

(a) Proof of the exponential order of growth. If $f\colon [0,\infty) \to (0,\infty)$ satisfies the equation $f'(x) = cf(x+1)$, put $\alpha(x) = f'(x)/f(x)$. Then

$$f(x) = f(0)\exp\left(\int_0^x \alpha(s)\,ds\right),$$

and

$$\alpha(x) = c\exp\left(\int_x^{x+1} \alpha(s)\,ds\right) > 0;$$

that is,

$$\ln\frac{\alpha(x)}{c} = \int_x^{x+1} \alpha(s)\,ds, \quad x \ge 0\,.$$

Choose k so that $k \ge \alpha(0)$ and $k \ge \ln(k/c)$. Begin to define the sequence $\{x_n\}_{n=0}^{\infty}$ in the following way:

$$x_0 = 0, \quad x_1 = \max\{x \in (0,1]\colon \alpha(x) \le k\}\,.$$

Since

$$\int_0^1 \alpha(s)\,ds = \ln\frac{\alpha(0)}{c} \le \ln\frac{k}{c} \le k\,,$$

x_1 is well defined. If $x_0, \dots, x_n$ are given, put

$$x_{n+1} = \max\{x \in (x_n, x_n+1]\colon \alpha(x) \le k\}\,.$$

Since

$$\int_{x_n}^{x_n+1} \alpha(s)\,ds = \ln\frac{\alpha(x_n)}{c} \le \ln\frac{k}{c} \le k\,,$$

x_{n+1} is well defined. Since $\alpha(x) > k$ on $(x_{n+1}, x_n + 1]$, it follows that $x_{n+2} > x_n + 1$. Hence

$$[0, n] \subset \bigcup_{l=1}^{2n} [x_{l-1}, x_l] .$$

Therefore, if $x \in [n-1, n)$, then

$$\begin{aligned}
\int_0^x \alpha(s)\, ds \le \int_0^n \alpha(s)\, ds &\le \sum_{l=1}^{2n} \int_{x_{l-1}}^{x_l} \alpha(s)\, ds \\
&\le \sum_{l=1}^{2n} \int_{x_{l-1}}^{x_{l-1}+1} \alpha(s)\, ds = \sum_{l=1}^{2n} \ln \frac{\alpha(x_{l-1})}{c} \\
&\le 2n \ln \frac{k}{c} \le 2nk \le 2xk + 2k .
\end{aligned}$$

Consequently,

$$f(x) = f(0) \exp\left(\int_0^x \alpha(s)\, ds \right) \le f(0) e^{2k} e^{2kx} \quad (x \ge 0).$$

(b) There is a function $\phi \in C^\infty[0,1]$, not identically zero, such that $\phi^{(n)}(0) = \phi^{(n)}(1) = 0$; $n = 0, 1, 2, \ldots$. For example, $\phi(x) = \exp(1(x(x-1)))$ if $x \in (0,1)$, and $f(0) = f(1) = 0$. Then the formula

$$f(x+n) = \frac{1}{c^n} \phi^{(n)}(x), \quad n = 0, 1, \ldots ; \quad x \in [0,1] ,$$

defines a solution of the equation $f'(x) = cf(x+1)$. Let $x \in (0,1)$ be such that $\phi(x) \neq 0$. By Taylor's theorem, there exists an $\eta \in (0, x)$ such that

$$\phi(x) = \sum_{l=0}^{n-1} \frac{\phi^{(l)}(0)}{l!} x^l + \frac{\phi^{(n)}(\eta)}{n!} x^n = \frac{\phi^{(n)}(\eta)}{n!} x^n .$$

Hence

$$|f(\eta + n)| = \frac{1}{c^n} \left| \phi^{(n)}(\eta) \right| = |\phi(x)| \frac{n!}{(cx)^n} ,$$

which increases faster than any $a \cdot e^{bn}$ as $n \to \infty$. Thus f has a nonexponential order of growth. □

3.4 GEOMETRY

Problem G.1. *Find the minimum possible sum of lengths of edges of a prism all of whose edges are tangent to a unit sphere.*

Solution. If the edges of a prism are tangent to a sphere, then each face of the prism intersects the sphere in a circle, tangent to the sides of the face. Hence, each face is a circumscribed polygon.

The translation along the generators of the prism that takes one base polygon to the other moves the inscribed circle of the first polygon to the inscribed circle of the second. These circles are congruent and have parallel planes since the bases of the prism are congruent and parallel. Since the translation that moves one of two congruent and parallel circles on the sphere to the other is perpendicular to the plane of the circles, the generators of the prism are perpendicular to the base, thus, the prism in question is a right prism.

Sides of a right prism are rectangles. Since they are also circumscribed, and only squares are circumscribed among rectangles, each side of the prism is a square. Consequently, the edges of the bases and the generators have the same length, the bases are regular, and the sum of the lengths of the edges of the prism is three times the perimeter of the base.

Let us cut the prism and the sphere by a plane through the center of the sphere, parallel to the base. The prism is cut in a polygon congruent to the base polygons; the sphere is cut in a great circle. The great circle contains the point of contact of every tangents of the sphere orthogonal to the plane of the great circle. Therefore, it contains a point from each generator. We get that the base polygon is inscribed in a unit circle. Since the base polygon has equal sides, it must be a regular polygon. We conclude that the prisms that satisfy the prescribed condition are regular prisms, with the length of the generators equal to the length of the sides of the base polygon scaled in such a way that the circumcircle of the base polygon has radius 1.

Conversely, every prism of this type satisfies the condition of the problem, since the center of such a prism is located at distance 1 from each edge. Indeed, the distance to the center from a generator is clearly equal to 1. On the other hand, the orthogonal projection of the center onto a lateral face, which is a square, is the center of the face. Thus the distances to the center from the sides of this square are equal.

To get the sum of the lengths of the edges of such a prism, we have to multiply by 3 the perimeter of a regular polygon inscribed in a unit circle. It remains to decide which of these regular polygons has the shortest perimeter.

The perimeter of a regular n-gon inscribed in a unit circle is equal to

$$2n \sin \frac{\pi}{n} = 2\pi \frac{\sin \alpha}{\alpha},$$

where $\alpha = \pi/n$. Since $\sin x$ is concave on the interval $(0, \pi)$, the slope of the chord connecting the origin to the point $(x, \sin x)$ is decreasing as x is increasing. Therefore, $\sin\alpha/\alpha$ is minimal when $\alpha = \pi/n$ is maximal, that is, n is minimal. Thus, the minimum is attained by a regular triangle.

As for the prisms, the minimum of the sum of the edge lengths is obtained for a regular prism over a triangular base, with lateral faces being squares, and the value of the minimum is $3 \cdot 6 \cdot \sin 60° = 9\sqrt{3}$. □

Problem G.2. *Show that the perimeter of an arbitrary planar section of a tetrahedron is less than the perimeter of one of the faces of the tetrahedron.*

Solution. Obviously, planar sections of a tetrahedron are triangles or quadrangles, since each face of the tetrahedron contains at most one edge of the intersection provided that the intersection does not coincide with the face.

The case of quadrangles can be reduced to the case of triangles. Namely, we show that if we translate the plane of a quadrangle intersection in both directions until it passes through the nearest vertex, then one of the translated planes intersects the tetrahedron in a triangle (possibly degenerated to an edge), the perimeter of which is not shorter than the perimeter of the quadrangle.

If both translated planes intersect the tetrahedron in a triangle, then introduce the notation shown in Figure G.1 and express the sides of the quadrangle with the help of the sides of the two triangles. Since the planes of the three sections are parallel, the ratio of the segments of a straight line cut off by these planes is the same for all straight lines. Set

$$\lambda = \overline{BP} : \overline{AB} = \overline{BQ} : \overline{BT} = \overline{FR} : \overline{UF} = \overline{GS} : \overline{AG}\,.$$

Then for the corresponding sides we have

$$p = \lambda a_1, \quad q = b_1 + (1-\lambda)(b_2 - b_1) = \lambda b_1 + (1-\lambda) b_2\,,$$
$$r = c_2 + \lambda(c_1 - c_2) = \lambda c_1 + (1-\lambda)c_2, \quad s = (1-\lambda)a_2\,.$$

From these, $p + q + r + s = \lambda(a_1 + b_1 + c_1) + (1-\lambda)(a_2 + b_2 + c_2)$, that is, $K = \lambda K_1 + (1-\lambda)K_2$, where $K = p + q + r + s$, $K_1 = a_1 + b_1 + c_1$, and $K_2 = a_2 + b_2 + c_2$. This implies

$$K \le \max\{K_1, K_2\}\,.$$

If one of the triangles $\triangle AUT$, or $\triangle BGF$, or both degeneratse to the edges $\overline{AD}$ or $\overline{BC}$, respectively, then this inequality remains valid, putting $K_1 = 2\overline{AD}$ or $K_2 = 2\overline{BC}$, respectively. Then we get that the perimeter of one of

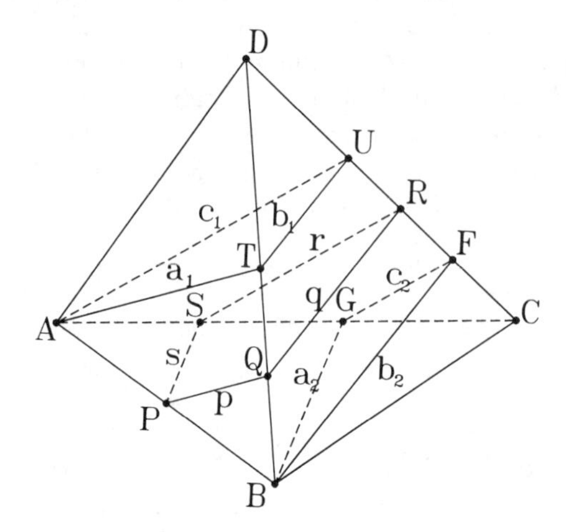

Figure G.1.

the faces of the tetrahedron is greater than the perimeter of the intersection quadrangle.

Hence, it is enough to consider the case of triangle intersections. We may suppose that the intersection triangle has a vertex in common with the tetrahedron. Otherwise we may translate the plane of the triangle to reach this situation (see Figure G.2). The translated plane meets the tetrahedron in a triangle similar to the original one, with ratio of similarity greater than 1 and it passes through that vertex of the nonintersected face of the tetrahedron that is nearest to the intersecting plane.

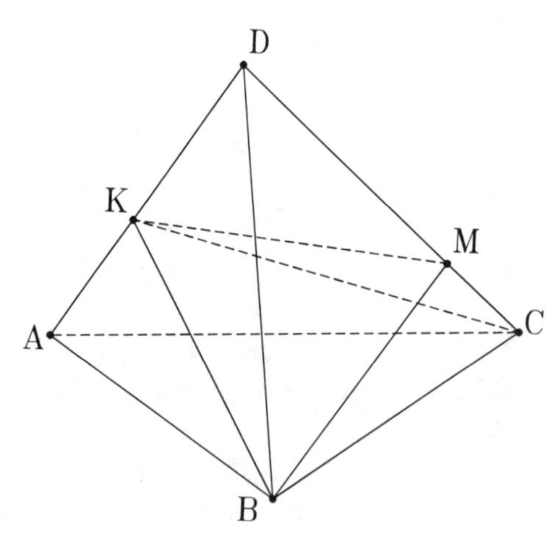

Figure G.2.

If the intersecting plane is parallel to the face not met by it, the assertion is trivial. Otherwise, we show that the perimeter of $\triangle KBM$ is smaller than the greater of the perimeters of $\triangle KBC$ and $\triangle KBD$. For this purpose, consider the smallest ellipsoid of revolution with foci K and B that contains both C and D. Since M is an internal point of it, we get

$$\overline{KM} + \overline{MB} < \max\{\overline{KC} + \overline{CB}, \overline{KD} + \overline{DB}\}.$$

But then, either $k_{KBM} < k_{KBD} < k_{ABD}$ or $k_{KBM} < k_{KBC}$, where k_{PQR} denotes the perimeter of $\triangle PQR$. In the first case, the assertion is proved, the second case can be finished by a repeated application of the previous arguments. □

Problem G.3. *Show that the center of gravity of a convex region in the plane halves at least three chords of the region.*

Solution. Let us denote the disc by T and its barycenter by S. If X is a point of the boundary curve G of T, then denote by $Y(X)$ the second intersection point of the straight line XS and the curve G. Let $f(X) = \overline{XS} - \overline{Y(X)S}$. We have $f(X) = -f(Y(X))$ for any $X \in G$. Since $f(X)$ changes continuously as X runs over the arc $\widehat{XY}(X)$, it attains any value between $f(X)$ and $-f(X)$. This implies the existence of a point $X_1 \in G$ such that $\overline{X_1S} = \overline{Y(X_1)S}$. If X_1S were the only straight line whose intersection with T is halved by S, then f would be positive along one of the arcs $\widehat{X_1Y}(X_1)$. Reflecting this arc through S, the reflected arc together with the other arc $\widehat{X_1Y}(X_1)$ of G would bound a domain T_1, the barycenter of which is different from S, since T_1 lies in a half-plane bounded by a straight line passing through S. Therefore, the barycenter of the union $T_2 = T_1 \cup T$ should be different from S. On the other hand, T_2 is centrally symmetric with respect to S, so its barycenter should be invariant under the reflection in S, which means that its barycenter should be S, which is a contradiction.

The same arguments show that if f has finite zeros, then it can not be nonnegative on a half-arc of G. Suppose that there are only two straight lines X_1S and X_2S for which the segment of the straight line cut off by the figure is halved by S $(X_1, X_2 \in G)$. By the previous remark, f must be negative (or positive) on the arcs $\widehat{X_2Y}(X_1)$ and $\widehat{X_1Y}(X_2)$, which is a contradiction, since the sign of f is opposite on arcs opposite to one another. We conclude that at least three chords of the figure are halved by S. The existence of four such chords cannot be proved in general, as it is shown by the case of an arbitrary triangle. □

Problem G.4. *Let $A_1, A_2, \ldots, A_n$ be the vertices of a closed convex n-gon K numbered consecutively. Show that at least $n-3$ vertices A_i have the property that the reflection of A_i with respect to the midpoint of $\overline{A_{i-1}A_{i+1}}$ is contained in K. (Indices are meant mod n.)*

Solution. We shall call a vertex P of K *reflectible* with respect to K if its reflection in the midpoint of the segment connecting the two neighbors of P belongs to K.

1. Let us begin with the first nontrivial case, with the case of quadrangles. For both pairs of opposite sides, the sum of angles lying on one of the two sides is at least π; let us denote by A a common vertex of two such sides and denote the other vertices in a cyclic order by B, C, D (see Figure G.3). We show that A is reflectible with respect to the quadrangle. We can construct the reflection A' of A by taking a parallel to AB through D, which intersects side BC in a point E because $\sphericalangle ADC + \sphericalangle BAD \geq \pi$, and then taking the parallel to AD through B, which intersects the segment DE at A' because $\sphericalangle DAB + \sphericalangle ABC \geq \pi$. Therefore, A' belongs to the quadrangle $ABCD$.

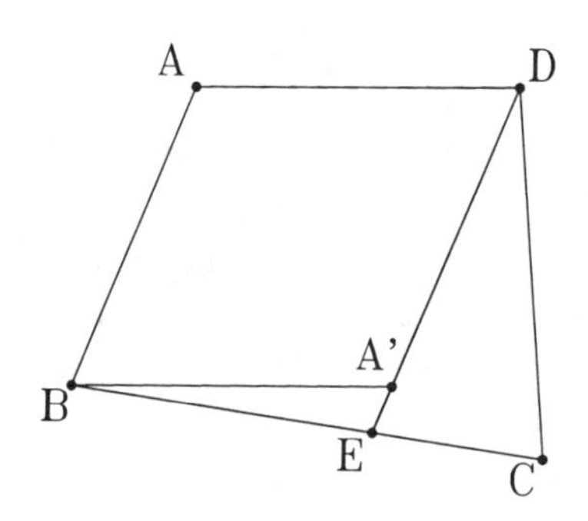

Figure G.3.

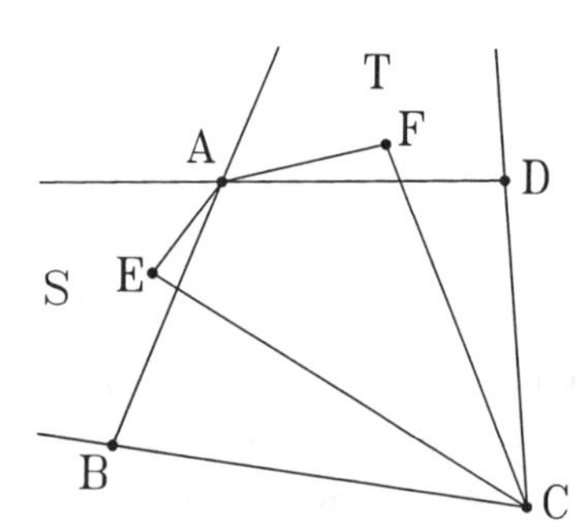

Figure G.4.

2. *If we show that among any four vertices of the convex polygon K, at least one is reflectible with respect to K, then the assertion of the problem is proved.*

We shall say that a vertex P of the polygon *precedes* the neighboring vertex R if going around K starting from P in the direction of R is a movement of positive orientation. Consider four arbitrary vertices of K and apply for them the previous notation, that is, let $\sphericalangle ADC + \sphericalangle BAD \geq \pi$, $\sphericalangle DAB + \sphericalangle ABC \geq \pi$ (see Figure G.4). If the vertex E of K that follows A is not B, then E is an internal point of the intersection S of three half-planes S_1, S_2, S_3, where S_1 is bounded by the straight line AD and contains $ABCD$, S_2 is bounded by BC and contains $ABCD$, and S_3 is bounded by AB and does not contain $ABCD$. Similarly, if the vertex F that precedes A is not D, then F is an internal point of the domain $T = T_1 \cap T_2 \cap T_3$, where the half-plane T_1 is bounded by the straight line AB and contains $ABCD$, T_2 is bounded by DC and contains $ABCD$, and T_3 is bounded by AD and does not contain $ABCD$. It is enough to prove that A is reflectible with respect to the quadrangle $AECF$, since then it is obviously reflectible with respect to K as well. For this purpose, it suffices to show that $\sphericalangle EAF + \sphericalangle AFC \geq \pi$ and $\sphericalangle FAE + \sphericalangle EAC \geq \pi$. It is enough to see the first inequality; the second can be shown in a similar way. But we have

$$\begin{aligned}\sphericalangle EAF + \sphericalangle AFC &= \sphericalangle EAC + (\sphericalangle CAF + \sphericalangle AFC)\\ &= \sphericalangle EAC + \pi - \sphericalangle ACF \geq \sphericalangle BAC + \pi - \sphericalangle ACD\\ &= \sphericalangle BAC + (\sphericalangle CAD + \sphericalangle ADC) = \sphericalangle BAD + \sphericalangle ADC \geq \pi\,,\end{aligned}$$

which proves the assertion of the problem. □

Remarks.

1. The assertion cannot be improved; that is, there exists for all n a convex n-gon having exactly $n-3$ reflectible vertices. Indeed, consider a convex quadrangle $ABCD$ with no parallel sides. It is easy to see that such a quadrangle has only one reflectible vertex; denote it by A. Now consider a convex n-gon K such that B, C, and D are vertices of K while the other vertices of K lie in a neighborhood of radius ε of A (obviously, such a convex n-gon exists for any n). It can be easily seen that if ε is sufficiently small, then B, C, and D are not reflectible with respect to K.

2. The assertion does not hold for concave polygons.

Problem G.5. *Is it true that on any surface homeomorphic to an open disc there exist two congruent curves homeomorphic to a circle?*

Solution. The answer is no: There exists a surface homeomorphic to a disc which does not contain two congruent curves homeomorphic to a circle. We show this by giving an example.

We start from a minimal surface, that is, from a surface for which $H = 1/2(g_1 + g_2) = 0$. (Here H denotes the Minkowski curvature, and g_1 and g_2 are the principal curvatures of the surface.) Although unnecessary, we explain why we are looking for counterexamples among minimal surfaces. If two congruent copies of a surface intersect one another in a closed curve, then by covering one copy with the other, the two positions of the intersection curve give two congruent curves on the surface, which are different in general. If two copies are tangent to one another at an isolated common point, then by a slight movement of one of them, they can be made to intersect one another along a curve. The choice of minimal surfaces is justified by the fact that if $H > 0$ at a point of a surface, then we can always find a congruent surface so that this point is an isolated common point of them. We can get such a surface by reflecting the original surface in the tangent plane at the point in question and then rotating it 90° about the normal of the surface. This follows from Euler's theorem, since by the inequality

$$g_1 \cos^2 \varphi + g_2 \sin^2 \varphi > -g_2 \cos^2 \varphi - g_1 \sin^2 \varphi\,,$$

the curvatures of the normal sections of the original surface are always greater than the curvatures of the corresponding normal sections of the transformed surface, and therefore the first surface is above the second one in a small neighborhood (looking at them from the direction of the normal vector).

Now we show that *in a sufficiently small domain of a minimal surface, any two congruent closed curves bound congruent domains of the surface.* For this purpose, take two copies of the surface and move one of them so that the congruent curves cover one another. Suppose that the piece of the minimal surface is small enough to ensure that both surfaces can be obtained as the graph of the single-valued functions $z = f_1(x,y)$ and $z = f_2(x,y)$. The areas of the domains bounded by the closed curve are obtained as integrals of the lengths of the vectors $m_1(-1, p_1, q_1)$ and $m_2(-1, p_2, q_2)$, respectively, over the same domain. (Here p_i and q_i denote the partial derivatives of f_i.)

We use the fact that the variation of the surface area of a minimal surface is 0, that is, if a minimal surface is embedded into a one-parameter family of surfaces all having the same boundary, then the derivative of the area of these surfaces with respect to the parameter is zero at the minimal surface. Consider the one-parameter family of surfaces defined by $z = \lambda f_1 + (1-\lambda) f_2$. The area of a member of this family is obtained as the integral of the magnitude of the vector $\lambda m_1 + (1-\lambda) m_2$. This magnitude is a strictly convex function of the parameter λ if $m_1 \neq m_2$ and does not depend on λ if $m_1 = m_2$. Consequently, the area, obtained by integrating convex functions of λ, is itself a convex function of λ. Its derivatives at $\lambda = 0$ and 1 can vanish only if it is a constant function, and this happens only if $m_1 = m_2$ at any point of the domain, that is, if the two pieces of surfaces coincide.

According to this observation, it is enough to find a minimal surface that does not contain different congruent pieces. Since two irreducible algebraic surfaces that have a domain in common coincide, it is enough to present an algebraic minimal surface that has no or only a finite number of automorphisms different from the identity. (An automorphism of a surface is a bijective mapping of the surface onto itself that can be extended to a congruence of the space.) If the surface has finite automorphisms, then a small neighborhood of a point different from its automorphic images has no automorphisms different from the identity.

Actually, any algebraic minimal surface would do, provided it is not a surface of revolution. However, we need to find only one of them. A simple computation shows that for the surface

$$x = u^3 - 3uv^2 + 3u\,,$$
$$y = v^3 - 3u^2v + 3v\,,$$
$$z = 6uv\,,$$

$H = 0$ and $K = -\frac{9}{4}(u^2 + v^2 + 1)^2$ (K is the Gauss curvature). The latter attains its minimum only at $u = v = 0$; thus this point and the Dupin indicatrix at this point must be fixed by any automorphism of the surface.

There may be only a finite number of such automorphisms, however, since the Dupin indicatrix is a hyperbola by $K < 0$. $\square$

Problem G.6. *The plane is divided into domains by n straight lines in general position, where $n \geq 3$. Determine the maximum and minimum possible number of angular domains among them. (We say that n lines are in general position if no two are parallel and no three are concurrent.)*

Solution. The minimal number of angular domains is three. Indeed, the convex hull of the intersection points of the straight lines is a convex polygon having at least three vertices. Each vertex is an intersection point of two straight lines and those half-lines of these lines, which go outside the convex hull, bound an angular domain of the considered subdivision, since there are no intersection points on them.

On the other hand, one can always position $n \geq 3$ straight lines in the plane so that the number of angular domains is exactly 3. Such a construction is given by n tangents to a quadrant (see Figure G.5).

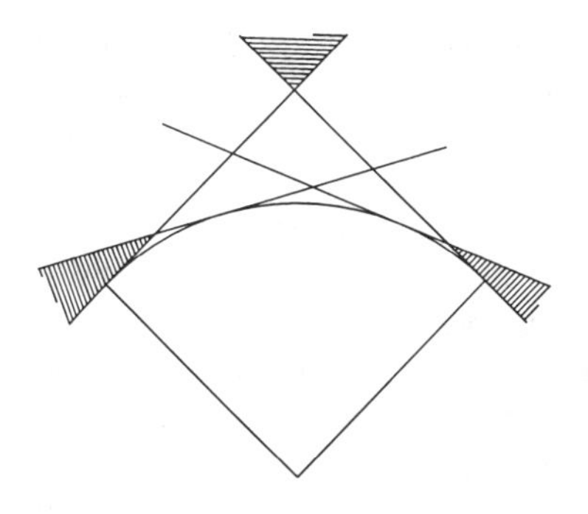

Figure G.5.

The maximal number of angular domains is

$$n + \frac{(-1)^{n+1} - 1}{2},$$

that is, n if n is odd and $n - 1$ if n is even. Indeed, each straight line is divided by the others into two half-lines and some segments. Angular domains can be bounded only by these half-lines, and each half-line bounds at most one angular domain (otherwise there would be a point that lies on three of the straight lines). It follows that $2n$ half-lines can bound no more than n angular domains.

For odd n, we can present an (essentially unique) construction with n straight lines and n angular domains: Consider the longest diagonals of a regular n-gon. These n straight lines are in general position. There is an

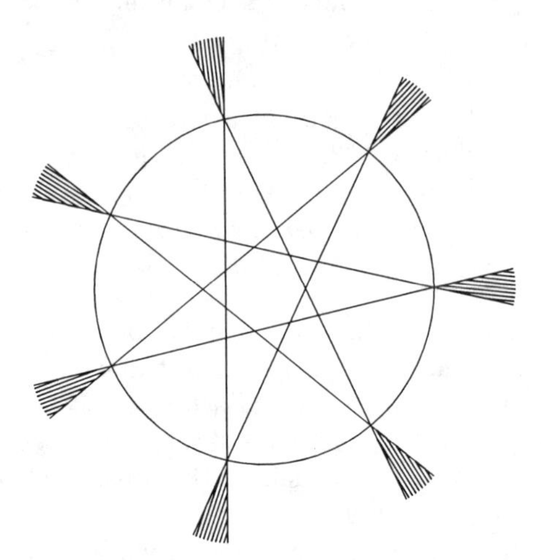

Figure G.6.

angular domain at each vertex of the polygon. Thus the number of angular domains is n (see Figure G.6).

We prove that if n straight lines in general position bound n angular domains, then n is odd. Removing the two half-lines considered above from each straight line gives n segments. Any endpoint of such a segment is shared by exactly two segments. If two of the segments have different endpoints, then they cross one another since the intersection point of their straight lines can lie neither outside nor at the end of the segments. Thus, the union of the segments yields a (number of) self-intersecting closed broken line(s). Fix an orientation on each of these broken lines, and consider three consecutive segments a, b, and c on one of them. Since a and c intersect, they lie on the same side of b. Omitting the endpoints of a and b (that is, three points), we can couple the remaining vertices, saying that the points P, Q form a pair if P is on the same side of b as the segments a and c, and Q is the neighboring vertex that comes after P according to the fixed orientation on the broken line containing P. Hence, the number, n, of vertices is odd.

In the case of even n, we can attain the maximal $(n-1)$ angular domains. This follows directly from Figure G.6; if we remove one straight line, the number of angular domains decreases by 2. □

Problem G.7. *Let $A = A_1A_2A_3A_4$ be a tetrahedron, and suppose that for each $j \neq k$, $[A_j, A_{jk}]$ is a segment of length ρ extending from A_j in the direction of A_k. Let p_j be the intersection line of the planes $[A_{jk}A_{jl}A_{jm}]$ and $[A_kA_lA_m]$. Show that there are infinitely many straight lines that intersect the straight lines p_1, p_2, p_3, p_4 simultaneously.*

Solution. It is natural to assume that we are in the projective space obtained from the Euclidean space by joining ideal elements.

Exclude first the singular cases, and assume the tetrahedron has no edges of length ρ. Let $\{j,k,l,m\} = \{1,2,3,4\}$. Denote the intersection point of the straight lines A_kA_l and $A_{jk}A_{jl}$ by B_{jm}. Obviously, B_{jm} is the intersection point of the plane $S_m = [A_jA_kA_l]$ opposite to A_m and the straight line p_j. Consider now the plane S_j (see Figure G.7). The points B_{kj}, B_{lj}, B_{mj} are adjacent to a straight line e_j, as is easy to see by a multiple application of Menelaos' theorem. This implies that e_j intersects p_1, p_2, p_3, p_4 simultaneously. (Indeed, the points B_{kj}, B_{lj}, B_{mj} belong to p_k, p_l, p_m, respectively, while p_j and e_j are coplanar.) The line e_j is not an edge of the tetrahedron, since it would result, the excluded case. The lines e_1, e_2, e_3, e_4 are all different since they lie in different planes of the tetrahedron, but none of them coincides with an edge.

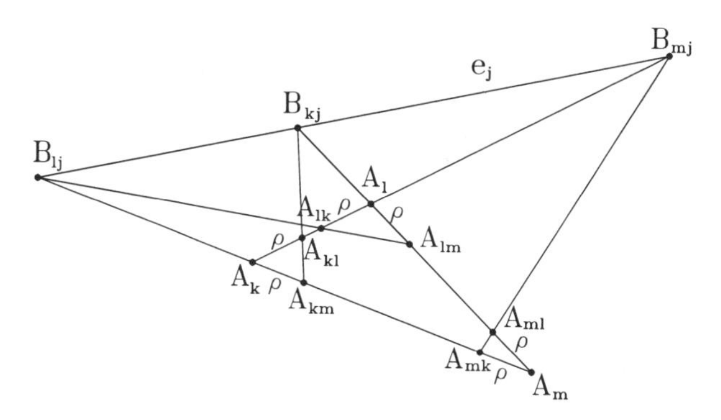

Figure G.7.

Therefore, the lines p_1, p_2, p_3, p_4 are intersected by four different straight lines simultaneously. If two of the lines p_i $(i = 1,2,3,4)$ are intersecting, then their point in common must be contained in the plane spanned by the two other straight lines; hence we can easily find infinitely many straight lines intersecting all four of them.

If p_1, p_2, p_3, p_4 are mutually skew, then take p_1, p_2, p_3. It is known that the straight lines that intersect all of these three lines sweep out a doubly ruled second-order surface. One family of straight lines on this surface is given by the straight lines intersecting p_1, p_2, p_3, while p_1, p_2, p_3 belong to the other family. The line p_4 has more than two points in common with the surface, hence it is also a generator of it and intersects every straight line in the first family.

Now let us discuss the singular cases.

(a) If the three edges starting from one of the vertices A_k have length ρ, then p_k is not properly defined, while the three other lines are adjacent to A_k. As p_k is not determined, the statement cannot be applied to this case. If, however, we define p_k as an arbitrary straight line in the two coinciding planes that should define it, the statement is clearly true.

(b) If there are two edges of length ρ starting from a vertex A_k, then denote by A_l the fourth vertex, not lying on these edges. It is easy to see that in this case p_k is the intersection of the planes S_k and S_l, $p_l \subset S_l$, while the two other lines, p_j, p_m, go through A_k. This means that any straight line e such that $A_k \in e \subset S_l$ intersects all the lines p_i.

(c) Assume now that the tetrahedron has exactly one edge of length ρ. In this case, one of the straight lines e_i coincides with this edge, and therefore we can only say that there are three different straight lines among e_1, e_2, e_3, e_4. However, to apply the arguments used in the generic case, it suffices to have three straight lines that intersect all of p_1, p_2, p_3, p_4. Thus, p_1, p_2, p_3, p_4 are intersected by infinitely many straight lines simultaneously.

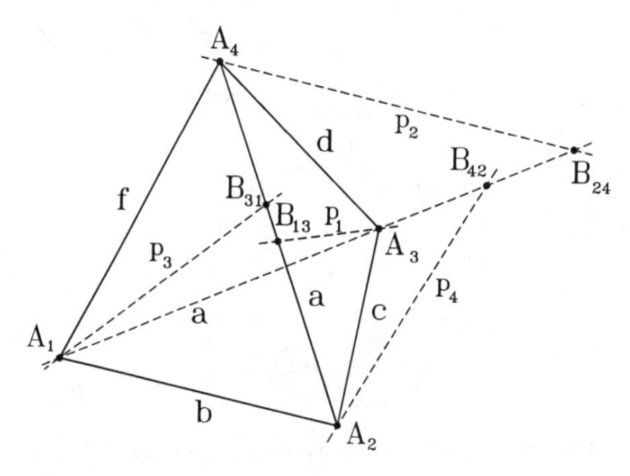

Figure G.8.

(d) Assume, finally, that the tetrahedron has two edges of length ρ, opposite to one another, say $A_1A_3 = A_2A_4 = \rho = a$. Introduce the notation $A_1A_2 = b$, $A_2A_3 = c$, $A_3A_4 = d$, $A_1A_4 = f$ (see Figure G.8). Applying Menelaos' theorem, we obtain the following divided ratios:

$$(A_1A_3B_{24}) = \frac{b-a}{a-c}, \qquad (A_1A_3B_{42}) = \frac{f-a}{a-d},$$

$$(A_2A_4B_{31}) = \frac{c-a}{a-d}, \qquad (A_2A_4B_{13}) = \frac{b-a}{a-f}.$$

Turning to cross ratios,

$$(A_1A_3B_{24}B_{42}) = \frac{(b-a)(a-d)}{(a-c)(f-a)} = (A_2A_4B_{13}B_{31}),$$

but then

$$(A_1A_3B_{24}B_{42}) = (B_{31}B_{13}A_4A_2).$$

Let us take into consideration that A_1 and B_{31} are on p_3; A_3 and B_{13} are on p_1; B_{24} and A_4 are on p_2; B_{42} and A_2 are on p_4. The equality we have just obtained means that p_1, p_2, p_3, p_4 cut the skew lines A_1A_3 and A_2A_4 in two 4-tuples having the same cross ratio. Thus these lines belong to the same family of generators on a doubly ruled second-order surface. This implies the statement of the problem. □

Problem G.8. *Consider the radii of normal curvature of a surface at one of its points P_0 in two conjugate directions (with respect to the Dupin indicatrix). Show that their sum does not depend on the choice of the conjugate directions. (We exclude the choice of asymptotic directions in the case of a hyperbolic point.)*

Solution. If P_0 is a parabolic point, then the indicatrix is a couple of parallel lines, and a direction not parallel to them is conjugate only to their (that is, the asymptotic) direction. The radius of normal curvature is infinite in the asymptotic direction and finite in any other direction, so the sum of the radii of normal curvature in two conjugate directions is always infinite.

It is known that the absolute value of the radius of normal curvature in a given direction is the square of the length of the segment from P_0 to the indicatrix point in the given direction. According to this, we need only to show that the sum (for the elliptic case) or the difference (for the hyperbolic case) of the square of the length of conjugate half-diameters of the Dupin indicatrix does not depend on the choice of the conjugate diameters.

A theorem of Apollonius states that the indicatrix is an ellipse at an elliptic point.

Now suppose that P_0 is a hyperbolic point, that is, the indicatrix is a pair of conjugate hyperbolas whose equation is of the form

$$\frac{x^2}{a^2} - \frac{y^2}{b^2} = \pm 1$$

in a properly chosen coordinate system. Let us denote by s_1 and s_2 the common asymptotes of them (see Figure G.9). Let S be a point on one of the hyperbolas and e_1 be the tangent of the hyperbola at S.

This tangent intersects s_1 at the point L, s_2 at the point M. Draw a tangent (different from s_2) from M to the other hyperbola. This tangent touches the hyperbola at the point T and crosses s_1 at N. It is known that the area of the triangle enclosed by the asymptotes and a tangent of the hyperbolas is equal to ab. Thus, $\triangle P_0LM$ and $\triangle P_0MN$ are of the same area, which, due to their common altitude MM^*, yields $\overline{P_0L} = \overline{P_0N}$. It is also known that S halves $\overline{LM}$ and T halves $\overline{MN}$. Hence, $\triangle P_0ST$ is the median triangle of $\triangle LMN$; consequently, $P_0S \parallel e_2$ and

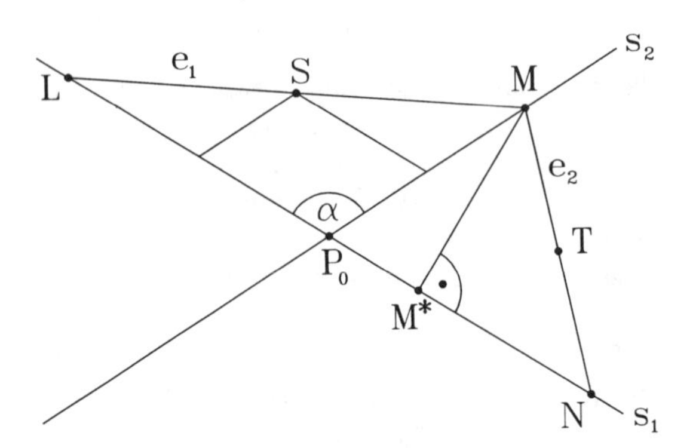

Figure G.9.

$P_0T \parallel e_1$. This means, however, that P_0S and P_0T are conjugate half-diameters. Applying the cosine law for the triangles $\triangle P_0LM$ and $\triangle P_0MN$, and using the equality $\overline{P_0L} = \overline{P_0N}$ shown above, we get

$$\begin{aligned} 4\left(\overline{P_0T}^2 - \overline{P_0S}^2\right) =& \overline{LM}^2 - \overline{NM}^2 \\ =& \overline{LP_0}^2 + \overline{P_0M}^2 - 2\overline{LP_0P_0M}\cos\alpha - \\ & \left[\overline{NP_0}^2 + \overline{P_0M}^2 - 2\overline{NP_0P_0M}\cos(\pi-\alpha)\right] \\ =& -4\overline{LP_0P_0M}\cos\alpha = -16\frac{\overline{LP_0}}{2}\frac{\overline{P_0M}}{2}\cos\alpha, \end{aligned}$$

where $\overline{LP_0}/2$ and $\overline{P_0M}/2$ are the contravariant coordinates of the vector $\overrightarrow{P_0S}$ with respect to a basis consisting of unit vectors pointing in the directions of the asymptotes. The product of them (as it is known) does not depend on the choice of the point S of the hyperbola. In such a way, $\left(\overline{P_0T}^2 - \overline{P_0S}\right)^2$ does not depend on S, that is, on the choice of the conjugate pair of diameters. □

Problem G.9. *Show that a segment of length h can go through or be tangent to at most $2[h/\sqrt{2}]+2$ nonoverlapping unit spheres. ([.] is integer part.)*

Solution. Let us denote by e the supporting straight line of the given segment of length h, project the centers of the spheres onto this straight line, and consider the open intervals of length $\sqrt{2}$ on e centered at the projections of the centers. We claim that no point of e belongs to more than two such intervals. If not, there would exist a cylinder of radius 1 and altitude $m < \sqrt{2}$ containing three points A, B, and C (the centers of the spheres) that are at least 2 units away from one another. We shall show that this is impossible.

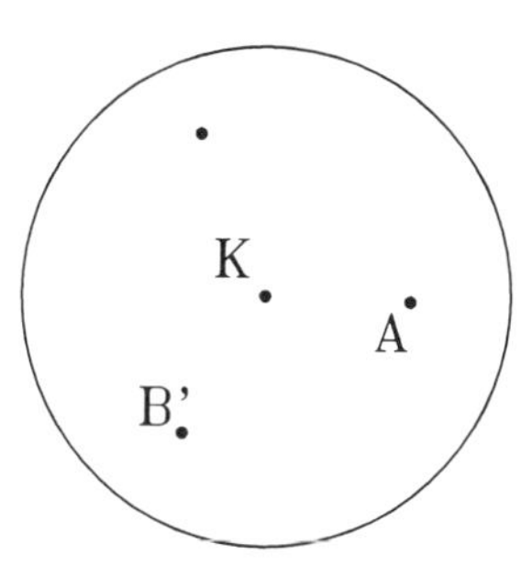

Figure G.10.

In the following, we shall use the term "the plane of a point" for the plane that is orthogonal to e and passes through the point. The plane of A intersects the cylinder in a circle centered at K. Let us project B and C onto the plane of A. Denoting the projections by B' and C', respectively (see Figure G.10), we have

$$m^2 + \overline{B'A}^2 \geq 2^2 ,$$

that is,

$$\overline{B'A}^2 \geq 2^2 - m^2 > 2 ,$$

hence, $\triangle B'KA$ is obtuse angled at K, since the radius of the circle equals 1. It follows that if we move A to the perimeter of the circle along the radius KA, $\overline{AB'}$ and hence $\overline{AB}$ increase (and similarly, so do $\overline{AC'}$ and hence $\overline{AC}$). The same is true for the other points, so we may suppose that A, B, and C are on the surface of the cylinder, at distance 1 from e. Since we required only $m < \sqrt{2}$ for the height of the cylinder, we may also suppose that one of the points, say A, lies in the upper base of the cylinder and one of the points, say C, lies in the lower base.

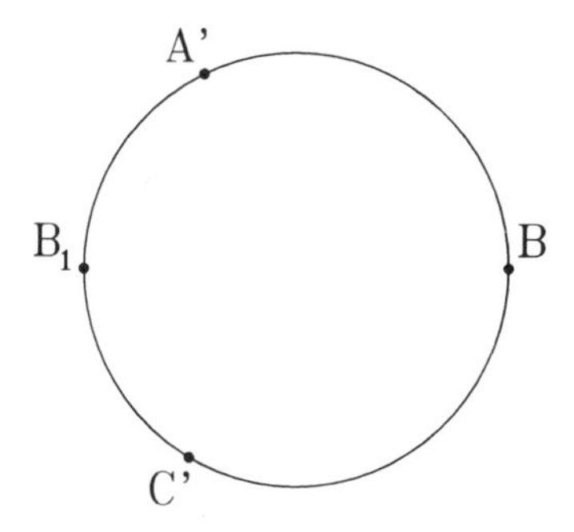

Figure G.11.

Consider the plane of B, and let B_1 be the reflection of B in e, A' and C' be the projections of A and C onto the plane of B, respectively (see Figure G.11). Since the distances between the planes of A, B and C are smaller than $\sqrt{2}$, while the distances between the points are not less than 2, the distances between the projections of these points are greater than $\sqrt{2}$. This implies that the endpoints A' and C' lie on different arcs determined by B and B_1. If we move both A' and C' toward B along the arc, and move A and C in a corresponding way, we can manage to have $\overline{AB} = \overline{CB} = 2$. But in this case $\overline{A'B_1}$ is just the distance of the planes of A and B, and $\overline{C'B_1}$ is just the distance of the planes of C and B. Therefore

$$\overline{A'B_1} + \overline{C'B_1} \leq m < \sqrt{2}\,.$$

Consequently,

$$\overline{A'C'} \leq \overline{A'B_1} + \overline{C'B_1} < \sqrt{2}\,.$$

Thus,

$$\overline{AC}^2 = m^2 + \overline{A'C'}^2 < 2 + 2\,,$$

and so $\overline{AC} < 2$. This is a contradiction that proves the proposition.

Let us elongate the segment of length h in both directions by a segment of length $\sqrt{2}$. Since the segment goes through or is tangent to the spheres, the intervals assigned to the spheres are all contained in this enlarged segment. Since every point of the enlarged segment is covered by at most two intervals, we get that the number of spheres is at most

$$2\left[\frac{h + 2\frac{1}{\sqrt{2}}}{2}\right] = 2\left[\frac{h}{\sqrt{2}}\right] + 2\,.$$

To see this, we put the intervals in an increasing order by their left endpoints (this ordering is not always unique). Then the intervals standing at odd (or even) positions are necessarily disjoint. Thus, the number of all intervals is at most

$$\left[\frac{h}{\sqrt{2}}\right] + 1 + \left[\frac{h}{\sqrt{2}}\right] + 1\,.$$

It is easy to see that the given estimation is exact. □

Problem G.10. *Characterize those configurations of n coplanar straight lines for which the sum of angles between all pairs of lines is maximum.*

Solution. We may suppose that all straight lines go through one point. Starting from an arbitrary straight line, let us number the straight lines in counterclockwise orientation (the order of coinciding straight lines is arbitrary): $e_1, \ldots, e_n$. For $1 \leq i \leq [n/2]$, we say that e_l is the ith neighbor of e_k if

$$l \equiv k + i \pmod{n}.$$

We shall denote the ith neighbor of e_k by e_k^i. Let us denote by (e, f) the angle between the straight lines e and f and denote by $< e, f >$ the angle of the counterclockwise rotation that takes e to f. For i fixed,

$$\begin{aligned}(e_1,e_1^i)+\cdots+(e_n,e_n^i) &\leq <e_1,e_1^i>+\cdots+<e_n,e_n^i> \\ &= i(<e_1,e_2>+<e_2,e_3>+\cdots+<e_n,e_1>) = i\pi,\end{aligned}$$

and equality holds if and only if

$$< e_k, e_k^i > \leq \frac{\pi}{2} \tag{1}$$

for all k. Thus, for odd n, the sum of angles between all pairs of straight lines can be estimated as follows:

$$\sum_{i=1}^{\frac{n-1}{2}} \left[(e_1, e_1^i) + \cdots + (e_n, e_n^i)\right] \leq \sum_{i=1}^{\frac{n-1}{2}} i\pi = \frac{n^2 - 1}{4}\frac{\pi}{2}, \tag{2}$$

and equality holds if and only if (1) holds for all $k = 1, 2, \ldots, n$ and $i = 1, 2, \ldots, (n-1)/2$. For this, it is enough to require that

$$< e_k, e_k^{(n-1)/2} > \leq \frac{\pi}{2} \tag{3}$$

hold for $k = 1, 2, \ldots, n$.

For even n, the sum of angles is

$$\begin{aligned}&\sum_{i=1}^{\frac{n}{2}-1} \left[(e_1, e_1^i) + \cdots + (e_n, e_n^i)\right] + \frac{1}{2}\left[(e_1, e_1^{n/2}) + \cdots + (e_n, e_n^{n/2})\right] \\ &\qquad \leq \sum_{i=1}^{\frac{n}{2}-1} i\pi + \frac{n}{2}\frac{\pi}{2} = \frac{n^2}{4}\frac{\pi}{2},\end{aligned}$$

and equality holds if and only if

$$< e_k, e_k^{n/2} > \leq \frac{\pi}{2}$$

holds for $k = 1, 2, \ldots, n$. For even n, these inequalities imply the requested characterization, since the inequalities

$$< e_k, e_k^{n/2} > \leq \frac{\pi}{2} \quad \text{and} \quad < e_{k+n/2}, e_{k+n/2}^{n/2} > \leq \frac{\pi}{2}$$

can be satisfied simultaneously only if

$$< e_k, e_k^{n/2} > = \frac{\pi}{2},$$

that is the $(n/2)$th neighbors are perpendicular to one another. To summarize, if n is even, the maximum is attained if the family of straight lines is composed of $n/2$ orthogonal pairs and any family of orthogonal pairs of straight lines yields maximal sum of angles.

Returning to the case of odd n, let us observe that orthogonal couples of straight lines can be removed from the family, since the omission of an orthogonal pair there decreases the sum of angles by $(n-1)\frac{\pi}{2}$ and

$$\frac{n^2}{4} - \frac{(n-2)^2}{4} = n - 1\,.$$

Suppose that after the removal of orthogonal pairs there remain $2m+1$ straight lines. These straight lines must be different, because if e were a double line then the mth neighbor of the mth neighbor f of e would be e, which would mean that e and f are perpendicular. If e is an arbitrary member of the $2m+1$ straight lines, then conditions (3) imply that each quadrant of the plane determined by e and the straight line perpendicular to e must meet exactly m straight lines. Adding to such a family of an odd number of straight lines an arbitrary number of orthogonal pairs, we get all the configurations that yield the maximum, and the maximum is equal to $[(n^2-1)/4] \cdot (\pi/2)$ by (2).

Finally, if we do not suppose that the straight lines go through one point, then taking a configuration characterized above and replacing each straight line by a parallel one, we can get all the requested configurations. □

Problem G.11. *Let $f(n)$ denote the maximum possible number of right triangles determined by n coplanar points. Show that*

$$\lim_{n\to\infty} \frac{f(n)}{n^2} = \infty \quad \textit{and} \quad \lim_{n\to\infty} \frac{f(n)}{n^3} = 0.$$

Solution.

Lower estimation. We shall show the following sharper result. There exists a constant $c > 0$ such that

$$f(n) > cn^2 \log n \quad (n \geq 3)\,.$$

Suppose first that $n = (3k+1)^2$, where k is a natural number. Consider those points of the lattice of points with integer coordinates, which lie in the square spanned by the vertices $(0,0)$, $(0,3k)$, $(3k,3k)$, $(3k,0)$. The number of these points is obviously $(3k+1)^2$. We shall count only those right triangles for which the right-angled vertex (q,r) satisfies $k \leq q \leq 2k$, $k \leq r \leq 2k$ and the two other vertices of which lie in the square of sides $2k$ centered at (q,r). Obviously, all these right triangles are good for us.

We have to determine the number of right triangles spanned by the lattice points of the square $(-k,-k)$, $(-k,k)$, (k,k), $(k,-k)$ so that the right-angled vertex lies in the origin $(0,0)$.

The perpendicular sides of such a right-angled triangle lie on straight lines given by equations of the form $ix = jy$ and $jx = -iy$, where $(i, j) = 1$, $-k \leq i, j \leq k$. Obviously, we estimate the number of right triangles from below if we take into consideration only those straight lines that satisfy $0 \leq i \leq j$ as well. In this case, each of the straight lines $ix = jy$ and $jx = -iy$ contains $2\,[k/j]$ lattice points different from $(0,0)$, so the number of right-angled triangles determined by them is

$$4\left[\frac{k}{j}\right]^2 \geq \frac{k^2}{j^2}\,.$$

For a fixed j, we find at least

$$\varphi(j)\frac{k^2}{j^2}$$

right triangles (φ denotes Euler's phi function), which gives

$$k^2\sum_{j=1}^{k}\frac{\varphi(j)}{j^2}$$

right-angled triangles altogether. Now returning to the original problem, the $(3k+1)^2$ points we constructed yield at least

$$(2k+1)^2k^2\sum_{j=1}^{k}\frac{\varphi(j)}{j^2}$$

right triangles.

Introducing the function

$$\Phi(x) = \sum_{j\leq x}\varphi(j)$$

and summing up partially, we get

$$f((3k+1)^2) \geq (2k+1)^2k^2\sum_{j=1}^{k-1}\frac{\Phi(j)(2j+1)}{j^2(j+1)^2}\,.$$

Combining this with the well-known inequality

$$\Phi(x) > c_1x^2, \tag{1}$$

we get

$$f((3k+1)^2) \geq c_2(2k+1)^2k^2\sum_{j=1}^{k-1}\frac{1}{j} \geq c_3k^4\log k\,.$$

Now let n be an arbitrary number. Set $k = [(\sqrt{n}-1)/3] > \sqrt{n}/4$. Then

$$f(n) \geq f((3k+1)^2) \geq c_3 k^4 \log k \geq cn^2 \log n\,.$$

Upper estimation. We shall prove a sharper result again: using the "descente infinie" method we show that

$$f(n) \leq n^2\sqrt{n}\,. \tag{2}$$

The statement is trivial for $n = 1, 2, 3, 4, 5$ (for these values $n^2\sqrt{n} > \binom{n}{3}$). If (2) were not true, then there would be a smallest natural number n such that

$$f(n) > n^2\sqrt{n}\,.$$

Let us take n points $P_1, \ldots, P_n$ on the plane that span $f(n)$ right-angled triangles. We claim that there must be a straight line that contains from the n points at least $2\sqrt{n}$ ones. Consider all possible ordered couples (P_i, P_j), and assign to each of them the number of right triangles $P_iP_jP_k$ such that the right angle is at P_i. This number is the number of points different from P_i on the straight line perpendicular to P_iP_j at P_i. The sum of these numbers is twice the number of right triangles and so it is at least $2n^2\sqrt{n}$. Since the number of summands is $n(n-1)(< n^2)$, one of the summands is at least $2\sqrt{n}$, and thus one of the perpendicular lines contains at least $2\sqrt{n}$ points.

Let us remove the points of this straight line and examine how many right triangles are destroyed by this at most. We divide the triangles destroyed into two groups:

(A) triangles for which the right-angled vertex is omitted;

(B) triangles with right-angled vertex lying off and one of the other vertices lying on the critical straight line.

We can give an upper bound for the number of triangles of type A in such a way that we have $n(n-1)/2$ choices for the hypotenuse of such a triangle, and a segment P_iP_j may serve as hypotenuse for at most two right triangles of type A (Thales' circle!), so the number of triangles of type A is at most $n(n-1) < n^2$.

In the case of triangles of type B, we have $n(n-1)/2$ choices for those vertices that are not required to lie on the critical straight line, and we can choose at most two points on the critical straight line which form a triangle of type B together with a fixed couple of points P_iP_j so that the right-angled vertex is either at P_i or at P_j. It follows that the number of triangles of type B is also at most $n(n-1) < n^2$.

In such a way, the removal of the points of the critical straight line destroys at most $2n^2$ right-angled triangles; consequently, the remaining, at most, $n-2$ points span at least $n^2\sqrt{n} - 2n^2$ right triangles. By the induction hypothesis,

$$n^2\sqrt{n} - 2n^2 \leq (n-2\sqrt{n})^2\sqrt{n-2\sqrt{n}} \leq (n-2\sqrt{n})^2\sqrt{n}\,,$$

from which

$$2n^2 \le 4n\sqrt{n}\,,$$

that is,

$$n \le 4\,,$$

for which cases we have already seen that the assertion is true. □

Remarks.

1. We could have avoided the use of (1) using only the elementary fact that $\varphi(p) = p-1$ for prime numbers. However, this would have given us only the estimation

$$f(n) \ge c_4 n^2 \log\log n\,,$$

where c_4 is a suitable positive constant.

2. László Lovász remarks that there is a positive constant c_5 such that the number of right triangles in the lattice construction described in the first part of the solution is at most

$$c_5 n^2 \log n\,.$$

3. Béla Bollobás proves the statement of the problem in the following more general form: Let us denote by $f(n,k)$ the maximal number of right-angled triangles spanned by n points of the $k(\ge 2)$-dimensional space, where the maximum is taken for all possible configurations of the points. Then

$$cn^2 \log n \le f(n,k) \le d_k n^{3-2^{1-k}}\,,$$

where c and d_k are positive constants. The lower bound comes from our previous arguments because of $f(n,k) \ge f(n,2)$. We prove the upper bound for $k = 3$. Denote by R a collection of n points in the three-dimensional space, and denote by p the number of those straight lines that contain more than $3\sqrt{n}$ points of R. Let $e_1, e_2, \ldots, e_p$ stand for these straight lines. e_1 contains at least $3\sqrt{n}$ points of R, e_1 and e_2 together contain at least $3\sqrt{n}(3\sqrt{n}-1)$ points, because two different straight lines may have at most one common point, and so on, $e_1, \ldots, e_{p'}$ cover at least

$$\sum_{i=1}^{p'-1} (3\sqrt{n}-1) \ge p' 3\sqrt{n} - p'^2$$

points of R for every $p' \le p$. Consequently, $p \le \sqrt{n}$. We prove that if we remove at most $\sqrt{n}$ "big" straight lines, the number of right triangles decreases at most by $4n^{5/2}$. Let e be a "big" straight line. Those right triangles that have at least one vertex on e can be divided into two classes:

i. *Triangles that have two vertices on e.* The third vertex of such a triangle can be chosen out of at most n points. This point, together with a point on e, leaves only two possibilities, and so the number of right triangles of this class is at most $2n^2$.

ii. *Triangles that have only one vertex on e.* We have $n(n-1)/2 < n^2/2$ choices for the two other vertices, which form a right-angled triangle together with at most 4 points on e, thus the number of right triangles of this type is less than $4n^2/2 = 2n^2$. Since we have at most $\sqrt{n}$ "big" straight lines, their removal causes the vanishing of not more than $\sqrt{n}4n^2 = 4n^{5/2}$ right triangles. After the removal, each straight line contains at most $3\sqrt{n}$ points. Let us denote by q the number of planes passing through the point P and least $6n^{3/4}$ points of the remaining family. By the idea used above, we get

$$q'6n^{3/4} - q'^2 3\sqrt{n} \leq n$$

for all $q' \leq q$, since two planes have at most $3\sqrt{n}$ common points. From here, $q \leq n^{1/4}$. There are less than $n^{1/4}3\sqrt{n}$ triangles satisfying that its right-angled vertex is at P, one of its perpendicular sides lies in one of these "big" planes, while the other is perpendicular to the plane. The number of right triangles with right-angled vertex at P, not satisfying these conditions, can be estimated by the sum $\sum_{i=1}^{l} f_i s_i$, where l denotes the number of orthogonal pairs of a plane and a straight line passing through P such that both the plane and the straight line contains at least one point different from P and f_i, s_i denote the number of points different from P lying on the plane and the straight line of the ith pair, respectively.
But then

$$\sum f_i s_i \leq 6n^{3/4} \sum f_i \leq 6n^{7/4}.$$

Thus, the number of right triangles in R is less than

$$4n^{5/2} + 6n^{7/4}n \leq 10n^{11/4}.$$

We remark that the case $k > 3$ requires the consideration of not only straight lines and l ($l < k$)-dimensional linear subspaces but also that of circles and spheres.

Problem G.12. *Suppose that a bounded subset S of the plane is a union of congruent, homothetic, closed triangles. Show that the boundary of S can be covered by a finite number of rectifiable arcs.*

Solution 1. Let us denote the boundary of S by H and let $P \in H$ be an arbitrary point. Then one can find a sequence of points $\{P_n\}$ converging to P such that each point P_n lies on the boundary of a triangle $\triangle_n$ constituting S.

We may suppose that each point P_n lies on the boundary of $\triangle_n$ at the same position (that is, the translation that takes $\triangle_n$to $\triangle_m$ takes P_n to P_m). (This can be shown by our assumptions. Translate the triangles $\triangle_n$ to a fixed triangle $\triangle$; denote by Q_n the point corresponding to P_n. The sequence $\{Q_n\}$ is bounded, and it thus contains a convergent subsequence

$\{Q_{n_k}\}$, which tends to Q. Translating $\triangle$ back to $\triangle_{n_k}$, Q corresponds to P'_{n_k}. Obviously, $P'_{n_k} \to P$ and the points P'_{n_k} lie at the same position on the boundary of $\triangle_{n_k}$). Obviously, each triangle $\triangle_n$ contains a homothetic triangle $\triangle'_n$ of half size such that P_n is a vertex of $\triangle'_n$ lying at the same position for all n. These triangles converge to a homothetic triangle $\triangle'_P$ with vertex P such that the interior of $\triangle'_P$ contains no point of H.

Let us number the three vertex positions of the triangles homothetic to $\triangle_n$. This gives rise to a natural ordering of the vertices of triangles homothetic to $\triangle_n$. Let us define the sets $H_i \subset H$ $(i = 1, 2, 3)$ in such a way that a point $P \in H$ belongs to H_i if and only if P is the ith vertex of $\triangle'_P$. Obviously, $H = H_1 \cup H_2 \cup H_3$.

Let us study H_1. Denote by α the angle lying at the first vertex of $\triangle'_P$. Let us introduce a coordinate system on the plane so that the y-axis shows in the direction of the bisectrix of the angle α, the direction of the x-axis coincides with the direction of the supplementary angle of α. Let us divide the plane into bands of equal width parallel to the x-axis. If the width of the bands is chosen to be small enough (for example, if it is smaller than the bisectrix of $\triangle'_P$ that passes through P), then the intersection of H_1 and one of the closed bands is the graph of a function $y = f(x)$, where f is defined on a bounded subset of the x-axis and satisfies Lipschitz condition with constant $k = \cot(\alpha/2)$, which gives the possibility to extend it to the closure of its original domain. The extension of f is also k-Lipschitz. After this, by extending f to the components of the open complement of its domain in a linear way, we can obtain a k-Lipschitz extension of f, defined on a sufficiently large bounded interval. The graph of this extension is a rectifiable curve that covers the intersection of H_1 and the considered band. Since S is bounded, it crosses only a finite number of bands; thus, H_1 can be covered by a finite number of rectifiable arcs. We can proceed similarly for H_2 and H_3, so the theorem is proved. □

Solution 2. Let $\mathbb{R}^2$ be the two-dimensional Euclidean vector space, and let $H \subset \mathbb{R}^2$ be a nonsingular closed triangle whose angles are the regions A_1, A_2, A_3, that is,

$$H = A_1 \cap A_2 \cap A_3 . \tag{1}$$

The statement of the problem can be reformulated as follows.

Statement. If M $(\subset \mathbb{R}^2)$ is bounded, than the boundary of $M + H$ can be covered by a finite number of rectifiable arcs. (From now on, we use operations on complexes.)

Since M is bounded, there exists a finite set $V \subset \mathbb{R}^2$ such that $M \subset V + \frac{1}{2}H$. Setting $M(v) = \left(v + \frac{1}{2}H\right) \cap H$, we get

$$M = \bigcup_{v \in V} M(v) ; \tag{2}$$

hence,

$$M + H = \bigcup_{v \in V} (M(v) + H) . \tag{3}$$

First we show that

$$M(v) + H = \bigcap_{i=1}^{3} (M(v) + A_i) \tag{4}$$

for all $v \in V$. It is clear from (1) that

$$M(v) + H \subseteq \bigcap_{i=1}^{3} (M(v) + A_i) .$$

Suppose now that

$$p \in \bigcap_{i=1}^{3} (M(v) + A_i) .$$

Then $p \in M(v) + A_i$ $(i = 1, 2, 3)$, but

$$(p - M(v)) \cap (A_i \setminus H) \subseteq (p - v - \frac{1}{2}H) \cap (A_i \setminus H) = \emptyset$$

must be satisfied at least for one i. Thus

$$(p - M(v)) \cap H \neq \emptyset ;$$

and this means that

$$p \in M(v) + H .$$

Hence,

$$M(v) + H \supseteq \bigcap_{i=1}^{3} (M(v) + A_i)$$

holds as well, and this proves (3).

Let us denote by fr (X) the boundary of a set X. Using that V is finite, (3) and (4) yield

$$\text{fr } (M + H) \subseteq \bigcup_{v \in V} \bigcup_{i=1}^{3} \text{fr } (M(v) + A_i) .$$

Now it is enough to show the following proposition.

Proposition. Let A be the angular domain of angle 2α $(0 < 2\alpha < \pi)$, the vertex of which is the origin and the bisectrix of which is the negative y-axis. Let $M \subset \mathbb{R}^2$ be a bounded set; furthermore,

$$T = \{(x, y) \in \mathbb{R}^2 \colon 0 \leq x \leq 1\} .$$

Then

$$G = \text{fr } (M + A) \cap T$$

is a rectifiable curve.

Proof. To show this, let $(t, g(t))$ denote for $0 \leq t \leq 1$ the point of G the whose abscissa is t, that is, set

$$g(t) = \sup\{y - |x - t| \cot \alpha \colon (x, y) \in M\} .$$

Since g satisfies the Lipschitz condition

$$|g(t) - g(t')| \leq |t - t'| \cot \alpha ,$$

the latter proposition is trivial. $\square$

Problem G.13. *Let F be a surface of nonzero curvature that can be represented around one of its points P by a power series and is symmetric around the normal planes parallel to the principal directions at P. Show that the derivative with respect to the arc length of the curvature of an arbitrary normal section at P vanishes at P. Is it possible to replace the above symmetry condition by a weaker one?*

Solution 1. Since the normal planes parallel to the principal directions are perpendicular, the symmetry of F around them implies the symmetry around their intersection, that is, around the normal of F at P. Since the plane of normal sections contains the normal line, normal sections at P are also symmetric around the normal. The analyticity of the surface ensures that the normal sections have arc length and curvature at each point, which is a differentiable function of the arc length. Let $g(s)$ be the curvature expressed as a function of the arc length. If P corresponds to the parameter s_0, then the symmetry of the normal section around the normal at P gives the identity $g(s_0+s)=g(s_0-s)$. Differentiating with respect to s at $s=0$, we get $g'(s_0)=-g'(s_0)$, that is, $g'(s_0)=0$, which proves the proposition.

As we see, symmetry around the normal planes parallel to the principal directions can be substituted by the symmetry around the normal of F at P. □

Solution 2. By our assumption, introducing a coordinate system with origin at P, the surface is given by an equation of the form

$$z=\sum_{i,k=0}^{\infty} a_{ik}x^i y^k \,,$$

where $a_{00}=0$. We may also suppose that the x- and y-axes show in the principal directions of F. In this case, $a_{10}=a_{01}=0$. With such a choice of coordinate system, according to our assumptions, $z(x,y)=z(-x,y)=z(-x,-y)=z(x,-y)$, which can be the case only if $a_{ik}=0$ whenever either i or k is odd. Then the equation of the surface has the form

$$z=\sum_{i,k=0}^{\infty} a_{2i,2k}x^{2i}y^{2k}\,.$$

A normal section at P can be parameterized as follows:

$$\mathbf{r}=\mathbf{r}(t):\ x=c_1t,\ y=c_2t\,,$$

$$z=\sum_{i,k=0}^{\infty} a_{2i,2k}(c_1t)^{2i}(c_2t)^{2k}==\sum_{n=1}^{\infty} b_nt^{2n}\,,$$

where the constants c_1, c_2 $(c_1^2+c_2^2\neq 0)$ depend on the plane of the normal section, and the coefficients b_n are functions of $a_{2i,2k}$.

The curvature of the curve $\mathbf{r} = \mathbf{r}(t)$ can be computed by the formula

$$g(t) = \frac{\sqrt{|\mathbf{r}'|^2 |\mathbf{r}''|^2 - (\mathbf{r}' \cdot \mathbf{r}'')^2}}{|\mathbf{r}'|^{3/2}}.$$

By our assumption, $g(0) \neq 0$, and instead of needing to prove the equation $dg/ds = 0$, it is enough to show the relation $g'(0) = 0$, since

$$\frac{dg}{ds} = \frac{dg}{dt}\frac{dt}{ds} = \frac{dg}{dt}\frac{1}{|\mathbf{r}'|}.$$

Taking into consideration that

$$\mathbf{r}' = \left(c_1, c_2, \sum_{n=1}^{\infty} 2n\ b_n t^{2n-1}\right),$$

$$\mathbf{r}'' = \left(0, 0, \sum_{n=1}^{\infty} 2n(2n-1)\ b_n t^{2n-2}\right),$$

$$\mathbf{r}''' = \left(0, 0, \sum_{n=1}^{\infty} 2n(2n-1)(2n-2)\ b_n t^{2n-3}\right),$$

the relation $g'(0) = 0$ is obtained by direct computation.

The above solution shows that in order to weaken the symmetry condition, it is enough to require that the equation $z = \sum_{i,k=0}^{\infty} a_{ik} x^i y^k$ of the surface in the coordinate system attached to the normal and principal directions does not contain terms of order three. □

Problem G.14. *Let $\sigma(S_n, k)$ denote the sum of the kth powers of the lengths of the sides of the convex n-gon S_n inscribed in a unit circle. Show that for any natural number greater than 2 there exists a real number k_0 between 1 and 2 such that $\sigma(S_n, k_0)$ attains its maximum for the regular n-gon.*

Solution. We shall show the following sharper result.

If $1 \leq k \leq (\tan(\pi/n))/(\pi/n)$, then $\sigma(S_n, k) \leq \sigma(S_n^*, k)$ for any convex n-gon inscribed in a unit circle, where S_n^* is the regular n-gon inscribed in a unit circle.

Proof. To show this statement, it is enough to consider only those polygons S_n that contain the center of the circle on the boundary or inside. Indeed, let S_n be a polygon not satisfying this condition, and let $A_1, \dots, A_n$ be the vertices of S_n. Suppose that A_1A_2 is the closest side of S_n to the center of the circle. Denote by A'_2 the antipodal pair of A_1, and consider the polygon S'_n with vertices $A_1, A'_2, A_3, \dots, A_n$. Then $\overline{A_1A'_2} > \overline{A_1A_2}$, and $\overline{A'_2A_3} > \overline{A_2A_3}$ (since the angle opposite to the side $A_3A'_2$ in $\triangle A'_2A_2A_3$ is obtuse). For $k \geq 1$, obviously $\sigma(S'_n, k) > \sigma(S_n, k)$.

Now let S_n be a convex n-gon inscribed in a unit circle, such that S_n contains the center of the circle on the boundary or inside. Let $x_1, \ldots, x_n$ denote half of the central angles corresponding to the sides of S_n. Then

$$\sigma(S_n, k) = 2^k \sum_{i=1}^{n} \sin^k x_i, \text{ where } 0 < x_i \leq \frac{\pi}{2}, \sum_{i=1}^{n} x_i = \pi.$$

Our proposition is well known for $k = 1$, so it suffices to deal with the case $k > 1$. Let us fix the exponent $1 < k \leq (\tan(\pi/n)/(\pi/n)$. Since the function $(\tan x)/x$ is continuous and strictly increasing in the interval $(0, \pi/2$, furthermore, $\lim_{x \to 0} (\tan x)/x = 1$, there exists a unique real number $a = a(k)$ such that $0 < a < \pi/n$ and $k = (\tan a)/a$.

Consider the function $f(x) = \sin^k x$. Since

$$f''(x) = k(k-1) \sin^{k-2} x \cos^2 x - k \sin^k x = k \sin^{k-2} x (k \cos^2 x - 1),$$

one can easily check that $f(x)$ is convex on the interval $\left(0, \arccos(1/\sqrt{k})\right)$ and concave on the interval $\left(\arccos(1/\sqrt{k}), \pi/2\right)$. Therefore, it is concave on the interval $(a, \pi/2)$ because

$$f''(a) = \frac{\tan a}{a} \sin^{k-2} a \left(\frac{\tan a}{a} \cos^2 a - 1 \right) < 0.$$

Set

$$g(x) = \begin{cases} k \sin^{k-1} a \cos ax, & \text{if } 0 \leq x \leq a, \\ \sin^k x, & \text{if } a < x \leq \pi/2. \end{cases}$$

Obviously, g is continuous in the interval $[0, \pi/2]$. One can see easily by a study of the derivative that the function $(\sin^k x)/x$ is strictly increasing on the interval $[0, a]$ and strictly decreasing on the interval $[a, \pi/2]$. One can derive from this that g is convex on $[0, \pi/2]$ and also that $f(x) \leq g(x)$ holds for $x \in [0, \pi/2]$. These facts, together with the inequality

$$\sum_{i=1}^{n} g(x_i) \leq n g\left(\frac{\pi}{n}\right),$$

obtained by an application of Jensen's inequality, yield

$$\sigma(S_n, k) = 2^k \sum_{i=1}^{n} f(x_i) \leq n 2^k g\left(\frac{\pi}{n}\right).$$

Since $a \leq \pi/n$, we have $f(\pi/n) = g(\pi/n)$, and thus

$$\sigma(S_n, k) = \leq n 2^k \sin^k \frac{\pi}{n} = \sigma(S_n^*, k).$$

The proposition is proved. □

Remark. Many contestants observed that if $3 \leq n' < n$ and $1 \leq k \leq (\tan(\pi/n))/(\pi/n)$, then we also have the inequality $\sigma(S_{n'}, k) \leq \sigma(S_n^*, k)$. Indeed, considering the derivative of the function $y = x\sin^k(\pi/x)$, we see that

$$y' = \sin^k \frac{\pi}{x} - \frac{k\pi}{x}\sin^{k-1}\frac{\pi}{x}\cos\frac{\pi}{x} = \sin^{k-1}\frac{\pi}{x}\left(\sin\frac{\pi}{x} - \frac{k\pi}{x}\cos\frac{\pi}{x}\right) > 0$$

provided that $x > 2$ and $k < (\tan(\pi/x))/(\pi/x)$. Using the relations $\pi/n' > \pi/n$ and

$$\frac{\tan\frac{\pi}{n'}}{\frac{\pi}{n'}} > \frac{\tan\frac{\pi}{n}}{\frac{\pi}{n}},$$

we obtain

$$\sigma(S_n^*, k) = 2^k n \sin^k \frac{\pi}{n} > 2^k n' \sin^k \frac{\pi}{n'} = \sigma(S_n^*, k) \geq \sigma(S_n, k).$$

Problem G.15. *Let h be a triangle of perimeter 1, and let H be a triangle of perimeter λ homothetic to h. Let $h_1, h_2, \ldots$ be translates of h such that, for all i, h_i is different from h_{i+2} and touches H and h_{i+1} (that is, intersects without overlapping). For which values of λ can these triangles be chosen so that the sequence $h_1, h_2, \ldots$ is periodic? If $\lambda \geq 1$ is such a value, then determine the number of different triangles in a periodic chain $h_1, h_2, \ldots$ and also the number of times such a chain goes around the triangle H.*

Solution. We may restrict ourselves to the case of regular triangles since any triangle can be transformed into a regular one by an affine transformation and affinities preserve homothetic and touching position of triangles and also the ratio of perimeters of homothetic triangles.

Let A, B, C be the vertices of the triangle H; let P_i, Q_i, R_i be the vertices of the triangle h_i; and let a, b, c and p_i, q_i, r_i be the sides of the triangles H and h_i opposite to the vertices A, B, C and P_i, Q_i, R_i, respectively. These sides are considered to be half-open segments excluding the endpoint C from side a, the endpoint R_i from side p_i, etc. It is assumed that the vertices A, B, and C correspond homothetically to the vertices P_i, Q_i, and R_i, respectively (see Figure G.12).

There are two ways a triangle h_i can touch the triangle H: either a vertex of H lies on the side of h_i opposite to the homothetically corresponding vertex of h_i, or a vertex of h_i lies on the side of H opposite to the homothetically corresponding vertex. According to our assumptions, we always have exactly one of the two cases. In the first case, we say that h_i touches a vertex of H; in the second, we say that h_i touches a side of H. The common point of the two triangles will be called the point of contact. We fix the orientations (P_i, Q_i, R_i) and (A, B, C) of the triangles h_i and H.

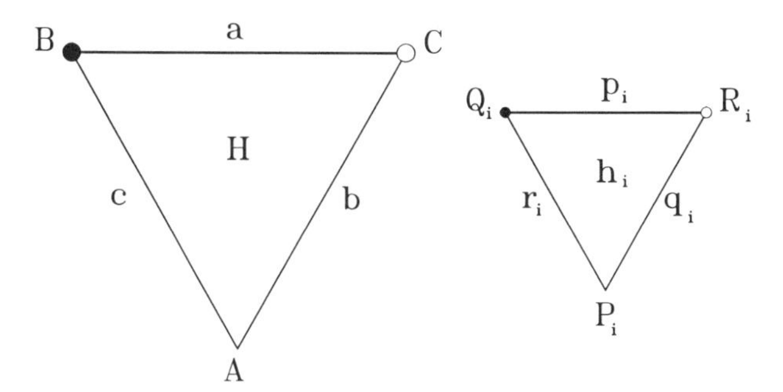

Figure G.12.

If h_i touches a side of H, then denote by $f(h_i)$ the distance between the point of contact and the vertex of H that comes next to the contact point according to the fixed orientation.

If h_i touches a vertex of H, then denote by $f(h_i)$ the distance between the point of contact and the vertex of h_i that comes next to the contact point according to the orientation opposite to the fixed one. Obviously, $0 < f(h_i) \leq \lambda/3$ in the first case, $0 \leq f(h_i) < 1/3$ in the second (see Figure G.13).

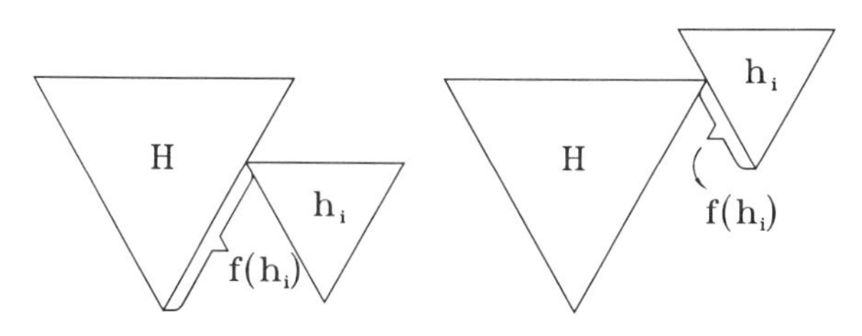

Figure G.13.

It is clear that the contact points of h_i and h_{i+1} are contained in one (closed) side of H. Thus, the fixed orientation of H gives rise to an ordering of the triangles h_i and h_{i+1}.

We may establish the following facts.

Let h' denote an arbitrary translate of h touching H. We want to describe the translates h'' of h that touch both h' and H and come after h' (according to the fixed orientation). (See Figures G.14.a–e).

(a) If h' touches a side of H and $f(h') > 1/3$, then h'' is unique, it touches the same side of H, and the distance of contact points is $1/3$ (thus $f(h'') = f(h') - 1/3$)) (Figure G.14.a).

(b) If h' touches a side of H and $f(h') \leq 1/3$, then h'' is unique, it touches the vertex of H next to the contact point of h' with respect to the

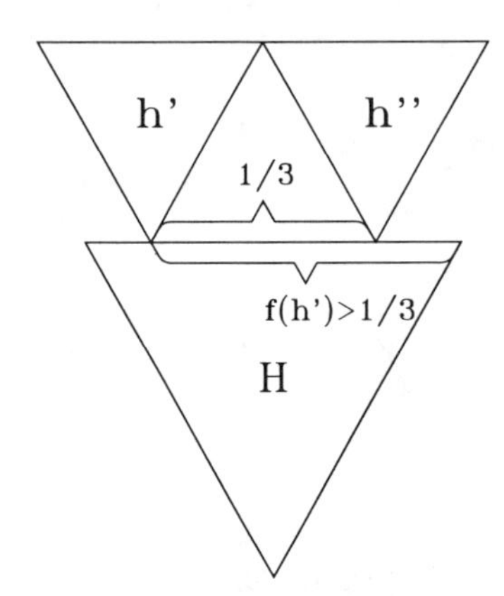

Figure G.14.a

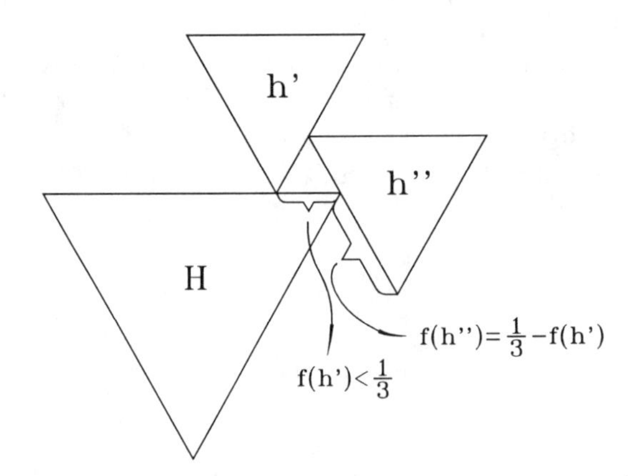

Figure G.14.b

fixed orientation, and $f(h'') = 1/3 - f(h')$ (Figure G.14.b).

(c) If h' touches a vertex of H and $f(h') < \lambda/3$, then h'' is unique, it touches the side of H that contains the contact point of h', and $f(h'') = \lambda/3 - f(h')$ (Figure G.14.c).

(d) If h' touches a vertex of H and $f(h') = \lambda/3$, then we can choose for h'' any triangle that touches the vertex of H next to the contact point of h' satisfying $f(h'') \leq (1-\lambda)/3$ (Figure G.14.d).

(e) Finally, if h' touches a vertex of H and $f(h') > \lambda/3$, then h'' is unique again, it touches the vertex of H next to the contact point of h', and it satisfies $f(h'') = (1-\lambda)/3$ (Figure G.14.e).

I. Suppose first that $\lambda < 1$. We show that in this case one can always find a periodic chain of triangles.

Let h_1 touch the vertex A of H so that $f(h') = 1 - \lambda$. Set $k = -(-\lambda/(1-\lambda))$. Starting from h_1 and advancing according to the fixed orientation, one can construct the sequence $h_2, \dots, h_{2k+1}$ uniquely by the

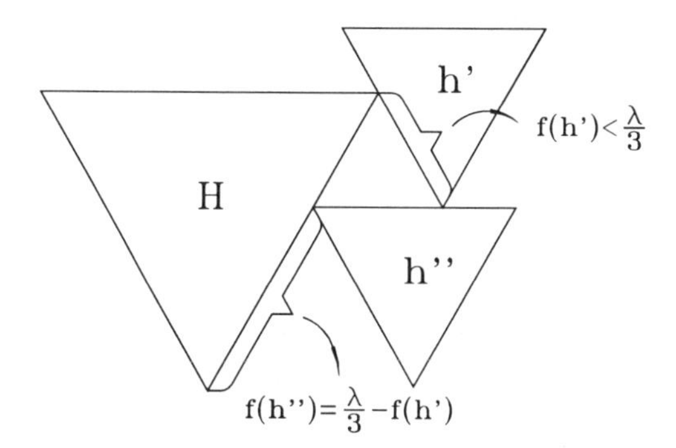

Figure G.14.c

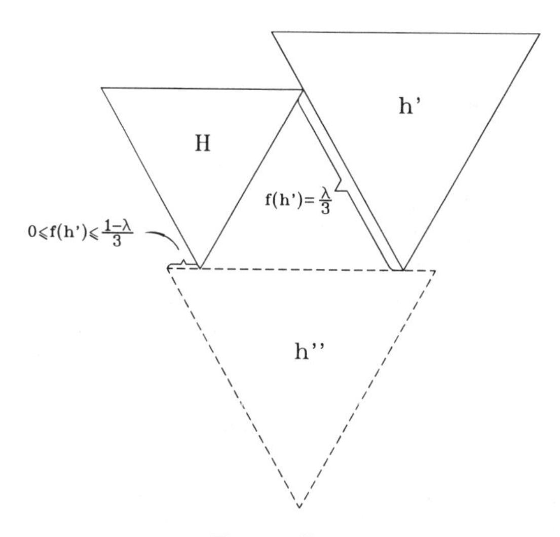

Figure G.14.d

previous observations (a)–(e), so that the even triangles $h_2, \ldots, h_{2k}$ touch a side of $H, h_1, \ldots, h_{2k+1}$ touch a vertex of H, and $f(h_{2k+1}) \geq \lambda$. Then we can choose h_{2k+2} according to (d) so that $f(h_{2k+2}) = 1 - \lambda$ is fulfilled. The triangle h_{2k+2} can be obtained from h_1 by a rotation of angle $2\pi i/3$ about the center of H. Therefore, continuing the construction from h_{2k+2}, choosing $f(h_{4k+3}) = 1 - \lambda$, we can close the chain by setting $h_{6k+4} = h_1$, which finishes the proof.

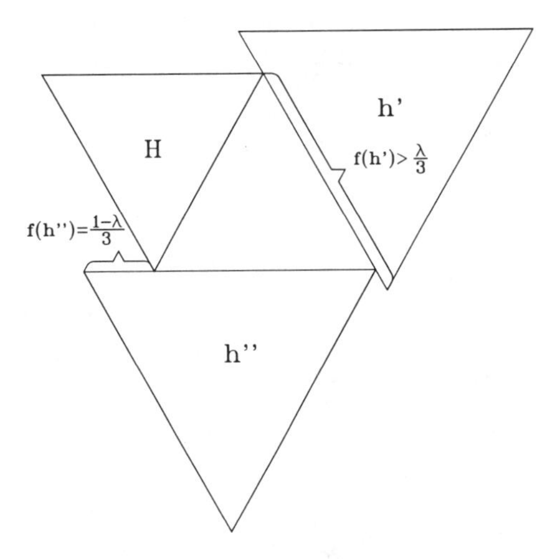

Figure G.14.e

II. Now let us examine the case $\lambda \geq 1$. Then only the cases (a), (b), and (c) are possible, and each h'' is uniquely determined by h'. The same would be true if we changed the orientation; hence, there are exactly two translates of h that touch H and a given h'. Consequently, given h_1 and h_2 that touches h_1 and H, the sequence $h_1, h_2, \ldots$ can be continued uniquely.

Suppose, we are given a periodic chain $h_1, h_2, \ldots$ with period $n (h_{n+1} = h_1)$ winding around H k times. The phrase "the chain winds around H k times" makes sense since the contact points of the triangles h_i follow one another according to a fixed orientation of the boundary of H. We may suppose that this orientation coincides with the orientation (A, B, C).

Observations (a), (b), and (c) make clear that the contact points of the triangles h_i that touch a side of H form a cyclic sequence of $n-3k$ elements running around H k times such that the peripheral distance between two consecutive contact points is 1/3. Thus,

$$\frac{1}{3}(n-3k) = \lambda k\,,$$

that is,

$$\lambda = \frac{n-3k}{3k}, \quad n = 3k(\lambda+1)\,. \tag{1}$$

One obtains from this equation that in the case $\lambda \geq 1$, the existence of a periodic chain implies the rationality of λ. In this case, the number of times this chain goes around the triangle H is the denominator in the reduced simple fraction form of 3λ, and the number of triangles contained in the chain is obtained as the product of this winding number and $3(\lambda+1)$ (of course, there is a periodic chain for any integer multiple of n and k with period n and winding number k).

We still have to show that if (1) holds for some natural numbers k and n, then there exists a periodic chain with period n. Then it will also prove that the rationality of λ is sufficient for the existence of a periodic chain.

In addition to this proposition, we show that no matter where we put the starting triangle h_1, the sequence of h_is will get closed at the nth step after k rounds around H. We may restrict ourselves to the case when h_1 touches a side of H, since either h_1 or h_2 touches a side of H, and if the chain starting from h_2 gets closed after the nth step, then so does the chain starting from h_1 because of the unique reconstructability of the sequence in both directions.

Let us construct the beginning part $h_1, \ldots, h_{m+1}$ of the sequence until it goes around H k times and the distance between the contact points of h_{m+1} and h_1 is less than $1/3$. This situation is surely arrived at sooner or later, because at most λ consecutive elements of the chain can touch the same side of H. Then, as above, we have

$$0 \le \left| k\lambda - \frac{m-3k}{3} \right| < \frac{1}{3}. \tag{2}$$

Expressing λ with the help of (1) and multiplying by 3, we get

$$0 \le |n-m| < 1.$$

The integers n and m can satisfy the latter inequality and the equivalent relation (2) only if equality holds on the left, that is, $n = m$ and $h_{n+1} = h_1$. $\square$

Problem G.16. *The traffic rules in a regular triangle allow one to move only along segments parallel to one of the altitudes of the triangle. We define the distance between two points of the triangle to be the length of the shortest such path between them. Put $\binom{n+1}{2}$ points into the triangle in such a way that the minimum distance between pairs of points is maximal.*

Solution. In the course of the solution, "distance" will refer to the ordinary distance of points while the italic "*distance*" will mean distance in the new metric. Let ABC be the given triangle, $AB = 1$. In order not to lose the idea of the solution among the precise verification of simple geometric facts, we shall omit technical details.

(a) *Construction for lower bound.* Let us divide each side of the triangle into $n-1$ equal parts, connect each node of BC to a corresponding node of CA by a segment parallel to AB, and divide these segments into $n-2, n-3, \ldots, 2, 1$ equal parts, respectively, in the order of their distances from the side AB. We obtain the same set of points if we take through all nodes on the sides AB, BC, and CA straight lines parallel to the two other sides and consider all possible intersections of these straight lines. One may expect that this system of $\binom{n+1}{2}$ points is optimal. Let us denote by r_n the minimum of *distances* for this system.

(b) *Theorem.* Let $\binom{n}{2} < N \leq \binom{n+1}{2}$. If we are given N points in $\triangle ABC$, then one can always find two among them with *distance* $\leq r_n$. If $N = \binom{n+1}{2}$ and the *distance* of any two points is at least r_n, then the system of points coincides with the one constructed in (a).

(c) For the proof, we need the proposition that a ball of radius r with center R in the metric space described in the problem is a regular hexagon centered at R, the vertices of which are obtained by moving R off to distance r parallel with one of the altitudes. The proof of this proposition is left to the reader.

(d) We show here that if the *distance* of any two points of a certain subset of the $\triangle ABC$ is greater than $r_n = 2c = \sqrt{3}/n-1$, then the subset contains at most $\binom{n}{2}$ points. Consider triangles of the triangular lattice constructed in (a). Triangles homothetic to $\triangle ABC$ with ratio $1/(n-1)$ will be called recumbent, and those homothetic to ABC with ratio $-1/(n-1)$ will be called standing. There are $\binom{n}{2}$ recumbent triangles. Let us place a *ball of radius* c that is a regular hexagon of side c around the centers of recumbent triangles. They cover the recumbent triangles and the standing ones as well, since each standing triangle is surrounded by three recumbent ones and the hexagons put around the centers of the neighboring triangles cover the standing triangle obviously. Thus, the given $\binom{n}{2}$ *balls* cover $\triangle ABC$, that is, every point of the considered subset belongs to one of these hexagons. Because the *diameter* of such a hexagon is $2c = r_n$, a system of points for which the *distance* between any two points is greater than r_n meets each hexagon in at most one point; hence, it contains at most $\binom{n}{2}$ points. This proves the first part of the theorem.

(e) We can show the uniqueness for $N = \binom{n+1}{2}$ points in the following way. If $q < 1$ and we shrink $\triangle ABC$ from A with ratio q, then we get a triangle $AB'C'$ that contains at most $\binom{n}{2}$ points. If $q \to 1$, we obtain that side BC contains at least n points. Of course, side BC cannot contain more than n points, and the points have to divide side BC into $n-1$ equal parts. Let AB^+C^+ be the dilatation of ABC with respect to A with ratio $(1-1/(n-1))$. The hexagons of side c around the points lying on BC cover the difference of the triangles ABC and AB^+C^+. Consequently, $\triangle AB^+C^+$ contains $\binom{n}{2}$ points, so that the distance of any two points is at least $r_n = r_{n-1}(1-1/(n-1)) = r_{n-1}AB^+$. Since the theorem is trivial for $n=2$, we may complete the proof by induction. □

Problem G.17. *Let C be a simple arc with monotone curvature such that C is congruent to its evolute. Show that under appropriate differentiability conditions, C is a part of a cycloid or a logarithmic spiral with polar equation $r = ae^{\vartheta}$.*

Solution. Let $\mathbf{r}_1 = \mathbf{r}(s)$, $s \in [0, \ell]$ be the parameterization of C by arc length for which the radius of curvature $\rho(s)$ is a monotone increasing function. The evolute E is parameterized then by $\mathbf{r}_2 = \mathbf{r}(s) + \rho(s)\mathbf{n}(s)$, where $\mathbf{n}$ is the principal normal vector. The length of the arc of the evolute bounded by $\mathbf{r}_2(0)$ and $\mathbf{r}_2(s)$ is $\sigma(s) = \rho(s) - \rho_0$, where $\rho_0 = \rho(0)$. Let Φ be the congruence that superimposes C on E. There are two possibilities:

1. Φ takes $\mathbf{r}(0)$ to $\mathbf{r}_2(0)$;
2. Φ takes $\mathbf{r}(0)$ to $\mathbf{r}_2(\ell)$.

The radius of curvature of the evolute at the point $\mathbf{r}_2(s)$ is $\rho(\rho(s)-\rho_0)$ in the first case and $\rho(\ell-\rho(s)+\rho_0)$ in the second. Since, by Frenet equations,

$$\frac{d\mathbf{r}_2}{d\sigma} = \mathbf{n}, \qquad \frac{d^2\mathbf{r}_2}{d\sigma^2} = -\frac{1}{\rho(s)\rho'(s)}\mathbf{t},$$

we have $\rho(\rho(s) - \rho_0) = \rho(s)\rho'(s)$ in the first case and $\rho(\ell - \rho(s) + \rho_0) = \rho(s)\rho'(s)$ in the second.

We show that $s = \rho(s) - \rho_0$ for all $s \in [0, \ell]$ in the first case. Otherwise, we could find an interval $[a, b]$ by the suitable differentiability conditions such that $a = \rho(a) - \rho_0$, $b = \rho(b) - \rho_0$, and $s > \rho(s) - \rho_0$ or $s < \rho(s) - \rho_0$ holds everywhere inside $[a, b]$. However, in this case, we would have

$$b - a = \rho(b) - \rho(a) = \int_a^b \rho'(s)ds = \int_a^b \frac{\rho(\rho(s) - \rho_0)}{\rho(s)}ds \neq b - a$$

since $\rho(s) - \rho_0 > s$ implies $\rho(\rho(s) - \rho_0) > \rho(s)$ and $\rho(s) - \rho_0 < s$ implies $\rho(\rho(s) - \rho_0) < \rho(s)$. We conclude that the solutions belonging to the first case have a natural equation of the form $s = \rho(s) - \rho_0$. These curves are logarithmic spirals the polar, equation of which has the form $r = ae^{\vartheta}$, and one easily checks that any arc of such a spiral is a solution.

In the second case, $\ell - s = \rho(\ell - \rho(s) + \rho_0)$ holds for all $s \in [0, \ell]$. To see this, we have to show first that Φ is an orientation-reversing transformation of the plane. We may suppose without loss of generality that while $\mathbf{r}(s)$ is running along the curve C, the unit tangent vector $\mathbf{t}(s)$ is rotating counterclockwise. Meanwhile $\mathbf{r}_2(s)$ is running along the evolute E, and its unit tangent vector $\mathbf{n}(s)$ is also rotating counterclockwise. Since Φ takes $\mathbf{r}(0)$ to $\mathbf{r}_2(\ell)$, the point $\Phi(\mathbf{r}(s))$ is running along E in the opposite direction and its unit tangent vector is rotating clockwise. Therefore, Φ reverses orientation. Orientation-reversing isometries of the plane are glide reflections, so the evolute F of the evolute of C is a translate of C. Let $\mathbf{r}_3(s)$ denote the center of curvature of E at $\mathbf{r}_2(s)$. The length of the arc of F lying between $\mathbf{r}_3(\ell)$ and $\mathbf{r}_3(s)$ is equal to $\rho(\ell - \rho(s) + \rho_0) - \rho_0$. Monotonicity of the curvature implies that the direction of the tangent vector uniquely characterizes the point at which the vector is tangent to the curve. As a consequence, we get that the translation above takes $\mathbf{r}(s)$ to $\mathbf{r}_3(s)$. For this reason, $\rho(\ell - \rho(s) + \rho_0) = \ell - s$ holds, as we wanted to show. From here, we obtain $\rho(s)\rho'(s) = \ell - s + \rho_0$ by our previous considerations. Thus, in the second case, the solutions are given by arcs of the curves with natural equation $\rho^2(s) + (\ell - s + \rho_0)^2 = c$. These curves are cycloids given by $x = a(u - \sin u)$, $y = a(1 - \cos u)$, where $a = \frac{1}{4}\sqrt{\rho_0^2 + (\ell + \rho_0)^2}$. We may conclude that, in the second case, the solutions are those subarcs $\widehat{P_1P_2}$ of the arc corresponding to the parameter domain $u \in [0, \pi]$ for which $a = \sqrt{\rho^2(P_1) + [\widehat{P_1P_2} + \rho(P_1)]^2}$ (for example, the whole arc fits the requirement). □

Remark. We may pose the more general problem of finding curves that are similar to their evolute. The solution of this problem involves epi- and hypocycloids and logarithmic spirals of the general polar equation $r = ae^{c\vartheta}$ as well.

Problem G.18. *Given four points A_1, A_2, A_3, A_4 in the plane in such a way that A_4 is the centroid of the $\triangle A_1A_2A_3$, find a point A_5 in the plane that maximizes the ratio*

$$\frac{\min_{1\le i<j<k\le 5} T(A_iA_jA_k)}{\max_{1\le i<j<k\le 5} T(A_iA_jA_k)}.$$

($T(ABC)$ denotes the area of the triangle $\triangle ABC$.)

Solution. The medians, which go through the centroid of the triangle, divide the plane into six angular regions. We claim that there is exactly one point in each of these regions for which the ratio in question attains its maximal value, namely $1/4$. Let F be the midpoint of A_2A_3, let E denote the reflection of A in F, let R denote the trisecting point of the segment A_3E that is closer to A_3, and finally, let A_5 denote the point on the half-line A_4R for which $A_4A_5 = 3/2 \cdot A_4R$. A_5 is the point in the angular region A_3A_4E, that yields the maximum (see Figure G.15).

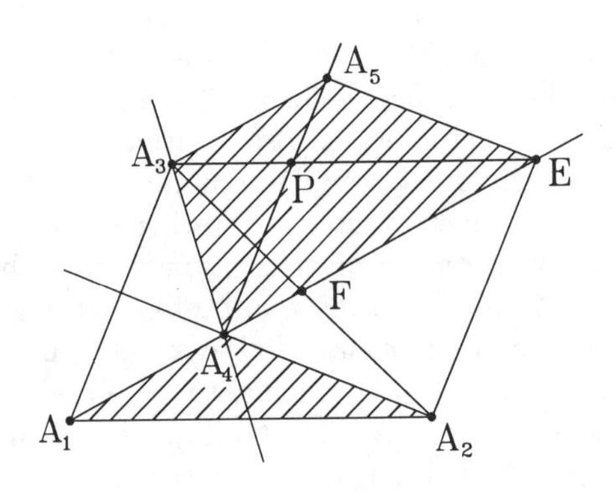

Figure G.15.

Suppose for a while that A_5 is an arbitrary point in the angle A_3A_4E. If this point is not in $\triangle A_3A_4E$, then using the notation $T(A_4A_5E) =$

$2T(A_1A_5A_4) = 2b$, $T(A_3A_5A_4) = a$, $T(A_1A_2A_4) = c$, $T(A_1A_2A_5) = d$, we can express the area of the hatched domain as

$$a + 2b + c = d\,, \tag{1}$$

since this area is the sum of the area of three triangles each having a side of length $A_1A_2 = A_3E$, and the sum of the corresponding altitudes equals the altitude of $\triangle A_1A_2A_5$ corresponding to the side A_1A_2. Formula (1) remains true also when A_5 is inside $\triangle A_3A_4E$, only we have to take signed areas. In both cases, the areas with no sign satisfy

$$a + 2b + c \le d\,.$$

From this,

$$\min(a, b, c) \le \frac{1}{4}d\,. \tag{2}$$

Therefore, the ratio in question is $\le 1/4$.

The necessary and sufficient condition of having equality in (2) is $a = b = c = 1/4$. On one hand, this means that A_5 is four times farther from the straight line A_1A_2 than A_4 is; on the other hand, it implies that $T(A_3A_4A_5) = T(A_1A_4A_5)$, that is, $A_4A_5 \perp A_1A_3$. These two conditions together are satisfied only by the point A_5 constructed above.

One can show by a direct computation that in this case $\triangle A_1A_2A_5$ has the largest area among all $\triangle A_iA_jA_k$ and that $T(A_iA_jA_k) \ge d/4$ for all i, j, k. $\square$

Problem G.19. *Let K be a compact convex body in the n-dimensional Euclidean space. Let $P_1, P_2, \ldots, P_{n+1}$ be the vertices of a simplex having maximal volume among all simplices inscribed in K. Define the points $P_{n+2}, P_{n+3}, \ldots$ successively so that P_k $(k > n+1)$ is a point of K for which the volume of the convex hull of $P_1, \ldots, P_k$ is maximal. Denote this volume by V_k. Decide, for different values of n, about the truth of the statement "the sequence $V_{n+1}, V_{n+2}, \ldots$ is concave."*

Solution. The statement makes no sense for $n = 0$ and is obviously true for $n = 1$. Consider the case $n = 2$.

It is clear that the vertices of the starting simplex lie on the boundary of K, otherwise we could move the point lying inside K farther from the opposite hyperplane. We show by general induction for the two-dimensional case that further points are also chosen from the boundary.

The base clue has already been proved. Suppose that the first n points lie on the boundary. In this case, they are vertices of a convex n-gon, each side of which cuts off a (possibly trivial) part of the convex figure. The new point is chosen from one of these parts, which results in the polygon "growing" by a triangle. Because of maximality, the new point must come from the boundary.

To prove concavity, we have to show that when taking two consecutive differences of the sequence, the second does not exceed the first. If the new points come from arcs belonging to different sides, then this is obvious, since otherwise we would have to add these points to the sequence in the opposite order. If two consecutive points are chosen from the same arc over a side of the convex hull of preceding points, then let A and B denote the endpoints of this side and C and D be the points added to the sequence. Since C is taken first, the area of $\triangle ABC$ is greater than or equal to the area of $\triangle ABD$. Denoting by E the intersections of the diagonals of the quadrangle spanned by the points $ABCD$, the statement follows from the inequality $T(EAB) \geq T(ECD)$, since the first difference is $T(ABC)$, while the second is $T(ABD) + T(ECD) - T(EAB)$. Observe that the supporting half-lines of the convex figure taken at A and B in the half-plane bounded by AB containing C (see Figure G.16) are parallel or intersect each other; otherwise the one from A and B that was chosen later into the sequence would not have the maximality property. Let us draw a straight line through C parallel to AB. This intersects the straight line BD at a point D' lying outside the segment $\overline{BD}$, so it suffices to prove the inequality with D' instead of D. The triangles ABE' and $CD'E'$ are similar, and by the above observation, the segment $\overline{AB}$ is not shorter than the segment $\overline{CD'}$, and thus the statement is implied.

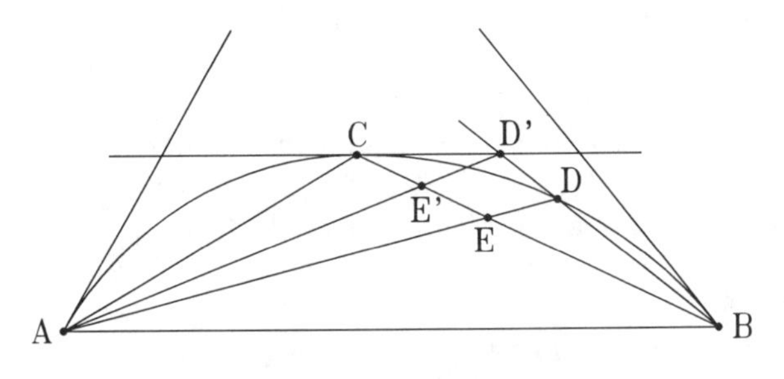

Figure G.16.

Now we are going to prove that the sequence is not necessarily concave for $n \geq 3$. For $n = 3$, the regular octahedron serves as a counterexample. Let us choose four arbitrary, noncoplanar vertices of the octahedron for the vertices of the first inscribed simplex. The volume of this simplex is maximal. Indeed, suppose that $ABCD$ is an inscribed simplex of maximal volume having as many common vertices with the octahedron as possible. If one of the vertices, say A, is not a vertex of the octahedron, then consider the plane through A parallel to the plane BCD. By the maximality assumption, this plane may not contain an internal point of the octahedron, but it has a nonempty intersection with the octahedron and thus contains at least one vertex of the octahedron, which could have been chosen instead of A. The contradiction proves our proposition. One can similarly show, that the fifth point can also be taken from the vertices of the octahedron. The point is that opposite faces of the octahedron are parallel, so the points lying at maximal distance from two neighboring faces form an edge of the octahedron. The sixth point then must be the sixth vertex of the octahedron. The second difference of this sequence is twice as much as the first one; hence the sequence is not concave.

Turning to the general case, consider a prism over the octahedron, that is, the convex hull of a three-dimensional octahedron and an $(n-3)$-dimensional simplex lying in a complementary subspace of the octahedron. We can show that this is a suitable counterexample in the same way as above, since due to maximality, the starting simplex must contain all the vertices not belonging to the octahedron and the further points must be chosen from the three-dimensional plane generated by the octahedron. □

Problem G.20. *Let us connect consecutive vertices of a regular heptagon inscribed in a unit circle by connected subsets (of the plane of the circle) of diameter less than 1. Show that every continuum (in the plane of the circle) of diameter greater than 4, containing the center of the circle, intersects one of these connected sets.*

Solution. Let $A_1, A_2, \ldots, A_7$ be the vertices of the heptagon, let O be the center of the circumscribed circle, and let $H_1, H_2, \ldots, H_7$ denote the connected sets that connects the pairs A_1, A_2, A_2, A_3, ... ,$A_7 A_1$, respectively. Denote by C a continuum in question, by D the circle of radius 2 centered at O.

By the assumption $d(H_i) < 1$ $(i = 1, 2, \ldots, 7)$, H_i is inside D and $O \notin H_i$. On the other hand, conditions $d(C) > 4$ and $0 \in C$ imply that there exists a point $Q \in C$ which is outside D. Suppose to the contrary that $\cup_i H_i$ and C are disjoint. In this case, the distance $\delta(x)$ of a point $x \in H_i$ from the closed set $C \cup D$ not covering x is positive. Consider the open disc $U(x, \delta(x))$ of radius $\delta(x)$ centered at x for all $x \in H_i$. The union $\Gamma_i = \cup_{x \in H_i} U(x, \delta(x))$ of these discs is open and connected, and the same holds for the union $\Gamma = \cup_i \Gamma_i$. The set Γ is inside D; furthermore, $\Gamma \cap (C \cup D) = \emptyset$. Since Γ is open and connected, one can draw a closed broken line T^* inside Γ that connects the points $A_1, A_2, \ldots, A_7$. Let T be the boundary of the connected component of $\mathbb{R}^2 \setminus T^*$ containing O. T is a simple, closed, broken line inside D such that T and C are disjoint, $O \in C$ is inside T, and $Q \in C$ is outside T.

We are going to show that the existence of such a T contradicts the fact that C is a continuum. Denote by B the set of points that lie inside T, by K the set of points lying outside T. By $C \cap D = \emptyset$, we have

$$C = (C \cap B) \cup (C \cap K),$$

where

(a) both $C \cap B$ and $C \cap K$ are open–closed in the subspace topology of C; furthermore,

b) none of them is empty because $O \in C \cap B$ and $Q \in C \cap K$. However, these two facts together contradict the connectedness of the continuum C. □

Remark. One can give a counterexample for the proposition of the problem if C is required only to be connected, dropping the assumption that it is also closed. Indeed, draw two disjoint, infinite, broken lines into a square $EFLJ$ so that one of them starts from E, the other starts from F, and both have closure that cover the segment $\overline{LJ}$. Thus, $M_1 = T_1 \cup J$ and $M_2 = T_2 \cup J$ are two disjoint connected sets. Let us shrink and move the square and the figures contained in it in such a way that the points $\overline{E}$, $\overline{F}$ that correspond to the vertices E and F get onto the segment A_7A_1 in the order A_7, $\overline{E}$, $\overline{F}$, A_1 while $\overline{L}$ gets inside the heptagon. (We denote the image of a figure under this similarity by putting a bar over the corresponding sign.) Let Q^* be a point on the half-line $O\overline{F}$ whose distance from O is greater than 4. Then the claim of the proposition does not hold for the connected sets $H_j = \overline{A_jA_{j+1}}$ ($j = 1, 2, \ldots, 6$), $H_7 = \overline{E_7\overline{E}_1} \cup \overline{M_1} \cup \overline{\overline{L}A_1}$, and the bounded and connected set $C = \overline{O\overline{J}} \cup \overline{M_2} \cup \overline{\overline{F}Q^*}$. (This remark is due to László Babai.)

Problem G.21. *What is the radius of the largest disc that can be covered by a finite number of closed discs of radius 1 in such a way that each disc intersects at most three others?*

Solution. Draw a closed disc of radius 1 around each vertex of a square of side $\sqrt{2}$. These four discs cover the disc of radius $\sqrt{2}$ drawn around the center of the square.

We are going to show that a circle of radius greater than $\sqrt{2}$ does not have a covering having the prescribed properties.

In the following, we shall denote by $'$ the boundary of a set (the boundary of A is A') and "disc" will always mean a closed disc.

Let K be a disc of radius greater than $\sqrt{2}$, and suppose that it has a covering with the prescribed properties. Let us consider a subsystem $A_1, A_2, \ldots, A_m$ of the covering in such a way that

(i) the system $A_1, A_2, \ldots, A_m$ covers K',

(ii) if any of the discs $A_1, A_2, \ldots, A_m$ are removed, the remaining discs fail to cover K'. Such a subsystem can be produced obviously by omitting unnecessary discs one by one. There are no single points and empty sets among the intersections $K' \cap A_i$, since the discs different from A_i cover a closed subset of K', so if $K' \cap A_i$ is empty or consists of one point, then A_i can be omitted.

We claim that $m \geq 5$. Obviously, $K' \cap A_i$ is an arc with diameter not greater than 2; however, the central angle of such an arc is less than $\pi/2$ in a circle of radius greater than $\sqrt{2}$. Hence one needs at least five discs to cover K'.

Now let X_1 and X_2 be the endpoints of the arc $K' \cap A_1$. We may suppose without loss of generality that A_2 covers X_1 and A_3 covers X_2 since if both endpoints of an arc $K' \cap A_i$ were contained in the disc $A_j (i \neq j)$, then the whole arc would lie in A_j and A_i could be omitted.

Now if we suppose that $((A'_1 \cap K) \setminus A_2) \setminus A_3$ is not empty, then it is an arc of positive length. Observe that this arc must be covered by one disc, otherwise A_1 would intersect more then three discs. Denote by A this disc, and consider the arc $A'_3 \cap K$. This arc must be covered by A and A_1 together. Otherwise, A would also have a point in common with the disc that covers the remaining part, that is, A would intersect at least four other discs. Therefore, $A \cup A_1 \supset A'_3 \cap K$. But then A covers the endpoint of $K' \cap A_3$ not contained in A_1, and since four circles can not cover K', A must intersect at least one more disc, which would be the fourth disc intersected by A, and this is impossible.

We conclude that $((A'_1 \cap K) \setminus A_2) \setminus A_3 = \emptyset$, that is, $A_2 \cup A_3 \supset A'_1 \cap K$. Similarly, we can find further two discs, A_4 and A_5, such that $A_1 \cup A_4 \supset A'_3 \cap K$, $A_1 \cup A_5 \supset A'_2 \cap K$, and $A_1, \ldots, A_5$ are different. However, this is a contradiction, since in this case A_1 would intersect more than four other circles.

Thus, a disc of radius $\sqrt{2}$ admits a covering with the prescribed properties, but larger discs do not. On the basis of the above proof, we can also see that the covering of the disc of radius $\sqrt{2}$ is "essentially" unique (that is, unique up to rotations). □

Problem G.22. *Assume that a face of a convex polyhedron P has a common edge with every other face. Show that there exists a simple closed polygon that consists of edges of P and passes through all vertices.*

Solution 1. Let S be the distinguished face of P, and suppose that P has n vertices off S. The proof goes by general induction with respect to n. The base clause is obviously true for $n = 1$; now let $n > 1$. It is easy to see that by omitting the edges of S from the graph of edges of P, we obtain a tree. (Indeed, imagine that P is a planet and its edges are dams, S is filled with water, while the other faces are empty basins and explode the dams that bound S.) For this reason, P has a vertex B not belonging to S such that the edges starting from B end on S with the exception of one edge. Let us denote by C the endpoint of the exceptional edge and by $A_1, A_2, \ldots, A_m$ the endpoints of the other edges ordered in correspondence with an orientation of S. Finally, let A_0 be the vertex of S that precedes the vertex $A_1, and let A_{m+1}$ be the vertex that follows the vertex A_m according to the above orientation of S. (Figure G.17 shows a case with $m = 3$.)

Now if the half-line $\overrightarrow{CB}$ intersects the plane of S in a point A, then construct a polyhedron P' by gluing the prism $AA_1 \ldots A_m B$ to P. P' is obviously convex itself and satisfies the conditions of the proposition. However, B is not a vertex of P', and thus P' has only $n - 1$ vertices off the plane of S. By the induction hypothesis, the edge skeleton of P' contains a simple closed polygon that passes through each vertex of P'. This polygon goes through A in one of the orders A_0AA_{m+1}, A_0AC, CAA_{m+1}.

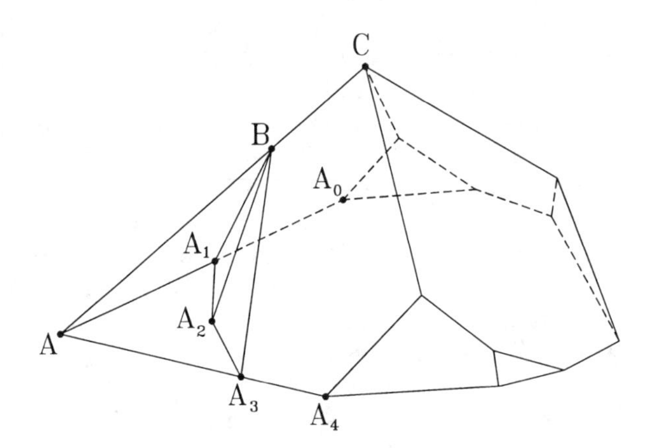

Figure G.17.

According to these three cases, replace vertex A of this polygon by the paths $A_1 \dots A_{m-1}BA_m$, $A_1A_2 \dots A_mB$, $BA_1A_2 \dots A_m$, respectively. The simple closed polygon we obtain this way passes through each vertex of P.

If CB is parallel to the plane of S, then choose a plane γ such that γ crosses the half-line $\overrightarrow{BC}$ but has empty intersection with P. Now if we take the projective augmentation of the Euclidean space by adding a plane ω of points at infinity and apply a projectivity Π that takes γ to ω, then the polyhedron ΠP will completely consist of proper points, and the intersection point of the half-line $\overrightarrow{(\Pi C)(\Pi B)}$ and the plane of ΠS will also be proper; consequently, we may apply the previous consideration.

We can proceed similarly if the half-line $\overrightarrow{BC}$ intersects the plane of S. □

Solution 2. Let S be the distinguished face, and let G be the graph obtained by removing the edges of S from the graph formed by the vertices and edges of the polyhedron. G is a tree, and the degree of a vertex of G not belonging to S is at least three.

Let us choose an arbitrary face, B_1, of P, and color the edges of G that lie on the boundary of B_1. Suppose that we have already chosen the faces $B_1, B_2, \dots, B_r$ in such a way that

(1) B_i and B_j have no vertex in common if $i \neq j$, $(i, j = 1, 2, \dots, r)$, and
(2) for all i $(i = 1, 2, \dots, r)$, there exists an edge of G that connects B_i to one of the faces $B_1, B_2, \dots, B_{i-1}$.

Using that G is connected, we get that if P has a vertex not belonging to S that is not covered by the faces $B_1, B_2, \dots, B_r$, then there is a vertex x among these that is connected to some B_i $(i \in \{1, 2, \dots, r\})$ by an edge xy. Since the degree of x is at least three, x is covered by a face B_{r+1} such that xy is not an edge of B_{r+1}. If some vertices of B_{r+1} were covered by the faces B_j $(j = 1, 2, \dots, r)$, then one of these vertices, say z, could be connected to x along edges of G without passing through points of the faces B_j. On the other hand, according to the induction hypothesis, we can join x and z by a path in G in such a way that all the vertices we go through belong to some B_j, which contradicts the fact that G is a tree. Consequently, the choice of faces $B_1, B_2, \ldots, B_{r+1}$ meets requirements (1) and (2). Now color the edges of G that lie on the boundary of B_{r+1}.

At the end of this process also color the edges of S that do not belong to any of the faces B_j. Thus, we obtain a system of colored edges that cover all vertices of P and form a polygon with the required properties. □

Problem G.23. *Let D be a convex subset of the n-dimensional space, and suppose that D' is obtained from D by applying a positive central dilatation and then a translation. Suppose also that the sum of the volumes of D and D' is 1, and $D \cap D' \neq \emptyset$. Determine the supremum of the volume of the convex hull of $D \cup D'$ taken for all such pairs of sets D, D'.*

Solution. The conditions imply that D and D' are bounded convex sets with nonempty interior. The problem is equivalent to the determination of the supremum of the quantity $V(\mathrm{co}(D \cup D'))//(V(D) + V(D'))$, where besides, having the properties just mentioned, D and D' have nonempty intersection and D' is obtained from D by a positive central dilatation and a translation. Let the ratio of the dilatation be a fixed positive number λ. We show that in the case of a fixed λ, the supremum in question is $(1 + \lambda + \cdots + \lambda^n)/(1 + \lambda^n) = f(\lambda)$, and this supremum is attained. The case $n = 1$ is trivial; let $n > 1$ in the following.

First, we show that the case $\lambda = 1$ can be derived from the case $\lambda \neq 1$. Indeed, if $\lambda = 1$, then stretch D' from a point $p \in D \cap D'$ with ratio $1 + \varepsilon$. Since D and the obtained D meet each other at p, we have

$$V(\mathrm{co}\,(D \cup D')) \leq V(\mathrm{co}\,(D \cup D'_\varepsilon)) \leq f(1+\varepsilon)(V(D) + V(D'_\varepsilon)) .$$

Taking the limit $\varepsilon \searrow 0$, the right-hand side tends to $f(1)(V(D) + V(D'))$.

Assume now that $\lambda \neq 1$. Changing the role of D and D' if necessary, we may also suppose that $0 < \lambda < 1$, since $f(\lambda) = f(\lambda^{-1})$. Then D' can be obtained from D by one central dilatation, the center O of which will be chosen for the origin.

$V(D) = \sup V(P)$, where supremum is taken for (closed) polyhedrons $P \subset D$. It is easy to see that $V(P)$ tends to $V(D)$ if and only if $r(P, D) = \sup_{x \in D} d(x, P)$ tends to 0, where $d(x, P)$ denotes the distance between x and P. It is also not difficult to see that if $r(P, D) \leq \varepsilon$, then $r(\mathrm{co}\,(P \cup P'), \mathrm{co}\,(D \cup D')) \leq \varepsilon$ (P' denotes the image of P in D'). Suppose that P contains a common point of D and D' and its preimage with respect to the similarity; then $P \cap P' \neq \emptyset$. Taking such polyhedrons P, if $V(P) \to V(D)$, then $V(\mathrm{co}\,(P \cup P')) \to V(\mathrm{co}\,(D \cup D'))$, and thus to prove that the supremum is $\leq f(\lambda)$, we may restrict ourselves to polyhedrons.

Now let D be a (closed) polyhedron, $D' = \lambda D$, $p \in D \cap D'$. We may assume that $0 \notin D$. It is easy to see that $\mathrm{co}\,(D \cup \lambda D) = \cup_{\lambda \leq \mu \leq 1} \mu D$. Set $E = (\cup_{0 \leq \mu \leq 1} \mu D) \setminus D$. Then

$$\bigcup_{\lambda \leq \mu \leq 1} \mu D = D \cup (E \setminus \lambda E),$$

and thus,

$$V\left(\bigcup_{\lambda\le\mu\le 1}\mu D\right) = V(D) + (1-\lambda^n)V(E)\,.$$

We show that $V(E) \le V(D)\cdot\lambda/(1-\lambda)$. Let $F_1, F_2, \ldots, F_r$ be those $(n-1)$-dimensional faces of D whose hyperplane separates O strictly from D. Then the segment that connects O to p intersects the hyperplane π_i at a point p_i. Therefore, using the notation $q = (1/\lambda)\cdot p$, we have

$$d(O,\pi_i) = \frac{\overline{Op_i}}{\overline{qp_i}}d(q,\pi_i) \le \frac{\overline{Op}}{\overline{qp}}d(q,\pi_i) = \frac{\lambda}{1-\lambda}d(q,\pi_i)\,.$$

Since $\overline{E}$ is the union of pyramids over the faces F_i with vertex O, and since the pyramids over the faces F_i with vertex q are contained in D, the previous inequality gives

$$\begin{aligned} V(E) &= \frac{1}{n}\sum_{i=1}^{r} d(O,\pi_i)V_{n-1}(F_i) \\ &\le \frac{\lambda}{1-\lambda}\sum_{i=1}^{r} d(q,\pi_i)V_{n-1}(F_i) \le \frac{\lambda}{1-\lambda}V(D) \end{aligned}$$

(V_{n-1} denotes the $(n-1)$-dimensional volume). From this,

$$\begin{aligned} \frac{V(\mathrm{co}\,(D\cup D'))}{V(D)+V(D')} &= \frac{V\left(\bigcup_{\lambda\le\mu\le 1}\mu D\right)}{V(D)(1+\lambda^n)} = \frac{V(D)+(1-\lambda^n)V(E)}{V(D)(1+\lambda^n)} \\ &\le \frac{1+(\lambda+\lambda^2+\cdots+\lambda^n)}{1+\lambda^n} = f(\lambda)\,. \end{aligned}$$

Now we show that $f(\lambda)$ is an exact maximum. Let D be a simplex with vertices $p_1, p_2, \ldots, p_{n+1}$; the corresponding vertices of D' are $p'_1, p'_2, \ldots, p'_{n+1}$, and let $p'_1 = p'_2$. (For $\lambda = 1$, we choose for D' the translate of D for which $p'_1 = p_2$.) In this case, $\mathrm{co}\,(D\cup D')$ is made up of the simplex D' and a truncated pyramid (or a prism if $\lambda = 1$), so its volume is $\lambda^n V(D) + (1+\lambda+\cdots+\lambda^{n-1})V(D) = f(\lambda)(V(D)+V(D'))$.

It remains to determine $\sup_{\lambda>0} f(\lambda)$. We claim that $f(\lambda) \le f(1) = (n+1)/2$, that is,

$$(n+1)(1+\lambda^n) - 2((1+\lambda+\cdots+\lambda^n) \ge 0\,.$$

However, this follows from summing up the inequalities

$$1+\lambda^n-\lambda^i-\lambda^{n-i} = (1-\lambda^i)(1-\lambda^{n-i}) \ge 0 \quad \text{for} \quad 0\le i\le n\,. \quad \square$$

Problem G.24. *Consider the intersection of an ellipsoid with a plane σ passing through its center O. On the line through the point O perpendicular to σ, mark the two points at a distance from O equal to the area of the intersection. Determine the loci of the marked points as σ runs through all such planes.*

Solution. Let B be an arbitrary body, O a point of it. Denote by $\mathcal{F}(\sigma; B, O)$ the pair of points $P = \{P^+, P^-\}$ that lie on the straight line perpendicular to σ at O and whose distance from O is equal to the area of the intersection of σ and B: Area $(\sigma \cap B)$. Let $\Phi(B)$ stand for the figure formed by the points P as σ is varied:

$$\Phi(B) = \bigcup_{\sigma} \mathcal{F}(\sigma; B, O).$$

Let Σ and t be a plane and a straight line intersecting one another orthogonally at O, $\mathcal{U}$ the stretching in the direction of t with ratio λ, and U* the stretching in the direction of Σ with ratio λ. Obviously, $\mathcal{U}$ and $\mathcal{U}^*$ are affinities.

Lemma. $\mathcal{U}^*(\Phi(B)) = \Phi(\mathcal{U}(B))$.

Proof. Set

$$\sigma' = \mathcal{U}(\sigma), \quad P = \mathcal{F}(\sigma; B, O),$$
$$P' = \mathcal{F}(\sigma'; \mathcal{U}(B), O) \equiv \mathcal{F}(\mathcal{U}(\sigma); \mathcal{U}(B), \mathcal{U}(O)).$$

The planes σ and σ' intersect each other in a straight line m, lying in the plane Σ. Thus P' belongs to the plane Λ spanned by t and P. (Figure G.18 shows the trace of the mentioned figures in the plane Λ viewed from the direction of m.) Consider the points $U \in \sigma$ and $U' \in \sigma'$ that lie over a given point $N \in \Sigma \cap \Lambda$ (that is, UN, $U'N \perp \Sigma$). They satisfy $\overline{NU'} = \lambda \overline{NU}$, and by elementary properties of affinities,

$$\frac{\overline{OP'}}{\overline{OP}} = \frac{\text{Area } (\sigma' \cap \mathcal{U}(B))}{\text{Area } (\sigma \cap B)} = \frac{\overline{OU'}}{\overline{OU}}.$$

This equation shows that $\triangle OPP'$ can be obtained from $\triangle OUU'$ by a dilatation and a right-angled rotation about O in the plane Λ. This similarity takes $\Sigma \cap \Lambda$ to t, the point N to the intersection M of t and PP', and since similarities preserve angles and the ratio between corresponding segments, we get that $PP' \perp t$ and $\overline{MP'} = \overline{MP}$, which means that $\mathcal{U}^*(P) = P'$, and this proves the lemma.

Now let a, b, c be the half-axes of the ellipsoid E $(E(a, b, c))$. The equation of E takes the canonical form

$$\frac{x^2}{a^2} + \frac{y^2}{b^2} + \frac{z^2}{c^2} = 1$$

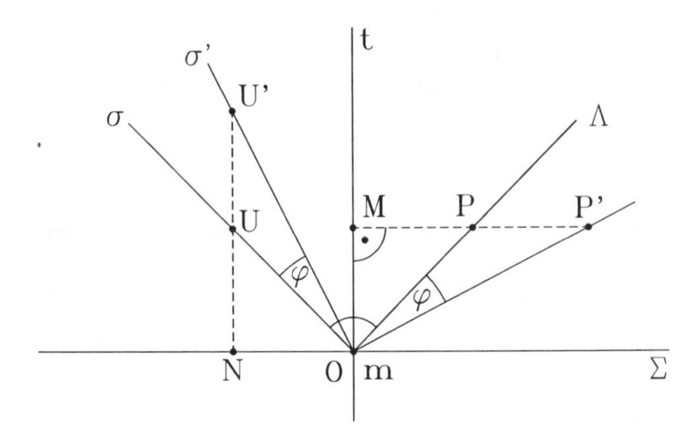

Figure G.18.

in a suitable coordinate system. Consider the sphere $S^2(a)$ of radius a around the origin. Then $\Phi(S^2(a))$ is also a sphere: $S^2(a^2\pi)$. Now let $\mathcal{U}_1$ be the stretching with ratio $\lambda = b/a$ in the direction of the y-axis. $\mathcal{U}_1(S^2(a))$ is the ellipsoid $E_1(a,b,a)$, and by the lemma,

$$\begin{aligned}\Phi(E_1) &= \Phi(\mathcal{U}_1(S^2(a))) = \mathcal{U}_1(\Phi(S^2(a))) \\ &= \mathcal{U}_1^*(S^2(a^2\pi)) = E_2(ab\pi, a^2\pi, ab\pi)\,.\end{aligned}$$

Now let $\mathcal{U}_2$ be the stretching with ratio c/a in the direction of the z-axis. Then $\mathcal{U}_2(E_1)$ is the given ellipsoid $E(a,b,c)$, and a repeated application of the lemma gives

$$\Phi(E) = \Phi(\mathcal{U}_2(E_1)) = \mathcal{U}_2^*(\Phi(E_1)) = \mathcal{U}_2^*(E_2) = E_3(bc\pi, ac\pi, ab\pi),$$

which was to be determined. □

Remarks.

1. We remark that the proposition of the lemma is true also for n-dimensional bodies B and $(n-1)$-dimensional hyperplanes. Thus, an application of the lemma gives an immediate answer to the n-dimensional version of the problem as well.

2. We also remark that the lemma, which lies in the base of the solution, concerns an arbitrary body B and an arbitrary point O of it. Hence, it is applicable to a much wider class of problems.

Problem G.25. *Construct on the real projective plane a continuous curve, consisting of simple points, which is not a straight line and is intersected in a single point by every tangent and every secant of a given conic.*

Solution. Removing a straight line from the projective plane we obtain a Euclidean plane. Let us introduce Cartesian coordinates (x, y) on it and denote by K the circle of radius 1 centered at $(0,-2)$. K is a conic of the projective plane. All conics of the projective plane are equivalent, that is, one can find a projective transformation $\mathcal{P}$ for any conic C such that $\mathcal{P}$ takes C to K. Therefore, if Γ is a solution of the problem with respect to the circle K, then $\mathcal{P}^{-1}(\Gamma)$ is a solution with respect to C since the projectivity $\mathcal{P}$ and its inverse $\mathcal{P}^{-1}$ are bijections that take straight lines into straight lines, and secants and tangents of a conic into a secant or tangent of the image conic, respectively. Thus if Γ meets the requirements of the problem with the conic K, then so does $\mathcal{P}^{-1}(\Gamma)$ with the conic C.

First, we construct on the Euclidean plane a continuous curve $\overline{\Gamma}$ consisting solely of simple points that is not a straight line and is intersected in a single point by every secant and tangent of K that is not parallel to the x-axis. Let $y = f(x)$ be a function defined on the whole x-axis such that

(a) f is continuous and differentiable,
(b) $0 \le f(x) \le 1$,
(c) the absolute value of the derivative of f is less than 1,
(d) f is increasing on the left of $x = 0$ and decreasing on the right,
(e) the limits $\lim_{x\to\infty} f(x)$ and $\lim_{x\to-\infty} f(x)$ exist.

We show that secants of the graph $\overline{\Gamma}$ of f do not intersect K. Indeed, suppose that $\ell(x)$ is a linear function such that

$$\ell(x_1) = f(x_1),\ \ell(x_2) = f(x_2); \quad x_1 < x_2 . \tag{1}$$

The graph $\overline{\ell}$ of $\ell(x)$ is a secant of $\overline{\Gamma}$. We have

$$|\ell'(x)| = |\tan\alpha| \le 1$$

because of (c), where α is the direction angle of $\overline{\ell}$. If $\alpha = 0$, then $\overline{\ell}$ either coincides with the x-axis or lies above it because of (b); hence it does not meet K. If $\alpha > 0$, then the zero x_0 of ℓ satisfies $x_0 \le x_1 < x_2$ by (b) and (1). In this case, the assumption $0 \le x_0$ together with $\alpha > 0$ and (1) implies $f(x_1) < f(x_2)$, $0 \le x_1 < x_2$, which is in contradiction with (d). If, however, $x_0 < 0$ and $|\tan\alpha| \le 1$, then $\overline{\ell}$ does not meet K. We can prove similarly that $\overline{\ell}$ has empty intersection with K also in the case $-1 \le \tan\alpha < 0$. We conclude that the secants of $\overline{\Gamma}$ do not meet K, or, in other words, the tangents and secants of K have at most one point in common with $\overline{\Gamma}$. Those secants and tangents of K that are parallel to the x-axis (denote the set of them by χ) obviously have empty intersection with $\overline{\Gamma}$. Other tangents and secants with nonzero direction angle intersect $\overline{\Gamma}$ since $\overline{\Gamma}$ divides the plane into two connected parts, one of which contains the half-plane $y < 0$ while the other contains the half-plane $y > 1$. A straight line with nonzero direction angle has points in both half-planes, hence it must cross the graph $\overline{\Gamma}$ of f. Thus, the curve $\overline{\Gamma}$ of the Euclidean plane has the required properties.

We augment the Euclidean plane to a projective plane by adding an ideal point to each family of parallel lines. Attaching the ideal point P_∞ of the x-axis to $\overline{\Gamma}$, we get a closed curve Γ on the projective plane, which is continuous by (e) and is intersected in a single point, namely in P_∞, by the straight lines of χ. The points of Γ are simple since the graph $\overline{\Gamma}$ of the continuous function f consists of simple points and we have closed $\overline{\Gamma}$ by adding one single boundary point.

Finally, we have to present a nonlinear function that has properties (a)–(e). For example, $y = e^{-x^2/2}$. This function gives a solution according to the previous considerations. But one can easily construct other suitable functions. The simple, though not differentiable, function whose graph is obtained from the x-axis by replacing the segment between the points $A(-2,0)$, $C(2,0)$ by the broken line ABC, where B has coordinates $(0,1)$, also does the job. □

Problem G.26. *Let T be a surjective mapping of the hyperbolic plane onto itself which maps collinear points into collinear points. Prove that T must be an isometry.*

Solution.

(A) First we show that T is injective. The proof (based on the idea of Nándor Simányi) consists of three steps.

1. We claim that the inverse image of a point is convex. For this purpose, we need to show that if $T(A) = T(B)$, then every point C of the segment $\overline{AB}$ is mapped to $T(A)$. Suppose it is not so: $T(C) \neq T(A)$ for some $C \in \overline{AB}$. Let e and f be the straight lines perpendicular to $T(A)T(C)$ at $T(A)$ and $T(C)$, respectively. Let us choose the points P and Q in such a way that their images are points of e and f, respectively, different from $T(A)$ and $T(C)$ (see Figure G.19). There are no three collinear points among A, B, P, Q since there are obviously no three collinear points among their images $T(A)$, $T(B)$, $T(P)$, $T(Q)$. According to this, the straight line CQ intersects side $\overline{AB}$ of $\triangle ABP$ in an inner point and thus must cross one of the two other sides. The image of this second intersection point must lie on both e and f, which is a contradiction since e and f do not intersect each other.

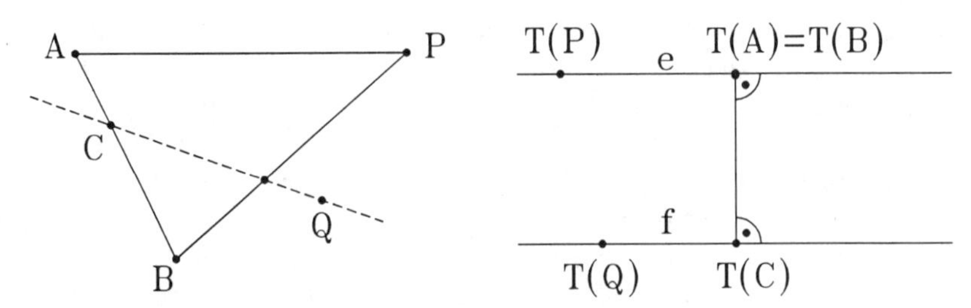

Figure G.19.

2. Now we show that the inverse image of a point X contains no interior point. Suppose to the contrary that the interior of $T^{-1}(X)$ is not empty. Draw a straight line f through X. We claim that int $(T^{-1}(f \setminus \{X\})) \neq \emptyset$. Indeed, consider a point Z of f different from X and denote by P a point of its pre-image (see Figure G.20). Then the images of straight lines that go through P and intersect $T^{-1}(X)$ lie in f. In other words, $T^{-1}(f)$ contains a pair of opposite angular domains with common vertex P. Since $T^{-1}(X)$ is completely contained in one of these angles, it is disjoint from the other; therefore, int $(T^{-1}(f \setminus \{X\})) \neq \emptyset$, as we claimed. Now rotate f about X and consider the sets $T^{-1}(f \setminus \{X\})$. They form an uncountable family of disjoint subsets of the hyperbolic plane such that each set has an interior point, which is a contradiction.

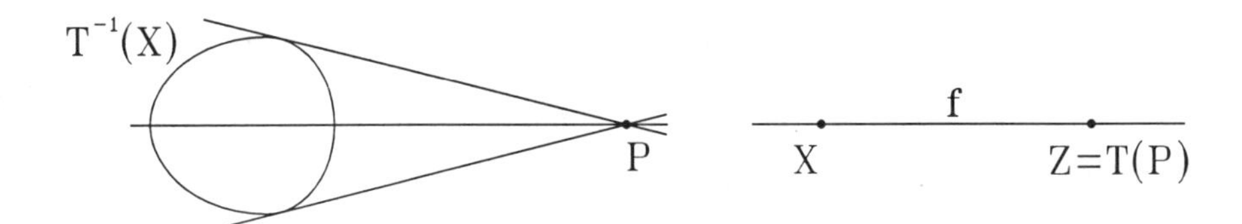

Figure G.20.

3. Now we are in the position to prove that the preimage $T^{-1}(X)$ of every point X consists of one single point. Again we use the indirect method. If $T^{-1}(X)$ is not a point, then it is a segment, half-line, or a straight line according to the previous considerations. In any case, $T^{-1}(X)$ is contained in a straight line, which will be denoted by e in the following. Let A and B be two points of e that are mapped to X (see Figure G.21). It is easy to see that $T(e)$ is a part of a straight line g. Draw a straight line f through X other than g. One can show as above that the interior of the set $T^{-1}(f \setminus \{X\})$ is not empty. Rotating f about X, we can again get a contradiction.

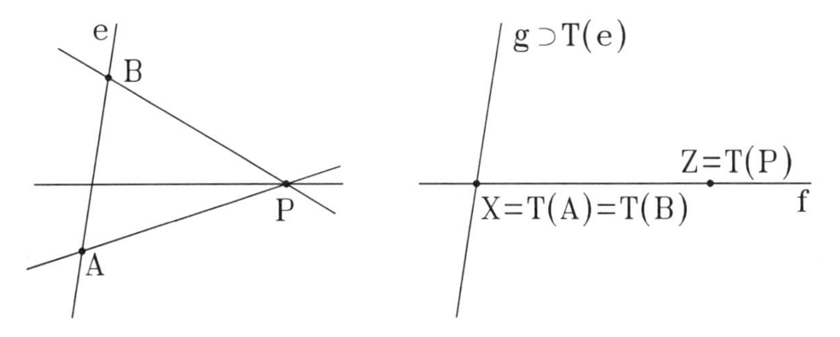

Figure G.21.

(B) The major difficulties of the proof are over. As a next step, we show that the images of noncollinear points are noncollinear. Indeed, if the noncollinear points P, Q, and R were mapped to a straight line e, then the images of the straight lines PQ, QR, and RP would also be contained in e. Consequently, every straight line that twice crosses the boundary of $\triangle PQR$ twice would be mapped into e. However, such a straight line passes through any point of the plane; therefore, the image of the whole plane should be contained in e, which contradicts the fact that T is surjective.

(C) From this it follows that T^{-1} also takes collinear points to collinear points; the image of a straight line (under T) is a straight line, and the images of intersecting straight lines are intersecting, the images of nonintersecting straight lines are nonintersecting.

(D) T preserves the ordering on straight lines. Indeed, if A, B, and C are three different points of a straight line e, then, as is known, C does not separate A and B if and only if one can find straight lines a, b, and c through A, B, and C, respectively, such that a and b intersect each other but c intersects neither a nor b (see Figure G.22). T preserves the latter property by (C), hence preserves ordering.

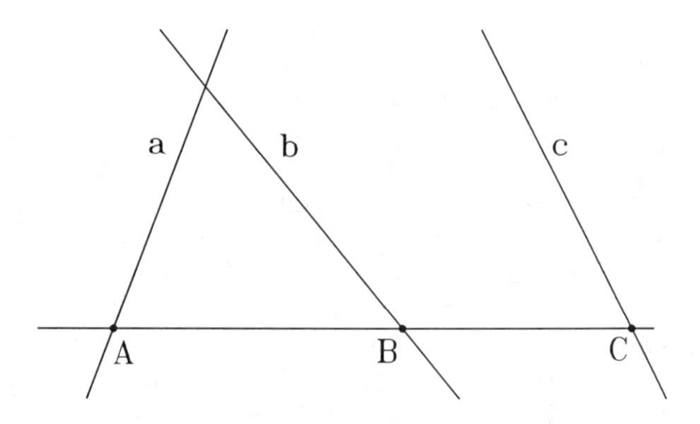

Figure G.22.

(E) Therefore, T preserves half-planes, half-lines, and segments. Furthermore, it preserves the ordering of pencils. Thus, the images of asymptotic half-lines are asymptotic.

(F) Now we show the existence of a length d such that T maps segments of length d onto segments of length d. Let us perform the following construction (see Figure G.23.a).
We denoted the two half-lines of e determined by P by f and g. Let us take a point X not lying on e and draw the half-lines h and k starting from X and asymptotic to f and g, respectively. Take a point Y on the elongation of k beyond X; let the segment $\overline{YP}$ intersect h at U. Draw an asymptotic half-line to g from U, and suppose that it intersects the segment $\overline{XP}$ in V. Denote by Q the intersection point of YV and g.

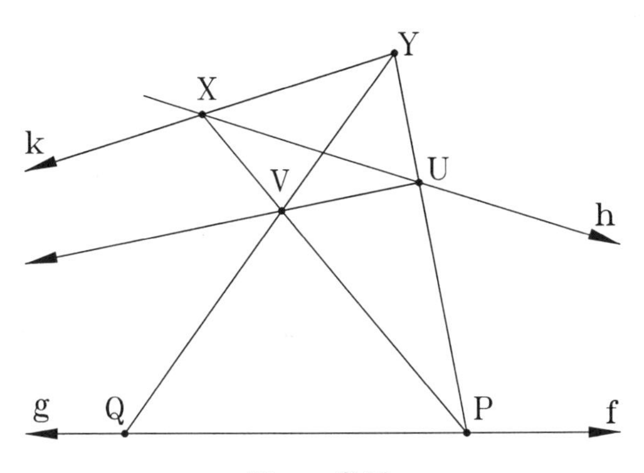

Figure G.23.a

We claim that Q *does not depend on the choice of* X *and* Y.
It is very easy to check this statement in the Cayley–Klein model of hyperbolic geometry (and it is also sufficient, since we know that if a proposition is true in every Cayley–Klein model, then it is a true theorem of hyperbolic geometry). The above construction is a construction of a complete quadrangle in the model (see Figure G.23.b); consequently, denoting by I and J the horizontal points of f and g, respectively, the pairs J, P and I, Q are conjugate, and the points P, J, and I determine the fourth harmonic Q uniquely.

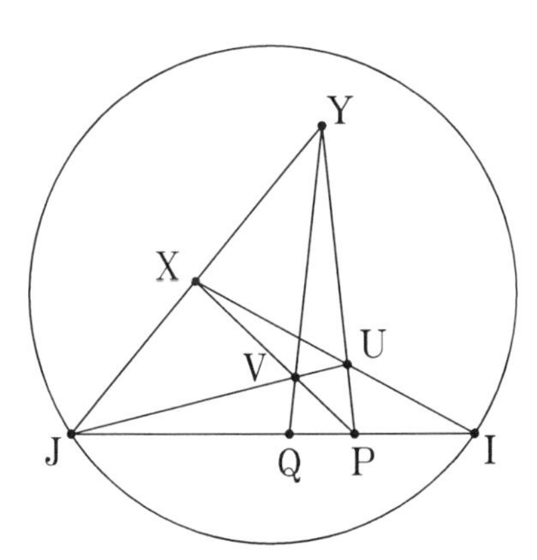

Figure G.23.b

It is clear that the length d of the segment PQ does not depend on the choice of P either. Since T transforms Figure G.23.a into a similar one, it moves segments of length d into segments of length d, as we stated.

(G) Of course, the segments of length nd are also invariant under T, as are the angles of regular triangles of side nd. Because the angle of these triangles tends to zero as n tends to infinity, there exist arbitrarily small invariant angles. As the sum of invariant angles is also invariant, a dense set of angles will be invariant under T. Taking into consideration that T preserves the ordering on pencils, it follows that T preserves angles. However, in hyperbolic geometry, a triangle is determined up to isometries by its angles, so the image of every triangle under T is a triangle congruent to the original one, hence T is an isometry. □

Problem G.27. *Let $X_1, \ldots, X_n$ be n points in the unit square ($n > 1$). Let r_i be the distance of X_i from the nearest point (other than X_i). Prove the inequality*

$$r_1^2 + \cdots + r_n^2 \leq 4.$$

Solution. We shall prove by general induction. The base clause is trivial for $n = 2$. Let $n > 2$, and suppose that the proposition is true for any system of points consisting of less than n points. This assumption means also that if the points $Y_1, \ldots, Y_k$ are enclosed in a square of side a and q_j denotes the distance of Y_j from the other points, then for $1 < k < n$ we have $\sum_{j=1}^{k} q_j^2 \leq 4a^2$.

Let us divide the square into four congruent squares by the medians. Denote the small squares by $N_1, \ldots, N_4$ (see Figure G.24). We shall separate some cases according to the possible distributions of the points X_i in the small squares and then study each case individually. If a point X_i lies on the common boundary of two squares, then decide which square it belongs to; it does not matter how.

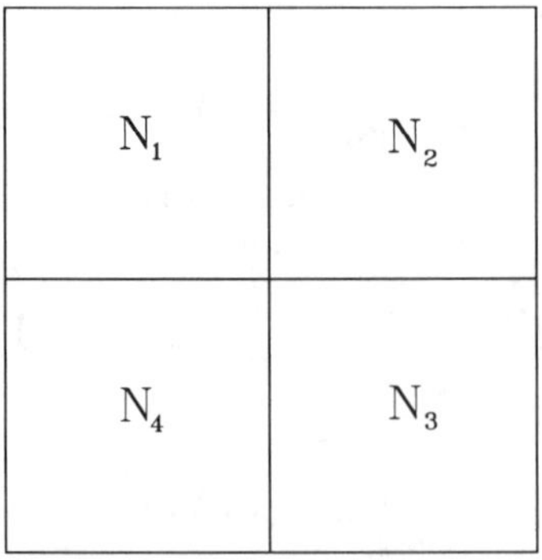

N_1	N_2
N_4	N_3

Figure G.24.

Case A. None of the small squares contains exactly one point. Here we may distinguish two further subcases: Either all points are in one small square or the points are distributed in more than one square. In the first case, applying the induction hypothesis for $X_1, \ldots, X_{n-1}$, we get

$$\sum_{i=1}^{n-1} r_i^2 \le 4\left(\frac{1}{2}\right)^2 = 1\,.$$

Since $r_n^2 \le 2$, the proposition of the problem is true in this case.

In the second subcase, however, each small square contains less than n points (and if it contains a point at all, then it certainly contains more than one). Therefore, by the induction hypothesis,

$$\sum_{X_i \in N_\nu} r_i^2 \le 4\left(\frac{1}{2}\right)^2 = 1 \tag{1}$$

for $1 \le \nu \le 4$, from which the proposition of the problem follows.

Case B. There is exactly one small square, say N_1, that contains one point. For example, let $X_1 \in N_1$. Inequality (1) holds also in this case for $\nu = 2, 3, 4$. Therefore, if $r_1^2 \le 1$, then we are ready. We may suppose for this reason that $r_1 > 1$. Of course, $r_1^2 \le 2$, and this estimation would be enough if one of the small squares were empty. Thus, we may suppose in the following that each small square contains a point; from this it follows that $r_1^2 \le 5/4$ because of the point in N_4.

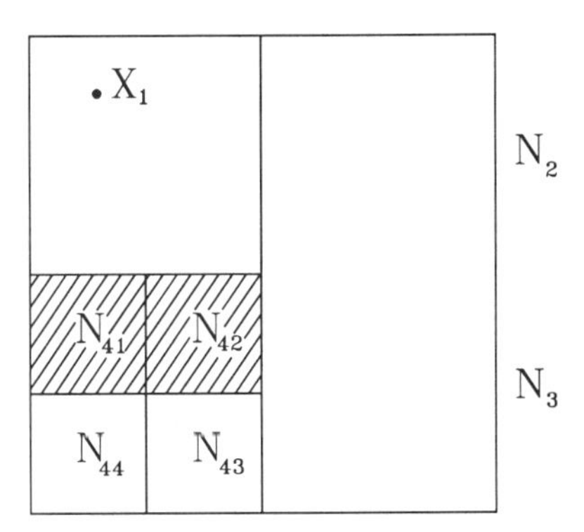

Figure G.25.

Let us divide N_4 into four smaller squares with the help of its medians, and we will denote these squares by N_{41}, N_{42}, N_{43}, N_{44} (see Figure G.25). As $r_1 > 1$, the shaded domain contains no point. We shall show that

$$\sum_{X_i \in N_{4j}} r_i^2 \le \frac{5}{16} \qquad (j = 3, 4)\,. \tag{2}$$

This follows from the induction hypothesis — applied for N_{4j} — if the number of points contained in N_{4j} is not exactly one. If, however, N_{4j}

contains exactly one point, then its distance from another point in N_4 is at most $\sqrt{5}/4$, that is, (2) holds in any case. Then

$$\sum_{i=1}^{n} r_i^2 = r_1^2 + \sum_{X_i \in N_2} r_i^2 + \sum_{X_i \in N_3} r_i^2 + \sum_{X_i \in N_4} r_i^2$$
$$\leq \frac{5}{4} + 1 + 1 + 2\frac{5}{16} < 4 .$$

Case C. The number of points is equal to one in exactly two small squares. In this case, the squares can be coupled in such a way that squares in a couple are neighbors and one of the squares in a couple contains exactly one point while the number of the points in the other is not one (see Figure G.26). Assume, for example, that N_1 and N_4 are in one couple and N_1 is the one that contains exactly one point. Obviously, it suffices to show that

$$\sum_{X_i \in N_1 \cup N_4} r_i^2 \leq 2.$$

This can be done in the same way as in case B.

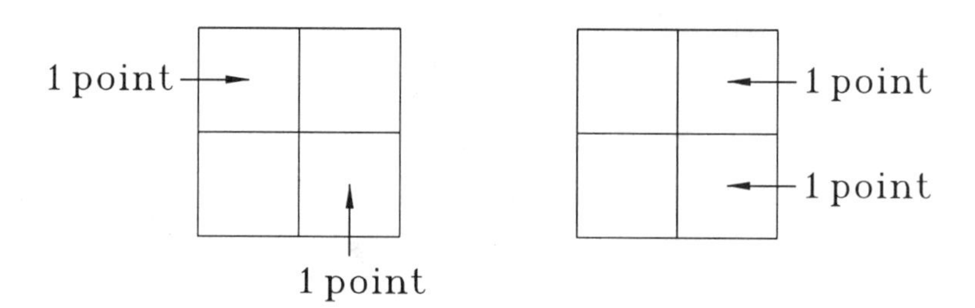

Figure G.26.

Case D. Exactly three squares, say N_1, N_2, and N_3, contain one point. Call these points X_1, X_2, and X_3, respectively. Let us divide N_4 into four small squares, as shown in Figure G.27. First of all, we remark that $r_2^2 \leq 5/4$.

We distinguish three subcases. Assume first that the rectangles $N_{41} \cup N_{42}$ and $N_{43} \cup N_{42}$ both contain a point. Then

$$r_1^2, r_3^2 \leq \left(\frac{3}{4}\right)^2 + \left(\frac{1}{2}\right)^2 = \frac{13}{16}.$$

Consequently,

$$\sum_{i=1}^{n} r_i^2 = r_1^2 + r_2^2 + r_3^2 + \sum_{X_i \in N_4} r_i^2 \leq \frac{13}{16} + \frac{5}{4} + \frac{13}{16} + 1 < 4 .$$

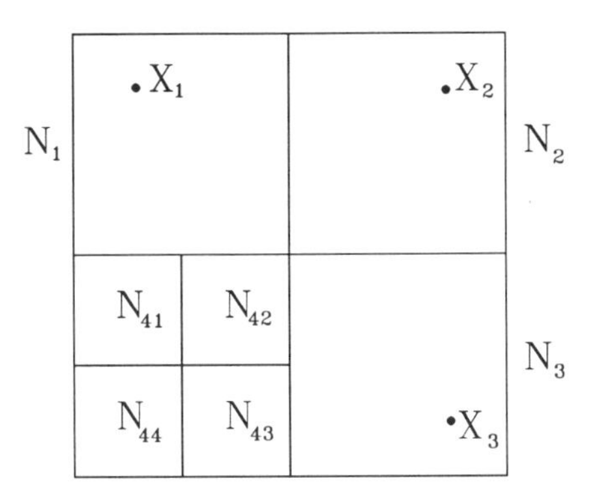

Figure G.27.

Second, suppose that, for example, $N_{41} \cup N_{42}$ contains no point but N_{43} and N_{44} do contain a point. In that case, $r_3^2 \leq 13/16$, $r_1^2 \leq 1 + 1/16$, and furthermore, (2) is now valid. Hence,

$$\sum_{i=1}^{n} r_i^2 = r_1^2 + r_2^2 + r_3^2 + \sum_{X_i \in N_4} r_i^2 \leq \frac{17}{16} + \frac{5}{4} + \frac{13}{16} + 2\frac{5}{16} < 4.$$

Third, it remains to study the case when neither N_{41} nor N_{42} contains a point and one of N_{43} and N_{44} is also empty. Applying the induction hypothesis to the other (nonempty) square we get

$$\sum_{i=1}^{n} r_i^2 = r_1^2 + r_2^2 + r_3^2 + \sum_{X_i \in N_4} r_i^2 \leq \frac{5}{4} + \frac{5}{4} + \frac{5}{4} + \frac{1}{4} = 4.$$

Case E. The last case that remains is when each of the small squares contains one point. Let $X_i \in N_i$. We prove that

$$d(X_1, X_2)^2 + d(X_2, X_3)^2 + d(X_3, X_4)^2 + d(X_4, X_1)^2 \leq 4\,. \tag{3}$$

($(d(P, Q)$ denotes the distance between the points P and Q.) This already implies the inequality of the problem.

Let us think of (3) as a strictly convex function defined on the set $\times_{\nu=1}^{4} N_\nu \subset \mathbb{R}^8$. Such a function can attain its maximum only at the extremal points of the boundary of its domain. An extremal point corresponds to such a configuration of the points X_i in which every X_i is positioned at a vertex of the corresponding square N_i. Since for these cases (3) can be easily verified, (3) holds in general. □

Remark. Contestants B. Brindza, V. Komornik, P.P. Pálfy, V. Totik, and Zs. Tuza described those configurations of the points for which the sum of squares in question is equal to 4. They found that there are only two possibilities: $n = 2$ and the points are positioned at opposite vertices of the square, or $n = 4$ and the points are put to the vertices of the square.

Problem G.28. *Give an example of ten different noncoplanar points $P_1, \ldots, P_5$, $Q_1, \ldots, Q_5$ in 3-space such that connecting each P_i to each Q_j by a rigid rod results in a rigid system.*

Solution. First, we describe two constructions.

I. Let g be a rigid structure. Take a further point P, and connect it by a rigid rod to three points Q_1, Q_2, Q_3 of g, for which P, Q_1, Q_2, Q_3 are not coplanar. Then a rigid system is obtained.

II. Let g be a rigid system and PQ a rod in it. Take a point T on this rod and connect it to two further points R, S of g, for which P, Q, R, S are not coplanar. The system obtained will also be rigid.

Putting it the other way around, these two statements mean that to get a rigid realization of a graph g, it is enough to have

(a) a rigid realization of a graph obtained from g by deleting a point of degree 3, or

(b) a rigid realization of a graph obtained from g by deleting a point of degree four and connecting two of its neighbors by an edge.

In our case, the graph to be realized is as shown in Figure G.28.

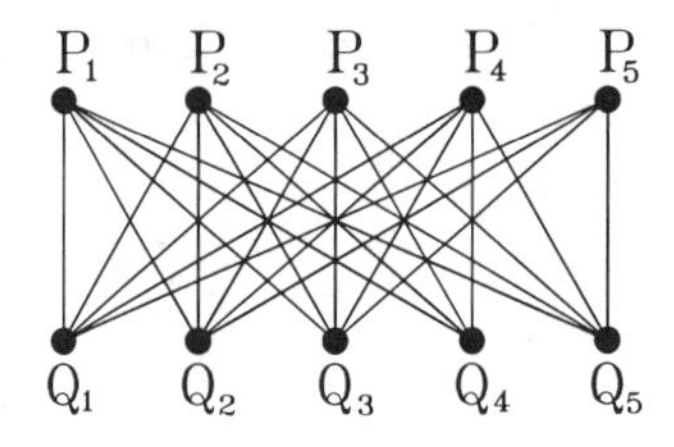

Figure G.28.

We omit the edge P_1Q_1. Without drawing edges between P_i's and Q_j's it is sufficient to find a realization of the following graphs:

by (b),

P_2 P_3 P_4 P_5

Q_1 Q_2 Q_3 Q_4 Q_5

again by (b),

P_2 P_3 P_4 P_5

Q_2 Q_3 Q_4 Q_5

again by (b),

P_3 P_4 P_5

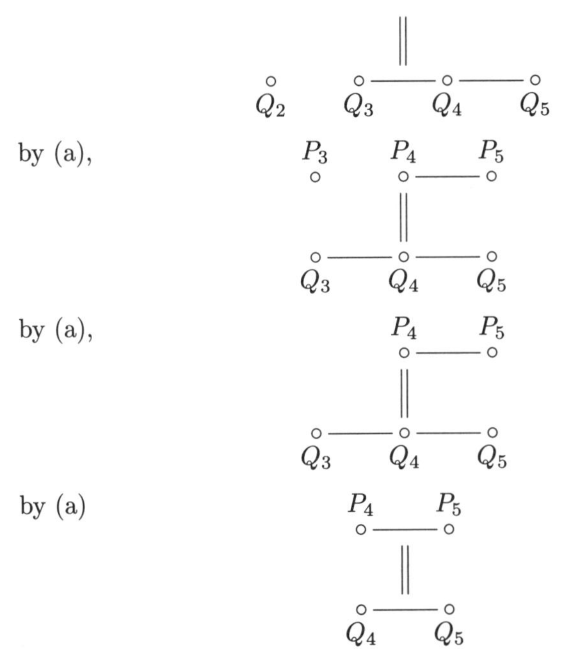

The last graph is a tetrahedron, which is obviously rigid. □

Problem G.29. *Let us define a pseudo-Riemannian metric on the set of points of the Euclidean space $\mathbb{E}^3$ not lying on the z-axis by the metric tensor*

$$\begin{pmatrix} 1 & 0 & 0 \\ 0 & 1 & 0 \\ 0 & 0 & -\sqrt{x^2+y^2} \end{pmatrix},$$

where (x, y, z) is a Cartesian coordinate system in $\mathbb{E}^3$. Show that the orthogonal projections of the geodesic curves of this Riemannian space onto the (x, y)-plane are straight lines or conic sections with focus at the origin.

Solution 1. Let α, β, and γ run through the indices 1,2, and denote by (x^1, x^2, x^3) the coordinates (x, y, z). We compute the differential equation of the geodesics. We can write the metric tensor and its inverse (g^{ik}) in the following form

$$g_{\alpha\beta} = \delta_{\alpha\beta}, \quad g^{\alpha\beta} = \delta^{\alpha\beta}, \quad g_{\alpha 3} = 0, \quad g^{\alpha 3} = 0,$$
$$g = -\sqrt{(x^1)^2 + (x^2)^2}, \quad g^{33} = -\frac{1}{\sqrt{(x^1)^2 + (x^2)^2}}.$$

Using the formula

$$\Gamma^i_{jk} = \frac{1}{2} \sum_{h=1}^{3} g^{ih} \left\{ \frac{\partial g_{jh}}{\partial x^k} + \frac{\partial g_{hk}}{\partial x^j} - \frac{\partial g_{jk}}{\partial x^h} \right\}$$

for the Christoffel symbols, we obtain

$$\Gamma^{\gamma}_{\alpha\beta} = \Gamma^{3}_{\alpha\beta} = \Gamma^{\gamma}_{3\beta} = \Gamma^{\gamma}_{\alpha 3} = 0\,;$$
$$\Gamma^{\gamma}_{33} = \frac{1}{2}\frac{x^{\gamma}}{\sqrt{(x^1)^2 + (x^2)^2}}\,;$$
$$\Gamma^{3}_{\gamma 3} = \Gamma^{3}_{3\gamma} = \frac{1}{2}\frac{x^{\gamma}}{(x^1)^2 + (x^2)^2}\,.$$

By these, the equation of the geodesics,

$$x^{j''} + \sum_{i,j} \Gamma^{i}_{jk} x^{j'} x^{k'} = 0$$

can be written in the form

$$x'' + \frac{1}{2}\frac{x}{\sqrt{x^2+y^2}} z'^2 = 0\,; \tag{1}$$
$$y'' + \frac{1}{2}\frac{y}{\sqrt{x^2+y^2}} z'^2 = 0\,; \tag{2}$$
$$z'' + \frac{x}{x^2+y^2} x'z' + \frac{y}{x^2+y^2} y'z' = 0\,. \tag{3}$$

This system is satisfied by the curves

$$x = as + b, \quad y = \bar{a}s + b, \quad z = c,$$

the projections of which onto the (x, y)-plane are straight lines. Furthermore, looking at the equations, we see that if $z' = 0$ at the initial point of an integral curve, then $z' \equiv 0$ along the curve.

Assume $z' \neq 0$. Then the third equation can be written in the form

$$\frac{z''}{z'} = -\frac{1}{2}\frac{d}{ds}\ln(x^2 + y^2)\,.$$

Therefore,

$$\ln z' = \ln\frac{c}{\sqrt{x^2+y^2}}, \qquad z' = \frac{c}{\sqrt{x^2+y^2}}$$

(where c is an arbitrary positive constant).

Substituting this into equations (1) and (2), we get

$$x'' + \frac{c^2}{2}\frac{x}{(x^2+y^2)^{3/2}} = 0\,;$$
$$y'' + \frac{c^2}{2}\frac{y}{(x^2+y^2)^{3/2}} = 0\,.$$

This system is non other than the well-known system of differential equations

$$x'' = \frac{\partial}{\partial x}\left(\frac{c^2}{2}\frac{1}{\sqrt{x^2+y^2}}\right),$$
$$y'' = \frac{\partial}{\partial y}\left(\frac{c^2}{2}\frac{1}{\sqrt{x^2+y^2}}\right)$$

of the Kepler problem, the solutions of which are conic sections with focus at the origin. □

Solution 2. The problem of finding geodesics of the given space is equivalent to the problem of finding stationary curves of the variational problem corresponding to the integral

$$\int \sqrt{\dot{x}^2 + \dot{y}^2 - \sqrt{x^2+y^2}\,\dot{z}^2}\, dt\,.$$

Since the variational function is positively homogeneous of degree one in the variables $\dot{x}, \dot{y}, \dot{z}$, it is well known that those stationary curves of it that satisfy the condition

$$\dot{x}^2 + \dot{y}^2 - \sqrt{x^2+y^2}\,\dot{z}^2 = 1 \tag{4}$$

coincide with the stationary curves of the problem corresponding to the integral

$$\int \dot{x}^2 + \dot{y}^2 - \sqrt{x^2+y^2}\,\dot{z}^2\, dt\,.$$

Changing over to the coordinate system (r, φ, z) with the coordinate transformation $x = r\cos\varphi$, $y = r\sin\varphi$, $z = z$, we obtain the variational problem

$$\int (\dot{r}^2 + r^2\dot{\varphi}^2 + r\dot{z}^2)\, dt\,.$$

The integrand function is independent of φ and z. Therefore, the Euler–Lagrange equations

$$\frac{\partial F}{\partial \varphi} = \frac{d}{dt}\frac{\partial F}{\partial \dot{\varphi}}, \quad \frac{\partial F}{\partial z} = \frac{d}{dt}\frac{\partial F}{\partial \dot{z}}$$

imply that

$$\frac{\partial F}{\partial \dot{\varphi}} = A, \quad \frac{\partial F}{\partial z} = B \qquad (A \text{ and } B \text{ are constant}),$$

that is, we obtain the first integrals

$$2r^2\dot{\varphi} = A, \quad -2r\dot{z} = B\,.$$

We introduce the new unknown function $u(\varphi) = 1/r(\varphi)$. For this, we have

$$\dot{r} = -\frac{1}{u}\dot{u} = -\frac{1}{u^2}\frac{du}{d\varphi}\dot{\varphi} = -\frac{1}{u^2}\frac{du}{d\varphi}\left(\frac{A}{2r^2}\right) = -\frac{A}{2}\frac{du}{d\varphi}.$$

Therefore, by the relation (4),

$$\begin{aligned} 1 &= \dot{r} + r^2\dot{\varphi}^2 - r\dot{z}^2 \\ &= \frac{A^2}{4}\left(\frac{du}{d\varphi}\right)^2 + \frac{1}{u^2}\left(\frac{A^2}{4}u^4\right) - \frac{1}{u}\left(\frac{B^2}{4}u^2\right) \\ &= \frac{A^2}{4}\left[\left(\frac{du}{d\varphi}\right)^2 + u^2\right] - \frac{B^2}{4}u\,. \end{aligned}$$

Let us differentiate with respect to φ:

$$\left\{\frac{A^2}{2}\left[\frac{d^2u}{d\varphi^2} + u\right] - \frac{B^2}{4}\right\}\frac{du}{d\varphi} = 0\,.$$

The equation $du/d\varphi = 0$ characterizes circles centered at the origin. Dividing by $du/d\varphi$, we get the equality

$$\frac{d^2u}{d\varphi^2} + u = \frac{B^2}{2A^2}\,,$$

the general solution of which can be written in the form

$$u(\varphi) = \frac{B^2}{2A^2}(1 + e\cos(\varphi + \omega))\,,$$

that is,

$$r(\varphi) = \frac{p}{1 + e\cos(\varphi + \omega)} \qquad \left(p = \frac{B^2}{2A^2}\right),$$

and this is the focal equation of a conic section with focus at the origin. □

Remark. Analogously, the trajectory of a point moving in a conservative force field can be obtained as the spacelike projection of a geodesic curve of a suitable Riemannian metric on the four-dimensional space-time plane.

Problem G.30. *Let us divide by straight lines a quadrangle of unit area into n subpolygons and draw a circle into each subpolygon. Show that the sum of the perimeters of the circles is at most $\pi\sqrt{n}$ (the lines are not allowed to cut the interior of a subpolygon).*

Solution. It is known that the area of an n-gon circumscribed about a circle of radius R is at least $R^2 n\tan(\pi/n)$. Cutting the quadrangle by one straight line after the other, we determine the number of the domains produced and the sum of the angles of the domains. When we draw a new straight line, the number of polygons increases by k, and the sum of the angles increases by at most $2k\pi$, depending on how many nodes of the previous decomposition the new straight line goes through. This way, if the number of sides of the polygons in the resulting decomposition: $k_1, k_2, \ldots, k_n$, then

$$\sum_{i=1}^{n}(k_i - 4) \le 0\,.$$

Denote by T_i the area of the ith domain and by K_i the perimeter of the inscribed circle. We know that

$$K_i^2 \le \frac{4\pi^2}{k_i \tan \frac{\pi}{k\,i}} T_i\,.$$

Then

$$\sum_{i=1}^{n} K_i \le \pi \sum_{i=1}^{n} \sqrt{\frac{4}{k_i \tan \frac{\pi}{k\,i}}}\sqrt{T_i} \le \pi \sqrt{\sum_{i=1}^{n} \frac{4}{k_i \tan \frac{\pi}{k\,i}}}\sqrt{\sum_{i=1}^{n} T_i}$$

by the Cauchy–Schwarz inequality. Therefore, it suffices to show that

$$\sum_{i=1}^{n} \frac{4}{k_i \tan \frac{\pi}{k\,i}} \ge n\,.$$

If $k_i = 4$, then the corresponding term in the sum is just equal to 1. If $k_i \ge 5$, then by $\sum(k_i - 4) \le 0$, this term comes together with $(k_i - 4)$ triangles, and for this reason, it is enough to see that for $k \ge 5$,

$$\frac{4}{k \tan \frac{\pi}{k}} + (k-4)\frac{4}{3 \tan \frac{\pi}{3}} \le k - 3\,.$$

Since

$$\frac{4}{3 \tan \frac{\pi}{3}} < 0.7699\,,$$

it suffices to show that for $k \ge 5$,

$$\frac{4}{k \tan \frac{\pi}{k}} < 0.23k + 0.079\,.$$

The left-hand side is less than $4/\pi$, and this is enough for $k \ge 6$. When $k = 5$, the left-hand side is less than 1.11 and the right-hand side is greater than 1.2. □

Problem G.31. *Let K be a convex cone in the n-dimensional real vector space $\mathbb{R}^n$, and consider the sets $A = K \cup (-K)$ and $B = (\mathbb{R}^n \setminus A) \cup \{0\}$ (0 is the origin). Show that one can find two subspaces in $\mathbb{R}^n$ such that together they span $\mathbb{R}^n$, and one of them lies in A and the other lies in B.*

Solution. First, we show that if $x \in \operatorname{int} \overline{K}$ then $x \in K$. Take a simplex with vertices in $\overline{K}$ that has x in its interior. The vertices can obviously be moved to K by a small perturbation in such a way that x remains in the simplex; thus $x \in K$.

Now we show that if $x \notin \overline{K}$, then there exists $u \in \mathbb{R}^n$ such that $(u, x) < 0$, but $(u, y) \geq 0$ for every $y \in \overline{K}$.

Let k be the point of $\overline{K}$ nearest to x. The perpendicular bisector hyperplane of the segment connecting x to k separates x from $\overline{K}$, that is, there exist $u \in \mathbb{R}^n$ and $b \in \mathbb{R}$ such that $(u, y) \geq b$ for every $y \in K$, but (u, x) <b. Since $O \in \overline{K}$, $b \leq 0$. If we had $(u, y) = \varepsilon < 0$ for some $y \in \overline{K}$, $(u, \lambda y) = \lambda\varepsilon < b$ would hold with a suitable $\lambda > 0$. Thus, $(u, y) \geq 0$ for every $y \in \overline{K}$.

We prove the proposition of the problem using induction. The statement is obvious for $n = 0$. Assume that $n \geq 1$ and the statement is true for $\mathbb{R}^s$ provided that $0 \leq s < n$. If the maximal number of linearly independent elements of K is s and $s < n$, then K is contained in an $(n-1)$-dimensional subspace H. In this case, let $E_0 \subset K \cup (-K)$ and $F_0 \subset (H \setminus A) \cup \{O\}$ be subspaces with which the proposition holds in H.

Taking E_0 for E and the span of F_0 and a vector $f \in \mathbb{R}^n \setminus H$ for F, we find that the proposition is true. We may thus assume that K contains n linearly independent vectors and, as a consequence, it has an interior point.

Set

$$N = \{u\colon (u, x) \geq 0 \text{ for every } x \in \overline{K}\}.$$

If $N = \{O\}$, then $\overline{K} = \mathbb{R}^n$, and thus $K = \mathbb{R}^n$. Therefore, we may assume that $N \neq \{O\}$. Choose a maximal linearly independent system $u_1, \dots, u_k$ consisting of vectors from N. Let H be the hyperplane

$$H = \{x\colon (u_1 + \cdots + u_k, x) = 0\}.$$

If $H \cap K$ is a cone,

$$(H \cap K) \cup (-(H \cap K)) = H \cap A, \quad (H \setminus (H \cap A)) \cup \{O\} = H \cap B;$$

therefore, the proposition can be applied to $H \cap A$ and $H \cap B$. We obtain subspaces $E_0 \subset H \cap A$ and $F_0 \subset H \cap B$, which span H. Let us observe that

$$E_0 \subset \{x\colon (u_1, x) = (u_2, x) = \cdots = (u_k, x) = 0\}.$$

Let $l_1, \dots, l_m$ be a basis of E_0, and let $f_1, \dots, f_l$ be a basis of F_0. We may obviously assume that $l_1, \dots, l_m \in K$; furthermore, $m + l \geq n - 1$ holds. $(K \setminus H)$ has at least one interior point, call any of them e, and let E be the subspace generated by E_0 and e. If we show that $E \subset A$, then we are done with the choice $F = F_0$.

First, we show that $E_0 \subset \overline{K}$. If $x \in E_0$ and $u \in N$, then $u = \lambda_1 u_1 + \cdots + \lambda_k u_k$, and thus

$$(u, x) = \lambda_1(u_1, x) + \cdots + \lambda_k(u_k, x),$$

from which $x \in \overline{K}$. Now, if $l_0 \in E_0$, then $l_0 \in \overline{K}$ and hence $l_0 + e \in \overline{K}$; moreover, $l_0 + e \in \operatorname{int} \overline{K} \subset K$. From this, $\lambda(l_0 + e) \in A$ for every $\lambda \in \mathbb{R}$, that is, $E \subset A$. □

Problem G.32. *Let V be a bounded, closed, convex set in $\mathbb{R}^n$, and denote by r the radius of its circumscribed sphere (that is, the radius of the smallest sphere that contains V). Show that r is the only real number with the following property: for any finite number of points in V, there exists a point in V such that the arithmetic mean of its distances from the other points is equal to r.*

Solution.

(a) Since V is bounded, it is easy to see that there exists a minimal sphere of radius r that contains V. Call this sphere G, and denote its boundary by S, its center by O (the origin). We show that $O \in \operatorname{co}(S \cap V)$. Suppose that this is not true. Then there exists a hyperplane H that separates O from $S \cap V$, since $S \cap V$ is compact. We may assume that

$$H = \{(x_1, \ldots, x_n) : x_n = c\}, \text{ where } 0 < c < r.$$

Since r is minimal, the spheres of radius less then r centered at $O_k = (0, \ldots, 0, 1/k)$ do not cover V. Therefore, we can find points $a_k \in V$ such that $|a_k - O| \geq r$. The sequence $\{a_k\}$ possesses a convergent subsequence $a_{k_i} \to a^*$. Obviously, $a^* \in V$ and $|a^*| = r$; therefore, $a^* \in S \cap V$ and thus $a^* = (a_1^*, \ldots, a_n^*)$, where $a_n^* > c$. Then, for $i > i_0$, $a_{k_i} = (a_1^i, \ldots, a_n^i)$, where $a_n^i > c$ and $\sum_{j=1}^n (a_j^i)^2 \leq r^2$ since $a_k^i \in V$. From this, we get

$$\begin{aligned} |a_{k_i} - O_{k_i}| &= \left(\sum_{j=1}^{n-1} (a_j^i)^2 + (a_n^i - 1/k_i)^2 \right)^{1/2} \\ &\leq (r^2 - 2c/k_i + 1/k_i^2)^{1/2} < r, \end{aligned}$$

but this is impossible.

This shows that $O \in \operatorname{co}(S \cap V)$, and therefore $O \in \operatorname{co}(V) = V$ as well. If $a_1, \ldots, a_m \in V$, then $|a_i| \leq r$, and thus the average of the distance between O and the points a_i is $\leq r$. On the other hand, $s = (1/m) \cdot \sum_{i=1}^m a_i \in V$, and by the minimality of r we can find a point y in V such that $|s - y| \geq r$. Thus,

$$\frac{1}{m} \sum |a_i - y| \geq \left| \frac{1}{m} \sum (a_i - y) \right| = |s - y| \geq r.$$

By the convexity of V, the segment $[O, y]$ lies in V, and one can obviously find a point z on this segment for which $(1/m) \cdot \sum |a_i - z| = r$.

(b) If $u > r$, then for $a_1 = O$, there is no such point $x \in V$ for which $|a_1 - x| = u$.

(c) Now let $u < r$. Since $O \in \operatorname{co}(S \cap V)$, there exist points $a_1, \dots, a_k \in S \cap V$ and numbers $t_1, \dots, t_k$ such that for $0 < t_i$, $\sum_{i=1}^k t_i = 1$ and $\sum_{i=1}^k t_i a_i = 0$. Set $\varepsilon = (r - u)/kr > 0$, and choose the natural numbers q and p_i in such a way that

$$\sum_{i=1}^{k} p_i = q, \text{ and } |p_i/q - t_i| < \varepsilon \qquad (i = 1, \dots, k).$$

Let the system of points $b_1, \dots, b_q$ consist of the points a_i, so that a_i is taken p_i times for each i. Then $|b_j| = r$ (since $b_j \in S$) and

$$\left| \frac{1}{q} \sum_{i=1}^{q} b_i \right| = \left| \frac{1}{q} \sum_{i=1}^{k} p_i a_i \right| = \left| \sum_{i=1}^{k} (p_i/q - t_i) a_i \right| < \varepsilon \sum_{i=1}^{k} |a_i| = kr\varepsilon.$$

We show that for any $x \in V$, $(1/q) \cdot \sum_{i=1}^q |b_i - x| > u$. Indeed, denoting by (a, b) the dot product, for $x \in V$, we have

$$\frac{1}{q} \sum_{i=1}^{q} |b_i - x| \geq \frac{1}{q} \sum_{i=1}^{q} \frac{1}{r} (b_i - x, b_i) = \frac{1}{qr} \left(\sum_{i=1}^{q} r^2 - \left(x, \sum_{i=1}^{q} b_i \right) \right)$$
$$= r - \frac{1}{qr} \left(x, \sum_{i=1}^{q} b_i \right) > r - \frac{1}{qr} r \left| \sum_{i=1}^{q} b_i \right| > r - kr\varepsilon > u \,. \quad \square$$

Problem G.33. *Show that for any natural number n and any real number $d > 3^n/(3^n - 1)$, one can find a covering of the unit square with n homothetic triangles with area of the union less than d.*

Solution. Let us circumscribe congruent homothetic triangles about the faces of a regular hexagonal tessellation such that the area of the triangles is $3/2$ times larger than the area of the hexagonal face. These triangles cover the plane with density $3/2$.

Let the radius of the circle inscribed in one of the triangles be equal to 1. Let us translate the sides of every triangle toward its center by ε. The density of the new triangles is equal to $(3/2) \cdot (1 - \varepsilon)^2$. There are holes produced in the covering: upside-down triangular holes circumscribed about a circle of radius ε. The relative number of the holes is twice the number of the triangles, so the density of the holes is $(3/2) \cdot 2\varepsilon^2 = 3\varepsilon^2$.

Let us cover each hole with some density $d > 1$. By this we mean that we cover each hole with some plates, whose total area is equal to d times the area of the hole. The joint density of the triangles and the plates is

$D = (3/2)\cdot(1-\varepsilon)^2 + 3d\varepsilon^2$. D is minimal at $\varepsilon = 1/(2d-1)$. The minimum is $D = 3d/(2d+1)$.

Let D_n denote the infimum of the densities of the coverings with homothetic triangles of n different sizes. As we have seen, $D_1 \leq 3/2$. Since the holes in the above construction can be covered by homothetic triangles of $(n-1)$ different sizes with density arbitrarily close to D_{n-1}, we have

$$D_2 \leq \frac{3D_1}{2D_1+1},\ D_3 \leq \frac{3D_2}{2D_2+1},\ldots,$$

that is,

$$D_n \leq \frac{3^n}{3^n-1}. \quad \square$$

Problem G.34. *Let R be a bounded domain of area t in the plane, and let C be its center of gravity. Denoting by T_{AB} the circle drawn with the diameter AB, let K be a circle that contains each of the circles T_{AB} $(A, B \in R)$. Is it true in general that K contains the circle of area $2t$ centered at C?*

Solution. We shall show that the answer to the question is negative. Let R be the semicircle in the plane with Cartesian coordinates (x,y) defined by the inequalities $x^2+y^2 \leq 1, x \geq 0$. The area of R is $t = \pi/2$, and the radius of the circle of area $2t$ is equal to 1. Assume that $P \in T_{AB}$ for some pair of points $A, B \in$ R. Denote the position vectors of the points P, A, B by p, a, b, respectively. The condition $P \in T_{AB}$ is equivalent to the inequality

$$|p-(a-b)/2| \leq |a-b|/2.$$

Since

$$|p-(a+b)/2| \geq |p| - |(a+b)/2|,$$

we have

$$\begin{aligned}|p| &\leq |a+b|/2 + |a-b|/2 \leq \left((|a+b|^2+|a-b|^2)/2\right)^{1/2} \\ &= \left((|a|^2+|b|^2+2(a,b)+|a|^2+|b|^2-2(a,b))/2\right)^{1/2} \\ &= \left(|a|^2+|b|^2\right)^{1/2} \leq \sqrt{2}.\end{aligned}$$

Therefore, if we choose for K the circle of radius $\sqrt{2}$ centered at the origin, then K will contain all the circles T_{AB} $(A, B \in R)$. Let us compute the coordinates (x_C, y_C) of the center of gravity C of the figure R. By the symmetry of the figure, $y_C = 0$. To compute x_C, we use the well-known method

$$x_C = \frac{1}{t}\int_0^1 x2\sqrt{1-x^2}\,dx = \frac{1}{t}\left[-\frac{2}{3}(1-x^2)^{3/2}\right]_0^1 = \frac{2}{3t} = \frac{4}{3\pi}.$$

We want to show that the circle of radius 1 centered at $(4/3\pi, 0)$ does not contain the circle K. For this purpose, we have only to show that $4/3\pi + 1 > \sqrt{2}$. Using $\pi < 3.2$, $2.4 \cdot 0.416 < 1$, and $(1.416)^2 > 2$, we obtain

$$\frac{4}{3\pi} + 1 > \frac{4}{3 \cdot 3.2} + 1 = \frac{1}{2.4} + 1 > 1.416 > \sqrt{2}.$$

Therefore, R is indeed a counterexample to the proposition in question. □

Remark. Competitors disproved the proposition with many different counterexamples. Most constructions resembled the set-theoretic union of the circle $x^2 + y^2 \leq 1$ and the ellipse $x^2/(1+\varepsilon)^2 + y^2/\varepsilon^2 \leq 1$ for some suitably small positive ε.

Problem G.35. *Let $M^n \subset \mathbb{R}^{n+1}$ be a complete, connected hypersurface embedded into the Euclidean space. Show that M^n as a Riemannian manifold decomposes to a nontrivial global metric direct product if and only if it is a real cylinder, that is, M^n can be decomposed to a direct product of the form $M^n = M^k \times \mathbb{R}^{n-k}$ $(k < n)$ as well, where M^k is a hypersurface in some $(k+1)$-dimensional subspace $E^{k+1} \subset \mathbb{R}^{n+1}$, $\mathbb{R}^{n-k}$ is the orthogonal complement of E^{k+1}.*

Solution. The solution rests upon the two fundamental equations of classical surface theory, namely, the Gauss equation

$$R(X,Y)Z = g(Z, A(X))A(Y) - g(Z, A(Y))A(X)$$

and the Codazzi–Mainardi equation

$$(\nabla_X A)(Y) = (\nabla_Y A)(X).$$

In these formulas, X, Y, Z are vector fields tangential to the hypersurface, $g(X,Y)$ is the first fundamental form, $A(X) = d_X\mathbf{m}$ is the Weingarten map ($\mathbf{m}$ is the unit normal vector field of the hypersurface, d_X is the directional derivative in $\mathbb{R}^{n+1}$), $\nabla_X Y$ is the covariant derivative, and, finally, $R(X,Y)Z$ is the Riemannian curvature tensor.

$A(X)$ is a self-adjoint linear mapping on each tangent space. It is also called the extrinsic geometrical fundamental form, since with its help one can describe the shape of the surface in the space. For example, the following statements are known and can simply be derived from the Codazzi–Mainardi equations.

• Let W_p^0 be the kernel of A_p at the point $p \in M^n$, and assume dim W_p^0 = constant = k on some open subset $U \subset M^n$. Then the distribution of the subspaces W_p^0 is integrable, and the integral manifolds are open subsets of a subspace $\mathbb{R}^k \subset \mathbb{R}^{n+1}$.

• The Riemannian curvature of the hypersurface M^n is 0 if and only if the rank of A is at most 1 at any point. For such a surface let U denote the open set of those points at which $A \neq 0$, and let $\mathbf{n}$ be a unit vector field on

U consisting of eigenvectors of A such that $A(\mathbf{n}) = \lambda\mathbf{n}$ holds with $\lambda \neq 0$. Then, by the previous proposition, integral manifolds perpendicular to $\mathbf{n}$ are open subsets of some subspace $\mathbb{R}^{n-1}$ in $\mathbb{R}^{n+1}$. Let $c(s)$ be a curve on such an integral manifold parameterized by arc length, and consider the function $\lambda(s) = \lambda(c(s))$. Codazzi–Mainardi equations yield by a simple computation that $\lambda' = \varphi(s)\lambda$, where $\varphi(s) := -g(\dot{c}(s), \nabla_{\mathbf{n}}\mathbf{n})$, and thus

$$\lambda(s) = \lambda(0)e^{\int_0^s \varphi(t)\,dt}.$$

Consequently, if $\lambda \neq 0$ at one of the points of the integral manifold perpendicular to $\mathbf{n}$, then $\lambda \neq 0$ along the whole integral manifold and also at its boundary points. For this reason, if M^n is complete, then the maximal integral manifolds perpendicular to $\mathbf{n}$ must fill the whole subspace $\mathbb{R}^{n-1}$. Consider a bending of such a complete hypersurface having Riemannian curvature 0 onto $\mathbb{R}^n$. (If M^n is not simply connected, then we bend its universal covering space.) The above integral manifolds will be bent onto parallel hyperplanes of $\mathbb{R}^n$; Thus the distance between the points of an integral manifold and another integral manifold is constant. This means, that the integral manifolds themselves are parallel, $(n-1)$-dimensional subspaces $\mathbb{R}^{n-1}$ in $\mathbb{R}^{n+1}$, since otherwise the above distance would not be constant. The orthogonal complement $\mathbb{R}^2$ of the subspaces $\mathbb{R}^{n-1}$ cuts M^n in a curve M^1, and we obviously have the cylinder decomposition $M^n = M^1 \times \mathbb{R}^{n-1}$. To sum up, we have the following propositions.

Proposition 1. A complete hypersurface of Riemannian curvature 0 is always a cylinder of the form $M^n = M^1 \times \mathbb{R}^{n-1}$, and M^1 is a curve in the orthogonal complement $\mathbb{R}^2$ of $\mathbb{R}^{n-1}$.

Proposition 2. If a complete hypersurface M^n decomposes into a metric direct product in the form $M^n = M^k \times M^{n-k}$, then either M^n has Riemannian curvature 0, or the Weingarten map vanishes at each point on the tangent spaces of one of the manifolds.

Proof. Denote by T_p the tangent space of M^n at $p \in M^n$ and by $T_p = T_p^1 \times T_p^2$ the decomposition of the tangent space corresponding to the direct product decomposition of M^n. Since for a metric direct product the Riemannian curvatures are also multiplied in a proper way, $R(T_p^1, T_p^2)X = 0$ holds. Combining this with the Gauss equation we get

$$g(X, A(T_p^1))A(T_p^2) = g(X, A(T_p^2))A(T_p^1).$$

This identity can hold for a self-adjoint mapping A only if A has rank 1 or A vanishes along one of the subspaces T_p^i.

We have to show that if the Riemannian curvature of the surface M^n is not identically equal to zero, then A vanishes along T^i at each point of the manifold.

Let $p \in M^n$ be a point at which $R(X,Y)Z \neq 0$. Then, as we have seen above, A vanishes on one of the subspaces T_p^i. Suppose that $A(T_p^2) = 0$ holds. By Gauss's equation, the Riemannian curvature $R^{(2)}(X,Y)Z$ of the manifold M^{n-k} vanishes, while the curvature $R^{(1)}(X,Y)Z$ of the manifold M^k is different from 0 at p. Since the tensor $R^{(1)}(X,Y)Z$ is constant along each copy of M^{n-k} in the direct product, $A(T^2) = 0$ along the copy of M^{n-k} that goes through p. This means that the Riemannian curvature of M^{n-k} is equal to 0 identically. It follows also that $A(T_q^2) = 0$ in every point $q = (q_1, q_2) \in M^k \times M^{n-k}$ such that the tensor $R^{(1)}$ is different from 0 at q_1.

It remains to show that $A(T_q^2) = 0$ also at points $q = (q_1, q_2)$ for which $R^{(1)}|_{q_1} \neq 0$.

Let $V^k \subset M^k$ be a maximal connected open subset such that $R^{(1)} = 0$ at each point of V^k. The Riemannian curvature vanishes on the part $U = V^k \times M^{n-k} \subset M^n$ of the hypersurface, and for this reason, the rank of A is at most 1 here. We have to show that the only eigenvector $\mathbf{n}$ of A corresponding to the nonzero eigenvalue lies in the tangent space T^1.

Suppose to the contrary that $\mathbf{n} \notin T^1$ at some point p. Let N be the maximal integral manifold perpendicular to $\mathbf{n}$ and going through p. Since the submanifolds N, V^k, and M^{n-k} are subspaces (totally geodesic submanifolds with 0 curvature) of the locally Euclidean space U, the angle between $\mathbf{n}$ and the subspace T^1 is constant and nonzero along N. Since the eigenvalue λ of $\mathbf{n}$ does not vanish on the boundary of N either, $A \neq 0$ on the boundary of N and $\mathbf{n} \notin T^1$ holds at these points as well. In any neighborhood of such a boundary point, one can find a point $q = (q_1, q_2)$ such that $R^{(1)}|_{q_1} \neq 0$, and thus the image $A(T_q)$ lies in the subspace T_q^1, while at the boundary points $A(T_p)$ is not contained in T_p^1. For this reason, the existence of such a boundary point contradicts the continuity of A. However, such a boundary point does exist obviously, otherwise N would be complete and its orthogonal projection onto the manifold V^k would give a complete open subset of V^k. This is impossible because of the definition of V^k. This proves Proposition 2.

The theorem follows immediately from the two propositions. Indeed, it is clear that cylinders decompose into a metric direct product, and conversely, if M^n decomposes into a metric direct product, then either $R(X,Y)Z \equiv 0$, and then by Proposition 1 M^n is a cylinder, or $A(T^2) = 0$, and then the copies of M^{n-k} are parallel $(n-k)$-dimensional subspaces. M^n is obviously a cylinder in the latter case as well. □

Problem G.36. *Among all point lattices on the plane intersecting every closed convex region of unit width, which one's fundamental parallelogram has the largest area?*

Solution. Let r^* be the lattice generated by the vertices of a regular triangle of altitude $\frac{1}{2}$. The area of the fundamental parallelogram is equal to $1/(2\sqrt{3})$.

Let r be another lattice such that r is not congruent to r^*, but the area of the fundamental parallelogram of r is also $1/(2\sqrt{3})$. We shall show that in this case, there exists a regular triangle of altitude 1 which shares no point in common with r. Let P^*, Q^* and P, Q be the two nearest points in the lattices r^* and r, respectively. Obviously,

$$a = \overline{PQ} < a = \overline{P^*Q^*} = \frac{1}{\sqrt{3}}.$$

Let the straight line PQ be horizontal. Place a regular triangle with altitude 1 onto the plane in such a way that one of its sides lies horizontally below PQ, while the other two sides go through P and Q, respectively. The distance of the straight line PQ from the upper vertex of the triangle is $(a\sqrt{3}/2)$, and its distance from the base side is $(1 - a\sqrt{3}/2)$. The distance between two horizontal ranges of points in r is equal to $1/(2\sqrt{3}a)$. Since

$$1 - \frac{a\sqrt{3}}{2} < \frac{1}{2a\sqrt{3}} < \frac{a\sqrt{3}}{2}, \qquad 0 < a < \frac{1}{\sqrt{3}},$$

the triangle contains no lattice points other than P and Q. Translating the triangle a little downward, we obtain a triangle of unit width that contains no lattice points.

It remains to show that r^* has a point on every closed convex figure k of width 1. For this purpose, let us remark that the radius of the maximal inscribed circle of a convex figure of width 1 is at least $1/3$. This can be proved from the facts that every convex figure is contained in a triangle or a band circumscribed about the inscribed circle of the figure and that among the triangles circumscribed about a given circle, the width of the regular triangle is maximal. As a consequence, the width of k is at most three times the radius of its inscribed circle. Now we need only remark that every circle of radius $1/3$ contains a lattice point from r^*. Therefore, r^* has a point in common not only with every convex figure of unit width, but also with the inscribed circle of such a figure. □

Problem G.37. *Let S be a given finite set of hyperplanes in $\mathbb{R}^n$, and let O be a point. Show that there exists a compact set $K \subseteq \mathbb{R}^n$ containing O such that the orthogonal projection of any point of K onto any hyperplane in S is also in K.*

Solution. We may assume without loss of generality that O is the origin of the space $\mathbb{R}^n$ and that the normal vectors of the hyperplanes in S generate $\mathbb{R}^n$ (if this latter condition is not fulfilled, then we may add further hyperplanes to S until we reach the desired property).

Denote by $\mathcal{P}_1$ the set of orthogonal projections onto the $(n-1)$-dimensional linear subspaces of $\mathbb{R}^n$ parallel to a hyperplane in S. For $i = 2, \ldots, n$, let $\mathcal{P}_i$ be the collection of all projection operators, the image of which is $(n-i)$-dimensional and can be obtained as the intersection of the images of some operators from $\mathcal{P}_1$. Then, by the finiteness of S and the assumption we made for the normal vectors of the hyperplanes, $\mathcal{P}_i$ is finite and nonempty for every $1 \leq i \leq n$. Thus, $\mathcal{P}_n$ has one element, $\mathcal{P}_n = \{P_n\}$, where P_n is the zero operator. For the sake of unified notation, set $\mathcal{P}_0 = \{P_0\}$, where P_0 is the identical transformation. If $P_i \in \mathcal{P}_i$ and $P_j \in \mathcal{P}_j$, and the image of P_i contains the image of P_j, or equivalently, if the kernel of P_i is contained in the kernel of P_j, then we write $P_i \geq P_j$.

The orthogonal projection onto a plane s from S can be given in the form $x \mapsto P_s x + p_s$, where $P_s \in \mathcal{P}_1$ is the operator of the orthogonal projection onto the $(n-1)$-dimensional linear subspace parallel to s, and p_s is the vector drawn from O perpendicular to s. Obviously, $P_s p_s = 0$ for every $s \in S$.

Lemma. One can find a sequence of numbers $0 = R_0 < R_1 < \cdots < R_n$ such that for any $1 \leq j \leq n$, $P_j \in \mathcal{P}_j$, and $R \geq R_j$, the set

$$G(P_j, R) := \{x \in \mathbb{R}^n \colon P_j x = 0 \text{ and } \|P_i x\|^2 \leq R^2 - R_i^2 \\ \text{for every } 0 \leq i \leq j-1, P_j \leq P_i \in \mathcal{P}_i\}$$

is mapped into itself by any projection $x \mapsto P_s x + p_s$ onto a plane $s \in S$ for which $P_s \geq P_j$.

Since $P_n = 0$, applying the lemma for $j = n$ and $R = R_n$, we obtain that the closed set $K = G(0, R_n)$ is mapped into itself by any projection onto a plane in S. On the other hand, this set is bounded, for it is contained in the sphere of radius R_n centered at the origin. Thus, it satisfies the requirements formulated in the problem.

Proof. We prove the lemma by induction on j. For $j = 1$, let $P_1 \in \mathcal{P}_1$ be an arbitrary projection. Then

$$G(P_1, R) = \{x \in \mathbb{R}^n \colon P_1 x = 0, \|x\| \leq R\}.$$

If, for some $s \in S$, $P_s \geq P_1$, then, since the images of the operators P_1 and P_s are $(n-1)$-dimensional, $P_s = P_1$. Therefore, $P_s x_0 + p_s = p_s$ for any $x_0 \in G(P_1, R)$. Thus, if we take $\sup_{s \in S} \|p_s\|$ for R_1, the claim of the lemma holds.

Now assume that the lemma has been proved for the values $1, 2, \ldots, j-1$ and that the numbers $R_0 < R_1 < \cdots < R_{j-1}$ have already been constructed. Let $P_j \in \mathcal{P}_j$ be an arbitrary projection and $s \in S$ such that $P_s \geq P_j$.

If $P_j x_0 = 0$, then $P_j(P_s x_0 + p_s) = P_j x_0 + P_j p_s = 0$. Thus, projection onto s preserves the first defining property of the elements of $G(P_j, R)$.

Second, we show that if $R \geq R_{j-1}$ and $x_0 \in G(P_j, R)$, then

$$\|P_i(P_s x_0 + p_s)\|^2 \leq R^2 - R_i^2 \tag{1}$$

holds for $0 \leq i \leq j-2$ and $P_j \leq P_i \in \mathcal{P}_i$.

If $P_s \geq P_i$, then $P_i(P_s x_0 + p_s) = P_i x_0$, and there is nothing to prove.

If $P_s \geq P_i$, then let us denote by P_{i+1} the projection operator whose image is the intersection of the images of P_s and P_i. Consider the set

$$G(P_{i+1}, R^*) = G(P_{i+1}, \sqrt{R^2 - \|P_{i-1} x_0\|^2}).$$

Then

$$R^{*2} = R^2 - \|P_{i+1}x_0\|^2 \geq R^2 - (R^2 - R_{i+1}^2) = R_{i+1}^2\,,$$

and therefore, $R^* \geq R_{i+1}$. Thus, $i+1 \leq j-1$, and by the induction hypothesis, the projection $x \mapsto P_s x + p_s$ maps $G(P_{i+1}, R^*)$ into itself (since $P_s \geq P_{i+1}$). However, $(P_0 - P_{i+1})x_0 \in G(P_{i+1}, R^*)$, since for $P_{i+1}(P_0 - P_{i+1})x_0 = 0$ and $P_{i+1} \leq P_k \in \mathcal{P}_k$ $(k = 0, \dots, i)$, we have

$$\begin{aligned}\|P_k(P_0 - P_{i+1})x_0\|^2 &= \|(P_k - P_{i+1})x_0\|^2 = \|P_k x_0\|^2 - \|P_{i+1}x_0\|^2 \\ &\leq R^2 - R_k^2 - \|P_{i+1}x_0\|^2 = R^{*2} - R_k^2\,.\end{aligned}$$

Therefore, $P_s(P_0 - P_{i+1})x_0 + p_s = (P_s - P_{i+1})x_0 + p_s$ is also contained in $G(P_{i+1}, R^*)$. Thus,

$$\|P_i[(P_s - P_{i+1})x_0 + p_0]\|^2 \leq R^{*2} - R_i^2\,,$$

that is,

$$\|P_i(P_s x_0 + p_s) - P_{i+1}x_0\|^2 \leq R^2 - \|P_{i+1}x_0\|^2 - R_i^2\,,$$

from which we get (1).

To prove the Lemma we have only to show that if $R \geq R_{j-1}$ is sufficiently large, then (1) holds for $i = j-1$ and any $x_0 \in G(P_j, R)$ and any $P_{j-1} \in \mathcal{P}_{j-1}$ such that $P_{j-1} \geq P_j$.

If $P_s \geq P_{j-1}$, then we are done because $P_{j-1}(P_s x_0 + p_s) = P_{j-1}x_0$.

If $P_s \geq P_{j-1}$, then the intersection of the images of P_s and P_{j-1} is $(n-j)$-dimensional and consequently coincides with the image of P_j.

We show that $P_{j-1}P_s$ is a contraction on the kernel of the operator P_j. For this purpose, it suffices to show that if $x \neq 0$, $P_j x = 0$, then $\|P_{j-1}P_s x\| < \|x\|$. The assumption that the two sides of the latter inequality are equal yields $\|P_s x\| = \|x\|$, from which we obtain $P_s x = x$. Repeating the same argumentat yields $P_{j-1}x = x$. Therefore, x is in the image of P_s and P_{j-1}, that is, in the image of P_j. However, this contradicts $x \neq 0$ and $P_j x = 0$. This means that $P_{j-1}P_s$ is indeed a contraction, that is, there exists a number $0 < q < 1$ such that $P_j x = 0$ implies $\|P_{j-1}P_s x\| \leq q\|x\|$. Now let us choose the numbers $r_j = r_j(P_s, P_{j-1}, P_j)$ in such a way that

$$qR + R_1 \leq \sqrt{R^2 - R_{j-1}^2}, \qquad \text{when} \qquad R \geq r_j\,. \tag{2}$$

(This is possible since if both sides are divided by R, the left-hand side tends to q and the right-hand side tends to 1 as $R \to \infty$.)

Now, if $R \geq r_j$ and $x_0 \in G(P_j, R)$, then

$$\|P_{j-1}(P_s x_0 + p_s)\| \leq q\|x_0\| + \|p_s\| \leq qR + R_1\,;$$

therefore, by (2), (1) holds for $i = j-1$. Now let R_j be the maximum of the numbers $r_j(P_s, P_{j-1}, P_j)$ for all P_s, P_{j-1}, P_j. Then we obtain that (1) holds for $R \geq R_j$, with $i = j-1$, if

$$\begin{gathered}P_j \in \mathcal{P}_j, \quad x_0 \in G(P_j, R), \quad P_j \leq P_{j-1} \in \mathcal{P}_{j-1}, \\ s \in S \quad \text{and} \quad P_s \geq P_j\,.\end{gathered}$$

This completes the proof of the lemma. □

Problem G.38. *Let k and K be concentric circles on the plane, and let k be contained inside K. Assume that k is covered by a finite system of convex angular domains with vertices on K. Prove that the sum of the angles of the domains is not less than the angle under which k can be seen from a point of K.*

Solution. Let O be the common center of the circles, and r and R the radius of k and K, respectively. Let us define a function F on the interior of k as follows. If the distance from O to P is ρ, then set

$$F(P) = f(\rho) = \frac{1}{\pi(R^2 - \rho^2)}\sqrt{\frac{R^2 - r^2}{r^2 - \rho^2}}.$$

If T is a measurable set in the interior of k, then let

$$m(T) = \int_T F(P)\, dP.$$

The function m is obviously a measure on the interior of k. We show that m has the following property:

If g and h are half-lines starting from a point A of K such that g is tangent to k, h intersects k, and the angle between h and g is α, then the angular domain bounded by g and h cuts the interior of k in a domain T, the measure m of which is α.

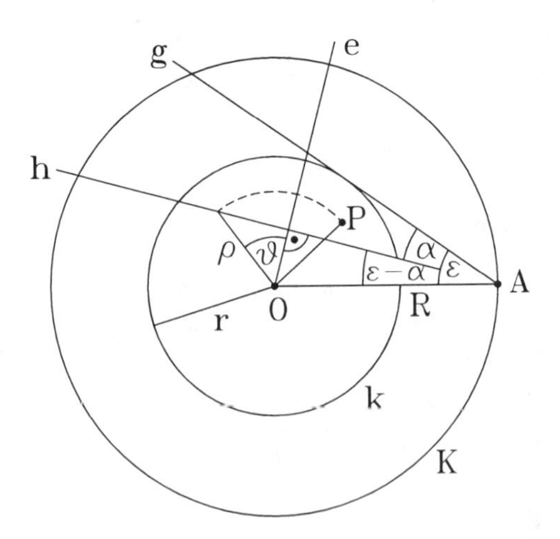

Figure G.29.

Let us compute $m(T)$ by successive integration. Let e be the half-line starting from O perpendicular to h. Let us introduce a polar coordinate system on the plane, with polar axis e. Then a point $P(\rho, \varphi)$ belongs to T

if and only if $d \le \rho \le r$, $-\vartheta \le \varphi \le \vartheta$, where $d = \rho\cos\vartheta = R\sin(\varepsilon - \alpha)$ is the distance of O from h, ε is the angle between the half-lines AO, and g (see Figure G.29). Based on these, we get

$$m(T) = \int_T F(P)\,dP = \int_d^r \int_{-\arccos(d/\rho)}^{\arccos(d/\rho)} f(\rho)\rho d\varphi\,d\rho$$

$$= \int_d^r 2f(\rho)\rho\arccos(\frac{d}{\rho})\,d\rho$$

$$= \int_{R\sin(\varepsilon-\alpha)}^r 2f(\rho)\rho\arccos\left(\frac{R\sin(\varepsilon-\alpha)}{\rho}\right)\,d\rho\,.$$

Therefore, we have to show that, for $0 \le \alpha \le 2\varepsilon$, we have

$$= \int_{R\sin(\varepsilon-\alpha)}^r 2f(\rho)\rho\arccos\left(\frac{R\cdot\sin(\varepsilon-\alpha)}{\rho}\right)\,d\rho = \alpha\,.$$

Substituting $R\cdot\sin(\varepsilon - \alpha) = t$, this means that

$$\int_t^r 2f(\rho)\rho\arccos(\frac{t}{\rho})\,d\rho = \varepsilon - \arcsin(\frac{t}{R})$$

if $-r \le t \le r$. This relationship is obviously true for $t = r$. Thus, it is enough to show that the derivatives of the two sides with respect to t are the same:

$$\int_t^r \frac{2f(\rho)\rho}{\sqrt{\rho^2 - t^2}}\,d\rho = \frac{1}{\sqrt{R^2 - t^2}}, \qquad -r \le t \le r$$

$$\int_t^r \frac{2f(\rho)\rho}{\sqrt{\rho^2 - t^2}}\,d\rho = \int_t^r \frac{2}{\pi}\frac{\rho}{R^2 - \rho^2}\sqrt{\frac{R^2 - r^2}{(r^2 - \rho^2)(\rho^2 - t^2)}}\,d\rho$$

$$= \left[\frac{2}{\pi}\frac{1}{\sqrt{R^2 - t^2}}\tan^{-1}\sqrt{\frac{R^2 - r^2}{R^2 - t^2}\frac{\rho^2 - t^2}{r^2 - \rho^2}}\right]_{\rho=t}^{\rho=r} = \frac{1}{\sqrt{R^2 - t^2}}\,.$$

The proposition proved, combined with the additivity of the measure m, yields immediately that if both half-lines bounding a convex angular domain of angular measure α intersect k, then the m measure of the intersection of the angle and k is equal to α (since the intersection can be obtained as the difference of two segments of k).

Now consider a covering of k with a finite number of convex angular domains. Let us denote by $\alpha_1, \dots, \alpha_n$ the angular measure of the angles, and by $T_1, \dots, T_n$ the intersections of the domains with k. Then, by the proposition just proved, $\alpha_i \ge m(T_i)$, $i = 1, \dots, n$. (Equality holds here if and only if the both boundary lines of the ith angle have a point in common with k.) Therefore, by the subadditivity of measures,

$$\alpha_1 + \dots + \alpha_n \ge m(T_1) + \dots + m(T_n) \ge m\left(\bigcup_{i=1}^{n} T_i\right) = 2\varepsilon$$

(since $\bigcup_{i=1}^n T_i$ fills the interior of k and the m measure of k is exactly 2ε). This completes the proof. $\square$

Problem G.39. *Let $\xi(E, \pi, B)$ $(\pi : E \to B)$ be a real vector bundle of finite rank, and let*

$$\tau_E = V\xi \oplus H\xi \tag{$*$}$$

be the tangent bundle of E, where $V\xi = \text{Ker } d\pi$ is the vertical subbundle of τ_E. Let us denote the projection operators corresponding to the splitting $(*)$ by v and h. Construct a linear connection ∇ on $V\xi$ such that

$$\nabla_X \vee Y - \nabla_Y \vee X = v[X, Y] - v[hX, hY].$$

(X and Y are vector fields on E, $[.,.]$ is the Lie bracket, and all data are of class $\mathcal{C}^\infty$).

Solution. Let $\mathcal{C}^\infty(E)$ be the ring of differentiable (of class $\mathcal{C}^\infty$) functions $E \to \mathbb{R}$, and denote by $\mathcal{X}(E)$ and $\mathcal{X}_V(E)$ the $\mathcal{C}^\infty(E)$-module of all vector fields on E and the submodule of vertical vector fields, respectively. A mapping

$$D \colon \mathcal{X}_V(E) \times \mathcal{X}_V(E) \to \mathcal{X}_V(E), \quad (Z, W) \mapsto D_Z W$$

will be called a *pseudoconnection on* $V\xi$ if it has the formal properties of linear connections, that is, it is $\mathcal{C}^\infty(E)$-linear in Z and an $\mathbb{R}$-linear derivation in W $(D_Z fW = (Zf)W + fD_Z W,\ f \in \mathcal{C}^\infty(E))$. Assume that $\mathring{D}$ is a pseudoconnection such that

$$\mathring{D}_Z W - \mathring{D}_W Z = [Z, W] \text{ for every } Z, W \in \mathcal{X}_V(E),$$

and define the mapping

$$\nabla \colon \mathcal{X}(E) \times \mathcal{X}_V(E) \to \mathcal{X}_V(E), \quad (X, Y) \mapsto \nabla_X Y$$

by the formula

$$\nabla_X Y := \mathring{D}_{vX} Y + v[hX, Y].$$

Then ∇ is a linear connection. It can be seen immediately that ∇ is $\mathbb{R}$-linear in X and Y and that for $f \in \mathcal{C}^\infty(E)$ we have

$$\begin{aligned}\nabla_X fY &= f\mathring{D}_{vX} Y + (vX)fY + fv[hX, Y] + v(hX)fY \\ &= f\nabla_X Y + (vX + hX)fY = f\nabla_X Y + (Xf)Y\end{aligned}$$

(applying linearity at each point and $vY = Y$). Finally, a similarly simple computation shows that $\nabla_{fX} Y = f\nabla_X Y$. We show that ∇ satisfies the required relationship. Indeed,

$$\begin{aligned}\nabla_X vY - \nabla_Y vX &= \mathring{D}_{vX} vY - \mathring{D}_{vY} vX + v[hX, vY] - v[hY, vX] \\ &= [vX, vY] + v[hX, vY] - v[hY, vX] \\ &= v([vX, vY] + [hX, vY] - [hY, vX]) \\ &= v([X, Y] - [hX, hY]) = v[X, Y] - v[hX, hY].\end{aligned}$$

It remains to show that an "auxiliary" pseudoconnection $\mathring{D}$ does exist. We shall prove this by a simple local construction. Assume $\dim B = n$, rank $\xi = r$. By the local triviality of the vector bundle, each point of B has a neighborhood U such that $\pi^{-1}(U) \to U$ is diffeomorphic to the trivial bundle $U \times \mathbb{R}^r \to \mathbb{R}^r$. If $\psi\colon \pi^{-1}(U) \to U \times \mathbb{R}^n$ is a vector bundle isomorphism, $(u^i)_{i=1}^n$ is a local coordinate system on U, and $(l^\alpha)_{\alpha=1}^r$ is the dual basis of the canonical basis of $\mathbb{R}^r$, then

$$\begin{aligned} x^i &:= u^i \circ \pi \qquad & (1 \le i \le n), \\ y^\alpha &:= l^\alpha \circ pr_2 \circ \psi \qquad & (1 \le \alpha \le r) \end{aligned}$$

is a local coordinate system on $\pi^{-1}(U)$, and a direct computation shows that the vector fields $\partial/\partial y^\alpha$ form a local basis of $\mathcal{X}_V(E)$. Let us define the pseudoconnection $\mathring{D}$ by the requirement

$$\mathring{D}_{\frac{\partial}{\partial y^\alpha}} \frac{\partial}{\partial y^\beta} = 0 \quad (1 \le \alpha, \beta \le r).$$

(It follows from standard methods of the theory of linear connections that $\mathring{D}$ can be defined this way.) Then for any two vector fields

$$Z = Z^\alpha \frac{\partial}{\partial y^\alpha}, \qquad W = W^\beta \frac{\partial}{\partial y^\beta}$$

over $\pi^{-1}(U)$, we have

$$\mathring{D}_Z W = Z^\alpha \frac{\partial W^\beta}{\partial y^\alpha} \frac{\partial}{\partial y^\beta},$$

and consequently, $\mathring{D}_Z W - \mathring{D}_W Z = [Z, W]$, which completes the proof. □

Problem G.40. *Consider a latticelike packing of translates of a convex region K. Let t be the area of the fundamental parallelogram of the lattice defining the packing, and let $t_{\min}(K)$ denote the minimal value of t taken for all latticelike packings. Is there a natural number N such that for any $n > N$ and for any K different from a parallelogram, $nt_{\min}(K)$ is smaller than the area of any convex domain in which n translates of K can be placed without overlapping? (By a latticelike packing of K we mean a set of nonoverlapping translates of K obtained from K by translations with all vectors of a lattice.)*

Solution. We show that such an N does not exist. Let h be the branch of the hyperbola defined by the equation $xy = 1$ that lies in the first quadrant of the plane. Let A be the intersection point of the x-axis and the tangent of h at $(a, 1/a)$, and let B be the intersection point of the y-axis and the tangent of h at $(1/a, a)$. Let us replace the arcs of the hyperbola "beyond" $(a, 1/a)$ and $(1/a, a)$ with the segments of the tangents between A, $(a, 1/a)$

and $B, (1/a, a)$, respectively. Denote by g the curve consisting of the two segments and the arc of the hyperbola between them. Then the area of any triangle bounded by the coordinate axes and a tangent of h equals 2. On the other hand, if $x \to \infty$, then the area of the domain between the coordinate axes and g tends to infinity as well. Therefore, we can choose a in such a way that the ratio of the two areas is equal to an arbitrarily small ε. Let us round two opposite corners of the unit square with the help of two suitably scaled copies of the arc g cutting off the square two domains of area δ. The area of the rounded square q equals $1-2\delta$. Since q is central symmetric, $t_{\min}(q)$ is the area of the circumscribed hexagon of q having minimal area, that is, $t_{\min}(q) = 1 - 2\varepsilon\delta$. On the other hand, n translates of q can be packed into a domain of area $n - 2\delta$. Thus, for any n, we can choose an $\varepsilon > 0$ such that $\varepsilon n < 1$, and then $nt_{\min}(q) = n-2\varepsilon\delta > n-2\delta$. □

Problem G.41. *Show that there exists a constant c_k such that for any finite subset V of the k-dimensional unit sphere there is a connected graph G such that the set of vertices of G coincides with V, the edges of G are straight line segments, and the sum of the kth powers of the lengths of the edges is less than c_k.*

Solution 1. Let G be a connected graph such that the set of vertices of G is V, the edges of G are straight line segments, and the sum of the length of the edges is minimal. By the characterization of G, G is a tree. Take an open ball about the midpoint of every edge of G with radius $r(\sqrt{3}-1)/4$, where r is the length of the edge. We prove that these spheres do not intersect one another. Assume, to the contrary, that the segments AB and CD are edges in G, $\rho(A, B) = R$, $\rho(C, D) = r$, and

$$S((A+B)/2,\ R(\sqrt{3}-1)/4) \cap S((C+D)/2, \quad r(\sqrt{3}-1)/4) \neq \emptyset,$$

where ρ is the distance and $S(a, r)$ is the open ball of radius r centered at a. By symmetry, we may assume $R \geq r$. Then

$$\begin{aligned}\rho(A, (C+D)/2) &\leq \rho(A, (A+B)/2) + \rho((A+B)/2, (C+D)/2)\\ &< R/2 + R(\sqrt{3}-1)/2 = R\sqrt{3}/2\,.\end{aligned}$$

Since at least one of the angles $(A, (C+D)/2, C)$ and $(A, (C+D)/2, D)$ is not obtuse if $A \neq (C+D)/2$, we get

$$\begin{aligned}\min\{\rho(A, C), \rho(A, D)\} &\leq (\rho^2(A, (C+D)/2) + r^2/4)^{1/2}\\ &< (3R^2/4 + R^2/4)^{1/2} = R\,.\end{aligned}$$

If the edge AB is removed from the graph G, then the subgraph is the union of two trees. By symmetry, we may assume that B is contained in the same component as C (and D as well). But then removing from G the edge AB and adding to it the shorter of AC and AD, the length of which is less then R, we obtain a connected graph G' satisfying all the requirements for G, but the sum of the length of its edges is less then the sum for G, and this is a contradiction.

Since the balls we constructed are disjoint and are contained in a sphere of radius 2, their total volume is less than the volume of the ball of radius 2. Thus,

$$2^k\omega_k \geq \sum_{AB \text{ is an edge in } G} \rho^k(A,B)\left(\frac{\sqrt{3}-1}{4}\right)^k \omega_k\,,$$

where ω_k is the volume of the k-dimensional unit ball. Therefore,

$$\sum_{AB \text{ is an edge in } G} \rho^k(A,B) \leq \left(\frac{4}{\sqrt{3}-1}\right)^k 2^k = c_k\,,$$

as was to be proved. □

Solution 2. It suffices to show the following claim:

Claim. If T is a box such that the ratio of any two edges of T is less than or equal to 3, and $V \subseteq T$ is a finite set of points, then there exists a connected graph $G = G(V)$ set of vertices V such that the sum of the kth power of the edges of G satisfies

$$S(G) \leq c_k \operatorname{Vol}(T),$$

where c_k is a constant depending only on k, $\operatorname{Vol}(T)$ is the volume of T.

Using induction on the number of points in V, we shall show that one may choose $c_k = 3^{k+1}k^{k/2}$. Let $|V| = n$. The statement is true for $n = 1, 2$, and assume it has already been proved for $1, 2, \dots, n-1$. Let us shrink T, that is, move the faces of T inside as long as the assumptions on T are not violated and T contains V. This way, we make the inequality to prove sharper. If T cannot be shrunk further and e is one of the longest edges of T, and furthermore, Q_1 and Q_2 are the faces perpendicular to e, then V has a point on both Q_1 and Q_2 (otherwise, we could shrink T further). Let T_1, T_2, and T_3 be the consecutive boxes obtained by cutting T into three equal parts by hyperplanes perpendicular to e. We distinguish two cases depending on whether T_2 contains a point of V or not.

Case 1. V has no point in T_2. Let $V_1 = V \cap T_2, V_3 = V \cap T_3$. Since $Q_i \cap V \neq \emptyset$ for $i = 1, 2$, we have $|V_i| < n$. Applying the induction hypothesis for T_1, V_1 and T_3, V_3, we obtain connected graphs $G_1 = G(V_1)$ and $G_3 = G(V_3)$, for which $S(G_i) \leq c_k \operatorname{Vol}(T_i) = (1/3)\operatorname{Vol}(T)$, $i = 1, 3$. Connecting a point of G_1 to a point of G_3, we obtain a connected graph G such that

$$\begin{aligned} S(G) &\leq S(G_1) + S(G_3) + (\sqrt{k})^k |e|^k \\ &\leq \frac{2}{3} c_k \operatorname{Vol}(T) + (\sqrt{k})^k 3^k \operatorname{Vol}(T) \leq c_k \operatorname{Vol}(T) \end{aligned}$$

by the choice of c_k.

Case 2. V has a point in T_2. Choose $x \in V \cap T_2$ and cut T into two closed boxes T_1^* and T_2^* by a hyperplane passing through x perpendicular to e. Then T_i^* also has the property that the ratio of any two of its edges is not more than 3. Thus, the induction hypothesis can be applied again to T_1^*, $T_1^* \cap V$ and T_2^*, $T_2^* \cap V$ since the sets $T_1^* \cap V$ and $T_2^* \cap V$ have less than n points. Therefore, one obtains the connected graphs $G_1^* = G(V \cap T_1^*)$ and $G_2^* = G(V \cap T_2^*)$ such that

$$S(G_i^*) \leq c_k \mathrm{Vol}(T_i^*) \qquad (i = 1, 2)\,.$$

Setting $G^* = G_1^* \cup G_2^*$ yields a connected graph on the set of vertices V and

$$S(G^*) = S(G_1^*) + S(G_2^*) \leq c_k \mathrm{Vol}(T_1^*) + c_k \mathrm{Vol}(T_2^*) = c_k \mathrm{Vol}(T)\,. \quad \square$$

Remarks.

1. We may choose for G a Hamiltonian circuit as well, but the proof of this fact is more difficult.
2. One can prove that $c_2 = 4$ is the smallest suitable constant and that it is good for the Hamiltonian circuit problem also. It is an interesting problem to find the exact constants for $k > 2$.

Problem G.42. *Let us draw a circular disc of radius r around every integer point in the plane different from the origin. Let E_r be the union of these discs, and denote by d_r the length of the longest segment starting from the origin and not intersecting E_r. Show that*

$$\lim_{r \to 0} (d_r - \frac{1}{r}) = 0.$$

Solution. Let the second endpoint (x, y) of one of the longest segments be on the circumference of the disc centered at (k, n). Reflecting the segment in the point $(k/2, n/2)$, we obtain that the segment connecting the points $(k - x, n - y)$ to (k, n) has a point in common only with the circle centered at (k, n). Since the points $(0, 0)$, $(k - x, n - y)$, (x, y), and (k, n) are vertices of a parallelogram, the two other sides of which have length r, this parallelogram contains no lattice points in its interior. The segment connecting the origin to (k, n) has a point in common only with the disc about its endpoint. The length of this segment differs from d_r by at most r.

Let D_r be the length of the longest segment connecting the origin to a lattice point and having a point in common only with the disc about its endpoint. As we observed, $| D_r - d_r | <$ r. According to this, it suffices to show that

$$\lim_{r \to 0+} (D_r - \frac{1}{r}) = 0\,.$$

For this, we shall apply the well-known theorem of lattice geometry saying that a triangle that has lattice points for its vertices but contains no further lattice points in its interior and on the boundary (except for the vertices) has area 1/2.

Let $V = (n, k)$ be the endpoint of a segment of length D_r, and let H be the domain containing those points of the plane whose orthogonal projection onto the straight line OV lies in the closed segment OV. Since translates of H with integer multiples of the vector (n, k) cover the whole plane, the distance of any lattice point $P \in H$ from the straight line OV is at least r if P is different from O and V. Let Q be a lattice point in H different from O and V such that the distance of Q from the straight line OV is minimal. Then there are no further lattice points on the segments OQ and OV, nor in the interior of triangle OVQ, because these points are closer to OV than Q. Furthermore, there are no lattice points in the interior of the segment OQ since every disc of radius r centered at a lattice point different from O and V is disjoint from the segment OV. Therefore, the area of triangle OVQ is equal to 1/2. Consequently, the distance of Q from the straight line OV, that is, the altitude of triangle OVQ, is $1/\mid OV\mid$. We find that the distance of Q from OV is at least r if and only if $\mid OV\mid\leq 1/r$.

This yields another characterization of D_r: D_r is the length of the longest segment connecting O to a lattice point without going through further lattice points (that is, the coordinates of the endpoint are relative primes) and having length at most $1/r$.

By this, $D_r \leq 1/r$, of course.

It is well known that for any $\varepsilon > 0$, if n is large enough, then there is a prime number between n and $n + \varepsilon n$.

Let $0 < \varepsilon < \sqrt{2} - 1$, and choose a prime p satisfying

$$\frac{1}{(1+\varepsilon)r} < p < \frac{1}{r} - 1\,.$$

Such a prime exists by the theorem mentioned above, provided that r is small enough.

Let q be the largest nonnegative integer such that

$$p^2 + q^2 \leq \frac{1}{r^2}\,.$$

Then $q < p$ since if we had $q \geq p$, then by

$$\frac{1}{\sqrt{2}r} < \frac{1}{(1+\varepsilon)r} < p, \quad p^2 + q^2 \geq 2p^2 > \frac{1}{r^2}$$

would hold. On the other hand, $q \geq 1$, as for $r < 1$, $p^2 + 1^2 < (1/r - 1)^2 + 1 < 1/r^2$. Thus, $1 \leq q < p$, and since p is a prime, p and q are relative primes. This implies that the segment connecting the origin to (p, q) is suitable. Therefore

$$D_r^2 \geq p^2 + q^2 > p^2 + \left(\sqrt{\frac{1}{r^2} - p^2} - 1\right)^2 = \frac{1}{r^2} - 2\sqrt{\frac{1}{r^2} - p^2} + 1$$

$$\geq \frac{1}{r^2} - \frac{2}{r}\sqrt{1 - \frac{1}{(1+\varepsilon)^2}} + 1 > \left(\frac{1}{r} - \sqrt{1 - \frac{1}{(1+\varepsilon)^2}}\right)^2,$$

that is,

$$D_r > \frac{1}{r} - \sqrt{1 - \frac{1}{(1+\varepsilon)^2}}\,.$$

We conclude that for any $0 < \varepsilon < \sqrt{2} - 1$, and for any sufficiently small r, we have

$$\frac{1}{r} - \sqrt{1 - \frac{1}{(1+\varepsilon)^2}} < D_r < \frac{1}{r}\,.$$

Since

$$\sqrt{1 - \frac{1}{(1+\varepsilon)^2}} \to 0 \quad \text{as} \quad \varepsilon \to 0\,,$$

the proposition follows immediately. □

Problem G.43. *We say that the point (a_1, a_2, a_3) is above (below) the point (b_1, b_2, b_3) if $a_1 = b_1$, $a_2 = b_2$ and $a_3 > b_3$ ($a_3 < b_3$). Let $e_1, e_2, \ldots, e_{2k}$ ($k \geq 2$) be pairwise skew lines not parallel with the z-axis, and assume that among their orthogonal projections to the (x, y)-plane no two are parallel and no three are concurrent. Is it possible that going along any of the lines the points that are below or above a point of some other line e_i alternately follow one another?*

Solution. Assume that it is possible. From this, we derive a contradiction.

Let $P_{i,j}$ be the point of e_i, which is above or below the line e_j ($1 \leq i,\ j \leq 2k,\ i \neq j$). In the first case, we call $P_{i,j}$ an upper point, and in the latter case we call it a lower point. In the text below, by a point of e_i we mean one of the points $P_{i,j}$. Those two points of e_i that have all points of e_i on one side will be called the "ends" of e_i. Two points of a line will be called neighbors if there is no point of the line between them.

Lemma. Every straight line contains at least two points that are above or below an end of another line.

Proof. Obviously, it is enough to consider the straight line e_1. We may also assume that e_1 is horizontal (that is, parallel with the (x, y)-plane), since this can always be managed applying a transformation of the form $(x, y, z) \mapsto (x, y, z + \alpha y + \beta z)$, which leaves the location of upper and lower points unchanged. Let S be a plane perpendicular to e_1, and denote by $f_2, f_3, \ldots, f_{2k}$ the orthogonal projections of the lines $e_2, e_3, \ldots, e_{2k}$ onto S (see Figure G.30). Let us introduce a coordinate system on S having horizontal and vertical axes and take from among $f_2, f_3, \ldots, f_{2k}$ the ones, say f_2 and f_3, whose slope is minimal and maximal, respectively. We claim that if e_2 is below (above) e_1, then each point of it lies on the negative (positive) side of e_1 with respect to the positive direction of the horizontal coordinate axis. Assume that this is not so and, for example, e_2 is below e_1 and has a point on the positive side of e_1 (the other cases can be treated analogously). Let $P_{i_0,2}$ be the nearest to $P_{2,1}$ among these points. This

is an upper point since it is a neighbor of a lower point. The point P_{2,i_0} below it lies on the straight line e_{i_0}. It is easy to see that one can find a sequence of indices $i_1, \dots, i_n$ such that
1. $i_n = 1$;
2. the points $P_{i_j,i_{j-1}}$ and $P_{i_j,i_{j+1}}$ are neighbors on e_{i_j} $(j = 1, 2, \dots, n-1)$;
3. $P_{i_j,i_{j+1}}$ is in the negative direction from $P_{i_j,i_{j-1}}$ $(j = 1, 2, \dots, n-1)$.

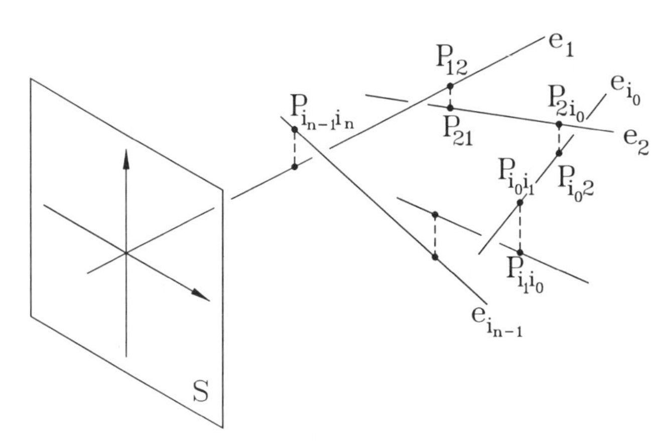

Figure G.30.

It is also easy to see, however, that in this case, the slope of one of the straight lines $f_{i_1}, \dots, f_{i_{n-1}}$ is less than the slope of f_2, and this is a contradiction.

This way, we proved that one of the ends of e_2 and e_3 is above or below e_1. Therefore, the lemma holds.

We remark that there cannot be more than two ends above or below a straight line, because it would mean that the $2k$ straight lines would have more than $4k$ ends. Therefore, there are exactly two points on every straight line that are above or below an end of another line, and these latter lines are the ones giving the minimal and maximal slopes after the above transformation and projection.

Let the ends of e_1 be above or below e_2 and e_3, respectively. Since the straight line contains $2k-1$, that is, an odd number of points, either both e_2 and e_3 are above e_1 or both of them are below e_1. We may assume that they are above e_1. We also suppose that e_2 and e_3 are horizontal since this can be reached by applying a suitable transformation of the form $(x, y, z) \mapsto (x, y, z + \alpha x + \beta y)$. Let S be a plane perpendicular to e_2, and introduce a coordinate system on S as shown in Figure G.31.

By the previous remark, among the orthogonal projections of the straight lines $e_j (j \neq 2)$ onto S, the projection of e_1 has maximal slope, e_1 has maximal slope, since e_1 goes below e_2 and has a point in the positive direction from $P_{1,2}$, and since the projection of e_3 is horizontal, that of e_1 has positive slope. This implies that $P_{1,3}$ is higher than $P_{1,2}$ (its z-coordinate is

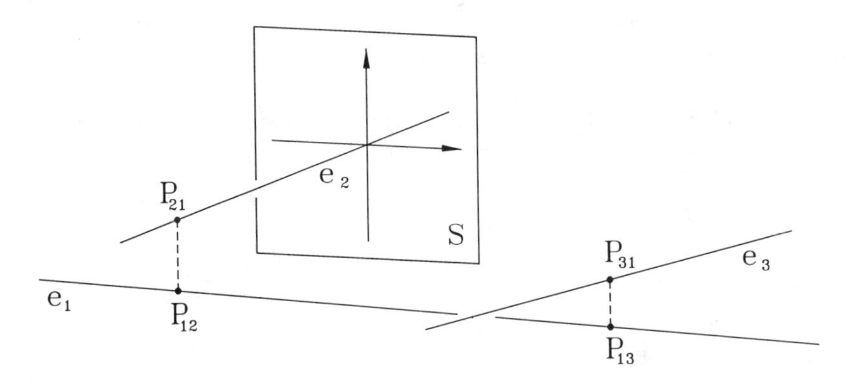

Figure G.31.

greater). Repeating the same argument but changing the role of e_2 and e_3, we obtain that $P_{1,2}$ is higher than $P_{1,3}$. However, these two implications contradict one another.

We conclude that the indirect assumption is false: upper and lower points cannot follow one another alternating on each straight line. □

Problem G.44. *Suppose that HTM is a direct complement to a vertical bundle VTM over the total space of the tangent bundle TM of the manifold M. Let v and h denote the projections corresponding to the decomposition $TTM = VTM \oplus HTM$. Construct a bundle involution $P : TTM \to TTM$ such that $P \circ h = v \circ P$ and prove that, for any pseudo-Riemannian metric given on the bundle VTM, there exists a unique metric connection ∇ such that*

$$\nabla_X PY - \nabla_Y PX = P \circ h[X, Y]$$

if X and Y are sections of the bundle HTM, and

$$\nabla_X Y - \nabla_Y X = [X, Y]$$

if X and Y are sections of the bundle VTM.

Solution. As usual, call the sections of the bundles $VTM \to TM$ and $HTM \to TM$ vertical and horizontal vector fields on TM, and denote their modules by $\mathfrak{X}_V TM$ and $\mathfrak{X}_H TM$, respectively. If $\mathfrak{X}(TM)$ denotes the module of vector fields in $TTM \to TM$, then

$$\mathfrak{X}(TM) = \mathfrak{X}_V TM \oplus \mathfrak{X}_H TM,$$

and, for simplicity, the projections corresponding to this decomposition are also denoted by v and h.

Let $(U,(u^1,\ldots,u^n))$ be a chart for M, and consider the induced chart

$$(\pi^{-1}(U),(x^1,\ldots,x^n;y^1,\ldots,y^n))$$

for TM, where $\pi : TM \to M$ is the projection of the tangent bundle, and

$$x^i = u^i \circ \pi, \quad y^i(v) = v(u^i) \quad (v \in TM,\ 1 \leq i \leq n).$$

When this chart is fixed, there exist unique smooth functions

$$N_i^k : \pi^{-1}(U) \to R \quad (1 \leq i,k \leq n)$$

such that the vector fields

$$\frac{\delta}{\delta x^i} = \frac{\partial}{\partial x^i} - N_i^k \frac{\partial}{\partial y^k} \quad (1 \leq i \leq n)$$

form a local basis in $\mathfrak{X}_H TM$. Then

$$\left(\frac{\partial}{\partial y^i}, \frac{\delta}{\delta x^i}\right) \quad (1 \leq i \leq n)$$

is a local basis in $\mathfrak{X}(TM)$, and on the coordinate patch $\pi^{-1}(U)$ every vector field $X \in \mathfrak{X}(TM)$ can be written in the form

$$X = X^i \frac{\delta}{\delta x^i} + Y^i \frac{\partial}{\partial y^i},$$

where the first term is horizontal and the second is vertical.

Define the map P by the formula

$$X^i \frac{\delta}{\delta x^i} + Y^i \frac{\partial}{\partial y^i} \mapsto Y^i \frac{\delta}{\delta x^i} + X^i \frac{\partial}{\partial y^i}.$$

An easy calculation using the above formulas shows that this map is independent of the choice of the chart. From the formula, it is immediate that $P^2 = 1$ and $P \circ h = v \circ P$. It is also clear that P is a bundle involution.

Since $\mathfrak{X}(TM) = \mathfrak{X}_V TM \oplus \mathfrak{X}_H TM$, and a connection $\nabla : \mathfrak{X}(TM) \times \mathfrak{X}_V TM \to \mathfrak{X}_V TM$ is linear over the function ring in its first variable, it suffices to define ∇ on the summands $\mathfrak{X}_V TM \times \mathfrak{X}_V TM$ and $\mathfrak{X}_H TM \times \mathfrak{X}_V TM$.

Case 1. Consider the map $F : (\mathfrak{X}_V TM)^3 \to \mathfrak{X}_V TM$ defined by

$$\begin{aligned} F(X,Y,Z) = {} & X\langle Y,Z\rangle + Y\langle Z,X\rangle - Z\langle X,Y\rangle \\ & - \langle X,[Y,Z]\rangle + \langle Y,[Z,X]\rangle + \langle Z,[X,Y]\rangle, \end{aligned}$$

where $\langle\,.\,,\,.\,\rangle$ denotes the pseudo-Riemannian metric given on the vertical bundle. For fixed X and Y, the map $Z \mapsto F(X,Y,Z)$ is linear over the function ring, and, since the form $\langle .\,,.\rangle$ is nondegenerate, there exists a unique $\nabla_X Y \in \mathfrak{X}_V TM$ such that

$$\langle 2\nabla_X Y, Z\rangle = F(X, Y, Z)$$

for all $Z \in \mathfrak{X}_V TM$. An argument analogous to the proof of the theorem on Levi–Civita connections shows that the map $(X, Y) \mapsto \nabla_X Y$ satisfies all requirements.

Case 2. The restriction of the connection ∇ to $\mathfrak{X}_H TM \times \mathfrak{X}_V TM$ is defined by the requirement that

$$\begin{aligned}\langle 2\nabla_U Y, Z\rangle = {} & U\langle Y, Z\rangle + PY\langle Z, PU\rangle - PZ\langle PU, Y\rangle \\ & - \langle PU, P\circ h[PY, PZ]\rangle + \langle Y, P\circ h[PZ, U]\rangle \\ & + \langle Z, P\circ h[U, PY]\rangle\end{aligned}$$

hold for all $U \in \mathfrak{X}_H TM$, $Y, Z \in \mathfrak{X}_V TM$. A simple but lengthy calculation shows that this restriction of the connection also has all required properties. In order to prove its uniqueness, suppose that the map

$$\nabla' : \mathfrak{X}_H TM \times \mathfrak{X}_V TM \to \mathfrak{X}_V TM$$

also fulfills the requirements of the problem. Then, for all $U \in \mathfrak{X}_H TM$, $Y, Z \in \mathfrak{X}_V TM$, we have

$$\begin{aligned}U\langle Y, Z\rangle + PY\langle Z, PU\rangle - PZ\langle PU, Y\rangle \overset{(1)}{=} {} & \langle \nabla'_U Y, Z\rangle + \langle Y, \nabla'_U Z\rangle \\ & + \langle \nabla'_{PY} Z, PU\rangle + \langle Z, \nabla'_{PY} PU\rangle \\ & - \langle \nabla'_{PZ} PU, Y\rangle - \langle PU, \nabla'_{PZ} Y\rangle \\ = {} & 2\langle \nabla'_U Y, Z\rangle + \langle \nabla'_{PY} PU - \nabla'_U Y, Z\rangle \\ & + \langle \nabla'_U Z - \nabla'_{PZ} PU, Y\rangle + \langle \nabla'_{PY} Z - \nabla'_{PZ} Y, PU\rangle \\ \overset{(2)}{=} {} & 2\langle \nabla'_U Y, Z\rangle - \langle P\circ h[U, PY], Z\rangle \\ & - \langle P\circ h[PZ, U], Y\rangle + \langle P\circ h[PY, PZ], PU\rangle,\end{aligned}$$

where at (1) compatibility of ∇ with the metric is used, and at (2) the assumption $\nabla'_X PY - \nabla'_Y PX = P\circ h[X, Y]$ is used. Therefore, it follows that

$$\langle \nabla'_U Y, Z\rangle = \langle \nabla_U Y, Z\rangle,$$

that is,

$$\langle \nabla'_U Y - \nabla_U Y, Z\rangle = 0 \quad \text{for all} \quad Z \in \mathfrak{X}_V TM.$$

Since the form $\langle .\,,.\rangle$ is nondegenerate, this implies that $\nabla' = \nabla$, which completes the proof. $\square$

Problem G.45. *Let A be a finite set of points in the Euclidean space of dimension $d \geq 2$. For $j = 1, 2, \ldots, d$, let B_j denote the orthogonal projection of A onto the $(d-1)$-dimensional subspace given by the equation $x_j = 0$. Prove that*

$$\prod_{j=1}^{d} |B_j| \geq |A|^{d-1}.$$

Solution. We use induction on the cardinality of the set A. If $|A| = 0$ or 1, then the statement is obvious.

If $|A| > 1$, then the projection of A onto a suitable coordinate axis contains at least two distinct points. Without loss of generality, we can assume that this coordinate is the dth. This means that a suitable hyperplane with equation $x_d = \alpha$ divides A into two disjoint nonempty subsets: $A = X \cup Y$. Let C_j and D_j denote the projections of X and Y, respectively, onto the hyperplane $x_j = 0$; then $|C_j| + |D_j| = |B_j|$ for each $j \in \{1, \ldots, d-1\}$, and $|C_d| \leq |B_d|$ and $|D_d| \leq |B_d|$.

By the induction hypothesis, we have

$$|X|^{d-1} \leq |B_d| \cdot \prod_{j=1}^{d-1} |C_j| \qquad \text{and}$$

$$|Y|^{d-1} \leq |B_d| \cdot \prod_{j=1}^{d-1} |D_j|.$$

Using the inequality

$$(a_1 \cdot \dots \cdot a_k)^{\frac{1}{k}} + (b_1 \cdot \dots \cdot b_k)^{\frac{1}{k}} \leq \left(\prod_{i=1}^{k} (a_i + b_i) \right)^{1/k},$$

which is an equivalent form of the usual inequality between arithmetic and geometric means, we have

$$\begin{aligned}
\prod_{j=1}^{d} |B_j|^{1/(d-1)} &= |B_d|^{1/(d-1)} \cdot \prod_{j=1}^{d-1} (|C_j| + |D_j|)^{1/(d-1)} \\
&\geq |B_d|^{1/(d-1)} \cdot \left(\prod_{j=1}^{d-1} |C_j|^{1/(d-1)} + \prod_{j=1}^{d-1} |D_j|^{\frac{1}{d-1}} \right) \\
&\geq |X| + |Y| = |A|,
\end{aligned}$$

that is,

$$\prod_{j=1}^{d} |B_j| \geq |A|^{d-1},$$

which is what we wanted to prove. $\square$

Problem G.46. *Let $A_1^{(0)}, \dots, A_n^{(0)}$ be a sequence of $n \geq 3$ points in the Euclidean plane $\mathbb{R}^2$. Define the sequence $A_1^{(i)}, \dots, A_n^{(i)}$ $(i = 1, 2, \dots)$ by induction as follows: let $A_j^{(i)}$ be the midpoint of the segment $A_j^{(i-1)} A_{j+1}^{(i-1)}$, where $A_{n+1}^{(i-1)} = A_1^{(i-1)}$. Show that, with the exception of a set of zero Lebesgue measure, for every initial sequence $(A_1^{(0)}, \dots, A_n^{(0)}) \in (\mathbb{R}^2)^n$, there exists a natural number N such that the points $A_1^{(N)}, \dots, A_n^{(N)}$ are consecutive vertices of a convex n-gon.*

Solution. If $\mathbb{R}^2$ is naturally identified with the complex plane $\mathbb{C}$, then our sequences can be considered as vectors in the space $\mathbb{C}^n$, and the induction step corresponds to the linear transformation

$$L(z_1, \dots, z_n) = \left(\frac{z_1 + z_2}{2}, \dots, \frac{z_n + z_1}{2} \right).$$

In $\mathbb{C}^n$ there exists a basis consisting of eigenvectors of L: we have

$$L(e_j) = \lambda_j e_j \qquad (j = 1, 2, \dots, n)$$

for the vectors $e_j = (1, \varepsilon^j, \varepsilon^{2j}, \dots, \varepsilon^{(n-1)j})$ and scalars $\lambda_j = (1 + \varepsilon_j)/2$ $(j = 1, \dots, n)$, where $\varepsilon = e^{2\pi i/n}$ is a primitive nth root of unity.

All vectors v in $\mathbb{C}^n$ can be written in the form $v = \sum_{j=1}^n a_j e_j$. Since $e_n = (1, 1, \dots, 1)$, adding the vector $a_n e_n$ means a translation of our system of points in the original plane $\mathbb{R}^2$ by a fixed vector. Therefore, it suffices to consider the problem restricted to the subspace generated by the vectors $e_1, \dots, e_{n-1}$. Suppose that

$$\left(A_1^{(0)}, \dots, A_n^{(0)} \right) = v = \sum_{j=1}^{n-1} a_j e_j.$$

Applying the induction step N times, we obtain

$$\left(A_1^{(N)}, \dots, A_n^{(N)} \right) = L^N(v) = \sum_{j=1}^{n-1} a_j \lambda_j^N e_j.$$

An elementary calculation shows that $|\lambda_j| < |\lambda_1|$ for $1 < j < n-1$, and $|\lambda_{n-1}| = |\lambda_1|$. So, for $1 < j < n-1$, we have $|\lambda_j|^N / |\lambda_1|^N \to 0$ when $N \to 0$. Therefore, it suffices to show that for almost all v the vector $a_1 e_1 + a_{n-1} e_{n-1}$ corresponds to successive vertices of a convex n-gon. This is indeed the case if $|a_1| \neq |a_{n-1}|$, because e_1 is a vector corresponding to successive vertices of a regular n-gon, and the vector $a_1 e_1 + a_{n-1} e_{n-1}$ corresponds to the sequence of points obtained from successive vertices of this regular n-gon by applying the linear map $z \mapsto a_1 z + a_{n-1} \bar{z}$ in $\mathbb{R}^2$. This transformation is nondegenerate, since $a_1 z + a_{n-1} \bar{z} = 0$ can only hold when $z = 0$ or $|a_1| = |a_{n-1}|$, and thus it takes convex n-gons into convex n-gons.

To summarize the property required in the problem is true for all vectors $\sum_{j=1}^n a_j e_j$ of $\mathbb{C}^n$ except when $|a_1| = |a_{n-1}|$, and this exceptional set does indeed have zero Lebesgue measure. $\square$

Problem G.47. *Suppose that n points are given on the unit circle so that the product of the distances of any point of the circle from these points is not greater than two. Prove that the points are the vertices of a regular n-gon.*

Solution. Let the circle be the unit circle around the origin in the complex plane. Let $z_1, z_2, \ldots, z_n$ be the complex numbers that represent the points. By a suitable rotation, we may assume that $z_1 \cdot z_2 \cdot \cdots \cdot z_n = 1$.

Consider the following polynomial:

$$\begin{aligned} P(w) &= (w - z_1)(w - z_2) \cdots (w - z_n) \\ &= w^n + a_1 w^{n-1} + \cdots + a_{n-1} w + 1 = w^n + Q(w) + 1\,. \end{aligned}$$

Then $|P(z)|$ is the product of the distances of the point represented by the complex number z from the given points. So, if z is a complex number of absolute value 1, then $|P(z)| \le 2$.

Let $w_1, w_2, \ldots, w_n$ denote the nth roots of unity. It is well known that $w_1^k + w_2^k + \cdots + w_n^k = 0$ for all $k = 1, 2, \ldots, n-1$. This implies that $Q(w_1) + \cdots + Q(w_n) = 0$. If $Q(w)$ is not identically zero, then, for some j, the complex number $Q(w_j)$ is different from zero and has nonnegative real part. Then $|P(w_j)| = |2 + Q(w_j)| > 2$, which contradicts the assumption. Therefore, the polynomial Q is identically zero, that is, $P(z) = z^n + 1$. Then the roots $z_1, z_2, \ldots, z_n$ of the polynomial $P(z)$ form a regular n-gon. □

3.5 MEASURE THEORY

Problem M.1. *Let f be a finite real function of one variable. Let $\overline{D}f$ and $\underline{D}f$ be its upper and lower derivatives, respectively, that is,*

$$\overline{D}f(x)=\limsup_{\substack{h,k\to 0\\ h,k\geq 0\\ h+k>0}}\frac{f(x+h)-f(x-k)}{h+k},\quad \underline{D}f(x)=\liminf_{\substack{h,k\to 0\\ h,k\geq 0\\ h+k>0}}\frac{f(x+h)-f(x-k)}{h+k}.$$

Show that $\overline{D}f$ and $\underline{D}f$ are Borel-measurable functions.

Solution. It is obviously sufficient to prove the assertion for $\overline{D}f$. It is also clear that

$$\{x : \overline{D}f(x) > c\} = \bigcup_{i=1}^{\infty}\bigcap_{k=1}^{\infty} A_{ik}$$

for all real values c, where A_{ik} denotes the union of intervals $[a, b]$ shorter than $1/k$ satisfying $(f(b) - f(a))/(b - a) > c + 1/i$. We show that A_{ik} is a set of type F_σ which implies the assertion.

More generally let $\{I_\gamma\}$ $(\gamma \in \Gamma)$ be a system of open intervals, and let $A = \cup_{\gamma\in\Gamma}\overline{I}_\gamma$ (where $\overline{I}_\gamma$ stands for the closure of I_γ). Obviously,

$$A = B \cup C \cup D,$$

where $B = \cup_{\gamma\in\Gamma}I_\gamma$, C is the set of all points that are left-hand endpoints of at least one I_γ but are not interior points of any of them, and D is the set of those points that are right-hand endpoints of at least one I_γ but do not lie in the interior of any I_γ. Clearly, B is an open set. If $x \in C$, let r_x be a rational point in the interior of the interval I_γ that has the left-hand endpoint x. Points r_x corresponding to distinct points x are distinct since in the case $x < y$, $r_x = r_y$, the point y would lie in the interior of the interval I_γ starting at x. Consequently, C, and similarly D, is countable. So each of the sets B, C, D, and therefore A as well is of type F_σ. □

Remark. It is interesting that for the left-hand and right-hand derivatives a similar theorem does not hold. For instance, if E is a nonmeasurable set that along with its complementary set is dense, and $f(x)$ is the characteristic function of E, then $\overline{D}^+ f(x) = 0$ or $+\infty$ if $x \in E$ or $x \notin E$, respectively, and so $\overline{D}^+ f(x)$ is nonmeasurable in this case.

It should be noted that in the book by *S. Saks (Theory of the Integral, Hafner, New York, 1937*, Vitali's covering theorem, *pp. 112–113)*, the author proves only Lebesgue measurability of $\overline{D}f$ and $\underline{D}f$ with much more powerful tools, although generalized to arbitrary dimension.

Problem M.2. *Let E be a bounded subset of the real line, and let Ω be a system of (nondegenerate) closed intervals such that for each $x \in E$ there exists an $I \in \Omega$ with left endpoint x. Show that for every $\varepsilon > 0$ there exist a finite number of pairwise nonoverlapping intervals belonging to Ω that cover E with the exception of a subset of outer measure less than ε.*

Solution. Let Δ be a bounded open interval containing E. Denote by S_n the set of all points of Δ having neighborhoods not containing points of E that are starting points of intervals belonging to Ω and longer than $1/n$. Obviously S_n is an open set, $S_{n+1} \subset S_n$ and, furthermore, $(\cap_{n=1}^{\infty} S_n) \cap E = 0$ since at each point of E there begins at least one nondegenerate interval belonging to Ω and if $\frac{1}{n}$ is already less than the length of the latter, then the point cannot belong to S_n. It is a well-known property of the outer measure that for any set A the outer measure $\lambda^*(X \cap A)$ as a function of X is a measure on the Lebesgue-measurable sets. Thus, taking $A = E$ and using the relation $S_{n+1} \subset S_n$ as well as the finiteness of the outer measure of E, we obtain

$$\lim_{n \to +\infty} \lambda^*(S_n \cap E) = \lambda^*((\bigcap_{n=1}^{\infty} S_n) \cap E) = \lambda^*(0) = 0.$$

Fix n_0 so that $\lambda^*(S_{n_0} \cap E) < \varepsilon/2$. If the set $E_1 = E - (S_{n_0} \cap E)$ is nonempty, which we may assume, then in each neighborhoood of any of its points there is a point of E that is also a point of E_1 and at which an interval longer than $1/n_0$ begins. There is a point $a_1 \in E_1$ such that

$$\lambda^*((-\infty, a_1) \cap E_1) < \frac{\varepsilon}{2(n_0\lambda(\Delta) + 1)},$$

and by the former arguments we may assume at once that some interval $[a_1, b_1]$ starting from a_1 and belonging to Ω is longer than $1/n_0$. Then we choose (if still possible) a point $a_2 \in E_1$ such that $a_2 > b_1$,

$$\lambda^*((b_1, a_2) \cap E_1) < \frac{\varepsilon}{2(n_0\lambda(\Delta) + 1)},$$

and

$$[a_2, b_2] \in \Omega, \qquad b_2 - a_2 > \frac{1}{n_0},$$

and so on. The procedure ends after a finite number of steps, namely we can construct at most $n_0\lambda(\Delta) + 1$ intervals $[a_k, b_k]$ in this way (since $b_k - a_k > 1/n_0$). What is left out from E by these intervals is covered by $S_{n_0} \cap E$ and the not more than $n_0\lambda(\Delta) + 1$ sets $(b_{k-1}, a_k) \cap E_1$ $(b_0 = -\infty)$ of outer measure not exceeding $\varepsilon/(2(n_0\lambda(\Delta) + 1))$. Their total outer measure is less than ε, which completes the proof. $\square$

Remarks.

1. In addition to being nonoverlapping, the intervals we have chosen are disjoint.

2. The assertion cannot be replaced by the stronger statement that it is possible to choose an at most countable number of nonoverlapping intervals such that the part not covered by them has measure 0. This is shown by the following simple example: $E = (0,1)$, $\Omega = \{[x,1] : x \in (0,1)\}$.

Problem M.3. *Let $f(t)$ be a continuous function on the interval $0 \leq t \leq 1$, and define the two sets of points*

$$A_t = \{(t,0)\colon t \in [0,1]\}, \quad B_t = \{(f(t),1)\colon t \in [0,1]\}.$$

Show that the union of all segments $\overline{A_tB_t}$ is Lebesgue-measurable, and find the minimum of its measure with respect to all functions f.

Solution. We first show that the set $A = \cup_{0 \leq t \leq 1} \overline{A_tB_t}$ is closed, hence Lebesgue-measurable. Let P_n be a convergent sequence of points in A, $P_n \to P_0$. By the definition of A, to every n there is a t_n such that $P_n \in \overline{A_{t_n}B_{t_n}}$. According to the Bolzano–Weierstrass theorem, the sequence t_n contains a convergent subsequence: $t_{n_k} \to t_0$. The distance from the point P_{n_k} to the segment $\overline{A_{t_0}B_{t_0}}$ is not greater than $\max\{|f(t_{n_k}) - f(t_0)|, |t_{n_k} - t_0|\}$, which tends to 0 by the continuity of $f(t)$. It follows that $P_0 \in \overline{A_{t_0}B_{t_0}} \subseteq A$. Since A contains the limit point of any convergent sequence of its points, A is closed.

A simple calculation shows that the point of the segment $\overline{A_tB_t}$ that lies on the straight line $y = c$ has abscissa $(1-c)t + cf(t)$, which, by the continuity of $f(t)$, is a continuous function of t. Consequently, if two points of the set A lie on the line $y = c$, then A contains the segment that joins these points.

Now we can determine the minimum of the measure of A. If $f(t)$ is constant, then A is a triangle of unit base and unit altitude, so it has measure 1/2. If the segments $\overline{A_0B_0}$ and $\overline{A_1B_1}$ do not intersect, then the trapezoid with vertices A_0, B_0, A_1, and B_1 is a subset of A, so the measure of A is not less than 1/2. If the segments $\overline{A_0B_0}$ and $\overline{A_1B_1}$ intersect at some point C, then the triangles A_0CA_1 and B_0CB_1 are subsets of A, so the measure of A is not smaller than

$$t(d) = \frac{1}{2}\left(\frac{1}{1+d} + \frac{d}{1+d}d\right) = \frac{1}{2}\frac{1+d^2}{1+d},$$

where d denotes the distance from B_0 to B_1. By a simple calculation, we obtain that the minimum of $t(d)$ on the positive half-axis is $\sqrt{2}-1$. Thus, the measure of A cannot be less than $\sqrt{2}-1$. On the other hand, for $f(t) = (\sqrt{2}-1)(1-t)$, the measure of A is exactly $\sqrt{2}-1$. □

Remark. I. N. Berstein (Doklady Acad. Nauk. SSSR 146 (1962), 11-13) refers to the result of the problem but he states erroneously that the minimum is $\frac{1}{2}$.

Problem M.4. *A "letter T" erected at point A of the x-axis in the xy-plane is the union of a segment AB in the upper half-plane perpendicular to the x-axis and a segment CD containing B in its interior and parallel to the x-axis. Show that it is impossible to erect a letter T at every point of the x-axis so that the union of those erected at rational points is disjoint from the union of those erected at irrational points.*

Solution 1. We call width of the letter T erected at A the minimum of the lengths CB and BD. We devide the irrational points into countably many classes $H_1, \ldots, H_k, \ldots$. The class H_k consists of all irrational points for which the widths of the letters T erected at them are greater than $1/k$. Baire's theorem implies that the set of all irrational points cannot be represented as the countable union of nowhere-dense sets, so some H_k is dense in a suitable interval I. Selecting an arbitrary rational point A of the interval I, denote by δ the width of the letter T erected at it. The class H_k contains an element A_1 whose distance from A is less than $\min\{\delta, 1/k\}$. The letters T erected at A and A_1 obviously intersect each other since they both have widths greater than the distance from A to A_1. The proof is complete. □

Solution 2. Project on the x-axis the segment CD parallel to the x-axis of each letter T, and denote the projection by I_x. The projection I_x is an interval containing x in its interior. We prove the existence of an irrational α and a rational r such that I_α and I_r contain r and α, respectively, in their interior. This already implies that the letters T erected at α and r intersect.

For each rational number $r = p/q$ (we shall write all rational numbers in irreducible form), we choose a closed subinterval J_r of I_r containing r in its interior and having length not greater than $1/q^2$. Consider a sequence of rational numbers where the denominators are strictly monotone increasing and each number lies in the intervals J_r corresponding to the previous terms:

$$r_1 = \frac{p_1}{q_1}, r_2 = \frac{p_2}{q_2}, \ldots, r_n = \frac{p_n}{q_n}, \ldots;$$
$$q_1 < q_2 < \cdots < q_n < \ldots;$$
$$r_n \in J_{r_1} \cap J_{r_2} \cap \cdots \cap J_{r_{n-1}}.$$

Such a sequence obviously exists and, moreover, it is a Cauchy sequence, so it has a limit $\lim_{n\to\infty} r_n = \alpha$. Since J_{r_n} contains the points $r_{n+1}, r_{n+2}, \ldots$ and is closed, it follows that $\alpha \in J_{r_n}$ and, in particular,

$$|\alpha - r_n| \leq \frac{1}{q_n^2}.$$

From the theory of numbers it is well known that in this case α can only be irrational. I_α contains α in its interior, so for sufficiently large n we have

$r_n \in I_\alpha$. On the other hand, $\alpha \in J_{r_n} \subset I_{r_n}$. Thus, for sufficiently large n, the letters T erected at the points α and r_n intersect each other. □

Solution 3. Assume that the assertion is false, that is, we have succeeded in erecting a letter T at each point of the real line so that none of the letters T erected at rational points intersects a T erected at an irrational point. We may assume that the letters T are symmetric, that is, B is the midpoint of the segment CD (let C always have the smaller abscissa among C and D).

We define a monotonically increasing sequence of numbers $\alpha_1, \alpha_2, \dots, \alpha_n, \dots$ by recursion. Let α_1 be any real number. If α_n is already defined as a rational number, then let α_{n+1} be irrational, while if α_n is irrational, then let α_{n+1} be rational, and choose $\alpha_{n+1} (> \alpha_n)$ so that $\alpha_{n+1} - \alpha_n$ is smaller than one-quarter of the length of B_nD_n. Denote the abscissa of D_n by δ_n. Making use of the fact that the letters T erected at α_n and α_{n+1} do not intersect, we obtain the following sequence of inequalities:

$$\alpha_{n+1} < \delta_{n+1} < \alpha_n + 2(\alpha_{n+1} - \alpha_n) < \alpha_n + 2\frac{\delta_n - \alpha_n}{4} = \frac{\alpha_n + \delta_n}{2} < \delta_n.$$

It follows that $\{\alpha_n\}$ is bounded, and so $\alpha = \lim \alpha_n$ exists, and also that $\alpha < \delta_n$ for every n. Then, however, for sufficiently large n, the letter T erected at α_n intersects the T erected at α, which contradicts the initial assumption. □

Remarks.

1. In addition to the assertion of the problem, a contestant has proved that in any interval there is a continuum of irrational numbers α such that the letter T erected at α intersects a T erected at a rational point. Really, from the reasoning of Solution 2 it turns out that if $r_1, r_2, \dots$ is a sequence of rational numbers such that $J_{r_1} \supset J_{r_2} \supset \dots$ and $q_1 < q_2 < \dots$, then $\{r_n\}$ is obviously convergent and its limit is a "suitable" irrational number. Let $r = p/q$ be a rational number belonging to the interval (a, b) such that $J_r \subset (a, b)$. Let $r_0 = p_0/q_0$ and $r_1 = p_1/q_1$ be rational numbers satisfying $q_0 > q$, $q_1 > q$, $J_{r_0} \subset J_r$, $J_{r_1} \subset J_r$, and $J_{r_0} \cap J_{r_1} = \emptyset$. The existence of r_0 and r_1 of this kind is evident. Once we have defined the rational numbers $r_{i_1,\dots,i_k}$ $(i_1 = 0, 1; \dots; i_k = 0, 1)$, let $r_{i_1,\dots,i_k,0}$ and $r_{i_1,\dots,i_k,1}$ be two rational numbers such that

$$\begin{aligned} &q_{i_1,\dots,i_k,0} > q_{i_1,\dots,i_k}, \quad q_{i_1,\dots,i_k,1} > q_{i_1,\dots,i_k}, \\ &J_{r_{i_1,\dots,i_k,0}} \subset J_{r_{i_1,\dots,i_k}}, \quad J_{r_{i_1,\dots,i_k,1}} \subset J_{r_{i_1,\dots,i_k}}, \\ &J_{r_{i_1,\dots,i_k,0}} \cap J_{r_{i_1,\dots,i_k,1}} = \emptyset. \end{aligned}$$

With each number $0 < x < 1$, we associate an irrational number $\alpha \in (a, b)$ such that the letter T erected at α surely intersects a T erected at some rational point. If the infinite dyadic form of x is

$$x = 0.i_1 i_2 \dots,$$

we associate with x the limit of the sequence

$$r_{i_1}, r_{i_1,i_2}, \ldots, r_{i_1,i_2,\ldots,i_k}, \ldots .$$

This will be a "suitable" irrational number, and the construction guarantees that α_1 and α_2 associated with distinct numbers x_1 and x_2 are distinct.

2. The union of a segment AB in the upper half-plane that is erected perpendicularly at point A of the x-axis and a segment BC that is parallel to the x-axis will be called a $\lceil$ or a $\rceil$ depending on whether the abscissa of C is greater than or less than that of B. Two contestants proved that if we erect a $\lceil$ at each rational point and a $\rceil$ at each irrational point, then the union of the $\lceil$ intersects the union of the $\rceil$. This assertion is obviously sharper than that of the problem. Two other contestants showed by examples that it is possible to erect a $\lceil$ at each point so that the $\lceil$ are pairwise disjoint. Indeed, if the points B of the $\lceil$ lie on the graph of a strictly monotone decreasing positive function and the length of BC is arbitrary, then clearly any two $\lceil$ are disjoint.

3. It is easy to see that the next theorem generalizes the statement of the problem: Let X be a complete metric space of second category, and let $X = P \cup Q$, where P and Q are disjoint dense sets. Then it is impossible to define a real function f on X such that each point $p \in P$ has a neighborhood V_p for which $x \in V_p \cap Q$ implies $f(x) < f(p)$ and such that each point $q \in Q$ has a neighborhood V_q for which $x \in V_q \cap P$ implies $f(x) < f(q)$.
This was proved by some contestants for complete metric spaces and compact metric spaces, respectively. It was also shown that the hypothesis of completeness cannot be dropped.

Problem M.5. *Let f and g be continuous positive functions defined on the interval $[0, \infty)$, and let $E \subset [0, \infty)$ be a set of positive measure. Prove that the range of the function defined on $E \times E$ by the relation*

$$F(x, y) = \int_0^x f(t)dt + \int_0^y g(t)dt$$

has a nonvoid interior.

Solution. Put

$$\phi(x) = \int_0^x f(t)dt, \qquad \psi(x) = \int_0^x g(t)dt.$$

The function ϕ and its inverse ϕ^{-1} (which obviously exists) are absolutely continuous, so they map sets of measure 0 onto sets of measure 0. Therefore, the image $\phi(E) = A$ of the set E of positive measure is measurable and has positive measure. Similarly, the set $\psi(E) = B$ has positive measure. Since the range of the function $F(x, y)$ considered on $E \times E$ is exactly the set $A+B=\{a+b: a\in A, b\in B\}$, it is sufficient to prove that for sets A and B of positive measure, the sum $A+B$ contains some interval.

Let u and v be points of density 1 of A and B, respectively, and let $\varepsilon>0$ be chosen so that the relations $0\le\delta\le 2\varepsilon$, $0\le\delta'\le 2\varepsilon$, $\delta+\delta'>0$ imply

$$\mu(A\cap[u-\delta,u+\delta'])>\frac{1}{2}(\delta+\delta'),\quad \mu(B\cap[v-\delta,v+\delta'])>\frac{1}{2}(\delta+\delta'),$$

where μ stands for Lebesgue measure. We show that $(u+v-\varepsilon,u+v+\varepsilon)\subset A+B$. Let $t\in(u+v-\varepsilon,u+v+\varepsilon)$ and

$$A^*=t-A=\{t-a: a\in A\}.$$

Then $A^*\cap[t-u,v+\varepsilon]$ is congruent with $A\cap[t-v-\varepsilon,u]$, so

$$\mu(A^*\cap[t-u,v+\varepsilon])>\frac{1}{2}(u+v+\varepsilon-t)\quad\text{and}$$
$$\mu(B\cap[t-u,v+\varepsilon])>\frac{1}{2}(u+v+\varepsilon-t),$$

from which it follows that $A^*\cap B\neq\emptyset$. Let $x\in A^*\cap B$; then $t-x\in A$, $x\in B$, and, consequently, $t=(t-x)+x\in A+B$. □

Remarks.

1. Two contestants showed that it is sufficient to assume that f and g are positive functions integrable in every finite interval.

2. A contestant remarked that if we know of E only that it has positive outer measure, then the statement becomes false.

3. Denote by A_1 and B_1 the set of all points of density 1 of A and B, respectively. If $u\in A_1$ and $v\in B_1$, then by the density theorem of Lebesgue u and v have density 1 also in A_1 and B_1, respectively, so by the solution above, A_1+B_1 contains some neighborhood of $u+v$. This means exactly that A_1+B_1 is an open set (see *J. B. H. Kemperman, A general functional equation, Trans. Am. Math. Soc. 86 (1957), 28–56*, Theorem 2.2).

Problem M.6. *In n-dimensional Euclidean space, the union of any set of closed balls (of positive radii) is measurable in the sense of Lebesgue.*

Solution. Let $\{G_\alpha\}$ $(\alpha\in A)$ be an arbitrary set of closed balls of positive radii. Put $H=\cup_{\alpha\in A}G_\alpha$. Denote by G the set of all closed balls (of positive radii) that are contained in some G_α. Clearly, the elements of G constitute a cover of H in the sense of Vitali. Therefore, by Vitali's covering theorem (see for example *S. Saks, Theory of the Integral, Hafner, New York, 1937, p. 109)*, we can choose an at most countable number of pairwise disjoint balls $S_1,S_2,\dots$ belonging to G such that $H\setminus\cup_i S_i$ has Lebesgue measure 0. Since $S_i\subset H$, the relation

$$H=\Big(H\setminus\bigcup_i S_i\Big)\cup\Big(\bigcup_i S_i\Big)$$

implies that H is Lebesgue-measurable. □

Remarks.

1. Most contestants solved the problem with the help of Vitali's covering theorem.
2. Several contestants remarked that H can be obtained as the union of at most countably many Jordan measurable sets.
3. One contestant proved the following generalization of the problem: The union of any set of convex sets with nonvoid interiors is measurable in the sense of Lebesgue.

Problem M.7. *In n-dimensional Euclidean space, the square of the two-dimensional Lebesgue measure of a bounded, closed, (two-dimensional) planar set is equal to the sum of the squares of the measures of the orthogonal projections of the given set on the n-coordinate hyperplanes.*

Solution 1. We solve the problem in the generalized form where a bounded, closed subset of H or an r-dimensional subspace L is projected to the r-dimensional coordinate subspaces.

H is the difference of two bounded open sets $N_2 \in N_1$. Each of them is the union of an at most countably infinite number of disjoint parallelepipeds t_{1i} and t_{2j}. Their images under a linear transformation T (the orthogonal projection) are parallelepipeds whose volume (Lebesgue measures, denoted by λ) satisfies

$$\lambda(Tt_{1i}) = c\lambda(t_{1i}) \quad \text{and} \quad \lambda(Tt_{2j}) = c\lambda(t_{2j}),$$

where c is a constant depending on the position of the two subspaces and the direction of projecting only. Further, since a projection takes the union and the difference of sets into the union and the difference, respectively, of the image sets, from the additivity of the measure it follows that TH is measurable and $\lambda(TH) = c\lambda(H)$.

If $T_{i_1 \dots i_r}$ is the orthogonal projection to the coordinate subspace $[x_{i_1}, \dots, x_{i_r}]$ and $c_{i_1 \dots i_r}$ denotes the corresponding constant, then the sum of the squared measures of the projections is

$$\sum_{i_1,\dots,i_r} \lambda^2(T_{i_1\dots i_r}H) = \sum_{i_1,\dots,i_r} c_{i_1\dots i_r}\lambda^2(H) = \lambda^2(H) \sum_{i_1,\dots,i_r} c_{i_1\dots i_r}, \qquad (1)$$

where the summation is extended to all combinations $i_1, \dots, i_r$ of order r of the numbers $1, 2, \dots, n$.

If $e_1, \dots, e_n$ are unit vectors of a Cartesian system of coordinates in the n-dimensional Euclidean space and H is the r-dimensional cube spanned by the orthogonal unit vectors

$$u_k = \sum_{i=1}^{n} \alpha_i^{(k)} e_i \qquad (k = 1, \dots, r)$$

belonging to L, then $\lambda(H) = 1$, and the value $c_{i_1\dots i_r} = \lambda(T_{i_1\dots i_r}H)$ is equal to the determinant $|A_{i_1\dots i_r}|$ formed of the columns $i_1, \dots, i_r$ of the matrix $A = \{\alpha_i^{(k)}\}$. Then, however, by the Binet–Cauchy formula,

$$\sum_{i_1,\dots,i_r} c^2_{i_1\dots i_r} = \sum_{i_1,\dots,i_r} |A_{i_1\dots i_r}|^2 = |AA^*| = 1,$$

which, owing to (1), verifies our assertion. □

Solution 2. We prove the assertion in the previous generalized form where a (not necessarily compact) Lebesgue-measurable set H lying in the r-dimensional subspace L is projected to the r-dimensional coordinate subspaces.

Let $e_1, \dots, e_n$ be the orthonormal basis vectors of a Cartesian system of coordinates in the n-dimensional Euclidean space E_n, and $u_1, \dots, u_r$ be the orthonormal basis vectors of a system of coordinates in the subspace L. We may assume that L contains the origin since this can be achieved by translation, and translation means translation of the projections, so the measure of H and the measures of its projections remain unchanged. Consequently, we may choose

$$u_k = \sum_{i=1}^{n} \alpha_i^{(k)} e_i \qquad (k = 1, \dots, r),$$

where the matrix $A = \{\alpha_i^{(k)}\}$ satisfies $|AA^*| = 1$.

Denote by $A_{i_1\dots i_r}$ the operator of orthogonal projection from L to the coordinate subspace $[x_{i_1}, \dots, x_{i_r}]$ of E_n. The operator $A_{i_1\dots i_r}$ sends the basis vectors u_k to the vectors of the latter subspace with components $\alpha_{i_1}^{(k)}, \dots, \alpha_{i_r}^{(k)}$. Let $(A_{i_1\dots i_r})$ be the matrix of this projection operator (that is, the square matrix formed of the columns $i_1, \dots, i_r$ of A), and let $|A_{i_1\dots i_r}|$ be the determinant of this matrix.

The Lebesgue measure of H is given by

$$\mu = \int_L \chi_H \, d\lambda,$$

where χ_H denotes the characteristic function of H. By the formula for transformation of integrals, the measure of $A_{i_1\dots i_r}H$ is

$$\int_{[x_{i_1},\dots,x_{i_r}]} \chi_{A_{i_1\dots i_r}H} \, d\lambda = |A_{i_1\dots i_r}| \int_L \chi_H \, d\lambda = |A_{i_1\dots i_r}|\mu.$$

(This is valid also when $|A_{i_1\dots i_r}| = 0$, since in this case $A_{i_1\dots i_r}H$ lies in an at most $(r-1)$-dimensional subspace, that is, a subspace of measure 0 with respect to λ.) Summing for all combinations $i_1, \dots, i_r$ of order r, by the Binet–Cauchy formula we obtain

$$\sum_{i_1,\dots,i_r} \{ \int_{[x_{i_1},\dots,x_{i_r}]} \chi_{A_{i_1\dots i_r}H} \, d\lambda\}^2 = \mu^2 \sum_{i_1,\dots,i_r} |A_{i_1\dots i_r}|^2 = \mu^2 |AA^*| = \mu^2,$$

which proves the assertion. □

Problem M.8. *Let us use the word N-measure for nonnegative, finitely additive set functions defined on all subsets of the positive integers, equal to 0 on finite sets, and equal to 1 on the whole set. We say that the system $\mathfrak{A}$ of sets determines the N-measure μ if any N-measure coinciding with μ on all elements of $\mathfrak{A}$ is necessarily identical with μ. Prove the existence of an N-measure μ that cannot be determined by a system of cardinality less than continuum.*

Solution. We shall need a definition and a lemma.

Definition. We say that the system A consisting of certain subsets of a set is independent, if for any distinct members $X_1, \ldots, X_j$, $Y_1, \ldots, Y_k$ of A, the set

$$X_1 \cap \cdots \cap X_j \cap \overline{Y_1} \cap \cdots \cap \overline{Y_k}$$

is infinite (here $\overline{U}$ denotes the complement of U).

Lemma. Any countable set has an independent system of subsets that is of power continuum.

Proof. Indeed, take the following set for the countable basic set. Let M be the set of all sets of real numbers that can be obtained as finite unions of intervals belonging to any type of closedness but having rational endpoints. Clearly, there is only a countable number of such sets. Denote by M_a the set of those sets of real numbers described above that contain the real number a. Then the system $\{M_a\}$ of subsets of M, for a running through the set of all real numbers, has power continuum. It is easy to see that $\{M_a\}$ is independent. Really, for given real numbers $a_1, \ldots, a_j$, $b_1, \ldots, b_k$ of A, the set

$$M_{a_1} \cap \cdots \cap M_{a_j} \cap M_{b_1} \cap \cdots \cap M_{b_k}$$

consists of those elements of M, that is, those sets of real numbers described above, that contain $a_1, \ldots, a_j$ but do not contain $b_1, \ldots, b_k$. The set of these sets, however, is countably infinite.

Next we construct an N-measure that takes the values 0 and 1 only. To this end, we shall need Zorn's lemma, so our proof is not purely constructive.

A system of subsets of the set N of all real numbers is called a filter if it is closed for taking finite intersections and if together with any set it contains all larger sets as well. To exclude the filter consisting of all subsets, we also require that a filter should not contain the empty set. Since the union of an increasing chain of filters is again a filter, to any filter with the help of Zorn's lemma we can find a maximal filter containing it. This is called an ultrafilter. If a filter contains neither the set A nor $\overline{A}$, then it can be extended by A. If it contained both, then it would also contain their intersection, that is, the empty set. Therefore, an ultrafilter contains exactly one of the sets A and $\overline{A}$. If we now associate with an ultrafilter

$\mathcal{M}$ the N-measure μ that takes the value 1 on elements of $\mathcal{M}$ and 0 on their complements, then we obtain a set function with values 0 and 1 that is defined on all subsets of N and is additive. Additiveness could fail only if neither of the sets has measure 0, but this is impossible: as the empty set does not belong to our ultrafilter, two disjoint sets of measure 1 do not exist. We shall prove for a measure of this kind that it cannot be determined by less than a continuum number of sets.

From the subsets of N, in the manner described in the lemma, we construct an independent system of power continuum, and denote its elements by N_a. If c is an infinite sequence consisting of distinct real numbers, we denote by S_c the union of the complements of the corresponding sets N_a. The filter generated by the N_a, the S_c, and the complements of finite sets consists of all sets containing finite intersections of these. To see that the latter is really a filter, we have to show that it does not contain the empty set. In the opposite case, we could find $a_1, \ldots, a_m$ and $c_1, \ldots, c_n$ so that the set

$$X = N_{a_1} \cap \cdots \cap N_{a_m} \cap S_{c_1} \cap \cdots \cap S_{c_n}$$

is finite. Here c_i is an infinite sequence of distinct real numbers, and consequently we can choose real numbers $b_1, \ldots, b_n$ from the elements of $c_1, \ldots, c_n$ so that $a_1, \ldots, a_m$, and $b_1, \ldots, b_n$ are distinct. Independence implies that the set

$$N_{a_1} \cap \cdots \cap N_{a_m} \cap \overline{N_{b_1}} \cap \cdots \cap \overline{N_{b_n}} = Y$$

is infinite. On the other hand, the definition of the S_c yields $Y \subseteq X$. This contradicts our assumption. It follows that the system of sets considered does not contain the empty set; hence it is a filter. We extend this filter to an ultrafilter $\mathcal{M}$ and define the N-measure μ with the help of the latter.

We have to prove that μ cannot be determined by a system $\mathfrak{A}$ of power less than continuum. If $\mathfrak{A}$ is a determining system containing, among others, sets of measure 0, then these can be replaced by their complements: it is sufficient to consider determining systems $\mathfrak{A}$ all of whose elements belong to $\mathcal{M}$. Taking a $\mathfrak{A}$ of this kind, we assume that it contains less than a continuum of sets.

If we consider the finite intersections of the sets in $\mathfrak{A}$, their power is still less than continuum. Therefore, we may assume from the beginning that $\mathfrak{A}$ is closed for taking finite intersections. If there were a $Z \in \mathcal{M}$ that contained no element of $\mathfrak{A}$, then $\overline{Z}$ would not be disjoint from any set in $\mathfrak{A}$ and, consequently, the sets containing sets of the form $\overline{Z} \cap A$ $(A \in \mathfrak{A})$ would constitute a filter. This filter could be extended to an ultrafilter, and the latter would define an N-measure ν. Then μ and ν would coincide on the elements of $\mathfrak{A}$, whereas $\mu(\overline{Z}) = 0$ and $\nu(\overline{Z}) = 1$. This is impossible since $\mathfrak{A}$ is a determining system. So each element of $\mathcal{M}$ contains a set from $\mathfrak{A}$. Then at least one element W of $\mathfrak{A}$ is contained in infinitely many elements of $\{N_a\}$; let c be a countable sequence of the corresponding subscripts. The set $\overline{N_a}$ is disjoint from W for each element of c, from which it follows that $S_c = \cup \overline{N_a}$ is also disjoint from W. But this means that $\mathcal{M}$ contains two disjoint sets, which contradicts our assumption. The contradiction has been caused by the assumption that μ can be determined by less than a continuum of sets. For the completion of the proof, it remains to show that μ vanishes on every finite set. This is evident since $\mu(D) = 1$ whenever D is the complement of a finite set. □

Remarks.

1. If the continuum hypothesis is assumed, the problem becomes trivial. In fact, it is easy to prove that a countable set cannot determine an N-measure.
2. It would be a sharpening of the problem to prove that no N-measure can be determined by less than a continuum number of sets. This, however, is undecidable within the usual Zermelo–Fraenkel system of axioms for set theory.
3. The problem can be generalized to other cardinal numbers. For any cardinal m, there exists a finitely additive 0–1 measure that vanishes on all sets of power less than m and cannot be generated by less than 2^m subsets. To prove this, it is sufficient to give an adequate generalization of the lemma. One contestant proved this generalization.
4. One may raise the question of whether there are relatively many or relatively few N-measures for the determination of which 2^m subsets are needed. There are altogether 2^{2^m} N-measures, and there are equally so many N-measures that cannot be determined by less than 2^m sets. In the original case where the basic set is countable, one contestant proved this.

Problem M.9. *Let $\{\phi_n(x)\}$ be a sequence of functions belonging to $L^2(0,1)$ and having norm less than 1 such that for any subsequence $\{\phi_{n_k}(x)\}$ the measure of the set*

$$\{x \in (0,1) : |\frac{1}{\sqrt{N}} \sum_{k=1}^{N} \phi_{n_k}(x)| \geq y\}$$

tends to 0 as y and N tend to infinity. Prove that ϕ_n tends to 0 weakly in the function space $L^2(0,1)$.

Solution. Suppose there is a function $f \in L^2(0,1)$ such that

$$\int_0^1 \phi_n(x) f(x)\, dx$$

does not tend to 0 as $n \to \infty$. Then in view of the relations

$$\left| \int_0^1 \phi_n(x) f(x)\, dx \right| \leq \|\phi_n\| \|f\| \leq \|f\| \qquad (n = 1, 2, \dots)$$

(replacing, if necessary, $f(x)$ by $e^{i\vartheta}f(x)$ with a suitable ϑ), there is an $\alpha > 0$ and a subsequence $\{\phi_{n_k}(x)\}_{k=1}^{\infty}$ such that

$$\operatorname{Re}\int_0^1 \phi_{n_k}(x)f(x)\,dx \geq \alpha$$

for every k ($\|f\|$ stands for the norm of the function $f(x)$ in $L^2(0,1)$).

Obviously, to every $\varepsilon > 0$ there is $\delta > 0$ such that the relations $m(A) < \delta$, $A \subseteq (0,1)$ imply $\|\chi_A f\|^2 < \varepsilon$ (here $m(A)$ denotes the measure of the set A, and $\chi_A(x)$ denotes the characteristic function of A). Let ε be a positive number less than α, let δ be a positive value corresponding to ε in the sense just indicated, and choose $y(\varepsilon)$ and $N(\varepsilon)$ so that for $N \geq N(\varepsilon)$ the measure of the set

$$E = \left\{x \in (0,1) : \left|\frac{1}{\sqrt{N}}\sum_{k=1}^{N}\phi_{n_k}(x)\right| > y(\varepsilon)\right\}$$

is smaller than δ. Then, for $N \geq N(\varepsilon)$, it follows that

$$\begin{aligned}
N\alpha &\leq \sum_{k=1}^{N}\operatorname{Re}\int_0^1 f(x)\phi_{n_k}(x)\,dx \\
&= \operatorname{Re}\int_{\overline{E}} f(x)\sum_{k=1}^{N}\phi_{n_k}(x)\,dx + \operatorname{Re}\sum_{k=1}^{N}\int_E f(x)\phi_{n_k}(x)\,dx \\
&\leq \sqrt{N}\int_{\overline{E}}|f(x)|\left|\frac{1}{\sqrt{N}}\sum_{k=1}^{N}\phi_{n_k}(x)\right|\,dx + \sum_{k=1}^{N}\int_E |f(x)\phi_{n_k}(x)|\,dx \\
&\leq \sqrt{N}y(\varepsilon)\|f\|^2 + N\|\chi_E f\|^2 < \sqrt{N}y(\varepsilon)\|f\|^2 + N\varepsilon,
\end{aligned}$$

that is, $\sqrt{N}(\alpha - \varepsilon) < y(\varepsilon)\|f\|^2$, which is impossible for sufficiently large N since $\alpha - \varepsilon > 0$. Consequently,

$$\int_0^1 \phi_n(x)f(x)\,dx \to 0 \qquad (n \to \infty)$$

for all functions $f \in L^2(0,1)$, as stated. □

Remarks. Essentially all acceptable solutions have followed this way.

1. Two contestants noted that it is sufficient to require the following: for any subsequence $\{\phi_{n_k}(x)\}$ and any number $c > 0$, the measure of the set

$$\left\{x \in (0,1) : \left|\sum_{k=1}^{N}\phi_{n_k}(x)\right| > cN\right\}$$

tends to 0 as $N \to \infty$.

2. One contestant showed by a counterexample that without boundedness of the sequence $\{\phi_n(x)\}$ in the norm of $L^2(0,1)$ the assertion is not always true. He also mentioned that the hypotheses of the problem do not imply strong convergence.

3. Another contestant pointed out that it is sufficient to make the following assumption: to any subsequence $\{n_k\}$ of the natural numbers, there is a function $f(N) \neq 0$ for which $f(N)/N \to 0$ as $N \to \infty$ and the measure of the set

$$\left\{x \in (0,1) : \left|\frac{1}{f(N)} \sum_{k=1}^{N} \phi_{n_k}(x)\right| > y\right\}$$

tends to 0 as $y \to \infty$, $N \to \infty$. He also noted that the sequence $\{\phi_n(x)\}$ does not necessarily converge either in measure or in mean.

4. A contestant observed that the hypothesis may be weakened as follows: from any subsequence of the sequence $\{\phi_n(x)\}$ one can choose a subsequence $\{\phi_{n_k}(x)\}$ for which the measure of set (1) tends to 0 as $y \to \infty$, $N \to \infty$.

5. One contestant called attention to the fact that if one assumes boundedness of the sequence $\{\phi_n(x)\}$ in the norm of $L^p(0,1)$, where $1 < p \leq \infty$, then it can be shown that, under the hypothesis of the problem,

$$\int_0^1 \phi_n(x) f(x)\, dx \to 0 \qquad (n \to \infty)$$

follows for all functions $f(x) \in L^q(0,1)$, where $1/p + 1/q = 1$.

6. Four contestants treated the problem in the complex case.

Problem M.10. *We say that the real-valued function $f(x)$ defined on the interval $(0,1)$ is approximately continuous on $(0,1)$ if for any $x_0 \in (0,1)$ and $\varepsilon > 0$ the point x_0 is a point of interior density 1 of the set*

$$H = \{x : \ |f(x) - f(x_0)| < \varepsilon\}.$$

Let $F \subset (0,1)$ be a countable closed set, and $g(x)$ a real-valued function defined on F. Prove the existence of an approximately continuous function $f(x)$ defined on $(0,1)$ such that

$$f(x) = g(x) \quad \textit{for all} \quad x \in F.$$

Solution 1. Let $F = \{c_1, c_2, \dots\}$. Denote by $I(x,n)$ the closed interval of center x and radius $1/n$. By the "central p-multiple" $(0 \leq p < 1)$ of the interval (a,b) we shall mean the closed interval

$$\left[a + \frac{1}{2}(b-a)(1-p), b - \frac{1}{2}(b-a)(1-p)\right].$$

We shall define a system J_n of intervals by recursion so that it has the following properties. J_n consists of countably many disjoint closed intervals, none of which intersects F. Further, c_n is the only accumulation point of the set of endpoints of the intervals forming J_n and, finally, K_m and K_n are disjoint if $m \neq n$, where K_n stands for the union of the intervals forming J_n.

Suppose that for $n' < n$ we have already defined $J_{n'}$ so that they have the properties required. Then the closure of the set $\cup_{k=1}^{n-1} K_k = L_n$ is

$$L_n \cup \{c_1, \dots, c_{n-1}\},$$

and it does not contain c_n. Therefore, it does not intersect a suitable neighborhood $I(c_n, r_n)$ of c_n either. We may assume that $r_n \geq n$. For each j $(j = 1, 2, \dots)$, consider the interior of the set $I(c_n, r_n + j) \setminus I(c_n, r_n + j + 1) \setminus F$. This is the union of a (finite or) countable number of open intervals. Obviously, the total length of these open intervals is equal to the measure of the set $I(c_n, r_n + j) \setminus I(c_n, r_n + j + 1) = M(n, j)$, so the union of a finite number of them has at least $(1 - 1/j)$ times this measure. Take the central $(1 - 1/j)$-multiples of these finitely many intervals, and let these new intervals constitute the system $J_{n,j}$. Put $J_n = \cup_{j=1}^{\infty} J_{n,j}$.

Clearly, the systems J_n of intervals so obtained possess the required properties (in particular, the recursion is correct). Further, the intervals forming $J_{n,j}$ fill at least $(1 - 1/j)^2$ times the measure of $M(n, j)$ and, therefore, K_n, the union of the intervals forming J_n, has density 1 at c_n.

Define f in the following way. Let $f(x) = g(c_n)$ if $x \in K_n$ or $x = c_n$. If we prove that the set $\cup_{n=1}^{\infty} K_n \cup F$ is closed, then f can be defined at points not belonging to this set so that on the countable number of disjoint open intervals forming the complement of this closed set, f is linear, and at the endpoints of the intervals it takes the values already fixed. It is obvious that the function f defined in this way is approximately continuous in points of F and continuous in all other points. To see the latter, observe that for $x \notin F$ the distance from x to F is positive, say greater than $1/m$, where m is a positive integer. Thus, by the assumption $r_n \geq n$, K_n does not intersect the $1/2m$-neighborhood of x if $n > 2m$. Consequently,

$$I(x, 2m) \cap \Big(\bigcup_{n=1}^{\infty} K_n \cup F \Big) = I(x, 2m) \cap \Big(\bigcup_{n=1}^{2m} K_n \Big).$$

The right-hand side, however, is the union of a finite number of closed intervals. This proves, on the one hand, that $\cup_{n=1}^{\infty} K_n \cup F$ is closed and, on the other hand, that the restriction to $I(x, 2m)$ of the function f constructed with the help of this set is a continuous polygon function. □

Solution 2. Let $F = \{r_1, r_2, \dots\}$. We start from the idea that the closer x is to r_i, the more $f(x)$ should "feel" the value taken at r_i. We try to achieve this by a definition of the following type:

$$f(x) = \frac{\sum_{i=1}^{\infty} \frac{g(r_i)}{u(i)|x - r_i|}}{\sum_{i=1}^{\infty} \frac{1}{u(i)|x - r_i|}}, \qquad \text{if} \quad x \notin F. \tag{1}$$

We now impose various conditions on the order of $u(i)$ to ensure convergence of this series and approximate continuity of the function obtained, and finally we show that these conditions can be fulfilled.

The series will always be convergent if, for instance,

$$u(i) > i^2|g(r_i)| \quad \text{and} \quad u(i) > i^2 \quad (i = 1, 2, \dots). \tag{2}$$

Really, for $x \notin F$, the closedness of F gives $\min_i |x - r_i| > 0$.

f will be continuous in all points of $(0,1) \setminus F$ (in a small neighborhood, both the numerator and the denominator are sums of uniformly convergent series of continuous functions), hence it is also approximatively continuous there.

Consider therefore an r_i. Let $d = \min_{j<i} |r_i - r_j|$, $\delta < d/2$. We show that in the interval $(r_i - \delta, r_i + \delta)$, for δ sufficiently small, the numerator in (1) will generally be around $g(r_i)/(u(i)|x - r_i|)$, and the denominator around $1/(u(i)|x - r_i|)$. In any case,

$$\sum_{j<i} \frac{g(r_j)}{u(j)|x - r_j|} = O(1) \tag{3}$$

and

$$\sum_{j<i} \frac{1}{u(j)|x - r_j|} = O(1)$$

(since $|x - r_j| > d - \delta/2 > d/2$). We divide the values x into two classes. One consists of those x that satisfy the relation

$$\frac{1 + |g(r_j)|}{u(j)|x - r_j|} < \frac{1}{j^2} \tag{4}$$

for all $j > i$. For these x, by (3) and (4),

$$f(x) = \frac{\frac{g(r_i)}{u(i)|x - r_i|} + O(1)}{\frac{1}{u(i)|x - r_i|} + O(1)} = \frac{g(r_i) + O(|x - r_i|)}{1 + O(|x - r_i|)},$$

which lies in the ε-neighborhood of $g(r_i)$ if δ is sufficiently small. Thus, we have to achieve that the measure of such x in $(r_i - \delta, r_i + \delta)$ be $(2 + o(1))\delta$. If, however, (4) does not hold, then

$$|x - r_j| < \frac{j^2}{u(j)}(1 + |g(r_j)|) = \varepsilon_j.$$

The measure of such "bad" x in $(r_i - \delta, r_i + \delta)$ is

$$\leq 2 \sum_{|r_i - r_j| < \delta + \varepsilon_j} \varepsilon_j.$$

If now $\varepsilon_j \leq |r_i - r_j|^2/j^2$, then this measure is

$$\leq 2 \sum_{|r_i - r_j| < \delta + \frac{|r_i - r_j|^2}{j^2}} \frac{|r_i - r_j|^2}{j^2} \leq 2(\frac{4}{3}\delta)^2 \sum_{j=1}^{\infty} \frac{1}{j^2} = O(\delta^2) = o(\delta)$$

as $\delta \to 0$ (since $|r_i - r_j|^2/j^2 \leq |r_i - r_j|/4$ and so $(3/4) \cdot |r_i - r_j| < \delta$). Consequently, together with (2), we have only a finite number of conditions for each j, so they can be realized. □

Problem M.11. *Let $\{f_n\}_{n=0}^{\infty}$ be a uniformly bounded sequence of real-valued measurable functions defined on $[0,1]$ satisfying*

$$\int_0^1 f_n^2 = 1.$$

Further, let $\{c_n\}$ be a sequence of real numbers with

$$\sum_{n=0}^{\infty} c_n^2 = +\infty.$$

Prove that some re-arrangement of the series $\sum_{n=0}^{\infty} c_n f_n$ is divergent on a set of positive measure.

Solution. We first show that

$$\sum_{n=0}^{\infty} c_n^2 f_n^2 \tag{1}$$

is divergent on a set of positive measure. Suppose that (1) yields a function g that is finite a.e. By Egorov's theorem there is a set $A \subseteq [0,1]$ on which the series (1) is uniformly convergent and for which $\lambda(A) > 1 - (/2K)$, where K is a common upper bound of the functions f_n^2. Then, by the uniform convergence, the limiting function $g(t)$ is integrable on A, and

$$\sum_{n=0}^{\infty} c_n^2 \int_A f_n^2(t)\,dt = \int_A g(t)\,dt < +\infty\,. \tag{2}$$

On the other hand,

$$\int_A f_n^2(t)\,dt = \int_0^1 f_n^2(t)\,dt - \int_{[0,1]\setminus A} f_n^2(t)\,dt \geq 1 - K\frac{1}{2K} = \frac{1}{2},$$

which, in view of (2), yields

$$\frac{1}{2}\sum_{n=0}^{\infty} c_n^2 < \infty,$$

a contradiction.

As a second step we show that the series $\sum_{n=0}^{\infty} c_n f_n$ has a subseries divergent on a set of positive measure. Let $\varepsilon_1, \dots, \varepsilon_n, \dots$ be independent random variables on some field (Ω, A, P) taking the values 0, 1 with probability 1/2, and let $0 < t < 1$ be a number for which (1) is divergent. Then for the random variables $\varepsilon_n c_n f_n(t)$ we have

$$\sum_{n=0}^{\infty} \text{Var}\,(\varepsilon_n c_n f_n(t)) = \sum_{n=0}^{\infty} c_n^2 f_n^2(t) = +\infty$$

and therefore, by Kolmogorov's three-series theorem, $\sum_{n=0}^{\infty} \varepsilon_n c_n f_n(t)$ is divergent with positive probability. Thus, the set of all elements (ω, t) in the probability field $(\Omega, A, P) \times [0,1]$ for which

$$\sum_{n=0}^{\infty} (\varepsilon_n(\omega) c_n f_n(t) \tag{3}$$

is divergent has positive measure. Consequently, there exists $\omega \in \Omega$ for which the series (3) is divergent on a set of positive measure.

It follows that there is a set D of positive measure and an infinite sequence $n_1 < n_2 < \dots$ such that

$$\sum_{n=0}^{\infty} c_{n_k} f_{n_k}(t) \tag{4}$$

is divergent for all $t \in D$; so there exists $\delta(t) > 0$ such that for arbitrarily large indices the partial sums of (4) have oscillation not less than $\delta(t)$.

Define next the numbers $N_1 = 1 < N_2 < \dots$ as follows. Let the number N_{k+1} be such that the measure of the set

$$D_k = \left\{ f \in D : \max_{N_k \le \nu < \mu < N_{k+1}} \left| \sum_{i=\nu}^{\mu} c_{n_i} f_{n_i}(t) \right| \ge \frac{\delta(t)}{2} \right\}$$

is greater than $(1 - (1/2^{k+1})) \cdot \lambda(D)$. Such an N_{k+1} exists since for sufficiently large N_{k+1}, all values $t \in D$ belong to D_k. The set

$$D^* = \bigcap_{k=1}^{\infty} D_k$$

has a positive measure, and

$$\max_{N_k \le \nu < \mu < N_{k+1}} \left| \sum_{i=\nu}^{\mu} c_{n_i} f_{n_i}(t) \right| \ge \frac{\delta(t)}{2}$$

if $t \in D^*$ and $k \ge 1$. Therefore, any rearrangement of the original series $\sum_{n=0}^{\infty} c_n f_n$, for which the sums $\sum_{i=N_k}^{N_{k+1}} c_{n_i} f_{n_i}$ consist of consecutive terms, is divergent at all points $t \in D^*$. □

Problem M.12. *Let $\{f_n\}$ be a sequence of Lebesgue-integrable functions on $[0,1]$ such that for any Lebesgue-measurable subset E of $[0,1]$ the sequence $\int_E f_n$ is convergent. Assume also that $\lim_n f_n = f$ exists almost everywhere. Prove that f is integrable and $\int_E f = \lim_n \int_E f_n$. Is the assertion also true if E runs only over intervals but we also assume $f_n \ge 0$? What happens if $[0,1]$ is replaced by $[0,\infty)$?*

Solution. Let $E = [0,1]$ or $E = [0,\infty)$, and $E = \cup_{k=1}^{\infty} E_k$, where E_k is a measurable set of finite measure and on E_k the convergence of the sequence

$\{f_n\}$ is uniform. By Egorov's theorem, such a sequence $\{E_k\}$ obviously exists. It is clear that $\int_{E_k} f$ exists for every k, and $\int_F f = \lim_n \int_F f_n$ if F is a measurable subset of E_k.

The formula $\mu(G) = \lim_n \int_G f_n$ defines a finitely additive signed measure on measurable subsets G of E. The theorem of Beppo–Levi shows that the existence of $\displaystyle\int_E f$ together with the relation $\mu(G) = \displaystyle\int_G f$ for each measurable subset G of E are equivalent to the σ-additivity of μ. But the latter follows from the well-known fact that the limit of a pointwise convergent sequence of finite signed measures is a finite signed measure (in particular, it is σ-additive; see *P. R. Halmos, Measure Theory, Springer, New York, 1974, p. 170*, relation (14)). It can also be seen that $\displaystyle\int_E |f_n - f| \to 0$ $(n \to \infty)$. Thus, we have answered the first and third questions of the problem. The answer to the second question is negative, as shown by the example $f_n(x) = n$ if $0 \le x \le \dfrac{1}{n}$, and $f_n(x) = 0$ if $\dfrac{1}{n} < x \le 1$. □

Problem M.13. *Let $0 \le c \le 1$, and let η denote the order type of the set of rational numbers. Assume that with every rational number r we associate a Lebesgue-measurable subset H_r of measure c of the interval $[0, 1]$. Prove the existence of a Lebesgue-measurable set $H \subset [0, 1]$ of measure c such that for every $x \in H$ the set*

$$\{r : x \in H_r\}$$

contains a subset of type η.

Solution. We give two solutions. Both make use of the following simple lemma:

Lemma. Let M be a system of sets that consists of certain subsets of the interval $(0, 1)$, and suppose that for each $A \in M$ there is an $x \in A$ such that $A \cap (0, x) \in M$ and $A \cap (x, 1) \in M$. Then every element of M has a subset of type η.

Proof. Arrange the rational points of $(0, 1)$ in a sequence $r_1, r_2, \ldots$. Let $A \in M$ be arbitrary. We shall define by recursion a sequence of points $x_n \in A$ having the following properties:

1. It is ordered in the same way as the sequence $\{r_n\}$;
2. $A \cap (0, x_n) \in M$, $A \cap (x_n, 1) \in M$ for every n, and $A \cap (x_i, x_j) \in M$ for every pair $x_i < x_j$.

By assumption, there is an $x_1 \in A$ such that $A \cap (0, x_1) \in M$ and $A \cap (x_1, 1) \in M$. Next suppose that $x_1, x_2, \ldots, x_n$ have already been chosen, and let the immediate neighbors of r_{n+1} from among $r_1, r_2, \ldots, r_n$ be r_i and r_j, $r_i < r_j$. By the induction hypothesis, we have $x_i < x_j$ and $A \cap (x_i, x_j) \in M$, so there is an $x_{n+1} \in A \cap (x_i, x_j)$ such that $A \cap (x_i, x_{n+1}) \in M$ and $A \cap (x_{n+1}, x_j) \in M$. A similar choice of x_{n+1} is possible in the cases where $r_{n+1} > r_i$ $(i \le n)$ or $r_{n+1} < r_i$ $(i \le n)$. The proof of the lemma is complete.

Solution 1. We need the following auxiliary theorem:

Auxiliary theorem. Let the sets $H_n \in [0,1]$ be measurable and have measure not less than c ($n = 1, 2, \dots$). Then the set H of all points x for which the sequence $\{n : x \in H_n\}$ has positive upper density is also measurable and has measure not less than c.

Proof. Denote by f_n the characteristic function of H_n, and put

$$g_n = \frac{1}{n}(f_1 + f_2 + \cdots + f_n).$$

Obviously, $x \in H$ if and only if $g_n(x)$ does not tend to 0 as $n \to \infty$, which yields the measurability of H. Since $0 \le g_n(x) \le 1$ for every n, we have

$$\begin{aligned}\lambda(H) &\ge \int_H g_n(x)\,dx = \int_0^1 g_n(x)\,dx - \int_{[0,1]\setminus H} g_n(x)\,dx \\ &= c - \int_{[0,1]\setminus H} g_n(x)\,dx.\end{aligned}$$

Now $x \in [0,1] \setminus H$ implies $\lim_{n\to\infty} g_n(x) = 0$, so by Lebesgue's theorem on the passage to the limit under the integral sign,

$$\lim_{n\to\infty} \int_{[0,1]\setminus H} g_n(x)\,dx = 0.$$

Thus, from the previous inequality, we obtain $\lambda(H) \ge c$, which proves the auxiliary theorem.

Let $\{r_n\}$ be an arrangement of the rational points of $(0,1)$ in a sequence of uniform distribution, that is, for any subinterval $(a,b) \in (0,1)$, the density of the sequence $\{n : r_n \in (a,b)\}$ is $b - a$. (A simple example of an arrangement of this kind is the following. We first fix an arbitrary arrangement in a sequence $\{s_n\}$, then from $\{s_n\}$ we consecutively choose the elements of minimal index lying respectively in the intervals

$$[0,1], \left[0, \frac{1}{2}\right], \left[\frac{1}{2}, 1\right], \dots, \left[0, \frac{1}{n}\right], \left[\frac{1}{n}, \frac{2}{n}\right], \dots, \left[\frac{n-1}{n}, 1\right], \dots;$$

meanwhile, we should make sure that each element is selected only once.)

Let $A \in (0,1)$, and assume that the sequence $\{n : r_n \in A\}$ has positive upper measure. We show that A contains a subset of type η. By the lemma we have presented in advance, it is sufficient to prove the existence of an $x \in A$ such that both $\{n : r_n \in A \cap (0,x)\}$ and $\{n : r_n \in A \cap (x,1)\}$ have positive upper density.

Denote by d the upper density of the sequence $\{n : r_n \in A\}$, and let $0 = x_0 < x_1 < \cdots < x_m = 1$ be a subdivision of $[0,1]$ finer than $d/3$. Since $\{r_n\}$ is uniformly distributed, the upper densities of the sequences $\{n : r_n \in A \cap (x_{i-1}, x_i)\}$ are smaller than $d/3$ for every i. Consequently, denoting by d_i the upper density of the sequence $\{n : r_n \in A \cap (0, x_i)\}$, we have $0 < d_i - d_{i-1} < d/3$ for $i = 1, 2, \ldots, m$. Since $d_m = d$, there is an i with $0 < d_{i-1} < d_i < d$. Then $d_{i-1} < d_i$ implies that the set $A \cap (x_{i-1}, x_i)$ is infinite, and it is easy to see that for any element $x \in A \cap (x_{i-1}, x_i)\}$, the sequences $\{n : r_n \in A \cap (0, x)\}$ and $\{n : r_n \in A \cap (x, 1)\}$ also have positive upper measures.

Now consider the sets H_r appearing in the statement, and let H_1 be the set of those x for which the sequence $\{n : x \in H_{r_n}\}$ has positive upper measure. By the previous observation, if $x \in H_1$, then $\{r : x \in H_r\}$ contains a subset of type η. Applying our auxiliary theorem to the sequence of the sets H_{r_n}, it follows that H_1 is measurable and $\lambda(H_1) \geq c$, as required. □

Solution 2. If the closure of a set A has positive measure, then A contains a subset of type η since the system $M = \{A \subset (0,1) : \lambda(\overline{A}) > 0\}$ obviously satisfies the condition of the lemma. It is therefore sufficient to prove that the set

$$H_2 = \{x : \lambda(\overline{\{r : x \in H_r\}}) > 0\}$$

is measurable and $\lambda(H_2) \geq c$.

Consider the function $f(x) = \lambda(\overline{\{r : x \in H_r\}})$. Clearly, $0 \leq f(x) \leq 1$; if we show that $f(x)$ is Lebesgue-measurable and $\int_0^1 f(x)\, dx \geq c$, then we shall be done.

For any $A \in [0,1]$, put

$$A_n = \bigcup_{[\frac{i-1}{2^n}, \frac{i}{2^n}] \cap A \neq \emptyset} \left[\frac{i-1}{2^n}, \frac{i}{2^n}\right].$$

Obviously, $A_1 \supset A_2 \supset \ldots$ and $\cap_{n=1}^{\infty} A_n = \overline{A}$; hence $\lim_{n\to\infty} \lambda(A_n) = \lambda(\overline{A})$. So let $f_n(x) = \lambda(\{r : x \in H_r\}_n)$; then $f_n \searrow f$, and therefore we need only to show that $f_n(x)$ is Lebesgue-integrable with $\int_0^1 f_n(x)\, dx \geq c$.

We have the relation

$$\lambda(A_n) = \frac{1}{2^n} \sum_{i=1}^{2^n} g_{n,i}(A),$$

where $g_{n,i}(A) = 1$ or 0 depending on whether A has or does not have a point in $[(i-1)/2^n, i/2^n]$. Clearly,

$$f_{n,i}(x) := g_{n,i}(\{r : x \in H_r\}) = \sup\left\{k_r : r \in \left[\frac{i-1}{2^n}, \frac{i}{2^n}\right]\right\},$$

where k_r is the characteristic function of H_r. The function k_r is integrable and $\int_0^1 k_r(x)\, dx = c$, so $f_{n,i}$ is integrable and $\int_0^1 f_{n,i}(x)\, dx \geq c$. Since $f_n = (1/2^n) \cdot \sum_{i=1}^{2^n} f_{n,i}$, it follows that f_n is integrable and $\int_0^1 f_n(x)\, dx \geq c$. □

Remark. Denote by H_0 the set of those x for which the set $\{r : x \in H_r\}$ contains a subset of type η. Let $\{r_n\}$ be a fixed arrangement of the rational points of $(0,1)$ in a uniformly distributed sequence, and denote by H_1 the set of those x for which the sequence $\{n : x \in H_{r_n}\}$ has positive upper density. Let H_2 be the set of those x for which the closure of the set $\{r : x \in H_r\}$ has positive measure. Finally, let H_3 be the set of those x for which $\{r : x \in H_r\}$ is dense in a subinterval of $(0,1)$. It is easy to see that for any set A of rational numbers, the upper density of the sequence $\{n : r_n \in A\}$ is at most $\lambda(\overline{A})$. Therefore, $H_3 \subset H_1 \subset H_2 \subset H_0$ holds for any system H_r (thus Solution 1 proves a deeper assertion than Solution 2, since it shows from a narrower set that its measure is at least c). On the other hand, the statement $\lambda(H_3) \geq c$ is false: it can be shown that, for any $0 < c < 1$, there exists a system H_r satisfying the conditions of the problem and such that the set $\{r : x \in H_r\}$ is nowhere dense for every $x \in [0,1]$, in particular, $H_3 = \emptyset$.

We also note that if the sets H_r are measurable, then so is H_0. Actually, it can be proved that if the H_r are Borel (or, more generally, analytic) sets, then H_0 is analytic, hence measurable. From this it follows easily that H_0 is also measurable in this general case.

Problem M.14. *Find a perfect set $H \subset [0,1]$ of positive measure and a continuous function f defined on $[0,1]$ such that for any twice differentiable function g defined on $[0,1]$, the set $\{x \in H : f(x) = g(x)\}$ is finite.*

Solution. Delete an interval of length $1/6$ from the center of $[0,1]$, intervals of length $1/18$ from the centers of the remaining two intervals, and, following the procedure, intervals of length $1/2 \cdot 3^{n+1}$ from the centers of the 2^n intervals remaining after the nth step. This is possible because the length of the remaining intervals is

$$2^{-n}\left(1 - \sum_{k=0}^{n-1} 2^k \cdot \frac{1}{2 \cdot 3^{n+1}}\right) > 2^{-n}\left(1 - \frac{1}{6}\sum_{k=0}^{\infty}\left(\frac{2}{3}\right)^k\right)$$

$$= 2^{-n}\left(1 - \frac{1}{2}\right) = 2^{-n-1} > \frac{1}{2 \cdot 3^{n+1}}.$$

Denote by $I_{n,i}$ $(n = 0, 1, \ldots; i = 1, \ldots, 2^n)$ the open intervals of length $1/2 \cdot 3^{n+1}$ deleted in the nth step. Put $H = [0,1] \setminus \cup I_{n,i}$. Then H is perfect and has measure $1/2$. Denote by h the function that vanishes on H and whose graph on $I_{n,i}$ is an isosceles triangle of altitude $1/n$ for $n = 0, 1, \ldots$ and $i = 1, \ldots, 2^n$. Then h is continuous on $[0,1]$, and so the function

$$f(x) = \int_0^x h(t)\, dt$$

is continuously differentiable on $[0,1]$ and satisfies the relation $f'(x) = 0$ for all $x \in H$. Let g be twice differentiable on $[0,1]$; we establish that the set $A = \{x \in H : f(x) = g(x)\}$ is finite. Suppose A is infinite. Then there is a convergent sequence $x_k \to x_0$ in A. Since $f'(x_0) = 0$, we have $g'(x_0) = 0$; hence by l'Hôpital's rule,

$$\lim_{x\to x_0} \frac{g(x)-g(x_0)}{(x-x_0)^2} = \lim_{x\to x_0} \frac{g'(x)}{2(x-x_0)} = \lim_{x\to x_0} \frac{g'(x)-g'(x_0)}{2(x-x_0)} = \frac{g''(x_0)}{2}.$$

It follows that

$$\lim_{k\to\infty} \frac{f(x_k)-f(x_0)}{(x_k-x_0)^2} = \frac{g''(x_0)}{2}.$$

The contradiction will arise from the fact that

$$\lim_{x\to x_0, x\in H} \left|\frac{f(x)-f(x_0)}{(x-x_0)^2}\right| = \infty.$$

Indeed, let $x \in H$, $x \neq x_0$, be arbitrary. Then there is an interval $I_{n,i}$ that separates x and x_0. Let n be the smallest subscript for which there exists an $I_{n,i}$ of this kind. It is easy to see that $|x-x_0| < 2^{-n}$. On the other hand, if, say, $x > x_0$, then

$$f(x)-f(x_0) = \int_{x_0}^{x} h(t)\,dt \geq \int_{I_{n,i}} h(t)\,dt = \frac{1}{2}\cdot\frac{1}{2\cdot 3^{n+1}}\cdot\frac{1}{n} = \frac{1}{12\cdot 3^n\cdot n}.$$

Hence

$$\left|\frac{f(x)-f(x_0)}{(x-x_0)^2}\right| \geq \frac{1}{12\cdot 3^n\cdot n\cdot 4^{-n}} = \left(\frac{4}{3}\right)^n\cdot\frac{1}{12n}.$$

If $x \to x_0$, then $n\to\infty$ and $(4/3)^n\cdot(1/n)\to\infty$, which proves the assertion. □

Problem M.15. *Prove that if $E \subset \mathbb{R}$ is a bounded set of positive Lebesgue measure, then for every $u < 1/2$, a point $x = x(u)$ can be found so that*

$$|(x-h, x+h)\cap E| \geq uh$$

and

$$|(x-h, x+h)\cap(\mathbb{R}\setminus E)| \geq uh$$

for all sufficiently small positive values of h.

Solution. We shall prove a stronger statement, namely, we verify the conclusion for $u = 1/2$. Let

$$F(x) = m\left((-\infty, x)\cap E\right),$$

where m denotes the Lebesgue measure. Then F is continuous, increasing, and satisfies $|F(x)-F(y)| \leq |x-y|$ for all $x, y \subset \mathbb{R}$. By Lebesgue's density theorem, there exists $x \in \mathbb{R}$, which is a point of density 1 of E, and hence there is a positive $k \in \mathbb{R}$ with $F(x+k) - F(x) > k/2$. Since for sufficiently small x we have $F(x+k) - F(x) = 0$, continuity ensures the existence of an x_0 satisfying $F(x_0 + k) - F(x_0) = k/2$. Consider the function $G(x) = F(x) - F(x_0) - (x - x_0)/2$. Then $G(x_0) = G(x_0 + k) = 0$. As almost every point of $\mathbb{R}$ is a point of density 1 of either E or $\mathbb{R} \setminus E$, we have $|G'(x)| = 1/2$ a.e. Thus, G is not identically 0 on $[x_0, x_0 + k]$. Consequently,

$$\text{either} \quad \max_{t \in [x_0, x_0+k]} G(t) > 0 \quad \text{or} \quad \min_{t \in [x_0, x_0+k]} G(t) < 0.$$

It is sufficient to consider the first possibility since the second is similar to it. Let $x_1 \in (x_0, x_0 + k)$ be a point at which G assumes its maximum just mentioned. Then for every

$$0 < h < \min\{x_1 - x_0, x_0 + k - x_1\},$$

we have $G(x_1 - h) \leq G(x_1) \geq G(x_1 + h)$, that is, $F(x_1 - h) + h/2 \leq F(x_1) \geq F(x_1 + h) - h/2$. Thus, $F(x_1 + h) - F(x_1 - h) \leq F(x_1) + h/2 - (F(x_1) - h) = 3h/2$ and $F(x_1 + h) - F(x_1 - h) \geq F(x_1) - (F(x_1) - h/2) = h/2$. From these two inequalities, we obtain

$$m\left((x_1 - h, x_1 + h) \cap E\right) \geq \frac{h}{2}$$

and

$$m\left((x_1 - h, x_1 + h) \cap (\mathbb{R} \setminus E)\right) \geq \frac{h}{2}.$$

The proof is complete. □

Remark. One contestant did not make use of the boundedness of E and for $u = 1/2$ proved the assertion in the case where E is measurable and neither E nor $\mathbb{R} \setminus E$ is a null set, and also in the case where $E \subseteq \mathbb{R}$, $m^*(E) \neq 0$, and $m_*(E) \neq 0$ (here m^* and m_* denote the exterior and interior Lebesgue measures, respectively).

Problem M.16. *Show that there exist a compact set $K \subset \mathbb{R}$ and a set $A \subset \mathbb{R}$ of type F_σ such that the set*

$$\{x \in \mathbb{R} : K + x \subset A\}$$

is not Borel-measurable (here $K + x = \{y + x : y \in K\}$). Show that there exist a compact set $K \subset \mathbb{R}$ and a set $A \subset \mathbb{R}$ of type F_σ such that the set

$$\{x \in \mathbb{R} : K + x \subset A\}$$

is not Borel-measurable (here $K + x = \{y + x : y \in K\}$).

Solution. Let P and K be bounded perfect sets such that the sums $x + y$ $(x \in P, y \in K)$ are pairwise distinct, that is, the relations $x_1, x_2 \in P$,

$y_1, y_2 \in K$, $x_1 + y_1 = x_2 + y_2$ imply $x_1 = x_2$ and $y_1 = y_2$. It is easy to see that the sets

$$P = \left\{ \sum_{i=1}^{\infty} \frac{a_i}{10^i}; \quad a_i = 0, 1; \quad i = 1, 2, \dots \right\}$$

and

$$K = \left\{ \sum_{i=1}^{\infty} \frac{a_i}{10^i}; \quad a_i = 0, 2; \quad i = 1, 2, \dots \right\}$$

are of this kind. It is well known that there exists a set $U \subset P \times K$ of type G_δ such that the set

$$B = \{x \in P : \text{ there exists } y \in K \text{ with } (x, y) \in U\}$$

(which is the projection of U onto P) is not Borel-measurable.

Put $V = (P \times K) \setminus U$ and $A = \{x + y : (x, y) \in V\}$. The mapping $\phi(x, y) = x + y \quad (x \in P, y \in K)$ is continuous and, by the choice of P and K, one-to-one. Therefore, ϕ is a homeomorphism of the compact set $P \times K$ onto $\{x + y : x \in P, y \in K\}$. Since V is of type F_σ, it follows that $A = \phi(V)$ is an F_σ-type subset of $\phi(P \times K) = P + K$. Since the latter set is closed in $\mathbb{R}$, the set A is a subset of type F_σ of $\mathbb{R}$ as well.

It is easy to see that $K + x \subset A$ if and only if $\phi^{-1}(K + x) \subset \phi^{-1}(A)$, that is, if $x \in P$ and

$$(\{x\} \times K) \cap U = \emptyset,$$

(that is, if $x \in P \setminus B$). So

$$\{x \in \mathbb{R} : K + x \subset A\} = P \setminus B,$$

which is not Borel-measurable. □

Problem M.17. *For which Lebesgue-measurable subsets E of the real line does a positive constant c exist for which*

$$\sup_{-\infty < t < \infty} \left| \int_E e^{itx} f(x) dx \right| \leq c \sup_{n = 0, \pm 1, \dots} \left| \int_E e^{inx} f(x) dx \right|$$

for all integrable functions f on E ?

Solution. We show that $(*)$ holds for those and only those Lebesgue-measurable sets that — apart from a subset of measure 0 — can be covered by a set consisting of the finite union of closed intervals and containing no pair of congruent points modulo 2π.

Denote by $\mathcal{E}$ the system of all sets satisfying the requirements of the problem and by $\mathcal{F}$ the system of all sets having the properties just described.

If $E \in \mathcal{E}$, then those points of E for which there is a congruent point in E (mod 2π) form a set of measure 0. Indeed, if, for example, the set

$$E' = E \cap (E + 2j\pi) \cap (2k\pi, 2(k+1)\pi)$$

had positive measure ($j \neq 0$ integer, $E + u$ denotes the translate by u of E), then with

$$f(x) = \begin{cases} 1 & \text{if } x \in E', \\ -1 & \text{if } x \in E' - 2j, \\ 0 & \text{otherwise}, \end{cases}$$

it would follow that

$$\int_E e^{inx} f(x)\, dx = 0 \quad \text{for every } n,$$

while, as $f \not\equiv 0$ a.e.,

$$\sup \left| \int_E e^{itx} f(x)\, dx \right| > 0.$$

The same property trivially holds also for every $E \in \mathcal{F}$.

Denote by * the reduction modulo 2 with range space $(-\pi, \pi]$. That is, u^* is the (signed) deviation of u from the nearest multiple of 2π (if there are two such multiples, then $u^* = \pi$). Furthermore, put

$$\begin{aligned} E^* &= \{u^* : u \in E\}, \\ h^*(x) &= h(u), \text{ where } x \in E^*, \ x = u^*, \ u \in E. \end{aligned}$$

By the property above, the latter definition is correct for a.e. $x \in E^*$ since there is only one such u. Therefore, we may write

$$\begin{aligned} \phi(t) &= \int_E e^{itx} f(x)\, dx = \int_E \left(e^{itx}\right)^* f^*(x)\, dx, \\ \phi(n) &= \int_E e^{inx} f(x)\, dx = \int_E \left(e^{inx}\right)^* f^*(x)\, dx. \end{aligned}$$

Here $f^*(x) \in \mathcal{L}_1(E^*)$, and every element of $\mathcal{L}_1(E^*)$ occurs among the $f^*(x)$, where $f(x) \in \mathcal{L}_1(E)$.

We show that the relation

$$\left| \int_{E^*} a(x) f^*(x)\, dx \right| \le c \max |\phi(n)| = c \max_n \int_{E^*} e^{inx} f^*(x)\, dx$$

is valid for every function $f^* \in \mathcal{L}_1(E^*)$ if and only if there exists a function $b(x) = \sum a_n e^{inx}$, $\sum |a_n| < +\infty$, such that $b(x) = a(x)$ for a.e. point $x \in E^*$. Then the choice $c = \sum_{n=-\infty}^{\infty} |a_n|$ is possible.

Suppose first that there exists a function $b(x)$ of this kind. Then

$$\left|\int_{E^*} a(x)f^*(x)\,dx\right| = \left|\int_{E^*} b(x)f^*(x)\,dx\right|$$
$$= \left|\sum_{n=-\infty}^{\infty} a_n \int_{E^*} e^{inx} f^*(x)\,dx\right|$$
$$= \left|\sum_{n=-\infty}^{\infty} a_n \phi(n)\right| \leq \sum_{n=-\infty}^{\infty} |a_n| \max_n |\phi(n)|.$$

Conversely, consider the sequences $\{j_n\}_{n=-\infty}^{+\infty}$ that tend to 0 as $|n| \to \infty$; they form a Banach space with respect to the norm $\max|j_n|$. The set of all sequences of the form $\{\phi(n)\}$ is a subspace of this space, and the functional $\int_{E^*} a(x)f^*(x)\,dx$ is linear and, by our assumption, bounded on it. According to the Hahn–Banach theorem, this functional can be boundedly extended to all sequences $\{j_n\}$. But the bounded linear functionals on the latter have the form $\sum a_n j_n$, where $\sum |a_n| < +\infty$. Consequently, restricted to the sequences $\{\phi(n)\}$,

$$\int_{E^*} a(x)f^*(x)\,dx = \sum a_n \phi(n) = \sum_{n=-\infty}^{\infty} a_n \int_{E^*} e^{inx} f^*(x)\,dx$$
$$= \int_{E^*} b(x)f^*(x)\,dx$$

(where $b(x) = \sum a_n e^{inx}$), for every $f^* \in \mathcal{L}_1(E^*)$. Hence, $a(x) = b(x)$ for a.e. $x \in E^*$.

Apply the proposition just proved to the function $a(x) = e^{itx}$ with fixed t. Let $E \in \mathcal{F}$. If the finite number of intervals covering E is somewhat augmented, then the set obtained will not contain any pair of points congruent modulo 2π either. Let $k(x)$ be a "sufficiently smooth" function that equals 1 on the original intervals, and 0 outside the augmented intervals. Finally, put

$$b(x) = \sum_{j=-\infty}^{\infty} k(x+2j)e^{it(x+2j\pi)}.$$

Since at each point at most one term of the series is different from 0, the definition of $b(x)$ makes sense. Furthermore,

$$b(x) = a(x)^* = (e^{itx})^* \quad \text{if } x \in E^*.$$

The function $b(x)$ is periodic with period 2π and Fourier coefficients

$$b_n = \frac{1}{2\pi}\int_{-\infty}^{\infty} e^{-inx} b(x)\,dx$$
$$= \frac{1}{2\pi}\int_{-\infty}^{\infty} e^{-inx} \sum_{j=-\infty}^{\infty} k(x+2j)e^{it(x+2j\pi)}\,dx$$
$$= \frac{1}{2\pi}\int_{-\infty}^{\infty} e^{itx-inx} k(x)\,dx.$$

Then

$$|b_n| \le \frac{1}{2\pi} \int_{-\infty}^{\infty} |k(x)|\, dx \le c_1,$$

or, after integrating by parts twice,

$$|b_n| \le \frac{1}{2\pi(n-t)^2} \int_{-\infty}^{\infty} |k''(x)|\, dx \le \frac{c_2}{(n-t)^2}$$

(for a suitable choice of $k(x)$). Hence,

$$\sum |b_n| \le c_1 + c_2 \sum_{n=-\infty}^{\infty} \frac{1}{(n-t)^2} = c_3.$$

Applying the statement we have proved previously, it follows that $E \in \mathcal{E}$.

Conversely, suppose that $E \in \mathcal{E}$. We establish the existence of $\delta > 0$ such that for any two Lebesgue density points u, v of E, the relation $|u-v| \ge \pi$ implies $|(u-v)^*| \ge \delta$. Otherwise, indeed, there are sequences $\{u_n\}$, $\{v_n\}$ of density points with $|u_n - v_n| \ge \pi$ but $(u_n - v_n)^* \to 0$. It can be shown that there is a number t such that $[t(u_n - v_n)]^* \not\to 0$. To see this, first let $u_n - v_n = w_n$ be a bounded sequence; then $t = 1/\sup|u_n - v_n|$ is an appropriate value. If w_n is unbounded, then we may assume that $w_n \to \infty$ and, what is more, $w_{n+1}/w_n \to \infty$. We form a sequence t_n as follows:

$$t_0 = 0, \quad t_n = t_{n-1} - \frac{(t_{n-1}w_n)^* - \pi/2}{w_n}, \quad n = 1, 2, \dots .$$

Here $t_n - t_{n-1} = O(1/w_n)$, and by the fast increase of w_n the limit of the sequence t_n exists. Put

$$t = \lim t_n = \sum_{n=1}^{\infty} (t_n - t_{n-1}).$$

Then

$$t - t_n = O\left(\frac{1}{w_{n+1}}\right).$$

Consequently,

$$tw_n = (t - t_n)w_n + t_n w_n = O\left(\frac{w_n}{w_{n+1}}\right) + t_{n-1}w_n - (t_{n-1}w_n)^* + \frac{\pi}{2},$$

and, since $t_{n-1}w_n - (t_{n-1}w_n)^*$ is a multiple of 2π,

$$(tw_n)^* = O\left(\frac{w_n}{w_{n+1}}\right) + \frac{\pi}{2} = \frac{\pi}{2} + o(1) \not\to 0.$$

So, this t is suitable.

Using this t, we form the function

$$a(x) = (e^{itx})^*, \quad x \in E^*,$$

and apply the statement proved earlier. Omitting a set of measure 0, $a(x)$ can be extended to a 2π-periodic continuous function (even a function with absolutely convergent Fourier series), to be denoted by $b(x)$. Since

$$\left(u_n - \frac{1}{n}, u_n + \frac{1}{n}\right) \cap E,$$

as well as its *, has positive measure, and similar statements hold for v_n, after the omission of the set of measure 0 mentioned above, there remain some points in these sets. Let u'_n, v'_n be such points. Then $u'_n - u_n \to 0$, $v'_n - v_n \to 0$, hence $(u'_n - v'_n)^* \to 0$, and $a\left(u'^*_n\right) = b\left(u'^*_n\right)$, $a\left(v'^*_n\right) = b\left(v'^*_n\right)$. Therefore, by the continuity of $b(x)$, it follows that $a\left(u'^*_n\right) - a\left(v'^*_n\right) \to 0$. But

$$a\left(u'^*_n\right) = e^{itu'_n}, \quad a\left(v'^*_n\right) = e^{itv'_n},$$

whence

$$0 = \lim[t(u'_n - v'_n)]^* = \lim[t(u_n - v_n)]^*,$$

and this contradicts the construction of t. Thus δ exists.

For each density point of E, consider its neighborhood of radius $\delta/4$. Their union S is open and almost covers E. For any two points u_1 and v_1 of S, the relation $|(u_1 - v_1)^*| < \delta/4$ implies $|u_1 - v_1| < \delta/4$; otherwise (assuming that $\delta < \pi$), $|u_1 - v_1| > (3/2)\pi$ would follow, and by the construction of S there would exist density points u and v in E with $|u - u_1| < \delta/4$, $|v - v_1| < \delta/4$, hence $|u - v| \geq \pi$, while $|(u - v)^*| < |(u_1 - v_1)^*| + 2\cdot\delta/4 < \delta$, which contradicts the property of δ.

Thus, the closure of S cannot have two congruent points modulo 2π. Hence, it also follows that the measure of S does not exceed 2π, and since each component of the open set S has length greater than or equal to $\delta/2$, the number of the components must be finite, that is, the closure of S is a finite union of closed intervals. Consequently, $E \in \mathcal{F}$. □

Problem M.18. *Show that any two intervals $A, B \subseteq \mathbb{R}$ of positive lengths can be countably disected into each other, that is, they can be written as countable unions $A = A_1 \cup A_2 \cup \dots$ and $B = B_1 \cup B_2 \cup \dots$ of pairwise disjoint sets, where A_i and B_i are congruent for every $i \in \mathbb{N}$.*

Solution 1. In $\mathbb{R}$, consider the equivalence relation

$$\sim = \left\{(x, y) \in \mathbb{R}^2 : x - y \in \mathbb{Q}\right\},$$

which induces the disjoint classification $\mathbb{R} = \cup_{\gamma\in\Gamma} Q_\gamma$. Obviously, each equivalence class Q_γ is dense in $\mathbb{R}$, so $A \cap Q_\gamma$ and $B \cap Q_\gamma$ are countably infinite sets. Consequently, assuming the axiom of choice, there exists a bijection

$$\phi_\gamma : A \cap Q_\gamma \longrightarrow B \cap Q_\gamma$$

between them for each γ. Then $\phi_\gamma(x) \sim x$ for every $\gamma \in \Gamma$, $x \in A \cap Q_\gamma$.

Let $\mathbb{Q} = \{q_1, q_2, \dots\}$ be a numbering of the rationals, and consider the sets

$$\begin{aligned} A_i &= \bigcup_{\gamma\in\Gamma} \{x \in A \cap Q_\gamma : \phi_\gamma(x) - x = q_i\}, \\ B_i &= \bigcup_{\gamma\in\Gamma} \{\phi_\gamma(x) : x \in A \cap Q_\gamma,\, \phi_\gamma(x) - x = q_i\}, \end{aligned}$$

for $i \in \mathbb{N}$. From the definitions, we see that

$$A = \bigcup_{i=1}^{\infty} A_i \quad \text{and} \quad B = \bigcup_{i=1}^{\infty} B_i$$

are partitions, and $B_i = A_i + q_i$ for every $i \in \mathbb{N}$. $\square$

Solution 2. Denote by σ the relation that holds between the subsets of $\mathbb{R}$ if and only if they can be countably dissected into each other. It is easy to verify that σ is a translation-invariant equivalence relation satisfying

$$\left(\bigcup_{n=1}^{\infty} S_n\right) \sigma \left(\bigcup_{n=1}^{\infty} T_n\right)$$

if S_n and T_n, respectively, are pairwise disjoint sets ($n \in \mathbb{N}$), and $S_n \sigma T_n$ for every n.

We now observe that any interval is the disjoint union of a countably infinite number of open intervals and a countably infinite set. So, by the above remarks, we may assume that $A =]0,a[$, $B =]0,b[$, $0 < a \le b$.

In $\mathbb{R}$, we now consider the equivalence relation

$$\sim = \{(x,y) \in \mathbb{R}^2 : x - y \in \mathbb{Q}\}.$$

Its equivalence classes are dense in $\mathbb{R}$, so — assuming the axiom of choice — there exists a set $X \subset]0, a/3[$ containing exactly one element of each class. Put

$$Q_A = \mathbb{Q} \cap]0, \frac{2a}{3}[, \quad Q_B = \mathbb{Q} \cap]0, b - \frac{a}{3}[$$

and

$$A^* = \bigcup_{q\in Q_A} (X+q), \quad B^* = \bigcup_{q\in Q_B} (X+q),$$

where $X + q$ stands for the set X translated by the number $q \in \mathbb{R}$. Obviously, A^* and B^* are countably infinite, disjoint unions of sets congruent with X, so $A^* \sigma B^*$. One also easily verifies the inclusions

$$\left[\frac{a}{3}, \frac{2a}{3}\right] \subset A^* \subset A, \quad \left[\frac{a}{3}, b - \frac{a}{3}\right] \subset B^* \subset B.$$

Put

$$A^- = \left]0, \frac{a}{3}\right[\setminus A^*, \quad A^+ = \left]\frac{2a}{3}, a\right[\setminus A^*,$$
$$B^- = \left]0, \frac{a}{3}\right[\setminus B^*, \quad B^+ = \left]b - \frac{a}{3}, b\right[\setminus B^*.$$

A simple calculation shows that $B^- = A^-$ and $B^+ = A^+ + (b-a)$. Consequently, the partitions

$$A = A^- \cup A^* \cup A^+ \quad \text{and} \quad B = B^- \cup B^* \cup B^+$$

prove the statement. □

Remarks.

1. A direct consequence of the statement is the significant fact that there is no σ-additive, translation-invariant measure defined on all subsets of $\mathbb{R}$ for which the unit interval has measure 1.

2. The first proof can immediately be applied to subsets with nonvoid interior of nondiscrete, separable, Hausdorff topological Abelian groups.

3. By the usual tools of measure theory, the statement can be extended to Lebesgue-measurable sets as follows. Any sets $A, B \subseteq \mathbb{R}$ of positive Lebesgue measures can almost be countably dissected into each other, that is, they have partitions

$$A = \bigcup_{i=0}^{\infty} A_i \quad \text{and} \quad B = \bigcup_{i=0}^{\infty} B_i$$

such that A_0 and B_0 have measure 0, while A_i is congruent with B_i, for every $i \in \mathbb{N}$.

4. The assertion of the previous remark cannot be sharpened by deleting the sets of measure 0, since — as noticed by a contestant — a set of first Baire category cannot be countably dissected into a set of second Baire category; however, among sets of positive Lebesgue measure, there are examples of this kind.

Problem M.19. *Let $H \subset \mathbb{R}$ be a bounded, measurable set of positive Lebesgue measure. Prove that*

$$\liminf_{t \to 0} \frac{\lambda((H+t) \setminus H)}{|t|} > 0,$$

where $H + t = \{x + t : x \in H\}$ and λ is the Lebesgue measure.

Solution. We show that

$$\liminf_{t \to 0} \frac{\lambda((H+t) \setminus H)}{|t|} \geq 1.$$

Lemma. For any $0 < \varepsilon < 1$, there is an interval $[a, b)$ such that $\lambda(H \cap [a, b)) > (1 - \varepsilon)(b - a)$.

Proof. Assume that the conclusion of the lemma is false. Let us cover H by a countably infinite number of intervals (closed from the left), so that the sum of the lengths of the intervals is less than $\lambda(H)/(1 - \varepsilon)$. By the definition of the Lebesgue measure, this is possible. Let the ith interval be $[a_i, b_i)$. Then by the indirect assumption,

$$\lambda(H \cap [a_i, b_i)) \leq (1 - \varepsilon)(b_i - a_i)$$

and therefore

$$\lambda(H) \leq \sum_{i=1}^{\infty} \lambda(H \cap [a_i, b_i)) \leq (1 - \varepsilon) \sum_{i=1}^{\infty} (b_i - a_i) < \lambda(H),$$

which is a contradiction. The proof of the lemma is complete.

Let $0 < \varepsilon < 1$, and let the interval $[a, b)$ be such that

$$\lambda(H \cap [a, b)) > (1 - \varepsilon)(b - a).$$

Let $0 < t < b - a$ and $n = [(b - a)/t] + 1$. Then

$$(1 - \varepsilon)(b - a) < \lambda(H \cap [a, b)) \leq \sum_{k=1}^{n} \lambda(H \cap [a + (k - 1)t, a + kt)).$$

Hence, there exists an integer k, $1 \leq k \leq n$, such that

$$\lambda(H \cap [a + (k - 1)t, a + kt)) \geq \frac{(1 - \varepsilon)(b - a)}{n} \geq \frac{(1 - \varepsilon)(b - a)}{\frac{b-a}{t} + 1} = \frac{(1 - \varepsilon)t}{1 + \frac{t}{b-a}}.$$

Put

$$A_i = H \cap [a + (k + i - 1)t, a + (k + i)t) \qquad (i = 0, 1, 2, \dots).$$

Since, H being bounded, A_i is empty if i is sufficiently large, we have

$$\lambda((H + t) \setminus H) \geq \sum_{i=0}^{\infty} \lambda((A_i + t) \setminus A_{i+1}) \geq \sum_{i=0}^{\infty} (\lambda(A_i) - \lambda(A_{i+1}))$$
$$= \lambda(A_0) = \lambda(H \cap [a + (k - 1)t, a + kt)) > \frac{(1 - \varepsilon)t}{1 + \frac{t}{b-a}}.$$

Quite similarly,

$$\lambda((H - t) \setminus H) \geq \frac{(1 - \varepsilon)t}{1 + \frac{t}{b-a}}.$$

It follows that

$$\liminf_{t \to 0} \frac{\lambda((H + t) \setminus H)}{|t|} \geq \lim_{t \to 0} \frac{1 - \varepsilon}{1 + \frac{t}{b-a}} = 1 - \varepsilon.$$

But this is true for every $0 < \varepsilon < 1$. Thus,

$$\liminf_{t \to 0} \frac{\lambda((H + t) \setminus H)}{|t|} \geq 1. \quad \square$$

3.6 NUMBER THEORY

Problem N.1. *Let f and g be polynomials with rational coefficients, and let F and G denote the sets of values of f and g at rational numbers. Prove that $F = G$ holds if and only if $f(x) = g(ax + b)$ for some suitable rational numbers $a \neq 0$ and b.*

Solution. By a "polynomial" we shall always mean a polynomial with rational coefficients, and by the "range" of a polynomial we shall mean the set of its values assumed at rational numbers.

We use two classifications of polynomials. We write $f \sim g$ if the ranges of these polynomials coincide, and $f \approx g$ if $f(x) = g(ax + b)$ for all x with suitable rational numbers $a \neq 0$, b. These are clearly equivalence relations, and they are compatible with multiplications by a constant, that is, if $f \sim g$ or $f \approx g$, then $cf \sim cg$ or $cf \approx cg$, respectively ($c \neq 0$ rational). Our aim is to show that these relations are identical. One implication is clear: if $f \approx g$, then $f \sim g$ holds obviously.

The converse will be proved in several steps.

1. It is sufficient to prove the statement for polynomials with integral coefficients. Indeed, assume $f \sim g$, and let c be an integer such that the coefficients of cf and cg are all integers. We have $cf \sim cg$, thus $cf \approx cg$, assuming the integer case, which yields $f \approx g$ as wanted.
2. If $f \sim g$ are polynomials with integral coefficients, we show that they must be of equal degree. Indeed, assume that

$$f(x) = ax^n + \dots, \qquad g(x) = bx^k + \dots,$$

where $a \neq 0$, $b \neq 0$, and $n \leq k$. Choose a prime $p \nmid ab$. We have $f(1/p) = c/p^n$, where $p \nmid c$. By assumption, g also takes on this value at some number u/v with $(u, v) = 1$. Now if $p \nmid v$, then p cannot divide the denominator of $g(u/v)$; consequently, p must divide v. Now, by $(p, b) = 1$, the exponent of p in the denominator of $b(u/v)^k$ is higher than in any other term of $g(u/v)$, thus the denominator (in the reduced form) of $g(u/v)$ contains p with an exponent $\geq k$. This yields $n \geq k$, hence $n = k$ as asserted.

3. It is sufficient to prove the statement for polynomials with integral cofficients where at least *one* of the leading cefficients is 1. Indeed, let $f \sim g$, $f(x) = ax^n + \dots$. Then we have

$$a^{n-1} f\left(\frac{x}{a}\right) \sim a^{n-1} f(x) \sim a^{n-1} g(x),$$

and here $a^{n-1} f(x/a)$ still has integral coefficients while its leading coefficient became 1. By assumption, we conclude $a^{n-1} f(x/a) \approx a^{n-1} g(x)$, which yields $f \approx g$.

4. Let $f \sim g$ be polynomials with integral coefficients, $f(x) = x^n + \dots$, $g(x) = bx^n + \dots$. We claim that every prime divisor of b has an exponent

$\geq n$. To show this, take a $p|b$ and consider the polynomials

$$f_1(x) = p^{n(n-1)} f\left(\frac{x}{p^{n-1}}\right) \sim p^{n(n-1)} f\left(\frac{x}{p^n}\right) \sim g_1(x)$$
$$= p^{n(n-1)} g\left(\frac{x}{p^n}\right).$$

We know that all coefficients of f_1 as well as all but the leading coefficient of g_1 are integral, and the leading coefficient of f_1 is 1. Hence every value of f_1 has the following property: if (in the reduced form) its denominator is divisible by p, it is divisible by p^n. By $f_1 \sim g_1$ the same must hold for g_1. Now $g_1(1) = (b/p^n) + c$ with an integer c, and by $p|b$ the exponent of p in the denominator is strictly less than n, thus it must be 0, that is, $p^n|b$.

5. It is sufficient to prove the statement for polynomials with integral cofficients where *both* leading cefficients are 1. Let $f \sim g$ be polynomials with integral coefficients, $f(x) = x^n + \ldots$, $g(x) = bx^n + \ldots$. Let c_n be the largest nth power dividing b. Consider the polynomials

$$f_1(x) = c^{n(n-1)} f\left(\frac{x}{c^{n-1}}\right) = x^n + \ldots,$$
$$g_1(x) = c^{n(n-1)} g\left(\frac{x}{c^{n-1}}\right) = \frac{b}{c^n} x^n + \ldots.$$

We have $f_1 \sim g_1$. By step 4, the exponent of any prime in the leading coefficient of g_1 must be either 0 or $\geq n$, and by the definition of c this coefficient must be 1 or -1. We can exclude the case -1 for even n, since then one domain would be bounded from below and the other from above. Finally, if n is odd, then replacing c by $-c$ we can change a -1 into 1. So if the statement is true under this restriction, then we obtain $f_1 \approx g_1$, which yields $f \approx g$.

6. We can also assume that the coefficient of x^{n-1} is 0 in both polynomials. Indeed, if

$$f(x) = x^n + ax^{n-1} + \cdots \sim g(x) = x^n + bx^{n-1} + \ldots,$$

then the polynomials

$$f_1(x) = n^n f\left(\frac{x}{n} - a\right) \sim g_1(x) = n^n g\left(\frac{x}{n} - a\right)$$

have this property, and $f_1 \approx g_1$ again implies $f \approx g$.

7. Combining the previous arguments, we find that to solve the problem it is sufficient to prove the following statement:

Statement. If $f \sim g$ are polynomials of the form

$$f(x) = x^n + ax^{n-2} + \ldots, \qquad g(x) = x^n + bx^{n-2} + \ldots$$

with integral coefficients, then $f \approx g$.

Proof. Observe that, since the leading coefficient is 1, the only rational numbers where f and g assume integral values are the integers.

Since $f(x+1)-g(x)=nx^{n-1}+\dots$, we have $g(x)<f(x+1)$ for large x, and similarly we obtain $g(x)>f(x-1)$. Thus, for a large positive integer x, the only positive rational y that can satisfy $f(y)=g(x)$ is $y=x$. If n is even, then there can also be a negative y, and similarly we find that its only possible value is $y=-x$. This means that one of the equations $f(x)=g(x)$ and $f(-x)=g(x)$ has infinitely many solutions; hence it must be an identity. This concludes the proof. □

Problem N.2. *Show that*

$$\prod_{1\le x<y\le\frac{p-1}{2}}(x^2+y^2)\equiv(-1)^{\left[\frac{p+1}{8}\right]}\pmod p$$

for every prime $p\equiv 3\pmod 4$. *([.] is integer part.)*

Solution 1. Write $p=4k+3$. Consider all the sums x^2+y^2, $1\le x,y\le 2k+1$. Let r denote the number of those pairs x,y for which this sum $\equiv 1$.

First, we show that the number of those pairs for which $x^2+y^2\equiv -1\pmod p$ is $r+1$. Since every quadratic residue has a unique representation in the form x^2, $1\le x\le 2k+1$, while every nonresidue has a unique representation in the form $-y^2$, the solutions of $x^2+y^2=x^2-(-y^2)\equiv 1$ count the consecutive numbers in the form (nonresidue, residue) within the sequence $1,2,\dots,p-2,p-1$. Similarly, the pairs with $x^2+y^2\equiv -1$ correspond to pairs of type (residue, nonresidue). Since this sequence starts with a residue and (recall that $p\equiv -1\pmod 4$) ends with a nonresidue, the second case must happen $r+1$ times.

Next, we prove that the number of solutions of $x^2+y^2\equiv a^2\pmod p$, $1\le x,y\le 2k+1$, is r for every a. Indeed, the mapping $x\equiv\pm ax_1$, $y\equiv\pm ay_1$, where the signs are chosen so that x,x_1,y,y_1 are all in $[1,2k+1]$, provides a one-to-one correspondence between solutions of $x^2+y^2\equiv a^2$ and $x_1^2+y_1^2\equiv 1$. Similarly, we obtain that the number of solutions of $x^2+y^2\equiv -1$ is $r+1$.

The above observations mean that these sums x^2+y^2 represent every quadratic residue r times and every nonresidue $r+1$ times. Since the total number of these pairs is $((p-1)/2)^2$ while the number of residues and the number of nonresidues are both equal to $(p-1)/2$, we find $r=(p-3)/4=k$.

Now, the product we want to compute is not over all these pairs but only over those with $x<y$. Consider a quadratic residue a. We know that the total number of solutions of $x^2+y^2\equiv a$ is k. If among them there are u with $x<y$, v with $x=y$, and w with $x>y$, then $u=w$ by symmetry and v is 0 or 1, hence $u=[k/2]$. Analogously, for a nonresidue the number of solutions with $x<y$ is $[(k+1)/2]$.

The product of all quadratic residues is

$$1^2\cdot 2^2\cdot\dots\cdot\left(\frac{p-1}{2}\right)^2\equiv\left(\left(\frac{p-1}{2}\right)!\right)^2\equiv 1\pmod p,$$

and the product of nonresidues is

$$(-1^2)\cdot(-2^2)\cdot\ldots\cdot\left(-\left(\frac{p-1}{2}\right)^2\right) \equiv (-1)^{\frac{p-1}{2}}\left(\left(\frac{p-1}{2}\right)!\right)^2$$
$$\equiv -1 \pmod p.$$

Hence, the original product is

$$1^{[k/2]}(-1)^{\left[\frac{k+1}{2}\right]} = (-1)^{\left[\frac{p-1}{8}\right]}. \quad \square$$

Solution 2. Denote this product by P. In the product, all sums of pairs of quadratic residues are multiplied. Let g be a primitive root mod p, and put $h = g^2$. The numbers $1, h, h^2, \ldots, h^{2k}$ represent every quadratic residue once; thus we have

$$P \equiv \prod_{0\le i<j\le 2k} (h^i + h^j) \qquad \pmod p.$$

This implies

$$P \cdot \prod_{0\le i<j\le 2k} (h^i - h^j) \equiv \prod_{0\le i<j\le 2k} (h^{2i} - h^{2j}) \qquad \pmod p.$$

Here each product is the value of a nonzero Vandermonde determinant. Thus, this equation can be rewritten as

$$P \cdot V(1, h, h^2, \ldots, h^{2k}) \equiv V(1, h^2, \ldots, h^{4k}) \qquad \pmod p.$$

Since $h^{2k+1} = g^{4k+2} \equiv 1 \pmod p$, the generators of the second Vandermonde determinant are congruent to

$$1, h^2, \ldots, h^{2k}, h, h^3, \ldots, h^{2k-1}.$$

Thus, the second determinant can be obtained from the first by $k+(k-1)+\cdots+2+1 = k(k+1)/2$ transpositions of rows. By the familiar properties of determinants, this means

$$P \equiv (-1)^{\frac{k(k+1)}{2}} \equiv (-1)^{\left[\frac{k+1}{2}\right]} \equiv (-1)^{\left[\frac{p+1}{8}\right]} \qquad \pmod p. \quad \square$$

Problem N.3. *Let p be a prime and let*

$$l_k(x,y) = a_k x + b_k y \quad (k = 1, \ldots, p^2),$$

be homogeneous linear polynomials with integral coefficients. Suppose that for every pair (ξ, η) of integers, not both divisible by p, the values $l_k(\xi, \eta)$, $1 \le k \le p^2$, represent every residue class mod p exactly p times.

Prove that the set of pairs $\{(a_k, b_k) : 1 \leq k \leq p^2\}$ *is identical mod* p *with the set* $\{(m, n) : 0 \leq m, n \leq p-1\}$.

Solution 1. Assume that the statement does not hold. Then there are numbers $i \neq j$ such that $a_i \equiv a_j$ and $b_i \equiv b_j$ (every congruence is meant mod p). Consider the number of those triplets (k, x, y), $(x, y) \neq (0, 0)$ that satisfy

$$l_k(x, y) \equiv l_i(x, y).$$

By the assumption, for every fixed pair $(x, y) \neq (0, 0)$, the number of solutions in k is p; thus the total number of these triplets is $p(p^2 - 1)$.

Now consider the solutions in x, y for a fixed k. If $k = i$ or j, this is an identity, which means $2(p^2 - 1)$ solutions. For any other k it is easy to see that the number of solutions is at least $p - 1$. This means that the total number of solutions is at least

$$(p^2 - 2)(p - 1) + 2(p^2 - 1) = p(p^2 - 1) + p(p - 1) > p(p^2 - 1),$$

a contradiction. $\square$

Solution 2. It is sufficient to prove that for every u and v there is exactly one k with $a_k \equiv u$, $b_k \equiv v$.

The uniformity of representations implies that for every $(\xi, \eta) \not\equiv (0, 0)$, we have

$$\sum_{k=1}^{p^2} e^{\frac{2\pi i}{p}(a_k \xi + b_k \eta)} = 0.$$

We multiply both sides by $e^{(2\pi i/p)\cdot(u\xi + v\eta)}$ to get

$$\sum_{k=1}^{p^2} e^{\frac{2\pi i}{p}((a_k - u)\xi + (b_k - v)\eta)} = 0,$$

for $(\xi, \eta) \not\equiv (0, 0)$. For $\xi \equiv \eta \equiv 0$, the same sum obviously gives p^2. Now summing these sums for all possible values of ξ and η, including $(0, 0)$, we obtain

$$S = \sum_{\xi=0}^{p-1} \sum_{\eta=0}^{p-1} \sum_{k=1}^{p^2} e^{\frac{2\pi i}{p}((a_k - u)\xi + (b_k - v)\eta)} = p^2.$$

Changing the order of the summations, we get

$$\begin{aligned} S &= \sum_{k=1}^{p^2} \sum_{\xi=0}^{p-1} \sum_{\eta=0}^{p-1} e^{\frac{2\pi i}{p}((a_k - u)\xi + (b_k - v)\eta)} \\ &= \sum_{k=1}^{p^2} \left(\left(\sum_{\xi=0}^{p-1} e^{\frac{2\pi i}{p}(a_k - u)\xi} \right) \left(\sum_{\eta=0}^{p-1} e^{\frac{2\pi i}{p}(b_k - v)\eta} \right) \right) = p^2. \end{aligned}$$

Now, a typical factor in this product vanishes, unless $a_k \equiv u$ (for the first) or $b_k \equiv v$ (for the second). Thus, the whole product is 0, unless $a_k \equiv u$ and $b_k \equiv v$, in which case it is equal to p^2. Since the sum of these products is p^2, this case must happen for exactly one value of k. □

Remarks.

1. The method of the second solution can be applied to prove the following generalization of the problem: Let p be a prime, r a positive integer, and the l_k's homogeneous linear polynomials in n variables of the form

$$l_k(x_1, \ldots, x_n) = a_1^k x_1 + \cdots + a_n^k x_n \qquad (k = 1, \ldots, rp^n).$$

Assume that for every k-tuple $(\xi_1, \ldots, \xi_n)$ of integers, not all divisible by p, the values of $l_k(\xi_1, \ldots, \xi_n)$ represent every residue class $\bmod\, p$ exactly rp^{n-1} times. Then every n-tuple $(m_1, \ldots, m_n)$ is represented $\bmod\, p$ among the n-tuples $(a_1^k, \ldots, a_n^k)$ of coefficients for exactly r values of k.

2. If we relax the requirement of uniform representation to those pairs (ξ, η) where neither ξ nor η is divisible by p, then the following can be asserted:

Assertion. Let $f(m, n)$ denote the number of those pairs (a_k, b_k) that satisfy $a_k \equiv m$, $b_k \equiv n$. Then the values $l_k(\xi, \eta)$ will be uniformly distributed for all of the $p^2 - 2p + 1$ admissible pairs (ξ, η) if and only if $f(m, n)$ can be represented as $f(m, n) = g(m) + h(n)$.

Problem N.4. *Let p be a prime, n a natural number, and S a set of cardinality p^n. Let $\mathbf{P}$ be a family of partitions of S into nonempty parts of sizes divisible by p such that the intersection of any two parts that occur in any of the partitions has at most one element. How large can $|\mathbf{P}|$ be?*

Solution 1. This maximum is $(p^n - 1)/(p - 1)$. Let H be the set to be partitioned, and let $C_1, \ldots, C_k$ be its partitions into sets whose cardinalities are multiples of p, such that any two classes may have at most one common element. Consider an $h \in H$, and let $C_i(h)$ be the class of C_i that contains h. By the assumptions, the sets

$$C_1(h) \backslash \{h\}, \ldots, C_k(h) \backslash \{h\}$$

are pairwise disjoint, and each has at least $p - 1$ elements. Consequently, we have $k(p - 1) \leq p^n - 1$, that is, $k \leq (p^n - 1)(p - 1)$.

Now we prove the corresponding lower estimate. Consider the n-dimensional projective space P_n over a finite field K of p elements, a hyperplane σ in it, and the affine space $P_n' = P_n \backslash \sigma$. Observe that P_n has $(p^{n+1}-1)/(p-1)$ points, σ has $(p^n - 1)/(p - 1)$ and P_n' has p^n.

For an arbitrary point $P \in \sigma$, consider those lines of P_n that contain P but do not lie in σ. If we omit the point P from each such line, the remaining affine lines (as sets of points) form a partition C_P of the affine space P_n' into sets of cardinality p. The number of these partitions is the same as the number of points of σ, that is, $(p^n - 1)/(p - 1)$. □

Solution 2. We keep the upper estimation from the first proof, and present a different construction to show the lower bound.

For H we take the set of all n digit numbers in base p. This set has p^n elements. The partitions will be given in forms of integer-valued functions; $f(x)$ will mean the number of the class containing a given $x \in H$. The functions will be indexed by pairs (j, a), where $j = 0, \dots, n-1$, and for a given value of j, the possible values of a are $a = 0, \dots, p^j - 1$. The number of possibilities for a is p^j; thus the total number of these functions is

$$1 + p + p^2 + \cdots + p^{n-1} = \frac{p^n - 1}{p - 1},$$

as wanted.

We define the function $f_{ja}(x)$ as follows. Let the representations of x and a to base p be

$$x = \xi_0 + \xi_1 p + \cdots + \xi_{n-1} p^{n-1},$$
$$a = \alpha_0 + \alpha_1 p + \cdots + \alpha_{j-1} p^{j-1}.$$

Let $[m]$ denote the (smallest nonnegative) residue of an integer $m \mod p$. Now we put

$$f_{ja}(x) = [\xi_0 + \alpha_0 \xi_j] + [\xi_1 + \alpha_1 \xi_j]p + \cdots + [\xi_{j-1} + \alpha_{j-1}\xi_j]p^{j-1}$$
$$+ \xi_{j+1} p^j + \cdots + \xi_{n-1} p^{n-2}.$$

In the case $j = 0$, $a = 0$, we interpret this as

$$f_{00}(x) = \xi_1 + \xi_2 p + \cdots + \xi_{n-1} p^{n-2}.$$

We show that every class contains p elements. Indeed, given the value of

$$f_{ja}(x) = y = \eta_0 + \cdots + \eta_{n-2} p^{n-2},$$

we have $\xi_k = \eta_{k-1}$ for $k = j+1, \dots, n$, we can choose the value of ξ_j that will give p possibilities, and after fixing ξ_j we can uniquely determine $\xi_0, \dots, \xi_{j-1}$ from the congruences

$$\xi_i + \alpha_i \xi_j \equiv \eta_i \pmod p \qquad (i = 0, \dots, j-1).$$

Next, we have to show that the intersection of two classes has at most one element, that is, the system of equations

$$f_{ja}(x) = y, \qquad f_{j'a'}(x) = y'$$

has at most one solution. Assume first that $j \neq j'$, say $j' < j$. Then we obtain $\xi_j = \eta'_j$ from the second equation, and we already know that $f_{ja}(x)$ and ξ_j determine x uniquely.

Finally, consider the case $j = j'$. We must have $a \neq a'$, say $\alpha_k \neq \alpha'_k$ for some digit $k \leq j-1$. Then the congruences

$$\xi_k + \alpha_k \xi_j \equiv \eta_k \pmod p,$$
$$\xi_k + \alpha'_k \xi_j \equiv \eta'_k \pmod p$$

determine ξ_j, which together with $f_{aj}(x)$ determines x uniquely. □

Problem N.5. *Let f be a complex-valued, completely multiplicative, arithmetical function. Assume that there exists an infinite increasing sequence N_k of natural numbers such that*

$$f(n) = A_k \neq 0 \quad \textit{provided} \quad N_k \leq n \leq N_k + 4\sqrt{N_k}.$$

Prove that f is identically 1.

Solution. We show a slightly stronger statement, where the 4 is replaced by $2+\varepsilon$. Assume that f is (nonzero and) constant on the intervals $I_k = [N_k, N_k + M_k]$, where $M_k = (2+\varepsilon)\sqrt{N_k}$.

First, we prove that $f(n) \neq 0$ for any n. Indeed, if $M_k > n$, then $nx \in I_k$ for some x; hence $f(n)f(x) = f(nx) \neq 0$. Next, we use an interval of constancy to create another. Let f be constant on $I = [N, N+M]$. If for an n we can find an integer x such that $nx, (n+1)x \in I$, then $f(n)f(x) = f(nx) = f((n+1)x) = f(n+1)f(x)$, hence $f(n) = f(n+1)$. Now this condition can be reformulated as

$$\frac{N}{n} \leq x \leq \frac{N+M}{n+1}.$$

Such an integer x can be found if

$$\frac{N+M}{n+1} - \frac{N}{n} \geq 1,$$

or, after rearranging, $n^2 + n(1-M) + N \leq 0$. This means that n lies between the roots of the corresponding quadratic equation, that is, $(M-1)^2 > 4N$ (which shows that our method does not work with $M = (2-\varepsilon)\sqrt{N}$) and

$$\frac{M-1-\sqrt{(M-1)^2-4N}}{2} \leq n \leq \frac{M-1+\sqrt{(M-1)^2-4N}}{2}.$$

If $M = (2+\varepsilon)\sqrt{N}$, then the above interval includes an interval of the form $I' = [c_1 M, c_2 M]$, where $0 < c_1 < c_2$ are constants depending on ε.

Now we repeat this argument for the interval I'. The result is an interval $I'' = [c_3, c_4 M]$ of constancy. As I runs over all the intervals I_k, these intervals I'' cover the whole half-line $[c_3, \infty]$; thus $f(n)$ is constant for $n > c_3$. Now take an arbitrary positive integer m, and select an $n > c_3$. We have $f(n) = f(mn)$, hence $f(m) = 1$ as asserted. □

Remark. The statement will not hold if the $4\sqrt{N_k}$ is reduced to exp $(c\sqrt{\log N_k \log\log\log N_k})$ with a suitably small positive c.

To see this, consider a sequence $I_k = [N_k, N_k + M_k]$ of intervals, $N_k > M_k > N_{k-1} + M_{k-1}$, with the property that every number $n \in I_k$ has a prime factor $p > M_k$. Choose the numbers $A_k \neq 0$ arbitrarily. We claim that there is a completely multiplicative function f that is identically A_k on I_k; if not, every A_k is 1, and then our function will not be identically 1.

We construct our function recursively. Assume that $f(p)$ is fixed for every prime $p \le N_{k-1} + M_{k-1}$; we define $f(p)$ up to $N_k + M_k$. For every $n \in I_k$, let $p_n > M_k$ be the largest prime divisor of n. These primes are distinct, since for $n \neq n'$ we have $(n, m) \le |n' - n| \le M_k$. Choose $f(p)$ arbitrarily for every prime that does not occur among the p_n's, and then set $f(p_n)$ to achieve $f(n) = A_k$.

Now we have to find such a sequence of intervals. Assume that $I_1, \dots,$ I_{k-1} are given. Take a large N, say $N > \exp N_{k-1}$; we try to find an I_k in $[N/2, N]$.

Let $R(x, y)$ denote the number of those integers $n \le x$ whose prime factors are all less than y. If M is such that $M(R(N, M) + 1) < N/2$, then the interval $[N/2, N]$ contains a subinterval of length M that is free of these numbers. This interval can serve as our I_k. The feasibility of taking $M = \exp(c\sqrt{\log N_k \log\log\log N_k})$ follows from Rankin's inequality:

$$R(x, y) < x \exp\left(-\frac{\log\log y}{\log y}\log x + O(\log\log y)\right).$$

Problem N.6. *If c is a positive integer and p is an odd prime, what is the smallest residue (in absolute value) of*

$$\sum_{n=0}^{\frac{p-1}{2}} \binom{2n}{n} c^n \pmod{p}?$$

Solution. Let $q = (p-1)/2$. With this notation, we have $2j+1 \equiv -2(q-j)$ for every j. Consequently, we have

$$\begin{aligned}\binom{2n}{n} = \frac{(2n)!}{n!^2} &= \frac{2^n \cdot n! \cdot 1 \cdot 3 \cdot \ldots \cdot (2n-1)}{n!^2} \\ &= \frac{2^n \cdot 1 \cdot 3 \cdot \ldots \cdot (2n-1)}{n!} \\ &\equiv \frac{2^n \cdot (-2)^n \cdot q \cdot (q-1) \cdot \ldots \cdot (q-n+1)}{n!} \\ &= (-4)^n \binom{q}{n}.\end{aligned}$$

Hence,

$$\sum_{n=0}^{q} \binom{2n}{n} c^n \equiv \sum_{n=0}^{q} \binom{q}{n} (-4c)^n = (1-4c)^q.$$

This is congruent to 0 if $1 - 4c$ is divisible by p. Otherwise, it is 1 if $1 - 4c$ is a quadratic residue mod p and -1 if it is a nonresidue. □

Problem N.7. *Find a constant $c > 1$ with the property that, for arbitrary positive integers n and k such that $n > c^k$, the number of distinct prime factors of $\binom{n}{k}$ is at least k.*

Solution. Write $t = [1, 2, \ldots, n]$. We shall show that the number of prime factors of $\binom{n}{k}$ is at least k for $n \geq t + k$.

Take a prime $p \leq k$, and let $p^s \leq k < p^{k+1}$. The exponent of p in the decomposition of $k!$ is $\sum_{i=1}^{s}[k/p^i]$, and in t it is s. Among the numbers n, $\ldots$, $n-k+1$, there are at least $[k/p^i]$ multiples of p^i. For $i \leq s$, these also enter the decomposition of the corresponding $(n-j, t)$; thus the exponent of p in the product $(n, t) \cdot (n-1, t) \cdot \ldots \cdot (n-k+1, t)$ is at least as high as in $k!$. Since this holds for every p, we conclude that

$$k! | (n, t) \cdot (n-1, t) \cdot \ldots \cdot (n-k+1, t).$$

This implies that

$$\prod_{i=0}^{k-1} \frac{n-i}{(n-i, t)} \,\Big|\, \binom{n}{k}.$$

The factors on the left side are pairwise coprime, since $(n-i, n-j)|(i-j)|t$ for $i \neq j$, and if $n \geq t + k$, they are all larger than 1. Taking a prime factor of each, we find k different prime factors of $\binom{n}{k}$.

Since $t = \exp(k + o(k))$ by the prime number theorem, every $c > e$ satisfies the requirement of the problem for large k. If we want to exhibit a concrete value of c that works for every k, we can argue as follows. Applying Chebyshev's estimate,

$$\prod_{p \leq x} p < 4^x,$$

we obtain

$$t = \prod_{p^s \leq k < p^{s+1}} p^s \leq \prod_{p \leq \sqrt{k}} k \prod_{\sqrt{k} < p \leq k} p < k^{\sqrt{k}-1} 4^k.$$

Therefore,

$$t + k \leq k^{\sqrt{k}} 4^k = \left(4k^{1/\sqrt{k}}\right)^k < 9^k,$$

since the function $x^{1/\sqrt{x}}$ assumes its maximum at e^2, and its value is $e^{2/e} < 2.1$. □

Remark. The value $c = 9$ we gave can be further reduced with a little extra effort; one of the contestants gave $c = 4.3$. It seems to be difficult to decide the improvability of $c = e + \varepsilon$ for large k.

If the "prime k-tuple conjecture" is true, then there are arbitrary large values of n for which $\binom{n}{k}$ has exactly k prime factors.

Problem N.8. *Let $f(n)$ be the largest integer k such that n^k divides $n!$, and let $F(n) = \max_{2\le m\le n} f(n)$. Show that*

$$\lim_{n\to\infty} \frac{F(n)\log n}{n\log\log n} = 1.$$

Solution. First, we find an upper estimate for $f(n)$. Let $n = \prod_{i=1}^k p_i^{\alpha_i}$ be the prime factorization of n. Since $n^{f(n)}|n!$, we have

$$\alpha_i f(n) \le \sum_{j=1}^{\infty} \left[\frac{n}{p_i^j}\right] < \frac{n}{p_i - 1},$$

that is,

$$\alpha_i \log p_i \le \frac{n}{f(n)} \frac{\log p_i}{p_i - 1},$$

for all i. Summing these inequalities, we obtain

$$\log n = \sum \alpha_i \log p_i \le \frac{n}{f(n)} \sum \frac{\log p_i}{p_i - 1}.$$

Let $q_1 < q_2 < \dots$ be the sequence of all primes. Since $(\log p)/(p-1)$ is decreasing, we have

$$\sum_{i=1}^{k} \frac{\log p_i}{p_i - 1} \le \sum_{i=1}^{k} \frac{\log q_i}{p_i - 1} = \log k + O(1).$$

Combining the last two inequalities, we get

$$f(n) \le (1 + o(1)) \frac{n\log k}{\log n}.$$

On the other hand,

$$n = p_1^{\alpha_1} \dots p_k^{\alpha_k} \ge q_1 \dots q_k \ge 2^k.$$

Consequently, $k \ll \log n$, $\log k \le \log\log n + O(1)$, and the previous estimate of $f(n)$ yields

$$f(n) \le (1 + o(1)) \frac{n\log\log n}{\log n},$$

as wanted.

Now we construct a number $m \le n$ with a large value of $f(m)$.

Define m_1 by $(m_1+1)! \le n < (m_1+2)!$. We have $m_1 \sim (\log n)/(\log\log n)$ as $n \to \infty$. Let p be the largest prime up to $(n/m_1!)^{1/3}$, and put $m = p^3 m_1!$. We have $m \le n$ obviously, and, since $n/m_1! > m_1 \to \infty$, we have $p \sim (n/m_1!)^{1/3}$, that is, $m \sim n$.

Since $(m_1!)^{m/m_1}|m!$, the exponent of any prime other than p is at least m/m_1 times as high in $m!$ as in m. We have to compare the exponents of p. The exponent of p in $m_1!$ is $\sum_{i=1}^{\infty}\left[m_1/p^i\right] < m_1/(p-1)$, hence in m it is less than $m_1/(p-1)+3$. On the other hand, the exponent of p in $m!$ is $\sum_{i=1}^{\infty}\left[m/p^i\right] \geq [m/p] > (m/p)-1$. Therefore, the quotient of these exponents is at least

$$\frac{\frac{m}{p}-1}{\frac{m_1}{p-1}+3} \sim \frac{m}{m_1},$$

because $p \to \infty$, but

$$p \leq (n/m_1!)^{1/3} < ((m_1+1)(m_1+2))^{1/3} = o(m).$$

Consequently, we have

$$F(n) \geq f(m) \geq (1+o(1))\frac{m}{m_1} = (1+o(1))\frac{n}{m_1} = (1+o(1))\frac{n\log\log n}{\log n},$$

as asserted. $\square$

Remark. It can be proved that

$$F(n) = \frac{n(\log\log n - \log\log\log n)}{\log n} + O\left(\frac{n}{\log n}\right).$$

Problem N.9. *Prove that a necessary and sufficient condition for the existence of a set $S \subset \{1,\ldots,n\}$ with the property that the integers $0,1,\ldots,n-1$ all have an odd number of representations in the form $x-y$, $x,y \in S$, is that $(2n-1)$ has a multiple of the form $2\cdot 4^k-1$.*

Solution. Let S be a set with this property, and write

$$f(x) = \sum_{a\in S} x^{a-1}, \quad g(x) = \sum_{a\in S} x^{n-a},$$

which we regard as polynomials over $GF(2)$. Then we have

$$f(x) = x^{n-1}g(\frac{1}{x}), \tag{1}$$

and

$$f(x)g(x) = \sum_{j=0}^{2n-2} x^j \sum_{\substack{a,b\in S\\ a-1+n-b=j}} 1 = 1+x+x^2+\cdots+x^{2n-2}, \tag{2}$$

by our assumptions on S. Conversely, if f and g are polynomials over $GF(2)$ that satisfy (1) and (2), then the set

$$S = \{a:\ \text{the coefficient of } x^{a-1} \text{ in } f \text{ is } 1\}$$

has the desired property.

Now put

$$1 + x + x^2 + \cdots + x^{2n-2} = p_1(x) \dots p_l(x), \tag{3}$$

where the factors $p_1, \dots, p_l$ are irreducible over $GF(2)$. With a suitable choice of the suffices, we have

$$f = p_1 \dots p_r, \quad g = p_{r+1} \dots p_l$$

by (2).

Observe that the roots of $1 + x + x^2 + \cdots + x^{2n-2}$ are the $(2n-1)$th roots of unity except 1. Furthermore, we have

$$(1 + x + x^2 + \cdots + x^{2n-2})' = 1 + x^2 + \cdots + x^{2n-4}$$

and

$$(1 + x + x^2 + \cdots + x^{2n-2}) - (x + x^2)(1 + x^2 + \cdots + x^{2n-4}) = 1,$$

and hence this polynomial has no multiple roots. Since the roots of f are the reciprocals of the roots of g by (1), no p_i can have two roots that are reciprocals of each other. Conversely, if no p_i has reciprocal roots, then the polynomials $p_1, \dots, p_l$ can be coupled so that the roots of any p_i are the reciprocals of the roots of its mate. We now multiply one element from each pair to get the polynomial f and the others to get g; these obviously satisfy (1) and (2). Thus, *such a set S exists if and only if for every root α of $1+x+x^2+\cdots+x^{2n-2}$, α and $1/\alpha$ belong to different irreducible factors.*

We construct this decomposition (3).

Lemma. Consider the permutation of the roots of the equation $x^{2n-1} - 1 = 0$ given by the operation of squaring. Let $C_1, \dots, C_l$ be the cycles of this permutation. The irreducible factors of $x^{2n-1} - 1$ are the polynomials

$$\prod_{\xi \in C_i} (x - \xi), \qquad i = 1, \dots, l.$$

We prove this lemma at the end of the solution; see also *E. R. Berlekamp, Algebraic Coding Theory, McGraw–Hill, New York, 1968*, Chapter 6.

Hence, our condition for the existence of the set S is satisfied if and only if the roots α and $1/\alpha$ cannot be tranformed into each other by repeated squarings for any $\alpha \neq 1$, that is,

$$\alpha^{2^k+1} \neq 1$$

for all $k \geq 0$ and $(2n-1)$th roots of unity $\alpha \neq 1$. This can be reformulated as

$$(x^{2n-1} - 1, x^{2^k+1} - 1) = x - 1$$

for all k. Since

$$(x^u - 1, x^v - 1) = x^{(u,v)} - 1,$$

the set S exists if and only if

$$(2n - 1, 2^k + 1) = 1 \tag{4}$$

for all k. Now let d be the smallest exponent for which $2n-1|2^d-1$. If d is even, then

$$2n-1|(2^{d/2}-1)(2^{d/2}+1),$$

but $2n-1\nmid(2^{d/2}-1)$, and thus $(2n-1,2^{d/2}+1)>1$, which contradicts (4). Conversely, if d is odd, then we have

$$(2n-1,2^k+1)=(2^d-1,2^k+1)=1.$$

Consequently, S exists if and only if d is odd, which is clearly equivalent to the condition given in the problem.

Proof of the lemma. Let

$$\phi(x)=\sum_{k=0}^{N}a_kx^k$$

be a polynomial over any field of characteristic 2. We have

$$(\phi(x))^2=\sum_{k=0}^{N}a_k^2x^{2k}.$$

Thus, the equation

$$(\phi(x))^2=\phi(x^2)$$

holds identically if and only if $a_k^2=a_k$ for all k, that is, every a_k is 0 or 1. On the other hand, the above equality is clearly equivalent to the assumption that for every root α of ϕ, α^2 is a root as well. □

Problem N.10. *Prove that the set of rational-valued, multiplicative arithmetical functions and the set of complex rational-valued, multiplicative arithmetical functions form isomorphic groups with the convolution operation $f\circ g$ defined by*

$$(f\circ g)(n)=\sum_{d|n}f(d)g(\frac{n}{d}).$$

(We call a complex number complex rational, if its real and imaginary parts are both rational.)

Solution. Let Q be either the field of rational numbers or the field of complex rational numbers, and let G_Q denote the group of Q-valued multiplicative functions. It is well known that G_Q is an Abelian group. Now we prove that for every natural number n and every $f\in G_Q$,the equation

$$\underbrace{g\circ g\circ\cdots\circ g}_{n\text{ times}}=g$$

is solvable and the solution is unique.

Indeed, we have $g(1) = 1$ by multiplicativity. For every prime power p^k, we have

$$f(p^k) = \sum_{k_1+\cdots+k_n=k} g(p^{k_1})\dots g(p^{k_n})$$
$$= \sum_{\substack{k_1+\cdots+k_n=k\\ k_i<n}} g(p^{k_1})\dots g(p^{k_n}) + ng(p^k).$$

This yields a recursion for $f(p^k)$, which shows the unicity. To show the existence, consider the function g whose values are defined by the above recursion at prime powers and that is extended multiplicatively to the other numbers. $g \circ \dots \circ g$ is a multiplicative function that coincides with f at prime powers; thus they must be identical.

Thus, both groups are divisible, torsion-free Abelian groups. By the fundamental theorem of divisible Abelian groups, both are isomorphic to some discrete direct power of the additive group of rational numbers.

Since we can prescribe the values of a multiplicative function at prime powers arbitrarily, we conclude that the cardinality of G_Q is $2^{\aleph_0}$ for both choices of Q. This implies that the direct power must have $2^{\aleph_0}$ factors in both cases. This proves the isomorphy of the two groups. □

Problem N.11. *Let H denote the set of those natural numbers for which $\tau(n)$ divides n, where $\tau(n)$ is the number of divisors of n. Show that*
(a) $n! \in H$ for all sufficiently large n,
(b) H has density 0.

Solution.
(a) We show that $\tau(n!)|n!$ for all $n \neq 3, 5$. We have

$$n! = \prod_{p\le n} p^{a_p}, \qquad a_p = \sum_{i=1}^{\infty}[n/p^i],$$

and consequently,

$$\tau(n!) = \prod_{p\le n}(a_p + 1).$$

To prove $\tau(n!)|n!$, it is sufficient to find, for every prime $p \le n$, a natural number $h(p) \le n$ such that $a_p + 1|h(p)$ and $h(p) \neq h(q)$ whenever $p \neq q$. If $p \le \sqrt{n}$, put $h(p) = a_p + 1$. We have

$$h(p) = 1 + a_p \le 1 + \sum_{i=1}^{\infty}[n/2^i] < 1 + \sum_{i=1}^{\infty} n/2^i = 1 + n,$$

hence $h(p) \le n$. Also, if $p < q \le \sqrt{n}$, then

$$\frac{n}{p} - \frac{n}{q} = \frac{(q-p)n}{pq} \ge \frac{n}{pq} > 1;$$

consequently, $[n/p] > [n/q]$. Since also $[p/p^j] \geq [n/q^j]$, for all j, we can infer that $h(p) > h(q)$. For the primes $\sqrt{n} < p \leq n$, we define $h(p)$ by recursion. Assume that $h(q)$ is already defined for every prime $q < p$. We need to find a multiple of $a_p + 1$ that is not among the already distributed numbers $h(q)$, $q < p$. Observe that $[n/p^j] = 0$, for $j \geq 2$; thus $a_p = [n/p]$. Hence, the number of multiples of $a_p + 1$ up to n is

$$\left[\frac{n}{a_p+1}\right] = \left[\frac{n}{1+[n/p]}\right] \geq \frac{n-[n/p]}{1+[n/p]} \geq \frac{n-(n/p)}{1+(n/p)}$$
$$= (p-1)\frac{n}{n+p} \geq \frac{p-1}{2},$$

with strict inequality unless $p = n$. The number of already occupied values is $\pi(p) - 1 \leq (p-1)/2$, with strict inequality unless $p = 3$, 5, or 7. Thus there is a free number for $h(p)$ except possibly in the cases $n = p = 3, 5, 7$. In these cases, the divisibility $\tau(n!)|n!$ holds for $n = 7$ and does not hold for $n = 3$ or 5.

(b) We split H into two parts. Choose a parameter $K > 1$. We put those numbers in which the exponent of every prime is $< K$ into H_1 and the rest into H_2. Consider first an $n \in H_1$. If $n = \prod_{j=1}^{s} p_j^{a_j}$, $a_j < K$, then $\tau(n) = \prod(a_j + 1)$ consists exclusively of primes $\leq K$, and by $\tau(n)|n$, the exponent of every prime is less than K. Consequently,

$$\tau(n) \leq \left(\prod_{p\leq K} p\right)^K < K^{K^2}.$$

On the other hand, we have $\tau(n) \geq 2^s$, and these inequalities together imply

$$s \leq [K^2 \log_2 K] = r.$$

It is well known that the sequence of numbers that have at most r prime factors has density 0, hence so does H_1.

Every element of H_2 is divisible by the Kth power of a prime. Hence, the number of elements of H_2 up to x is at most

$$\sum_p [\tfrac{x}{p^K}] < x\sum_{i=2}^{\infty} i^{-K} < x\int_1^{\infty} t^{-k}\,dt = \frac{x}{K-1}.$$

Hence, the density of H is at most $1/(K-1)$. Since K was arbitrary, H must have density 0. □

Remark. Let $H(x)$ denote the number of elements of H up to x. It can be shown that

$$x(\log x)^{-(1/2+\varepsilon)} \ll H(x) \ll x(\log x)^{-1/2},$$

for $x > x_0(\varepsilon)$. We outline a proof of these inequalities.

We start with the upper estimate. Write n in the form

$$n = 2^l p_1 p_2 \dots p_k m^2,$$

where $p_1, \dots, p_k$ are odd primes and m is an odd integer.

The number of possibilities for m is $\leq \sqrt{x}$, for l it is $O(\log x)$. Hence, the number of those n for which $p_1 \dots p_k < x^{1/3}$ is $O(x^{5/6} \log x) = o(x(\log x)^{-1/2})$. In what follows, we assume that $p_1 \dots p_k \geq x^{1/3}$.

The exponent of each p_j in n is odd, hence $2^k | \tau(n)$, so $k \leq l$ if $n \in H$. For fixed l and m we have

$$p_1 \dots p_k \leq y = \frac{x}{2^l m^2},$$

and $y \geq x^{1/3}$ by the previous assumption. Now by a classical theorem of Hardy and Ramanujan, the number of integers $\leq y$ that are products of k distinct primes is

$$\ll \frac{y}{\log y} \frac{(c + \log \log y)^{k-1}}{(k-1)!}$$

with an absolute constant c. Since $\log x > \log y > (1/3) \log x$ in our case, this is

$$\ll \frac{y}{\log x} \frac{(c + \log \log x)^{k-1}}{(k-1)!}.$$

We have to sum this for all possible values of l and m. Write $l = k + j$; we know that $j \geq 0$. With this substitution, our sum is

$$\sum_{k,j,m} 2^{-k-j} m^{-2} \frac{x}{\log x} \frac{(c + \log \log x)^{k-1}}{(k-1)!}.$$

Here the sum over m and j contributes only a constant factor, while the sum over k is just the power series expansion of

$$\exp \frac{c + \log \log x}{2} = c'(\log x)^{1/2}.$$

This concludes the proof of the upper estimate.

The lower estimate can be obtained by considering numbers of the form

$$n = 2^{q-1} q p_1 \dots p_k,$$

where $q, p_1, \dots, p_k$ are distinct odd primes. This number has $\tau(n) = 2^{k+1} q$ divisors, thus $n \in H$ if $k \leq q - 2$. For a fixed q, $q < (1 - \varepsilon) \log \log x$, the number of such $n \leq x$ is approximately

$$q^{-1} 2^{-q} \frac{x}{\log x} \frac{(\log \log x)^{q-3}}{(q-3)!}.$$

The lower estimate follows by taking $q \sim \frac{1}{2} \log \log x$.

Problem N.12. *Prove that if a, x, y are p-adic integers different from 0 and $p|x$, $pa|xy$, then*

$$\frac{1}{y}\,\frac{(1+x)^y-1}{x} \equiv \frac{\log(1+x)}{x} \pmod{a}.$$

Solution. Every p-adic number α can be uniquely represented in the form $\alpha = p^{\nu(\alpha)}\varepsilon$, where ε is a unit and $\nu(\alpha)$ is an integer; p-adic integers are characterized by $\nu(\alpha) \geq 0$. Hence, a divisibility $\alpha|\beta$ is equivalent to $\nu(\alpha) \leq \nu(\beta)$. In particular, the condition $pa|xy$ means $1+\nu(a) \leq \nu(x)+\nu(y)$, and the statement to be proved means $\nu(A) \geq \nu(a)$, where

$$A = \frac{1}{y}\frac{(1+x)^y-1}{x} - \frac{\log(1+x)}{x}.$$

By the previous observation, it is sufficient to show that $\nu(A) \geq \nu(x) + \nu(y) - 1$.

The series

$$(1+x)^y = 1 + \binom{y}{1}x + \dots \binom{y}{n}x^n + \dots$$

and

$$\log(1+x) = x - \frac{x^2}{2} + \dots + (-1)^{n-1}\frac{x^n}{n} + \dots$$

are convergent since $p|x$, $\binom{y}{n}$ is a p-adic integer, and $n - \nu(n) \to \infty$. Substituting these expansions into A, we obtain

$$A = \sum_{n=2}^{\infty} B_n,$$

where

$$\begin{aligned} B_n &= \left(\frac{1}{y}\binom{y}{n} - (-1)^{n-1}\frac{1}{n}\right)x^{n-1} \\ &= \left\{(y-1)(y-2)\dots(y-(n-1)) - (-1)^{n-1}(n-1)!\right\}\frac{x^{n-1}}{n!}. \end{aligned}$$

Here the first factor is a polynomial in y whose constant term vanishes; thus it is a multiple of y. Consequently, $B_n = cyx^{n-1}/n!$ with a p-adic integer c, thus

$$\nu(B_n) \geq \nu(y) + (n-1)\nu(x) - \nu(n!).$$

To prove that $\nu(A) \geq \nu(x)+\nu(y)-1$, it is sufficient to show that $\nu(B_n) \geq \nu(x) + \nu(y) - 1$ for every n; by the previous inequality, this would follow from $\nu(n!) \leq (n-2)\nu(x)+1$. Since $\nu(x) \geq 1$ by assumption, we need only to show $\nu(n!) \leq n-1$. But the exponent of p in $n!$ is

$$\nu(n!) = \sum_{j=1}^{\infty}\left[\frac{n}{p^j}\right] < \sum_{j=1}^{\infty}\frac{n}{p^j} = \frac{n}{p-1} \leq n.$$

Consequently, $\nu(n!) \leq n-1$ since it must be an integer. This concludes the proof. □

Problem N.13. *Let $1 < a_1 < a_2 < \cdots < a_n < x$ be positive integers such that $\sum_1^n 1/a_i \le 1$. Let y denote the number of positive integers smaller than x not divisible by any of the a_i. Prove that*

$$y > \frac{cx}{logx}$$

with a suitable positive constant c (independent of x and the numbers a_i).

Solution. The number of multiples of a_j up to x is $[x/a_j]$. Hence, the total number of multiples is

$$\le \sum \left[\frac{x}{a_j}\right] \le \sum \frac{x}{a_j} \le x.$$

We have to improve on this trivial bound.

Choose an x_0 such that the number of primes between $x/2$ and x is at least $x/(3\log x)$ for all $x \ge x_0$; such an x_0 exists by the prime number theorem. For $x < x_0$, we use the trivial estimate $y \ge 1$.

For $x \ge x_0$, we distinguish two cases.

If the number of $a_j > x/2$ is less than $x/(6\log x)$, then there are at least $x/(6\log x)$ primes $x/2 < p < x$ that are not contained among the a_j. These primes cannot be divisible by any a_j; thus in this case we obtain $y \ge x/(6\log x)$.

If the number of $a_j > x/2$ is at least $x/(6\log x)$, then we have

$$\sum_{a_j \le x/2} \frac{1}{a_j} \le 1 - \sum_{a_j > x/2} \frac{1}{a_j} \le 1 - \frac{1}{x}\frac{x}{6\log x} = 1 - \frac{1}{6\log x}.$$

Hence, the number w of integers up to $x/2$ that are divisible by some a_j satisfies

$$w \le \sum_{a_j \le x/2} \left[\frac{x/2}{a_j}\right] \le \frac{x}{2} - \frac{x}{12\log x}.$$

This implies

$$y \ge \left[\frac{x}{2}\right] - w \ge \frac{x}{12\log x} - 1. \quad \square$$

Problem N.14. *Let $T \in SL(n,\mathbb{Z})$, let G be a nonsingular $n \times n$ matrix with integer elements, and put $S = G^{-1}TG$. Prove that there is a natural number k such that $S^k \in SL(n,\mathbb{Z})$.*

Solution. Let d be the absolute value of the determinant of G, a positive integer by the assumption. Every element of G^{-1} is a fraction with denominator d. Since we have

$$S^k = G^{-1}T^kG = G^{-1}(T^k - I)G + I,$$

it is sufficient to prove that there is a positive integer k for which every element of $T^k - I$ is a multiple of d (the determinant of S^k is automatically 1).

By the box principle, there are two matrices among $T^0, T^1, \ldots, T^{d^{n^2}}$, say T^l and T^m, $l > m$, such that every element of T^l is congruent to the corresponding element of T^m. This means that every element of $T^l - T^m = T^m(T^{l-m} - I)$ is a multiple of d. This property is preserved if we multiply it by the matrix T^{-m}, which has integer elements by $T \in SL(n, \mathbb{Z})$. □

Problem N.15. *Let*

$$f(n) = \sum_{\substack{p \mid n \\ p^\alpha \le n < p^{\alpha+1}}} p^\alpha.$$

Prove that

$$\limsup_{n\to\infty} f(n)\,\frac{\log\log n}{n \log n} = 1.$$

Solution. It is well known that the number $\omega(n)$ of different prime divisors of n is at most $(1+o(1)) \cdot (\log n)/(\log\log n)$. Since obviously $f(n) \le n\omega(n)$, we conclude immediately that

$$\limsup_{n\to\infty} f(n)\frac{\log\log n}{n\log n} \le 1.$$

To prove the other inequality, we select a parameter k and try to compose an n from primes below k. The number of these primes is $\sim k/\log k$ by the prime number theorem. Now take another number m, which we shall later specify in terms of k. Every prime $p \le k$ has a power between m and km. We cover the interval $[m, km]$ with intervals of the type $[m(1+\varepsilon)^{i-1}, m(1+\varepsilon)^i]$, where $i = 1, \ldots, [c \log k]$, c depends only on ε. For a suitable i, there are at least $l = [c_1 k/(\log k)^2]$ primes $p \le k$ that have a power in this interval. We select l of them; let P be the product of these l primes. If $P < \varepsilon m$, then there is a multiple of P in the interval $[m(1+\varepsilon)^i, m(1+\varepsilon)^{i+1}]$; let n be such a multiple. We have

$$f(n) \ge lm(1+\varepsilon)^{i-1} > ln(1-\varepsilon)^2.$$

We have to estimate n in terms of l. We have $P \le k^l$ anyway; thus the choice $m = k^{l+1}$ guarantees $P < \varepsilon m$ if $k > 1/\varepsilon$. Further, we have

$$n \le m(1+\varepsilon)^{i+1} < km(1+\varepsilon)^2 < k^{l+2},$$

hence $l \ge (\log n)/(\log k) - 2$. Also $n < k^k$ from the above inequality. Thus $k \gg (\log n)/\log\log n$, and we obtain

$$l \ge (1-\varepsilon)\frac{\log n}{\log\log n}, \qquad f(n) > (1-\varepsilon)^3\frac{n\log n}{\log\log n},$$

for large k. □

Remark. It can be shown that

$$\max_{n \le x} f(n) \sim \frac{n \log n}{\log\log n}.$$

Problem N.16. *Let a and b be positive integers such that when dividing them by any prime p, the remainder of a is always less than or equal to the remainder of b. Prove that $a = b$.*

Solution. By taking a $p > \max(a, b)$, we immediately see that $a \leq b$. Assume that $a < b$, and write

$$k = \min(a, b - a).$$

By a theorem of Sylvester and Schur, there is a prime $p > k$ that divides $\binom{b}{k}$. We show that this prime contradicts the assumption.

We know

$$p | b(b-1)\dots(b-k+1),$$

and thus the residue of b is a number $r \leq k - 1$. Now if $a \leq b/2$, then $k = a < p$, and thus the residue of a is $a = k > r$, as claimed.

If $a > b/2$, then we have $k = b - a$. Write $b = qp + r$. We have

$$a - (q-1)p = (b-k) - (q-1)p = p + r - k > r,$$

while $p + r - k \leq p - 1$, and thus this is the residue of a and it was larger than r. □

Problem N.17. *Call a subset S of the set $\{1, \dots, n\}$ exceptional if any pair of distinct elements of S are coprime. Consider an exceptional set with a maximal sum of elements (among all exceptional sets for a fixed n). Prove that if n is sufficiently large, then each element of S has at most two distinct prime divisors.*

Solution. We prove the following slightly more general statement.

Statement. We call a subset S of the set $\{1, \dots, n\}$ k-exceptional if any k distinct elements of S are coprime. Consider a k-exceptional set with a maximal sum of elements (among all exceptional sets for a fixed n). If n is sufficiently large, then each element of S has at most two distinct prime divisors.

By maximality, every prime $p \leq n$ must have a multiple in S. We divide the primes $p \leq n$ into two classes: into P we put those for which there is an $m > 1$ such that $pm \in S$, and into Q those for which the only multiple of p in S is p itself.

We can estimate the number of $p \in P$, $p > x$, as follows. For every such p, take an $m > 1$, $pm \in S$, and then a prime $q | m$. These primes satisfy $q \leq n/x$ and any prime is represented at most $k - 1$ times among them; thus the number of our p's is at most $(k-1)\pi(n/x)$. Consequently, if $1 < x < y \leq n$, then the number of primes $q \in Q$, $x < q \leq y$, is at least

$$\pi(y) - \pi(x) - (k-1)\pi\frac{n}{x}. \tag{1}$$

Now assume that there is a $t \in S$ that has at least three distinct prime divisors, say $t = u^\alpha v^\beta w^\gamma r$, where $u < v < w$ are primes, and α, β, γ, and r are positive integers. Our plan is to find two suitable primes $p, q \in Q$ and replace t, p, q by up and vq. We must have $p \leq n/u$, $q \leq n/v$; thus the loss is

$$t + p + q \leq n + \frac{n}{u} + \frac{n}{v} \leq n + \frac{n}{2} + \frac{n}{3} = \frac{11}{6}n.$$

If we can achieve $up > cn$ and $vq > cn$, where $c = 11/12$, then the inclusion of up and vq offsets this loss. If there are at least two elements of Q in both $(cn/u, n/u]$ and $(cn/v, n/v]$, then this is possible. (We need two elements to guarantee that $p \neq q$.) By (1), for this it is sufficient that

$$\pi\frac{n}{u} - \pi\frac{cn}{u} \geq (k-1)\pi\frac{u}{c} + 2$$

and

$$\pi\frac{n}{v} - \pi\frac{cn}{v} \geq (k-1)\pi\frac{v}{c} + 2.$$

By the prime number theorem, these inequalities hold if $v \leq \delta\sqrt{n}$ with a suitable constant δ depending on k (say, $\delta = 1/(3k)$ is a good choice).

If $v > \delta\sqrt{n}$, so that the previous procedure fails, we argue as follows. We have

$$u \leq \frac{n}{vw} < \frac{n}{v^2} < \frac{1}{\delta^2}.$$

We try to find a prime $p \in Q$ and replace t and p by $u^j p$ and vw with a suitable exponent j. The new number is admissible if $p \leq n/u^j$. The loss is at most $p + n$, while the sum of the new terms is $pu^j + vw \geq pu^j + \delta^2 n$. Thus the process yields a profit if $p(u^j - 1) > (1 - \delta^2)n$, that is, we need to find a $p \in Q$ satisfying

$$\frac{1-\delta^2}{u^j - 1}n < p \leq \frac{n}{u^j}.$$

By (1), such a p exists if

$$\pi\left(\frac{n}{u^j}\right) - \pi\left(\frac{1-\delta^2}{u^j-1}n\right) > (k-1)\pi\left(\frac{u^j-1}{1-\delta^2}\right). \tag{2}$$

Recall that $u < 1/\delta^2$, hence there is a power of u in the interval $(\delta^{-4}, \delta^{-6}]$. With this as u^j, the left side of (2) is of order $n/\log n$, while the right side stays bounded, so (2) will hold for every sufficiently large n. $\square$

Problem N.18.

(a) *Prove that for every natural number k, there are positive integers $a_1 < a_2 < \cdots < a_k$ such that $a_i - a_j$ divides a_i for all $1 \leq i, j \leq k$, $i \neq j$.*

(b) *Show that there is an absolute constant $C > 0$ such that $a_1 > k^{Ck}$ for every sequence $a_1, \ldots, a_k$ of numbers that satisfy the above divisibility condition.*

Solution.

(a) For $k = 1$, we can put $a_1 = 1$. Now we show the inductive step from k to $k+1$. Assume that $0 < a_1 < \cdots < a_k$ is such a set. Write $b = \prod_{i=1}^{k} a_i$. Then the numbers $b < b + a_1 < \cdots < b + a_k$ form a suitable collection of $k+1$ numbers. Indeed, we have

$$(b + a_i) - (b + a_j) = a_i - a_j | a_i | b + a_i,$$

for $1 \le i, j \le k$, $i \ne j$, and also $(b + a_i) - b = a_i | b$.

(b) Consider any prime $p \le k$. If $a_i \equiv a_j \pmod{p}$, then $a_i \equiv 0 \pmod{p}$ by the divisibility condition. Consequently, there are at most $p - 1$ numbers a_i that are not divisible by p. Consider now the divisible ones. Dividing them by p, we again get an admissible system. Thus there can be at most $p - 1$ numbers not divisible by p among them. Repeating this argument, we find that the exponent of p in $A = \prod_{i=1}^{k} a_i$ is at least

$$\big(k - (p-1)\big) + \big(k - 2(p-1)\big) + \cdots + \left(k - \left[\frac{k}{p-1}\right](p-1)\right)$$
$$= \left[\frac{k}{p-1}\right]\left(k - \frac{p-1}{2}\left(1 + \left[\frac{k}{p-1}\right]\right)\right) \ge \frac{k^2}{3p}$$

if $p \le \sqrt{k}$. Consequently, we have (with $c = 1/3$)

$$A \ge \prod_{p \le \sqrt{k}} p^{ck^2/p},$$

and hence

$$a_k = \max a_j \ge A^{1/k} \ge \prod_{p \le \sqrt{k}} p^{ck/p}$$
$$= \exp ck \sum_{p \le \sqrt{k}} \frac{\log p}{p} \ge \exp c' k \log k = k^{c'k}$$

for $k \ge 4$. From the relation $a_k - a_1 | a_1$, we infer $a_1 \ge a_k/2 \ge 2$, hence

$$a_1 = \sqrt{a_1^2} \ge \sqrt{2a_1} \ge \sqrt{a_k} > k^{Ck}$$

with $C = c'/2$. We have proved the inequality $a_1 > k^{ck}$ for $k \ge 4$. For $k = 3$ it is easy to see that the minimal possible value of a_1 is 2, hence it holds if $C < (\log 2)/(3 \log 3)$. For $k = 1$ and 2, it fails by the examples $\{1\}$ and $\{1, 2\}$. □

Problem N.19. *Let $n_1 < n_2 < \dots$ be an infinite sequence of natural numbers such that $n_k^{1/2^k}$ tends to infinity monotone increasingly. Prove that $\sum_{k=1}^{\infty} 1/n_k$ is irrational. Show that this statement is best possible in a sense by giving, for every $c > 0$, an example of a sequence $n_1 < n_2 < \dots$ such that $n_k^{1/2^k} > c$ for all k but $\sum_{k=1}^{\infty} 1/n_k$ is rational.*

Solution. Assume indirectly that $\sum 1/n_k = p/q$, where p and q are positive integers. Then, for arbitrary k, we have

$$\sum_{i=1}^{k} \frac{1}{n_i} + \sum_{i=k+1}^{\infty} \frac{1}{n_i} = \frac{p}{q}.$$

Multiplying both sides of this equality by $qn_1 \dots n_k$, we see that

$$qn_1 \dots n_k \sum_{i=k+1}^{\infty} \frac{1}{n_i}$$

is a positive integer for all k. Now we show that

$$n_1 \dots n_k \sum_{i=k+1}^{\infty} \frac{1}{n_i} \to 0 \quad (k \to \infty),$$

and this contradiction will disprove the indirect assumption.

By the monotonicity condition, we have

$$n_{k+1}^{2^{-(k+1)}} \geq n_k^{2^{-k}} \geq n_{k-1}^{2^{-(k-1)}} \geq \dots \geq n_1^{2^{-1}},$$

which yields

$$n_1 \dots n_k \leq n_{k+1}^{2^{-1}+\dots+2^{-k}}.$$

On the other hand, by applying the inequalities

$$n_{k+1}^{2^{-(k+1)}} \leq n_{k+2}^{2^{-(k+2)}} \leq n_{k+3}^{2^{-(k+3)}} \leq \dots$$

and obvious estimations, we obtain

$$\begin{aligned} \frac{1}{n_{k+1}} + \frac{1}{n_{k+2}} + \frac{1}{n_{k+3}} + \dots &\leq \frac{1}{n_{k+1}} + \frac{1}{n_{k+1}^2} + \frac{1}{n_{k+1}^{2^2}} + \dots \\ &< \frac{1}{n_{k+1}} + \frac{1}{n_{k+1}^2} + \frac{1}{n_{k+1}^3} + \dots \\ &= \frac{1}{n_{k+1} - 1}. \end{aligned}$$

These two estimations together imply

$$n_1 \dots n_k \sum_{i=k+1}^{\infty} \frac{1}{n_i} \leq \frac{n_{k+1}^{1-2^{-k}}}{n_{k+1} - 1} = \frac{n_{k+1}}{n_{k+1} - 1} \left(n_{k+1}^{-2^{-(k+1)}} \right)^2.$$

Here the first term tends to 1 and the second to 0 as $k \to \infty$; thus the whole expression tends to 0.

To solve the second part of the problem, we are going to construct a sequence such that

$$n_k^{2^{-k}} > c \quad (k = 1, 2, \dots) \text{and} \sum_{k=1}^{\infty} \frac{1}{n_k} \text{ is rational.}$$

Take an arbitrary integer $n_1 > c^2 + 1$, and let $n_{k+1} = n_k^2 - n_k + 1$ for $k = 1, 2, \dots$. The definition implies

$$n_{k+1} - 1 > (n_k - 1)^2 > \cdots > (n_1 - 1)^{2^k} > c^{2^{k+1}}.$$

Consequently, $n_k^{2^{-k}} > c$ for all k. On the other hand, an induction shows that

$$\sum_{i=1}^{k} \frac{1}{n_i} = \frac{1}{n_1 - 1} - \frac{1}{n_k(n_k - 1)}.$$

Thus, the sum of the series is the rational number $1/(n_1 - 1)$. □

Problem N.20. *Prove that for every positive number K, there are infinitely many positive integers m and N such that there are at least $KN/\log N$ primes among the integers $m + 1, m + 4, \dots, m + N^2$.*

Solution. The basic idea of the proof is to average for several values of m while excluding divisibility by small primes.

Let $Q = 3 \cdot 5 \cdot \ldots \cdot p_l$, the product of the first l odd primes, and let c be a natural number $\leq Q$ such that $-c$ is a quadratic nonresidue for the moduli $3, 5, \dots, p_l$; such a c exists by the Chinese remainder theorem. We try to find m in the form $m = c + kQ$, where $1 \leq k \leq N^2$.

Consider an $1 \leq i \leq N$; by the choice of c, we have $(i^2 + c, Q) = 1$. By the prime number theorem for arithmetical progressions, the number of primes among the first N^2 elements of this progression is asymptotically equal to

$$\frac{1}{\phi(Q)} \frac{N^2 Q}{\log N^2 Q};$$

thus for sufficiently large N (for a fixed value of Q), it exceeds

$$\frac{Q}{3\phi(Q)} \frac{N^2}{\log N}.$$

Thus, the total number of primes (counted with multiplicity) among the integers

$$i^2 + c + kQ, \qquad 1 \leq i \leq N, \quad 1 \leq k \leq N^2$$

is at least

$$\frac{Q}{3\phi(Q)} \frac{N^3}{\log N}.$$

Consequently, for a suitable value of $m = c + kQ$, there are at least

$$\frac{Q}{3\phi(Q)} \frac{N}{\log N}$$

primes among the integes $m+1, m+4, \ldots, m+N^2$.

Since

$$\frac{Q}{\phi(Q)} = \prod_{i=1}^{l} \frac{p_i}{p_i - 1} > \prod_{i=1}^{l} \left(1 + \frac{1}{p_i}\right) > \sum_{i=1}^{l} \frac{1}{p_i}$$

and the sum $\sum 1/p$ is divergent, we can select Q so that $Q/3\phi(Q) > K$. Then the previous arguments yield that for every sufficently large N, the integer m can be chosen so that there are more than $K \cdot (N/\log N)$ primes in the sequence $m+1, m+4, \ldots, m+N^2$. □

3.7 OPERATORS

Problem O.1. *Let $a, b_0, b_1, \ldots, b_{n-1}$ be complex numbers, A a complex square matrix of order p, and E the unit matrix of order p. Assuming that the eigenvalues of A are given, determine the eigenvalues of the matrix*

$$B = \begin{pmatrix} b_0E & b_1A & b_2A^2 & \cdots & b_{n-1}A^{n-1} \\ ab_{n-1}A^{n-1} & b_0E & b_1A & \cdots & b_{n-2}A^{n-2} \\ ab_{n-2}A^{n-2} & ab_{n-1}A^{n-1} & b_0E & \cdots & b_{n-3}A^{n-3} \\ & & \ddots & & \\ ab_1A & ab_2A^2 & ab_3A^3 & \cdots & b_0E \end{pmatrix}.$$

Solution. We show that if the eigenvalues of A are $\lambda_1, \ldots, \lambda_p$ then the eigenvalues of B will be the numbers

$$\phi(\alpha_k\lambda_j) \quad (k = 1, \ldots, n;\ j = 1, \ldots, p),$$

where $\alpha_1, \ldots, \alpha_n$ denote the roots of the equation $z^n - a = 0$, and

$$\phi(x) = b_0 + b_1x + \cdots + b_{n-1}x^{n-1}.$$

In fact, if $A_0, A_1, \ldots, A_{n-1}$ are quadratic matrices of order p having complex entries, a is an arbitrary complex number and

$$C = \begin{pmatrix} A_0 & A_1 & A_2 & \cdots & A_{n-1} \\ aA_{n-1} & A_0 & A_1 & \cdots & A_{n-2} \\ \vdots & \vdots & \vdots & \ddots & \vdots \\ aA_1 & aA_2 & aA_3 & \cdots & A_0 \end{pmatrix},$$

then

$$\det C = \det M(\alpha_1) \ldots \det M(\alpha_n), \tag{1}$$

where

$$M(x) = A_0 + A_1x + \cdots + A_{n-1}x^{n-1}.$$

For $a = 0$ the assertion of this theorem is trivial. For $a \neq 0$, the validity of the theorem follows from the simple fact that

$$C = W \begin{pmatrix} M(\alpha_1) & \cdots & (0) \\ \vdots & \ddots & \vdots \\ (0) & \cdots & M(\alpha_n) \end{pmatrix} W^{-1},$$

where W is the Kronecker product with E of the Vandermonde matrix built from the roots of the equation $z^n = a$, that is,

$$W = V \otimes E = \begin{pmatrix} E & \cdots & E \\ \alpha_1 E & \cdots & \alpha_n E \\ \vdots & \ddots & \vdots \\ \alpha_1^{n-1} E & \cdots & \alpha_n^{n-1} E \end{pmatrix}.$$

Apply (1) to the matrix $B - \lambda\mathcal{E}$ (where $\mathcal{E}$ is the unit matrix of order np):

$$\det(B - \lambda\mathcal{E}) = \prod_{k=1}^{n} \det\left(\phi(\alpha_k A) - \lambda E\right).$$

On the other hand it is well known that the eigenvalues of the matrix

$$\phi(\alpha_k A) = b_0 E + b_1 \alpha_k A + \cdots + b_{n-1}\alpha_k^{n-1} A^{n-1}$$

are

$$\phi(\alpha_k \lambda_1), \ldots, \phi(\alpha_k \lambda_p).$$

This proves the assertion. $\square$

Remark. Another group of solutions is based on the theorem stated, but not proved, by one participant in the following general form:

Theorem. If the eigenvalues of the quadratic matrix A of order p are $\lambda_1, \ldots, \lambda_p$, further $\max_j |\lambda_j| = \lambda$, and $f_{ij}(z)$ $(i, j = 1, \ldots, n)$ are regular functions on the disc $|z| \le \lambda$, then the eigenvalues of the matrix

$$\begin{pmatrix} f_{11}(A) & \cdots & f_{1n}(A) \\ \vdots & \ddots & \vdots \\ f_{n1}(A) & \cdots & f_{nn}(A) \end{pmatrix}$$

are given by the eigenvalues of the matrices

$$\begin{pmatrix} f_{11}(\lambda_i) & \cdots & f_{1n}(\lambda_i) \\ \vdots & \ddots & \vdots \\ f_{n1}(\lambda_i) & \cdots & f_{nn}(\lambda_i) \end{pmatrix} \qquad (i = 1, \ldots, p).$$

One participant proved this theorem in the less general case where the functions $f_{ij}(z)$ are polynomials; this proof can be applied in the more general case as well.

Problem O.2. *Let U be an $n \times n$ orthogonal matrix. Prove that for any $n \times n$ matrix A, the matrices*

$$A_m = \frac{1}{m+1} \sum_{j=0}^{m} U^{-j} A U^j$$

converge entrywise as $m \to \infty$.

Solution 1. For an arbitrary $n \times n$ matrix A, let $\|A\|$ denote the norm of A, that is,

$$\|A\| = \sup_{\|x\|=1} \|Ax\|,$$

where x varies in the n-dimensional Euclidean space. If U is orthogonal, then U^{-1} is also orthogonal, and $\|UA\| = \|AU\| = \|A\|$.

Entrywise convergence and convergence in norm of a sequence of matrices are equivalent. Thus, by the Bolzano–Weierstrass theorem, from a norm-convergent sequence of matrices it is possible to select a norm-convergent subsequence.

For a given orthogonal matrix U, natural number m, and arbitrary matrix B, put

$$B_m = \frac{1}{m+1} \sum_{i=0}^{m} U^{-i} B U^i.$$

We shall need the following lemma:

Lemma. For any natural number m,

$$\lim_{p \to \infty} \|(A_m)_p - A_p\| = 0.$$

Proof. Let $p > m$. Then

$$(A_m)_p = \frac{1}{p+1} \frac{1}{m+1} \sum_{i=0}^{p} \sum_{j=0}^{m} U^{-i-j} A U^{i+j}$$

$$= \frac{1}{p+1} \frac{1}{m+1} \sum_{k=0}^{p+m} s(k) U^{-k} A U^k,$$

where

$$s(k) = \sum_{0 \le i \le p, 0 \le j \le m} 1 \le m+1 \quad (k = 0, \dots, p+m);$$

if $m \le k \le p$, then $s(k) = m+1$. Hence

$$\|(A_m)_p - A_p\| = \left\| \frac{1}{p+1} \frac{1}{m+1} \sum_{k=0}^{p+m} s(k) U^{-k} A U^k - \frac{1}{p+1} \sum_{k=0}^{p} U^{-k} A U^k \right\|$$

$$= \frac{1}{p+1} \frac{1}{m+1} \left\| \sum_{k=0}^{m-1} (s(k) - m - 1) U^{-k} A U^k + \sum_{k=p+1}^{p+m} s(k) U^{-k} A U^k \right\|$$

$$\le \frac{2m}{p+1} \|A\| \to 0$$

as $p \to \infty$. This proves the lemma.

Now the assertion of the problem can be proved as follows. Since $\|A_m\| \le \|A\|$ for all natural number m, the sequence $\{A_m\}$ contains a subsequence $\{A_{m_k}\}$ that is convergent in norm to a matrix H. Then for $k \to \infty$ we have

$$\begin{aligned}\|H - U^{-1}HU\| &\le \|H - A_{m_k}\| + \|A_{m_k} - U^{-1}A_{m_k}U\| \\ &\quad + \|U^{-1}(A_{m_k} - H)U\| \\ &\le 2\|H - A_{m_k}\| + \frac{2}{m_k+1}\|A\| \to 0,\end{aligned}$$

whence $H = U^{-1}HU$. Thus, $H_p = H$ for every natural number p.

We prove that $\lim_{p\to\infty} \|A_p - H\| = 0$. Actually, for every natural number k we have

$$\begin{aligned}\|A_p - H\| = \|A_p - H_p\| &\le \|A_p - (A_{m_k})_p\| + \|(A_{m_k})_p - H_p\| \\ &\le \|A_p - (A_{m_k})_p\| + \|A_{m_k} - H\|,\end{aligned}$$

whence by the lemma we obtain

$$\limsup_{p\to\infty} \|A_p - H\| \le \|A_{m_k} - A\|.$$

For $k \to \infty$, the assertion follows. □

Solution 2. Consider the natural embedding of the n-dimensional real space in the n-dimensional complex space. In the latter, the n by n real orthogonal matrices are unitary matrices. Thus, we prove more than required if in the statement of the problem we replace the orthogonal matrix by a unitary matrix, and real matrices A by matrices with complex entries.

So let $U = (U_{ij})$ be a unitary matrix in the n-dimensional complex space. Then, as we know, in a suitable basis the matrix U has the form $(U_{ij}) = (\varepsilon_i \delta_{ij})$, where δ_{ij} is the Kronecker symbol and $|\varepsilon_1| = \cdots = |\varepsilon_n| = 1$. Obviously,

$$(U^j)_{ik} = \varepsilon_i^j \delta_{ik}, \qquad (U^{-j})_{ik} = \varepsilon_i^{-j} \delta_{ik}.$$

Now, if $A = (a_{ik})$ is an arbitrary matrix, then

$$(U^{-j}AU^j)_{ik} = \sum_{r,s} \varepsilon_i^{-j} \delta_{ir} a_{rs} \varepsilon_s^j \delta_{sk} = \varepsilon_i^{-j} a_{ik} \varepsilon_k^j = \varepsilon_i^{-j} \varepsilon_k^j a_{ik}.$$

Therefore,

$$(A_m)_{ik} = \frac{1}{m+1} \sum_{j=0}^{m} \varepsilon_i^{-j} \varepsilon_k^j a_{ik} = \frac{a_{ik}}{m+1} \sum_{j=0}^{m} \varepsilon_i^{-j} \varepsilon_k^j.$$

If $\varepsilon_i = \varepsilon_k$, then $\varepsilon_i^{-j}\varepsilon_k^j = 1$, so $\lim_{m\to\infty} (A_m)_{ik} = a_{ik}$.

On the other hand, if $\varepsilon_i \neq \varepsilon_k$, then

$$(A_m)_{ik} = \frac{a_{ik}}{m+1}\sum_{j=0}^{m}\left(\overline{\varepsilon_i}\varepsilon_k\right)^j = \frac{a_{ik}}{m+1}\frac{\left(\overline{\varepsilon_i}\varepsilon_k\right)^{m+1}-1}{\overline{\varepsilon_i}\varepsilon_k - 1} \to 0$$

as $m \to \infty$, since

$$\left|\frac{\left(\overline{\varepsilon_i}\varepsilon_k\right)^{m+1}-1}{\overline{\varepsilon_i}\varepsilon_k - 1}\right| \leq \frac{2}{\left|\overline{\varepsilon_i}\varepsilon_k - 1\right|}.$$

We see that in this case $\lim_{m\to\infty}(A_m)_{ik} = 0$.

It remains to note that if for a sequence of matrices $\left(a_{ik}^{(m)}\right)_{m=1}^{\infty}$ we know that for each pair of indices i, k the numbers $a_{ik}^{(m)}$ tend to some a_{ik} as $m \to \infty$ then the sequence of matrices, evidently, is entrywise convergent to the matrix (a_{ik}). □

Solution 3. As a generalization of the problem, we prove the following theorem.

Theorem. Let H be a complex Hilbert space, and U a unitary operator in H. Then, for any compact operator A of H, the sequence of operators

$$\Phi_m(A) = \frac{1}{m+1}\sum_{j=0}^{m} U^{-j}AU^j \qquad (m = 0, 1, \dots)$$

is weakly convergent, that is, for every pair of elements f, g in H

$$\lim_{m,n\to\infty}((\Phi_m(A) - \Phi_n(A))f, g) = 0,$$

where (f, g) denotes the inner product of the elements $f, g \in H$.

(This statement obviously contains the statement of the problem. Indeed, the n by n real orthogonal matrix appearing in the problem induces a unitary operator of the n-dimensional complex Hilbert space; in this n-dimensional space all linear operators, in particular those induced by n by n real matrices, are compact, and in finite-dimensional spaces weak and entrywise — so-called strong — convergences of operators coincide.)

Proof. For proving the theorem, we first observe that the mappings $A \to \Phi_m(A)$ $(m = 1, 2, \dots)$, defined for all bounded operators A of H, have the following properties:

a. For any pair of bounded linear operators A, B of H and any pair of complex numbers α and β,

$$\Phi_m(\alpha A + \beta B) = \alpha\Phi_m(A) + \beta\Phi_m(B) \qquad (m = 0, 1, \dots).$$

b. For any bounded linear operator A of H,

$$\|\Phi_m(A)\| \leq \|A\| \qquad (m = 0, 1, \dots),$$

where $\|A\|$ stands for the norm of A.

c. If the sequence $\{A_k\}_{k=1}^{\infty}$ of bounded linear operators is uniformly convergent to a bounded linear operator A, that is, $\|A_k - A\| \to 0$ as $k \to \infty$, and the sequence $\{\Phi_m(A_k)\}_{m=0}^{\infty}$ is weakly convergent for each k, then the sequence $\{\Phi_m(A)\}_{m=0}^{\infty}$ is also weakly convergent.

From these three properties only c. needs to be verified. So let $\{A_k\}_{k=1}^{\infty}$ be a sequence of operators with the properties above, and let f and g be two elements of the Hilbert space H. Without loss of generality, we may assume that $\|f\| = \|g\| = 1$. Then

$$\begin{aligned}
&|((\Phi_m(A) - \Phi_n(A))f, g)| \\
&= |([\Phi_m(A) - \Phi_m(A_k) + \Phi_m(A_k) - \Phi_n(A_k) + \Phi_n(A_k) - \Phi_n(A)]\, f, g)| \\
&\le |(\Phi_m(A - A_k)f, g)| + |((\Phi_m(A_k) - \Phi_n(A_k))\, f, g)| + |(\Phi_n(A_k - A)f, g)| \\
&\le 2\|A - A_k\| + |((\Phi_m(A_k) - \Phi_n(A_k))\, f, g)|
\end{aligned}$$

(we have made use of properties a and b and applied the Schwarz inequality).

Now let $\varepsilon > 0$ be arbitrary. Then, by one of the assumptions on the sequence $\{A_k\}_{k=1}^{\infty}$, there is an index $k = k(\varepsilon)$ such that

$$\|A_{k(\varepsilon)} - A\| < \frac{\varepsilon}{4}.$$

Next, let $N = N(k(\varepsilon), \varepsilon)$ be a positive integer satisfying

$$\left|\left(\left(\Phi_m(A_{k(\varepsilon)}) - \Phi_n(A_{k(\varepsilon)})\right) f, g\right)\right| < \frac{\varepsilon}{2}$$

for $m, n > N$. By the assumptions, such N exists. Then, in view of the foregoing, we obtain

$$|((\Phi_m(A) - \Phi_n(A))\, f, g)| < \varepsilon$$

for $m, n > N$. This proves property c.

It is well known that every compact operator is the uniform limit of finite rank operators, further every finite rank operator is a finite linear combination of rank 1 operators. Consequently, by properties a. and c., it is sufficient to prove our assertion for operators of rank 1. So let A be an operator of rank 1. Then there are two elements ϕ and ψ in the space H such that

$$Af = (f, \phi)\psi$$

for all f in H. Now, if h and g are any two elements of H, then

$$(\Phi_m(A)h, g) = \frac{1}{m+1} \sum_{j=0}^{m} (AU^j h, U^j g) = \frac{1}{m+1} \sum_{j=0}^{m} (U^j h, \phi)\left(U^{-j}\psi, g\right).$$

Denote by $H \otimes H$ the tensor product of H with itself. (The definition of the Hilbert space $H \otimes H$ is the following. Denote by H_0 the algebraic tensor product of H with itself. As is well known, this is the free module generated by the symbols $x \otimes y\,(x, y \in H)$ and having the complex field for operator domain. By the inner product of two elements

$$\hat{\psi} = \sum_{i=1}^{n} x_i \otimes y_i, \qquad \hat{\psi}' = \sum_{j=1}^{m} x_j' \otimes y_j'$$

of H_0, we mean the number

$$\langle \hat{\psi}, \hat{\psi}' \rangle = \sum_{i=1}^{n} \sum_{j=1}^{m} (x_i, x_j')(y_i, y_j'). \tag{1}$$

Two elements $\hat{\psi}, \hat{\psi}' \in H_0$ are considered to be identical if $\langle \hat{\psi} - \hat{\psi}', \hat{\psi} - \hat{\psi}' \rangle = 0$. On the factor space $\tilde{H}_0$ that arises after this identification, (1) defines a norm. Completing H_0 in the metric induced by this norm, we obtain the Hilbert space $\tilde{H} = H \otimes H$.)

Let U_0 be the operator of $\tilde{H}$ satisfying the relation

$$U_0(x \otimes y) = \big(Ux \otimes U^{-1}y\big) \qquad (x, y \in H).$$

U_0 is uniquely determined in this way and is unitary in H. On the other hand it is clear that

$$(\Phi_m(A)h, g) = \frac{1}{m+1} \sum_{j=0}^{m} \Big(U_0^j(h \otimes \psi), \phi \otimes g\Big).$$

By a classic theorem of ergodic theory (see for example *F. Riesz and B. Sz. Nagy, Functional Analysis, Blackie, London, 1956*, §144), the right-hand side is convergent for every pair of elements $\hat{\psi}, \hat{\psi}' \in \tilde{H}$ and so, in particular, for the pair $\hat{\psi} = h \otimes \psi$, $\hat{\psi}' = \phi \otimes g$. The proof is complete. □

Problem O.3. *Prove that if a sequence of Mikusiński operators of the form $\mu e^{-\lambda s}$ (λ and μ nonnegative real numbers, s the differentiation operator) is convergent in the sense of Mikusiński, then its limit is also of this form.*

Solution. Let $\mu_n e^{-\lambda_n s}$ be the convergent sequence of operators considered. We may assume that $\mu_n \to \mu$ and $\lambda_n \to \lambda$ $(0 \le \mu, \lambda \le +\infty)$ as $n \to \infty$; in fact, passing to a suitable subsequence, this can always be achieved without any influence on convergence and limiting operator.

By the definition of convergence, there exists an operator $\{p\}/\{q\}$ ($\{p\}$, $\{q\} \in C[0, \infty)$, $\{p\}, \{q\} \ne \{0\}$) such that

$$\mu_n e^{-\lambda_n s} \frac{\{p\}}{\{q\}} \tag{1}$$

is an almost uniformly (that is, uniformly on each finite interval) convergent sequence of functions in $C[0,\infty)$. We may assume that $p(0) = 0$, since otherwise $\{p\}/\{q\}$ in (1) can be replaced by the equal expression $(\{p\}\{1\})/(\{q\}\{1\})$. Define p to be zero on the negative half-line; then p will be a continuous function on the whole real line and not identically zero on the positive half-line. From (1) it follows that the sequence of functions

$$\mu_n e^{-\lambda_n s}\{p\} = \{\mu_n p(t-\lambda_n)\} \tag{2}$$

is almost uniformly convergent.

If $\lambda = +\infty$, then we see that (2) is almost uniformly convergent to the zero function; thus

$$\mu_n e^{-\lambda_n s} \to 0 = 0 \cdot e^{-s} \quad (n \to \infty).$$

If $\lambda < +\infty$, then we show that μ is also finite. Let t_0 be a point with $p(t_0 - \lambda) \neq 0$. Then $p(t_0 - \lambda_n) \to p(t_0 - \lambda)$ $(n \to \infty)$, so $\mu_n p(t_0 - \lambda_n)$ can only be convergent if $\mu < +\infty$. Then, however, $\mu_n p(t - \lambda_n)$ tends almost uniformly to the function $\{\mu p(t-\lambda)\} = \mu e^{-\lambda s}\{p\}$. Consequently,

$$\mu_n e^{-\lambda_n s} \to \mu e^{-\lambda s} \quad (n \to \infty). \quad \square$$

Problem O.4. *Prove that an idempotent linear operator of a Hilbert space is self-adjoint if and only if it has norm 0 or 1.*

Solution 1. If T is self-adjoint, then it is an orthogonal projection operator; it is well known that such operators have norm 1 or 0. Conversely, let $\|T\| \leq 1$. For any element x of the space and any number μ, we have

$$T\left(\mu Tx - (x - Tx)\right) = \mu Tx$$

and therefore

$$\begin{aligned} |\mu|^2\|Tx\|^2 &\leq \|\mu Tx - (x-Tx)\|^2 \\ &= -2\mathrm{Re}\left[\mu(Tx, x-Tx)\right] + |\mu|^2\|Tx\|^2 + \|x-Tx\|^2. \end{aligned}$$

Consequently,

$$\mathrm{Re}\left[\mu(Tx, x-Tx)\right] \leq \frac{1}{2}\|x-Tx\|^2.$$

But this is possible for every μ only if

$$(Tx, x-Tx) = 0.$$

Hence, it follows in a simple and well-known manner that $T^* = T$.

The proof applies to real as well as complex spaces. $\square$

Solution 2. The kernel (null space) of T, in view of the property $T^2 = T$, coincides with the range of $I - T$, that is, with the set of all elements of the form $(I - T)x$ where x runs through all elements of the space. It is a well-known fact, and easy to verify, that the orthogonal complement of the range of $I - T$ is equal to the kernel of $I - T^*$. By a well-known result of Nagy (see, for example, *B. Sz. Nagy and C. Foiaş, Harmonic Analysis of Operators on Hilbert Space, Akadémiai Kiadó and North-Holland Publ. Co., 1970*), from the inequality $\|T\| \leq 1$ it follows that the invariant elements of T and T^* are the same. So, the kernel of $I - T^*$ coincides with the kernel of $I - T$, which, in turn, coincides with the range of T (the latter fact follows from the property $(I - T)^2 = I - T$). Thus

$$(Tx, x - Tx) = 0$$

for every x, which implies the statement. $\square$

Problem O.5. *Let T be a bounded linear operator on a Hilbert space H, and assume that $\|T^n\| \leq 1$ for some natural number n. Prove the existence of an invertible linear operator A on H such that $\|ATA^{-1}\| \leq 1$.*

Solution. Let $(.,.)$ denote scalar multiplication in H. It is easy to verify that the formula

$$[x, y] = \sum_{i=0}^{n-1} (T^i x, T^i y)$$

defines a new scalar product on H. Let $\hat{H}$ denote the space equipped with the scalar product $[.,.]$. Obviously,

$$(x, x) \leq [x, x] \leq \left(\sum_{i=0}^{n-1} \|T^i\|^2 \right) (x, x)$$

for all $x \in H$, that is, the norm in H is equivalent to that in $\hat{H}$. Consequently, $\hat{H}$ is also a Hilbert space.

Since the dimension of a Hilbert space is the minimal power of complete sets, that is, sets whose closed linear span is the whole space, the dimensions of H and $\hat{H}$ are equal. But Hilbert spaces of equal dimension are isomorphic. So, consider a Hilbert space isomorphism $A : \hat{H} \to H$. Since the underlying vector space for H and $\hat{H}$ is the same, A is an invertible linear operator on H. Let $x \in H$ be arbitrary, and put $y = A^{-1}x$. Then

$$\begin{aligned}(x, x) - (ATA^{-1}x, ATA^{-1}x) &= (Ay, Ay) - (ATy, ATy) = [y, y] - [Ty, Ty] \\ &= \sum_{i=0}^{n-1} (T^i y, T^i y) - \sum_{i=0}^{n-1} (T^{i+1} y, T^{i+1} y) = (y, y) - (T^n y, T^n y),\end{aligned}$$

which is nonnegative since $\|T^n\| \leq 1$. Thus, we have obtained

$$(x, x) \geq (ATA^{-1}x, ATA^{-1}x)$$

for all $x \in H$, that is, $\|ATA^{-1}\| \leq 1$. Therefore, the operator A satisfies the conditions of the problem. $\square$

Remark. One contestant showed that the positive square root of the positive, bounded, self-adjoint, linear operator $\sum_{i=0}^{n-1} T^{*i}T^i$ can be chosen for A.

Problem O.6. *Is it true that if A and B are unitarily equivalent, self-adjoint operators in the complex Hilbert space $\mathcal{H}$, and $A \leq B$, then $A^+ \leq B^+$? (Here A^+ stands for the positive part of A.)*

Solution. We first show that there are self-adjoint operators $\hat{A}$ and $\hat{B}$ in $\mathbb{C}^2$ with eigenvalues λ_1, λ_2 and λ_3, λ_4, respectively, such that $\hat{A} \leq \hat{B}$ but $\hat{A}^+ \not\leq \hat{B}^+$. For instance, let $\hat{A}$ and $\hat{B}$ have matrices

$$\begin{pmatrix} 0 & -1 \\ -1 & -1 \end{pmatrix} \quad \text{and} \quad \begin{pmatrix} 1 & 0 \\ 0 & 0 \end{pmatrix},$$

respectively. Then for $(x, y) \in \mathbb{C}^2$, we have

$$\langle \hat{B}(x,y), (x,y) \rangle - \langle \hat{A}(x,y), (x,y) \rangle = |x|^2 + x\bar{y} + \bar{x}y + |y|^2 = |x+y|^2 \geq 0,$$

that is, $\hat{A} \leq \hat{B}$. From the matrices, it is clear that $\hat{A}$ and $\hat{B}$ are self-adjoint, $\hat{B}$ positive semidefinite, $\hat{A}$ indefinite. Thus $\hat{B}^+$ is equal to $\hat{B}$, and $\hat{A}^+$ is positive semidefinite with $\hat{A}^+(0,1) \neq (0,0)$, since $(0,1)$ is not an eigenvector of $\hat{A}$. Therefore,

$$\langle \hat{A}^+(0,1), (0,1) \rangle > 0 \quad \text{whereas} \quad \langle \hat{B}^+(0,1), (0,1) \rangle = 0.$$

So $\hat{A}^+ \not\leq \hat{B}^+$.

Now let $\mathcal{H} = l_2$ (complex). For simplicity, we write each element of $\mathcal{H}$ as a series $\sum_{n=1}^{\infty} a_n e_n$ convergent in l_2, where

$$e_n = (0, \ldots, 0, \overset{n.}{\breve{1}}, 0, \ldots).$$

Let

$$A\left(\sum_{n=1}^{\infty} a_n e_n\right) = \hat{A}(a_1, a_2) + \sum_{n=3}^{\infty} \lambda_n a_n e_n,$$

where $\lambda_n = \lambda_s$ if $n = 4k + s$, $1 \leq s \leq 4$, and $\lambda_1, \lambda_2, \lambda_3, \lambda_4$ are described above. Here e_1 and e_2 are identified with $(1,0)$ and $(0,1)$. Further, let

$$B\left(\sum_{n=1}^{\infty} a_n e_n\right) = \hat{B}(a_1, a_2) + \sum_{n=3}^{\infty} \lambda_n a_n e_n.$$

Both A and B are sums of two self-adjoint operators, so they are self-adjoint. Since $B - A = \hat{B} - \hat{A}$, we have

$$\left\langle (B-A)\sum_{n=1}^{\infty} a_n e_n, \sum_{n=1}^{\infty} a_n e_n \right\rangle = \left\langle (\hat{B}-\hat{A})(a_1,a_2),(a_1,a_2)\right\rangle \geq 0,$$

that is, $A \leq B$. Similarly, $B^+ - A^+ = \hat{B}^+ - \hat{A}^+$ and, consequently, $A^+ \not\leq B^+$.

We now prove that A and B are unitarily equivalent. Let

$$U_1\left(\sum_{n=1}^{\infty} a_n e_n\right) = \hat{U}_1(a_1,a_2) + \sum_{n=3}^{\infty} a_n e_n,$$

where $\hat{U}_1$ denotes the unitary transformation of coordinates that reduces the matrix of $\hat{B}$ to diagonal form:

$$(\hat{U}_1\hat{B}\hat{U}_1^{-1})(a,b) = (\lambda_3 a, \lambda_4 b).$$

U_1 is obviously unitary, and

$$(U_1 B U_1^{-1})\left(\sum_{n=1}^{\infty} a_n e_n\right) = \lambda_3 a_1 e_1 + \lambda_4 a_2 e_2 + \sum_{n=3}^{\infty} \lambda_n a_n e_n .$$

Let

$$U_2\left(\sum_{n=1}^{\infty} a_n e_n\right) = \sum_{n=1}^{\infty} a_{f(n)} e_n,$$

where f denotes the following permutation of $\mathbb{N}$: $f(n) = n-4$ for $n = 4k+3$ or $n = 4k+4$ with $k \geq 1$. Further, $f(3) = 1$, $f(4) = 2$, and $f(n) = n+4$ for $n = 4k+1$ or $n = 4k+2$ with $k \geq 0$. Then

$$\begin{aligned}\langle U_2\left(\sum_{n=1}^{\infty} a_n e_n\right), \sum_{n=1}^{\infty} b_n e_n\rangle &= \sum_{n=1}^{\infty} a_{f(n)}\bar{b}_n \\ &= \sum_{n=1}^{\infty} a_n \bar{b}_{f^{-1}(n)} = \langle \sum_{n=1}^{\infty} a_n e_n, U_2^{-1}\left(\sum_{n=1}^{\infty} b_n e_n\right)\rangle,\end{aligned}$$

which implies that U_2 is also unitary. Further, we have

$$(U_2 U_1 B U_1^{-1} U_2^{-1})\left(\sum_{n=1}^{\infty} a_n e_n\right) = \sum_{n=1}^{\infty} \lambda_n a_n e_n.$$

Finally, let

$$U_3\left(\sum_{n=1}^{\infty} a_n e_n\right) = \hat{U}_3(a_1,a_2) + \sum_{n=3}^{\infty} a_n e_n,$$

where $(\hat{U}_3^{-1}\hat{A}\hat{U}_3)(a,b) = (\lambda_1 a, \lambda_2 b)$, and $\hat{U}_3$ is unitary. With this third unitary transformation, we have

$$(U_3 U_2 U_1 B U_1^{-1} U_2^{-1} U_3^{-1})\left(\sum_{n=1}^{\infty} a_n e_n\right) = A\left(\sum_{n=1}^{\infty} a_n e_n\right),$$

and so A and B are unitarily equivalent.

Consequently, the assertion is false. □

Problem O.7. *Let K be a compact subset of the infinite-dimensional, real, normed linear space $(X, \|\cdot\|)$. Prove that K can be obtained as the set of all left limit points at 1 of a continuous function $g : [0, 1[\to X$, that is, x belongs to K if and only if there exists a sequence $t_n \in [0, 1[$ $(n = 1, 2, \dots)$ satisfying $\lim_{n\to\infty} t_n = 1$ and $\lim_{n\to\infty} \|g(t_n) - x\| = 0$.*

Solution. The compact set $K \subset X$ has, for every $\varepsilon = 1/n$ $(n \in \mathbb{N})$, a finite ε-net $z_1(n), z_2(n), \dots, z_{j_n}(n) \in K$, that is, if $x \in K$, then $\|x - z_i(n)\| \le \varepsilon = 1/n$ for some $1 \le i \le j_n$. Denote by $x_0, x_1, x_2, \dots$ the sequence obtained by writing the finite $(1/n)$-nets $z_1(1), z_2(1), \dots, z_{j_1}(1), z_1(2), z_2(2), \dots, z_{j_2}(2), z_1(3), \dots$ one after the other. It would already be possible to define a piecewise linear function g: $[0, 1[\to X$ such that $g(1 - 1/k) = x_k$ $(k \in N)$, and therefore each point of K is a left-hand limit point at 1 of the function g. The problem is that g may have further limit points that do not belong to K. So, we have to improve the procedure by interpolating a new sequence of function values $y_0, y_1, y_2, \dots$. To define the latter, we shall need the following proposition (which is essentially the same as the well-known Riesz lemma on almost orthogonal vectors):

If $L \subset X$ is a linear subspace of finite dimension, then there exists a vector $y \in X \setminus L$ such that $\|y\| = 1$ and $\operatorname{dist}(y, L) = 1$, where

$$\operatorname{dist}(A, B) = \inf\{\|a - b\| \colon a \in A, b \in B\}$$

denotes the distance between the subsets $A, B \subset X$, and the singleton $\{a\}$ is replaced by a. Indeed, since X is infinite dimensional, it contains a vector $v \in X \setminus L$, and the subset

$$D = \{u \colon u \in L, \|v - u\| \le \|v\|\}$$

of the finite-dimensional subspace L is bounded and closed. Thus D is compact, so it contains a vector $u_0 \in D$, which lies closest to v. Then, setting $v_0 = v - u_0$,

$$\|v_0\| = \min\{\|v - u\| \colon u \in D\} = \operatorname{dist}(v, L) = \operatorname{dist}(v_0, L).$$

Consequently, the vector $y = v_0/\|v_0\|$ has the required properties.

Next, we define a sequence in X by induction. Let $L_0 = \operatorname{lin}\{x_0, x_1\}$, and using the proposition just proved, choose a vector $y_0 \in X \setminus L_0$ such that

$$\|y_0\| = 1 = \operatorname{dist}(y_0, L_0).$$

If $y_0, y_1, \dots, y_{k-1}$ $(k \in \mathbb{N})$ are already defined, then set

$$L_k = \operatorname{lin}\{x_0, y_0, x_1, y_1, \dots, x_{k-1}, y_{k-1}, x_k, x_{k+1}\},$$

and using the preceding proposition again, choose the vector $y_k \in X \setminus L_k$ so that

$$\|y_k\| = 1 = \operatorname{dist}(y_k, L_k).$$

For simplicity, we first define a continuous function $\phi\colon [0,\infty[\to X$ that has K for the set of limit points at ∞. For $k \in \mathbb{N}_0 = \mathbb{N}\cup\{0\}$ and $0 \le r \le 1$, put

$$\begin{aligned}\phi(3k+r) &= (1-r)x_k + r(x_k+y_k),\\ \phi(3k+1+r) &= (1-r)(x_k+y_k) + r(x_{k+1}+y_k),\\ \phi(3k+2+r) &= (1-r)(x_{k+1}+y_k) + rx_{k+1}.\end{aligned}$$

Let us verify the desired properties.

a. By the construction of the sequence $x_0, x_1, x_2, \ldots$, for any $x \in K$ there is a subsequence $x_{k_n} \to x$; setting $\tau_n = 3k_n$, we have

$$\lim_{n\to\infty} \phi(\tau_n) = \lim_{n\to\infty} x_{k_n} = x.$$

b. On the other hand, we show that no $x \notin K$ can be a limit point of ϕ at ∞. In fact, the closedness of K yields

$$\delta = \min\{\operatorname{dist}(x,K), 1\} > 0,$$

and if for some $\tau_0 \in [3k_0, 3k_0+3]$ $(k_0 \in \mathbb{N}_0)$ we have $\operatorname{dist}(x,\phi(\tau_0)) < \delta/3$, then for every $\tau > 3k_0+3$ the relation $\operatorname{dist}(x,\phi(\tau)) \ge \delta/3$ is valid. Now there exists a natural number $k_0 < k \in \mathbb{N}$ such that

b1. either $3k+(2\delta/3) < \tau < 3k+3-(2\delta/3)$, in which case the relation $\phi(\tau_0) \in L_k$ and the definition of ϕ imply

$$\operatorname{dist}(\phi(\tau_0),\phi(\tau)) > \frac{2\delta}{3}$$

and therefore

$$\operatorname{dist}(x,\phi(\tau)) \ge \operatorname{dist}(\phi(\tau_0),\phi(\tau)) - \operatorname{dist}(x,\phi(\tau_0)) > \frac{\delta}{3},$$

b2. or $|3k-\tau| \le 2\delta/3$, in which case for $x_k \in K$, by the definition of ϕ, we have

$$\operatorname{dist}(x_k,\phi(\tau)) \le \frac{2\delta}{3},$$

and so

$$\operatorname{dist}(x,\phi(\tau)) \ge \operatorname{dist}(x,K) - \operatorname{dist}(K,\phi(\tau)) \ge \frac{\delta}{3}.$$

Thus, there is indeed no sequence $\tau_n \to \infty$ with $\phi(\tau_n) \to x$. Finally, the required function $g\colon [0,1[\to X$ is provided by the transformation

$$g(t) = \phi\left(\frac{t}{1-t}\right), \qquad t \in [0,1[. \quad \square$$

Problem O.8. *Denote by $B[0,1]$ and $C[0,1]$ the Banach space of all bounded functions and all continuous functions, respectively, on the interval $[0,1]$ with the supremum norm. Is there a bounded linear operator*

$$T : B[0,1] \to C[0,1]$$

such that $Tf = f$ for all $f \in C[0,1]$?

Solution. Suppose the existence of such an operator T.

For any $0 < a < 1$, consider the function

$$f_a(x) = \begin{cases} 0 & \text{if } x < a, \\ 1 & \text{if } x \geq a. \end{cases}$$

The function $h_a = f_a - Tf_a$ is, apart from the jump of size 1 at point a, continuous and satisfies $T(h_a) = T(f_a - Tf_a) = 0$.

Replacing h_a by $-h_a$ is necessary. We may assume that $h_a > 1/3$ in a small, punctured, right-hand or left-hand neighborhood of a.

Let $a_1 = 1/2$, and define the sequence $a_1, a_2, \ldots$ so that $a_i \in (0,1)$ and a_i belongs to a corresponding small neighborhood of a_{i-1} for all $i \geq 2$. Then for the function $g_n = \sum_{i=1}^{n} h_{a_i}$, we have $g_n \in B[0,1]$, g_n is continuous apart from the jumps of size 1 at the points $a_1, \ldots, a_n$, and $g_n > n/3$ in a small, one-sided, punctured neighborhood of a_n. It is easy to construct a continuous function f_n satisfying the relation

$$\|f_n - g_n\| \sup_{x \in [0,1]} |f_n(x) - g_n(x)| \leq \frac{1}{2}.$$

(To speak exactly, the sign of equality is valid.) Then, for the norm of the operator T, we obtain the estimate

$$\|T\| \geq \frac{\|T(f_n - g_n)\|}{\|f_n - g_n\|} \geq 2\|Tf_n - Tg_n\| = 2\|f_n\| \geq \frac{2}{3}n - 1,$$

valid for any natural number n. Consequently, T cannot be a bounded operator. It follows that there is no operator T satisfying the requirements of the problem. □

Problem O.9. *Does there exist a bounded linear operator T on a Hilbert space H such that*

$$\bigcap_{n=1}^{\infty} T^n(H) = \{0\} \quad \textit{but} \quad \bigcap_{n=1}^{\infty} T^n(H)^- \neq \{0\},$$

where $^-$ denotes closure?

Solution. There exists an operator T of this kind. Let H be a Hilbert space of countably infinite dimension, and let $\{e_n\}_{n=1}^{\infty}$ be an orthonormal basis in H. We define T on the basis vectors first. Let

$$Te_n = \begin{cases} 0 & \text{if } n = 1, \\ e_{n+1} & \text{if } n \geq 2 \text{ is not a square,} \\ \alpha_j e_1 + \frac{\alpha_j}{j} e_{n+1} & \text{if } n = j^2 \text{ for an integer } j \geq 2. \end{cases}$$

For the sequence $\{\alpha_j\}_{j=2}^{\infty}$, we assume that $0 < |\alpha_j| \leq 1$ for each j, and

$$\alpha = \sum_{j=2}^{\infty} |\alpha_j|^2 < \infty.$$

We next extend the definition of T to all of H. For an arbitrary vector $x \in H$, put

$$Tx := \sum_{n=1}^{\infty} \xi_n Te_n, \quad \text{where} \quad \xi_n = \langle x, e_n \rangle.$$

First, we have to prove that the series that defines Tx is convergent, that is, the sequence of the partial sums is a Cauchy sequence. Let $k < l$ be positive integers. Then

$$\left\| \sum_{n=k}^{l} \xi_n Te_n \right\|^2 \leq \sum_{n=k}^{l} |\xi_n|^2 + \left| \sum_{k \leq j^2 \leq l} \alpha_j \xi_{j^2} \right|^2,$$

since, substituting the defining expressions of the vectors Te_n in the sum $\sum_{n=k}^{l} \xi_n Te_n$, in the linear combination of basis vectors so obtained the coefficients of the vectors e_n with $k+1 \leq n \leq l+1$ have absolute value less than or equal to 1, the coefficient of e_1 is $\sum_{k \leq j^2 \leq l} \alpha_j \xi_{j^2}$, and the remaining basis vectors do not appear in the linear combination. Using the Cauchy inequality, it follows that

$$\begin{aligned} \left\| \sum_{n=k}^{l} \xi_n Te_n \right\|^2 &\leq \sum_{n=k}^{l} |\xi_n|^2 + \left(\sum_{k \leq j^2 \leq l} |\alpha_j|^2 \right) \left(\sum_{k \leq j^2 \leq l} |\xi_{j^2}|^2 \right) \\ &\leq (1 + \alpha) \sum_{n=k}^{l} |\xi_n|^2. \end{aligned}$$

Since $\sum_{n=1}^{\infty} |\xi_n|^2 = \|x\|^2 < \infty$, the series $\sum_{n=1}^{\infty} \xi_n Te_n$ is convergent. The linearity of T is now obvious, and if in the preceding inequality we write $k = 1$ and let l tend to infinity, then we obtain the boundedness of T, namely, $\|T\| \leq 1 + \alpha$.

We show that $e_1 \in T^l(H)^-$ for every positive integer l. In fact,

$$e_{j^2} = T^{2j-2} e_{(j-1)^2+1} \qquad \text{for} \quad j \geq 2$$

and therefore

$$e_1 + \frac{1}{j}e_{j^2+1} = \frac{1}{\alpha_j}Te_{j^2} = T^{2j-1}\left(\frac{1}{\alpha_j}e_{(j-1)^2+1}\right) \in T^{2j-1}(H).$$

Thus,

$$e_1 + \frac{1}{j}e_{j^2+1} \in T^{2j-1}(H) \subset T^l(H) \quad \text{if} \quad 2j - 1 \geq l,$$

and letting j tend to infinity, we obtain

$$e_1 = \lim_{j\to\infty}\left(e_1 + \frac{1}{j}e_{j^2+1}\right) \in T^l(H)^- .$$

It remains to show that $\cap_{l=1}^{\infty} T^l(H) = \{0\}$. Suppose $x \in \cap_{l=1}^{\infty} T^l(H)$. From the definition of T it is clear that, for any l, the vectors e_n with $2 \leq n \leq l+1$ are orthogonal to $T^l(H)$. Thus $\langle x, e_n\rangle = 0$ for all integers $n \geq 2$. On the other hand, if $Ty \neq 0$ for some $y \in H$, then there necessarily exists an integer $n \geq 2$ with $\langle y, e_n\rangle \neq 0$. Then, however, $\langle Ty, e_{n+1}\rangle \neq 0$ and, by the foregoing, $x = T0 = 0$. □

3.8 PROBABILITY THEORY

Problem P.1. *From a given triangle of unit area, we choose two points independently with uniform distribution. The straight line connecting these points divides the triangle, with probability one, into a triangle and a quadrilateral. Calculate the expected values of the areas of these two regions.*

Solution. First, observe that an affine transformation does not change the ratio of the areas, therefore the expectation of the ratio does not change. We will use this fact by choosing the type of the triangle in the most convenient way for our purposes.

Denote the vertices of the triangle by A_1, A_2, A_3, and the chosen points by X and Y. The probability that the line e connecting the points X and Y crosses one of the points A_i $(i = 1, 2, 3)$ is equal to zero because of the independence of X and Y. So the probability that we will get a triangle and a quadrilateral by dividing the triangle by the line e is equal to one. Denote by $\mathcal{A}_i$ the event that the point A_i is a vertex of the small triangle. The events $\mathcal{A}_i$ $(i = 1, 2, 3)$ form a complete system of events, so by the theorem of "complete expectations,"

$$E(\frac{t}{T}) = \sum_{i=1}^{3} E(\frac{t}{T}|\mathcal{A}_i)P(\mathcal{A}_i),$$

where t denotes the area of the small triangle, and T denotes the area of the original triangle.

If the original triangle is equilateral, then $E(t/T|\mathcal{A}_i)$ is independent of i (so does $P(\mathcal{A}_i)$ for $i = 1, 2, 3$), hence

$$E(\frac{t}{T}) = E(\frac{t}{T}|\mathcal{A}_1)\sum_{i=1}^{3} P(\mathcal{A}_i) = E(\frac{t}{T}|\mathcal{A}_1).$$

Suppose now that we have a right-angled isosceles triangle, A_i is the right-angled vertex, and the equal sides have unit length. Denote by $F(a, b)$ the (conditional) probability, that for the line segments ξ and η of the sides cut by e, we have $\xi < a$ and $\eta < b$, assuming that $\xi < 1$ and $\eta < 1$. That is,

$$F(a, b) = P(\xi < a, \eta < b|\xi < 1, \eta < 1).$$

Since $a < 1$ and $b < 1$,

$$F(a, b) = \frac{P(\xi < a, \eta < b)}{P(\xi < 1, \eta < 1)}.$$

If $X = (x_1, x_2)$ and $Y = (y_1, y_2)$, then

$$P(\xi < a, \eta < b) = c\int_{\triangle(a,b)} dx_1dx_2dy_1dy_2,$$

where c is a constant, independent of a and b, and $\triangle(a,b)$ is the domain of the four-dimensional space such that, for the corresponding line, $\xi < a$ and $\eta < b$. By substitutions $x_i = ax_i'$ and $y_i = by_i'$, the domain $\triangle(a,b)$ is transformed to $\triangle(1,1)$, and therefore

$$P(\xi < a, \eta < b) = ca^2b^2 \int_{\triangle(1,1)} dx_1' dx_2' dy_1' dy_2' = a^2b^2 P(\xi < 1, \eta < 1),$$

that is, $F(a,b) = a^2b^2$. So the density function is

$$f(a,b) = \frac{\partial^2 F(a,b)}{\partial a \partial b} = 4ab,$$

and thus the ratio of the areas of the triangles is given by

$$g(a,b) = \frac{\frac{1}{2}ab}{\frac{1}{2}\cdot 1 \cdot 1} = ab.$$

The expectation of the ratio is

$$E(\frac{t}{T}) = E\ g(a,b) = \int_0^1 \int_0^1 g(a,b) f(a,b)\ da\ db = \int_0^1 \int_0^1 4a^2b^2\ da\ db = \frac{4}{9}.$$

So the expectations of the areas of the triangle and the quadrilateral are 4/9 and 5/9, respectively. □

Problem P.2. *Select n points on a circle independently with uniform distribution. Let P_n be the probability that the center of the circle is in the interior of the convex hull of these n points. Calculate the probabilities P_3 and P_4.*

Solution 1. Assume that the circumference of the circle is one. Fix a point on the circle, and introduce the arclength parameter. Let $\tau_1, \dots, \tau_n$ be the parameter values associated with the chosen points. Denote by A_n the event that the center of the circle is not inside the convex hull of the points. Furthermore, denote by B_i the event that there is no point on the arc $(\tau_i, \tau_i + 1/2)$. By the independence and uniform distribution of the points, it is obvious, that, except for an event of probability zero, the events B_i are mutually disjoint and $A_n = B_1 + \dots + B_n$, $P(B_i) = 1/2^{n-1}$, and therefore $P_n = 1 - P(A_n) = 1 - n/2^{n-1}$. Hence $P_3 = 1/4$, $P_4 = 1/2$. □

Solution 2. Denote by $\widehat{\tau_i\tau_j}$ the length of the shorter arc with endpoints τ_i and τ_j and by B_{ij} $(1 \le i < j \le n)$ the event that all of the points are on the shorter arc with endpoints τ_i and τ_j. (If τ_i and τ_j are the endpoints of a half-circle, then B_{ij} denotes the event that all of the points are on one of the half-circles defined by them.) It is obvious that $A_n = \sum_{i,j} B_{ij}$. If the points are distinct, the events B_{ij} are mutually disjoint. Since the points are independent and uniformly distributed, they do not coincide with probability one, and so $P(A_n) = \sum_{i,j} P(B_{ij})$. Furthermore, $P(\widehat{\tau_i\tau_j} < x) = 2x$, so $(P(\widehat{\tau_i\tau_j} < x))' = 2 \quad (0 < x < 1/2)$ and $P(B_{ij}|\widehat{\tau_i\tau_j} < x) = x^{n-2}$. Because of the theorem of total probability,

$$P(B_{ij}) = 2\int_0^{1/2} x^{n-2}dx = \frac{1}{(n-1)2^{n-2}}.$$

Hence,

$$P_n = 1 - P(A_n) = 1 - \binom{n}{2}\frac{1}{(n-1)2^{n-2}} = 1 - \frac{n}{2^{n-1}}\ . \quad \square$$

Problem P.3. *Let $\varepsilon_1, \varepsilon_2, \ldots, \varepsilon_{2n}$ be independent random variables such that $P(\varepsilon_i = 1) = P(\varepsilon_i = -1) = 1/2$ for all i, and define $S_k = \sum_{i=1}^k \varepsilon_i$, $1 \le k \le 2n$. Let N_{2n} denote the number of integers $k \in [2,\, 2n]$ such that either $S_k > 0$, or $S_k = 0$ and $S_{k-1} > 0$. Compute the variance of N_{2n}.*

Solution. It is known (see, for example, *W., Feller, An Introduction to Probability Theory and Its Applications, Wiley, New York, 1957, Vol. I, p. 77*) that the distribution of N_{2n} is given as

$$P(N_{2n} = 2k) = \frac{\binom{2k}{k}\binom{2(n-k)}{n-k}}{2^{2n}} \qquad (k = 0, 1, \ldots, n), \tag{1}$$

$$P(\mathbb{N}_{2n} = 2k+1) = 0 \qquad (k = 0, 1, \ldots, n-1). \tag{2}$$

The symmetry of the distribution implies that $E(N_{2n}) = n$, and from (1) and (2) we conclude that

$$E(N_{2n}^2) = \sum_{k=0}^{n} (2k)^2 \frac{\binom{2k}{k}\binom{2(n-k)}{n-k}}{2^{2n}}.$$

Assume that $|x| < 1$, and consider the functions

$$F(x) = \sum_{k=1}^{\infty} (2k)^2 \frac{\binom{2k}{k}}{2^{2k}} x^{2k} \text{ and } G(x) = (1-x^2)^{-1/2}.$$

It is easy to see that

$$G(x) = \sum_{k=0}^{\infty} \frac{\binom{2k}{k}}{2^{2k}} x^{2k}. \tag{3}$$

The generating function of $E(N_{2n}^2)$ is $F(x)G(x)$, that is,

$$F(x)G(x) = \sum_{k=0}^{\infty} E(N_{2n}^2)x^{2k}.$$

Using (3), it is easy to verify that

$$F(x) = x(xG'(x))',$$

which implies that

$$F(x) = x(x^2(1-x^2)^{-3/2})' = 2x^2(1-x^2)^{-3/2} + 3x^4(1-x^2)^{-5/2}.$$

Therefore,

$$F(x)G(x) = 2x^2(1-x^2)^{-2} + 3x^4(1-x^2)^{-3}. \tag{4}$$

Simple calculations show that

$$(1-x^2)^{-2} = \sum_{k=1}^{\infty} kx^{2k-2}$$

and

$$(1-x^2)^{-3} = \sum_{k+1}^{\infty} \frac{k(k-1)}{2} x^{2k-4}.$$

So from (4), we get

$$F(x)G(x) = \sum_{k=1}^{\infty} 2kx^{2k} + \sum_{k=1}^{\infty} \frac{3}{2}k(k-1)x^{2k} = \sum_{k=1}^{\infty} \left(\frac{3}{2}k^2 + \frac{1}{2}k\right) x^{2k}.$$

Therefore,

$$E(N_{2n}^2) = \frac{3}{2}n^2 + \frac{1}{2}n\ .$$

Hence,

$$\operatorname{Var}(N_{2n}) = E(N_{2n}^2) - E^2(N_{2n}) = \left(\frac{3}{2}n^2 + \frac{1}{2}n\right) - n^2 = \frac{n(n+1)}{2}\ . \quad \square$$

Problem P.4. *A gambler plays the following coin-tossing game. He can bet an arbitrary positive amount of money. Then a fair coin is tossed, and the gambler wins or loses the amount he bet depending on the outcome. Our gambler, who starts playing with x forints, where $0 < x < 2C$, uses the following strategy: if at a given time his capital is $y < C$, he risks all of it; and if he has $y > C$, he only bets $2C - y$. If he has exactly $2C$ forints,* *he stops playing. Let $f(x)$ be the probability that he reaches $2C$ (before going bankrupt). Determine the value of $f(x)$.*

Solution 1. We are going to prove that $f(x) = x/2C \quad (0 < x < 2C)$. First, we show that $f(x)$ is nondecreasing. Let $0 < x_1 < x_2 < 2C$, and suppose there is a sequence of tossings such that he can reach $2C$ from x_1. We show that by the same sequence of tossings he can reach $2C$ from x_2 (maybe earlier), so $f(x_2) \geq f(x_1)$.

Let us see how the amount of the player's money changes after one toss:

(a) if $x_1 < x_2 < C$, then $2x_1 < 2x_2$ for a win, and $0 \leq 0$ for a loss;

(b) if $x_1 < C < x_2$, then $2x_1 < 2C$ for a win, and $0 < 2(x_2 - C)$ for a loss;

(c) if $C < x_1 < x_2$, then $2C \leq 2C$ for a win, and $2(x_1 - C) < 2(x_2 - C)$ for a loss.

Thus if $x_1 < x_2$, then, for the new amounts x_1' and x_2', we have $x_1' \leq x_2'$.

Define $x = (a/2^n) \cdot 2C \quad (1 \leq a \leq 2^n - 1$, a is odd, $n = 1, 2, \ldots)$. We will prove by induction that for such an x, $f(x) = a/2^n = x/2C$. For $n = 1$, $f((1/2) \cdot 2C) = f(C) = 1/2$, since starting from C the player will have $2C$ with probability $1/2$. Suppose the formula holds for $n = k \quad (k \geq 1)$, that is, $f(x) = f((a/2^k) \cdot 2C) = a/2^k$. If $x = (a/2^{k+1}) \cdot 2C$ and $a < 2^k$ (that is, $x < C$), then the player will win $2x$ forints or lose everything with probabilities $1/2$, $1/2$, respectively. So

$$f(x) = \frac{1}{2} f(\frac{a}{2^k} 2C) + \frac{1}{2} f(0) = \frac{a}{2^{k+1}} = \frac{x}{2C}.$$

If $x = (a/2^{k+1}) \cdot 2C$ and $2^k < a < 2^{k+1}$ $(a \neq 2^k$ since it is odd), then the inequality $x > C$ implies

$$f(x) = \frac{1}{2} f(2C) + \frac{1}{2} f\left(\frac{a - 2^k}{2^k} - 2C\right) = \frac{1}{2} + \frac{1}{2}\frac{a - 2^k}{2^k} = \frac{a}{2^{k+1}} = \frac{x}{2C}.$$

Suppose now $x_0 \neq (a/2^n) \cdot 2C$ $(1 \leq a \leq 2^n - 1$, a is odd), and suppose $f(x_0) \neq x_0/2C$, but, for example, $f(x_0) > x_0/2C$ .Then there should exist a number having the form $a/2^n$ such that

$$\frac{x_0}{2C} < \frac{a}{2^n} < f(x_0), \tag{1}$$

and it contradicts the fact that $f(x)$ is nondecreasing. Namely, because of (1), $x_0 < (a/2^n) \cdot 2C$, and, on the other hand,

$$\frac{a}{2^n} = f(\frac{a}{2^n} 2C) < f(x_0).$$

We can obtain a similar contradiction for $f(x_0) < x_0/2C$, therefore $f(x) = x/2C \, (0 < x < 2C)$. □

Solution 2. Introduce the notation $g(x) = f(2Cx)$ $(0 \le x \le 1)$. Since $f(0) = 0$ and $f(2C) = 1$, we get $g(0) = 0$ and $g(1) = 1$. It is easy to verify that

$$g(x) = \begin{cases} \frac{1}{2}g(2x) & \text{if } x \in [0, \frac{1}{2}], \\ \frac{1}{2} + \frac{1}{2}g(2x-1) & \text{if } x \in (\frac{1}{2}, 1]. \end{cases} \tag{2}$$

To see this, observe that if $x \in (0, 1/2]$, the gambler wins or loses $2Cx$ forints with probability $1/2$, so

$$g(x) = \frac{1}{2}g(2x) + \frac{1}{2}g(0) = \frac{1}{2}g(2x).$$

If $x \in (1/2, 1]$, the gambler will have 2C or $(2x-1)2C$ forints with equal probabilities $1/2$, so

$$g(x) = \frac{1}{2}g(1) + \frac{1}{2}g(2x-1) = \frac{1}{2} + \frac{1}{2}g(2x-1).$$

We will next show that the only bounded solution of the functional equation (2) is the identity function $g(x) = x$ (from which we get the same solution as before).

Put $h(x) = g(x) - x$. Then $h(x)$ is bounded and

$$h(x) = \begin{cases} \frac{1}{2}h(2x) & \text{if } x \in [0, \frac{1}{2}], \\ \frac{1}{2}h(2x-1) & \text{if } x \in (\frac{1}{2}, 1]. \end{cases} \tag{3}$$

We are going to show that $h(x) \equiv 0$. Since $h(x)$ is bounded, both $M = \sup_{x\in[0,1]} h(x)$ and $m = \inf_{x\in[0,1]} h(x)$ are finite, and

$$h(x) = \begin{cases} \frac{1}{2}h(2x) \le \frac{M}{2} & \text{if } x \in [0, \frac{1}{2}], \\ \frac{1}{2}h(2x-1) \le \frac{M}{2} & \text{if } x \in (\frac{1}{2}, 1]. \end{cases}$$

The definition of M implies that $M \le M/2$, that is, $M \le 0$. A similar argument shows that $m \ge 0$, that is, $h(x) \equiv 0$. □

Problem P.5. *For a real number x in the interval $(0, 1)$ with decimal representation*

$$0.a_1(x)a_2(x)\dots a_n(x)\dots,$$

denote by $n(x)$ the smallest nonnegative integer such that

$$\overline{a_{n(x)+1}a_{n(x)+2}a_{n(x)+3}a_{n(x)+4}} = 1966.$$

Determine $\int_0^1 n(x)dx$. ($\overline{abcd}$ denotes the decimal number with digits a, b, c, d.)

Solution 1. The integral is understood as a Lebesgue integral. Introduce the following notation:

$$L_n = \{x : \ x \in (0,1), \ \ n(x) = n\},$$
$$A_n = \{x : \ x \in (0,1), \ \ \overline{a_{n+1}(x)a_{n+2}(x)a_{n+3}(x)a_{n+4}(x)} = 1966\}.$$

λ is the Lebesgue measure, $\lambda_n = \lambda(L_n)$, $\alpha_n = \lambda(A_n)$, where $(n = 0, 1, 2, \dots)$. Observe that L_n and A_n are the unions of finitely many intervals, so they are measurable sets.

First, we show that $\sum_{n=0}^{\infty} \lambda_n = 1$, that is, the integrand is defined almost everywhere. If

$$H = \{x : x \in (0,1),\ \overline{a_{k+1}(x)a_{k+2}(x)a_{k+3}(x)a_{k+4}(x)} \neq 1966, k = 1, 2, \dots\}$$

and

$$C_n = \{x : x \in (0,1),\ \overline{a_{4k+1}(x)a_{4k+2}(x)a_{4k+3}(x)a_{4k+4}(x)} \neq 1966, k = 1, 2, \dots\},$$

then $H \subset C_n \quad (n = 0, 1, 2, \dots)$ and

$$\lambda(C_n) = \left(\frac{10^4 - 1}{10^4}\right)^{n+1}$$

Therefore, $\lim_{n\to\infty} \lambda(C_n) = 0$, $\lambda(H) = 0$. The lengths of the intervals in A_n are equal to $10^{-(n+4)}$, the number of these intervals is 10^n (we can choose the first n decimal digits in 10^n different ways), and therefore $\alpha_n = 10^{-4}$. Obviously, $A_n \cap L_n = L_n$ $(n = 0, 1, \dots)$, and if $n - 3 \leq k \leq n - 1$, then $A_n \cap L_k = 0$ since two sequences 1966 cannot overlap. If $0 \leq k \leq n - 4$, then

$$\lambda(A_n \cap L_k) = \lambda(A_{n-k-4})\lambda(L_k) = 10^{-4}\lambda_k,$$

since the assumption that x is in L_k restricts only the first $(k+4)$ decimal digits of x. So

$$10^{-4} = \lambda(A_n) = \sum_{k=0}^{n} \lambda(A_n \cap L_k) = \sum_{k=0}^{n-4} 10^{-4}\lambda_k + \lambda_n,$$

that is,

$$10^4\lambda_n = 1 - \sum_{k=0}^{n-4} \lambda_k.$$

By the equality $\sum_{k=0}^{\infty} \lambda_k = 1$, we get

$$\sum_{k=n+1}^{\infty} \lambda_k = 10^4\lambda_{n+4}. \tag{1}$$

For the integral of $n(x)$,

$$\int_0^1 n(x)dx = \sum_{n=0}^{\infty} n\lambda_n = \sum_{n=0}^{\infty} \sum_{k=n+1}^{\infty} \lambda_k.$$

Hence, by (1),

$$\int_0^1 n(x)dx = \sum_{n=0}^{\infty} 10^4\lambda_{n+4} = 10^4 \sum_{n=4}^{\infty} \lambda_n$$
$$= 10^4(1 - \lambda_0 - \lambda_1 - \lambda_2 - \lambda_3),$$

and since $\lambda_0 = \lambda_1 = \lambda_2 = \lambda_3 = 10^{-4}$, we clearly get

$$\int_0^1 n(x)dx = 10^4 - 4 = 9996. \quad \square$$

Solution 2. We solve the problem in a more general form. Let $\sigma = \overline{s_1, s_2, \dots, s_k}$ be an integer given in the decimal system, and for $x \in (0,1)$ define the function $n(x)$ as follows: let $n(x)$ be the smallest nonnegative integer such that

$$\overline{a_{n(x)+1}a_{n(x)+2}\dots a_{n(x)+k}} = \overline{s_1 s_2 \dots s_k}.$$

We are going to calculate the value of $\int_0^1 n(x)dx$. Let $(\Omega, \mathcal{A}, P)$ be the probability space of the Lebesgue-measurable subsets of the interval $(0,1)$ endowed with the Lebesgue measure. Then on the set $\Omega = (0,1)$, the digits $a_1, a_2, \dots, a_k, \dots$ are independent and

$$P(a_k = i) = \frac{1}{10} \quad (k = 1, 2, \dots;\ i = 0, 1, \dots, 9).$$

On this space, $n(x)$ is a random variable, and its expectation is

$$\int_0^1 n(x)\, dx.$$

Put $P_n = P(n(x) = n)$ and $Q_n = P(\overline{a_{n+1}a_{n+2}\dots a_{n+k}} = \sigma) = 1/10^k$ $(n = 0, 1, \dots)$.

Let l be the smallest natural number such that shifting σ by l positions, the digits on the same positions coincide. (That is $s_i = s_{i+l}$ for all $i = 1, 2, \dots, k-l$). Obviously, $1 \le l \le k$. Rewrite k in the form $k = pl + r$, where $0 \le p$, $1 \le r \le l$. The σ consists of p blocks $\tau = \overline{s_1 s_2 \dots s_l}$ of length l, and the first r digits of this block τ. Notice that

$$\begin{aligned} Q_n = {} & (P_0 Q_{n-k} + P_1 Q_{n-k-1} + \dots + P_{n-k} Q_0) \\ & + (P_{n-k+r} 10^{-(k-r)} + P_{n-k+r+l} 10^{-(k-r-l)} + \dots + P_{n-l} 10^{-l} + P_n) \\ & (n = k, k+1, \dots) \end{aligned} \tag{2}$$

since the event

$$\overline{a_{n+1}a_{n+2}\dots a_{n+k}} = \sigma$$

can occur in two different cases, as follows:

(a) Block σ appears only from the $(i+1)$th index $(0 \leq i \leq n-k)$, and after $n-k-i$ digits it occurs again. (The probability of this case is $P_i Q_{n-k-i}$.)

(b) Block σ appears from the $(n-k+r+il)$th index $(i = 0, 1, \ldots, p)$, thus

$$\overline{a_{n-k+r+il+1} a_{n-k+r+il+2} \cdots a_{n+r+il}} = \sigma,$$

and it appears again starting from the $(n+1)$th index. Note that besides the previous assumption, it is also necessary that the values of the undefined $(k-r-il)$ digits are given. The probability of this case is $P_{n-k+r+il} 10^{k-l-il}$.

Assume that $P_n = Q_n = 0$ if $n = -1, -2, \ldots$. Then (2) holds also for $0 \leq n < k$. Multiply (2) by z^n, add up the resulting equations for all values of n, and introduce the notation

$$P(z) = \sum_{n=0}^{\infty} P_n z^n, \qquad Q(z) = \sum_{n=0}^{\infty} Q_n z^n$$

to get

$$Q(z) = z^k P(z) Q(z) + \sum_{i=0}^{p} z^{k-r-il} P(z) 10^{-(k-r-il)}.$$

Using the obvious formula

$$Q(z) = \sum_{n=0}^{\infty} \frac{1}{10^k} z^n = \frac{1}{10^k(1-z)},$$

we have

$$P(z) = \frac{1}{z^k + 10^k(1-z)\sum_{i=0}^{p} z^{k-r-il} 10^{-(k-r-il)}}.$$

From $P(1) = 1$ we can see that $n(x)$ is defined almost everywhere, and

$$\int_0^1 n(x)dx = P'(1) = -k + \sum_{i=0}^{p} 10^{r+il} .$$

This last equality shows that the integral is always an integer. If $\sigma = 1966$, then $k = l = r = 4$, $p = 0$, and thus

$$\int_0^1 n(x)dx = 10^4 - 4 = 9996.$$

If $\sigma = 1961$, then $k = 4$, $l = 3$, $p = r = 1$, and thus

$$\int_0^1 n(x)dx = 10 + 10^4 - 4 = 10,006.$$

If $\sigma = 1919$, then $k = 4$, $l = r = 2$, $p = 1$, thus

$$\int_0^1 = 10^2 + 10^4 - 4 = 10,096,$$

and, finally, if $\sigma = 1111$, then $k = 4$, $l = r = 1$, $p = 3$, thus

$$\int_0^1 n(x)dx = 10 + 10^2 + 10^3 + 10^4 - 4 = 11,106. \quad \square$$

Remarks.

1. The problem can obviously be generalized for any number system not just the decimal one.

2. One can reduce the problem to compute the expectation of the first return time of a recurrent sequence of events. In this case, the value of the integral can be determined using the theorem by Erdős, Feller, and Pollard.

Problem P.6. *Let f be a continuous function on the unit interval* $[0, 1]$. *Show that*

$$\lim_{n\to\infty} \int_0^1 \cdots \int_0^1 f\left(\frac{x_1 + \cdots + x_n}{n}\right) dx_1 \dots dx_n = f\left(\frac{1}{2}\right)$$

and

$$\lim_{n\to\infty} \int_0^1 \cdots \int_0^1 f(\sqrt[n]{x_1 \dots x_n}) dx_1 \dots dx_n = f\left(\frac{1}{e}\right).$$

Solution 1. Let k be an arbitrary positive integer and $n \geq k$. Consider the multinomial expansion of $(\sum_{i=1}^n x_i)^k$. The number of terms that contain all variables with powers not exceeding one is

$$n(n-1)\dots(n-k+1) = n^k + O(n^{k-1}).$$

The integrals of these terms on the unit cube

$$R_n = \{(x_1, \dots, x_n) : \; 0 \leq x_i \leq 1\} \quad (i = 1, \dots, n)$$

are equal to $1/2^k$. The number of terms containing at least one x_i with a power higher than 1 is not greater than $n \cdot n^{k-2} = n^{k-1}$. The integrals of these terms on R_n are not greater than 1. From this observation, we get

$$\int_0^1 \cdots \int_0^1 \left(\frac{x_1 + \cdots + x_n}{n}\right)^k dx_1 \dots dx_n \;\to\; \left(\frac{1}{2}\right)^k$$

as $n \to \infty$.

An easy calculation shows that

$$\int_0^1 \cdots \int_0^1 (\sqrt[n]{x_1 \dots x_n})^k dx_1 \dots dx_n = \prod_{i=1}^n \int_0^1 x_i^{k/n} dx_i = \frac{1}{(1+\frac{k}{n})^n} \to \left(\frac{1}{e}\right)^k$$

as $n \to \infty$.

Hence, we have shown that the statements are valid for $f(x) = x^k$. It is also obvious that the statements hold for constant functions, and if they are valid for two continuous functions, they are valid for their linear combinations as well. So we have proved the theorem for polynomials.

If the function f is continuous on $[0,1]$, then by the Weierstrass approximation theorem, for any positive ε there is a polynomial $p(x)$ such that $|f(x) - p(x)| < \varepsilon/3$ $(0 \le x \le 1)$. Since the statements are valid for polynomials, there is a K such that

$$\left|\int_0^1 \cdots \int_0^1 p\left(\frac{x_1 + \dots + x_n}{n}\right) dx_1 \dots dx_n - p\left(\frac{1}{2}\right)\right| < \frac{\varepsilon}{3},$$
$$\left|\int_0^1 \cdots \int_0^1 p(\sqrt[n]{x_1 \dots x_n}) dx_1 \dots dx_n - p\left(\frac{1}{e}\right)\right| < \frac{\varepsilon}{3},$$

for any $n > K$. Therefore, for any $n > K$,

$$\begin{aligned}
&\left|\int_0^1 f\left(\frac{x_1 + \dots + x_n}{n}\right) dx_1 \dots dx_n - f\left(\frac{1}{2}\right)\right| \\
&\le \int_0^1 \cdots \int_0^1 \left|f\left(\frac{x_1 + \dots + x_n}{n}\right) - p\left(\frac{x_1 + \dots + x_n}{n}\right)\right| dx_1 \dots dx_n \\
&\quad + \left|\int_0^1 \cdots \int_0^1 p\left(\frac{x_1 + \dots + x_n}{n}\right) dx_1 \dots dx_n - p\left(\frac{1}{2}\right)\right| \\
&\quad + \left|p\left(\frac{1}{2}\right) - f\left(\frac{1}{2}\right)\right| < \varepsilon,
\end{aligned}$$

and similarly,

$$\left|\int_0^1 \cdots \int_0^1 f(\sqrt[n]{x_1 \dots x_n}) dx_1 \dots dx_n - f\left(\frac{1}{e}\right)\right| < \varepsilon.$$

Now the proof is completed. □

Solution 2. Let $\xi_1, \dots, \xi_n, \dots$ be mutually independent, uniformly distributed random variables on the interval $(0,1)$. Then the variables $\eta_n = \log \xi_n$ $(n = 1, 2, \dots)$ are also mutually independent, and they have the same distribution. Since $E(\xi_n) = 1/2$ and $E(\eta_n) = -1$ $(n = 1, 2, \dots)$, a theorem by Kolmogorov implies that

$$\varphi_n = \frac{\xi_1 + \dots + \xi_n}{n} \to \frac{1}{2}$$

and

$$\varphi'_n = \frac{\eta_1 + \cdots + \eta_n}{n} \to -1$$

with probability 1.

By the assumptions the function f is bounded on the interval $[0, 1]$ and continuous at $x = 1/2$, and the function $g(x) = f(e^x)$ is also bounded on $(-\infty, 0]$ and continuous at $x = -1$. Using the Lebesgue theorem, we get

$$E(f(\varphi_n)) \to E(f(\frac{1}{2})) = f(\frac{1}{2}),$$
$$E(g(\varphi'_n)) \to E(g(-1)) = E(f(\frac{1}{e})) = f(\frac{1}{e}),$$

and, finally, the definitions of φ_n, φ'_n, and g imply that

$$f\left(\frac{1}{2}\right) = E(f(\varphi_n)) = \int_0^1 \cdots \int_0^1 f\left(\frac{x_1 + \cdots + x_n}{n}\right) dx_1 \ldots dx_n$$

and

$$f\left(\frac{1}{e}\right) = E(g(\varphi'_n)) = \int_0^1 \cdots \int_0^1 f(\sqrt[n]{x_1 \ldots x_n}) dx_1 \ldots dx_n. \quad \square$$

Remark. The statements also hold in the more general case when f is bounded, integrable, and continuous at $x = 1/2$ and $x = 1/e$, respectively. In the first statement we can also drop the condition of boundedness (as was pointed out by László Lovász).

Problem P.7. *Let $A_1, \ldots, A_n$ be arbitrary events in a probability field. Denote by C_k the event that at least k of $A_1, \ldots, A_n$ occur. Prove that*

$$\prod_{k=1}^{n} P(C_k) \leq \prod_{k=1}^{n} P(A_k).$$

Solution. We start with three remarks:

Comment A. The statement is true for $n = 2$, that is,

$$P(A + B)P(AB) \leq P(A)P(B).$$

To see this, use the notation $AB = x$, $A\overline{B} = y$, and $\overline{A}B = z$. Then our inequality becomes

$$(P(x) + P(y) + P(z))P(x) \leq (P(x) + P(y))(P(x) + P(z)),$$

which obviously holds.

Comment B. If $A_1 \supseteq \cdots \supseteq A_n$, then $C_k = A_k$.

Comment C. C_k is identical to the event that at least k occur from the events A_1+A_2, A_1A_2, $A_3,\dots,A_n$, that is, from the events A_1+A_2, A_1A_2, $A_3,\dots,A_n$, exactly as many occur as from the events $A_1,A_2,\dots,A_n$. This statement is trivial.

Our proof is based on the above comments. Put $A_i^0 = A_i$ $(i=1,\dots,n)$, and assume that the events $A_1^\mu, \dots, A_n^\mu$ are already defined. Next, define the events $A_1^{\mu+1}, \dots, A_n^{\mu+1}$ as follows: choose a pair A_i^μ, A_j^μ of events, $i<j$, for which $A_i^\mu \not\supseteq A_j^\mu$, $A_j^\mu \not\supseteq A_i^\mu$ (that is, A_i^μ and A_j^μ are incomparable) and define $A_i^{\mu+1} = A_i^\mu + A_j^\mu$, $A_j^{\mu+1} = A_i^\mu A_j^\mu$ and $A_k^{\mu+1} = A_k^\mu$ for $k \neq i,j$. Continue this procedure until there exists at least one pair of incomparable events. This procedure will terminate in finitely many steps, since in each step the number of incomparable pairs decreases. This observation follows from the facts that $A_i^{\mu+1}$ and $A_j^{\mu+1}$ become comparable, and any event is comparable to at least as many of the events $A_i^\mu + A_j^\mu$, $A_i^\mu A_j^\mu$, as of A_i^μ and A_j^μ.

By comment A,

$$\prod_{k=1}^{n} P(A_k^{\mu+1}) \leq \prod_{k=1}^{n} P(A_k^\mu). \tag{1}$$

Observe that comment C implies that the events C_k remain unchanged, and because of comment B they are identical to the final events A_i^μ. Therefore, (1) implies the original statement.

From the proof of comment A, we see that if $\prod_{k=1}^n P(A_k^\mu) \neq 0$ and $P(A_i^\mu \overline{A}_j^\mu) P(\overline{A}_i^\mu A_j^\mu) \neq 0$, then strict inequality holds in (1). Hence, equality holds iff either $\prod_{k=1}^n P(A_k) = 0$ or the original system of events $\{A_k\}$ are already "ordered." □

Remarks.

1. If multiplication is replaced by addition, then in the statement we always have equality. That is,

$$\sum_{i=1}^{n} P(C_i) = \sum_{i=1}^{n} P(A_i).$$

2. Let $S_k(x_1,\dots,x_n)$ be the kth elementary symmetric polynomial of the variables $x_1,\dots,x_n$. Then

$$S_k(P(C_1),\dots,P(C_n)) \leq S_k(P(A_1),\dots,P(A_n)).$$

Slight modifications of the above proof can be applied to show the validity of this more general statement.

Problem P.8. *Let A and B be nonsingular matrices of order p, and let ξ and η be independent random vectors of dimension p. Show that if ξ, η and $\xi A + \eta B$ have the same distribution, if their first and second* *moments exist, and if their covariance matrix is the identity matrix, then these random vectors are normally distributed.*

Solution. Put $\zeta = \xi A + \eta B$. Without loss of generality, we can suppose that $E(\xi) = E(\eta) = E(\zeta) = 0$. Since ξ and η are independent,

$$\mathrm{Var}\,(\xi) = E[(\xi A + \eta B)^*(\xi A + \eta B)] = A^* \,\mathrm{Var}\,(\xi) A + B^* \,\mathrm{Var}\,(\eta) B,$$

where $\mathrm{Var}\,(\xi)$, $\mathrm{Var}\,(\eta)$, and $\mathrm{Var}\,(\zeta)$ denote the covariance matrices of ξ, η, and ζ, respectively. They are equal to the identity matrix, so

$$A^*A + B^*B = I.$$

Thus, for arbitrary vector t,

$$(tA^*, tA^*) + (tB^*, tB^*) = (t, t),$$

and by the regularity of A and B,

$$|tA^*| < |t| \quad |tB^*| < |t| \quad (t \neq 0), \tag{1}$$

where $|.|$ denotes the Euclidean length of vectors.

Since ξ and η are independent,

$$E(e^{i(t, \xi A + \eta B)}) = E(e^{i(tA^*, \xi)}) E(e^{i(tB^*, \eta)}).$$

If $\phi(t)$ denotes the common characteristic function of ξ, η, and ζ, then

$$\varphi(t) = \varphi(tA^*)\varphi(tB^*). \tag{2}$$

We are going to show that $\varphi(t) \neq 0$. Assume that $\varphi(t)$ has real roots. Then its continuity implies that there is a root t_0 with smallest absolute value. Since $\varphi(0) = 1,\ t_0 \neq 0$. Thus, (1) implies that

$$\varphi(t_0 A^*) \neq 0, \quad \varphi(t_0 B^*) \neq 0,$$

and this contradicts equation (2).

Consider the function

$$\psi(t) = \log \varphi(t) + \frac{1}{2}(t, t),$$

and choose the branch of the logarithmic function for which $\psi(0) = 0$.

The assumptions imply the existence of vectors

$$\varphi'(t) = \frac{d}{dt}\varphi(t), \quad \varphi'(0) = iE(\xi) = 0$$

and matrices

$$\varphi''(t) = \frac{d^2}{dt^2}\varphi(t), \quad \varphi''(0) = -\mathrm{Var}\,(\xi) = -E.$$

Therefore,

$$\left[\frac{d}{dt}\psi(t)\right]_{t=0} = 0, \quad \left[\frac{d^2}{dt^2}\psi(t)\right]_{t=0} = 0,$$

and hence

$$\psi(t) = o((t,t)) \quad \text{if} \quad |t| \to 0. \tag{3}$$

We are going to show that $\psi(t) \equiv 0$. Suppose $\psi(t) \not\equiv 0$. Then there exists a positive ε and a $t \neq 0$ such that $|\psi(t)| \geq \varepsilon(t,t)$. Because of the continuity of $\varphi(t)$, there exists such a t_1 with minimal absolute value , and (3) implies that $t_1 \neq 0$. From inequality (1), we conclude that

$$|\psi(t_1A^*)| < \varepsilon(t_1A^*, t_1A^*)$$

and

$$|\psi(t_1B^*)| < \varepsilon(t_1B^*, t_1B^*).$$

From (2),

$$\psi(t) = \psi(tA^*) + \psi(tB^*),$$

so

$$|\psi(t_1)| \leq |\psi(t_1A^*)| + |\psi(t_1B^*)| < \varepsilon[(t_1A^*, t_1A^*) + (t_1B^*, t_1B^*)] = \varepsilon(t_1, t_1).$$

This inequality contradicts the definition of t_1. Consequently, $\psi(t) \equiv 0$, and so

$$\varphi(t) = e^{-\frac{1}{2}(t,t)}. \quad \square$$

Problem P.9.. *Let $\xi_1, \xi_2, \dots$ be independent random variables such that $E\xi_n = m > 0$ and* Var$(\xi_n) = \sigma^2 < \infty$ $(n = 1, 2, \dots)$. *Let $\{a_n\}$ be a sequence of positive numbers such that $a_n \to 0$ and $\sum_{n=1}^{\infty} a_n = \infty$. Prove that*

$$P\left(\lim_{n\to\infty} \sum_{k=1}^{n} a_k\xi_k = \infty\right) = 1.$$

Solution. We can solve the problem using the Kolmogorov inequality, but in the solution given below we only use the Chebyshev inequality.

Instead of $a_n \to 0$ and $\mathrm{Var}\,(\xi_n) = \sigma^2$, it is sufficient to assume that $\{a_n\}$ is bounded and $\mathrm{Var}\,(\xi_n) = \sigma^2$. For instance, suppose $0 < a_n < 1$. Put $S_n = \sum_{k=1}^{n} a_k$, and define the sequence $\{r_n\}$ by $S_{r_n} < n^2 \leq S_{r_n+1}$. Then

$$n^2 - 1 < S_{r_n} < n^2. \tag{1}$$

If $\lim_n \sum_{k=1}^{n} a_k\xi_k = \infty$ does not hold, then there is a K such that

$$\sum_{k=1}^{n} a_k\xi_k < K \tag{2}$$

occurs infinitely many times.

It is sufficient to show that for any fixed K, this event has probability 0 since the union of these events for $n = 1, 2, \ldots$ includes all event sequences for which $\lim_n \sum a_k\xi_k = \infty$ does not hold.

Let A_k be the event that there is an n between r_k and r_{k+1} such that (2) holds. Note that (2) holds infinitely many times if infinitely many A_k occur. Denote the probability of A_k by P_k. We have to show that with probability one only finitely many A_k occur. We prove this by verifying that $\sum_{k=1}^{\infty} P_k$ is convergent. In this case, $\sum_{k\geq m} P_k \to 0$, and $\sum_{k\geq m} P_k$ is greater than the probability that infinitely many A_k occur. Therefore, this latest probability is equal to 0. (Actually, here we have proved and applied the Borel–Cantelli lemma.)

Put $\eta_k = \sum_{i=1}^{r_k} a_i\xi_i$ and $\psi_k = \sum_{i=r_k+1}^{r_{k+1}} a_i|\xi_i|$. If (2) holds for some $r_k \leq z < r^{k+1}$ with z replacing n, then either

$$\sum_{i=1}^{r_k} a_i\xi_i \leq k^2\frac{m}{2}$$

or

$$\sum_{i=r_k+1}^{z} a_i\xi_i \leq -k^2\frac{m}{3},$$

supposing $k^2m > 6K$.

In the last case, $\psi_k = \sum_{i=r_k+1}^{r_{k+1}} a_i|\xi_i| \geq k^2m/3$, so at least one of the inequalities $\psi_k \geq k^2m/3$ and $\eta_k \leq k^2m/2$ is valid. Thus,

$$P_k \leq P\left(\eta_k \leq \frac{k^2m}{2}\right) + P\left(\psi_k \geq \frac{k^2m}{3}\right). \tag{3}$$

We have to show that on the right side both terms are small.

It is obvious that $E(\eta_k) = mS_{r_k}$ and $\mathrm{Var}\,(\eta_k) \leq \sigma^2 S_{r_k}$ since $a_i < 1$ implies that

$$\mathrm{Var}\,(\eta_k) \leq \sigma^2 \sum_{i=1}^{r_k} a_i^2 \leq \sigma^2 \sum_{i=1}^{r_k} a_i \quad = \sigma^2 S_{r_k}.$$

Furthermore,

$$E(|\xi_i|) < E(1+\xi_i^2) = 1 + E^2(\xi_i) + \mathrm{Var}\,(\xi_i) = 1 + m^2 + \sigma^2 = A$$

and

$$\mathrm{Var}\,(|\xi_i|) = E(\xi_i^2) - E^2(|\xi_i|) \leq E(\xi_i^2) < A.$$

Since the absolute values of independent random variables are independent,

$$E(\psi_k) \le (S_{r_{k+1}} - S_{r_k}) \max_{r_k < i \le r_{k+1}} E(|\xi_i|) < (S_{r_{k+1}} - S_{r_k})A = O(k)$$

and

$$\mathrm{Var}\,(\psi_k) \le \sum_{i=r_k+1}^{r_{k+1}} a_i^2 \max_{r_k < i \le r_{k+1}} \mathrm{Var}\,(|\xi_i|) \le (S_{r_{k+1}} - S_{r_k})A = O(k).$$

Now we can use the Chebyshev inequality and (1) to see that

$$E(\eta_k) \sim k^2 m, \qquad \mathrm{Var}\,(\eta_k) \sim k^2 \sigma,$$

$$P(\eta_k \le \frac{k^2 m}{2}) \le P(|\eta_k - E(\eta_k)| \ge \frac{k^2 m}{3}) = O(\frac{1}{k^2}),$$

$$P(\psi_k \ge \frac{k^2 m}{3}) \le P(|\psi_k - E(\psi_k)| \ge \frac{k^2 m}{6}) = O(\frac{1}{k^2})$$

if k is large enough. From these relations, the convergence of $\sum P_k$ follows. □

Remarks.

1. One can sharpen the proof to see that $\sum_{k \le n} a_k \xi_k$ tends to infinity as fast as $\sum_{k \le n} a_k$ with probability one.
2. The condition $a_n = O(S_n)$ is weaker than boundedness, and it does not guarantee the convergence to infinity with probability one, but the slightly stronger condition

$$a_n = O\left(S_n^{1-(1+\varepsilon)\frac{\log \log S_n}{\log S_n}}\right)$$

(which is still weaker than boundedness) does.

P.10. *Let $\vartheta_1, \ldots, \vartheta_n$ be independent, uniformly distributed, random variables in the unit interval $[0,1]$. Define*

$$h(x) = \frac{1}{n} \#\{k : \vartheta_k < x\}.$$

Prove that the probability that there is an $x_0 \in (0,1)$ such that $h(x_0) = x_0$, is equal to $1 - \frac{1}{n}$.

Solution 1. Let us denote by I_k the interval $((k-1)/n, k/n)$. Consider the elementary events that for every i, ϑ_i belongs to I_{k_i}, where k_i is given for every i. So we decomposed the original sample space to the union of n^n disjoint events of probability $1/n^n$ (events with probability zero are omitted). These events will be called atoms.

Since the value $(1/n) \cdot \sum_{\vartheta_i < x} 1 = h(x)$ is an integer multiple of $1/n$, in order to check whether $h(x_0) = x_0$ holds for some x_0, it is sufficient to know which intervals I_{k_i} contain ϑ_i, that is, which atom will be the result of the experiment series. They have the same probability; therefore, it is sufficient to determine the number of atoms for which there is no k such that exactly k ϑ_i occur until k $(1 \le k < n)$. Denote by A_n $(1 \le k < n)$ the number of atoms for which there is no k such that the first k intervals contain exactly k ϑ_i values.

Let B_k denote the number of atoms for which there are exactly k ϑ_i values in $\cup_{j \le k} I_j$ but for any $t < k$, $\cup_{j \le t} I_j$ does not contain exactly t ϑ_i values. So every atom that has exactly k ϑ_i values in the first k intervals belongs to exactly one B_k. Next, we determine the value of B_k. We can choose the ϑ_is belonging to $\cup_{j \le k} I_j$ in $\binom{n}{k}$ different ways. We can choose the first k intervals I_{k_i} containing these ϑ_i values in A_k different ways, and we can arrange the remaining $n-k$ ϑ_i values in the remaining $n-k$ intervals in $(n-k)^{n-k}$ different ways. So we have the recursion

$$n^n - A_n = \sum_{k=1}^{n-1} \binom{n}{k} A_k (n-k)^{n-k}, \qquad A_1 = 1. \tag{1}$$

We are going to verify that $A_n = n^{n-1}$ by mathematical induction. It is easy to check that $A_1 = 1^0$ and $A_2 = 2^1$ or $A_3 = 3^2$. Suppose that we have already verified this relation up to $n-1$. In order to show this equation for n, we have to show that

$$n^n - n^{n-1} = \sum_{k=1}^{n-1} \binom{n}{k} k^{k-1} (n-k)^{n-k}. \tag{2}$$

It is easy to see by interchanging k and $n-k$ that this equation is equivalent to

$$n^{n-1}(n-1) = \sum_{k=1}^{n-1} \binom{n}{k} k^2 (n-k) k^{k-2} (n-k)^{n-k-2}. \tag{3}$$

We will prove this equality by a well-known theorem of Cayley. Consider the trees having the numbers $1, 2, \ldots, n$ as vertices. (A tree is a connected graph not containing a circle.) Two graphs are different if they do not have the same (i, j) pairs as edges. By the Cayley theorem, there are n^{n-2} different such trees. Now choose in every tree in all possible different ways one vertex and one edge. We will call these graphs vertex-edge-signed trees, and consider these graphs different if they are different before our vertex-edge selection, or the selected vertex-edge pair is different. Since a tree has $n-1$ edges, we have $n^{n-1}(n-1)$ different vertex-edge-signed trees, and this number is exactly the left-hand side of (3). Let us count the number of these vertex-edge-signed trees in another way. By omitting the signed edge, we get a tree with k vertices with one signed vertex and a normal tree with $(n-k)$ vertices $(k = 1, 2, \ldots, n-1)$. On the other hand, if we choose

k vertices out of n, define a tree on them, select a vertex and define a tree from the remaining $n-k$ vertices, and connect the two trees by an edge, which will be called the signed edge, then we get a vertex-edge-signed tree. Thus, the two procedures are the reverse of each other. In this way, we get

$$\sum_{k=1}^{n-1}\binom{n}{k}k\cdot k^{k-2}(n-k)^{n-k-2}k(n-k)$$

different graphs, which is the right-hand side of (3). □

Solution 2. The first part of the solution is the same, but here we give an algebraic proof of (2).

Lemma 1. If $0<j<m$, then

$$\sum_{k=1}^{m}\binom{m}{k}(-1)^k k^j=0.$$

Proof. Denote by D the operation of differentiation followed by a multiplication by x. Thus $D^m(x^k)=k^m x^k$. It is easy to see by induction that if x_0 is a root of the polynomial $p(x)$ with multiplicity m_0, then x_0 will also be a root of $D^m(p(x))$ with multiplicity at least m_0-m. Apply this observation to the polynomial $p(x)=(1-x)^m$ and the operator D^j, $j<m$. Then we get that $x=1$ is a zero of the polynomial

$$D^j((1-x)^m)=\sum_{k=1}^{m}\binom{m}{k}(-1)^k k^j x^k.$$

This completes the proof of Lemma 1.

Lemma 2.

$$\sum_{k=1}^{n-1}\binom{n}{k}k^{k-1}(z-k)^{n-k}=nz^{n-1}-n^{n-1}$$

Proof.

$$\sum_{k=1}^{n-1}\binom{n}{k}k^{k-1}(z-k)^{n-k}$$

$$\sum_{k=1}^{n-1}\binom{n}{k}k^{k-1}\sum_{j=0}^{n-k}\binom{n-k}{n-k-j}z^j(-k)^{n-k-j}$$

$$=\sum_{k=1}^{n-1}\sum_{j=0}^{n-k}\frac{n!}{k!j!(n-k-j)!}z^j(-1)^{n-k-j}k^{n-1-j}$$

$$= \sum_{k=1}^{n-1} \sum_{j=0}^{n-k} \binom{n}{j} \binom{n-j}{k} z^j (-1)^{n-k-j} k^{n-1-j}$$

$$= \sum_{k=1}^{n} \sum_{j=0}^{n-k} \binom{n}{j} \binom{n-j}{k} z^j (-1)^{n-k-j} k^{n-1-j} - n^{n-1}$$

$$= \sum_{j=0}^{n-1} \sum_{k=1}^{n-j} \binom{n}{j} \binom{n-j}{k} z^j (-1)^{n-k-j} k^{n-1-j} - n^{n-1}$$

$$= \sum_{j=0}^{n-2} \sum_{k=1}^{n-j} \binom{n}{j} \binom{n-j}{k} z^j (-1)^{n-k-j} k^{n-1-j} n^{n-1} + nz^{n-1}$$

$$= \sum_{j=0}^{n-2} \binom{n}{j} (-1)^{n-j} z^j (\sum_{k=1}^{n-j} \binom{n-j}{k} (-1)^k k^{n-j-1}) - n^{n-1} + nz^{n-1}$$

$$= nz^{n-1} - n^{n-1}$$

since, by Lemma 1, the inner sum is equal to zero.

Substitute $z = n$ to get (2). □

Problem P.11. *We throw N balls into n urns, one by one, independently and uniformly. Let $X_i = X_i(N, n)$ be the total number of balls in the ith urn. Consider the random variable*

$$y(N, n) = \min_{1 \le i \le n} |X_i - \frac{N}{n}|.$$

Verify the following three statements:

(a) If $n \to \infty$ and $N/n^3 \to \infty$, then

$$P\left(\frac{y(N,n)}{\frac{1}{n}\sqrt{\frac{N}{n}}} < x \right) \to 1 - e^{-x\sqrt{2/\pi}} \quad \text{for all } x > 0.$$

(b) If $n \to \infty$ and $N/n^3 \le K$ (K constant), then for any $\varepsilon > 0$ there is an $A > 0$ such that

$$P(y(N, n) < A) > 1 - \varepsilon.$$

(c) If $n \to \infty$ and $N/n^3 \to 0$ then

$$P(y(N, n) < 1) \to 1.$$

Solution. The basic idea of the proof is the following: X_i is a $B(p, N)$ (binomial distribution with parameters p, N), where $p = 1/n$. If the X_i's were independent, the statements could be proved easily. But of course they are not, since $X_1 + \cdots + X_n = N$. However, we will show that k of them can be considered independent for sufficiently small values of k.

Suppose that the X_i's are independent and their distribution is $B(p, N)$. Let $a_1, \ldots a_k$ be natural numbers in the interval $[0, N]$,

$$\alpha = P(X_i = a_i,\, i = 1, \ldots k)\,, \quad \alpha' = P(X_i' = a_i,\, i = 1, \ldots, k).$$

Introduce some additional notations:

$$a = \sum_{i=1}^{k} a_i\,, \quad a_i = pN + b_i\,, \quad b = \sum_{i=1}^{k} b_i \quad .$$

We will show that $\alpha \sim \alpha'$ under the conditions $k = O(1)$, $b_i = o(\sqrt{N})$, and $N \geq n$.

Obviously,

$$\alpha = \frac{N!}{(\prod_{i=1}^{k} a_i!)(N-a)!} p^a (1-kp)^{N-a}$$

and

$$\alpha' = \left(\prod_{i=1}^{k} \binom{N}{a_i} \right) p^a (1-p)^{kN-a},$$

so

$$\beta = \frac{\alpha'}{\alpha} = \frac{N!^{k-1}(N-a)!(1-p)^{kN-a}}{(\prod_{i=1}^{k}(N-a_i)!)(1-kp)^{N-a}}.$$

The value of β will be estimated by the Stirling formula:

$$t! = \sqrt{2\pi t}\; t^t e^{-t+O(1/t)}.$$

The error of $\log t$, that is, the difference of the logarithms of the true and estimated β values can be determined from the $O(1/t)$ terms as

$$O\left(\sum_{i=1}^{k} \frac{1}{N-a_i} + \frac{k-1}{N} = \frac{1}{N-a} \right) = O(\frac{1}{N}) = o(1),$$

since

$$a_i = pN + o(\sqrt{N}),\, a = kpN + o(\sqrt{N}),\; p = \frac{1}{n} \to 0$$

and $N \geq n$ implies $N \to \infty$. The terms e^{-t} and $\sqrt{2\pi}$ are cancelled, and the value of the $\sqrt{t}$ terms can be given as

$$\sqrt{\frac{N^{k-1}(N-a)}{\prod_{i=1}^{k}(N-a_i)}} = \sqrt{\frac{1-\frac{a}{N}}{\prod_{i=1}^{k}(1-\frac{a_i}{N})}} = 1 + o(1).$$

So, finally,

$$\begin{aligned}\beta &\sim \frac{N^{N(k-1)}(N-a)^{N-a}}{\prod_{i=1}^{k}(N-a_i)^{n-a_i}}\,\frac{(1-p)^{kN-a}}{(1-kp)^{N-a}}\\ &= \frac{(1-\frac{a}{N})^{N-a}}{(1-kp)^{N-a}} : \frac{\prod_{i=1}^{k}(1-\frac{a_i}{N})^{N-a_i}}{(1-p)^{kN-a}}\\ &= \left(1-\frac{b}{N(1-kp)}\right)^{N-a} : \prod_{i=1}^{k}\left(1-\frac{b_i}{N(1-p)}\right)^{N-a_i}\end{aligned}$$

Since each term converges to 1, we can apply the formula $\log(1+\varepsilon) = c\varepsilon + O(\varepsilon^2)$. The value of the second-order term is

$$O\left(\frac{b^2+\sum_{i=1}^{k} b_i^2}{N}\right) = o(1).$$

The value of the first-order term is

$$\frac{-b(N-kpN-b)}{N(1-kp)} + \sum_{i=1}^{k}\frac{b_i(N-pN-b_i)}{N(1-p)} =$$

$$\frac{b^2}{N(1-kp)} - \sum_{i=1}^{k}\frac{b_i^2}{N(1-p)} = o(1)\,,$$

which implies that $\log\beta = o(1)$.

Solution to part (a). Put

$$\begin{aligned}\alpha_k &= P(|X_i - pN| < x\sqrt{\frac{N}{n^3}};\ i=1,\dots,k),\\ c &= P\left(y(N,n) < x\sqrt{\frac{N}{n^3}}\right).\end{aligned}$$

Recall that if $A_1,\dots,A_n$ are arbitrary events and

$$S_j = \sum_{1\le i_1<\dots<i_j\le n} P(A_{i_1},\dots,A_{i_j}),$$

then

$$\sum_{j=0}^{2s+1}(-1)S_j \le P(\overline{A_1}\dots\overline{A_n}) \le \sum_{j=0}^{2s}(-1)^j S^j$$

(see, for example, *K. Bognárné, J. Mogyoródi, A. Prékopa, A. Rényi, D. Szász, Exercises in Probability Theory (in Hungarian), Tankönyvkiadó, Budapest, 1971*, Problem I.2.11.c). So

$$\sum_{j=0}^{2s+1}(-1)^j\binom{n}{j}\alpha_j \le 1-c \le \sum_{j=0}^{2s}(-1)^j\binom{n}{j}\alpha_j.$$

Note that

$$\alpha_k \sim \alpha'_k = P\left(|X'_i - pN| < x\sqrt{\frac{N}{n^3}};\ i = 1, \dots, k\right),$$

since we can apply the previous asymptotic relation for any system $X_i - pN = b_i$, when $|b_i| < x\sqrt{N/n^3}$. Since $\alpha'_k = {\alpha'}^k$, with

$$\alpha' = P\left(|X'_1 - pN| < x\sqrt{\frac{N}{n^3}}\right),$$

$$1 - c = \sum_{j=0}^{k} (-1)^j \binom{n}{j} {\alpha'}^j + O\left(\binom{n}{k+1} {\alpha'}^{k+1}\right) + o\left(\sum_{j=0}^{k} \binom{n}{j} {\alpha'}^j\right).$$

The first sum is the first k terms of $(1-\alpha')^n$, and substituting them by $(1-\alpha')^n$ itself, the error will be only $O(\binom{n}{k+1}{\alpha'}^{k+1})$. From the local De Moivre–Laplace theorem we conclude that

$$\alpha' \sim \sqrt{\frac{2}{\pi}}\frac{x}{n}$$

(here we use the fact that $N/n^3 \to \infty$).

Select an arbitrary, small $\varepsilon > 0$. If k is sufficiently large, then the first error term is less than ε, the second term is $o((1+\alpha')^n) = o(1)$, and finally,

$$\log(1-\alpha') = -\alpha' + O({\alpha'}^2), \quad n \log(1-\alpha') = -\sqrt{\frac{2}{\pi}}x + o(1),$$

which proves the assertion.

Solution to part (c). We are going to prove for an arbitrary, small $\varepsilon > 0$ that there exist δ and n_0 such that for $n > n_0$ and $N/n^3 < \delta$, we have $P(y(N,n) < 1) > 1 - \varepsilon$.

If we wish to apply the previous reasoning, we face the difficulty that $\binom{n}{k+1}{\alpha'}^{k+1}$ is not small enough, and $(1+\alpha')^n$ is not bounded if α' is too large. We can, however, overcome this difficulty since these quantities can be replaced by

$$\binom{[A\sqrt{\frac{N}{n}}]}{k+1}\left(c\sqrt{\frac{n}{N}}\right)^{k+1} \quad \text{and} \quad \left(1 + c\sqrt{\frac{n}{N}}\right)^{[A\sqrt{\frac{N}{n}}]},$$

respectively. Consider the probability that for any of the first $[A\sqrt{N/n}]$ urns, $|X_i - N/n| < 1$. (If δ is small enough, then $A\sqrt{N/n} < n$). Here

$$P\left(|X'_i - \frac{N}{n}| < 1\right) = c\sqrt{\frac{n}{N}},$$

where c remains between positive bounds for $N \geq n$. (We can assume this since otherwise the assertion is obvious.) Then the proof can be completed as it was demonstrated in the proof of (a).)

Solution to part (b). We can assume that with the δ given in the previous part $N/n^3 \geq \delta$, otherwise, the previous result implies the desired inequality. Now $\alpha' = cA/n$, where c is between positive bounds (they depend on δ and K). The proof then can be completed with this α' in the same way as was shown at the end of the solution to part (a). □

Problem P.12. *Determine the value of*

$$\sup_{1 \leq \xi \leq 2} [\log E\xi - E \log \xi],$$

where ξ is a random variable and E denotes expectation.

Solution. Since $\log \xi$ is concave in ξ, and $1 \leq \xi \leq 2$,

$$\log \xi = \log[(2-\xi) + (\xi - 1)2] \geq (\xi - 1)\log 2,$$

which implies that $E(\log \xi) \geq [E(\xi) - 1]\log 2$. Note that $1 \leq E(\xi) \leq 2$, and therefore

$$\begin{aligned}\log E(\xi) - E(\log \xi) &\leq \log E(\xi) - [E(\xi) - 1]\log 2 \\ &\leq \max_{1 \leq t \leq 2} [\log t - (t-1)\log 2] \\ &= \log t_0 - (t_0 - 1)\log 2 = K \quad ,\end{aligned}$$

where $t_0 = 1/\log 2 \in [1, 2]$. We will next show that K is the least upper bound. Consider the random variable ξ_0 defined by $P(\xi_0 = 1) = 2 - t_0$ and $P(\xi_0 = 2) = t_0 - 1$. Then $\log E(\xi_0) = \log t_0$ and $E(\log \xi_0) = (t_0 - 1)\log 2$, therefore, $K = -\log\log 2 - 1 + \log 2$. □

Remark. A similar proof shows that if $\varphi(t)$ is a concave and continuous function on the closed interval $[a, b]$, then

$$\sup_{a \leq \xi \leq b} \{\varphi(E(\xi)) - E(\varphi(\xi))\} = \max_{a \leq t \leq b} [\varphi(t) - \frac{b-t}{b-a}\varphi(a) - \frac{t-a}{b-a}\varphi(b)].$$

A further generalization is also possible for continuous but not necessarily concave functions φ.

Problem P.13. *Find the limit distribution of the sequence η_n of random variables with distribution*

$$P\left(\eta_n = \arccos(\cos^2 \frac{(2j-1)\pi}{2n})\right) = \frac{1}{n} \qquad (j = 1, , \ldots, n).$$

(arccos(.) *denotes the main value.)*

Solution. Let ξ_n be defined as

$$P(\xi_n = \frac{2j-1}{2n}) = \frac{1}{n} \qquad (j = 1, 2, \ldots, n).$$

Then $\eta_n = \arccos(\cos^2 \pi\xi_n)$. Since $\xi_n \in (0,1)$ $(n = 1, 2, \ldots)$, and the function $\arccos(\cos^2 \pi x)$ maps the interval (0,1) into $(0, \pi/2)$, $P(\eta_n < x) = 0$ for $x \leq 0$ and $P(\eta_n < x) = 1$ for $x \geq \pi/2$. Suppose $x \in [0, \pi/2]$. Since $\cos x$ is strictly decreasing in the interval $[0, \pi]$, $\arccos x$ is strictly decreasing in the interval $[-1, 1]$. The distribution of ξ_n is symmetric with respect to 1/2, therefore,

$$\begin{aligned}
P(\eta_n < x) &= P(\arccos(\cos^2 \pi\xi_n) < x) = P(\cos^2(\pi\xi_n) > \cos x) \\
&= P(\cos(\pi\xi_n) > \sqrt{\cos x}) + P(\cos(\pi\xi_n) < -\sqrt{\cos x}) \\
&= P(\pi\xi_n > \arccos(-\sqrt{\cos x})) + P(\pi\xi_n < \arccos(\sqrt{\cos x})) \\
&= 2P(\pi\xi_n < \arccos\sqrt{\cos x}) = 2P(\xi_n < \frac{\arccos\sqrt{\cos x}}{\pi}).
\end{aligned}$$

For $x \epsilon (0, \pi/2)$, $\arccos\sqrt{\cos x} \in (0, \pi/2)$; thus only the value of $P(\xi_n < y)$ for $y \epsilon (0, 1/2)$ is to be determined.

Let k be the largest integer such that $k/n \leq y$. Then

$$P(\xi_n < \frac{k}{n}) = \sum_{\frac{2j-1}{2n} < \frac{k}{n}} \frac{1}{n} = \frac{k}{n} \leq P(\xi_n < y) \leq P(\xi_n < \frac{k+1}{n}) = \frac{k+1}{n}.$$

Therefore, $\lim_{n\to\infty} P(\xi_n < y) = y$, and consequently,

$$\lim_{n\to\infty} P(\eta_n < x) = \begin{cases} 0 & \text{if } x \leq 0, \\ \frac{2}{\pi} \arccos\sqrt{\cos x} & \text{if } x \in (0, \frac{\pi}{2}), \\ 1 & \text{if } x \geq \frac{\pi}{2}. \quad \square \end{cases}$$

Remarks.

1. The random variable η_n may take the same value twice.
2. The numbers $\lambda_j = \cos((2j-1)/2n)\pi$ $(j = 1, 2, \ldots, n)$ are the zeros of the Chebyshev polynomial $T_n = \cos(n \arccos x)$, and the corresponding Cotes numbers are $A_{n,j} = \pi/n$. Using the characteristic function of η_n, one can easily prove the following.

Theorem. Let f be a continuous function in $[-1, 1]$, and let ξ_n be a random variable defined by

$$P(\xi_n = f(\lambda_j)) = \frac{1}{n} \quad (j = 1, \ldots, n), \ (n = 1, 2, \ldots).$$

Assume that ξ is a random variable with density function $1/(\pi\sqrt{1-x^2})$ in the interval $(-1, 1)$ and zero otherwise. Then the sequence $\{\xi_n\}$ converges weakly to the random variable $f(\xi)$.

In the problem, select $f(x) = \arccos x^2$ so that the limit distribution of the sequence $\{\eta_n\}$ is the distribution of $\eta = \arccos \xi^2$. Simple calculations show that the density function of η is

$$\frac{\sqrt{2}}{\pi} \frac{1}{\sqrt{1 - \tan^2 \frac{x}{2}}} \quad \text{if} \quad x\epsilon(0, \frac{\pi}{2}) \quad \text{and} \quad 0 \text{ otherwise.}$$

Problem P.14. *Let μ and ν be two probability measures on the Borel sets of the plane. Prove that there are random variables $\xi_1, \xi_2, \eta_1, \eta_2$ such that*

(a) *the distribution of (ξ_1, ξ_2) is μ and the distribution of (η_1, η_2) is ν,*
(b) *$\xi_1 \leq \eta_1$, $\xi_2 \leq \eta_2$ almost everywhere, if and only if $\mu(G) \geq \nu(G)$ for all sets of the form $G = \cup_{i=1}^k (-\infty, x_i) \times (-\infty, y_i)$.*

Solution. The necessity part of the assertion is obvious, since if $\xi = (\xi_1, \xi_2)$ and $\eta = (\eta_1, \eta_2)$, then $\eta\epsilon G$ implies $\xi\epsilon G$, so

$$\nu(G) = P(\eta\epsilon G) \leq P(\xi\epsilon G) = \mu(G).$$

First, we prove the sufficiency part of the assertion for the special case where μ and ν are concentrated to the finite set $x_1, \ldots ,x_m$ and $y_1, \ldots ,y_n$, respectively. Define the graph $\mathcal{G}$ with vertices $x_1, \ldots, x_m$, $y_1, \ldots, y_n$ and where the vertices x_i and y_j are connected by an edge if and only if $x_i \leq y_j$ ($(a, b) \leq (c, d)$ means $a \leq c$ and $b \leq d$). Define a random variable (ξ, η) that takes the values (x_i, y_j) with probability $a_{i,j}$. These numbers $a_{i,j}$ satisfy the following relations:

(a) $a_{i,j} \geq 0$;
(b) $\sum_{i=1}^m a_{i,j} = \nu(y_j)$;
(c) $\sum_{j=1}^n a_{i,j} = \mu(x_i)$;
(d) If $a_{i,j} > 0$, then there is an edge between x_i and y_j.

By the Kőnig–Egerváry theorem, such numbers $a_{i,j}$ exist if and only if for any set $Y \subseteq \{y_1, \ldots, y_n\}$ the μ measure of the set X consisting of all points x_i connected with the elements of Y is at least $\nu(Y)$. Assume that

$$Y = \{y_1, \ldots, y_k\} \quad \text{with} \quad y_i = (y'_i, y''_i)$$

and

$$G = \cup_{i=1}^k (-\infty, y'_i) \times (-\infty, y''_i).$$

Then a point x_i is connected to Y if and only if $x_i \in G$. Therefore,

$$\sum_{x_i \epsilon X} \mu(x_i) = \mu(X \cap G) = \mu(G) \geq \nu(G) \geq \nu(Y).$$

Consequently, there are numbers $a_{i,j}$ having properties (a)–(d), and any vector variable (ξ, η) that has the values (x_i, y_j) with probabilities $a_{i,j}$ satisfies the requirements of the theorem.

Now consider the general case. Denote by $F(x, y)$ the distribution function of the underlying variable (ξ, η), and put

$$F_\mu(x', x'') = \mu((-\infty, x') \times (-\infty, x'')),$$
$$F_\nu(y', y'') = \nu((-\infty, y') \times (-\infty, y'')).$$

The existence of a distribution function F satisfying

$$\text{(i)} \qquad F(x, \infty) = F_\mu(x)$$
$$\text{(ii)} \qquad F(\infty, y) = F_\nu(y)$$
$$\text{(iii)} \qquad F(y, y) = F_\nu(y)$$

solves the problem, since for a random variable (ξ, η) having distribution F, $\xi \le \eta$ holds with probability one because $F(y, y) = F(\infty, y)$.

Divide the square $[-n, n] \times [-n, n]$ into n^4 small squares of sides $1/n$. Define μ_n as follows: concentrate the μ measure of every point into the closest node. (More precisely, consider the set of all points of the plane that are closest to a given vertex, and concentrate the μ measure of this set to the given vertex. If a point is in the same distance from two or more vertices, then select the one that has the smallest abscissa among the vertices with smallest ordinate.)

Let ν_n be defined analogously. Then (μ_n, ν_n) satisfies the requirements of the theorem, since they are measures concentrated in finitely many points. Therefore, there is a variable (ξ_n, η_n) such that $\xi_n \le \eta_n$, the distribution of ξ_n is μ_n, and the distribution of η_n is ν_n. Let F_n be the distribution function of (ξ_n, η_n). Then

$$F_n(y, y) = F_n(\infty, y),$$

furthermore,

$$F_n(x', x'', \infty, \infty) = \mu_n((-\infty, x') \times (-\infty, x'')),$$
$$F_n(\infty, \infty, y', y'') = \nu_n((-\infty, y') \times (-\infty, y'')).$$

Let S be a countable and everywhere-dense set of the continuity points of F_ν and F_μ. Select a subsequence n_i such that $F(a, b) = \lim_{i\to\infty} F_{n_i}(a, b)$ exists for all $a, b \in S$. Extend the definition of F by the relation

$$F(x, y) = \sup\{\lim_{i\to\infty} F_{n_i}(a, b) : a \le x,\, b \le y,\, a, b \in S\}.$$

It is easy to verify that F is a distribution function and that it satisfies (i), (ii), (iii). Thus, the proof is completed. □

Problem P.15. *Let $X_1, X_2, \ldots, X_n$ be (not necessarily independent) discrete random variables. Prove that there exist at least $n^2/2$ pairs (i, j) such that*

$$H(X_i + X_j) \geq \frac{1}{3} \min_{1 \leq k \leq n} \{H(X_k)\},$$

where $H(X)$ denotes the Shannon entropy of X.

Solution. Consider the graph G with vertices $\{1, 2, \ldots, n\}$, and assume that i and j are connected with an edge if and only if the inequality

$$H(X_i + X_j) \geq \frac{1}{3} \min_{1 \leq k \leq n} H(X_k) \tag{1}$$

does not hold. There is no loop in G, since for all i

$$H(X_i + X_i) = H(X_i) \geq \frac{1}{3} \min_{1 \leq k \leq n} H(X_k).$$

We are going to show that there is no triangle in G. Let i, j, and l be different elements of $\{1, 2, \ldots, n\}$. Obviously,

$$X_i = \frac{1}{2}([X_i + X_j] + [X_i + X_l] - [X_j + X_l]),$$

that is,

$$\begin{aligned} H(X_i) &= H([X_i + X_j] + [X_i + X_l] - [X_j + X_l]) \\ &\leq H(X_i + X_j) + H(X_i + X_l) + H(X_j + X_l) \\ &\leq 3 \max\{H(X_i + X_j), H(X_i + X_l), H(X_j + X_l)\}. \end{aligned}$$

So

$$\begin{aligned} \frac{1}{3} \min_{1 \leq k \leq n} H(X_k) &\leq \frac{1}{3} H(X_i) \\ &\leq \max\{H(X_i + X_j), H(X_i + X_l), H(X_j + X_l)\}. \end{aligned}$$

Therefore, there are two vertices among i, j, and l that are not connected.

By a well known theorem of Pál Turán (see, *Mat. Fiz. Lapok 48, (1941)*, pp. 436–452), the number of edges of such a graph is not larger than $n^2/4$. Consequently, the number of pairs (i, j) such that (1) holds is at least $n^2 - 2 \cdot (n^2/4) = n^2/2$. □

Remark. The bound $n^2/2$ is the best possible. Let X be a random variable such that $H(X) \neq 0$, and define

$$\begin{aligned} X_1 = X_2 = \cdots = X_{[\frac{n}{2}]} = X, \\ X_{[\frac{n}{2}]+1} = \cdots = X_n = -X. \end{aligned}$$

If either $1 \leq i \leq n/2 < j \leq n$ or $1 \leq j \leq n/2 < i \leq n$, then $H(X_i + X_j) = 0$. The number of such pairs (i, j) is equal to $[n^2/2]$.

Problem P.16. *Let $\xi_1, \xi_2, \dots$ be independent, identically distributed random variables with distribution*

$$P(\xi_1 = -1) = P(\xi_1 = 1) = \frac{1}{2}.$$

Write $S_n = \xi_1 + \xi_2 + \dots + \xi_n$ $(n = 1, 2, \dots)$, $S_0 = 0$, and

$$T_n = \frac{1}{\sqrt{n}} \max_{0 \le k \le n} S_k \; .$$

Prove that $\liminf_{n\to\infty} (\log n)\, T_n = 0$ with probability one.

Solution. Denote $\max_{a \le k \le b} S_k$ by $(S_b - S_a)^*$, and $\max_{0 \le k \le n} S_k$ by S_n^*. First, suppose that we have two sequences $g(n)$ and $f(n)$ such that

$$\sum_{n=1}^{\infty} P\{(f(n))^{-1/2} S_{f(n)}^* > g(f(n))\} < \infty \tag{a}$$

and

$$\sum_{n=1}^{\infty} P\left\{ \frac{(S_{f(n+1)} - S_{f(n)})^* + \sqrt{f(n)}g(f(n))}{\sqrt{f(n+1)}} < \frac{\varepsilon}{\log f(n+1)} \right\} = \infty \tag{b}$$

for all $\varepsilon > 0$. If such sequences $g(n)$, $f(n)$ are found, then by the Borel–Cantelli lemma with probability one, only finitely many of the events

$$\frac{S_{f(n)}^*}{\sqrt{f(n)}} > g(f(n))$$

and infinitely many of the events

$$\frac{(S_{f(n+1)} - S_{f(n)})^* + \sqrt{f(n)}g(f(n))}{\sqrt{f(n+1)}} < \frac{\varepsilon}{\log f(n+1)}$$

occur. Consequently, infinitely many of the events

$$\frac{S_{f(n+1)}^*}{\sqrt{f(n+1)}} \le \frac{(S_{f(n+1)} - S_{f(n)})^* + S_{f(n)}^*}{\sqrt{f(n+1)}} < \frac{\varepsilon}{\log f(n+1)}$$

will occur with probability one. Hence, it is sufficient to find sequences $g(n)$, $f(n)$ satisfying (a) and (b).

We are going to apply the approximation

$$P\left(\frac{S_n^*}{\sqrt{n}} < x\right) \approx \sqrt{\frac{2}{\pi}} \int_0^x e^{-u^2/2} du$$

(see, for example, *A. Rényi, Foundation of Probability, Holden Day, San Francisco, 1970*). This approximation holds if x is replaced by a sequence $x_n = o(n^{\frac{1}{3}})$.

So

$$P\left(\frac{(S_{f(n+1)} - S_{f(n)})^* + \sqrt{f(n)}g(f(n))}{\sqrt{f(n+1)}} < \frac{\varepsilon}{\log f(n+1)}\right)$$
$$\geq P\left(\frac{(S_{f(n+1)} - S_{f(n)})^*}{\sqrt{f(n+1) - f(n)}} < \frac{\varepsilon}{\log f(n+1)} - \frac{\sqrt{f(n)}g(f(n))}{\sqrt{f(n+1)}}\right)$$
$$\approx \sqrt{\frac{2}{\pi}} \int_0^A e^{-u^2/2} du \geq \frac{1}{2}A,$$

where

$$A = \frac{\varepsilon}{\log f(n+1)} - \frac{\sqrt{f(n)}g(f(n))}{\sqrt{f(n+1)}}$$

is small enough. Therefore, we have to find sequences $f(n)$ and $g(n)$ such that

$$\text{(i)} \qquad \sum_{n=1}^{\infty} \frac{1}{\log f(n+1)} = \infty$$

$$\text{(ii)} \qquad \sum_{n=1}^{\infty} \sqrt{\frac{f(n)g^2(f(n))}{f(n+1)}} < \infty$$

$$\text{(iii)} \qquad \sum_{n=1}^{\infty} \int_{g(f(n))}^{\infty} e^{-u^2/2} du < \infty.$$

It is easy to see that, for example, $g(f(n)) = n^3$ and $f(n+1) = (n!)^{14}$ satisfy these relations. □

Problem P.17. *Let the sequence of random variables $\{X_m,\ m \geq 0\}$, $X_0 = 0$, be an infinite random walk on the set of nonnegative integers with transition probabilities*

$$p_i = P(X_{m+1} = i+1 \,|\, X_m = i) > 0, \quad i \geq 0,$$

$$q_i = P(X_{m+1} = i-1 \,|\, X_m = i) > 0, \quad i > 0.$$

Prove that for arbitrary $k > 0$ there is an $\alpha_k > 1$ such that

$$P_n(k) = P\left(\max_{0 \leq j \leq n} X_j = k\right)$$

satisfies the limit relation

$$\lim_{L \to \infty} \frac{1}{L} \sum_{n=1}^{L} P_n(k)\alpha_k^n < \infty\ .$$

Solution. Introduce the notation

$$P_n^*(k) = P(\max_{0\le j\le n} X_j \le k),$$
$$P_{i,n}^*(k) = P(\max_{0\le j\le n} X_j \le k \quad \text{and} \quad X_n = i) \quad (0 \le i \le k).$$

Obviously, $P_n(k) \le P_n^*(k) = \sum_{i=0}^k P_{i,n}^*(k)$. It is easy to see that the following inequality holds:

$$\sum_{i=0}^{k} P_{i,n}^*(k) p_i p_{i+1} \dots p_{i+k+1} + P_{n+k+1}^*(k) \le P_n^*(k).$$

Define

$$\varepsilon_k = \min_{0\le i\le k} \{p_i p_{i+1} \dots p_{i+k+1}\}.$$

Then $0 < \varepsilon_k < 1$ and

$$P_{n+k+1}^*(k) \le (1-\varepsilon_k) P_n^*(k).$$

Using mathematical induction, one can easily show that for all positive integers n,

$$P_n^*(k) \le (1-\varepsilon_k)^{1/(k+1)-2}.$$

If

$$1 < \alpha_k < [\frac{1}{1-\varepsilon_k}]^{1/(k+1)-2},$$

then

$$P_n^*(k)\alpha_k^n \le (1-\varepsilon_k)^{2n-2},$$

and this inequality implies the statement. □

Problem P.18. *Let Y_n be a binomial random variable with parameters n and p. Assume that a certain set H of positive integers has a density and that this density is equal to d. Prove the following statements:*
(a) $\lim_{n\to\infty} P(Y_n \in H) = d$ if H is an arithmetic progression.
(b) The previous limit relation is not valid for arbitrary H.
(c) If H is such that $P(Y_n \in H)$ is convergent, then the limit must be equal to d.

Solution.
(a) By the assumptions of the problem,

$$P(Y_n = k) = \binom{n}{k} p^k q^{n-k} \qquad (q = 1-p).$$

For any given k, $\lim_{n\to\infty} P(Y_n = k) = 0$; consequently, $\lim_{n\to\infty} P(Y_n \in H)$ does not change if finitely many elements of H are changed. We can therefore assume that H is a complete residue class with respect to a module D:

$$H = \{k : k \equiv a \bmod D\}.$$

Then

$$P(Y_n \in H) = \sum_{k \equiv a(D)} \binom{n}{k} p^k q^{n-k}. \tag{1}$$

Denote by ε a primitive Dth unit root. It is known that

$$\sum_{\nu=0}^{D-1} \varepsilon^{(k-a)\nu} = \begin{cases} D & \text{if } k \equiv a(D), \\ 0 & \text{if } k \not\equiv a(D). \end{cases}$$

We can write (1) in the following form:

$$\begin{aligned} P(Y_n \in H) &= \sum_{k=0}^{\infty} \sum_{\nu=0}^{D-1} \frac{1}{D} \binom{n}{k} p^k q^{n-k} \varepsilon^{(k-a)\nu} \\ &= \frac{1}{D} \sum_{\nu=0}^{D-1} \varepsilon^{-a\nu} \sum_{k=0}^{\infty} \binom{n}{k} (p\varepsilon^{\nu})^k q^{n-k} \\ &= \frac{1}{D} \sum_{\nu=0}^{D-1} \varepsilon^{-a\nu} (q + \varepsilon^{\nu} p)^n \to \frac{1}{D} = d, \end{aligned}$$

since in the last sum, for $\nu > 0$, the absolute values of the corresponding terms are less than one, so they converge to zero as $n \to \infty$.

(b) We construct a sequence H with zero density such that $\limsup_{n\to\infty} P(Y_n \in H) = 1$. The expectation and variance of Y_n are np and npq, respectively, so the Chebyshev inequality implies that

$$P(|Y_n - np| \geq \lambda\sqrt{npq}) \leq \frac{1}{\lambda^2}.$$

If λ_n is a sequence such that $\lim_{n\to\infty} \lambda_n = \infty$, then

$$\lim_{n\to\infty} P(|Y_n - np| < \lambda_n \sqrt{n}) = 1.$$

Let n_k be an arbitrary increasing sequence of natural numbers, and let

$$H = \bigcup_{k=1}^{\infty} (n_k - [n_k^{3/4}], n_k + [n_k^{3/4}]).$$

By the previous observation,

$$\lim_{k\to\infty} P(Y_{n_k} \in H) = 1,$$

and so

$$\limsup_{n\to\infty} P(Y_n \in H) = 1.$$

Finally, if n_k increases sufficiently fast (for example, $n_k = 10^k$), then H has zero density.

(c) The majority of the contestants verified that if H has density, then $\lim P(Y_n \in H) = d$ in the sense of Cesáro, that is,

$$\frac{S_n}{n} = \frac{\sum_{\nu=1}^{n} P(Y_\nu \epsilon H)}{n} \to d \qquad (n \to \infty).$$

From this property, the original statement follows, since for arbitrary convergent sequence, the Cesáro limit is the same as the usual limit. One can also prove that we do not need to assume the existence of the density of H, since if $P(Y_n \in H)$ is convergent, then H automatically has a density that equals $\lim_{n\to\infty} P(Y_n \in H)$.

A more general result can also be verified: for an arbitrary set H of natural numbers, define $h(x) = |\{n \in H : n < x\}|$. Then

$$\lim_{n\to\infty} \frac{S_n}{n} - \frac{h(np)}{np} = 0.$$

Thus H has a density if and only if $P(Y_n \in H)$ is convergent in the sense of Cesáro.

After these remarks, we can start to prove part (c). We are going to prove that if the density of H is d, then $p_n = P(Y_n \in H) \to d$ in the sense of Abel, that is, $\sum_{n=0}^{\infty}(p_n - p_{n-1}) = d$ in the sense of Abel, namely, $\lim_{x\to 1-0} \sum_{n=0}^{\infty}(p_n - p_{n-1})x^n = d$. If we prove this result, then we are ready, since $\lim_{n\to\infty} P(Y_n \in H) = c$ implies $P(Y_n \in H) \to c$ in the sense of Abel, that is, $c = d$.

For $|x| < 1$,

$$\begin{aligned}
G(x) &= \sum_{n=0}^{\infty}(p_n - p_{n-1})x^n = (1-x)\sum_{n=0}^{\infty} p_n x^n \\
&= (1-x)\sum_{n=0}^{\infty}\sum_{x\in H}\binom{n}{k} p^k q^{n-k} x^n \\
&= (1-x)\sum_{k\in H}(px)^k \sum_{n=0}^{\infty} \frac{n(n-1)\dots(n-k+1)}{k!}(qx)^{n-k} \\
&= (1-x)\sum_{k\in H}(px)^k \frac{1}{(1-qx)^{k+1}} = \frac{1-x}{1-qx}\sum_{k\in H}(\frac{px}{1-qx})^k.
\end{aligned}$$

Substitute $z = px/(1-qx)$. If $x \to 1-0$, then $z \to 1-0$, and furthermore $(1-x)/(1-qx) \sim 1-z$. So

$$\lim_{x\to 1-0} G(x) = \lim_{z\to 1-0}(1-z)\sum_{k\in H} z^k \tag{2}$$

if this last limit exists.

Let a_n be the indicator sequence of H, that is, a_n equals 1 if $n \in H$, and equals 0 if $n \notin H$. The limit of a_n in the sense of Cesáro is the density of H, which is d. Then, by a theorem of Frobenius, the limit of the sequence a_n in the sense of Abel also exists and equals d. This Abel limit is the right-hand side of (2). Consequently, $\lim_{x\to 1-0} G(x) = d$, which was to be proved. □

Problem P.19. *Let* $\{\xi_{kl}\}_{k,l=1}^{\infty}$ *be a double sequence of random variables such that*

$$E\xi_{ij}\xi_{kl} = O((\log(2|i-k|+2)\log(2|j-l|+2))^{-2}) \qquad (i,j,k,l = 1,2,\dots).$$

Prove that with probability one,

$$\frac{1}{mn}\sum_{k=1}^{m}\sum_{l=1}^{n}\xi_{kl} \to 0 \quad \text{as } \max(m,n)\to\infty.$$

Solution. Let

$$S(b,c;m,n) = \sum_{k=b+1}^{b+m}\sum_{l=c+1}^{c+n}\xi_{kl} \quad \text{where} \quad b,c\geq 0, \quad \text{and} \quad m,n\geq 1.$$

Using assumption (1) and the identity thickmuskip=1.4mu

$$\begin{aligned}E\big(S^2(b,c,m,n)\big) = &\sum_{k=b+1}^{b+m}\sum_{l=c+1}^{c+n}E(\xi_{kl}^2)+2\Big\{\sum_{k=b+1}^{b+m}\sum_{l=c+1}^{c+n-1}\sum_{j=1}^{c+n-l}E(\xi_{kl}\xi_{k,l+j})\\ &+\sum_{l=c+1}^{c+n}\sum_{k=b+1}^{b+m-1}\sum_{i=1}^{b+m-k}E(\xi_{kl}\xi_{k+i,l})\Big\}\\ &+2\sum_{k=b+1}^{b+m-1}\sum_{l=c+1}^{c+n-1}\sum_{i=1}^{b+m-k}\sum_{j=1}^{c+n-l}\Big(E(\xi_{kl}\xi_{k+i,l+j})\\ &+E(\xi_{k,l+j}\xi_{k+i,l})\Big),\end{aligned}$$

it is easy to see that

$$\begin{aligned}E(S^2(b,c,m,n)) &= O\{mn\sum_{i=0}^{m-1}\sum_{j=0}^{n-1}\frac{1}{(\log(1+2i)\log(1+2j))^2}\}\\ &= O\{mn\frac{mn}{(\log 2m)^2(\log 2n)^2}\} \qquad (m,n=1,2,\dots).\end{aligned} \tag{2}$$

Put

$$M(b,c;m,n) = \max_{1\leq k\leq m}\max_{1\leq l\leq n}|S(b,c,k,l)|,$$

and use the following result (Theorem 4 in *F. Móricz, Momemt inequalities for the maximum of partial sums of random fields, Acta. Sci. Math. 39 (1977)*, pp. 353–366).

Assume that the nonnegative function $f(b,c;m,n)$ $(b,c,m,n \in N)$ satisfies the following inequalities:

$$f(b,c;h,n)+f(b+h,c;m-h,n) \leq f(b,c,m,n),$$
$$f(b,c,m,n)+f(b,c+i;m,n-i) \leq f(b,c;m,n),$$

for $b \geq 0$, $c \geq 0$, and $1 \leq h \leq m$, $1 \leq i < n$. Let $\chi(m)$ and $\lambda(n)$ be two nondecreasing sequences, and define K and Λ as follows:

$$K(1)=\chi(1), \quad \Lambda(1)=\lambda(1).$$

For $m \geq 2$ and $n \geq 2$,

$$K(m)=\chi(h)+K(h-1), \qquad h=[\frac{1}{2}(m+2)],$$
$$\Lambda(n)=\lambda(i)+\Lambda(i-1), \qquad i=[\frac{1}{2}(n+2)]$$

($[\,.\,]$ denotes the integer part of real numbers). Assume that for some $\gamma \geq 1$, and for all $b,c \geq 0$ and $m,n \geq 1$,

$$E(|S(b,c;m,n)|^{\gamma}) \leq K^{\gamma}(m)\lambda^{\gamma}(n)\,f(b,c;m,n).$$

Then

$$E(M^{\gamma}(b,c;m,n)) \leq K^{\gamma}(m)\Lambda^{\gamma}(n)\,f(b,c;m,n)$$

for all $b,c \geq 0$ and $m,n \geq 1$.

In our case, choose $f(b,c;m,n)=mn$ and $\chi(m)=\sqrt{m}/\log 2m$, $\lambda(n)=\sqrt{n}/\log 2n$, $\gamma=2$. Note that $f(b,c;m,n)$ is an additive set function on the rectangle $[b+1,b+m]\times[c+1,c+n]$. Furthermore, for $2^p \leq m \leq 2^{p+1}$,

$$\begin{aligned}K(m) \leq K(2^{p+1}-1) &= \sum_{k=0}^{p}\chi(2^p)=\sum_{k=0}^{p}\frac{2^{k/2}}{k+1}\\ &= O\left(\frac{2^{p/2}}{p+1}\right)=O\left(\frac{\sqrt{m}}{\log 2m}\right).\end{aligned}$$

Consequently,

$$E(M^2(b,c;m,n)=O\{mn\frac{mn}{(\log 2m)^2(\log 2n)^2}\}. \tag{3}$$

In the case $b=c=0$, define $S(m,n)=S(0,0;m,n)=\sum_{k=1}^{m}\sum_{l=1}^{n}\xi_{kl}$. Since (2) holds,

$$\sum_{k=0}^{\infty}\sum_{l=0}^{\infty}\frac{1}{2^{2k}2^{2l}}E(S^2(2^k,2^l))=O(1)\sum_{k=0}^{\infty}\sum_{l=0}^{\infty}\frac{1}{2^{2k}2^{2l}}\frac{2^{2k}2^{2l}}{(k+1)^2(l+1)^2}<\infty.$$

A theorem of B. Levi implies that with probability one,

$$\frac{1}{2^k 2^l} S(2^k 2^l) \to 0 \quad \text{as} \quad \max(k,l) \to \infty. \tag{4}$$

To complete the proof, we have to show that

$$\frac{1}{2^k 2^l} \max_{2^k < m \leq 2^{k+1}} \max_{2^l < n \leq 2^{l+1}} |S(m,n) - S(2^k, 2^l)| \to 0 \tag{5}$$

with probability one, as $\max(k,l) \to \infty$. Consider the following decomposition:

$$\begin{aligned} S(m,n) = {} & S(2^k, 2^l) + S(2^k, 0; m - 2^k, 2^l) \\ & + S(0, 2^l; 2^k, n - 2^l) + S(2^k, 2^l; m - 2^k, n - 2^l), \end{aligned}$$

from which we see that the left-hand side of (5) is not greater than

$$\frac{1}{2^k 2^l} \{M(2^k, 0; 2^k 2^l) + M(0, 2^l; 2^k, 2^l) + M(2^k, 2^l; 2^k, 2^l)\}.$$

By (3),

$$\sum_{k=0}^{\infty} \sum_{l=0}^{\infty} \frac{1}{2^{2k} 2^{2l}} E(M^2(2^k, 2^l; 2^k, 2^l)) < \infty,$$

and the theorem of B. Levi again implies that

$$\frac{1}{2^k 2^l} M(2^k, 2^l; 2^k, 2^l) \to 0 \tag{6}$$

with probability one, as $\max(k,l) \to \infty$. Notice that

$$M(2^k, 0; 2^k, 2^l) \leq \sum_{q=-1}^{l-1} M(2^k, 2^q; 2^k, 2^q),$$

where, by definition, $M(2^k, 2^{-1}; 2^k, 2^{-1}) = M(2^k, 0; 2^k, 1)$. A Toeplitz lemma and (6) imply that with probability one

$$\frac{1}{2^k 2^l} M(2^k, 0; 2^k, 2^l) \leq \sum_{q=-1}^{l-1} \frac{1}{2^{l-q}} \frac{1}{2^k 2^q} M(2^k, 2^q; 2^k, 2^q) \to 0$$

as $\max(k,l) \to \infty$. A similar argument shows that

$$\frac{1}{2^k 2^l} M(0, 2^l; 2^k, 2^l) \to 0,$$

as $\max(k,l) \to \infty$, with probability one.

Combining (4) and (5), we complete the proof. □

Problem P.20. *Let P be a probability distribution defined on the Borel sets of the real line. Suppose that P is symmetric with respect to the origin, absolutely continous with respect to the Lebesgue measure, and its density function p is zero outside the interval $[-1,1]$ and inside this interval it is between the positive numbers c and d $(c < d)$. Prove that there is no distribution whose convolution square equals P.*

Solution. Assume indirectly that there exists such a distribution Q. Then $Q([-1/2,1/2]) = 1$, and its moments satisfy the relations

$$|M_k| = \left|\int_{-\infty}^{\infty} x^k dQ(x)\right| = \left|\int_{-1/2}^{1/2} x^k dQ(x)\right| \leq 1.$$

Thus,

$$\limsup_{k\to\infty} \sqrt[k]{\frac{|M_k|}{k!}} = 0.$$

Therefore, the characteristic function φ_Q of Q is analytic on the real line. Since P is symmetric, φ_P is real. The equality $\varphi_P(0) = 1$ and the continuity of φ_P imply that, for any x with sufficiently small absolute value, $\varphi_P(x) > 0$. For such x values $\varphi_P(x) = (\varphi_Q(x))^2$, so $\varphi_Q(x)$ is also real. Since φ_Q is analytic, it is real everywhere, which implies that $\varphi_P(x) \geq 0$ for all x. By a well-known theorem (see for example *E. Hewitt, and K. Stromberg, Real and Abstract Analysis, Springer, 1965, p. 409*), we know that if a density function is bounded and its characteristic function is nonnegative, then its characteristic function is integrable. Therefore, φ_P is integrable, so the density function p should be equal to a continuous function almost everywhere, but because of its "jumps" at -1 and 1, this is impossible. □

Remark. For many related negative and positive decomposition results see *I. Z. Ruzsa and G. J. Székely, Algebraic Probability Theory, Wiley, New York, 1988* and *G. J. Székely, Paradoxes in Probability Theory and Mathematical Statistics, Reidel (Kluwer), Dordrecht, 1986*.

Problem P.21. *Let $p_0, p_1, \ldots$ be a probability distribution on the set of nonnegative integers. Select a number according to this distribution and repeat the selection independently until either a zero or an already selected number is obtained. Write the selected numbers in a row in order of selection without the last one. Below this line, write the numbers again in increasing order. Let A_i denote the event that the number i has been selected and that it is in the same place in both lines. Prove that the events A_i $(i = 1, 2, \ldots)$ are mutually independent, and $P(A_i) = p_i$.*

Solution. We will show that for any k $(k = 1, 2, \ldots)$, and any sequence $1 \leq i_1 < i_2 < \cdots < i_k$,

$$P(A_{i_1} A_{i_2} \ldots A_{i_k}) = p_{i_1} p_{i_2} \ldots p_{i_k}.$$

Modify the experiment in such a way that in addition to zero and a repetition of a number, we also stop at selecting any of the numbers $i_1, i_2, \ldots, i_k$. Consider now a (finite) outcome of the original experiment in which $A_{i_1} A_{i_2} \ldots A_{i_k}$ occurs. (The probability of infinite outcome is equal to zero.) From the original sequence, omit the values $i_1, i_2, \ldots, i_k$. In this way, an arbitrary outcome of the modified experiment can be uniquely obtained. On the other hand, it is easy to see that there is only one way to insert the numbers $i_1, i_2, \ldots, i_k$ in such way that event $A_{i_1} A_{i_2} \ldots A_{i_k}$ occurs, that is, the numbers $i_1, i_2, \ldots, i_k$ are placed in their increasing locations. This one-to-one correspondence implies that the probability of the event $A_{i_1} A_{i_2} \ldots A_{i_k}$ is equal to $p_{i_1} p_{i_2} \ldots p_{i_k} \cdot 1$. $\square$

Problem P.22. *Let $X_1, \ldots, X_n$ be independent, identically distributed, nonnegative random variables with a common continuous distribution function F. Suppose in addition that the inverse of F, the quantile function Q, is also continuous and $Q(0) = 0$. Let $0 = X_{0:n} \leq X_{1:n} \leq \cdots \leq X_{n:n}$ be the ordered sample from the above random variables. Prove that if EX_1 is finite, then the random variable*

$$\Delta = \sup_{0 \leq y \leq 1} \left| \frac{1}{n} \sum_{i=1}^{[ny]+1} (n+1-i)(X_{i:n} - X_{i-1:n}) - \int_0^y (1-u)dQ(u) \right|$$

tends to zero with probability one as $n \to \infty$.

Solution. Introduce the notation

$$H_n(y) = \frac{1}{n} \sum_{i=1}^{[ny]+1} (n+1-i)(X_{i:n} - X_{i-1:n})$$

and

$$H_F(y) = \int_0^y (1-u)dQ(u).$$

Thus, $H_F(1) = E(X_1)$ and

$$H_n(1) = \lim_{y \uparrow 1} H_n(y) = \frac{1}{n} \sum_{i=1}^{n} X_i, \tag{1}$$

that is, the problem generalizes the strong law of large numbers for nonnegative random variables.

Because of the continuity of F, the independent random variables $Y_i = F(X_i)$, $1 \leq i \leq n$, are uniformly distributed on the interval $(0,1)$. Denote by $E_n(y) = n^{-1} \times \{k : \ 1 \leq k \leq n \ , \ Y_k \leq y\}$ the empirical distribution function of $Y_1, Y_2, \ldots, Y_n$, and let

$$U_n(y) = \begin{cases} Y_{k:n} & \text{if } \frac{k-1}{n} \leq y \leq \frac{k}{n} \ k = 1, \ldots, n, \\ Y_{n:n} & \text{if } y = 1 \end{cases}$$

be their empirical quantile function. Further, let

$$F_n(x) = n^{-1} \times \{k \quad : \quad 1 \leq k \leq n, \quad X_k \leq x\} \quad (0 \leq x < \infty)$$

and

$$Q_n(y) = \begin{cases} X_{k:n} & \text{if } \frac{k-1}{n} \leq y < \frac{k}{n} \; k = 1, \dots, n, \\ X_{n:n} & \text{if } y = 1 \end{cases}$$

be the empirical distribution and quantile function of the original sample, respectively.

Consider the random function

$$G_n(x) = \int_0^{U_n(y)} (1 - E_n(u))dQ(u).$$

By the continuity of F and Q,

$$G_n(y) = \int_0^{Q(U_n(y))} (1 - E_n(F(x)))dx = \int_0^{Q_n(y)} (1 - F_n(x)dx$$

with probability one, where the exceptional set (where the original sample elements coincide) is independent of y. If $(k-1)/n \leq y < k/n$ for some integer $1 \leq k \leq n$, then this last integral is the following:

$$\int_0^{X_{k:n}} (1 - F_n(x))dx = \sum_{i=1}^{k} \int_{X_{i-1:n}}^{X_{i:n}} (1 - F_n(x))dx,$$

where

$$\sum_{i=1}^{k} (1 - \frac{i-1}{n})(X_{i:n} - X_{i-1:n}) = H_n(y).$$

Since $G_n(1) = H_n(1)$ with the common value given in (1), we conclude that for $n = 1, 2, \dots,$

$$P\{\sup_{0 \leq y \leq 1} |H_n(y) - G_n(y)| = 0\} = 1.$$

It is sufficient to show that for $n \to \infty$,

$$\Delta_n^* = \sup_{0 \leq y \leq 1} |G_n(y) - H_F(y)| \to 0$$

with probability one. Obviously,

$$\begin{aligned} \Delta_n^* &\leq \sup_{0 \leq y \leq 1} |G_n(y) - H_F(U(y))| + \sup_{0 \leq y \leq 1} |H_F(U_n(y)) - H_F(y)| \\ &= \Delta_n^{(1)} + \Delta_n^{(2)}. \end{aligned} \tag{2}$$

Notice that for $0 \le y \le 1$, $U_n(y)$ has only n different values from the interval $[0, 1]$; therefore,

$$\Delta_n^{(1)} \le \Delta_n^{(3)} = \sup_{0 \le y \le 1} |\int_0^y (1 - E_n(u))dQ(u) - \int_0^y (1-u)dQ(u)|.$$

Let $0 < \varepsilon < 1$ be an arbitrary value. Then

$$\Delta_n^{(3)} \le \int_{1-\varepsilon}^1 (1 - E_n(u))dQ(u) + \int_{1-\varepsilon}^1 (1-u)dQ(u)$$
$$+ Q(1-\varepsilon) \sup |y - E_n(y)|.$$

Since $Q(1-\varepsilon) < \infty$, the third term tends to zero with probability one (Glivenko–Cantelli theorem, see, for example, *A. Rényi, Probability Theory, Akadémiai Kiadó, Budapest, 1970*, VII. §8). The first term ($I(A)$ denotes the indicator function of the event A) can be written as

$$\int_{1-\varepsilon}^1 (1 - E_n(u))dQ(u) = \frac{1}{n} \sum_{i=1}^n \int_{1-\varepsilon}^1 (1 - I(\{U_i \le u\}))dQ(u),$$

and because of the strong law of large numbers, this quantity tends to $\int_{1-\varepsilon}^1 (1-u)dQ(u)$ with probability one.

In summary,

$$P\{\limsup_{n\to\infty} \Delta_n^{(3)} \le 2 \int_{1-\varepsilon}^1 (1-u)dQ(u)\} = 1,$$

where the upper bound can be made arbitrarily small by choosing a sufficiently small ε since $E(X_1) < \infty$. Thus, we have proved that the first term of (2) tends to zero ($\Delta_n^{(1)} \to 0$) with probability one as $n \to \infty$.

It is known (see, for example, *M. Csörgő and P. Révész, Strong Approximations in Probability and Statistics, Akadémiai Kiadó, Budapest, 1981*, p. 162) that the Glivenko–Cantelli theorem is also valid for the quantile function of a uniformly distributed sample. Therefore,

$$\sup_{0 \le y \le 1} |U_n(y) - y| \to 0$$

with probability one, as $n \to \infty$. Using the fact that the integral $\int_0^y (1$ $u)dQ(u)$ is a continuous function of y, for the second term in (2) we get

$$P\{\lim_{n\to\infty} \Delta_n^{(2)} = 0\} = 1,$$

which completes the proof. □

Remark. Suppose n machines start working at time $t = 0$, and let $X_{1:n} \le X_{2:n} \le \dots$ be the failure times for these machines. Then $nH_n(y)$ is the total time until the $([ny]+1)$th failure. The best published result (*N. A. Langberg, R. V. Leon, and F. Proschan, Characterization of nonparametric classes of life distribution, Annals of Probability 8(1980), pp. 1163–1170*, Theorem 3.2) shows only pointwise convergence, that is, for all fixed $0 \le y \le 1$, $H_n(y) \to H_F(y)$ with probability one, as $n \to \infty$.

Problem P.23. *Let $X_0, X_1, \dots$ be independent, identically distributed, nondegenerate random variables, and let $0 < \alpha < 1$ be a real number. Assume that the series*

$$\sum_{k=0}^{\infty} \alpha^k X_k$$

is convergent with probability one. Prove that the distribution function of the sum is continuous.

Solution. Define

$$Z = \sum_{k=1}^{\infty} \alpha^{k-1} X_k \qquad \text{and} \qquad Y = \sum_{k=0}^{\infty} \alpha^k X_k.$$

Then $Y = X_0 + \alpha Z$, where Y and Z have the same distribution, and X_0 and Z are independent.

Assume that the distribution function of Y is not continuous. Define

$$p = \max_a P(Y = a),$$

and denote by $a_1, \dots, a_n$ the points for which

$$P(Y = a_j) = p.$$

Then

$$\begin{aligned} p = P(Y = a_j) &= P(X_0 + \alpha Z = a_j) \\ &= \sum_{P(X_0 = x) > 0} P(X_0 = x) P(Z = \frac{a_j - x}{\alpha}) \\ &\leq \sum_{P(X_0 = x) > 0} P(X_0 = x) \max_i P(Z = \frac{a_i - x}{\alpha}) \leq p. \end{aligned}$$

Therefore, we have equality everywhere. This implies $\sum_{x \in R} P(X_0 = x) = 1$, that is, X_0 has a discrete distribution, and

$$P(Z = \frac{a_i - x}{\alpha}) = p \qquad \text{if} \qquad P(X_0 = x) > 0.$$

Put

$$\begin{aligned} K &= \{x : \quad P(X_0 = x) > 0\}, \\ P_a &= \{x : \quad P(Z = \frac{a - x}{\alpha}) = p\}. \end{aligned}$$

Then

$$K \subset \bigcap_{j=1}^{n} P_{a_j}.$$

Since $|P_a| = n$ and $P_a = P_0 + a$,

$$K \subset P_0 + a_j,$$

that is,

$$K - a_j \subset P_0 \quad , \quad j = 1, \dots, n.$$

This relation can hold only if $|K| = 1$, that is, if X_0 is degenerate. □

Remark. For another proof and some related results see *I. Z. Ruzsa and G. J. Székely, Algebraic Probability Theory, Wiley, New York, 1988,* Section 5.5.

Problem P.24. *Let $X_1, X_2, \dots$ be independent random variables with the same distribution:*

$$P(X_i = 1) = P(X_i = -1) = \frac{1}{2} \qquad (i = 1, 2, \dots).$$

Define

$$S_0 = 0, \qquad S_n = X_1 + X_2 + \cdots + X_n \quad (n = 1, 2, \dots),$$
$$\xi(x, n) = |\{k : 0 \le k \le n, \ S_k = x\}| \qquad (x = 0, \pm 1, \pm 2, \dots),$$

and

$$\alpha(n) = |\{x : \xi(x, n) = 1\}| \qquad (n = 0, 1, \dots).$$

Prove that

$$P(\liminf \alpha(n) = 0) = 1$$

and that there is a number $0 < c < \infty$ such that $P(\limsup \alpha(n)/\log n = c) = 1$.

Solution. Represent the values of $S_0, S_1, \dots$ in a two-dimensional coordinate system as follows: starting from the origin, make a step to the right, and move up or down if $X_n = +1$ or $X_n = -1$. Note that $\alpha(n)$ counts the points whose second coordinates are visited exactly once by this random walk.

First, we show that

$$P(\liminf \alpha(n) = 0) = 1,$$

which means that the events $A_n = \{\alpha(n) = 0\}$ occur infinitely many times with probability one. In the moments when the random walk returns to zero, there are infinitely many such points with probability one, the value of $\alpha(n)$ is at most one, since only the extreme values can occur once. However, the probability that between two returns the random walk visits the extreme values at least twice is positive. Therefore, among the infinitely many instants, there will be one, and thus there will be infinitely many ones, when the value of $\alpha(n)$ is zero.

Let us consider the second statement. We will proceed via several lemmas. Denote by $\mathcal{A}_n$ the set of second coordinates visited only once, that is, $\mathcal{A}_n = \{x :$ there is a k, $1 \le k \le n$, such that $S_k = x$, and for $j \neq k$, $S_j \neq x\}$. Then $\alpha(n) = |\mathcal{A}_n|$.

Lemma 1. $\lim_{n\to\infty} P(S_j > 0,\, 0 < j \le n,\, S_n - S_j > 0,\, 0 \le j < n) = c^* > 0$.

Proof. Define the following conditional distribution:

$$\mu_n(dx, dy) = P\left(\frac{1}{\sqrt{n}} S_n = dx, \frac{1}{\sqrt{n}} \sup_{0\le k<n} S_k = dy \quad | S_j > 0 \qquad 1 \le j \le n\right).$$

If $n = 2m+1$, then

$$\begin{aligned}
&P(S_j > 0,\, 0 < j \le n,\, S_n - S_j > 0,\, 0 \le j < n) \\
&\quad = P(S_j > 0,\, 0 < j \le m,\, S_m > \sup_{m<k\le n} (S_n - S_k) - (S_n - S_m), \\
&S_n - S_j > 0,\, m \le j < n,\, S_n - S_m > \sup_{0\le k\le m} (S_k - S_m)) \\
&\quad = P(S_j > 0,\, 0 < j \le m)\, P(S_n - S_j > 0,\, m \le j \le n) \mu_m * \mu_m(A),
\end{aligned}$$

where

$$A = \{(x_1, y_1, x_2, y_2) : x_1 > y_2 - x_2,\, x_2 > y_1 - x_1\}.$$

We know that the sequence μ_n of measures tends weakly to a measure μ^* without atoms and has a positive value on the open subsets of the set $\{(x, y) : x > 0, y > 0\}$. On the other hand,

$$P(S_j > 0,\, 0 < j \le m) P(S_n - S_j > 0,\, m \le j < n) \sim \frac{9}{4\pi} \cdot \frac{1}{n},$$

so the statement of the lemma holds for integers of the form $2m+1$. For integers of the form $2m$, the proof is similar.

Lemma 2. Suppose $k \sim \alpha \log n$, $\alpha_k(n) = \binom{\alpha(n)}{k}$. Then for arbitrary $\eta > 0$, there is an $n_0 = n_0(k, \eta)$ such that for any $n > n_0$,

$$[(c^* - \eta) \log n]^k < E\alpha_k(n) < [(c^* + \eta) \log n]^k.$$

Proof. Put

$$\begin{aligned}
C(r,t) &= \{S_r < S_j < S_t\,,\ r < j < t\}, \\
D_1(t) &= \{S_j < S_t\,, \qquad 0 \le j < t\}, \\
D_2(t) &= \{S_j > S_t\,, \qquad t < j \le n\}.
\end{aligned}$$

Then

$$\begin{aligned}
P(C(r,t)) &= P(C(0, t-r)) = \frac{c^*}{t-r}(1 + o(1)), \\
P(D_1(t)) &= \frac{K}{\sqrt{t}}(1 + o(t)) \qquad (K = \frac{3}{2\sqrt{2\pi}}), \\
P(D_2(t)) &= \frac{K}{\sqrt{n-t}}(1 + o(1)).
\end{aligned}$$

Put

$$\mathcal{A}_n^+ = \mathcal{A}_n \cap \{z : z \geq 0\},$$
$$\alpha^+(n) = |\mathcal{A}_n^+|, \quad \alpha_k^+(n) = \binom{\alpha^+(n)}{k},$$

and for $0 \leq j_1 < j_2 < \cdots < j_k \leq n$, use the notation

$$B_{j_1,\ldots,j_k} = D_1(j_1)C(j_1,j_2)\ldots C(j_{k-1},j_k)D_2(j_k).$$

Then

$$E\alpha_k^+(n) = \sum_{0 \leq j_1 < \cdots < j_k \leq n} P(B_{j_1,\ldots,j_k}),$$

since the event $B_{j_1,\ldots,j_k}$ means that

$$(j_1,\ldots,j_k) \subset \mathcal{A}_n^+$$

and the event $\alpha(n) = l$ contains exactly $\binom{l}{k}$ such events. Put

$$U(j,l) = \sum_{j=j_1<\cdots<j_k=l} P(C(j_1,j_2)) \cdot \ldots \cdot P(C(j_{k-1},j_k)) = U(0,l-j).$$

So

$$\begin{aligned}
E\alpha_k^+(n) &= \sum_{r=0}^{\infty} U(0,r) \sum_{j=0}^{n-r} P(D_1(j))P(D_2(r+j)) \\
&\leq \text{const} \cdot \sum_{r=0}^{n} U(0,r) \sum_{j=1}^{n-r} \frac{1}{\sqrt{j}} \frac{1}{\sqrt{n-j-r}} \leq \text{const} \cdot \sum_{r=0}^{n} U(0,r) \\
&\leq \text{const} \left[\sum_{j=0}^{n} P(C(0,j))\right]^{k-1} \leq [(c^* + \eta)\log n]^k.
\end{aligned}$$

On the other hand,

$$\begin{aligned}
E\alpha_k^+(n) &\geq \sum_{\substack{0 \leq j_1 < n/3 \\ n/3k \geq j_t - j_{t-1} > 0}} P(D_1(j_1)C(j_1,j_2)\ldots C(j_{k-1},j_k)D_2(j_k)) \\
&\geq \frac{\text{const}}{\sqrt{n}} \sum_{j=1}^{n/3} P(D_1(j)) \left[\sum_{j=1}^{n/3} P(C(0,j))\right]^{k-1} \\
&\geq \text{const} \left[(c^* - \frac{\eta}{2}) \log \frac{n}{3k}\right]^{k-1} \geq [(c^* - \eta)\log n]^k
\end{aligned}$$

since $k \sim \alpha \log n$, so $[(c^* - \eta/2)/(c^* - \eta)]^{\alpha \log n} > \log n$ if n is large enough.

We can get the same result for $\mathcal{A}_n^-$ in a similar way. So

$$\mathcal{A}_n = \begin{cases} \mathcal{A}_n^+ & \text{plus maybe one more point, if } S_n \geq 0, \\ \mathcal{A}_n^- & \text{plus maybe one more point, if } S_n \leq 0. \end{cases}$$

So the statement of the lemma is valid for $\alpha(n)$.

Lemma 3. For all $K > 0$ and $\varepsilon > 0$, there is an $n_0 = n_0(K, \varepsilon)$ such that for all $n > n_0$,

$$n^{-(K/c^*)-\varepsilon} \leq P(\alpha(n) < K \log^2 n) \leq n^{-(K/c^*)+\varepsilon}.$$

Proof. Put $k = (K \log n)/c^*$. Then

$$P\left(\alpha(n) \geq K \log^2 n\right) = P\left(\alpha_k(n) \geq \binom{K \log^2 n}{k}\right) \leq \frac{E\alpha_k(n)}{\binom{K \log^2 n}{k}}.$$

If n is large enough, then we can use Lemma 2 to get

$$P(\alpha(n) \geq K \log^2 n) \leq n^{-(K/c^*) \log((c^* e)/(c^* + \eta))} \leq n^{-(K/c^*)+\varepsilon}.$$

For the proof of the lower bound, put $q_m = P(\alpha(n) = m)$. Then

$$E\alpha_k(n) = \sum_m q_m \binom{m}{k}.$$

Put $k' = (K/c^*) \cdot (1 + \varepsilon^2) \log n$. Then for any sufficiently small ε,

$$\sum_{m > (k+\varepsilon) \log^2 n} q_m \binom{m}{k'} < \frac{1}{3} E\alpha_{k'}(n)$$

and

$$\sum_{m \leq K \log^2 n} q_m \binom{m}{k'} < \frac{1}{3} E\alpha_{k'}(n)$$

because if $k'' = ((K + \varepsilon)/c^*) \cdot \log n$, then $k'' > k'$ and

$$\begin{aligned} \sum_{m > (K+\varepsilon) \log^2 n} q_m \binom{m}{k'} &= \sum_m q_m \binom{m}{k''} \frac{\binom{m}{k'}}{\binom{m}{k''}} \\ &\leq \frac{\binom{(K+\varepsilon) \log^2 n}{k'}}{\binom{(K+\varepsilon) \log^2 n}{k''}} \sum_m q_m \binom{m}{k''} \\ &\leq \frac{\binom{(K+\varepsilon) \log^2 n}{k'}}{\binom{(K+\varepsilon) \log^2 n}{k''}} E\alpha_{k''}(n). \end{aligned}$$

Again using Lemma 2 with $\eta = \varepsilon^3$, we get

$$\sum_{m>(K+\varepsilon)\log^2 n} q_m \binom{m}{k'} \leq (c^* \log n)^{k'} \exp\{(-\frac{\varepsilon^2}{2Kc^*} + O(\varepsilon^2)) \log n\}$$

$$\leq \frac{1}{3} E\alpha_{k'}(n)$$

if n is sufficiently large. On the other hand,

$$\sum_{m \leq K \log^2 n} q_m \binom{m}{k'} \leq \frac{\binom{K \log^2 n}{k'}}{\binom{K \log^2 n}{k}} \sum_{m \leq K \log^2 n} q_m \binom{m}{k}.$$

Using Lemma 2 again, we get

$$\sum_{m \leq K \log^2 n} q_m \binom{m}{k'} \leq \frac{1}{3} E\alpha_{k'}(n).$$

These inequalities imply

$$\sum_{K \log^2 n < m < (K+\varepsilon) \log^2 n} q_m \binom{m}{k'} \geq \frac{1}{3} E\alpha_{k'}(n).$$

So

$$P(\alpha(n) \geq K \log^2 n) \geq \sum_{K \log^2 n < m < (K+\varepsilon) \log^2 n} q_m \binom{m}{k'} \frac{1}{\binom{(K+\varepsilon)\log^2 n}{k'}}$$

$$\geq \frac{1}{3} \frac{E(\alpha_{k'}(n))}{\binom{(K+\varepsilon)\log^2 n}{k'}} \geq \frac{1}{3} n^{-\frac{K}{c^*}(1+\varepsilon^3)\log[e \frac{c^*}{c^*-\eta} \frac{K+\varepsilon}{K} (1+\varepsilon^3)^{-2}]}.$$

This last expression tends to 1 for $\varepsilon > 0$ and $\eta > 0$, so we have checked the lower bound. Using the above lemmas, we are able to prove that

$$P\left(\limsup \frac{\alpha(n)}{\log^2 n} = c\right) = 1.$$

Put

$$M_n = \inf\{j : S_j \geq n\}.$$

Then

$$\lim \frac{M_n}{n^4} = 0 \qquad \text{and} \qquad M_{(n+1)^2} - M_{n^2} \geq 2n + 1.$$

Consider the instant when the random walk reaches the height k^2 for the first time. Start a k-step walk from this point, and denote it by $\mathcal{U}_k$, that is,

$$\mathcal{U}_k = \{0, S_{M_{k^2}+1} - S_{M_{k^2}}, \ldots, S_{M_{k^2}+k} - S_{M_{k^2}}\}.$$

These walks are independent from each other. Put

$$B_k = \left\{ \begin{array}{c} \text{the walk } \mathcal{U}_k \text{ visits exactly once at least} \\ (c^* - \varepsilon)\log^2 k \text{ points with positive coordinates} \end{array} \right\}.$$

Then $P(B_k) > k^{-1+\varepsilon/2}$ if k is large enough. So $\sum P(B_k) = \infty$, hence $P(\limsup A_k) = 1$. But on the event $\mathcal{A}_k$ $\alpha(M_{k^2} + k) \geq (c^* - \varepsilon)\log^2 k$, and since $M_{k^2} + k < k^8 + k$ for sufficiently large values of k,

$$\frac{\alpha(n)}{\log^2 n} > \frac{c^* - \varepsilon}{64}$$

for infinitely many n with probability one.

The upper bound of Lemma 3 with $K = c^* + \varepsilon$ shows

$$\sum P(\alpha(n) \geq (c^* + \varepsilon)\log^2 n) < \infty,$$

that is,

$$\limsup \frac{\alpha(n)}{\log^2 n} \leq c^* + \varepsilon,$$

for large values of n with probability one. Recall that the Kolmogorov 0–1 law implies that $\limsup \alpha(n)/\log^2 n$ is constant with probability one; therefore,

$$P\left(\limsup \frac{\alpha(n)}{\log^2 n} = c\right) = 1$$

with a suitable $c^*/64 \leq c \leq c^*$. □

Remark. Unfortunately, the original formulation of the second statement of the problem was incorrect. It contained a log divisor instead of $\log^2$. The proof of the correct statement appeared in *P. Major, On the set visited once by a random walk, Probab. Th. Rel. Fields 77, (1988), pp. 117–128.* The present proof follows this work.

Problem P.25. *Let $(\Omega, \mathcal{A}, P)$ be a probability space, and let $(X_n, \mathcal{F}_n)$ be an adapted sequence in $(\Omega, \mathcal{A}, P)$ (that is, for the σ-algebras $\mathcal{F}_n$, we have $\mathcal{F}_1 \subseteq \mathcal{F}_2 \subseteq \cdots \subseteq \mathcal{A}$, and for all n, X_n is an $\mathcal{F}_n$-measurable and integrable random variable). Assume that*

$$E(X_{n+1}|\mathcal{F}_n) = \frac{1}{2}X_n + \frac{1}{2}X_{n-1} \qquad (n = 2, 3\ldots).$$

Prove that $\sup_n E|X_n| < \infty$ implies that X_n converges with probability one as $n \to \infty$.

Solution. Put $Y_n = X_n + (1/2)\cdot X_{n-1}$ $(n = 2, 3, \ldots)$. Then Y_n is an $\mathcal{F}_n$-measurable and integrable random variable. Furthermore,

$$E(Y_{n+1}|\mathcal{F}_n) = E\left(X_{n+1} + \frac{1}{2}X_n|\mathcal{F}_n\right) = \frac{1}{2}X_n + \frac{1}{2}X_{n-1} + \frac{1}{2}X_n = Y_n$$

for $n \geq 2$. That is, $(Y_n, \mathcal{F}_n,\quad n = 2, 3\ldots)$ is a martingale.

Since $\sup_n E(|Y_n|) \le (3/2) \cdot \sup_n E(|X_n|) < \infty$, the martingale convergence theorem implies that Y_n is convergent with probability one, and

$$\lim_{n\to\infty} Y_n(\omega) = Y(\omega)$$

if $\omega \in \Omega'$, where $P(\Omega') = 1$. Here Y is a random variable that is finite with probability one, and therefore we may assume that it is finite for $\omega \in \Omega'$.

We will show that for $\omega \in \Omega'$, the sequence $X_n(\omega)$ is also convergent. Let $\omega \in \Omega'$ be given, and $a_n = X_n(\omega)$. We will prove that if the sequence $b_n = a_n + (1/2)\cdot a_{n-1}$ converges to c, then a_n converges to $(2/3)\cdot c$. Assume first that $c = 0$. Then for any $\varepsilon > 0$, there is an N such that $|b_n| \le \varepsilon/2$ as $n > N$. So

$$|a_{n+1}| = |b_{n+1} - \frac{1}{2}a_n| \le |b_{n+1}| + \frac{1}{2}|a_n| \le \frac{\varepsilon + |a_n|}{2}$$

for all $n \ge N$. Using this inequality for $n = N, \dots, N+k-1$, we get

$$|a_{N+k}| \le \frac{\varepsilon}{2} + \frac{\varepsilon}{4} + \cdots + \frac{\varepsilon}{2^k} + \frac{|a_N|}{2^k} < \varepsilon + \frac{|a_N|}{2^k} < 2\varepsilon$$

if k is large enough. However, for all $n > N+k$, $|a_n| < 2\varepsilon$. That is, the sequence converges to zero.

If $c \ne 0$, then the same proof applies for the sequence $a'_n = a_n - (2/3)\cdot c$. □

Problem P.26. *Let $X_1, X_2, \dots$ be independent, identically distributed random variables such that $X_i \ge 0$ for all i. Let $EX_i = m$, $\mathrm{Var}(X_i) = \sigma^2 < \infty$. Show that, for all $0 < \alpha \le 1$,*

$$\lim_{n\to\infty} n \,\mathrm{Var}\left(\left[\frac{X_1 + \cdots + X_n}{n}\right]^\alpha\right) = \frac{\alpha^2\sigma^2}{m^{2(1-\alpha)}}.$$

Solution. Put $\overline{X}_n = (X_1 + \cdots + X_n)/n$ and $S_n = X_1 + \cdots + X_n$. Notice first that

$$n\mathrm{Var}\left[\left(\frac{X_1 + \cdots + X_n}{n}\right)^\alpha\right] - nE[(\overline{X}_n^\alpha - m^\alpha + m^\alpha - E(\overline{X}_n^\alpha))^2]$$
$$= nE(\overline{X}_n^\alpha - m^\alpha)^2 - n(E(\overline{X}_n^\alpha - m^\alpha))^2.$$

First we show that the second term tends to zero as $n \to \infty$. The quadratic Taylor polynomial of $(1+z)^\alpha$ with remainder term implies that for $z \ge -1$, $1-(1+z)^\alpha \le (1-\alpha)z^2 - \alpha z$. From the Jensen inequality and this relation, we get

$$0 \le m^\alpha - E(\overline{X}_n^\alpha) = m^\alpha E\left(1 - \left(\frac{\overline{X}_n}{m}\right)^\alpha\right)$$
$$\le m^\alpha\left[(1-\alpha)E\left(\frac{\overline{X}_n}{m} - 1\right)^2 - \alpha E\left(\frac{\overline{X}_n}{m} - 1\right)\right] = (1-\alpha)\frac{m^\alpha}{m^2}\frac{\sigma^2}{n},$$

which implies the assertion. Consider next the first term. Introduce the notation $Z_i = (X_i/m) - 1$ and $\overline{Z}_n = (Z_1 + Z_2 + \cdots + Z_n)/n$. Then $Z_i \geq -1$, $E(Z_i) = 0$, $\mathrm{Var}(Z_i) = \sigma^2/m^2$, furthermore for arbitrary $0 < \varepsilon < 1$,

$$\begin{aligned} nE(\overline{X}_n^\alpha - m^\alpha)^2 &= m^{2\alpha} nE((1+\overline{Z}_n)^\alpha - 1)^2 \\ &= m^{2\alpha}[nE(((1+\overline{Z}_n)^\alpha - 1)^2 I_{|\overline{Z}_n| \leq \varepsilon}) \\ &+ nE(((1+\overline{Z}_n)^\alpha - 1)^2 I_{|\overline{Z}_n| > \varepsilon})]. \end{aligned}$$

In the next step, we will use the following inequalities. In the first term, $\alpha|z|/(1+|z|) \leq |(1+z)^\alpha - 1| \leq \alpha|z|/(1-|z|)$ for $|z| < 1$. In the second term, $|(1+z)^\alpha - 1| \leq |z|^\alpha$ for $z \geq -1$. Observe, furthermore, that if $|z| \leq \varepsilon$, then $|z|/(1+|z|) \geq |z|/(1+\varepsilon)$ and $|z|/(1-|z|) \leq |z|/(1-\varepsilon)$, therefore,

$$\begin{aligned} m^{2\alpha}\left[n\frac{\alpha^2}{(1+\varepsilon)^2}E(\overline{Z}_n^2 I_{|\overline{Z}_n| \leq \varepsilon})\right] &\leq nE(\overline{X}_n^\alpha m^\alpha)^2 \\ &\leq m^{2\alpha}\left[n\frac{\alpha^2}{(1-\varepsilon)^2}E(\overline{Z}_n^2 I_{|\overline{Z}_n| \leq \varepsilon})\right. \\ &\left. c + nE(|\overline{Z}_n|^{2\alpha} I_{|\overline{Z}_n| > \varepsilon})\right]. \end{aligned}$$

Therefore, it is sufficient to show that

$$nE(\overline{Z}_n I_{|\overline{Z}_n| \leq \varepsilon}) \to \frac{\sigma^2}{m^2} \quad \text{and} \quad nE(\overline{Z}_n^{2\alpha} I_{|\overline{Z}_n| > \varepsilon}) \to 0.$$

Let ϕ be the standard normal distribution function. Then for large n,

$$\begin{aligned} \frac{\sigma^2}{m^2} &= nE(\overline{Z}_n^2) \geq nE(\overline{Z}_n^2 I_{|\overline{Z}_n| \leq \varepsilon}) = \frac{\sigma^2}{m^2}E\left[\left(\frac{\sqrt{n}\overline{Z}_n}{\sigma/m}\right)^2 I_{(\frac{\sqrt{n}\overline{Z}_n}{\sigma/m}) \leq \frac{\varepsilon\sqrt{n}}{\frac{\sigma}{m}}}\right] \\ &\geq \frac{\sigma^2}{m^2}E\left[\left(\frac{\sqrt{n}\overline{Z}_n}{\frac{\sigma}{m}}\right)^2 I_{\frac{\sqrt{n}\overline{Z}_n}{\sigma/m} \leq a}\right] \to \frac{\sigma^2}{m^2}\int_{-a}^{a} y^2 d\phi(y) \end{aligned}$$

since $\sqrt{n}\overline{Z}_n/(\sigma/m) \to \phi$ in probability. In addition,

$$\begin{aligned} nE(|\overline{Z}_n|^{2\alpha} I_{|Z_n| > \varepsilon}) &= \varepsilon^{2\alpha} nE\left(\left|\frac{\overline{Z}_n}{\varepsilon}\right|^{2\alpha} I_{|\frac{\overline{Z}_n}{\varepsilon}| > 1}\right) \\ &\leq n\varepsilon^{2\alpha} E\left(\left|\frac{\overline{Z}_n}{\varepsilon}\right|^{2} I_{|\frac{\overline{Z}_n}{\varepsilon}| > 1}\right) \\ &= n\varepsilon^{2\alpha-2} E(\overline{Z}_n^2 I_{|\overline{Z}_n| > \varepsilon}) \to 0. \end{aligned}$$

Since

$$\frac{\sigma^2}{m^2} = nE(\overline{Z}_n^2) = nE(\overline{Z}_n^2 I_{|\overline{Z}_n| \leq \varepsilon}) + nE(\overline{Z}_n^2 I_{|\overline{Z}_n| > \varepsilon}),$$

we get

$$nE(\overline{Z}_n^2 I_{|\overline{Z}_n| \leq \varepsilon}) \to \frac{\sigma^2}{m^2}$$

as $n \to \infty$. $\square$

Problem P.27. *Let F be a probability distribution function symmetric with respect to the origin such that $F(x) = 1 - x^{-1}K(x)$ for $x \geq 5$, where*

$$K(x) = \begin{cases} 1 & \text{if } x \in [5, \infty) \setminus \cup_{n=5}^{\infty}(n!, 4n!), \\ \frac{x}{n!} & \text{if } x \in (n!, 2n!], \quad n \geq 5, \\ 3 - \frac{x}{2n!} & \text{if } x \in (2n!, 4n!), \quad n \geq 5. \end{cases}$$

Construct a subsequence $\{n_k\}$ of natural numbers such that if $X_1, X_2, \ldots$ are independent, identically distributed random variables with distribution function F, then for all real numbers x

$$\lim_{x \to \infty} P\left\{ \frac{1}{n_k} \sum_{j=1}^{n_k} X_j < \pi x \right\} = \frac{1}{2} + \frac{1}{\pi} \arctan x.$$

Solution. Observe first that the independence of $X_1, X_2, \ldots$ implies that for any $n_k \geq 1$ and $t \neq 0$ the characteristic function of the random variable

$$Y_{n_k} = \frac{1}{\pi n_k} \sum_{j=1}^{n_k} X_j$$

can be given as

$$E(e^{itY_{n_k}}) = (1 - \frac{(1/\tau) \cdot h_{n_k}(t/\pi)|t|}{n_k})^{n_k}.$$

The assumed symmetry implies that, for any $s \neq 0$,

$$\begin{aligned} h_{n_k}(s) &= \frac{n_k}{|s|}(1 - E(e^{i\frac{s}{n_k}X_1}) = \frac{n_k}{|s|} \int_{-\infty}^{\infty} \left(1 - \cos\frac{s}{n_k}x\right) dF(x) \\ &= \frac{n_k}{|s|} \int_{-5}^{5} \left(1 - \cos\frac{s}{n_k}x\right) dF(x) - 2 \int_{\frac{5|s|}{n_k}}^{\infty} (1 - \cos y) d\left(\frac{1}{y} K\left(\frac{y}{|s|} n_k\right)\right) \end{aligned}$$

The above form of the characteristic function is useful since by a suitable choice of subsequence $\{n_k\}$ we can guarantee that these functions converge to $e^{-|t|}$. Note that this is the characteristic function of the Cauchy distribution, which is the limit distribution given in the problem. (See *A. Rényi, Probability Theory, Akadémiai Kiadó, Budapest, 1970*, IV. §10, VI. §2.)

The first term of the above representation of h_{n_k} tends to zero if for $k \to \infty, n_k \to \infty$. It is easy to see that if $K \equiv 1$, then the second term (with the choice $n_k = k$) tends to the limit

$$2 \int_0^{\infty} \frac{1 - \cos y}{y^2} dy = \pi.$$

Thus, in this case $n_k = k$ would be a suitable choice. Therefore, it is sufficient to choose the sequence $\{n_k\}$ so that for any $y > 0$ and $s \neq 0$, $K((y/|s|) \cdot n_k) \to 1$ as $k \to \infty$.

Let $\{a_k\}$ be an arbitrary sequence of positive integers such that $a_k \to \infty$ and $a_k/k \to 0$ as $k \to \infty$. Then for arbitrary $x > 0$ and sufficiently large k, $4k! < a_k k! x < (k+1)!$ and for any such k, $K(a_k k! x) = 1$. That is, if we choose $n_k = a_k k!$, $k = 1, 2, \ldots$, then the continuity theorem of P. Lévy, implies that $h_{n_k}(s) \to \pi$ for $k \to \infty$. Therefore, $E(e^{itY_{n_k}}) \to e^{-|t|}$ for arbitrary $t \neq 0$ as $k \to \infty$, and the repeated application of the continuity theorem completes the proof. □

Problem P.28. *Let $a \in \mathbb{C}$, $|a| \leq 1$. Find all values of $b \in \mathbb{C}$ for which there exist probability measures with characteristic function ϕ satisfying $\phi(2) = a$ and $\phi(1) = b$.*

Solution. If ϕ is a characteristic function with the required properties, then $\phi(0) = 1$, $\phi(-1) = \bar{b}$, and $\phi(-2) = \bar{a}$, and the self-adjoint matrix

$$\begin{pmatrix} \phi(0) & \phi(1) & \phi(2) \\ \phi(-1) & \phi(0) & \phi(1) \\ \phi(-2) & \phi(-1) & \phi(0) \end{pmatrix} = \begin{pmatrix} 1 & b & a \\ \bar{b} & 1 & b \\ \bar{a} & \bar{b} & 1 \end{pmatrix}$$

is positive semidefinite. So its determinant is real and nonnegative:

$$1 + \bar{a}b^2 + a\bar{b}^2 - 2b\bar{b} - a\bar{a} \geq 0. \tag{1}$$

If $a = u + vi$ and $b = x + yi$, then (1) can be rewritten as

$$(2 - 2u)x^2 - 4vxy + (2 + 2u)y^2 \leq 1 - u^2 - v^2. \tag{2}$$

Let a' be a complex number such that its square equals a. It is easy to see that if $|a| = 1$, then (1) holds for the points of the interval between a' and $-a'$ (since $|b| \leq 1$, it does not hold for the other points of the connecting line). If $|a| < 1$, then (1) holds for the points in the interior and on the curve of the ellipse with major axis $\sqrt{2 + 2|a|}$ and foci a' and $-a'$. These statements are shown next. Assume first that $|a| = 1$. Let the random variable X be defined as $\arg a'$. Then $\phi_X(1) = a'$ and $\phi_X(2) = a'^2 = a$, so $b = a'$ is suitable. A similar proof shows that $b = -a'$ also satisfies the required properties.

Next assume that $|a| < 1$, and let b be a boundary point of the ellipse. In this case, we have equality in (1). If $b^2 - a = re^{2\alpha i}$, then (1) implies that

$$\begin{aligned} r^2 &= (b^2 - a)(\bar{b}^2 - \bar{a}) = |b|^4 - a\bar{b}^2 - \bar{a}b^2 + |a|^2 \\ &= |b|^4 - 2|b|^2 + 1 = (|b|^2 - 1)^2, \end{aligned}$$

that is, $r = 1 - |b|^2$.

Put $c = \operatorname{Re} e^{-\alpha i} b$, $\omega_1 = \alpha + \arccos c$, and $\omega_2 = \alpha - \arccos c$. Furthermore, let

$$p_1 = \frac{1}{2} + \frac{\operatorname{Im} e^{-\alpha i} b}{2\sqrt{1-c^2}},$$
$$p_2 = \frac{1}{2} - \frac{\operatorname{Im} e^{-\alpha i} b}{2\sqrt{1-c^2}}.$$

Since $|b| < 1$, p_1 and p_2 are positive. Define X as follows:

$$P(X = \omega_1) = p_1 \qquad \text{and} \qquad P(= \omega_2) = p_2.$$

Then the characteristic function ϕ of X satisfies

$$\begin{aligned}
\phi(1) &= p_1 e^{\omega_1 i} + p_2 e^{\omega_2 i} = e^{\alpha i}(p_1 e^{\arccos c} + p_2 e^{\arccos c}) \\
&= e^{\alpha i}((p_1 + p_2)\cos(\arccos c) + i(p_1 - p_2)\sin(\arccos c)) \\
&= e^{\alpha i}((p_1 + p_2)c + 2i(p_1 - p_2)\sqrt{1-c}) \\
&= e^{\alpha i}((\operatorname{Re} e^{-\alpha i} b) + i(\operatorname{Im} e^{-\alpha i} b)) = b
\end{aligned}$$

and

$$\begin{aligned}
\phi(2) &= p_1 e^{2\omega_1 i} + p_2 e^{2\omega_2 i} = e^{2\alpha i}(p_1 e^{2\arccos c} + p_2 e^{-2\arccos c}) \\
&= e^{2\alpha i}((p_1 + p_2)\cos(2\arccos c) + i(p_1 - p_2)\sin(2\arccos c)) \\
&= e^{2\alpha i}((p_1 + p_2)(2c^2 - 1) + i(p_1 - p_2)(2c\sqrt{1-c^2})) \\
&= e^{2\alpha i}(2(\operatorname{Re} e^{-\alpha i} b)^2 - 1 + 2i(\operatorname{Re} e^{-\alpha i} b)(\operatorname{Im} e^{-\alpha i} b)) \\
&= e^{2\alpha i}((e^{-\alpha i} b)^2 + |e^{-\alpha i} b|^2 - 1) = b - e^{2\alpha i}(1 - |b|^2) \\
&= b^2 - r\, e^{2\alpha i} = b^2 - (b^2 - a) = a\,.
\end{aligned}$$

Hence $\phi(1) = b$ and $\phi(2) = a$. Finally, we show that the set of the suitable b values is convex, and consequently, the interior points are also good. Let b_1 and b_2 be two suitable points, and let Q_1 and Q_2 be two probability distributions such that $\phi_{Q_1}(2) = \phi_{Q_2}(2) = a$, $\phi_{Q_1}(1) = b_1$, and $\phi_{Q_2}(1) = b_2$. Define $Q_3 = \lambda Q_1 + (1-\lambda)Q_2$ with $0 \le \lambda \le 1$; then Q_3 is also a probability measure, $\phi_{Q_3}(2) = a$, and $\phi_{Q_3}(1) = \lambda b_1 + (1-\lambda)b_2$. $\square$

Problem P.29. *Let $Y(k)$, $k = 1, 2, \ldots$ be an m-dimensional stationary Gauss–Markov process with zero expectation, that is, suppose that*

$$Y(k+1) = A\,Y(k) + \varepsilon(k+1), \qquad k = 1, 2, \ldots$$

Let H_i denote the hypothesis $A = A_i$, and let $P_i(0)$ be the a priori probability of H_i, $i = 0, 1, 2$. The a posteriori probability $P_1(k) = P(H_1 | Y(1), \ldots, Y(k))$ of hypothesis H_1 is calculated using the assumptions $P_1(0) > 0$, $P_2(0) > 0$, $P_1(0) + P_2(0) = 1$.

Characterize all matrices A_0 such that $P\{\lim_{k\to\infty} P_1(k) = 1\} = 1$ if H_0 holds.

Solution. Let $Y(1)$ be an m-dimensional, normally distributed random vector with zero expectation and covariance matrix D. The so-called "white noise" $\varepsilon(k)$, $k = 1, 2, \ldots$, is independent from the past of the process $Y(k)$, that is the $\varepsilon(k)$'s are m-dimensional, normally distributed random vectors independent from $Y(1)$ and from each other, and they have zero expectation and covariance matrix Q.

The background of the problem is as follows: from hypothesis H_1 and H_2, we accept the one for which the posterior probability converges to 1. We examine the robustness of our procedure, that is, if H_0 holds, then our decision must result in the hypothesis that is "closer" to H_0.

Suppose that Q is regular. Since the process is stationary, each $Y(k)$ has the same covariance matrix D, $k = 1, 2, \ldots$. From (1) we obtain the following matrix equation:

$$\begin{aligned} D &= E(Y(k+1)Y^*(k+1)) = E([A\,Y(k) + \varepsilon(k+1)][A\,Y(k) + \varepsilon(k+1)]^*) \\ &= E(A\,Y(k)Y^*(k)A^*) + E(\varepsilon(k+1)\varepsilon^*(k+1)) = ADA^* + Q\,. \end{aligned}$$

Hence $Q = D - ADA^*$; therefore, D is also regular. Since Q is positive definite and D is at least semidefinite, for any $x \in R^n$ such that $Dx = 0$,

$$0 \le x^*Qx = x^*Dx - x^*ADA^*x = -(A^*x)^*D(A^*x) \le 0,$$

that is, $x = 0$.

Since $P_1(0) + P_2(0) = 1$, $P_1(k) + P_2(k) = 1$ for all $k = 1, 2, \ldots$, and therefore $\lim_{k\to\infty} P_1(k) = 1$ if and only if $\lim_{k\to\infty} P_1(k)/P_2(k) = \infty$. The Bayes theorem implies that, for $i = 1, 2$,

$$P(H_i|Y(1) = y_1, \ldots, Y(k) = y_k) = \frac{p_{Y(1),\ldots,Y(k)|H_i}(y_1, \ldots, y_k)}{p_{Y(1),\ldots,Y(k)}(y_1, \ldots, y_k)} P_i(0),$$

where $p_{Y(1),\ldots,Y(k)}$ (or $p_{Y(1),\ldots,Y(k)|H_i}$) denotes the probability density function of variables $Y(1), \ldots, Y(k)$ (or the corresponding conditional density function under H_i). This relation implies that

$$\frac{P(H_1|Y(1) = y_1, \ldots, Y(k) = y_k)}{P(H_2|Y(1) = y_1, \ldots, Y(k) = y_k)} = \frac{p_{Y(1),\ldots,Y(k)|H_1}(y_1, \ldots, y_k)P_1(0)}{p_{Y(1),\ldots,Y(k)|H_2}(y_1, \ldots, y_k)P_2(0)}\,. \qquad (2)$$

Under the hypothesis H_i, the random variables

$$Y(1) \approx N(0, D), \qquad Y(k+1) - A_iY(k) \approx N(0, D)$$

are independent. (Here $N(M, S)$ denotes the normal distribution with expectation M and covariance matrix S. In the following, we will denote the density function of $N(M, S)$ by $p_{N(M,S)}$.)

In this case,

$$\begin{aligned}
&p_{Y(1),\dots,Y(k)|H_i}(y_1,\dots,y_k)\\
&= p_{N(0,D)}(y_1)p_{N(0,Q)}(y_2-A_iy_1)\dots p_{N(0,Q)}(y_k-A_iy_{k-1})\\
&= \frac{1}{\sqrt{(2\pi)^m|D|}}e^{-\frac{1}{2}y_1^*D^{-1}y_1}\cdot\frac{1}{\sqrt{(2\pi)^m|Q|}}e^{-\frac{1}{2}(y_2-A_iy_1)^*Q^{-1}(y_2-A_iy_1)}\\
&\times\cdot\times\frac{1}{\sqrt{(2\pi)^m|Q|}}e^{-\frac{1}{2}(y_k-A_iy_{k-1})^*Q^{-1}(y_k-A_iy_{k-1})}\\
&= \frac{1}{\sqrt{(2\pi)^{km}|D||Q|^{k-1}}}e^{-\frac{1}{2}\left[y_1^*D^{-1}y_1+\sum_{j=2}^k(y_j-A_iy_{j-1})^*Q^{-1}(y_j-A_iy_{j-1})\right]}.
\end{aligned}$$

Substituting this relation into (2) and using the variables $Y(1),\dots,Y(k)$, we get the following ratio of the posterior probabilities:

$$\begin{aligned}
\frac{P_1(k)}{P_2(k)}=&\frac{P_1(0)}{P_2(0)}\exp(\frac{1}{2}\sum_{j=2}^k[Y(j)-A_2Y(j-1)]^*Q^{j-1}[Y(j)-A_2Y(j-1)]\\
&-\frac{1}{2}\sum_{j=2}^k[Y(j)-A_1Y(j-1)]^*Q^{-1}[Y(j)-A_1Y(j-1)]).
\end{aligned}$$

That is, if the hypothesis H_0 holds, that is, the process is based on the matrix A_0. Then with the random variable

$$\begin{aligned}
&L(Y(1),\dots,Y(k))\\
&=\sum_{j=1}^{k-1}[(A_0-A_2)Y(j)+\varepsilon(j+1)]^*Q^{-1}[(A_0-A_2)Y(j)+\varepsilon(j+1)]-\\
&\quad-\sum_{j=1}^{k-1}[(A_0-A_1)Y(j)+\varepsilon(j+1)]^*Q^{-1}[(A_0-A_1)Y(j)+\varepsilon(j+1)]
\end{aligned}$$

we get

$$\frac{P_1(k)}{P_2(k)}=\frac{P_1(0)}{P_2(0)}\exp\left(\frac{1}{2}L(Y(1),\dots,Y(k))\right).$$

The question to be answered is as follows: when does the limit relation

$$\lim_{k\to\infty}L(Y(1),\dots,Y(k))=\infty$$

hold with probability one?

The process $(Y(k),\ \varepsilon(k+1)\quad k=1,2,\dots)$ is an ergodic Gauss–Markov process. By the ergod theorem,

$$\begin{aligned}
&\lim_{k\to\infty}\frac{1}{k}L(Y(1),\dots,Y(k))\\
&=E([(A_0-A_2)Y(1)+\varepsilon(2)]^*Q^{-1}[(A_0-A_2)Y(1)+\varepsilon(2)])-\\
&\quad-E([(A_0-A_1)Y(1)+\varepsilon(2)]^*Q^{-1}[(A_0-A_1)Y(1)+\varepsilon(2)])\\
&=\operatorname{tr}[(A_0-A_2)^*Q^{-1}(A_0-A_2)D]-\operatorname{tr}[(A_0-A_1)^*Q^{-1}(A_0-A_1)D]=L
\end{aligned}$$

with probability one. If $L > 0$, then $\lim_{k\to\infty} P_1(k)/P_2(k) = \infty$ with probability 1, and this is the same as $\lim_{k\to\infty} P_1(k) = 1$. $\square$

Remark. For $L < 0$, we get $\lim_{k\to\infty} P_2(k) = 1$ with probability 1. One can also prove that for $L = 0$, neither $P_1(k)$ nor $P_2(k)$ tends to 1 with probability one, that is, no appropriate decision between H_1 and H_2 can be made.

Problem P.30. *Let X and Y be independent identically distributed, real-valued random variables with finite expectation. Prove that*

$$E|X+Y| \geq E|X-Y|.$$

Solution 1. For any real numbers α and β,

$$|\alpha+\beta| - |\alpha-\beta| = 2\,\mathrm{sign}(\alpha\beta)\min\{|\alpha|,|\beta|\}.$$

Therefore,

$$\begin{aligned}E(|X+Y|) - E(|X-Y|) =& E(|X+Y| - |X-Y|)\\ =& 2E(\min\{|X|,|Y|\}\mathrm{sign}(XY))\\ =& 2\int_0^\infty \{P(|X|\geq t, |Y|\geq t, XY \geq 0)\\ &- P(|X|>t,|Y|>t,XY<0)\}dt.\end{aligned}$$

Since X and Y are independent, identically distributed variables, and $P(Z \geq t) = P(Z > t)$ almost everywhere,

$$\begin{aligned}&E(|X+Y| - |X-Y|)\\ &\quad = 2\int_0^\infty \{P(X\geq t)]^2 + [P(X\leq -t)]^2 - 2\,P(X\geq t)P(X\leq -t)\}dt\\ &= 2\int_0^\infty [P(X\geq t) - P(X\leq -t)]^2 dt \geq 0.\end{aligned}$$

hence,

$$E(|X+Y|) \geq E(|X-Y|). \quad \square$$

Solution 2. This problem has a simple solution, if we use the Fourier transform of generalized functions. We may assume that X has a nice density function $f(x)$, say an infinitely many times differentiable function with finite support. Then the statement of Problem P.30 can be rewritten as

$$\int |x|(f*f)(x)\,dx \geq \int |x|(f*f^-)(x)dx,$$

where $*$ denotes convolution, and $f^-(x) = f(-x)$. This formula can be rewritten by means of the Plancherel formula if $|x|$ is considered as a generalized function. The Fourier transform of $|x|$, when it is considered as a generalized function, equals $-2\sigma^{-2}$. (See, for example, *I. M. Gelfand, G. E. Shilow, Verallgemeinerte Funtionen (Distributionen) I. VEB Deutscher Verlag der Wissenschaften, Berlin (1960) Vol. 1*). The Plancherel formula, which is actually the definition of the Fourier transform of generalized functions, and the definition of σ^{-2} as a generalized function (see formula (5) on page 60 of Gelfand and Shilow's book) state that the last formula can be rewritten as

$$-2\int_0^\infty \frac{1}{\sigma^2}\left[\tilde{f}^2(\sigma)+\tilde{f}^2(-\sigma)-2\tilde{f}^2(0)\right]d\sigma$$
$$\geq -2\int_0^\infty \frac{1}{\sigma^2}\left[\tilde{f}(\sigma)\tilde{f^-}(\sigma)+\tilde{f}(-\sigma)\tilde{f^-}(-\sigma)-2\tilde{f}(0)\tilde{f^-}(0)\right]d\sigma,$$

where $\tilde{}$ denotes the Fourier transform. A simple calculation shows that the last formula is equivalent to the relation

$$\int_0^\infty \frac{1}{\sigma^2}\left[\tilde{f}(\sigma)-\tilde{f}(-\sigma)\right]^2 d\sigma \leq 0\,.$$

This relation clearly holds, since $\tilde{f}(\sigma)-\tilde{f}(-\sigma)$ is a purely imaginary number. □

Remark. The second solution was suggested by Péter Major.

Problem P.31. *Let $X_1, X_2, \ldots$ be independent, identically distributed random variables such that, for some constant $0<\alpha<1$,*

$$P\left\{X_1 = 2^{k/\alpha}\right\} = 2^{-k}, \quad k=1,2,\ldots$$

Determine, by giving their characteristic functions or any other way, a sequence of infinitely divisible, nondegenerate distribution functions G_n such that

$$\sup_{-\infty<x<\infty}\left|P\left\{\frac{X_1+\cdots+X_n}{n^{1/\alpha}}\leq x\right\}-G_n(x)\right|\to 0 \quad \text{as} \quad n\to\infty.$$

Solution 1. Put $S_n = X_1+\cdots+X_n$, $n=1,2,\ldots$. The characteristic function

$$\varphi(t) = E\left(e^{itX_1}\right) = \sum_{k=1}^{\infty} e^{it2^{(k/\alpha)}}\frac{1}{2^k}\ , \qquad t\in\mathbb{R}.$$

Introduce the sequence $\gamma_n = n/2^{\lceil \log n\rceil}$, $n=1,2,\ldots$, where $\lceil u\rceil = \min\{n\in N\ :\ u\leq n\}$. Obviously, $1/2<\gamma_n\leq 1$ for all n. The independence assumption implies that for all t the value of the characteristic function of the random variable $S_n/n^{1/\alpha}$ is

$$\varphi_n(t) = E(e^{it(S_n/n^{1/\alpha})}) = \varphi^n(1/n^{1/\alpha}) = \left(1+\sum_{k=1}^{\infty}(e^{it2^{(1/\alpha)k}/n^{1/\alpha}}-1)\frac{1}{2^k}\right)^n$$

$$= \left(1+\frac{1}{2^{\lceil \log n\rceil}}\sum_{k=1}^{\infty}(e^{it2^{(1/\alpha)(k-\lceil \log n\rceil)}/\gamma_n^{1/\alpha}}-1)\frac{1}{2^{k-\lceil \log n\rceil}}\right)^n$$

$$= \left(1+\frac{1}{n}\sum_{r=-\lceil \log n\rceil+1}^{\infty}(e^{it2^{r/\alpha}/\gamma_n^{1/\alpha}}-1)\frac{\gamma_n}{2^n}\right)^n .$$

For a given $1/2 \leq \gamma \leq 1$, define

$$h_\gamma(t) = \sum_{r=-\infty}^{\infty}\left(e^{it2^{r/\alpha}/\gamma^{1/\alpha}}-1\right)\frac{\gamma}{2^r} \qquad t \in \mathbb{R}$$

and put

$$\xi_\gamma(T) = e^{h_\gamma(t)}, \qquad t \in R \quad .$$

Since for all $t \in \mathbb{R}$,

$$|h_\gamma(t)| \leq 2\gamma\sum_{r=1}^{\infty}2^{-r} + \gamma^{1-1/\alpha}|t|\sum_{r=0}^{-\infty}2^{(1/\alpha-1)r} \leq 2\gamma + \frac{\gamma^{1-1/\alpha}}{1-2^{1-1/\alpha}}|t|,$$

the above definition is correct, and $\xi_\gamma(.)$ is the characteristic function of the random variable

$$Z_\gamma = \sum_{r=-\infty}^{\infty}\frac{2^{r/\alpha}}{\gamma^{1/\alpha}}Y_r(\gamma),$$

where the $Y_r(\gamma)$'s are independent Poisson-distributed random variables with expectation $E(Y_r(\gamma)) = \gamma/2^r$, $r = 0, \pm 1, \pm 2, \dots$. For the sake of simplicity, Z_γ can be defined as the limit in distribution of the partial sums

$$\sum_{r=-n}^{n}\frac{2^{r/\alpha}}{\gamma^{1/\alpha}}Y_r(\gamma).$$

By the continuity theorem of P. Lévy it is easy to see that this is possible. Obviously, Z_r is an infinitely divisible random variable.

Observe that, for any fixed $t \in \mathbb{R}$, $\xi_\gamma(t)$ is continuous as the function of γ on the interval $[1/2, 1]$. ($\xi_\gamma(t)$ as a bivariable function is continuous on the domain $[1/2] \times \mathbb{R}$).

We are going to prove that the distribution function H_γ of Z_γ, which is uniquely defined by the relation

$$\xi_\gamma(t) = \int_{-\infty}^{\infty} e^{itx}dH_\gamma(x), \qquad t \in R,$$

is continuous on the whole real line for all $\gamma \in [1/2, 1]$. Simple calculations show that

$$\begin{aligned}
\frac{\gamma^{-1/\alpha}}{2}\int_{-\infty}^{\infty}|\xi_\gamma(t)|\,dt &= \frac{1}{2}\int_{-\infty}^{\infty}\left|\exp\left(\gamma\sum_{r=-\infty}^{\infty}\left(e^{is2^{r/\alpha}}-1\right)\frac{1}{2^r}\right)\right|ds\\
&= \int_0^{\infty}\exp\left(-\gamma\sum_{r=-\infty}^{\infty}\left(1-\cos(s2^{r/\alpha})\right)\frac{1}{2^r}\right)ds\\
&\le \int_0^{\infty}\exp\left(-\gamma\sum_{k=0}^{\infty}(1-\cos(s2^{-k/\alpha}))2^{\alpha}\right)ds\\
&\le \int_0^{\infty}\exp\left(-\gamma\sum_{k=\kappa(s)}^{\infty}(1-\cos(s2^{-k/\alpha}))2^{k}\right)ds, \qquad (1)
\end{aligned}$$

where for arbitrary $s > 0$, $\kappa(s)$ denotes the smallest integer $k \ge 1$ such that $\log(s/\pi) < k - 1$. Since $\kappa(s) \le 2 + \log(s/\pi)$ and $s2^{-k} < \pi/2$, for all $k \ge \kappa(s)$, $s2^{-k/\alpha} < \pi/2$. Thus, using the inequality

$$1-\cos x \ge \frac{4}{\pi^2}x^2 \qquad \left(0 \le x \le \frac{\pi}{2}\right),$$

we get

$$\begin{aligned}
\sum_{k=\kappa(s)}^{\infty}\left(1-\cos(s2^{-k/\alpha})\right)2^k &\ge \frac{4}{\pi^2}s^2\sum_{k=\kappa(s)}^{\infty}2^{(1-2/\alpha)k} \ge \frac{4}{\pi^2}s^2\sum_{k=\kappa(s)}^{\infty}\frac{1}{2}^k\\
&= \frac{4}{\pi^2}s^2 2^{1-\kappa(s)} \ge \frac{2}{\pi}s. \qquad (2)
\end{aligned}$$

Hence

$$\int_{-\infty}^{\infty}|\xi_\gamma(t)|dt \;\le\; 2\gamma^{1/\alpha}\int_0^{\infty}e^{-2\gamma/\pi s}ds \;<\; \infty.$$

The continuity of H_γ follows from the inversion formula of characteristic functions. (See, for example, *A. Rényi, Probability Theory, Akadémiai Kiadó, Budapest, 1970.*)

Consider the random variables Z_{γ_n} with distribution functions $G_n(x) = H_{\gamma_n}(x)$, $x \in \mathbb{R}$, and characteristic functions $\Psi_n(t) = \xi_{\gamma_n}(t)$, $t \in \mathbb{R}$, $n = 1, 2, \dots$.

Since for $n \to \infty$

$$h_{\gamma_n}(t) - \sum_{r=-\lceil\log n\rceil+1}^{\infty}\left(e^{it2^{r/\alpha}/\gamma_n^{1/\alpha}}-1\right)\frac{\gamma_n}{2^r} \to 0$$

for arbitrary $t \in \mathbb{R}$, it is easy to see that

$$\phi_n(t) - \Psi_n(t) \to 0 \qquad \text{as} \qquad n \to \infty. \qquad (3)$$

Finally, introduce

$$\triangle(F_n, G_n) = \sup_{-\infty<x<\infty} |F_n(x) - G_n(x)|,$$

where

$$F_n(x) = P\left\{\frac{S_n}{n^{1/\alpha}} \le x\right\}, \qquad x \in \mathbb{R},$$

and consider a sequence $\{n'\}$ of positive integers tending to ∞. Since $1/2 < \gamma_{n'} \le 1$, the Bolzano–Weierstrass theorem implies that there is a subsequence $\{n''\} \subset \{n'\}$ such that for some $\gamma \in [1/2, 1]$, $\gamma_{n''} \to \gamma$ as $n'' \to \infty$. In this case,

$$\triangle(F_{n''}, G_{n''}) \le \triangle(F_{n''}, H_\gamma) + \triangle(G_{n''}, H_\gamma)$$

and $\Psi_{n''}(t) = \xi_{\gamma_{n''}}(t) \to \xi_\gamma(t)$ for all $t \in \mathbb{R}$ as $n'' \to \infty$.

From the continuity theorem of Lévy, we know that $G_{n''}(x) = H_{\gamma_{n''}}(x) \to H_\gamma(x)$ at any continuity point of the limit function. Since the limit function is continuous everywhere, it converges uniformly by a well-known theorem of Pólya. Hence

$$\triangle(G_{n''}, H_\gamma) \to 0 \qquad \text{as} \qquad n'' \to \infty.$$

The relation (3) and a similar argument imply that

$$\triangle(F_{n''}, H_\gamma) \to 0 \qquad \text{as} \qquad n'' \to \infty.$$

Since the sequence $\{n'\}$ was arbitrary,

$$\triangle(F_n, G_n) \to 0 \qquad \text{as} \qquad n \to \infty. \quad \square$$

Solution 2. Define the sequence Z_{γ_n} of random variables in the same way as above:

$$\gamma_n = \frac{n}{2^{\lceil \log n \rceil}}, \qquad L_n = \lceil \log n \rceil, \qquad Z_{\gamma_n} = n^{-1/\alpha} \sum_{r=-\infty}^{\infty} 2^{(L_n + r)/\alpha} Y_r(\gamma_n),$$

where the $Y_r(\gamma_n)$'s are independent Poisson-distributed random variables with expectation $\gamma_n/2^r = n2^{-L_n - r}$. The "three-series theorem" implies that the sum defining Z_{γ_n} is convergent. We will use estimations (1) and (2), which guarantee that the absolute value of the characteristic function $\xi_n(t)$ of the distribution function $G_n(x)$ of the random variable Z_{γ_n} is integrable, and this integral remains bounded independently of n. The distribution functions $G_n(x)$ are obviously infinitely divisible, and we will show that they satisfy the requirements of the problem. This assertion will follow from the following two statements.

Put $S_n = n^{-1/\alpha} \sum_{k=1}^{n} X_k$, and let $F_n(x)$ denote its distribution function.

(a) There exists a sequence $\varepsilon_n \to 0$ of positive numbers such that for all $x \in (-\infty, \infty)$,

$$G_n(x-\varepsilon_n) - \varepsilon_n < F_n(x) < G_n(x+\varepsilon_n) + \varepsilon_n.$$

(b) For all n, the density function $g_n(x)$ of $G_n(x)$ exists, and there is a real M that is independent of n, and for all $x \in (-\infty, \infty)$, $g_n(x) < M$.

First, we show that the requirements of the problem follow from (a) and (b). Indeed,

$$F_n(x)-G_n(x) = F_n(x)-G_n(x+\varepsilon_n)+G_n(x+\varepsilon_n)-G_n(x) \le \varepsilon_n + M\,\varepsilon_n, \quad (4)$$

$$G_n(x)-F_n(x) = G_n(x-\varepsilon_n)-F_n(x)+G_n(x)-G_n(x-\varepsilon_n) \le \varepsilon_n + M\,\varepsilon_n. \quad (5)$$

Since $\varepsilon_n \to 0$, from (4) and (5) we have the limit $\sup_x |G_n(x) - F_n(x)| \to 0$, which was to be proved.

Statement (b) follows from estimations (1) and (2).

Proof of statement (a). Let d_n be a sequence of real numbers that tends to infinity slowly enough. Put

$$Z_n^{(1)} = n^{-1/\alpha} \sum_{r=-\infty}^{L_n-d_n} 2^{(L_n+r)/\alpha} Y_r(\gamma_n),$$

$$Z_n^{(2)} = n^{-1/\alpha} \sum_{r=L_n+d_n}^{\infty} 2^{(L_n+r)/\alpha} Y_r(\gamma_n),$$

$$Z_n^{(3)} = Z_{\gamma_n} - Z_n^{(1)} - Z_n^{(2)},$$

$$S_n^{(1)} = n^{1/\alpha} \sum_{k=1}^{n} X_k I_{(X_k < 2^{(L_n-d_n)/\alpha})} Y_k(\gamma_n),$$

$$S_n^{(2)} = n^{-1/\alpha} \sum_{k=1}^{n} X_k I_{(X_k > 2^{L_n+d_n)/\alpha})} Y_k(\gamma_n),$$

$$S_n^{(3)} = S_n - S_n^{(1)} - S_n^{(2)},$$

where I_A is the indicator function of the set A.

The relations $P(S_n^{(2)} \neq 0) \to 0$, $P(Z_n^{(2)} \neq 0) \to 0$, $E(S_n^{(1)}) \to 0$, $E(Z_n^{(1)}) \to 0$, and simple calculations imply that

$$S_n - S_n^{(3)} \Rightarrow 0 \qquad \text{and} \qquad Z_{\gamma_n} - Z_n^{(3)} \Rightarrow 0, \quad (6)$$

where $\Rightarrow$ denotes the stochastic convergence.

Observe that $S_n^{(3)}$ can be written in the form

$$S_n^{(3)} = \sum_{l=L_n-d_n}^{L_n+d_n} 2^{l/\alpha} \nu_l n^{-1/\alpha},$$

where $\nu_l = \sharp\{j,\ 1 \le j \le n\ , x_j = 2^{l/\alpha}\}$, $l = 1, 2, \dots$, and $\sharp A$ denotes the cardinality of the set A. Denote the distribution function of $S_n^{(3)}$ by $F_n^{(3)}(x)$, and that of $Z_n^{(3)}$ by $G_n^{3)}(x)$. We are going to prove that

$$\operatorname{Var}(F_n^{(3)}(x), G_n^{(3)}(x)) \to 0, \tag{7}$$

where $\operatorname{Var}(\{p_i\}, \{q_i\})$ denotes the distance $\sum_i |p_i - q_i|$ of the discrete distributions $\{p_i\}$ and $\{q_i\}$.

From relations (6) and (7), statement (a) follows immediately. We only have to prove the convergence in (7).

Denote by $\operatorname{distr}\{Y_j(\gamma_n),\ |j| \le d_n\}$ and $\operatorname{distr}\{\nu_{l+L_n}, |l| \le d_n\}$ the joint distr ibutions of the random variables $\{Y_j(\gamma_n)\}$ and $\{\nu_{l+L_n}\}$, respectively, $(|j| \le d_n, |l| \le d_n)$. The limit relation (7) follows from the convergence

$$\operatorname{Var}(\operatorname{distr}\{Y_j(\gamma_n),\quad |j| \le d_n\},\quad \operatorname{distr}\{\nu_{l+L_n}\},\quad |l| \le d_n). \tag{8}$$

This last statement can be proved as follows. Since d_n tends to infinity slowly enough, by restricting the range $\{p_l\ , |l| \le d_n\}$ of the random variables appearing in (8) by $\log^2 n$, the resulting error does not affect the validity of the following estimation:

$$\begin{aligned}
P(\nu_{l+L_n} = p_l\ ,\ |j| \le d_n) &= \frac{n!}{\prod_{l=-d_n}^{d_n} p_l! \left(n - \sum_{l-d_n}^{d_n} p_l\right)!} \cdot \\
&\quad \prod_{l=-d_n}^{d_n} 2^{-p_l(L_n+l)} \cdot \left(1 - \sum_{l=-d_n}^{d_n} 2^{(-L_n+l)}\right)^{n-\sum_{l=-d_n}^{d_n} p_l} \\
&= \prod_{l=-d_n}^{d_n} \frac{(n2^{-L_n-l})^{p_l}}{p_l!} e^{-n2^{-L_n-l}} \cdot \left(1 + O\left(\frac{\log^3 n}{n}\right)\right) \\
&= \left(1 + O\left(\frac{\log^3 n}{n}\right)\right) \prod_{l=-d_n}^{d_n} P(Y_l(\gamma_n) = p_l) \\
&= \left(1 + O\left(\frac{\log^3 n}{n}\right)\right) P(Y_n(\gamma_n) = p_l\ ,\ |l| \le d_n).
\end{aligned}$$

The proof is complete. $\square$

3.9 SEQUENCES AND SERIES

Problem S.1. *Let the Fourier series*

$$\frac{a_0}{2} + \sum_{k\geq 1}(a_k \cos kx + b_k \sin kx)$$

of a function $f(x)$ be absolutely convergent, and let

$$a_k^2 + b_k^2 \geq a_{k+1}^2 + b_{k+1}^2 \quad (k = 1, 2, \dots).$$

Show that

$$\frac{1}{h}\int_0^{2\pi} (f(x+h) - f(x-h))^2 \, dx \qquad (h > 0)$$

is uniformly bounded in h.

Solution. By the Denjoy–Lusin theorem, $\sum_{k=1}^{\infty}(|a_k| + |b_k|) < \infty$. Hence

$$\sum_{k=1}^{\infty} \varrho_k < \infty \quad (\varrho_k = \sqrt{a_k^2 + b_k^2};\ k = 1, 2, \dots). \tag{1}$$

So we obtain $\sum_{k=1}^{\infty} \varrho_k^2 < \infty$. By the Riesz–Fischer theorem and the completeness of trigonometric functions, it follows that $(f(x))^2$ and thereby $(f(x+h) - f(x-h))^2$ is integrable. Easy calculation shows that

$$f(x+h) - f(x-h) \sim 2\sum_{k=1}^{\infty} (b_k \sin kh \cos kx - a_k \sin kh \sin kx) \, .$$

The Parseval formula implies that

$$\int_0^{2\pi} (f(x+h) - f(x-h))^2 \, dx = 4\pi \sum_{k=1}^{\infty} \varrho_k^2 \sin^2 kh. \tag{2}$$

By the condition $\varrho_k^2 \geq \varrho_{k+1}^2 \ (k = 1, 2, \dots)$, we obtain $k\varrho_k \leq \sum_{l=1}^{k} \varrho_l \ (k = 1, 2, \dots)$. So by (1), $k\varrho_k = O(1)$. Using $\sin^2 x \leq |x|$, by (1) we obtain

$$\sum_{k=1}^{\infty} \varrho_k^2 \sin^2 kh \leq |h| \sum_{k=1}^{\infty} k\varrho_k^2 = O(h).$$

By this and (2), we conclude the theorem. □

Remark. Instead of the absolute convergence of the Fourier series and $\varrho_k \geq \varrho_{k+1}$ $(k = 1, 2, \dots)$, it is enough to suppose that $\sum_{k=1}^{\infty} k\varrho_k < \infty$. Gábor Halász and Tibor Nemetz remarked that by the conditions of the theorem $(1/h) \cdot \int_0^{2\pi} (f(x+h) - f(x-h))^2 \, dx = o(h)$ holds.

Problem S.2. *Let $y_1(x)$ be an arbitrary, continuous, positive function on $[0, A]$, where A is an arbitrary positive number. Let*

$$y_{n+1}(x) = 2\int_0^x \sqrt{y_n(t)}\,dt \quad (n = 1, 2, \dots).$$

Prove that the functions $y_n(x)$ converge to the function $y = x^2$ uniformly on $[0, A]$.

Solution. We will not only prove the theorem but also approximate the rate of the convergence.

Let $y_1^*(x)$ and $y_1^{**}(x)$ be given continuous functions on [0,A] such that $0 < y_1^*(x) \le y_1^{**}(x)$, for all $0 \le x \le A$. Denote by $y_1^*(x), y_2^*(x), \dots, y_n^*(x), \dots$ and by $y_1^{**}(x), y_2^{**}(x), \dots, y_n^{**}(x), \dots$ the functions obtained from $y_1^*(x)$ and $y_1^{**}(x)$ by the above iteration. Obviously, $y_n^*(x) \le y_n^{**}(x)$ for all $0 \le x \le A$. So it is enough to prove that the claim of the problem holds for a constant function of value $C > 0$, since if we choose C to be the maximum (minimum) of $y_1(x)$, then the nth element of the function series received from the constant function will be larger (smaller) than $y_n(x)$ at each x.

So let $Y_1(x) \equiv C$, where $C > 0$, and let $Y_2(x), Y_3(x), \dots, Y_n(x)$ be the functions obtained by the iteration. We can show by induction that

$$Y_n(x) = c_n x^{2-(1/2^{n-2})},$$

where c_n is defined by the recursion

$$c_{n+1} = \frac{\sqrt{c_n}}{1 - \frac{1}{2^n}} \quad (n = 1, 2, \dots;\ c_1 = C). \tag{1}$$

Take the 2^{n+1}th power of each side of (1):

$$c_{n+1}^{2n+1} = c_n^{2n} \frac{1}{(1 - \frac{1}{2^n})^{2^{n+1}}}.$$

The multiplier of C_n^{2n} is positive and converges to a finite positive value (to e^2) as n goes to infinity. So the quotient of the neighboring elements of the series $c_1^2, c_2^{2^2}, \dots c_n^{2^n}, \dots$ remains between two positive constants. Therefore, by appropriate constants m and M,

$$m^n < c_n^{2^n} < M^n \quad (n = 1, 2, \dots).$$

Hence, by

$$m^{n/2^n} < c_n < M^{n/2^n} \quad (n = 1, 2, \dots),$$

it follows that

$$c_n = e^{o(n/2^n)}. \tag{2}$$

Obviously,

$$Y_n(x) - x^2 = x^{2-(1/2^{n-1})} \cdot (c_n - 1) + (x^{2-(1/2^{n-1})} - x^2).$$

By (2), for all $0 \le x \le A$,

$$|x^{2-(1/2^{n-1})} \cdot (c_n - 1)| \le A^2 |c_n - 1| = O\left(\frac{n}{2^n}\right).$$

It is easy to show by derivation that the function $x^{2-(1/2^{n-1})} - x^2$ achieves its maximum on the interval (0,1) at $x_0 = (1 - \frac{1}{2^n})^{2^{n-1}}$, and

$$x_0^{2-(1/2^{n-1})} - x_0^2 = x_0^{2-(1/2^{n-1})} \cdot \left(1 - x_0^{(1/2^{n-1})}\right) < \left[1 - \left(1 - \frac{1}{2^n}\right)\right] = \frac{1}{2^n}.$$

For $A \ge 1$, $1 \le x \le A$,

$$|x^{2-\frac{1}{2^{n-1}}} - x^2| = x^{2-\frac{1}{2^{n-1}}} |1 - x^{\frac{1}{2^{n-1}}}| \le A^2 |1 - A^{\frac{1}{2^{n-1}}}| = O\left(\frac{1}{2^n}\right).$$

Thus,

$$Y_n(x) - x^2 = O\left(\frac{n}{2^n}\right).$$

Hence, it follows that, starting from the function $y_1(x)$, which satisfies the conditions of the problem, we get

$$y_n(x) - x^2 = O\left(\frac{n}{2^n}\right),$$

where the constant included in the sign O depends only on A, and on the maximum and minimum of $y_1(x)$. So the claim of the problem is proved. □

Remarks.

1. Gábor Halász proved the following generalization of the problem. If the function $G(y)$ is monotone increasing and continuous for all $y \ge 0$, $G(0) = 0, G(y) > 0$ for $y > 0$, and

$$\int_0^1 \frac{1}{G(y)}\, dy < +\infty,$$

but

$$\int_0^\infty \frac{1}{G(y)}\, dy = +\infty,$$

then obtaining the functions $y_n(x)$ by the recursion

$$y_{n+1}(x) = \int_0^x G(y_n(t))dt,$$

it follows that the functions $y_n(x)$ uniformly converge to the unique solution of the differential equation $y'(x) = G(y)$, which goes through (0,0) and does not agree with the constant function ($x = 0$) in any neighborhood of the origin.

2. If we suppose that $y_1(x)$ is nonnegative instead of positive, we can state the following:
Let δ be the upper bound of those x values for which $y_1(u) = 0$ for all $0 \leq u \leq x$, and let $f(x, \delta)$ be the function such that $f(x, \delta) = 0$ for $0 \leq x \leq \delta$ and $f(x, \delta) = (x - \delta)^2$ for $\delta \leq x \leq A$; then the functions $y_n(x)$ uniformly converge on [0,A] to the function $f(x, \delta)$.

Problem S.3. *Let E be the set of all real functions on $I = [0, 1]$. Prove that one cannot define a topology on E in which $f_n \to f$ holds if and only if f_n converges to f almost everywhere.*

Solution. Let

$$I_{2^m+k} = \left[\frac{k-1}{2^m}, \frac{k}{2^m}\right] \quad (m = 0, 1, 2, \ldots ; k = 1, 2, \ldots 2^m),$$

and denote by f_n the characteristic function of I_n. Then $\{f_n\}$ will nowhere converge to 0; but if $\{f_{n_k}\}$ is an arbitrary subsequence and $x_{n_k} \in I_{nk}$, where $\{x_{n_{kl}}\}$ converges to a number $x \in I$, then obviously $\{f_{n_{kl}}\} \to 0$ everywhere but at x, so in particular, almost everywhere. So if T is a topological space and $x_n \not\to x$, $x_n \in T$, $x \in T$, then there exist, a neighborhood of x such that for some subsequence x_{n_k}, $x_{n_k} \notin V$, and hence for any subsequence $x_{n_{kl}}$, $x_{n_{kl}} \not\to x$. □

Remarks.

1. The solution shows that the same statement is true if in the definition of almost-everywhere convergence we give the role of the zero-measurable sets to a set system $\Re$, where for any $x \in I$, $\{x\} \in \Re$ but $I \notin \Re$.

2. With a slight modification of the solution, one can show that the statement holds when I is the set of real numbers or a closed interval.

Problem S.4. *Let the continuous functions $f_n(x)$, $n = 1, 2, 3, \ldots$, be defined on the interval $[a, b]$ such that every point of $[a, b]$ is a root of $f_n(x) = f_m(x)$ for some $n \neq m$. Prove that there exists a subinterval of $[a, b]$ on which two of the functions are equal.*

Solution 1. Denote by E_{nm} the zero set of $f_n(x) - f_m(x)$ for $n \neq m$. These sets can be enumerated. Let

$$M_1, M_2, M_3, \ldots, M_k, \ldots$$

be an enumeration of these sets. If M_1 is a nowhere-dense set on $[a, b]$, then there exists a subinterval $[a_1, b_1]$ of $[a, b]$ that is disjoint from M_1. If M_2 is a nowhere-dense set on $[a, b]$, then there exists a subinterval $[a_2, b_2]$ of $[a_1, b_1]$ that is disjoint from M_2. If all of the M_k are nowhere dense on $[a, b]$, then we can continue this method. The intersection of all the intervals

$$[a, b] \supseteq [a_1, b_1] \supseteq \cdots \supseteq [a_k, b_k] \supseteq \dots$$

is not empty and does not contain a root of $f_n(x) = f_m(x)$ for any $n \neq m$. This is a contradiction. So there is an $M_{\overline{k}}$ that is dense on some $[c, d]$ part of $[a, b]$, that is, the roots of the corresponding equality $f_{\overline{n}}(x) = f_{\overline{m}}(x)$ are dense on $[c, d]$. By the continuity of the functions, it follows that $f_{\overline{n}}(x) = f_{\overline{m}}(x)$ on $[c, d]$. $\square$

Solution 2. The union of the sets E_{nm} of the previous solution is the interval $[a, b]$. If none of the sets E_{nm} is dense on any subinterval of $[c, d]$, then their union cannot be an interval, since an interval cannot be a union of countably many nowhere dense sets. So there is a set $E_{\overline{nm}}$ that is dense on some $[c, d]$ subinterval of $[a, b]$; that is, the set of the roots of the corresponding $f_{\overline{n}}(x) - f_{\overline{m}}(x)$ is dense on $[c, d]$. Since the functions are continuous, each point of $[a, b]$ is a root. $\square$

Remark. Many participants used the Baire category theorem to generalize the problem.

Problem S.5. *If* $\sum_{m=-\infty}^{+\infty} |a_m| < \infty$, *then what can be said about the following expression?*

$$\lim_{n\to\infty} \frac{1}{2n+1} \sum_{m=-\infty}^{+\infty} |a_{m-n} + a_{m-n+1} + \cdots + a_{m+n}|$$

Solution. Since $\sum_{m=-\infty}^{+\infty} |a_m| = \sigma < \infty$, $\sum_{m=-\infty}^{+\infty} a_m = S$ exists. Let

$$C_n = \frac{1}{2n+1} \sum_{m=-\infty}^{+\infty} |a_{m-n} + a_{m-n+1} + \cdots + a_{m+n}|.$$

We will show that $\lim_{n\to\infty} C_n = |S|$.

Let ε be an arbitrary positive number. Then there is a natural number M such that

$$\sum_{|m|>M} |a_m| < \varepsilon.$$

Hence, for $n \geq M$,

$$(2n+1)C_n = \sum_{|m|>n+M} |\dots| + \sum_{n+M\geq|m|>n-M} |\dots| + \sum_{|m|\leq n-M} |\dots|.$$

So

$$\sum_{|m|>n+M} |a_{m-n} + \cdots + a_{m+n}| \le \sum_{|m|>n+M} a_{m-n} + \cdots + a_{m+n}$$
$$\le (2n+1) \sum_{|m|>M} |a_m| \le (2n+1)\varepsilon,$$

and

$$\sum_{n+M \ge |m| > n-M} |a_{m-n} + \cdots + a_{m+n}| \le \sum_{n+M \ge |m| > n-M} \sigma \le 4M\sigma.$$

Since $m - n \le -M \le M \le m + n$ if $|m| \le n - M$, it follows that if $|m| \le n - M$, then

$$||a_{m-n} + \cdots + a_{m+n}| - |S|| \le \sum_{|m|>M} |a_m| < \varepsilon,$$

so

$$\left| \sum_{|m| \le n-M|} |a_{m-n} + \cdots + a_{m+n}| - (2n - 2M + 1)|S| \right|$$
$$\le \sum_{|m| \le n-M} ||a_{m-n} + \cdots + a_{m+n}| - |S|| \le (2n - 2M + 1)\varepsilon.$$

Therefore,

$$|(2n+1)C_n - (2n - 2m + 1)|S|| \le (2n+1)\varepsilon + 4M\sigma + (2n - 2M + 1)\varepsilon.$$

Dividing by $(2n+1)$ and taking the limit as $n \to \infty$,

$$\limsup_{n\to\infty} |C_n - |S|| \le \varepsilon + \varepsilon = 2\varepsilon.$$

Since ε is an arbitrary positive number, we conclude that

$$\lim_{n\to\infty} C_n = |S|. \quad \square$$

Remark. Let (M, σ, μ) be a commutative measurable group. Namely, M is an additive group, σ is an invariant σ-algebra of the subsets of M under addition and μ is an invariant measure, where $\mu(M) > 0$. Suppose that the function $S : M \times M \to M \times M$ defined by $S(x, y) = (x, x + y)$ maps measurable sets of $M \times M$ into measurable sets of $M \times M$.

Let $A_1, A_2, \ldots (A_i \in \sigma)$ be a sequence of sets symmetric about 0 (that is, $A_i = -A_i$) such that

$$\cup_1^\infty A_i = M; \quad A_i \subset A_{i+1}; \quad 0 < \mu(A_i) < \infty;$$
$$\mu(A_{i+1} - A_i) = o(\mu(A_i)); \quad A_i + A_j \subset A_{i+j}.$$

Claim. If $f(x) \in L_1(M, \mu)$, then

$$\lim_{n\to\infty} \frac{1}{\mu(A_n)} \int_M \left| \int_{A_t} f(x+y)\, d\mu_y \right| d\mu_x = \left| \int_M f\, d\mu \right|.$$

This can be proved by similar methods. Miklós Simonovits gave this generalization.

Problem S.6. *Let $a_1, a_2, \ldots, a_n$ be nonnegative real numbers. Prove that*

$$\left(\sum_{i=1}^{n} a_i\right)\left(\sum_{i=1}^{n} a_i^{n-1}\right) \le n\prod_{i=1}^{n} a_i + (n-1)\sum_{i=1}^{n} a_i^n .$$

Solution 1. If one of the a_i is zero, then the statement follows immediately from the known inequality between the nth power mean and lower power means. It also follows that equality holds iff all the other a_j numbers are equal. So we can suppose $a_i > 0$ for $i = 1, 2, \ldots, n$.

Denote by Σ_k the sum of the kth power of the numbers $a_1, a_2, \ldots, a_n$, and by S_k their kth elementary symmetrical polynomial. We generalize the inequality:

$$\binom{n-1}{k-1}\Sigma_1\Sigma_{k-1} \le nS_k + (k-1)\binom{n}{k}\Sigma_k, \quad 1 \le k \le n. \tag{1}$$

The problem is the special case of inequality (1) for $k = n$.

For $n = 1, 2$, $k = 1$, equality holds. So we can suppose $n \ge 3$, $k \ge 2$. We introduce the following notation. If $\alpha_1, \ldots, \alpha_n \ge 0$ are real numbers, then denote by $[\alpha_1, \ldots, \alpha_n]$ the sum $(1/n!) \cdot \sum a_{i_1}^{\alpha_1} \ldots a_{i_n}^{\alpha_n}$, where we sum over all permutations $i_1, \ldots, i_n$ of $1, \ldots, n$.

Lemma. If $n \ge 3$ and ν, $\alpha_4, \cdots \ge 0$, $\delta > 0$, are real numbers, then

$$[\nu + 2\delta, 0, 0, \alpha_4, \ldots] - 2[\nu + \delta, \delta, 0, \alpha_4, \ldots] + [\nu, \delta, \delta, \alpha_4, \ldots] \ge 0. \tag{2}$$

Equality holds iff all of the a_i are equal.

Proof. We cut the sum (2) into $\binom{n}{3}(n-3)!$ terms, so it is enough to prove that for arbitrary real numbers $b_1, b_2, b_3 > 0$,

$$\begin{aligned} b_1^{\nu+2\delta} + b_2^{\nu+2\delta} + b_3^{\nu+2\delta} - (b_1^{\nu+\delta}b_2^{\delta} + b_2^{\nu+\delta}b_1^{\delta} + b_1^{\nu+\delta}b_3^{\delta} + b_3^{\nu+\delta}b_1^{\delta} \\ + b_2^{\nu+\delta}b_3^{\delta} + b_3^{\nu+\delta}b_2^{\delta}) + b_1^{\nu}b_2^{\delta}b_3^{\delta} + b_1^{\delta}b_2^{\nu}b_3^{\delta} + b_1^{\delta}b_2^{\delta}b_3^{\nu} \ge 0, \end{aligned}$$

where equality holds iff $b_1 = b_2 = b_3$. This is straightforward by the following obvious inequality:

$$x^{\mu}(x-y)(x-z) + y^{\mu}(y-x)(y-z) + z^{\mu}(z-x)(z-y) \ge 0,$$

for $x, y, z > 0, \mu > 0$. Equality holds iff $x = y = z$. We conclude the lemma by changing x, y, z, and μ into $a_1^{\delta}, a_2^{\delta}, a_3^{\delta}$, and ν/δ, respectively. The ideas in the proof of the lemma appears in *G. H. Hardy, J. E. Littlewood, Gy. Pólya, Inequalities, Cambridge University Press, Cambridge, 1959.*

To prove inequality (1), write it in the following form:

$$\binom{n-1}{k-1}\sum_{i \ne j} a_i a_j^{k-1} \le nS_k + \left((k-1)\binom{n}{k} - \binom{n-1}{k-1}\right)\Sigma_k. \tag{3}$$

The following equalities hold by the definition of $[\alpha_1, \dots, \alpha_n]$:

$$\frac{1}{n(n-1)} \sum_{1 \neq j} a_i a_j^{k-1} = [k-1, 1, \underbrace{0, \dots 0}_{n-2}],$$

$$\frac{1}{n} \sum_k = [k, \underbrace{0, \dots 0}_{n-1}],$$

$$\frac{1}{\binom{n}{k}} S_k = [\underbrace{1, \dots 1}_{k}, \underbrace{0, \dots 0}_{n-k}]. \tag{4}$$

If we substitute the inequalities (4) for (3), and making use of the relation $\binom{n}{k} = \frac{n}{k} \cdot \binom{n-1}{k-1}$, we can write (3) in the following form:

$$(n-1)\frac{k}{n}[k-1, 1, \underbrace{0, \dots 0}_{n-2}] \leq [\underbrace{1, \dots 1}_{k}, \underbrace{0, \dots 0}_{n-k}] + \left(k - 1 - \frac{k}{n}\right)[k, \underbrace{0, \dots 0}_{n-1}]. \tag{5}$$

By $(n-1) \cdot (k/n) = k - 1 - (k/n) + 1$, we can write (5) as

$$[k-1, 1, \underbrace{0, \dots 0}_{n-2}] - [\underbrace{1, \dots 1}_{k}, \underbrace{0, \dots 0}_{n-k}]$$
$$\leq (k - 1 - \frac{k}{n})[k, \underbrace{0, \dots 0}_{n-1}] - [k-1, 1, \underbrace{0, \dots 0}_{n-2}]. \tag{6}$$

We introduce the following notation:

$$\Delta_t^k = [t+1, 1, \underbrace{1, \dots 1}_{k-t-1}, \underbrace{0, \dots 0}_{n-k+t-1}] - [t, 1, \underbrace{1, \dots 1}_{k-t}, \underbrace{0, \dots 0}_{n-k+t-1}]$$
$$n \geq 3, \quad k \geq 2, \quad t = 0, 1, \dots, k-1.$$

By substituting $\nu = t - 1, \delta = 1$ into the lemma, it is straightforward that

$$0 = \Delta_0^k \leq \Delta_1^k \leq \dots \leq \Delta_{k-1}^k. \tag{7}$$

Equality holds iff all of the a_is are equal.

We can write (6) in the following form:

$$\sum_{t=0}^{k-2} \Delta_t^k \leq \left(k - 1 - \frac{k}{n}\right) \Delta_{k-1}^k.$$

This inequality holds, since by (7),

$$\sum_{t=0}^{k-2} \Delta_t^k \leq (k-2)\Delta_{k-1}^k \quad \text{and} \quad k - 2 \leq k - 1 - \frac{k}{n}.$$

It is straightforward that if $n \geq 3$ and $k \geq 2$, then equality holds iff $k = n$ and all of the a_is are equal. $\square$

Remark. Inequality (1) was first proved by László Lovász. He also proved it for the case when some of the a_is are zero, but he did not examine the cases of equality. The previous proof is the generalization of Péter Gács's proof. He used this method for proving the case $k = n$, that is, for proving Problem S.6.

Solution 2. If one of the a_is is zero, then the statement follows immediately from the Chebyshev inequality for uniformly ordered sequences. It also follows that equality holds iff all the other numbers a_j are equal. So we can suppose $a_i > 0$ for $i = 1, 2, \ldots, n$. Let

$$F(a_1, \ldots, a_n) = n \prod_{i=1}^{n} a_i + (n-1) \sum_{i=1}^{n} a_i^n - \sum_{i=1}^{n} a_i \sum_{i=1}^{n} a_i^{n-1}.$$

By the homogeneity of F, it is enough to prove that

$$F(a_1, \ldots, a_n) \geq 0 \quad \text{if} \quad a_1 + \cdots + a_n = n. \tag{1}$$

$F(a_1, \ldots, a_n) = 0$ holds for $n = 1, 2$. Let us suppose $n \geq 3$. It is enough to prove (1) in those cases when F has a local minimum at $a_1 + \cdots + a_n = n$, that is, when

$$\frac{1}{n} a_i \frac{dF}{da_i} = (n-1)(a_i^n - a_i^{n-1}) + \prod_{i=1}^{n} a_i = \lambda a_i \quad (i = 1, \ldots, n) \tag{2}$$

for some real λ.

By the Descartes sign rule and $\Pi_{i=1}^{n} a_i > 0$, the polynomial $(n-1)x - (n-1)x^{n-1} + \lambda x + \prod_{i=1}^{n} a_i$ cannot have more than two positive roots. So by (2) it is enough to prove that

$$F(\underbrace{x, \ldots x}_{k}, \underbrace{y, \ldots, y}_{n-k}) \geq 0 \quad \text{for} \quad x, y > 0.$$

By the homogeneity, it is enough to prove that $g(x) \geq 0$, where

$$g(x) = F(\underbrace{x, \ldots x}_{k}, \underbrace{1, \ldots, 1}_{n-k}), \quad 0 < x \leq 1. \quad g(1) = 1.$$

If we show $g(x) > 0$ for $0 < x < 1$, $k \neq 0$, $k \neq 1$, then it also follows that in case $n \geq 3$, $a_i > 0$, equality holds iff all of the a_is are equal. Let $0 < x < 1$, $0 < k < n$. By rearranging, we get

$$g(x) = k(n-k)(x^n - x^{n-1} - x + 1) - kx^n + nx^k - n + k.$$

Hence,

$$\begin{aligned}\frac{g(x)}{1-x} &= k(n-k)(1-x^{n-1}) - n(1+\cdots+x^{k-1}) + k(1+\cdots+x^{n-1}) \\ &= k(n-k) - (n-k)(1+\cdots+x^{k-1}) + k(x^k+\cdots+x^{n-1}) \\ &\quad - k(n-k)x^{n-1} \geq 0,\end{aligned}$$

since

$$k(n-k) \geq (n-k)(1+\cdots+x^{k-1}) \quad \text{and} \quad k(x^k+\cdots+x^{n-1}) \geq k(n-k)x^{n-1}.$$

Equality holds in both inequalities if $k = 1,\ k = n-1$, that is, iff $n = 2$. □

Solution 3.

$$n\sum_{i=1}^{n} a_i^t - \sum_{i=1}^{n} a_i - \sum_{i=1}^{n} a_i^{t-1} = \frac{1}{2}\sum_{i=1}^{n}\sum_{j=1}^{n}(a_i-a_j)^2(a_i^{t-2}+a_i^{t-3}a_j+\cdots+a_j^{t-2})$$

for arbitrary t, since

$$(a_i-a_j)^2(a_i^{t-2}+\cdots+a_j^{t-2}) = a_i^t + a_j^t - a_i a_j^{t-1} - a_j a_i^{t-1} \quad \text{for} \quad i,j = 1,2,\ldots n. \tag{1}$$

If

$$0 \leq a_{n+1} \leq a_n \leq \cdots \leq a_1 \quad \text{and} \quad \sum_{i=1}^{n} a_i = 1,$$

then

$$\begin{aligned}\prod_{i=1}^{n} a_i &= \prod_{i=1}^{n}(a_{n+1}+a_i-a_{n+1}) \geq a_{n+1}^n + a_{n+1}^{n-1}\sum_{i=1}^{n}(a_i-a_{n+1}) \\ &= a_{n+1}^n(1-n) + a_{n+1}^{n-1}.\end{aligned} \tag{2}$$

We shall use induction. Equality holds for $n = 1, 2$. Suppose that $n \geq 2$, $a_{n+1} \leq a_n \leq \cdots \leq a_1$, and that the inequality of the problem holds for n. We are going to prove it for $n+1$. By the homogeneity, we can suppose that $\sum_{i=1}^{n} a_i = 1$. By the induction,

$$(n-1)\sum_{i=1}^{n} a_i^n + n\prod_{i=1}^{n} a_i - \sum_{i=1}^{n} a_i^{n-1} \geq 0. \tag{3}$$

We have to prove

$$n\sum_{i=1}^{n+1} a_i^{n+1} + (n+1)\prod_{i=1}^{n+1} a_i - \sum_{i=1}^{n+1} a_i \sum_{i=1}^{n+1} a_i^n \geq 0. \tag{4}$$

The left side of (4) can be written in the following form:

$$n\sum_{i=1}^{n} a_i^{n+1} + na_{n+1}^{n+1} + na_{n+1}\prod_{i=1}^{n} a_i + a_{n+1}\prod_{i=1}^{n} a_i$$
$$- (a + a_{n+1})(\sum_{i=1}^{n} a_i^n + a_{n+1}^n)\,. \tag{5}$$

By (3), it is enough to prove that (5) is at least as large as a_{n+1} times the left side of (3). So, by rearrangement, we have to prove that

$$(n\sum_{i=1}^{n} a_i^{n+1} - \sum_{i=1}^{n} a_i^n) - a_{n+1}(n\sum_{i=1}^{n} a_i^n) - \sum_{i=1}^{n} a_i^{n-1})$$
$$+ a_{n+1}(\prod_{i=1}^{n} a_i - (1-n)a_{n+1}^n - a_{n+1}^{n-1}) \geq 0. \tag{6}$$

By (1) and $a_i \geq a^{n+1}$ $(i = 1, 2, \ldots, n)$, the difference of the first two terms is

$$\frac{1}{2}\sum_{i=1}^{n}\sum_{j=1}^{n}(a_i - a_j)^2[a_i^{n-1} + a_i^{n-2}a_j + \cdots + a_j^{n-2}$$
$$- a_{n+1}(a_i^{n-2} + a_i^{n-3}a_j + \cdots + a_j^{n-2})] \geq 0\,. \tag{7}$$

By (2), the third term is also nonnegative.

By (7), if $n \geq 2$, then equality holds iff $a_1 = a_2 = \cdots = a_n$. If $a_{n+1} > 0$ and $a_1 = a_2 = \cdots = a_n$, then the third term of (6) can be zero only in the case $a_i - a_{n+1} = 0$; otherwise, strict equality would hold in (2). □

Problem S.7. *Let $f(x) \geq 0$ be a nonzero, bounded, real function on an Abelian group G, $g_1, \ldots, g_k$ are given elements of G and $\lambda_1, \ldots, \lambda_k$ are real numbers. Prove that if*

$$\sum_{i=1}^{k} \lambda_i f(g_i x) \geq 0$$

holds for all $x \in G$, then

$$\sum_{i=1}^{k} \lambda_i \geq 0.$$

Solution 1. We can suppose that $f(g_1) \geq 0$. Denote by A_n the set of those elements that can be written in the form $g_1^{\alpha_1}, \ldots, g_k^{\alpha_k}$, where the maximum absolute value of the numbers $\alpha_1, \ldots, \alpha_k$ is n, where $n \geq 0$ is an integer. Denote by $S(H)$ the sum $\sum_{x \in H} f(x)$, where H is a finite set.

$$\inf_{n>0} \frac{S(A_{n+1}) - S(A_{n-1})}{S(A_n)} = 0$$

holds, since if for some $\varepsilon > 0$ and for all $n > 0$,

$$\frac{S(A_{n+1}) - S(A_{n-1})}{S(A_n)} > \varepsilon$$

would hold, then

$$S(A_{n+1}) > S(A_{n-1}) + \varepsilon S(A_n) \geq (1+\varepsilon) S(A_{n-1})$$

and so

$$S(A_{2n+1}) \geq (1+\varepsilon)^n S(A_1)$$

would hold, which is a contradiction, since

$$S(A_n) \leq \sup_{x \in G} f(x) (2n+1)^k.$$

By (1),

$$\sum_{i=1}^{k} \lambda_i \frac{S(g_i A_n)}{S(A_n)} = \frac{1}{S(A_n)} \sum_{x \in A_n} \sum_{i=1}^{k} \lambda_i f(g_i x) \geq 0,$$

that is,

$$\sum_{i=1}^{k} \lambda_i \geq \sum_{i=1}^{k} \lambda_i \frac{S(A_n) - S(g_i A_n)}{S(A_n)},$$

hence

$$\left| \frac{S(A_n) - S(g_i A_n)}{S(A_n)} \right| \leq \frac{S(A_{n+1}) - S(A_{n-1})}{S(A_n)},$$

and therefore

$$\sum_{i=1}^{k} \lambda_i \geq -\sum_{i=1}^{k} |\lambda_i| \frac{S(A_{n+1}) - S(A_{n-1})}{S(A_n)}.$$

It follows that

$$\sum_{i=1}^{k} \lambda_i \geq -\sum_{i=1}^{k} |\lambda_i| \inf_{n>0} \frac{S(A_{n+1}) - S(A_{n-1})}{S(A_n)} = 0. \quad \square$$

Solution 2. Let us again suppose $f(g_1) > 0$. It is enough to consider the subgroup G' generated by the elements $g_1, \ldots, g_k$. We are going to define a discrete translation-invariant measure on G'. Let $\varepsilon > 0$, and let $\alpha_1 \dots \alpha_k$ be integers. For $x = g_1^{\alpha_1} \dots g_k^{\alpha_k}$, denote by $\mu_\varepsilon(x)$ the number $(1+\varepsilon)^{-(|\alpha_1|+\dots+|\alpha_k|)}$. The measure of G' is finite since

$$\begin{aligned}\sum_{x\in G} \mu_\varepsilon(x) &\le 2^k \sum_{\alpha_1=0}^{\infty} \cdots \sum_{\alpha_k=0}^{\infty} (1+\varepsilon)^{-(\alpha_1+\dots+\alpha_k)} \\ &= 2^k \sum_{\alpha_1=0}^{\infty} (1+\varepsilon)^{-\alpha_1} \cdots \sum_{\alpha_k=0}^{\infty} (1+\varepsilon)^{-\alpha_k},\end{aligned}$$

and so

$$0 < S = \int_{G'} f(x)\, d\mu_\varepsilon(x) < \infty.$$

Furthermore,

$$\mu_\varepsilon(g_i x) = (1+\varepsilon)^{\pm 1} \mu_\varepsilon(x),$$

and so

$$|\mu_\varepsilon(g_i x) - \mu_\varepsilon(x)| \le \varepsilon \mu_\varepsilon(g_i x).$$

Integrating inequality (1) on G', we get

$$\begin{aligned}0 \le \int_{G'} \sum_{i=1}^{k} \lambda_i f(g_i x)\, d\mu_\varepsilon(x) &\le \sum_{i=1}^{k} \lambda_i \int_{G'} f(x)\, d\mu_\varepsilon(x) \\ + \sum_{i=1}^{k} |\lambda_i| \int_{G'} \varepsilon f(g_i x)\, d\mu_\varepsilon(g_i x) &= S \sum_{i=1}^{k} \lambda_i + S\varepsilon \sum_{i=1}^{k} |\lambda_i|.\end{aligned}$$

So by $\varepsilon \to 0$,

$$S \sum_{i=1}^{k} \lambda_i \ge 0$$

it follows that

$$\sum_{i=1}^{k} \lambda_i \ge 0. \quad \square$$

Remarks.

1. Péter Gács and András Simonovits remarked that the statement of the problem holds for an arbitrary group if $k = 2$.

2. Péter Gács and László Lovász showed that the statement generally does not hold for a noncommutative group.

3. László Lovász, Endre Makai, and Imre Ruzsa showed that $f(x)$ has to be bounded.

4. László Lovász discussed situations where the converse theorem holds.

Problem S.8. *Show that the following inequality holds for all $k \geq 1$, real numbers $a_1, a_2, \ldots, a_k$, and positive numbers $x_1, x_2, \ldots, x_k$.*

$$\ln \frac{\sum\limits_{i=1}^{k} x_i}{\sum\limits_{i=1}^{k} x_i^{1-a_i}} \leq \frac{\sum\limits_{i=1}^{k} a_i x_i \ln x_i}{\sum\limits_{i=1}^{k} x_i}$$

Solution 1. By the known inequality relating weighted arithmetic and geometric means,

$$\frac{\sum_{i=1}^{k} x_i y_i}{\sum_{i=1}^{k} x_i} \geq \left(\prod_{i=1}^{k} y_i^{x_i}\right)^{\frac{1}{\sum_{i=1}^{k} x_i}} \qquad (x_i > 0,\ y_i > 0,\ i = 1, 2, \ldots k;\ k \geq 1).$$

If we substitute $y_i = x_i^{-a_i}$ for $i = 1, \ldots, k$ and take -1 times the logarithm of each side, then we get the claim of the problem. □

Solution 2. We are going to prove the following generalization of the problem.

Generalization. Let (X, S, μ) be a measure space, where $\mu(x) > 0$. Let $x(s) > 0$ and let $a(s)$ be measureable functions on X, and $x(s)$, $x(s)^{1-a(s)}$, $a(s)x(s)\ln(x(s))$ are integrable functions on X. In this case,

$$\ln \frac{\int\limits_X x(s) d\mu}{\int\limits_X x(s)^{1-a(s)} d\mu} \leq \frac{\int\limits_X a(s)x(s) \ln x(s) d\mu}{\int\limits_X d(s) d\mu}. \tag{1}$$

Equality holds iff $x(s)^{-a(s)}$ is a constant for almost all $s \in X$.

Proof. The following inequality holds for all positive d:

$$d - 1 \geq \ln d. \tag{2}$$

Put

$$u = \exp \frac{\int\limits_X a(s)x(s) \ln x(s) d\mu}{\int\limits_X x(s) d\mu}.$$

By substituting $d = u \cdot x(s)^{-a(s)}$, multiplying the inequality by $x(s)$, and integrating each side on X, we get the following equality:

$$u \int\limits_X x(s)^{1-a(s)}\, d\mu - \int\limits_X x(s) d\mu \geq 0. \tag{3}$$

This is equivalent to the claim of the generalization. Equality holds in (2) iff $d = 1$. So equality holds in (3) (that is, in (1)) iff $u \cdot x(s)^{-a(s)}$ is a constant for almost all $s \in X$, that is, iff $x(s)^{-a(s)}$ is a constant for almost all $s \in X$. □

Problem S.9. *Construct a continuous function $f(x)$, periodic with period 2π, such that the Fourier series of $f(x)$ is divergent at $x = 0$, but the Fourier series of $f^2(x)$ is uniformly convergent on $[0, 2\pi]$.*

The background of the problem. The following statement is a special case of the Wiener–Lévy theorem. If a positive function $f(x)$ is periodic by 2π and the Fourier series of $f^2(x)$ is absolutely convergent, then the Fourier series of $f(x)$ is also absolutely convergent. The following question arose. Can the absolute convergence be replaced by uniform convergence? The problem above states that it cannot be replaced without assuming the positivity. (R. Salem constructed an $f(x)$ whose Fourier series is uniformly convergent, but the Fourier series of $f^2(x)$ is not.) There are more ways to solve the problem. Lászó Lovász slightly modified Lipót Fejér's known construction: By adding the Fejér polynomials under a relatively strong gap condition, he received a continuous function $f(x)$, whose Fourier series is divergent at $x = 0$, and the Fourier series of $f^2(x)$ is uniformly convergent. It is also nontrivial to prove the uniform convergence of the Fourier series of $f^2(x)$. Lajos Pósa and Péter Gács solved the problem in an essentially similar way: They considered the function $f(x)$, instead of its Fourier series and used the observation that if the following inequality holds for a continuous, 2π-periodic function $g(x)$, then the Fourier series of $g(x)$ is uniformly convergent:

$$|g(x) - g(y)| \leq \frac{1}{(\log \frac{1}{|x-y|})^{1+\varepsilon}}$$

(See the Dini–Lipschitz condition in *I. P. Natanson, Constructive Function Theory 1–3, 1964–65, VII. 3§.*)

Now we prove the claim.

Solution 1. Denote by $s_n(f; x)$ the nth partial sum of the Fourier series of $f(x)$, where $f(x)$ is periodic by 2π and integrable, that is,

$$s_n(f; x) = \frac{1}{\pi} \int_0^{\pi} f(2\vartheta) \frac{\sin(2n+1)\vartheta}{\sin \vartheta} \, d\vartheta. \tag{0}$$

If, on $[0, 2\pi]$,

$$f_n(x) = \sin \frac{(2n+1)x}{2}, \tag{1}$$

then

$$\begin{aligned} s_n(f_n; 0) &= \frac{1}{\pi} \int_0^{\pi} \frac{\sin^2(2n+1)\vartheta}{\sin \vartheta} \, d\vartheta > \frac{1}{\pi} \int_0^{\pi/2} \frac{\sin^2(2n+1)\vartheta}{\sin \vartheta} \, d\vartheta \\ &> \frac{1}{\pi} \int_0^{\pi/2} \frac{\sin^2(2n+1)\vartheta}{\vartheta} \, d\vartheta > \frac{1}{\pi} \int_0^{n\pi} \frac{\sin^2 \vartheta}{\vartheta} d\vartheta > A_1 \log n, \end{aligned} \tag{2}$$

where A_1 and later $A_2, A_3, \dots$ are positive constants. Note that if $|g(x)| \leq 1$ is integrable, then for all x,

$$|s_n(g; x)| \leq A_2 \log n. \tag{3}$$

The function we are going to construct will have the form

$$F(x) = \sum_{\nu=1}^{\infty} c_\nu \sin \frac{2n_\nu + 1}{2} x, \tag{4}$$

where c_ν, n_ν are still undefined. Outside $[0, 2\pi]$ we will get the function by periodic continuation. The restriction

$$|c_\nu| \leq 4^{-\nu} \quad (\nu = 1, 2, \dots) \tag{5}$$

and $F(0) = F(2\pi) = 0$ will guarantee the continuity of $F(x)$ and the absolute convergence of (4). Let $c_1 = 1$, $c_2 = 1/4$, $n_1 = 1$, and suppose that $c_1, c_2, \dots, c_{\mu-1}, c_\mu$ and $n_1, n_2, \dots, n_{\mu-1}$ are already defined.

Since the functions $f_n(x)$ — defined in (1) — are twice differentiable and their Fourier series are uniformly convergent, the following inequality holds at each x, for $m > A_3$

$$|s_m(f_{n_\nu}; x)| \leq 2, \quad \mu = 1, 2, \dots, \nu - 1. \tag{6}$$

Denote by n_ν the smallest positive integer for which

$$c_\mu \log n_\mu > \frac{\mu}{A_1} \tag{7}$$

and

$$n_\mu > 2 + \max\{A_3, 3n_{\mu-1}\}. \tag{8}$$

If

$$c_{\mu+1} = \min\left\{\frac{c_\mu}{4}, \frac{1}{\log n_\mu}\right\}, \tag{9}$$

then (5) is straightforward. So $F(x)$ is defined.

We will use the fact that the following numbers are pairwise different:

$$n_i + n_k + 1 \text{ and } n_j - n_m \quad (1 \leq i \leq k, \ 1 \leq m < j). \tag{10}$$

By $n_1 = 1$ and by (8), the numbers above are different if $1 \leq i \leq k \leq 2$, $1 \leq m < j \leq 2$. If they are different for n_i, n_k, n_j, n_m, where i, k, j, m are less then μ, then by joining n_μ to n_i, n_k, n_j, n_m, we get the following sums and differences:

$$\begin{aligned} n_\mu + n_\mu + 1 > n_\mu + n_{\mu-1} + 1 > \cdots > n_\mu + n_1 + 1 \\ > n_\mu - n_2 > \cdots > n_\mu - n_{\mu-1}. \end{aligned}$$

Even the smallest of them, $n_\mu - n_{\mu-1}$, is larger than the largest of the previous ones, $n_{\mu-1}+n_{\mu-1}+1$. So the numbers at (10) are indeed pairwise different.

By (4) and (5),

$$F^2(x) = \frac{1}{2}\sum_{\mu,\nu} c_\mu c_\nu \{\cos(n_\mu - n_\nu)x - \cos(n_\mu + n_\nu + 1)x\}$$

$$= \frac{1}{2}\sum_{i=1}^{\infty} c_\nu^2 + \sum_{1\le\nu<\mu} c_\mu c_\nu \cos(n_\mu - n_\nu)x$$

$$- \frac{1}{2}\sum_{\nu=1}^{\infty} c_\nu^2 \cos(2n_\nu + 1)x - \sum_{1\le\nu<\mu} c_\mu c_\nu \cos(n_\mu + n_\nu + 1)x.$$

Since the numbers at (10) are pairwise different, this gives the Fourier series of $F^2(x)$. By (5), this is absolutely convergent. We still have to prove that the Fourier series of $F(x)$ is divergent at $x = 0$. To see this, divide the series into three parts:

$$s_n(F;0) = c_\mu s_n(f_n;0) + \sum_{j=1}^{\mu-1} c_j s_n(f_{n_j};0) + \sum_{j=\mu+1}^{\infty} c_j s_{n_\mu}(f_{n_\mu};0)$$

$$= U_1 + U_2 + U_3\,. \tag{11}$$

By (2) and (7),

$$U_1 > \mu. \tag{12}$$

By (6) and (8),

$$|U_2| \le 2\sum_{j=1}^{\mu-1} c_j < \frac{8}{3}. \tag{13}$$

Finally, by (3) and (5),

$$|U_3| < A_2 \log n_\mu \sum_{j=\mu+1}^{\infty} c_j, \qquad A_4 c_{\mu+1} \log n_\mu \le A_4. \tag{14}$$

So by (11), (12), (13), and (14), the Fourier series of $F(x)$ is divergent at $x = 0$. □

Solution 2. We will define the function $f(x)$ instead of its Fourier series, such that (0) will hold for $f^2(x)$, $f(0) = 0$, and

$$\limsup_{n\to\infty} \int_0^\pi \frac{\sin nt}{t} f(t)dt > 0. \tag{15}$$

$f(x)$ will be an even function. Its nth subsum $S_n(0)$ at 0 thus has the following form:

$$S_n(0) = \frac{2}{\pi} \int_0^{\pi} f(2t) \frac{\sin(2n+1)t}{\sin t} dt.$$

If a continuous function is convergent, then its Fourier series and the Fejér means will converge to the same value, to the value of the function. So to see the divergence of $S_n(0)$, it is enough to prove that $S_n(0) \not\to 0$. To prove this, we modify the kernel:

$$\frac{\sin(2n+1)t}{\sin t} = \sin 2nt \cot t + \cos 2nt = \frac{\sin 2nt}{t} + \sin 2nt \left(\cot t - \frac{1}{t} \right) + \cos 2nt.$$

$\cot t - (1/t)$ can be modified to be continuous at 0, so by applying the Riemann lemma twice, we get

$$\begin{aligned} \limsup_{n\to\infty} \int_0^{\pi} f(2t) \frac{\sin(2n+1)t}{\sin t} dt &= \limsup_{n\to\infty} \int_0^{\pi} f(2t) \frac{\sin 2nt}{t} dt \\ &= \limsup_{n\to\infty} \int_0^{\frac{\pi}{2}} f(2t) \frac{\sin 2nt}{t} dt \\ &= \limsup_{n\to\infty} \int_0^{\pi} f(t) \frac{\sin nt}{t} dt > 0. \end{aligned}$$

Thus $S_n(0) \not\to 0$, that is, $S_n(0)$ is divergent if (15) holds. Since $f(n)$ is even, it is enough to construct it on $[0, \pi]$.

Let

$$a_n = \pi e^{-n^3}, \quad \delta_n = n^{-2}, \quad p_n = e^{n^2} \quad (n \geq 0)$$

and

$$I_n = [a_n, a_{n-1}], \quad g_n^*(x) = \delta_n \sin p_n x\,.$$

Define $g_n(x) = g_n^*(x)$ between the second and the last roots of $g^*(x)$ on I_n, and $g_n(x) = 0$ elsewhere. We will define $f(x)$ as a sum of a subsequence of the sequence $\{g_n(x)\}$.

Construct a sequence $n_1 < n_2 < \ldots$ of natural numbers such that

$$\left| \int_0^{\pi} \frac{\sin p_{n_k+1} x}{x} f_k(x)\, dx \right| < \varrho \quad \text{if} \quad f_k = \sum_{i=1}^{k} g_{n_i}, \tag{16}$$

$$\left| \int_0^{\frac{\pi}{2}} \frac{\sin p_{n_j} x}{x} g_{n_k+1}(x)\, dx \right| < \varrho 2^{-k} \quad \text{for all} \quad j \leq k, \tag{17}$$

where ϱ is a sufficiently small constant, say $\varrho = 10^{-5}$. To see that such a sequence exists, observe that, on the one hand, by $p_n \to \infty$, by the continuity of $g_n(x)/x$, and by the Riemann lemma, all terms of the integral in (16) converge to zero; on the other hand, by the continuity of $(\sin \varrho_n x)/x$, the integral in (17) converges to 0 for any fixed j. Define $f(x) = \sum_{k=0}^{\infty} g_{n_k}(x)$. Obviously, $f(x)$ is everywhere continuous, even at 0. We will show that the Fourier series of $f(x)$ does not converge to 0 at 0, that is, it is divergent. For the reader's convenience, denote by $S_n \approx T_n$ if $S_n/T_n \to 1$. By (16) and (17), the difference of $\int_0^\pi (\sin p_{n_k} t)/t\, dt$ and $\int_{a_{n_k}}^{a_{n_k-1}} g_{n_k}(t) \cdot (\sin p_{n_k} t)/t\, dt$ is a maximum of 2ϱ.

$$\begin{aligned}
\int_{a_{n_k}}^{a_{n_k-1}} g_{n_k}(t) \frac{\sin p_{n_k} t}{t}\, dt &\approx \int_{\pi e^{-n_k^3}}^{\pi e^{-(n_k-1)^3}} \delta_{n_k} \sin^2 e^{n_k^3 t}\, dt \\
&= \delta_{n_k} \int_{\pi}^{\pi e^{n_k^3-(n_k-1)^3}} \frac{\sin^2 u}{u} du \\
&\approx \delta_{n_k} \left(\sum_{m=1}^{e^{3n_k^2-3n_k-1}} \frac{1}{m} \right) \int_0^\pi \sin^2 u\, du \\
&\approx \frac{3\pi}{2} n_k^{-2} n_k^2 = \frac{3\pi}{2}.
\end{aligned}$$

So the Fourier series of $f(x)$ is divergent at 0.

Now we are going to show that

$$|f^2(x) - f^2(y)| = O\left(\left(\frac{1}{\log(|x-y|)} \right)^2 \right), \tag{18}$$

that is, the Fourier series of $f^2(x)$ is uniformly convergent. Let $x > y$ and $x \in I_n$, $y \in I_m$ for some $n \geq m$. We move y to the sine curve $\rho_n \sin \rho_n t$ of I_m such that y lies on the same quarter period of the sine curve as x, but $f^2(x)$ remains the same. $|x-y|$ does not increase if $n = m$ or $n > m$. So we have to prove (18) only in the case when x and y are on the same I_n and $|x-y| \leq \pi/p_n$. By the Lagrange mean value theorem,

$$\begin{aligned}
|f^2(x) - f^2(y)| &= \delta_n^2 |\sin p_n^2 x - \sin p_n^2 y| \leq 2\delta_n^2 p_n |x-y| \\
&= 2\delta_n^2 p_n \frac{|x-y|(\log|x-y|)^2}{(\log \frac{1}{|x-y|})^2}.
\end{aligned} \tag{19}$$

Since $h(\log h)^2$ monotonically increases if $h < e^{-2}$, by (19) we have to prove (18) only in the case $|x-y| = \pi/p_n$; but this case is trivial. This finishes the solution. □

Remark. Solution 1 proves more, as it proves that the Fourier series of $f^2(x)$ is absolutely convergent.

Problem S.10. *Prove that for every ϑ, $0 < \vartheta < 1$, there exist a sequence λ_n of positive integers and a series $\sum_{n=1}^{\infty} a_n$ such that*

(i) $\lambda_{n+1} - \lambda_n > (\lambda_n)^{\vartheta}$,

(ii) $\lim\limits_{r\to 1-0} \sum\limits_{n=1}^{\infty} a_n r^{\lambda_n}$ *exists,*

(iii) $\sum\limits_{n=1}^{\infty} a_n$ *is divergent.*

The background of the problem. András Simonovits noted that $\lambda_{n+1} > c\lambda_n$ $(c > 1)$ and the (C,1) summability together guarantee the convergence. Moreover, if we substitute $x = e^{-y}$ in *G. H. Hardy, Divergent Series, Clarendon Press, Oxford, 1949,* §7.13, then Theorem 114 can be stated as follows. Given $c > 1$ constant and a sequence of natural numbers, such that $\lambda_{n+1} > c\lambda_n$ $(c > 1)$, $\sum_{n=1}^{\infty} a_n r^{\lambda_n}$ convergent in the unit circle and its limit exists at $r \to 1 + 0$, then $\sum_{n=1}^{\infty} a_n$ is convergent.

So this theorem, conjectured by Littlewood and proved by Hardy and Littlewood, states that the Abel summability and the convergence are equivalent notions in the case of sequences satisfying the Hadamard gap condition $\lambda_{n+1} > c\lambda_n$ $(c > 1)$. Our problem states that the Hadamard gap condition cannot be substituted with a much weaker condition. The participants gave two kinds of generalization for the problem. Some of them substitute the Abel summability with the stronger (C,1) summability; others showed that for any sequence λ_n satisfying $\liminf(\lambda_{n+1} - \lambda_n)/\lambda_n = 0$, there exists a sequence $\sum_{n=1}^{\infty} a_n$ that satisfies the conditions of the problem. This latter statement is also a generalization of the theorem, since if $\lambda_n = e^{\sqrt{n}}$, then

$$\lambda_{n+1} - \lambda_n = e^{\sqrt{n+1}} - e^{\sqrt{n}} = (e^{\sqrt{n+1}-\sqrt{n}} - 1)e^{\sqrt{n}}$$
$$= e^{\sqrt{n}}(e^{\frac{1}{\sqrt{n}+\sqrt{n+1}}} - 1) = (1 + (o(1))\frac{e^{\sqrt{n}}}{2\sqrt{n}},$$

and therefore,

$$\frac{\lambda_{n+1} - \lambda_n}{\lambda_n} \to 0, \qquad \frac{\lambda_{n+1} - \lambda_n}{\lambda_n^{\vartheta}} \to \infty \text{ for } \vartheta < 1.$$

Solution. We will prove the following statement, which is slightly weaker than the previous two generalizations.

Theorem. For any series $\{\lambda_k\}$ that satisfies

$$\liminf \frac{\lambda_{k+1} - \lambda_k}{\lambda_k} = 0,$$

there exists a divergent series $\sum_{n=1}^{\infty} a_n$ that is (C,1) summable and $a_n = 0$ only if n is not in $\{\lambda_k\}$.

Proof. Let $1 < m_1 < m_2 < m_3 < \dots$ be a subsequence of $\lambda_1, \dots, \lambda_k, \dots$ such that

$$k \sum_{n=1}^{2k-1} m_n < m_{2k} \quad \text{and} \quad \frac{m_{2k} - m_{2k-1}}{m_{2k}} < \frac{1}{k}.$$

Choose a_{m_k} to be $(-1)^{k+1}$, the others to be zero. Since $\sum_{j=1}^n a_j$ is 0 infinitely many times and 1 infinitely many times, $\sum_{j=1}^n a_j$ is divergent. Denote $\sum_{j=1}^n a_j$ by s_n. Since

$$0 \leq t_n = \frac{s_1 + s_2 + \cdots + s_n}{n} < 1,$$

t_n is monotone decreasing in $[m_{2k}, m_{2k+1})$ and monotone increasing in $[m_{2k-1}, m_{2k})$. Therefore, to prove $t_m \to 0$, it is enough to show that $t_{m_{2k}} \to 0$:

$$t_{m_{2k}} = \frac{(m_2 - m_1) + (m_4 - m_3) + \cdots + (m_{2k} - m_{2k-1})}{m_{2k}} < \frac{1}{k} + \frac{1}{k} \to 0.$$

Hence, $t_m \to 0$, that is, $\sum_{j=1}^n a_j$ is (C,1) summable. □

Problem S.11. *Let $0 < a_k < 1$ for $k = 1, 2, \dots$. Give a necessary and sufficient condition for the existence, for every $0 < x < 1$, of a permutation π_x of the positive integers such that*

$$x = \sum_{k=1}^{\infty} \frac{a_{\pi_x(k)}}{2^k}.$$

Solution. Obviously, the following condition is sufficient:

$$\inf_k a_k = 0, \quad \sup_k a_k = 1.$$

We are going to prove that this condition is also necessary. Let $x \in (0,1)$ be fixed. For a permutation π, let $a_{\pi(k)} = b_k$ and

$$d_k = 2^k \left[x - \left(\frac{b_1}{2} + \frac{b_2}{4} + \cdots + \frac{b_k}{2^k} \right) \right] \quad (d_0 = x).$$

If we choose the permutation π such that $0 < d_k < 1$ holds for all k, then obviously

$$x = \sum_{k=1}^{\infty} \frac{b_k}{2^k}.$$

Let us suppose that $\pi(1), \pi(2), \dots, \pi(k)$ are already defined such that $0 < d_0, d_1, \dots, d_k < 1$. We want to define $\pi(k+1)$ such that $0 < d_{k+1} < 1$. This is equivalent to

$$2d_k - 1 < b_{k+1} < 2d_k, \tag{1}$$

and also equivalent to

$$2(1-d_k)-1<1-b_{k+1}<2(1-d_k). \tag{2}$$

We have to take care that a_k will be used exactly once in the construction.

Let a_l be the element with the smallest index such that $a_l \notin \{b_1, b_2, \dots, b_k\}$. If $2d_k - 1 < a_l < 2d_k$, then we can choose $b_{k+1} = a_l$. If $a_l \geq 2d_k$, then let n be the smallest natural number such that $2^{n+1}d_k > a_l$; then $2^{n+1}d_k \leq 2a_l < 1 + a_l$, so

$$2^{n+1}d_k - 1 < a_l < 2^{n+1}d_k.$$

Since $\inf_k a_k = 0$, $b_{k+1}, b_{k+2}, \dots, b_{k+n}$ can be chosen sufficiently small, such that by $d_{j+1} = 2d_j - b_{j+1}$, d_{j+n} will be close enough to $2^n d_k$. So we can achieve

$$2d_{k+n} - 1 < a_l < 2d_{k+n}.$$

Thus, we can choose $b_{k+n+1} = a_l$.

If $a_l < 2d_k - 1$, or equivalently $1 - a_l \geq 2(1-d_k)$, then we repeat the above process by substituting $1-a_l, 1-a_j, 1-d_j$ in the role of a_l, a_j, d_j, respectively, and by using (2) instead of (1); and in the meantime, we use the equality $1 - d_{j+1} = 2(1-d_j) - (1-b_{j+1})$. By $\sup_k a_k = 1$, we can choose the numbers $1 - b_j$ to be sufficiently small, that is, b_j is sufficiently close to 1.

Repeating the process above, we obtain the required permutation. □

Problem S.12. *Let $\lambda_1 \leq \lambda_2 \leq \dots$ be a positive sequence and let K be a constant such that*

$$\sum_{k=1}^{n-1} \lambda_k^2 < K\lambda_n^2 \quad (n = 1, 2, \dots). \tag{1}$$

Prove that there exists a constant K' such that

$$\sum_{k=1}^{n-1} \lambda_k < K'\lambda_n \quad (n = 1, 2, \dots). \tag{2}$$

Solution 1. We can suppose that K is an integer. We are going to prove by induction that (2) holds if $K' = 8K$. The theorem is straightforward if $n > 8K$, since $K > 1$. Let $n > 8K$.

We observe that by the monotonicity of $\{\lambda_n\}$ and by (1),

$$4K\lambda_{n-4K}^2 \leq \sum_{k=1}^{n-1} \lambda_k^2 < K\lambda_n^2,$$

so $\lambda_{n-4K} \geq \lambda_n/2$. By this, by the induction, and by the monotonicity of $\{\lambda_n\}$,

$$\sum_{k=1}^{n-1} \lambda_k = \sum_{k=1}^{n-4K-1} \lambda_k + \sum_{k=n-4K}^{n-1} \lambda_k$$
$$\leq 8K\lambda_{n-4K} + \sum_{k=n-4K}^{n-1} \lambda_k \leq 4K\lambda_n + 4K\lambda_n = 8K\lambda_n.$$

Thus, the solution is complete. □

Solution 2. The monotonicity of $\{\lambda_n\}$ is superfluous. We are going to prove the following. If $\mu_i > 0$, then the necessary and sufficient condition of

$$\sum_{t=1}^{n-1} \mu_i < K\mu_n, \quad n = 1, 2, \dots \tag{$*$}$$

is that there exist $c > 0$ real and r natural numbers such that, for each n,

$$\mu_{n+1} > c\mu_n \quad \text{and} \tag{a}$$
$$\mu_{n+r} > 2\mu_n. \tag{b}$$

This implies the statement of the theorem, since if $\lambda_{n+1}^2 > c\lambda_n^2$ and $\lambda_{n+r}^2 > 2\lambda_n^2$, then $\lambda_{n+1} > \sqrt{c}\lambda_n$ and $\lambda_{n+2r}^2 > \lambda_{n+r}^2 > 4\lambda_n^2$, so $\lambda_{n+2r} > 2\lambda_n$.

(a) is necessary: by $(*)$ and by $\mu_i > 0$, trivially $\mu_{n-1} < K\mu_n$, so (a) holds if $c = 1/K$.

(b) is necessary: by $(*)$ again,

$$\mu_{n+1} > \frac{1}{K}\mu_n,\ \mu_{n+2} > \frac{1}{K}\mu_n,\ \dots, \mu_{n+r-1} > \frac{1}{K}\mu_n.$$

Adding them together, we get

$$\mu_{n+1} + \cdots + \mu_{n+r-1} > \frac{r-1}{K}\mu_n,$$

so by $(*)$,

$$\mu_{n+r} > \frac{1}{K}(\mu_{n+1} + \cdots + \mu_{n+r-1}) > \frac{r-1}{K^2}\mu_n.$$

Thus, (b) holds if $r > 2K^2 + 1$.

(a) and (b) are sufficient: define $\mu_0 = \mu_{-1} = \cdots = 0$.

$$\sum_{i=1}^{n-1} \mu_i = \sum_{i=1}^{r}\sum_{j=0}^{\infty} \mu_{n-i-jr} < \sum_{i=1}^{r}\sum_{j=0}^{\infty} \frac{\mu_{n-i}}{2^j}$$
$$\leq 2\sum_{i=1}^{r} \mu_{n-i} < 2\sum_{i=1}^{r} \frac{\mu_n}{c^i} = K\mu_n \qquad \left(K = 2\sum_{i=1}^{r} c^{-i}\right).$$

In the last step, we used that, by (a), $\mu_{m+l} > c^l\mu_m$ holds.

This finishes the solution. Moreover, a similar argument shows that the theorem remains true if the exponent 2 is replaced by any exponent $\alpha > 0$. □

Problem S.13. *Given a positive, monotone function $F(x)$ on $(0, \infty)$ such that $F(x)/x$ is monotone nondecreasing and $F(x)/x^{1+d}$ is monotone nonincreasing for some positive d, let $\lambda_n > 0$ and $a_n \geq 0$, $n \geq 1$. Prove that if*

$$\sum_{n=1}^{\infty} \lambda_n F\left(a_n \sum_{k=1}^{n} \frac{\lambda_k}{\lambda_n}\right) < \infty, \tag{1}$$

or

$$\sum_{n=1}^{\infty} \lambda_n F\left(\sum_{k=1}^{n} a_k \frac{\lambda_k}{\lambda_n}\right) < \infty, \tag{2}$$

then $\sum_{n=1}^{\infty} a_n$ is convergent.

Solution 1. If $x \geq 1$, then $F(x)/x \geq F(1)$. If $x \leq 1$, then $F(x)/x^{1+d} \geq F(1)$. We can suppose that $F(1) = 1$.

$$F(x) \geq x \quad \text{if} \quad x \geq 1\,, \tag{3}$$

and

$$F(x) \geq x^{1+d} \quad \text{if} \quad x \leq 1\,. \tag{4}$$

First, we prove the statement of the problem when (1) holds. Denote by $\sum_n'$ the sum over those numbers n that satisfy

$$a_n \sum_{k=1}^{n} \frac{\lambda_k}{\lambda_n} \geq 1\,.$$

Denote by $\sum_n''$ the sum over the other numbers n. Then, by (1) and (3),

$$\infty > \sum_n{}' \lambda_n F\left(a_n \sum_{k=1}^{n} \lambda_k/\lambda_n\right) \geq \sum_n{}' a_n \sum_{k=1}^{n} \lambda_k \geq \sum_n{}' a_n \lambda_1.$$

Since $\lambda_1 > 0$,

$$\sum_n{}' a_n < \infty. \tag{5}$$

By (1) and (4),

$$\infty > \sum_n{}'' \lambda_n F\left(a_n \sum_{k=1}^{n} \lambda_k/\lambda_n\right) \geq \sum_n{}'' a_n^{1+d} \lambda_n^{-d} \left(\sum_{k=1}^{n} \lambda_k\right)^{1+d}. \tag{6}$$

Divide the sum $\sum_n''$ into two parts:

$$\sum_n{}'' = \sum_n{}''_{(1)} + \sum_n{}''_{(2)}\,,$$

where the first sum runs over those numbers n that satisfy

$$a_n < \lambda_n / \left(\sum_{k=1}^{n} \lambda_k \right)^{(1+d)/d},$$

and the second runs over the other numbers n.

Then by one of the theorems of Abel and Dini, and by $(1+d)/d > 1$,

$$\sum_n{}''_{(1)} a_n < \sum_{n=1}^{\infty} \lambda_n / \left(\sum_{k=1}^{n} \lambda_k \right)^{(1+d)/d} < \infty. \tag{7}$$

By (6),

$$\infty > \sum_n{}''_{(2)} a_n a_n^d \lambda_n^{-d} \left(\sum_{k=1}^{n} \lambda_k \right)^{1+d}$$
$$\geq \sum_n{}''_{(2)} a_n \left(\lambda_n / \left(\sum_{k=1}^{n} \lambda_k \right)^{(1+d)/d} \right)^d \lambda_n^{-d} (\sum_{k=1}^{n} \lambda_k)^{1+d} = \sum_n{}''_{(2)} a_n.$$

Comparing this with (5) and (7), we conclude that $\sum_{n=1}^{\infty} a_n$ is convergent.

Now, we prove the statement of the problem when (2) holds. If $a_n = 0$ for all but finitely many n, then the theorem is trivial; otherwise, we can delete those λ_n, a_n pairs for which $a_n = 0$. So we can suppose that $a_n > 0$. Let $X = \{n : a_n > 1\}$. If $n \in X$, then

$$\sum_{k=1}^{n} \frac{a_k \lambda_k}{\lambda_n} \geq \frac{a_n \lambda_n}{\lambda_n} > 1,$$

and so by (2) and (3),

$$\infty > \sum_{n \in X} \lambda_n \sum_{k=1}^{n} \frac{a_k \lambda_k}{\lambda_n} = \sum_{n \in X} \sum_{k=1}^{n} a_k \lambda_k \geq \sum_{n \in X} a_{n_0} \lambda_{n_0},$$

where n_0 is the smallest element of X. So X is a finite set and the sequence a_n is bounded. Let $\mu_n = a_n \lambda_n$. Then (2) can be written in the following way:

$$\sum_{n=1}^{\infty} \frac{\mu_n}{a_n} F \left(a_n \sum_{k=1}^{n} \frac{\mu_k}{\mu_n} \right) < \infty.$$

By the boundedness of the sequence a_n, (1) is satisfied with the substitution $\lambda_n = \mu_n$ and, as we proved before, from this it follows that $\sum_{n=1}^{\infty} a_n$ is convergent. □

Solution 2. We prove the statement of the problem when (2) holds, but we will not use (1). Solution 2 is the same as Solution 1 through formula (4). Denote by $\sum_n'$ the sum over those numbers n that satisfy

$$\sum_{k=1}^{n} \frac{a_k \lambda_k}{\lambda_n} \geq 1. \tag{5}$$

By (2) and (3), we obtain

$$\infty > \sum_n{}' \lambda_n F\left(\sum_{k=1}^{n} a_k \lambda_k / \lambda_n\right) \geq \sum_n{}' \sum_{k=1}^{n} a_k \lambda_k$$
$$\geq \sum_{n \leq n_0}{}' a_{n_0} \lambda_{n_0} = a_{n_0} \lambda_{n_0} \sum_{n \leq n_0}{}' 1.$$

By choosing n_0 such that $a_{n_0} \neq 0$, it follows that (5) holds only for finitely many n. So, by (2) and (4),

$$\infty > \sum_{n=1}^{\infty} \left(\sum_{k=1}^{n} \frac{a_k \lambda_k}{\lambda_n}\right)^{1+d} \lambda_n = \sum_{n=1}^{\infty} \lambda_n^{-d} \left(\sum_{k=1}^{n} a_k \lambda_k\right)^{1+d}. \tag{6}$$

Hence, assuming that not all of the a_n's are zero, it follows that

$$\sum_{n=1}^{\infty} \lambda_n^{-d} < \infty,$$

and therefore

$$\lambda_n \to \infty. \tag{7}$$

Let n_q be the largest natural number n such that $2^q \leq \lambda_n < 2^{q+1}$; if such an n does not exist, then n_q can be chosen arbitrarily. We will see that in the latter case the multiplier of λ_{n_q} is zero. Continuing (6),

$$\infty > \sum_{n=1}^{\infty} \lambda_n^{-d} \left(\sum_{k=1}^{n} a_k \lambda_k\right)^{1+d} \geq \sum_{q=1}^{\infty} \lambda_n^{-d} \left(\sum_{k: 2^q \leq \lambda_k < 2^{q+1}} a_k \lambda_k\right)^{1+d}$$
$$\geq \sum_{q=1}^{\infty} 2^{-d(q+1)} 2^{q(1+d)} S_q^{1+d} = 2^{-d} \sum_{q=1}^{\infty} 2^q S_q^{1+d},$$

where

$$S_q = \sum_{k: 2^q \leq \lambda_k < 2^{q+1}} a_k.$$

So there is a positive K for which

$$\sum_{q=1}^{\infty} 2^q S_q^{1+d} < K.$$

Hence, for any natural number q,

$$2^q S_q^{1+d} < K.$$

Therefore

$$S_q < K 2^{-q/(1+d)},$$

which trivially implies

$$\sum_{q=1}^{\infty} S_q < \infty.$$

By the definition of S_q and by (7), there is an n_0 such that $\sum_{q=1}^{\infty} S_q \geq \sum_{n=n_0}^{\infty} a_n$. Therefore, $\sum_{n=1}^{\infty} a_n < \infty$, which finishes the solution. □

Remarks.

1. Originally, László Leindler placed part (1) of this problem at the competition committee's disposal. Atilla Máté noticed that the statement also follows from (2).

2. László Babai states a similar theorem for integrals: Let $F(x)$ be a positive monotone function on $(0,\infty)$ such that $F(x)/x$ is monotone nondecreasing and there is a positive d for which $F(x)/x^{1+d}$ is monotone nonincreasing. Furthermore, let $\Lambda(t)$ and $A(t)$ be absolutely continuous functions on $(0,\infty)$, $A(t)$ monotone nondecreasing, $\Lambda'(t) > 0$ almost everywhere, and $\lim_{t\to 0} A(t) = \lim_{t\to 0} \Lambda(t) = 0$; then any of

$$\int_0^{\infty} \Lambda'(t) F\Big(\frac{A(t)}{\Lambda'(t)} \Lambda(t)\Big)\, dt < \infty$$

and

$$\int_0^{\infty} \Lambda'(t) F\Big(\frac{1}{\Lambda'(t)} \int_o^t A'(x)\Lambda'(x)\, dx\Big)\, dt < \infty$$

implies $\lim_{t\to\infty} A(t) < \infty$. (By the monotonicity of $F(x)/x$, the functions after the $\int$ signs are measurable.)

Problem S.14. *Let $f(x) = \sum_{n=1}^{\infty} a_n/(x+n^2)$, $(x \geq 0)$, where $\sum_{n=1}^{\infty} |a_n| n^{-\alpha} < \infty$ for some $\alpha > 2$. Let us assume that for some $\beta > 1/\alpha$, we have $f(x) = O(e^{-x^{\beta}})$ as $x \to \infty$. Prove that a_n is identically 0.*

Solution. Obviously, the series defining $f(x)$ is convergent at each $x \neq -n^{\alpha}$ $(n = 1, 2, \dots)$, so $f(x)$ is meromorphic on the whole plane. Let

$$F(x) = \prod_{n=1}^{\infty} \Big(1 + \frac{x}{n^{\alpha}}\Big), \quad h(x) = F(x) f(x).$$

Since $\alpha > 2$, this product is convergent for all x. Easy calculation shows that $F(x)$ is an entire function, so for any $\varepsilon > 0$,

$$f(x) = O(e^{|x|^{1/\alpha+\varepsilon}}) \quad .$$

The function $h(x)$ is also entire, and there is a sufficiently small ε such that, for $x > 0$,

$$|h(x)| = |f(x)| \cdot |F(x)| = O(e^{-x^\beta})O(e^{-x^{1/\alpha+\varepsilon}}) = o(1). \tag{1}$$

Furthermore, for any x,

$$|h(x)| \le \sum_{n=1}^{\infty} \frac{|a_n|}{n^\alpha} \prod_{m \ne n} \left(1 + \frac{|x|}{m^\alpha}\right) \le \left(\sum_{n=1}^{\infty} \frac{|a_n|}{n^\alpha}\right) F(|x|) = O(e^{|x|^{1/\alpha+\varepsilon}}).$$

So if $M(r) = \max |F(x)|$, then

$$\frac{\log M(r)}{\sqrt(r)} = O\left(\frac{r^{1/\alpha+\varepsilon}}{\sqrt{r}}\right) = O(1)\,. \tag{2}$$

So by a Phragmèn–Lendelöf type of theorem (*Gy. Pólya and G. Szegő, Problems and Theorems in Analysis, Springer, Berlin, 1976, vol. I,* III.6.332), (1) and (2) can hold for an entire function $h(x)$ only if $h(x)$ is constant. So by (1), $h(x) = 0$. Hence $f(x) = 0$. Therefore, if for some n, $a_n \ne 0$, then $f(z)$ has a pole at $z = -n^\alpha$. But this is a contradiction, so $a_1 = a_2 = \cdots = 0$. □

Problem S.15. *Given a positive integer m and $0 < \delta < \pi$, construct a trigonometric polynomial $f(x) = a_0 + \sum_{n=1}^{m}(a_n \cos nx + b_n \sin nx)$ of degree m such that $f(0) = 1$, $\int_{\delta \le |x| \le \pi} |f(x)|\, dx \le c/m$, and $\max_{-\pi \le x \le \pi} |f'(x)| \le c/\delta$, for some universal constant c.*

Solution 1. Let $\vartheta = \max\{\delta/2, 1/m\}$. Extend the following function periodically to all real numbers:

$$\varphi(x) = \begin{cases} 1 - |\frac{x}{\delta}|, & \text{if } |x| \le \vartheta, \\ 0, & \text{if } \vartheta \le |x| \le \pi\,. \end{cases}$$

This nearly satisfies our purpose, it just is not a trigonometric polynomial. Let us examine the Fejér kernel of its Fourier series. Let

$$f(x) = c_0 \int_{-\pi}^{\pi} \varphi(x-t) K_m(t)\, dt,$$

where

$$K_m(t) = \frac{\sin^2 \frac{m+1}{2} t}{2(m+1) \sin^2 \frac{t}{2}}$$

and c_0 is chosen such that $f(0) = 1$. Obviously,

$$1 = f(0) \geq c_0 \frac{1}{2} \int_{-\vartheta/2}^{\vartheta/2} K_m(t)\, dt \geq \frac{c_0}{2} \int_{-1/2m}^{1/2m} K_m(t)\, dt$$

trivially if $|t| < 1/2m$. Then

$$|K_m(t)| \geq \frac{1}{10} m.$$

Hence,

$$1 = f(0) \geq \frac{c_0}{20}, \quad c_0 \leq 20.$$

Since φ is absolutely continuous,

$$f'(x) = c_0 \int_{-\pi}^{\pi} \varphi'(x-t) K_m(t)\, dt.$$

So

$$|f'(x)| \leq c_0 \frac{1}{\vartheta} \int_{x-\vartheta}^{x+\vartheta} K_m(t) \leq \frac{c_0}{\vartheta} \int_{-\pi}^{\pi} K_m(t)\, dt < \frac{80}{\vartheta}.$$

Estimate $\int_{\delta \leq |x| \leq \pi} |f(x)|\, dx$ in cases $\vartheta = \delta/2$ and $\vartheta = 1/m$. If $\vartheta = 1/m$, then

$$\int_{\delta \leq |x| \leq \pi} |f(x)|\, dx = 2 \int_{\delta}^{\pi} |f(x)|\, dx \leq 2 \int_{\delta}^{\pi} \int_{x-\delta}^{x+\delta} K_m(t)\, dt\, dx$$

$$\leq 4\vartheta \int_{\delta-\vartheta}^{\pi+\vartheta} K_m(t)\, dt \leq 4\vartheta \int_{-\pi}^{\pi} K_m(t)\, dt = \frac{4\pi}{m}.$$

If $\vartheta = \delta/2$, then

$$\int_{\delta \leq |x| \leq \pi} |f(x)|\, dx \leq 4\vartheta \int_{\delta-\vartheta}^{\pi+\vartheta} K_m(t)\, dt = 2\delta \int_{\delta/2}^{\pi} + \frac{\delta}{2} K_m(t)\, dt.$$

Using elementary estimates on the interval $0 \leq t \leq \pi + \delta/2$ $(\leq (3/2)\pi)$,

$$K_m(t) \leq \frac{30}{mt^2},$$

and so

$$2\delta \int_{\delta/2}^{\pi+\delta/2} K_m(t)\,dt \le 2\delta \int_{\delta/2}^{\infty} \frac{30}{mt^2}\,dt = \frac{240}{m}.$$

This finishes the solution. □

Solution 2. Start from the function

$$h(x) = \begin{cases} \frac{1}{\Delta^4}(\Delta^2 - x^2)^2, & \text{if } x \le \Delta, \\ 0, & \text{if } \Delta \le |x| \le \pi, \end{cases}$$

where the parameter Δ will be defined later.

Let

$$h(x) = \sum_{n=-\infty}^{\infty} c_n e^{inx}$$

and

$$s(x) = \sum_{n=-l}^{l} c_n e^{inx}.$$

We will show that Δ can be chosen in such way that $f(x) = s^2(x)/s^2(0)$ satisfies the conditions of the problem. Obviously, $f(x)$ is a trigonometric polynomial of degree not more than m, and $f(0) = 1$. We will estimate the difference of $s(x)$ and $h(x)$. The Cauchy–Schwartz inequality yields

$$|s(x)-h(x)| \le \sum_{|n|>l} |c_n n| \frac{1}{|n|} \le \sqrt{\sum_{|n|\le l} |c_n n|^2} \sqrt{\sum_{|n|>l} \frac{1}{n^2}} \le \sqrt{\sum_{n=-\infty}^{+\infty} (c_n n)^2} \sqrt{\frac{2}{l}}.$$

By Parseval's formula,

$$\sum_{n=-\infty}^{+\infty} (c_n n)^2 = \frac{1}{2\pi} \int_{-\Delta}^{\Delta} |h'(x)|^2\,dx < \frac{1}{\Delta};$$

hence

$$|s(x) - h(x)| < \sqrt{\frac{2}{l\Delta}}.$$

Similarly,

$$|s'(x) - h'(x)| < \sqrt{\frac{9}{l\Delta^3}}.$$

So if

$$l\Delta \ge 9, \tag{1}$$

then

$$|s(x)| \le |h(x)| + \sqrt{\frac{2}{l\Delta}} \le \frac{3}{2}.$$

Moreover,

$$|s'(x)| \le |h'(x)| + \sqrt{\frac{9}{l\Delta^3}} \le \frac{4}{\Delta} + \frac{1}{\Delta} = \frac{5}{\Delta},$$

so

$$|f'(x)| \frac{2}{s^2(0)} |s(x)||s'(x)| \le \frac{60}{\Delta}$$

suffices if

$$\Delta \ge \delta. \tag{2}$$

This inequality and (1) also hold if we choose $\Delta = \max\{\delta, 9/l\}$. The integral of $|f(x)|$ can be estimated in two ways. If $\Delta = \delta$, then

$$\begin{aligned}\int\limits_{\delta \le |x| \le \pi} |f(x)|\,dx &= \frac{1}{s^2(0)} \int\limits_{\delta \le |x| \le \pi} |s^2(x)|\,dx = \frac{1}{s^2(0)} \int\limits_{\delta \le |x| \le \pi} (s(x) - h(x))^2\,dx \\ &\le \frac{1}{s^2(0)} \int\limits_{-\pi}^{\pi} (s(x) - h(x))^2\,dx = \frac{2\pi}{s^2(0)} \sum_{|n|>l} |c_n|^2 \\ &\le \frac{2\pi}{l^2} \sum_{n=-\infty}^{\infty} n^2 c_n^2 < \frac{2\pi}{\Delta l^2} < \frac{1}{l}.\end{aligned}$$

If $\Delta = 9/l \pm \delta$, then

$$\begin{aligned}\int\limits_{\delta \le |x| \le \pi} |f(x)|\,dx &\le \int\limits_{-\pi}^{\pi} \frac{s^2(x)}{s^2(0)}\,dx = \frac{2\pi}{s^2(0)} \sum_{n=-l}^{l} |c_n|^2 \\ &\le 2\pi \sum_{-\infty}^{\infty} |c_n|^2 = 4 \int\limits_{-\pi}^{\pi} h^2(x)\,dx \le 8\Delta = \frac{72}{l}.\end{aligned}$$

This finishes the solution. ⊔

Problem S.16. *Prove that if*

$$\sum_{n=1}^{m} a_n \le N a_m \quad (m = 1, 2, \dots)$$

holds for a sequence $\{a_n\}$ *of nonnegative real numbers with some positive integer* N, *then* $\alpha_{i+p} \ge p\alpha_i$ *for* $i, p = 1, 2, \dots,$ *where*

$$\alpha_i = \sum_{n=(i-1)N+1}^{iN} a_n \quad (i = 1, 2, \dots).$$

Solution 1. Since

$$\alpha_{i+p} = \alpha_{(i+p-1)N+1} + \cdots + \alpha_{(i+p)N} \geq \frac{1}{N}\left(\sum_{n=1}^{(i+p-1)N+1} a_n + \cdots + \sum_{n=1}^{(i+p)N} a_n\right)$$

$$\geq \frac{1}{N} N \sum_{n=1}^{(i+p-1)N} a_n = \sum_{n=1}^{N} a_n + \sum_{n=N+1}^{2N} a_n + \cdots +$$

$$+ \sum_{n=(i+p-2)N+1}^{(i+p-1)N} a_n = \alpha_1 + \alpha_2 + \cdots + \alpha_{i+p-1},$$

it follows that

$$\alpha_{i+p} \geq \alpha_1 + \alpha_2 + \cdots + \alpha_{i+p-1}.$$

Hence $\alpha_{i+1} \geq \alpha_i$, and therefore

$$\alpha_{i+p} \geq \alpha_i + \alpha_{i+1} + \cdots + \alpha_{i+p-1} \geq p\alpha_i\,. \quad \square$$

Solution 2. Similar to the previous solution,

$$\alpha_i \geq \sum_{k=1}^{i-1} \alpha_k. \tag{1}$$

We are going to prove that

$$\alpha_{i+p} \geq 2^{p-1} \sum_{k=1}^{i} \alpha_k \tag{2}$$

for any i and p.

Indeed, (2) is the same as (1) for $p = 1$, and if (2) holds for p, then by

$$\alpha_{i+p+1} \geq 2^{p-1} \sum_{k=1}^{i+1} \alpha_k = 2^{p-1}\left(\alpha_{i+1} + \sum_{k=1}^{i} \alpha_k\right) \geq 2^{p-1} 2 \sum_{k=1}^{i} \alpha_k$$

it also holds for $p+1$. Since $\alpha_i \geq 0$, from (2) it follows that $\alpha_{i+p} \geq 2^{p-1}\alpha_i$, which finishes the solution. $\square$

Remark. We can similarly prove the following, more general statement. Let $\{a_n\}$ be a sequence of nonnegative real numbers such that for some c,

$$a_n \geq c \sum_{i=1}^{n-1} a_i.$$

Define α_i as we did in the problem. What is the largest constant d_p such that for all such sequences the following holds?

$$\alpha_{i+p} \geq d_p \alpha_i$$

In the problem, $c = 1/(N-1)$. First, we will find a c_p such that

$$a_{n+p} \geq e_p \sum_{i=1}^{n} a_i.$$

$e_1 = c$ suffices for $p = 1$. Since

$$a_{n+p+1} \geq e_p \sum_{i=1}^{n-1} a_i \geq e_p(1+c) \sum_{i=1}^{n} a_i,$$

it is not a bad idea to choose $e_{p+1} = e_p(1+c)$, that is, $e_p = c(1+c)^{p-1}$. Then

$$\alpha_{i+p} = \sum_{n=(i+p-1)N+1}^{(i+p)N} a_n \geq \sum_{j=1}^{N} e_{(p-1)N+j} \sum_{n=1}^{iN} a_n \geq \left(\sum_{j=1}^{N} e_{(p-1)N+j} \right) \alpha_i.$$

So we can choose d_p such that

$$d_p = \sum_{j=1}^{N} c(1+c)^{(p-1)N+j-1} = (1+c)^{(p-1)N}[(1+c)^N - 1].$$

It is easy to see that if we define $a_1 = 1$, $a_n = c(1+c)^{n-2}$ $(n \geq 2)$, then for any N and p, $\alpha_{p+1} = d_p \alpha_1$ holds, so the inequality cannot be sharpened.

As stated, in the problem $c = 1/(N-1)$, so

$$d_p = \left(\frac{N}{N-1} \right)^{(p-1)N} \left[\left(\frac{N}{N-1} \right)^N - 1 \right] \geq e^{p-1}(e-1),$$

but even for $c = 1/N$ it is true that $d_p \geq 2^{p-1}$.

Problem S.17. *Let $S_\nu = \sum_{j=1}^{n} b_j z_j^\nu$ $(\nu = 0, \pm 1, \pm 2, \dots)$, where the b_j are arbitrary and the z_j are nonzero complex numbers. Prove that*

$$|S_0| \leq n \max_{0<|\nu|\leq n} |S_\nu|.$$

Solution. Let $\prod_{j=1}^{n}(z - z_j) = \sum_{k=0}^{n} a_k z^k$ and $\max_{0 \geq k \geq n} |a_k| = |a_m|$. Obviously, $|a_m| \geq 1$. Then

$$\sum_{k=0}^{n} a_k S_{k-m} = \sum_{k=0}^{n} a_k b_j z_j^{k-m} = \sum_{j=1}^{n} b_j z_j^{-m} \sum_{k=0}^{n} a_k z_j^k = 0.$$

Hence

$$|S_0| = \left| \sum_{k=0;k\neq m}^{n} (-\frac{a_k}{a_m}) S_{k-m} \right| \leq \sum_{k=0;k\neq m}^{n} |S_{k-m}| \leq n \max_{0<|\nu|\leq n} |S_\nu|. \quad \square$$

Remark. One can show that equality holds iff $b_1 = b_2 = \cdots = b_n$ and the set of numbers z_k is the same as $\{a \cdot e^{(2\pi i)/((n+1)j)}\ (j = 1, 2, \dots, n)\}$, where a is an arbitrary constant of absolute value 1.

Problem S.18. *Let $p \geq 1$ be a real number and $\mathbb{R}_+ = (0, \infty)$. For which continuous functions $g : \mathbb{R}_+ \to \mathbb{R}_+$ are the following functions all convex?*

$$M_n(x) = \left[\frac{\sum_{i=1}^n g(\frac{x_i}{x_{i+1}}) x_{i+1}^p}{\sum_{i=1}^n g(\frac{x_i}{x_{i+1}})} \right]^{\frac{1}{p}},$$

$$x = (x_1, \ldots, x_{n+1}) \in \mathbb{R}_+^{n+1}, \quad n = 1, 2, \ldots$$

Solution. We prove that M_n $(n = 2, 3, \ldots)$ are all convex iff g is a constant function. If g is a constant function, then by Minkovsky's inequality,

$$M_n(x + y) \leq M_n(x) + M_n(y) \quad (x, y \in \mathbb{R}_+^{n+1}), \tag{1}$$

and by $M_n(\alpha x) = \alpha M_n(x)$ $(\alpha \in \mathbb{R}_+, x \in \mathbb{R}_+^{n+1})$, it follows that M_n is convex.

Conversely, if all of the M_n are convex, then we will show the following. If $M_2(x, u, 1)$ is a convex function of x for all $u \in (1-\delta, 1+\delta)$, $\delta > 0$, then the function g is a constant.

Taking each side of the inequality

$$M_2(\alpha x_1 + (1-\alpha)x_2, u, 1) \leq \alpha M_2(x_1, u, 1) + (1-\alpha) M_2(x_2, u, 1)$$
$$(x_1, x_2 \in \mathbb{R}_+, \quad u \in (1-\delta, 1+\delta), \alpha \in (0, 1))$$

to the power p and using the inequality

$$\alpha t + (1-\alpha)s \leq (\alpha t^p + (1-\alpha)s^p)^{1/p} \quad (t, s > 0, \ \alpha \in (0, 1))$$

between the arithmetic mean and the n-th power mean, we obtain

$$M_2^p(\alpha x_1 + (1-\alpha)x_2, u, 1) \leq \alpha M_2^p(x_1, u, 1) + (1-\alpha) M_2^p(x_2, u, 1).$$

Deleting $u^p = \alpha u^p + (1-\alpha)u^p$ from both sides, we get

$$\frac{g(u)(1-u^p)}{g(\frac{\alpha x_1 + (1-\alpha)x_2}{u}) + g(u)} \leq \alpha \frac{g(u)(1-u^p)}{g(\frac{x_1}{u}) + g(u)} + (1-\alpha) \frac{g(u)(1-u^p)}{g(\frac{x_2}{u}) + g(u)},$$

for $x_1, x_2 \in \mathbb{R}_+, u \in (1-\delta, \ 1+\delta), \alpha \in (0, 1)$.

If $u < 1$, then divide by $g(u)(1-u^p)$ and take the limit $u \to 1-0$. Since g is continuous, we get

$$\frac{1}{g(\alpha x_1 + (1-\alpha)x_2) + g(1)} \leq \alpha \frac{1}{g(x_1) + g(1)} + (1-\alpha) \frac{1}{g(x_2) + g(1)}.$$

Using the same method for $u < 1$, we get the opposite inequality. Therefore, equality holds. So the function $f = 1/(g(x) + g(1))$ satisfies the following Jensen's equality:

$$f(\alpha x_1 + (1-\alpha)x_2) = \alpha f(x_1) + (1-\alpha) f(x_2) \quad (x_1, x_2 \in \mathbb{R}_+, \ \alpha \in (0, 1)).$$

Since f is continuous and positive, $f(x) = Cx + D$, where C and D are nonnegative constants. So

$$Cx + D = \frac{1}{g(x) + g(1)} < \frac{1}{g(1)} = 2f(1) = 2(C + D).$$

This does not hold for large values of x when $C > 0$. Therefore, $C = 0$ and g is constant. □

Remark. Some of the participants remarked that we do not have to suppose the continuity of g, since it follows from the convexity of M_n. Moreover, g is continuous if $M_2(x, u, 1)$ is a convex function of x for $u \in (1 - \delta, 1 + \delta)$.

Problem S.19. *Suppose that $R(z) = \sum_{n=-\infty}^{\infty} a_n z^n$ converges in a neighborhood of the unit circle $\{z : |z| = 1\}$ in the complex plane, and $R(z) = P(z)/Q(z)$ is a rational function in this neighborhood, where P and Q are polynomials of degree at most k. Prove that there is a constant c independent of k such that*

$$\sum_{n=-\infty}^{\infty} |a_n| \leq ck^2 \max_{|z|=1} |R(z)|.$$

Solution. Denote by D the unit disc of the complex plane $D = \{z : |z| < 1\}$, and by ∂D its boundary. We can suppose that $|R(z)| < 1$ on the perimeter of the unit circle. So we have to prove that $\sum_{n=-\infty}^{\infty} |a_n| < ck^2$. If we prove

$$\sum_{n=-\inf}^{-1} |a_n| \leq ck^2,$$

then we can apply it to $R(1/z)/z$, so that we obtain $\sum_{n=0}^{\infty} |a_n| < ck^2$, which finishes the solution.

We know that

$$a_n = \frac{1}{2\pi i} \oint_{\Gamma} \frac{R(z)}{z^{n+1}} dz, \tag{1}$$

where we can integrate along ∂D, or along such a curve Γ, such that $R(z)$ does not have a pole in the region bounded by ∂D and Γ. Now we will try to choose skillfully $\Gamma \subset D$. By the previous results,

$$\begin{aligned} \sum_{n=-\infty}^{-1} |a_n| &\leq \frac{1}{2\pi} \oint_{\Gamma} |R(z)|(1 + |z| + \cdots + |z|^i + \ldots)\, |dz| \\ &= \frac{1}{2\pi} \oint_{\Gamma} |R(z)| \frac{|dz|}{1 - |z|}. \end{aligned}$$

We are looking for a Γ for which the integrand is not too big in the rightmost integral. Because the multiplier is $1/(1-|z|)$, we will have to bring Γ as far from the unit circle as possible. The only problem is that we do not know anything about the behavior of $R(z)$ far away from ∂D.

Denote the poles of R in the unit disc by $q_1, q_2, \ldots, q_h$, where the points are included in the sequence with appropriate multiplicity. Obviously, $h \le k$. The Blaschke product

$$B(z) = \prod_{\nu=1}^{h} \frac{z - q_\nu}{1 - \overline{q_\nu} z}$$

is holomorphic in $\overline{D}$, vanishes at the points q_ν, and has absolute value not more than 1 in the points of the unit circle. So $S(z) = R(z)Q(z)$ is holomorphic in D and has absolute value not more than 1 in the boundary of the unit circle. Thus, by the maximum modulus principle, $|S(z)| \le 1$, for $z \in \overline{D}$, and so

$$|R(z)| \le \frac{1}{\prod_{\nu=1}^{h} \frac{z-q_\nu}{1-\overline{q_\nu} z}} \quad (|z| \le 1). \tag{2}$$

This is the approximation we were seeking.

Lct

$$\Gamma = \partial\left\{z \in D \ : \ \left|\frac{z - q_\nu}{1 - \overline{q_\nu} z}\right| \ge 1 - \frac{1}{k+1},\ 1 \le \nu \le h\, z \ : |z| = 1\right\}.$$

If $z \in \Gamma$, then by (2),

$$|R(z)| \le \frac{1}{(1 - \frac{1}{k+1})^k} = \left(1 + \frac{1}{k+1}\right)^k < e. \tag{3}$$

The curve defined above is not necessarily a Jordan curve, but it consists of finitely many closed Jordan curves. Nonetheless, $R(z)$ will be holomorphic in the region bordered by Γ and ∂D. So all of the consequences of (1) will hold.

Obviously, Γ consists of some subcurves of the curves

$$\Gamma_\nu = \left\{z \ : \ \left|\frac{z - q_\nu}{1 - \overline{q_\nu} z}\right| = \frac{k}{k+1}\right\}.$$

If $q_\nu = 0$, then Γ_ν is a circular arc. But it is also a circular curve when $q_\nu \ne 0$, since Γ_ν is obtained from the circular arc $|w| = k/(k+1)$ by the transformation

$$w = \frac{z - q_\nu}{1 - \overline{q_\nu} z}, \tag{4}$$

and such a transformation maps a circle onto a circle. (Of course, q_ν is not the centerpoint of Γ_ν. Moreover, if we are considering Q as the

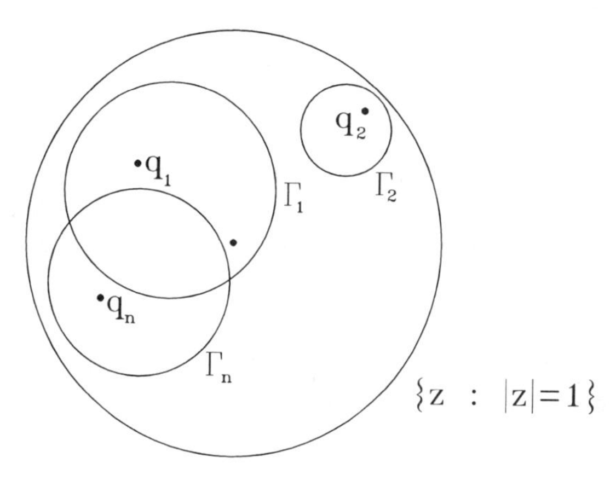

Figure S.1.

Poincaré model of the hyperbolic geometry, then the transformation (4) is a conformal transformation of the hyperbolic space, so q_ν will be the non-Euclidean center of Γ_ν. The non-Euclidean radius of Γ_ν is approximately $\log k$, see Figure S.1.)

Since $\Gamma \subset \cup_1^k \Gamma_\nu$, by (3),

$$\sum_{n=-\infty}^{-1} |a_n| \leq \frac{e}{2\pi} \oint_{\Gamma} \frac{|dz|}{1-|z|} \leq \frac{e}{2\pi} \sum_{\nu=1}^{h} \oint_{\Gamma_\nu} \frac{|dz|}{1-|z|}.$$

The latter sum can be approximated by substituting (4) and some calculations as

$$\oint_{\Gamma_\nu} \frac{|dz|}{1-|z|} \leq 2 \oint_{\Gamma_\nu} \frac{|dz|}{1-|z|^2} = 2 \oint_{|w|=\frac{k}{k+1}} \frac{|dw|}{1-|w|^2} \leq \frac{2}{1-\left(\frac{k}{k+1}\right)^2} 2\pi < 8\pi k.$$

(Here we had to do "some calculations" to prove the equality above. The equality also shows that curve $|dz|/(1-|z|^2)$ is invariant under the conformal transformation of the hyperbolic plane.) So

$$\sum_{n=-\infty}^{-1} |a_n| < 4ek^2,$$

and as we noted above it follows that

$$\sum_{n=-\infty}^{\infty} |a_n| < 8ek^2. \quad \square$$

Remark. The inequality above is not sharp. Gábor Somorjai showed with a better curve Γ that

$$\sum_{n=-\infty}^{\infty} |a_n| < ck \log k.$$

Problem S.20. *Let $K_n (n = 1, 2, \dots)$ be periodical continuous functions of period 2π, and write*

$$k_n(f; x) = \int_0^{2\pi} f(t) K_n(x - t) dt.$$

Prove that the following statements are equivalent:
(i) $\int_0^{2\pi} |k_n(f; x) - f(x)|\, dx \to 0$ $(n \to \infty)$ for all $f \in L_1[0, 2\pi]$.
(ii) $k_n(f; 0) \to f(0)$ for all continuous, 2π-periodic functions f.

Solution 1.

(i)→(ii): Denote by $C[0; 2\pi]$ the space of the continuous functions under the supremum norm. Let us examine the sequence of operators $k_n(\cdot; \cdot) : L_1 \to L_1$:

$$\begin{aligned} \|k_n(\cdot;\cdot)\| &= \sup_{\|f\|_{L_1} \le 1} \|k_n(f;\cdot)\|_{L_1} \\ &\le \sup_{\|f\|_{L_1} \le 1} \int_0^{2\pi} |f(t)|\, dt \int_0^{2\pi} |K_n(x-t)|\, dt \le \int_0^{2\pi} |K_n(t)|\, dt. \end{aligned} \tag{1}$$

On the other hand, the functions

$$f_\delta(x) = \begin{cases} \frac{1}{\delta}, & \text{if } x \in [0; \delta], \\ 0, & \text{if } x \in (\delta; 2\pi), \end{cases}$$

by the continuity of $K_n(t)$, satisfy

$$\|k_n(f_\delta; \cdot)\|_{L_1} \to \int_0^{2\pi} |K_n(t)|\, dt \quad (\delta \to 0),$$

and so

$$\|k_n(\cdot;\cdot)\| = \int_0^{2\pi} |K_n(t)|\, dt.$$

By applying the Banach–Steinhaus theorem to the sequence $\{k_n(\cdot;\cdot)\}$, we obtain $\int_0^{2\pi} |K_n(t)|\, dt \le C$, where C is the same constant for all n.

If f is continuous, then

$$\begin{aligned} |k_n(f; x) - k_n(f; x+h)| &= \left| \int_0^{2\pi} (f(t) - f(t-h)) K_n(x-t) dt \right| \\ &\le C \sup_t |f(t) - f(t-h)|, \end{aligned}$$

and so the functions $k_n(f; x)$ are equicontinuous.

Let us suppose that (ii) does not hold. Then there is a continuous f, an $\varepsilon > 0$, and a sequence $\{\nu_n\}$ such that $k_{\nu_n}(f;0) > f(0)+\varepsilon$ for $n = 1, 2, \dots$. By the uniform continuity of $\{k_{\nu_n}(f;x)\}$, there is a δ such that, for all n, $k_{\nu_n}(f;x) > f(x) + \frac{\varepsilon}{2}$ if $0 \le x \le \delta$ $(n = 1, 2, \dots)$. Hence,

$$\int_0^{2\pi} |k_{\nu_n}(f;x) - f(x)|\, dx \ge \int_0^{\delta} (k_{\nu_n}(f;x) - f(x))\, dx \ge \delta\frac{\varepsilon}{2} \quad (n = 1, 2, \dots),$$

but this contradicts (i).

(ii)→(i): It is well known that the norm of the linear functional $k_n(\cdot;0) : C[0;2\pi] \to \mathbb{R}$ is $\int_0^{2\pi} |K_n(t)|dt$. So by the Banach–Steinhaus theorem, $\int_0^{2\pi} |K_n(t)|dt \le C$ $(n = 1, 2, \dots)$, where C is a constant. Since

$$\begin{aligned} k_n(f, x_0) &= \int_0^{2\pi} f(t)k_n(x_0 - t)\, dt \\ &= \int_0^{2\pi} f(x_0 + t)K_n(-t)\, dt = k_n(f(x_0 + t); 0) \end{aligned}$$

is continuous for all f, $k_n(f;x) \to f(x)$ holds everywhere. But $k_n(f;x) \le C \sup(f)$, so by Lebesgue's theorem,

$$\int_0^{2\pi} |k_n(f,x) - f(x)|\, dx \to 0 \quad (n \to \infty,\ f \in C[0, 2\pi]).$$

Let $f \in L_1[0, 2\pi]$ and let $\varepsilon > 0$ be arbitrary. Choose a continuous function g in such a way that $\|g - f\|_{L_1} < \varepsilon$ holds. Then, by (1),

$$\begin{aligned} \int_0^{2\pi} |k_n(f,x) - f(x)|\, dx &\le \int_0^{2\pi} |k_n(f,x) - k_n(g;x)|\, dx \\ &\quad + \int_0^{2\pi} |k_n(g;x) - g(x)|\, dx, \\ \int_0^{2\pi} |g(x) - f(x)|\, dx &\le C\|g - f\|_{L_1} \\ &\quad + \int_0^{2\pi} |k_n(g;x) - g(x)|\, dx + \|g - f\|_{L_1}. \end{aligned}$$

By the previous argument, the middle term above goes to 0 if $n \to \infty$, so

$$\int_0^{2\pi} |k_n(f,x) - f(x)|\, dx$$

becomes arbitrarily small for sufficiently large n. This finishes the solution. □

Solution 2. We are going to use the following version of the Banach–Steinhaus theorem. Let X and Y be Banach spaces, and let $A, A_n : X \to Y$

$(n = 1, 2, \dots)$ be linear bounded operators. The sequence $\{A_n\}$ pointwise converges to A iff there is a constant C and a closed system $S \subseteq X$ such that

$$\|A_n\| \le C \quad (n = 1, 2, \dots), \tag{1}$$
$$A_n x \to Ax \quad (n \to \infty) \tag{2}$$

hold for all $x \in S$. ($S \subseteq X$ is called a closed system if the linear hull of S is dense in X.)

Let $A_n^1 : L_1 \to L_1$ be the operator $k_n(\cdot;\cdot)$, A^1 be an identity on L_1, $A_n^2 : C[0, 2\pi] \to \mathbb{R}$ be the functional $k_n(\cdot; 0)$, and $A^2 : C[0, 2\pi] \to \mathbb{R}$ be the mapping $f \to f(0)$. As we showed in the previous solution, $\|A_n^1\| = \|A_n^2\| = \int_0^{2\pi} |K_n(t)|\, dt$, so (1) holds for $\{A_n^1\}$ and $\{A_n^2\}$ at the same time.

Let $S = \{e_m = e^{imx}\}_{m=1}^{\infty}$. Since S is closed in L_1 and in $C[0, 2\pi]$, and

$$\begin{aligned}(A_n^1 e_m)(x) &= \int_0^{2\pi} e^{imt} K_n(x - t)\, dt = \int_0^{2\pi} e^{im(x+t)} K_n(-t)\, dt \\ &= e^{imx} \int_0^{2\pi} e^{imt} K_n(-t)\, dt = e_m(x) A_n^2 e_m,\end{aligned}$$

it follows that

$$\begin{aligned}\|A_n^1 e_m - A^1 e_m\|_{L_1} &= \int_0^{2\pi} |e_m(x) A_n^2 e_m - e_m(x)|\, dx \\ &= |A_n^2 e_m - A^2 e_m| \int_0^{2\pi} |e_m(x)|\, dx \\ &= 2\pi \|A_n^2 e_m - A^2 e_m\|_{\mathbb{R}},\end{aligned}$$

so (2) holds for $\{A_n^1\}$ and A^1. Furthermore, (2) also holds for $\{A_n^2\}$ and A^2 simultaneously. So we proved the equivalence of the pointwise convergences $A_n^1 \to A^1$ and $A_n^2 \to A^2$. □

Problem S.21. *Let us assume that the series of holomorphic functions* $\sum_{k=1}^{\infty} f_k(z)$ *is absolutely convergent for all* $z \in \mathbb{C}$. *Let* $H \subseteq \mathbb{C}$ *be the set of those points where the above sum function is not regular. Prove that* H *is nowhere dense but not necessarily countable.*

Solution. Let $g_n(z) = \sum_{k=1}^{n} f_k(z)$ and $g(z) = \sum_{k=1}^{\infty} f_k(z)$. Then $g_n(z)$ is holomorphic and $g_n(z) \to g(z)$ for all $z \in \mathbb{C}$. To prove the first part of the statement of the problem, we have to show that inside each circle K of $\mathbb{C}$, there is a circle k in which $g(z)$ is holomorphic. Let $\varphi(z) = \sup_n |g_n(z)|$. This exists by the convergence and is finite for all $z \in \mathbb{C}$. Let $A_k = \{z \in K | k - 1 \le \varphi(z) < k\}$ $(k = 1, 2, \dots)$. By Baire's theorem, there exists an N such that A_N is dense inside some circle $k \subseteq K$. If $\varphi(z_0) > N$ for some $z_0 \in k$, then $|g_n(z_0)| > N$ for some n, and by the continuity of $g_n(z)$ there is a neighborhood of z_0 where also $\varphi(z) > N$, which contradicts the fact that A_N is dense in k. So $\varphi(z) \le N$ if $z \in k$, and so $|g_n(z)| \le N$ if $z \in k$. By the Vitali–Montel theorem, there is a subsequence of $\{g_n(z)\}$ that is equiconvergent inside k. Since $g_n(z)$ is convergent, this subsequence converges to $g(z)$. Hence, by Weierstrass's theorem, $g(z)$ is holomorphic inside k. So we proved that H is nowhere dense. □

Remarks.

1. In the solution, we did not use the fact that the series of functions $\sum_{k=1}^{\infty} f_k(z)$ is absolutely convergent. To prove the second part of the statement of the problem, we will need the following result.

Mergelyan's theorem. Let K be a compact set of $\mathbb{C}$, such that its complement is connected, and let f be is a continuous complex function on K, such that it is holomorphic inside K. Then for any $\varepsilon > 0$ there exists a polynomial P, for which $|f(z) - P(z)| < \varepsilon$ for all $z \in K$. (See *W. Rudin, Real and Complex Analysis, McGraw-Hill, London, 1970.*)

Let $A = \{z | \text{Im } z < 0\}$, $B = \{z | \text{Im } z > 0\}$, $C = \{z\colon z \text{ is real}\}$. There are series of compact sets $\{A_n\}$, $\{B_n\}$, and $\{C_n\}$ $(n = 1, 2, \dots)$, such that

$$A_n \subseteq A_{n+1}, \quad B_n \subseteq B_{n+1}, \quad C_n \subseteq C_{n+1},$$
$$C \backslash (A_n \cup B_n \cup C_n) \text{ is connected for } n = 1, 2, \dots$$

and

$$\bigcup_1^{\infty} A_n = A, \quad \bigcup\nolimits_1^{\infty} B_n = B, \quad \bigcup\nolimits_1^{\infty} C_n = C.$$

So $A_n \cup B_n \cup C_n$ is compact, its complement is connected, and

$$f(z) = \begin{cases} 1, & \text{if } z \in A_n \cup B_n, \\ 0, & \text{if } z \in C_n. \end{cases}$$

Hence, by Mergelyan's theorem, there is a polynomial $P_{\varepsilon,n}$ for any $\varepsilon > 0$ and n, such that $|1 - P_{\varepsilon,n}| < \varepsilon$, if $z \in A_n \cup B_n$, and $|P_{\varepsilon,n}(z)| < \varepsilon$, if $z \in C_n$ $(n = 1, 2, \dots)$. Let

$$f_1(z) = Q_1(z),\ f_2(z) = Q_2(z) - Q_1(z), \dots, f_k(z) = Q_k(z) - Q_{k-1}(z), \dots,$$

where $Q_n(z) = P_{(1/2^n)\cdot n}(z)\,(n = 1, 2, \dots)$ is a series of polynomials. Then $\sum_{k=1}^{n} f_k(z) = Q_n(z)$, so

$$\sum_{k=1}^{\infty} f_k(z) = \begin{cases} 1, & \text{if } z \text{ is not real}, \\ 0, & \text{if } z \text{ is real}. \end{cases}$$

Here H is the set of real numbers, and is hence uncountable. On the other hand, for any $z \in \mathbb{C}$ there exists an N such that $z \in A_n \cup B_n \cup C_n$ if $n \ge N$, so in this point z,

$$\sum_{k=1}^{\infty} |f_{N+k}(z)| = \sum_{k=1}^{\infty} |Q_{N+k}(z) - Q_{N+k-1}(z)| \le \sum_{k=1}^{\infty} 2 \frac{1}{2^{N+k-1}} < \infty;$$

hence, $\sum_{k=1}^{\infty} f_k(z)$ is convergent at z.

2. We can prove a stronger statement: the measure of H can be positive. Let R be the real axis, I the imaginary axis. It is enough to show that if S is a nowhere-dense subset of I, then there is a sequence $\{f(n)\}_{n=1}^{\infty}$ of holomorphic functions that satisfies the conditions of the problem and $R \times S$ contained in the H set obtained from this sequence $\{f(n)\}_{n=1}^{\infty}$. So let S be a nowhere-dense, open set of I. Then $I \backslash S^{-}$ is a dense set of I, so it is a union of countably many disjoint intervals; let these be $i_k = (a_k, b_k)$ $(k = 1, 2, \dots)$. (We can suppose that a_k and b_k are finite.) Let $\{r_l^{(k)}\}_{l=1}^{\infty}$ be monotone decreasing, $\{r_l^{(k)}\}_{l=1}^{\infty}$ monotone increasing, such that $r_1^{(k)} < l_1^{(k)}$ and $r_l^{(k)} \to a_k$, $s_l^{(k)} \to b_k$ if $l \to \infty$. Let $\{\alpha_n\}_{n=1}^{\infty}$ be a sequence of natural numbers, such that it contains each natural number infinitely many times. Let n be an arbitrary number. Let $\alpha_n = k$, denote by l the cardinality of $\{m \colon m \leq n, \alpha_m = k\}$, and let

$$
\begin{aligned}
K_n &= \{z \big| r_{2l-1}^{(k)} < \operatorname{Im} z \leq s_{2l-1}^{(k)},\ |\operatorname{Re} z| \leq n\}, \\
A_n &= \{z \big| \operatorname{Im} z = r_{2l}^{(k)},\ |\operatorname{Re} z| \leq n\}, \\
B_n &= \{z \big| \operatorname{Im} z = s_{2l}^{(k)},\ |\operatorname{Re} z| \leq n\}, \\
L_n &= \{z \big| r_{2l-1}^{(k)} \geq \operatorname{Im} z \geq \min(a_k - n, -n),\ |\operatorname{Re} z| \leq n\}, \\
M_n &= \{z \big| s_{2l-1}^{(k)} \leq \operatorname{Im} z \leq \max(b_k + n, n),\ |\operatorname{Re} z| \leq n\}.
\end{aligned}
$$

The set $P_n = K_n \cup A_n \cup B_n \cup L_n \cup M_n$ is compact, and its complement is connected in C^{-}. So by Runge's theorem, there is a polynomial f_n such that

$$|f_n(z)| < \frac{1}{2^n} \quad \text{if} \quad z \in K_n \cup L_n \cup M_n$$

and

$$|f_n(z) - n| < \frac{1}{2^n} \quad \text{if} \quad z \in A_n \cup B_n.$$

So

(i) The functions f_n are polynomials, and hence holomorphic on $\mathbb{C}$.

(ii) For any $z \in \mathbb{C}$, there exists an N_0 such that $z \in K_n \cup L_n \cup M_n$ for any $n > N_0$. So $|f_n(z)| < 1/2^n$ for any $n > n_0$, and $\sum_{n=1}^{\infty} |f_n(z)| < \infty$.

(iii) Let $z_0 \in R \times S$ arbitrary. Let $\varepsilon > 0$ and $U_\varepsilon = \{z \big| |z - z_0| < \varepsilon\}$. There is an index k_0 such that $U_{\varepsilon/2} \cap (\mathbb{R} \times i_{k_0}) \neq 0$; hence, $U_\varepsilon \cap \{z \colon \operatorname{Im} z = a_{k_0}\} \neq 0$ or $U_\varepsilon \cap \{z \colon \operatorname{Im} z = b_{k_0}\} \neq 0$.

Let us suppose that $U_\varepsilon \cap \{z \colon \operatorname{Im} z = a_{k_0}\} \neq 0$ and $z_1 \in U_\varepsilon \cap \{z \colon \operatorname{Im} z = a_{k_0}\}$. (The other case can be discussed similarly.) Let $n_0 \geq \max(|\operatorname{Re} z_0| + \varepsilon, |\operatorname{Im} z_0| + \varepsilon)$ such that if $n \geq n_0$ and $\alpha_n = k_0$, then $K_n \cap U_\varepsilon \neq 0$. Let $g_{n_0} = \sum_{n=1}^{n_0} g_n$. Since g_{n_0} is continuous at z_1, there is a $0 < \delta < \varepsilon$ such that for all $z \in V_\delta = \{z \colon |z - z_1| < \delta\}$, $|g_{n_0}(z) - g_{n_1}| < 1$. Let $n_1 > n_0$ such that $\alpha_{n_1} = k_0$, $n_1 > 5 + 2|g_{n_0}(z_1)|$ and $K_{n_1} \cap V_\delta \neq 0$. Let $\lambda_1 \in A_{n_1} \cap U_\delta$ and $\lambda_2 \in \partial K_{n_1} \cap U_\delta$, where ∂K_{n_1} denotes the boundary of K_{n_1}.

If $n > n_0$ and $n \neq n_1$, then $|f_n(\lambda_1)| < 1/2^n$ and $|f_{n_1}(\lambda_1) - n_1| < 1/a^{n_1}$. So

$$|f(\lambda_1) - g_n(z_1) - n_1| \leq |g_{n_0}(\lambda_1) - g_{n_0}(z_1)| + |f_{n_1}(\lambda_1) - n_1| + \sum_{n>n_0, n\neq n_1} f_n(\lambda_1) \leq 1 + \sum_{n=n_0+1}^{\infty} \frac{1}{2^n} \leq 2.$$

On the other hand, $|f_n(\lambda_2)| < 1/2^n$ if $n \geq n_0$, so

$$|f(\lambda_2) - g_{n_0}(z_1)| \leq |g_{n_0}(\lambda_2) - g_n(z_1)| + \sum_{n=n_0+1}^{\infty} |f_n(\lambda_2)| \leq 2.$$

Hence,

$$|f(\lambda_1)| \geq |g_{n_0}(z_1) + n_1| - |f(\lambda_1) - g_{n_0}(z_1) - n_1| \geq n_1 - |g_{n_0}(z_1)| - 2$$

and

$$|f(\lambda_2)| \leq |f(\lambda_2) - g_{n_0}(z_1)| + |g_{n_0}(z_1)| \leq 2 + |g_{n_0}(z_1)|.$$

So

$$|f(\lambda_1)| - |f(\lambda_2)| \geq n_1 - 4 - 2g_{n_0}(z_1) > 1.$$

Therefore, the total variation of $|f|$ is more than 1 in U_ε. Since $\varepsilon > 0$ is arbitrary, $|f|$ is not continuous at z_0. So $|f|$ is not continuous in the points of $R \times S$. That is, $\{f_n\}_{n=1}^{\infty}$ satisfies the conditions of the problem, and the set H, obtained from $\{f(n)\}_{n=1}^{\infty}$, contains $R \times S$. By this, the statement of the remark is proved.

Problem S.22. *Let $f(x)$ be a nonnegative, integrable function on $(0, 2\pi)$ whose Fourier series is $f(x) = a_0 + \sum_{k=1}^{\infty} a_k \cos(n_k x)$, where none of the positive integers n_k divides another. Prove that $|a_k| \leq a_0$.*

Solution. Let $K_n(x) = 1/2 + (1 - \frac{1}{n+1}) \cos x + \dots$ be the nth Fejér kernel. It is known that $K_n(x) \geq 0$. By integrating, we get $0 \leq \int_0^{2\pi} f(x) K_n(n_k x) = \pi(a_0 + (1 - 1/(n+1))a_{n_k})$, so if $n \to \infty$, then $a_{n_k} \leq a_0$. $K_n(x - \pi) = 1/2 - (1 - 1/(n+1)) \cos x + \dots$, so

$$0 \leq \int_0^{2\pi} f(x) K_n(n_k x - \pi) = \pi \left(a_0 - \left(1 - \frac{1}{n+1}\right) a_{n_k} \right).$$

Therefore, if $n \to \infty$, then $a_{n_k} \leq -a_0$. □

Problem S.23. *We are given an infinite sequence of 1's and 2's with the following properties:*
(1) The first element of the sequence is 1.
(2) There are no two consecutive 2's or three consecutive 1's.
(3) If we replace consecutive 1's by a single 2, leave the single 1's alone, and delete the original 2's, then we recover the original sequence.
How many 2's are there among the first n elements of the sequence?

Solution. Since, by the above conditions, the first $n-1$ elements of the sequence determine the nth one, there can be only one sequence. It is easy to see that the following sequence of numbers 1 and 2 satisfies the conditions of the problem: in this sequence, the index of the nth 2 is larger by 2 or 3 than the index of the $(n-1)$th 2 iff the nth element of the sequence is 1 or 2, respectively.

The following condition is equivalent to (3):

(3′) If we change all 1's into 12 and all 2's into 112, then we obtain the original sequence.

Denote by $f(n)$ the number of 2's among the first n elements. We are going to prove that, for all n,

$$f(f(n)+2n) = n, \tag{a}$$
$$f(f(n)+2n-1) = n-1, \tag{b}$$
$$f(f(n)+2n+1) = n. \tag{c}$$

To prove this, change the first n elements of the sequence as described in (3′). Hence, we obtain $3 \cdot f(n) + 2 \cdot (n - f(n)) = f(n) - 2n$ elements, such that the number of 2's is n, so we proved (a). By (3′), this $f(n) - 2n$ ends with 12, and by (2), the $(f(n) - 2n + 1)$th element is 1, so we obtain (b) and (c). Obviously,

$$f(1) = 0 \quad \text{and} \quad f(2) = 1. \tag{d}$$

Now we are going to prove that if function $g : \mathbb{N} \to \mathbb{N}$ satisfies the conditions (a)–(d), then $g = f$.

$f(n) = g(n)$ follows from (d) for $n = 1, 2$. Let us suppose that we proved $f(k) = g(k)$ for all $k < n$. Since $h(k) = f(k) + 2k$, $h(k+1) - h(k) \leq 3$, and so there is a $k < n$ natural number for which $n = f(k) + 2k + \varepsilon_n$, where $\varepsilon_n = 0, \pm 1$. Then, by (a), (b), or (c) and the induction,

$$f(n) = f(f(k) + 2k + \varepsilon_n) = \begin{cases} k, & \text{if } \varepsilon_n = 0, 1, \\ k-1, & \text{if } \varepsilon_n = -1. \end{cases}$$

Moreover,

$$g(n) = g(f(k) + 2k + \varepsilon_n) = g(g(k) + 2k + \varepsilon_n) = \begin{cases} k, & \text{if } \varepsilon_n = 0, 1, \\ k-1, & \text{if } \varepsilon_n = -1. \end{cases}$$

Hence, $f(n) = g(n)$.

Finally, we are going to show that $g(n) = \left[(\sqrt{2}-1)n + 1 - 1/\sqrt{2}\right]$ satisfies the conditions (a)–(d), so $f(n) = \left[(\sqrt{2}-1)n + 1 - 1/\sqrt{2}\right]$ holds. Obviously, $g(1) = 0$ and $g(2) = 1$. To verify (a), write $g(n)$ in the form $g(n) = (\sqrt{2}-1)n+1-1/\sqrt{2}]-\varepsilon(n)$, where $\varepsilon(n) \in [0,1]$. (However, $\varepsilon(n) = 0$ cannot hold, but we do not need this fact.) Hence,

$$g(g(n)+2n) = \left[(\sqrt{2}-1)\Big(\sqrt{2}-1)n + 1 - \frac{1}{\sqrt{2}} - \varepsilon(n) + 2n\Big) + 1 - \frac{1}{\sqrt{2}}\right]$$
$$= [n + (1-\varepsilon(n))(\sqrt{2}-1)] = n\,.$$

We can similarly verify (b) and (c). □

Remark. After writing down the first few elements of $f(n)$, it becomes plausible that $f(n)$ is the integer part of some linear function. If we substitute the function $f(n) = [an+b]$ into (a) and (b), we obtain that (a) and (b) hold only for the pair $a = \sqrt{2}-1$, $b = 1 - 1/\sqrt{2}$ obtained in the solution.

Problem S.24. *Let $a_0, a_1, \ldots$ be nonnegative real numbers such that*

$$\sum_{n=0}^{\infty} a_n = \infty\,.$$

For arbitrary $c > 0$, let

$$n_j(c) = \min\left\{k \,:\, c\cdot j \le \sum_{i=0}^{k} a_i\right\}, \quad j = 1, 2, \ldots\,.$$

Prove that if $\sum_{i=0}^{\infty} a_i^2 < \infty$, then there exists a $c > 0$ for which $\sum_{j=1}^{\infty} a_{n_j(c)} < \infty$, and if $\sum_{i=0}^{\infty} a_i^2 = \infty$, then there exists a $c > 0$ for which $\sum_{j=1}^{\infty} a_{n_j(c)} = \infty$.

Solution. Let

$$s_k = \sum_{i=0}^{k} a_i\,,$$

and denote by $\chi_{k,j}$ the characteristic function of the interval $I_{k,j} = \left[(1/j)\cdot s_{k-1}, (1/j)\cdot s_k\right]$. Let

$$f(c) = \sum_{j=1}^{\infty} a_{n_j(c)}; \quad f\colon (0,\infty) \to (0,+\infty]\,.$$

Since $\chi_{k,j}(c) = 1$ iff $n_j(c) = k$, the function $f = \sum_{k,j} a_k \chi_{k,j}$ is Lebesgue-measurable. So

$$\int_a^b f(c)\,dc = \sum_{k=\infty, j=p}^{\infty} a_k \lambda(I_{k,j} \cup [a,b])\,.$$

We are going to give a lower and upper estimate for the value of the integral.

Lower estimate. We sum over only those pairs k, j for which

$$I_{k,j} \subset [a, b],$$

so if

$$a \leq \frac{1}{j} s_{k-1} \leq \frac{1}{j} s_k \leq b,$$

or, equivalently,

$$\frac{s_k}{b} \leq j \leq \frac{s_{k-1}}{a}.$$

Note that

$$\lambda(I_{k,j}) = \frac{a_k}{j}.$$

So

$$\int_a^b f(c)\, dc \geq \sum_{k=0}^{\infty} \sum_j \frac{1}{j} a_k^2 \geq \sum_{k=0}^{\infty} a_k^2 \int_{(s_k/b)+1}^{s_{k-1}/a} \frac{1}{x}\, dx$$

$$= \sum_{k=0}^{\infty} a_k^2 \ln \frac{s_{k-1}/a}{s_k/b+1} = \sum_{k=0}^{\infty} a_k^2 \ln \frac{b s_{k-1}}{a s_k + ab}.$$

Upper estimate. We sum over only those pairs k, j for which

$$I_{k,j} \cap [a, b] \neq \emptyset,$$

so if

$$\frac{1}{j} s_{k-1} < b \text{ and } a \leq \frac{s_k}{j},$$

or, equivalently,

$$\frac{s_{k-1}}{b} < j \leq \frac{s_k}{a},$$

then

$$\int_a^b f(c)\, dc \leq \sum_{k=0}^{\infty} a_k^2 \left(\sum_j \frac{1}{j} \right) \leq \sum_{k=0}^{\infty} a_k^2 \left(1 + \frac{1}{2} + \ln \frac{b s_k}{a s_{k-1} + ab} \right).$$

If $\limsup a_n = \alpha > 0$, then $a_n > \alpha/2$ for infinitely many n. So $\sum a_k^2 = \infty$. Since infinitely many terms of the sum $\sum_{j=1}^{\infty} a_{n_j(\alpha/2)}$ are larger than a_n, $f(\alpha/2) = \infty$.

So let us suppose $\lim a_n = 0$. Then $s_n/s_{n-1} \to 1$. Therefore,

$$\lim \ln \frac{b s_{k-1}}{a s_k + ab} = \ln \frac{b}{a} > 0,$$

$$\lim \ln \frac{b s_k}{a s_{k-1} + ab} = \ln \frac{b}{a}.$$

So there exist real numbers $A > 0$, $B > 0$, $C > 0$, and D such that

$$\sum_{k=0}^{\infty}\left(\ln\frac{bs_{k-1}}{as_k+ab}\right)a_k^2 \geq \left(\sum_{k=0}^{\infty}a_k^2\right)A+B,$$

$$\sum_{k=0}^{\infty}\left(\frac{3}{2}+\ln\frac{bs_k}{as_{k-1}+ab}\right)a_k^2 \leq \left(\sum_{k=0}^{\infty}a_k^2\right)C+D.$$

So if $\sum a_k^2 < \infty$, then $\int_b^a f(c)\,dc < \infty$. Therefore, $f(c) < \infty$ almost everywhere, and if $\sum a_k^2 = \infty$, then $\int_b^a f(c)\,dc = \infty$. So for any $a < b$, the set

$$\left\{\sum_{j=1}^{\infty}a_{n_j(c)} > m\right\}$$

is a dense set for an any m.

However, the function $a_{n_j(c)}$ is a left-upper-semicontinuous function of c, so the following sum is also left-upper-semicontinuous:

$$\sum_{j=1}^{\infty}a_{n_j(c)}.$$

Therefore, the set $\{a_{n_j(c)} > m\}$ is dense and open. So the set $\{f = \infty\}$ is of second category and therefore is not empty. □

Problem S.25. *Let $2/(\sqrt{5}+1) \leq p < 1$, and let the real sequence $\{a_n\}$ have the following property: for every sequence $\{e_n\}$ of 0's and ±1's for which $\sum_{n=1}^{\infty}e_np^n = 0$, we also have $\sum_{n=1}^{\infty}e_na_n = 0$. Prove that there is a number c such that $a_n = cp^n$ for all n.*

Solution. We start with a definition.

Definition. Let L denote the set of all positive and strictly decreasing sequences $\{l_n\}$ for which $L = \sum_{n=1}^{\infty}l_n < \infty$. We call such a sequence interval filling if for every $x \in [0, L]$ there is a sequence $\varepsilon_n \in \{0,1\}$ with $x = \sum_{n=1}^{\infty}\varepsilon_nl_n$.

We need three lemmas.

Lemma 1. If a sequence $\{l_n\} \in L$ satisfies

$$\lambda_n \leq \sum_{i=n+1}^{\infty}\lambda_i \quad (n = 1, 2, \ldots),$$

then it is interval filling.

Proof. For $x \in [0, L]$ we define the numbers ε_n inductively as follows:

$$\varepsilon_n = \begin{cases} 1, & \text{if } \sum_{i=1}^{n-1}\varepsilon_i\lambda_i + \lambda_n < x; \\ 0, & \text{if } \sum_{i=1}^{n-1}\varepsilon_i\lambda_i \geq x. \end{cases}$$

For every n for which $\varepsilon_n = 0$,

$$0 \leq x - \sum_{i=1}^{\infty} \varepsilon_i \lambda_i \leq x - \sum_{i=1}^{n-1} \varepsilon_i \lambda_i \leq \lambda_n,$$

hence if there are infinitely many such n, then $x = \sum_{i=1}^{\infty} \varepsilon_i \lim_{n \to 0} l_i$. If, however, there are only finitely many such n's, then for the largest one

$$x - \sum_{i=1}^{n-1} \varepsilon_i \lambda_i \leq \lambda_n \leq \sum_{i=n+1}^{\infty} \lambda_i = \sum_{i=n+1}^{\infty} \varepsilon_i \lambda_i,$$

from which

$$x \leq \sum_{n=1}^{\infty} \varepsilon_n l_n,$$

so our claim holds even in this case.

Lemma 2. For $2/(\sqrt{5}+1) \leq p < 1$, the sequence

$$p,\ p^2,\ p^3,\ \ldots$$

is interval filling, and for any natural number N, the sequence

$$p,\ p^2,\ \ldots,\ p^{N-2},\ p^{N-1},\ p^{N+1},\ p^{N+2},\ \ldots$$

is interval filling too.

Proof. By Lemma 1, it is enough to verify that

$$p^n \leq \sum_{i=n+1}^{\infty} p^i \quad \text{for all} \quad n \in N$$

and

$$p^{n-1} \leq \sum_{i=n+1}^{\infty} p^i \quad \text{for all} \quad n \in N$$

are satisfied. The first of these is equivalent to $1 \leq p/(1-p)$ and the second one to $1 \leq p^2/(1-p)$, hence each of these is true if $2/(\sqrt{5}+1) \leq p < 1$.

Lemma 3. Let A and B be nonempty disjoint sets such that their union is the set of the natural numbers, and let $2/(\sqrt{5}+1) \leq p < 1$. Then there exist nonempty sets $A' \subseteq A$, $B' \subseteq B$ with the property

$$\sum_{t \in A'} p^i = \sum_{i \in B'} p^i. \tag{1}$$

Proof. Let $L = p/(1-p)$ and $x = \sum_{i \in A} p^i$. Let us choose an $N \in A$ for which $x < L - p^N$. If A is finite, then N can be the largest element in

A; otherwise, we can choose any element in A with sufficiently large index. Since Lemma 2 says that the sequence

$$p,\ p^2,\ \dots,\ p^{N-2},\ p^{N-1},\ p^{N+1},\ p^{N+2},\ \dots$$

is interval filling, there is a set $C \subseteq \mathbb{N} \setminus \{N\}$ for which

$$\sum_{i\in C} p^i = x = \sum_{i\in A} p^i.$$

Let $A' = A \setminus C$, $B' = C \setminus A$. Then (1) clearly holds, and because $N \in A'$, the set A' is not empty. From this and (1), it follows that B' is not empty, either.

After these preparations, we turn to the proof of the statement in the problem. For every $c \in \mathbb{R}$, the sequence $a'_n = a_n - cp^n$ also satisfies the assumptions of the theorem. Suppose, on the contrary, that the claim is not true. Then we can choose a real c for which there is an n with $a'_n > 0$ and also another one with $a'_n < 0$. Let

$$A = \{n \in N : a'_n > 0\}, \quad B = \{n \in N : a'_n \le 0\}.$$

On applying Lemma 3, we get two sets A' and B' with the properties stated there, and with the aid of these we define

$$e_n = \begin{cases} 1, & \text{if } \ n \in A'; \\ -1, & \text{if } \ n \in B'; \\ 0, & \text{otherwise.} \end{cases}$$

Then

$$\sum_{i=1}^{\infty} e_i p^i = \sum_{i\in A'} p^i - \sum_{i\in B'} p^i = 0,$$

but

$$\sum_{i=1}^{\infty} e_i a'_i = \sum_{i\in A'} a'_i - \sum_{i\in B'} a'_i > 0,$$

and this contradicts the hypothesis of the problem. The contradiction obtained proves the claim. □

Remark. The result is true for any $0 < p < 1$ (see *Z. Daróczi, I. Kátai, T. Szabó, On Completely Additive Functions Related to Interval-filling Sequences, Arch. Math. 54 (1990), 173–179*).

Problem S.26. *Let S be the set of real numbers q such that there is exactly one 0–1 sequence $\{a_n\}$ satisfying*

$$\sum_{n=1}^{\infty} a_n q^{-n} = 1.$$

Prove that the cardinality of S is $2^{\aleph_0}$.

Solution. Let $\{b_n\}$ be a sequence such that $b_1 = b_2 = 1$, $b_{3n} = 1$, $b_{3n+1} = 0$, and $b_{3n+2} = 0$ or 1 $(n = 1, 2, \dots)$. There is exactly one number $q > 1$ such that

$$\sum_{n=1}^{\infty} \frac{b_n}{q^n} = 1.$$

Such a q exists, since the series $\sum_{n=1}^{\infty} b_n x^n$ is convergent if $|x| < 1$, continuous, it takes the value 0 at $x = 0$, and goes to infinity if $x \to 1-$.

We will prove that if there is a sequence $\{a_n\}$ with elements 0 or 1 and

$$\sum_{n=1}^{\infty} \frac{a_n}{q^n} = 1,$$

then $a_n = b_n$ for all n.

Let us suppose that this is false. Let k be the smallest number such that $a_k \neq b_k$. Then

$$\sum_{n=k}^{\infty} \frac{a_n}{q^n} = \sum_{n=k}^{\infty} \frac{b_n}{q^n}.$$

Since

$$1 = \sum_{n=1}^{\infty} \frac{b_n}{q^n} \geq \frac{1}{q} + \frac{1}{q^2} + \sum_{n=1}^{\infty} \frac{1}{q^{3n}} = \frac{q^4 + q^3 + q^2 - q - 1}{q^5 - q^2},$$

that is, $q^5 - q^4 - q^3 - 2q^2 + q + 1 \geq 0$ holds,

$$\begin{aligned} \frac{q^2 + q}{q^3 - 1} &= 1 - \frac{q^3 - q^2 - q - 1}{q^3 - 1} = 1 - \frac{(q^3 - q^2 - q - 1)(q^3 - 1)}{(q^3 - 1)^2} \\ &= 1 - \frac{q(q^5 - q^4 - q^3 - 2q^2 + q + 1) + 1}{(q^3 - 1)^2} < 1. \end{aligned}$$

So

$$\sum_{n=1}^{\infty} \left(\frac{1}{q^{3n-2}} + \frac{1}{q^{3n-1}} \right) < 1.$$

Suppose that $b_k = 0$ and $a_k = 1$. If $n > 1$, then at least one of b_n, b_{n+1}, and b_{n+2} is zero, so

$$\frac{b_n}{q^n} + \frac{b_{n+1}}{q^{n+1}} + \frac{b_{n+2}}{q^{n+2}} \leq \frac{1}{q^n} + \frac{1}{q^{n+1}}.$$

Hence

$$\begin{aligned} \sum_{n=k}^{\infty} \frac{b_n}{q^n} &= \sum_{n=k+1}^{\infty} \frac{b_n}{q^n} \leq \sum_{m=0}^{\infty} \left(\frac{1}{q^{k+3m+1}} + \frac{1}{q^{k+3m+2}} \right) \\ &= \frac{1}{q^k} \sum_{m=1}^{\infty} \left(\frac{1}{q^{3m-2}} + \frac{1}{q^{3m-1}} \right) < \frac{1}{q^k} \leq \sum_{n=k}^{\infty} \frac{a_n}{q^n}, \end{aligned}$$

and this is a contradiction.

Otherwise, suppose that $b_k = 1$ and $a_k = 0$. Since at least one of b_n, b_{n+1}, and b_{n+2} is 1,

$$\frac{b_n}{q^n} + \frac{b_{n+1}}{q^{n+1}} + \frac{b_{n+2}}{q^{n+2}} \geq \frac{1}{q^{n+2}},$$

so

$$\begin{aligned}\sum_{n=k}^{\infty} \frac{b_n}{q^n} &\geq \frac{1}{q^k} + \sum_{m=1}^{\infty} \frac{1}{q^{n+3m}} = \frac{1}{q^k}\left(1 + \sum_{m=1}^{\infty} \frac{1}{q^{3m}}\right) \\ &> \frac{1}{q^k}\left(\sum_{m=1}^{\infty}\left(\frac{1}{q^{3m-2}} + \frac{1}{q^{3m-1}}\right) + \sum_{m=1}^{\infty} \frac{1}{q^{3m}}\right) = \frac{1}{q^k}\sum_{n=1}^{\infty} \frac{1}{q^n} \\ &\geq \sum_{n=k+1}^{\infty} \frac{a_n}{q_n} = \sum_{n=k}^{\infty} \frac{a_n}{q_n},\end{aligned}$$

and this is also a contradiction. So only the sequence $\{b_n\}$ satisfies the conditions.

The cardinality of different sequences $\{b_n\}$ is $2^{\aleph_0}$. If two $\{b_n\}$ sequences are different, then the corresponding numbers q are also different. So the cardinality of S is maximum $2^{\aleph_0}$. But it cannot be more than this, since the cardinality of the set of all real numbers is $2^{\aleph_0}$. □

Problem S.27. *Given $a_n \geq a_{n+1} > 0$ and a natural number μ, such that*

$$\limsup_n \frac{a_n}{a_{\mu n}} < \mu,$$

prove that for all $\varepsilon > 0$ there exist natural numbers N and n_0 such that, for all $n > n_0$ the following inequality holds:

$$\sum_{k=1}^{n} a_k \leq \varepsilon \sum_{k=1}^{Nn} a_k\,.$$

Solution. We have to find a natural number n_0 such that for any $\varepsilon > 0$ there exists a natural number N such that

$$\frac{1}{\varepsilon} \leq \frac{\sum_{k=1}^{nN} a_k}{\sum_{k=1}^{N} a_k}$$

holds for all $n > n_0$, where condition (1) holds.

Choose a real number ν in the interval $(\limsup_n (a_n/a_{n\mu}), \mu)$. Define a natural number M such that $a_n/a_{n\mu} \leq \nu$, for all $n \geq M$. Therefore, for all natural numbers $l \geq 0$ and $n \geq M$, the inequality $a_n/\nu^l \leq a_{n\mu^l}$ holds.

First, we show that the series $\sum_{k=1}^{n} a_k$ is divergent. Obviously,

$$\sum_{k=1}^{M\mu^l} a_k \geq \sum_{k=1}^{M\mu^l} a_{M\mu^l} \geq \sum_{k=1}^{M\mu^l} \frac{a_M}{\nu^l} \geq M\mu^l \frac{a_M}{\nu^l} = M a_M \left(\frac{\mu}{\nu}\right)^l .$$

Here M and a_M are fixed, so when $l \to \infty$, then from $\nu < \mu$ it follows that the right side of the inequality also goes to infinity, that is, we estimated the series $\sum_{k=1}^{n} a_k$ from below by a divergent series. This divergence provides us with a number K, for which $\sum_{k=1}^{M} a_k \leq \sum_{k=M+1}^{K} a_k$; so for any $n \geq K$, $\sum_{k=1}^{n} a_k \leq 2\sum_{k=M+1}^{n} a_k$.

For given ε, choose l such that $1/\varepsilon \leq (\mu/\nu)^l/2$. Therefore,

$$\begin{aligned} \frac{1}{\varepsilon} \leq \frac{1}{2}(\mu/\nu)^l &= \frac{\mu^l \sum_{k=M+1}^{n} a_k/\nu^l}{2\sum_{k=M+1}^{n} a_k} = \frac{\sum_{k=M\mu^l}^{n\mu^l-1} a_{(\lfloor k/\nu^l \rfloor+1)}/\nu^l}{2\sum_{k=M+1}^{n} a_k} \\ &\leq \frac{\sum_{k=M\mu^l}^{n\mu^l-1} a_{(\lfloor k/\nu^l \rfloor+1)\mu^l}}{2\sum_{k=M+1}^{n} a_k} \leq \frac{\sum_{k=M\mu^l}^{n\mu^l-1} a_k}{2\sum_{k=M+1}^{n} a_k} \leq \frac{\sum_{k=1}^{nN} a_k}{\sum_{k=1}^{n} a_k}. \quad \square \end{aligned}$$

3.10 TOPOLOGY

Problem T.1. *Prove that any uncountable subset of the Euclidean n-space contains an uncountable subset with the property that the distances between different pairs of points are different (that is, for any points $P_1 \neq P_2$ and $Q_1 \neq Q_2$ of this subset, $\overline{P_1P_2} = \overline{Q_1Q_2}$ implies either $P_1 = Q_1$ and $P_2 = Q_2$, or $P_1 = Q_2$ and $P_2 = Q_1$). Show that a similar statement is not valid if the Euclidean n-space is replaced with a (separable) Hilbert space.*

Solution. For the proof of the first statement of the problem, we say that a subset of the Euclidean n-space E_n has property T if the distances between all pairs of points of this subset are different. By induction on n, we prove that if all subsets of a set $H(\subseteq E_n)$ that have property T are countable, then H is countable. The case $n = 0$ is trivial. Suppose that the statement is true for $n - 1(\geq 0)$. Let $H(\subseteq E_n)$ be a set such that all subsets with property T are countable. By Tukey's lemma, there exists a maximal subset M of H with property T. By maximality of M, any point of $H \setminus M$ either has equal distance from two distinct points of M or has a distance from a point of M that equals the distance between some pair of points in M. Therefore, H is covered by the perpendicular bisector hyperplanes of pairs of points of M, together with the spheres centered at and passing through points of M. The set of these hyperplanes F_i and spheres S_i is countable ($i = 1, 2, \dots$), since M is countable. Each $H \cap S_j$ is a countable set. Indeed, the preceding argument can be applied to $H \cap S_j$ instead of H (since $H \cap S_j$ satisfies all assumptions made on H); therefore, the set of S_i''s and F_i''s that replace the preceding S_i's and F_i's and cover $H \cap S_j$ is countable; by the induction hypothesis $H \cap F_i'$ and (since $S_j \cap S_i'$ is contained in a hyperplane of E_n) $H \cap S_j \cap S_i'$ are countable sets. Thus, $H \cap S_j$ is countable since it can be covered by a countable family of countable sets. Further, by the induction hypothesis, each $H \cap F_j$ is a countable set. Therefore, H is a countable set since it is covered by a countable family of countable sets.

In order to prove the second statement of the problem, it suffices to construct an uncountable subset of the Hilbert space spanned by the orthonormal basis e_i ($i = 1, 2, \dots$), such that the set of distances between all pairs of points in this subset is countable. To this end, consider an uncountable set $\mathfrak{A}$ of infinite sets of natural numbers with pairwise finite intersection. It is well known that such a set $\mathfrak{A}$ exists. (For instance, the set of bounded, monotone sequences of rational numbers has this property.) Each set $A(\in \mathfrak{A})$ determines a vector $\sum_{i=1}^{\infty}(1/2^i) \cdot e_{\alpha_i}$ in the Hilbert space, where α_i runs through A in increasing order. The set K of these vectors is uncountable. On the other hand, the distance between any pair of elements of K is the square root of a rational number, for if the vectors $a = \sum_{i=1}^{\infty}(1/2^i) \cdot e_{\alpha_i}$ and $a' = \sum_{i=1}^{\infty}(1/2^i) \cdot e'_{\alpha_i}$ are determined by the sets A and A' ($\in \mathfrak{A}$), then

$$\|a - a'\|^2 = \|a\|^2 = \|a'\|^2 - 2(a, a') = \frac{2}{3} - 2 \sum_{a_i = a'_j} \frac{1}{2^{i+j}},$$

where in the last term we have a finite sum since $A \cap A'$ is finite.

Remark. István Juhász and Béla Bollobás pointed out that a similar argument proves the following generalization of the first statement of the problem: If $\mathfrak{m}$ is a regular cardinal number, then any subset of cardinality $\mathfrak{m}$ of the Euclidean n-space has a subset of cardinality $\mathfrak{m}$ in which all distances between pairs of points are different; further, they considered generalizations of the statement to other metric spaces.

Problem T.2. *A sentence of the following type is often heard in Hungarian weather reports: "Last night's minimum temperatures took all values between −3 degrees and +5 degrees." Show that it would suffice to say, "Both −3 degrees and +5 degrees occurred among last night's minimum temperatures." (Assume that temperature as a two-variable function of place and time is continuous.)*

Remark. The formulation of the problem allows for various models. The proof is simplest when the country is assumed to be compact; not assuming compactness requires a different proof and yields a more general theorem. The argument needs some modification if the time interval is replaced with a (not necessarily metrizable) compact space. In the following solutions, the space is connected and the time is compact. The proof is carried out for metric spaces in the first one and for arbitrary topological spaces in the second one.

Solution 1. Since any continuous image of a connected space is connected, and in the real line only the intervals are connected sets, it is sufficient to prove that the function that assigns to each point the minimum temperature attained there during the night is continuous.

Let E be a connected metric space, I be a compact metric space, $\mathbb{R}$ be the real line, and $f : I \to \mathbb{R}$ be a continuous function. For any fixed $x \in E$, the function $f(x,t)$ is continuous on the compact space I; therefore, $g(x) = \min_{t \in I} f(x,t)$ exists. We show that $g(x)$ is continuous. If this were not the case, then there would exist an $\varepsilon > 0$ and a sequence $\{x_k\}$ ($x_k \in E, k = 1, 2, \dots$) such that $x_k \to x$, but

$$|g(x_k) - g(x)| \geq \varepsilon.$$

We consider two cases:

(a) There is a subsequence of $\{x_n\}$ such that

$$g(x_{n_k}) \leq g(x) - \varepsilon.$$

If $f(x_{n_k}, t_k) = g(x_{n_k})$, then the sequence $\{t_k\}$ has an accumulation point $t \in I$, and we can assume that $t_k \to t$. Then

$$f(x_{n_k}, t_k) = g(x_{n_k}) \leq g(x) - \varepsilon \leq f(x,t) - \varepsilon,$$

and f cannot be continuous.

(b) There is a subsequence of $\{x_n\}$ such that

$$g(x_{n_k}) \geq g(x) + \varepsilon.$$

If $f(x,t) = g(x)$, then

$$f(x_{n_k}, t) \geq g(x_{n_k}) \geq g(x) + \varepsilon \geq f(x,t) + \varepsilon,$$

and f cannot be continuous.

Since at least one of the cases must hold, we have a contradiction, which proves the statement. □

Solution 2. Now let E be a connected topological space and I be a compact topological space. We keep other notations of Solution 1.

We prove again that $g(x)$ is continuous on E. For $\varepsilon > 0$, $u \in E$, and $v \in I$, let $U(u,v)$ and $V(u,v)$ be respective neighborhoods of u and v such that for all $(x,t) \in U(u,v) \times V(u,v)$, we have

$$|f(u,v) - f(x,t)| < \varepsilon.$$

For u fixed, the neighborhoods $V(u,v)$ $(v \in I)$ cover I. Then, by its compactness, I is covered by a finite number of them:

$$V(u,v_1), V(u,v_2), \ldots, V(u,v_k).$$

Then $U = \cap_{i=1}^{k} U(u,v_i)$ is a neighborhood of u, and if $x \in U$, then $|g(x) - g(u)| < \varepsilon$, for if $t \in I$, then $t \in V(u,v_j)$ for some j, and then, since

$$(x,t), (u,t) \in U(u,v_j) \times V(u,v_j),$$

we have

$$|f(x,t) - f(u,t)| < \varepsilon. \qquad \square$$

Remarks.

1. Assuming space and time to be compact topological spaces, Attila Máté considered temperature to take values in a metric space and proved, under these conditions, that the function $g(x)$ is continuous. Compactness of space is not essential in this model.

2. György Vesztergombi proved that the statement of the problem remains true if the definition of night depends on place (astronomical night), provided that the beginning and the end of the night are continuous functions of place.

3. Miklós Simonovits gave an example to show that it is essential to assume compactness of night. (If the night is not compact, then, of course, we have to speak of local infimum instead of local minimum.) Let $E = [-1. +1]$, $I = (-\infty, +\infty)$, $f(x,t) = e^{-t^2x^2}$. Then

$$g(x) = \inf_{t \in I} f(x,t) = \begin{cases} 1 & \text{for } x = 0, \\ 0 & \text{for } x \neq 0, \end{cases}$$

and $g(x)$ is not continuous.

4. Juhász and Pósa gave examples that show that it is not sufficient to assume partial continuity of $f(x,t)$. Juhász's example is the following: put $E = I = [-1, +1]$, and

$$f(x,t) = \begin{cases} 1 & \text{if } x \le 0, \text{ or } x > 0 \text{ and } t \ge 0, \\ 1 + \frac{t}{x} & \text{if } x > 0 \text{ and } -x \le t < 0, \\ 1 + \frac{x}{t} & \text{if } x > 0 \text{ and } -1 \le t < -x. \end{cases}$$

In this case,

$$g(x) = \begin{cases} 1 & \text{for } x \le 0, \\ 0 & \text{for } x > 0 \end{cases}$$

is not continuous, and the statement of the problem is not true.

Problem T.3. *Let A be a family of proper closed subspaces of the Hilbert space $H = l^2$ totally ordered with respect to inclusion (that is, if $L_1, L_2 \in A$, then either $L_1 \subset L_2$ or $L_2 \subset L_1$). Prove that there exists a vector $x \in H$ not contained in any of the subspaces L belonging to A.*

Solution 1. More generally, we prove the statement for separable Banach spaces. Let B be a separable Banach space, and let R be a system of subspaces L of B that satisfies the requirements of the problem. Suppose that $\cup_{L \in R} L = B$. Consider a countable, everywhere-dense subset $\{x_1, x_2, \dots\}$ of B. By recursion, define a countable increasing sequence of elements of R as follows. Let L_1 be an element of B that contains x_1. Suppose that the subspaces $L_1, \dots, L_n$ ($\in R$) have already been defined. Let $L^{(n+1)}$ be an element of R that contains x_{n+1}. (Such an element exists by the assumptions.) One of the subspaces $L_1, \dots, L_n, L^{(n+1)}$ contains all others; let this one be denoted by L_{n+1}. Then $L_n \subseteq L_{n+1}$ and $x_n \in L_n$ $(n = 1, 2, \dots)$. Now, let L be an arbitrary element of R. Since L is a proper closed subspace of B, it cannot contain all elements of the dense set $\{x_1, x_2, \dots\}$. Suppose, say, that $x_k \notin L$. By definition, $x_k \in L_k$, so $L \subseteq L_k$ since the system B is ordered with respect to inclusion. This means that $\cup_{i=1}^{\infty} L_i = B$.

Now, for all natural numbers n, let f_n be a continuous linear functional on B for which $\|f_n\| = n$ and $f_n(x) = 0$ if $x \in L_n$. (Such functionals exist. For example, take an arbitrary element y_n in the complement of L_n. Since L_n is closed, the distance d of y_n from the subspace L_n is positive. Consider now the linear subspace $[y_n] + L_n$, where $[y_n]$ denotes the one-dimensional subspace generated by y_n, and let

$$f_n^{(0)}(\lambda y_n - x) = n\lambda d,$$

where $x \in L_n$ and λ is a complex number. $f_n^{(0)}$ is obviously linear on $[y_n] + L_n$ and $f_n^{(0)}(x) = 0$ for $x \in L_n$, and

$$\|f_n^{(0)}\| = \sup \frac{|n\lambda d|}{\|\lambda y_n - x\|} = \sup_{x \in L_n} \frac{nd}{\|y_n - x\|} = \frac{nd}{\inf_{x \in L_n} \|y_n - x\|} = n.$$

By the Hahn–Banach theorem, $f_n^{(0)}$ is extendable to a functional f_n defined on the whole B that has the required properties.) Now, if $x \in B$, then $\lim_{n\to\infty} f_n(x) = 0$, since, by $\cup_{i=1}^{\infty} L_i = B$, x is contained in some L_i, and the sequence $\{L_n\}$ of subspaces is increasing with respect to inclusion. So, the sequence $\{f_n\}$ of functionals is pointwise convergent; therefore, by the Banach–Steinhaus theorem, $\sup_n \|f_n\| < \infty$. But this contradicts the choice of f_n. □

Solution 2. We prove the following generalization of the problem.

Generalization. Let M be a separable topological space of second Baire category, and let $\mathfrak{A}$ be a system of nowhere-dense closed subsets of M ordered by inclusion. Then $\cap_{L\in\mathfrak{A}} L \neq M$.

(It is obvious that the space l^2 and the family of subspaces in the problem and, more generally, separable Hilbert and Banach spaces with a similar family of subspaces all satisfy the hypotheses of this statement; indeed, these spaces equipped with the norm topology are separable topological spaces of second Baire category and, in any of these spaces, a proper closed subspace is nowhere dense.)

Proof. Suppose that, to the contrary, $\cup_{L\in\mathfrak{A}} L = M$, and consider a countable, everywhere-dense set $R = \{x_1, x_2, \dots\}$ in M. Put

$$M_k = \bigcup\{L \in \mathfrak{A} : x_k \in L\} \quad (k = 1, 2, \dots).$$

Then M_k is a nonempty, nowhere-dense subset of M for all k. Therefore, $\cup_{k=1}^{\infty} M_k$ is of first category, and

$$\bigcup_{k=1}^{\infty} M_k \neq M.$$

On the other hand, if $L \in \mathfrak{A}$, then $L \neq M$. L cannot contain R, that is, there is a natural number k such that $x_k \notin L$. Since $\mathfrak{A}$ is ordered with respect to inclusion, $L \subseteq M_k$ follows from the definition of M_k. Thus, $\cup_{L\in\mathfrak{A}} L \subseteq \cup_{k=1}^{\infty} M_k$, a contradiction. □

Remark. Several contestants remarked that the statement of the problem is not true for nonseparable Hilbert spaces. Attila Máté gave the following example. Let H be a nonseparable Hilbert space. Then H is isomorphic to a Hilbert space $K \oplus L^2(\omega_1)$, where K is a suitably chosen Hilbert space, and ω_1 is the set of countable ordinals considered as the discrete measure space in which the measure of all singletons is 1. Then

$$\bigcup_{\xi\in\omega_1} \left(K \oplus L^2(\xi)\right) = K \oplus L^2(\omega_1).$$

Problem T.4. *Let K be a compact topological group, and let F be a set of continuous functions defined on K that has cardinality greater than continuum. Prove that there exist $x_0 \in K$ and $f \neq g \in F$ such that*

$$f(x_0) = g(x_0) = \max_{x \in K} f(x) = \max_{x \in K} g(x).$$

Solution. In the proof, we use the following theorem by Paul Erdős.

Theorem. If the edges of a complete graph of cardinality greater than continuum are labeled with natural numbers, then there exists an uncountable complete subgraph with all edges labeled with the same number. See, for example, *P. Erdős, A. Hajnal, and R. Radó, Partition Relations for Cardinal Numbers, Acta Math. Acad. Sci. Hung. 16 (1965)*, Theorem 1.

If F_x denotes the system of functions in F whose maximum is x, then it is clear that the cardinality of F_x is greater than continuum for some real number x. Suppose that $f_{=x} \cap g_{=x} = \emptyset$ for all pairs of functions $f \neq g \in F_x$. Consider the complete graph with the functions in F_x as vertices. If $f \neq g \in F_x$, then there exists a natural number n such that $f_{\geq x-(1/n)} \cap g_{\geq x-(1/n)} = \emptyset$; otherwise, the closed sets $\left\{f_{\geq x-(1/n)}, g_{\geq x-(1/n)}\right\}_{n=1,2,\ldots}$ would form a family with the finite intersection property whose intersection, by compactness, would be nonempty, but this intersection is contained in the set $f_{=x} \cap g_{=x}$. By the theorem quoted above, there is an n and an uncountable $F' \subseteq F_x$ such that all edges of F' are labeled with n. Thus, the sets $f_{x>1/n}$, $f \in F'$ are nonempty, open, and pairwise disjoint. This contradicts the well-known fact that K admits a finite Haar measure, and therefore any family of pairwise disjoint, nonempty, open sets in K is countable. So, the statement of the problem is true for some pair of functions $f \neq g \in F$. $\square$

In fact, we proved the following theorem.

Theorem. The statement of the problem holds for a compact space K if K satisfies the following condition: any family of pairwise disjoint, nonempty, open sets in K is countable.

Problem T.5. *Prove that two points in a compact metric space can be joined with a rectifiable arc if and only if there exists a positive number K such that, for any $\varepsilon > 0$, these points can be connected with an ε-chain not longer than K.*

Solution. If A and B are two points of the metric space, then let $t(A, B)$ denote their distance. By a rectifiable arc joining A and B we mean a homeomorphic image of the real interval $[a, b]$, where a and b are mapped to A and B, and for any subdivision $a = t_0 < t_1 < t_2 < \cdots < t_{n=1} < t_n = b$, denoting the image of t_i by T_i, we have

$$\sum_{i=0}^{n-1} t(T_i, T_{i+1}) \leq K$$

for some fixed K. The infimum of such numbers K is called the length of the arc. Therefore, the half of the problem stating that if two points can be joined with a rectifiable arc, then for any $\varepsilon > 0$ they can be joined with an ε-chain of length at most K, is obvious.

Now suppose that, for any $\varepsilon > 0$, A and B are joinable with an ε-chain not longer than K. Let $L = \{H_0, \dots, H_n\}$ be a sequence of points in the metric space. We use the following notation:

$$q(L) = \max_{1 \le i \le n} t(H_{i-1}, H_i)$$

and

$$K(L) = \sum_{i=1}^{n} t(H_{i-1}, H_i).$$

For $0 \le h < 1$, let $L(h)$ denote the H_k for which

$$\sum_{i=1}^{k} t(H_{i-1}, H_i) \le h \cdot K(L) < \sum_{i=1}^{k+1} t(H_{i-1}, H_i),$$

and let $L(1) = H_n$. Then, obviously,

$$t(L(H_1)L(H_2)) \le |h_1 - h_2| \cdot K(L) + q(L).$$

Now choose an indefinitely refining sequence S_0 of chains connecting A and B, where the lengths of these chains do not exceed a fixed constant K. The hypotheses in the problem guarantee the existence of such a sequence of chains. Arrange the rational points of the interval $[0, 1]$ into a sequence. Then let S_k be an indefinitely refining sequence S_0 of chains connecting A and B such that for $k = 1, 2, \dots,$

(a) S_k is a subsequence of S_{k-1}, and

(b) $L(h)$ is convergent in the first k rational points when the chains are taken from S_k and k is kept fixed.

(More precisely, instead of $L(h)$, we should write $L_{n,k}(h)$, where this denotes $L(h)$ for the nth chain of S_k. In (b), k and h are fixed while $n \to \infty$.) The sequence of chains S_k can be defined for all k by compactness of the space. Now define the function $f(h)$ on the set of rational numbers in $[0, 1]$ by the formula

$$f(h) = \lim_{n \to \infty} L_{k,n}(h)$$

if h is one of the first k rational numbers. This definition is correct by (a). Let h_1 and h_2 be rational numbers in $[0, 1]$. Choose a sufficiently large k so that h_1 and h_2 occur among the first k rational numbers. Then

$$t(f(h_1), f(h_2)) \le K \cdot |h_1 - h_2|.$$

Since a uniformly continuous function defined on a dense subset of the interval $[0, 1]$ can always be continuously extended over the whole interval

$[0,1]$, provided the target space is compact, the function $f(h)$ is extendable, and the last estimate remains valid. The image of the interval $[0,1]$ is a continuous curve connecting the points A and B, and from the estimate it immediately follows that all approximating chains have length at most K. Therefore, the curve is rectifiable of length $\leq K$.

In general, it is not true that we obtain an arc. But if we choose the above K to be the smallest constant such that, for any $\varepsilon > 0$, A and B are joinable with an ε-chain not longer than K, then we get an arc. Indeed, otherwise there would exist u and v in $[0,1]$ $(u \neq v)$, with $f(u) = f(v)$. Then we could delete the image of the open interval (u, v) from the curve constructed above, and we would get a rectifiable curve from A to B whose length K^+ is less than K. Then, by our initial observations, for any $\varepsilon > 0$, A and B would be joinable by an ε-chain shorter than K^+. This contradicts the minimality of K, proving the statement of the problem. □

Remarks.

1. László Lovász proved that, instead of compactness, assuming only local compactness and completeness of the space, the statement remains valid. By deleting a chord from a disk in the plane, we obtain a locally compact space in which two points on different sides of the chord cannot be joined with an arc although they are joinable with an arbitrarily fine ε-chain. Therefore, it does not suffice to assume only local compactness of the space. Completeness alone is not sufficient either. To show this, Lovász gave the following example.

Example. Let $e_0, e_1, \ldots, e_n, \ldots$ be an orthonormal basis in the Hilbert space l^2. Consider the segments connecting 0 with the endpoints of the vectors $e_1, e_2, \ldots$, and similarly the segments connecting e_0 with the points $e_1+e_0, e_2+e_0, \ldots$. Divide the segment between e_n and e_n+e_0 into n equal parts, that is, consider the points $e_n + (i/n) \cdot e_0$ when $i = 1, 2, \ldots, n-1$. The segments and points just defined form a closed set in the Hilbert space; therefore, they define a complete metric space. In this space the points 0 and e_0 for any $\varepsilon > 0$ are joinable with an ε-chain not longer than 3, but they cannot be joined with an arc. This space can be made into a connected counterexample by connecting the points $e_n + ((i-1)/n) \cdot e_0$ and $e_n + (i/n) \cdot e_0$ with suitably chosen, nonrectifiable arcs.

2. A sketch of the proof of Lovász's generalization is the following.

Proof. We call a point accessible if, for some t, B and the point are joinable with an arc of length t and, for any $\varepsilon > 0$, the point and A are joinable with an ε-chain not longer than $K - t$, where K is the minimal constant used above. Of the accessible points x and y we say that x is finer than y if, in the definition of accessibility of x, the arc from B can be chosen through y. Then, using Zorn's lemma, we define a "finest" point x_0, which we show to be necessarily A. Assuming the contrary, take a compact neighborhood U of x_0 that does not contain A. Then define a point z on the boundary of U that is a limit point of certain points of chains from x_0 to A in the definition of accessibility of x_0. We show that z is accessible and finer than x_0. To this end, it suffices to see that x_0 and z are joinable with a sufficiently short arc. But we can apply the result already obtained for compact spaces to the points x_0 and z, and this proves the statement.

Problem T.6. *Let a neighborhood basis of a point x of the real line consist of all Lebesgue-measurable sets containing x whose density at x equals 1. Show that this requirement defines a topology that is regular but not normal.*

Solution.

(1) Let $m(A)$ denote the measure of the Lebesgue-measurable set A. Let A and B be basis neighborhoods of the point x, we show that $A \cap B$ is as well. Since $A \cap B$ is measurable and $x \in A \cap B$, we only have to show that its density at the point x is 1, that is,

$$\frac{m((A \cap B) \cap I)}{m(I)} \to 1$$

when the interval I shrinks to x:

$$\begin{aligned}\left|\frac{m(A \cap B \cap I)}{m(I)} - 1\right| &= \left|\frac{m(I) - m(I - (A \cap B))}{m(I)} - 1\right| \\ &= \left|\frac{m((I - A) \cup (I - B))}{m(I)}\right| \\ &\leq \frac{m(I - A)}{m(I)} + \frac{m(I - B)}{m(I)}.\end{aligned}$$

The last expression converges to 0 when I shrinks to x. So $A \cap B$ is indeed a neighborhood of x. In order to obtain a topology, we show that any neighborhood A of x contains a neighborhood B of x that is a neighborhood of all of its points. Let B be the set of points in A where the density of A is 1. Obviously, $x \in B$, and by the Lebesgue density theorem, $A - B$ has measure 0. Therefore, B is also measurable, and at points of B, B has the same density as A, that is, 1. So, we have a topology indeed. Let E denote this new topology, and call its open sets E-open while keeping the adjective "open" for open sets in the usual topology. Similarly, we distinguish closed and E-closed sets. It is not obvious but it is true that the E-open sets are measurable. We can prove this as follows.

Let H be an E-open set. We can assume that H is bounded; then the outer measure $\overline{m}(H)$ of H is finite. Now put

$$\alpha = \sup\{m(C) : C \subseteq H,\ C \text{ is measurable}\},$$

and for all n choose C_n so that $C_n \subseteq H$ and $m(C_n) > \alpha - (1/n)$. Then $S = \cup_n C_n$ is a subset of H with the property that every measurable subset of $H-S$ has measure 0. (S is usually called the measurable core of H.) If $V \subseteq H$ is measurable and $Q = H-S$, then, by $(V\cap Q)\cup(V\cap S) = V$, $V \cap Q$ is measurable, and therefore it has measure 0. So,

$$m(V) = m(V \cap S).$$

Q decomposes as the union of two sets Z and X, where the former has measure 0, and the density of Q at all points of the latter is 1. In order to prove that H is measurable, we show that Q has measure 0, and to this end, we need X to be empty. This will imply measurability of H. Arguing by contradiction, we suppose that $p \in X$. Let I denote an interval shrinking to p. By definition of X, $\overline{m}(Q \cap I) = (1 + o(1))m(I)$. Now let U be an E-neighborhood of p contained in H. Then

$$m(U \cap I \cap S) = m(U \cap I) = (1 + o(1))m(I),$$

so $U \cap S$ has density 1 at p. Therefore, by $I - (U \cap S) \supseteq Q \cap I$, we have

$$\overline{m}(Q \cap I) \leq m(I - (U \cap S)) = o(m(I)),$$

which contradicts our assumption.

(2) We show that the topology E is regular. Suppose that K is an E-closed set, and $x \notin K$. Then x has a neighborhood A that does not meet K. So, $m(A \cap I) = (1 + o(1))m(I)$, where the interval I shrinks to x. Thus, $m(\overline{A} \cap I) = o(m(I))$. (Here $\overline{A}$ denotes the complement of the set A.) $K \subseteq \overline{A}$, so $m(K \cap I) \leq m(\overline{A} \cap I)$, and therefore

$$\frac{m(K \cap I)}{m(I)} \to 0.$$

Put $x_n = x - (1/n)$ and $y_n = x + (1/n)$, $n = 1, 2, \ldots$. It is well known that, for any measurable set X and $\varepsilon > 0$, there exists an open set G such that $X \subseteq G$ and $m(G) \leq m(X) + \varepsilon$. Choose the open set G_n in the interval $[x_{n-1}, x_{n+2}]$ so that $K \cap [x_n, x_{n+1}] \subseteq G_n$ and

$$m(G_n) \leq m(K \cap [x_n, x_{n+1}]) + \frac{1}{2^n}$$

hold. For $n = 1$, we require $G_1 \subseteq (-\infty, x_2]$. Choose the sets H_n in a similar way for the intervals $[y_{n+1}, y_n]$. Put

$$B = (-\infty, x_1) \cup G_1 \cup G_2 \cup \cdots \cup H_1 \cup H_2 \cup \cdots \cup (y_1, \infty);$$

then B is open and contains K. Since open sets are E-open (because they have density 1 at every point), it will suffice to prove that the density of B at x is 0. Then the complement of the E-closure of B and B will be disjoint E-neighborhoods of x and K, respectively. It is enough to show that $m(I \cap B)/m(I) \to 0$ for $I = [x, x + k]$ $(k \to 0)$, because a similar argument shows the corresponding estimate for the left-hand intervals, and these two together imply the same for arbitrary intervals I shrinking to x. Assume $y_{m+1} \leq x + k < y_m$. Then $m(I) = k \geq 1/(m+1)$. We have $m(B \cap I) \leq \sum_{i=m-2}^{\infty} m(H_i)$, since $H_i \cap [x, y_m] = \emptyset$ for $i < m-2$. Therefore

$$m(B \cap I) \leq \sum_{i=m-2}^{\infty} m(K \cap [y_i + 1, y_i]) + \sum_{i=m-2}^{\infty} \frac{1}{2^i}$$
$$= m(K \cap [x, y_n - 2]) + \frac{1}{2^{m-3}},$$

and so

$$\frac{m(B \cap I)}{m(I)} \leq \frac{m(K \cap [x, y_{m-2}])}{\frac{1}{m+1}} + \frac{\frac{1}{2^{m+3}}}{\frac{1}{m+1}}.$$

If I shrinks to x, then $m \to \infty$, so both terms in the right-hand side of the last inequality converge to 0. This is obvious for the second term; for the first one, use the already established convergence $m(K \cap I)/m(I) \to 0$ and the fact that $(m+1)/(m-2) \leq 4$ if $m \geq 3$. So, B indeed has density 0 at x, and the space is regular.

(3) We show that the space is not normal; namely, that an E-closed set K_1 of second Baire category and an E-closed everywhere-dense set K_2 cannot be separated by E-open sets. Since a countable set has measure 0, and thus is E-closed, the existence of such a set K_2 is clear. The existence of such a K_1 is also well known: for every n, there exists a nowhere-dense set of measure greater than $1 - 1/n$ in $[0, 1]$; the union of these is a first category set of measure 1; and the complement of this in $[0, 1]$ is of second category and has measure 0 and is therefore E-closed. Since K_1 is chosen to have measure 0, K_2 can be chosen in the complement of K_1, then K_1 and K_2 are disjoint.

Suppose that there exist two disjoint E-open sets C_1 and C_2 with $K_1 \subseteq C_1$ and $K_2 \subseteq C_2$. Then, for any $x \in C_1$, there is a neighborhood $A \subseteq C_1$ of x for which

$$\frac{M(A \cap I)}{m(I)} \to 1, \quad \text{and thus} \quad \frac{m(C_1 \cap I)}{m(I)} \to 1$$

when I shrinks to x. Therefore, by $K_1 \subseteq C_1$, for any $x \in K_1$, there exists an n_0 such that for $n > n_0$,

$$\frac{m\left(C_1 \cap [x - (1/n), x + (1/n)]\right)}{2/n} \geq \frac{1}{2},$$

that is,

$$m\left(C_1 \cap \left[x - \frac{1}{n}, x + \frac{1}{n}\right]\right) \geq \frac{1}{n} \quad (n > n_0).$$

If we divide the elements of K_1 into a countable number of subsets depending on whether the last inequality holds from a certain index, then at least one of these subsets is dense in an interval $[a, b]$. Thus, there exists an n_0 such that the last inequality holds from this n_0 on an everywhere-dense subset of the interval (a, b).

Since K_2 is everywhere dense, we can choose an element y of K_2 in (a, b). Since $y \in C_2$, there is an $n > n_0$ with

$$\frac{m\left(C_2 \cap [y - (1/n), y + (1/n)]\right)}{2/n} \geq \frac{3}{4},$$

that is,

$$m\left(C_2 \cap \left[y - \frac{1}{n}, y + \frac{1}{n}\right]\right) \geq \frac{3}{4n}.$$

K_1 is dense in (a, b), so there exists an x in the interval $(y, y + (1/8n))$ for which

$$m\left(C_1 \cap \left[x - \frac{1}{n}, x + \frac{1}{n}\right]\right) \geq \frac{1}{n}$$

holds. The length of the interval $I = [y - (1/n), x + (1/n)]$ is at most $(2/n) + (1/8n)$. C_1 and C_2 are disjoint and

$$m(C_1 \cap I) \geq \frac{1}{n} \quad \text{and} \quad m(C_2 \cap I) \geq \frac{3}{2n}.$$

Therefore,

$$\left(2 + \frac{1}{8}\right)\frac{1}{n} \geq m(I) \geq m(C_1 \cap I) + m(C_2 \cap I) \geq \left(1 + \frac{3}{2}\right)\frac{1}{n},$$

a contradiction, which proves that the topology E is not normal. □

Remarks.

1. In order to solve the problem, it is not necessary to prove that E-open sets are measurable, but this simplifies the proof at various points.

2. Lajos Pósa proved the nonnormality statement using a cardinality argument. The sketch of his proof is as follows.

Proof. There exists a set of cardinality $\mathfrak{c} = 2^{\aleph_0}$ and of measure 0. All subsets of this set are E-closed; therefore it can be partitioned into two disjoint E-closed sets in $2^{\mathfrak{c}}$ different ways. If the space were normal, then for each such partition we could find two disjoint E-open sets that separate the two parts. It can be proved that the pairs consisting of the E-closures of the separating E-open sets corresponding to different partitions are different. It is also easy to see that the E-closure of every E-open set equals the E-closure of an F_σ-subset, which only differs by a set of measure 0. Since the cardinality of the set of F_σ-sets is only $\mathfrak{c}$, the cardinality of the set of all pairs of closures of F_σ-sets is also only $\mathfrak{c}$. Therefore, the cardinality of the set of pairs of separating E-open sets is only $\mathfrak{c}$, a contradiction.

3. László Lovász proved complete regularity (which is stronger than regularity). He noticed that the above technique yields that if A and B are disjoint E-closed sets, then they can be separated by an open set containing B and an E-open set containing A. Then the proof of the Urysohn lemma applies.

Problem T.7. *Suppose that V is a locally compact topological space that admits no countable covering with compact sets. Let $\mathbf{C}$ denote the set of all compact subsets of the space V and $\mathbf{U}$ the set of open subsets that are not contained in any compact set. Let f be a function from $\mathbf{U}$ to $\mathbf{C}$ such that $f(U) \subseteq U$ for all $U \in \mathbf{U}$. Prove that either*

(i) *there exists a nonempty compact set C such that $f(U)$ is not a proper subset of C whenever $C \subseteq U \in \mathbf{U}$,*

(ii) *or for some compact set C, the set*

$$f^{-1}(C) = \bigcup \{U \in \mathbf{U} : f(U) \subseteq C\}$$

is an element of $\mathbf{U}$, that is, $f^{-1}(C)$ is not contained in any compact set.

Solution. The statement is trivial. Indeed, if $f^{-1}(\emptyset) = V$, then (ii) holds. If not, then for an arbitrary $x \in V - f^{-1}(\emptyset)$, (i) holds with $C = \{x\}$. (We do not use that V is not σ-compact, only noncompactness of V is necessary.) □

Remark. László Babai proves that if, under the hypotheses of the problem, (ii) is not true, then (i) holds with a set C of two elements. The proof of this is fairly difficult.

Problem T.8. *Let $\mathcal{T}_1$ and $\mathcal{T}_2$ be second-countable topologies on the set E. We would like to find a real function σ defined on $E \times E$ such that*

$$0 \le \sigma(x,y) < +\infty, \quad \sigma(x,x) = 0,$$
$$\sigma(x,z) \le \sigma(x,y) + \sigma(y,z) \quad (x,y,z \in E),$$

and, for any $p \in E$, the sets

$$V_1(p,\varepsilon) = \{x : \sigma(x,p) < \varepsilon\} \quad (\varepsilon > 0)$$

form a neighborhood base of p with respect to $\mathcal{T}_1$, and the sets

$$V_2(p,\varepsilon) = \{x : \sigma(p,x) < \varepsilon\} \quad (\varepsilon > 0)$$

form a neighborhood base of p with respect to $\mathcal{T}_2$. Prove that such a function σ exists if and only if, for any $p \in E$ and $\mathcal{T}_i$-open set $G \ni p$ $(i = 1,2)$, there exist a $\mathcal{T}_i$-open set G' and a $\mathcal{T}_{3-i}$-closed set F with $p \in G' \subset F \subset G$.

Solution. Suppose that there exists a function $\sigma(x,y)$ with the required properties. Let $p \in E$, G be open in $\mathcal{T}_i$, $p \in G$. Then there exists $\varepsilon > 0$ with $V_i(p,\varepsilon) \subset G$. Let $0 < \delta < \varepsilon$, $G' = V_i(p,\delta)$, and let F be the $\mathcal{T}_{3-i}$-closure of G'. Using the triangle inequality, it easily follows that G', F, G

indeed satisfy the hypotheses. (The assumption on second countability is not needed in this direction.)

In what follows, G always denotes an open set and F a closed set. Superscripts indicate which topology this means. For example, $\overline{G^{(1)}}^{(2)}$ denotes the $\mathcal{T}_2$-closure of the $\mathcal{T}_1$-open set $G^{(1)}$. We prove that the hypothesis given in the problem is sufficient for the existence of σ. The proof is a modification of the well known-proof of the Urysohn metrization theorem.

Tikhonov lemma. If $F^{(1)} \cap F^{(2)} = \emptyset$, then there exist $G^{(1)}$ and $G^{(2)}$ such that $G^{(1)} \supset F^{(2)}$, $G^{(2)} \supset F^{(1)}$, and $G^{(1)} \cap G^{(2)} = \emptyset$.

Proof. Let $p \in F^{(i)}$. Then $p \in E - F^{(3-i)}$, this set is $\mathcal{T}_{3-i}$-open, so by the assumption there exists a $G' = G'^{(3-i)}(p)$ such that $p \in G'^{(3-i)}(p)$ and $\overline{G'^{(3-i)}(p)}^{(i)} \subset E - F^{(3-i)}$, that is, $\overline{G'^{(3-i)}(p)}^{(i)} \cap F^{(3-i)} = \emptyset$.

To each such $G'^{(j)}(p)$, assign an element $U^{(j)}$ of the countable basis in $\mathcal{T}_j$: $p \in U^{(j)} \subset G'^{(j)}(p)$. We have countably many sets $U^{(j)}$; these can be arranged as $U_n^{(j)}$, $n = 1, 2, \ldots$, $(j = 1, 2)$. We also have

$$F^{(3-i)} \subset \bigcup_{n=1}^{\infty} U_n^{(i)}, \qquad \overline{U_n^{(i)}}^{(3-i)} \cap F^{(i)} = \emptyset. \tag{1}$$

Put

$$U_{n*}^{(i)} = U_n^{(i)} - \bigcup_{k=1}^{n} \overline{U_k^{(3-i)}}^{(i)}.$$

These $U_{n*}^{(i)}$ are $\mathcal{T}_i$-open, they also satisfy (1), and obviously $U_{n*}^{(i)} \cap U_{m*}^{(3-i)} = \emptyset$. Therefore, the sets

$$G^{(i)} = \bigcup_{n=1}^{\infty} U_{n*}^{(i)} \quad (i = 1, 2)$$

clearly satisfy the requirements of the lemma.

Urysohn lemma. If $F^{(1)} \subset G^{(2)}$, then there exists a real function $\rho(x, y)$ on $E \times E$ such that

$$0 \le \rho(x, y) \le 1, \quad \rho(x, x) = 0,$$
$$\rho(x, z) \le \rho(x, y) + \rho(y, z) \quad (x, y, z \in E); \tag{2}$$

the sets

$$V_\rho^{(1)}(p, \varepsilon) = \{x : \rho(x, p) < \varepsilon\} \quad (\varepsilon > 0) \tag{3}$$

are $\mathcal{T}_1$-open; the sets

$$V_\rho^{(2)}(p, \varepsilon) = \{x : \rho(p, x) < \varepsilon\} \quad (\varepsilon > 0) \tag{4}$$

are $\mathcal{T}_2$-open; and

$$\rho(x, y) = 1 \quad \text{if} \quad x \in F^{(1)},\ y \notin G^{(2)}. \tag{5}$$

Proof. Introduce the notations $G_0^{(2)} = \emptyset$, $F_0^{(1)} = F^{(1)}$, $G_1^{(2)} = G^{(2)}$, $F_1^{(1)} = E$. Let D be a countable dense subset of the closed interval $[0,1]$: $D = \{r_0, r_1, r_2, \dots\}$, where $r_0 = 0$, $r_1 = 1$.

By recursion on n, we define the sets $G_n^{(2)}$ and $F_n^{(1)}$ so that $G_n^{(2)} \subset F_n^{(1)}$ and, for $r_n < r_m$, $F_n^{(1)} \subset G_m^{(2)}$ hold. These inclusions are true in the cases already defined $(n,m = 0,1)$. If the sets with subscripts less than n $(n \geq 2)$ have already been defined, then find the neighbors of r_n among $r_0, r_1, \dots, r_{n-1}$. Let them be r_k and r_l: $r_k < r_n < r_l$, $0 \leq k,l \leq n-1$. Apply the Tychonoff lemma to the sets $F_k^{(1)}$ and $E - G_l^{(2)}$: by $r_k < r_l$ and the induction hypothesis, $F_k^{(1)} \cap (E - G_l^{(2)}) = \emptyset$, so by the the Tychonoff lemma, there exist $G_n^{(2)}$ and $G_{n*}^{(1)}$ with $G_n^{(2)} \cap G_{n*}^{(1)} = \emptyset$ and

$$G_n^{(2)} \supset F_k^{(1)}, \quad G_{n*}^{(1)} \supset E - G_i^{(2)},$$

that is, using the notation $F_n^{(1)} = E - G_{n*}^{(1)}$,

$$F_n^{(1)} \subset G_l^{(2)},$$

and $G_n^{(2)} \cap G_{n*}^{(1)} = \emptyset$ means that

$$G_n^{(2)} \subset F_n^{(1)}.$$

It is clear that the validity of the induction hypothesis is inherited to these sets, that is, the recursive definition is correct.

For $x \in E$, put

$$g(x) = \inf(\{r_n : x \in G_n^{(2)}\} \cup \{1\}),$$
$$f(x) = \sup(\{r_n : x \notin F_n^{(1)}\} \cup \{0\}),$$

$(0 \leq f(x), g(x) \leq 1)$. It is clear that

$$f|F^{(1)} = g|F^{(1)} \equiv 0, \quad f|E - G^{(2)} = g|E - G^{(2)} \equiv 1.$$

Further, if $n \geq 2$, then

$$g(x) < r_n \quad \Rightarrow \quad x \in G_n^{(2)} \quad \Rightarrow \quad x \in F_n^{(1)} \quad \Rightarrow \quad f(x) \leq r_n,$$

so $f(x) \leq g(x)$ for all $x \in E$. But if $f(x) < g(x)$ for some $x \in E$, then there would exist $r_n, r_m \in D$ such that

$$f(x) < r_n < r_m < g(x).$$

Then $x \in F_n^{(1)}$ and $x \notin G_m^{(2)}$, but $F_n^{(1)} \subset G_m^{(2)}$, a contradiction. Therefore, $f(x) = g(x)$ on E.

For $x, y \in E$, put

$$\rho(x,y) = \begin{cases} 0, & \text{if } f(x) \geq f(y), \\ f(y) - f(x), & \text{if } f(x) < f(y). \end{cases}$$

Obviously, $0 \leq \rho(x,y) \leq 1$, $\rho(x,x) = 0$, and (2) and (5) hold. To prove (3) and (4), by (2) it suffices to show that $p \in \operatorname{int}^{(i)} V_\rho^{(i)}(p,\varepsilon)$ ($p \in E$, $\varepsilon > 0$) ($i = 1, 2$).

For $i = 1$:

Let $\varepsilon > 0$, $p \in E$. Then

$$V_\rho^{(1)}(p,\varepsilon) = \{x : \rho(x,p) < \varepsilon\} = \{x : f(p) - f(x) < \varepsilon\} \supset E - F_n^{(1)} \ni p$$

if n is chosen so that $f(p) - \varepsilon < r_n < f(p)$. In case $f(p) = 0$, $V_\rho^{(1)}(p,\varepsilon) = E \ni p$. $E - F_n^{(1)}$ and E are $\mathcal{T}_1$-open.

For $i = 2$:

$$\begin{aligned} V_\rho^{(2)}(p,\varepsilon) &= \{x : \rho(x,p) < \varepsilon\} = \{x : f(x) - f(p) < \varepsilon\} \\ &= \{x : g(x) - g(p) < \varepsilon\} \supset G_n^{(2)} \ni p \end{aligned}$$

if $g(p) < r_n < g(p) + \varepsilon$. In case $g(p) = 1$, $V_\rho^{(2)}(p,\varepsilon) = E$.

This proves the Urysohn lemma.

We turn now to the proof of the theorem.

Let P be the set of all pairs $(U^{(i)}, V^{(i)})$, where $U^{(i)}$ and $V^{(i)}$ are elements of the countable basis for the topology $\mathcal{T}_i$ and

$$\overline{U^{(i)}}^{(3-i)} \subset V^{(i)} \quad (\text{for } i = 1, 2).$$

Then P is countable:

$$P = \{(U_1, V_1), (U_2, V_2), \dots\}.$$

Now for $k = 1, 2, \dots$ we define a function $\rho_k(x,y)$.

Put $(U_k, V_k) = (U_k^{(i)}, V_k^{(i)})$. If $i = 2$, then apply the Urysohn lemma to the pair of sets $F^{(1)} = \overline{U_k^{(2)}}^{(1)}$, $G^{(2)} = V_k^{(2)}$, and call ρ_k the function ρ given by the lemma. If $i = 1$, then put $F^{(2)} = \overline{U_k^{(1)}}^{(2)}$, $G^{(1)} = V_k^{(1)}$, and apply the version of the Urysohn lemma where the superscripts (1) and (2) are interchanged and $\rho(a,b)$ is replaced with $\rho(b,a)$ everywhere. (Then (3) and (4) simply interchange and (2) remains the same.) Call ρ_k the function ρ thus obtained.

Put

$$\sigma(x,y) = \sum_{k=1}^{\infty} \frac{1}{2^k} \rho_k(x,y).$$

It is clear that $0 \le \sigma(x,y) \le 1$, $\sigma(x,x) = 0$ and, by (2), $\sigma(x,z) \le \sigma(x,y) + \sigma(y,z)$ holds for all $x, y \in E$.

We prove that for $\varepsilon > 0$ and $p \in E$

$$p \in \operatorname{int}^{(i)} V_i(p,\varepsilon) \quad (i = 1,2).$$

Let N be sufficiently large so that $\sum_{k=N}^{\infty} \frac{1}{2^k} < \frac{\varepsilon}{2}$. Then

$$V_i(p,\varepsilon) \supset \bigcap_{k=1}^{N} V_{\rho_k}^{(i)}\left(p, \frac{\varepsilon}{2}\right).$$

But, by (3) and (4), the right-hand side is a $\mathcal{T}_i$-open set containing p.

It only remains to show that if $p \in \operatorname{int}^{(i)} U$, then $p \in V_i(p,\varepsilon) \subset U$ for some $\varepsilon > 0$. Since $p \in \operatorname{int}^{(i)} U$, there is a $G^{(i)}$, then, by the assumptions, there is a $G'^{(i)}$ in the countable basis for $\mathcal{T}_i$, such that

$$p \in G'^{(i)}, \quad \overline{G'^{(i)}}^{(3-i)} \subset G^{(i)} \subset U.$$

Therefore, $(G'^{(i)}, G^{(i)}) = (U_k^{(i)}, V_k^{(i)}) \in P$ for some k. So, using $p \in G'^{(i)}$ and (5),

$$V_i\left(p, \frac{1}{2^k}\right) \subset V_{\rho_k}^{(i)}(p,1) \subset G^{(i)}. \qquad \square$$

Problem T.9. *Prove that there exists a topological space T containing the real line as a subset, such that the Lebesgue-measurable functions, and only those, extend continuously over T. Show that the real line cannot be an everywhere-dense subset of such a space T.*

Solution.

1. Let $\{f_i : i \in I\}$ be the set of all Lebesgue-measurable functions, X_i be a copy of the real line with the usual topology ($i \in I$), and

$$T - \prod_{i \in I} X_i.$$

If $h : \mathbb{R} \to T$ is the map defined by

$$p_i(h(x)) = f_i(x),$$

where p_i is the projection onto X_i, then h is injective and $h(\mathbb{R})$ can be considered as a copy of the real line. We prove that T is as required.

Let f be a measurable function, say, $f = f_i$. Then $f = p_i | h(\mathbf{R})$ and a continuous extension of this over T is p_i. Conversely, if $k : T \to \mathbb{R}$ is continuous, then $k|h(\mathbb{R})$ is measurable. Indeed, by a well-known theorem (see, for example, *R. Engelking, Outline of General Topology, PWN–Polish Sci. Publ, Warsaw, 1968, p. 98*, Problem R), k only depends on countably many coordinates, that is, $k = k' \circ p$, where $p : T \to T'$ is the projection onto the product $T' = \prod_{i \in I'} X_i$, where $I' \subseteq I$ is countable and $k' : T' \to \mathbb{R}$ is continuous. Now,

$$\{x : k(h(x)) \geq c\} = \{x : k'(p(h(x))) \geq c\} = \{x : p(h(x)) \in U_c\}$$

where U_c is open in T'. The sets of the form

$$\prod_{i \in I'} Y_i,$$

where $Y_i = X_i$ with a finite number of exceptions when Y_i is an interval with rational endpoints, form a countable basis in T'. Therefore, the set $\{x : k(h(x)) \geq c\}$ can be written as a countable union of sets of the form $\{x : p(h(x)) \in \prod_{i \in I'} Y_i\}$. But

$$\left\{x : p(h(x)) \in \prod_{i \in I'} Y_i\right\} = \{x : f_i(x) \in Y_i \ (i \in I)\},$$

being a union of countably many measurable sets, is measurable. Thus, $\{x : k(h(x)) \geq c\}$ is measurable.

2. Suppose that T has the required properties and $\mathbb{R} \subset T$ is dense. Since all measurable functions are continuous, it is obvious that the topology of T induces the discrete topology on $\mathbb{R}$. Let $p \in T$. Let f^* denote the extension of the function $f(x) = x$ over T (this is unique since $\mathbb{R}$ is dense in T). Further, consider the function

$$g(x) = \begin{cases} 0, & \text{if } x = f^*(p), \\ \frac{1}{x - f^*(p)}, & \text{if } x \neq f^*(p), \end{cases}$$

and let g^* be its continuous extension over T. Put

$$U_p = \left\{q : |g^*(p) - g^*(q)| < 1, \ |f^*(p) - f^*(q)| < \frac{1}{g^*(q) + 1}\right\}.$$

Let $x \in U_p \cap \mathbb{R}$ (since $\mathbb{R}$ is dense, such an x exists); we show that $x = f^*(p)$. Indeed, if $x \neq f^*(p)$, then

$$|g(x)| = \left|\frac{1}{x - f^*(p)}\right| > |g^*(p)| + 1,$$

which contradicts that $|g^*(p) - g^*(x)| < 1$. So $U_p \cap \mathbb{R} = \{f^*(p)\}$. From this it also follows that $f^*(q) = f^*(p)$ for every $q \in U_p$; indeed, $U_q \cap U_p \cap \mathbb{R}$ is nonempty (since $\mathbb{R}$ is everywhere dense), but its only element must equal both $f^*(q)$ and $f^*(p)$. So $f^*(p) = f^*(q)$.

Now for an arbitrary, nonmeasurable, real function φ, the function

$$\varphi^* = \varphi \circ f^*$$

is a continuous extension of φ, because this function, being constant on the neighborhood U_p of any point $p \in T$, is continuous. Thus, nonmeasurable functions also have continuous extensions over T, and this is a contradiction. $\square$

Problem T.10. *Let A be a closed and bounded set in the plane, and let C denote the set of points at a unit distance from A. Let $p \in C$, and assume that the intersection of A with the unit circle K centered at p can be covered by an arc shorter than a semicircle of K. Prove that the intersection of C with a suitable neighborhood of p is a simple arc of which p is not an endpoint.*

Solution. Let ab be the minimal arc of K containing $K \cap A$ (possibly $a = b$). Introduce Cartesian coordinates in the plane so that p is the origin, a and b lie in the left half-plane symmetrically with respect to the horizontal axis, and in case $a \neq b$, b lies in the upper half-plane.

We claim that with a suitable $\delta > 0$, the intersection C_1 of C with the upper half-plane and with the disc of radius δ centered at p is a simple arc starting from p that only meets the horizontal axis at p. Then a similar statement is true for the lower half-plane, and this proves the theorem. In order to prove our claim, it suffices to show that, for any $r \leq \delta$, the set C meets the upper semicircle K_r of radius r centered at p in a single interior point. In this case, assigning to points of C_1 their distance from p is a topological map from the compact set C_1 onto the interval $[0, \delta]$, since it is continuous and bijective. Let T be a convex plane sector, containing a and b in its interior, with vertex at p, and with the horizontal coordinate axis as axis of symmetry. Then, obviously, $\varrho(p, A - T) = 1 + \varepsilon_1 > 1$. Further, there is an $\varepsilon_2 > 0$ such that if $x \in T$, say, in the upper half-plane, and $\varrho(x, p) < \varepsilon_2$, then the triangle pxb has an obtuse angle at p; Therefore, $\varrho(x, b) < \varrho(p, b) = 1$, and so $\varrho(x, A) < 1$.

Now let $\delta < \min(\varepsilon_1, \varepsilon_2)$. If $r < \delta$, then $K_r \cap C_1$ is nonempty. Indeed, let c and d be the endpoints of K_r in the left and right half-planes, respectively. Then $\varrho(d, A) < 1$ by the above. On the other hand, if $x \in A - T$, then

$$\varrho(x, c) \geq \varrho(p, x) - \varrho(p, c) \geq 1 + \varepsilon - \delta > 1;$$

and if $x \in A \cap T$, then the triangle xpc has an obtuse angle at p, so

$$\varrho(c, x) > \varrho(p, x) > 1.$$

Thus, $\varrho(c, A) > 1$. By continuity of ϱ, there is a point y on the arc K_r for which $\varrho(y, A) = 1$, that is, $y \in C_1$. It also follows that $c, d \notin C_1$.

On the other hand, suppose that $y_1, y_2 \in C_1 \cap K_r$. By the above, $y_1, y_2 \notin T$. Let y_1 be the one of y_1, y_2 that is closer to d, and let $t \in A$ be a point such that $\varrho(y_2, t) = 1$. Then $t \in T$, because

$$\varrho(p, t) \leq \varrho(p, y_2) + \varrho(y_2, t) \leq 1 + \delta < 1 + \varepsilon_1.$$

Since $\varrho(y_1, t) \geq 1 = \varrho(y_2, t)$, the point t is on the same side of the perpendicular bisector of the segment y_1y_2 (which goes through p) as y_2. Therefore, the angle $y_2pt\measuredangle$, measured in positive orientation, is less than 180 degrees and obviously is greater than 90 degrees. Then the triangle y_2pt has an obtuse angle at p, and so $\varrho(y_2, t) > \varrho(p, t) \geq 1$, a contradiction. $\square$

Problem T.11. *Suppose that τ is a metrizable topology on a set X of cardinality less than or equal to continuum. Prove that there exists a separable and metrizable topology on X that is coarser than τ.*

Solution 1. We say that a set A is separated by a family of subsets if for any $a, b \in A$, $a \neq b$, there exists a set in the family that contains a but does not contain b. If the cardinality of A is at most continuum, then A is always separated by a countable family of subsets. Indeed, embed A into R and take the subsets corresponding to intervals with rational endpoints.

First, we show that there exists a sequence of closed subsets in X that separates X. It is well known that every metrizable space admits a σ-discrete basis. Let $B = \cup_{n=1}^{\infty} B_n$ be a basis in X such that B_n consists of pairwise disjoint, nonempty sets. Then the cardinality of B_n is less than or equal to continuum, so we can take a family $\{B_n^i : i = 1, 2, \dots\}$ (a family of families of sets) that separates B_n. Let U_n^i be the union of the elements of B_n^i; then U_n^i is an open set with respect to the topology τ.

The family $\{U_n^i\}$ separates X. Indeed, let $x, y \in X$, $x \neq y$. Since B is a basis, there is an n and $V \in B_n$ such that $x \in V$ and $y \notin V$. There may or may not exist a $V' \in B_n$ with $y \in V'$, but there is at most one such V', since B_n is a disjoint family. Choose a B_n^i such that $V \in B_n^i$ and $V' \notin B_n^i$ (or, an arbitrary B_n^i containing V if there is no V'). Then, obviously, $x \in U_n^i$ and $y \notin U_n^i$. As the family $\{U_n^i\}$ separates X, so does the family consisting of the complements of the sets U_n^i. Arrange these complements into a sequence $F_1, F_2, \dots$. Then $\{F_n\}$ is a family that separates X and consists of closed sets.

Let d be a bounded metric on X that induces the topology τ. Consider the following map $f : X \to l^1$:

$$f(x) = \left(\frac{d(F_1, x)}{2^1}, \frac{d(F_2, x)}{2^2}, \dots \right).$$

Then f is injective, because if $x \neq y$ there would exist an F_n with $x \in F_n$ and $y \notin F_n$, and then $d(F_n, x) = 0 \neq d(F_n, y)$. The map f is continuous, for if D denotes the distance in l^1, then

$$D(f(x), f(y)) = \sum_{n=1}^{\infty} 2^{-n} |d(F_n, x) - d(F_n, y)| \leq \sum_{n=1}^{\infty} 2^{-n} d(x, y) = d(x, y).$$

Consider now the topology on X determined by f as the inverse image topology. Since l^1 is a separable metric space (and separability is a hereditary property of metric spaces) and f is injective, this topology is separable and metrizable and, since f is continuous, it is coarser than τ. $\square$

Solution 2. For any cardinal number $\mathfrak{m} > 0$, we denote by $J(\mathfrak{m})$ the following metric space: the points of $J(\mathfrak{m})$ are the pairs (i, x), where $i \in I$

(I is an arbitrary set of cardinality $\mathfrak{m}$) and $x \in [0,1]$, with the points of the form $(i, 0)$ identified. The metric d of $J(\mathfrak{m})$ is defined as

$$d((i,x),(j,y)) = \begin{cases} x+y, & \text{if } i \neq j, \\ |x-y|, & \text{if } i = j. \end{cases}$$

In the proof, we use the theorem that asserts that every metric space of weight $\mathfrak{m}$ is topologically embeddable into the topological product of countably many copies of the space $J(\mathfrak{m})$. (See, for example, *R. Engelking, Outline of General Topology, PWN–Polish Sci. Publ, Warsaw, 1968, p. 197*, Theorem 7.)

Let P denote the following property of topological spaces (X, τ): there exists on X a separable metrizable topology coarser than τ. It is obvious that all subspaces of a space with property P and topological products of countable families of spaces with property P also have property P. It is also clear that metric spaces satisfying the hypotheses of the problem are of weight less than or equal to $\mathfrak{c}$ (continuum). Since, by the theorem cited above, all metric spaces of weight $\leq \mathfrak{c}$ are embeddable into the topological product of countably many copies of $J(\mathfrak{c})$, it suffices to show that $J(\mathfrak{c})$ has property P.

It is obvious that $J(\mathfrak{c})$ is homeomorphic to the metric space whose underlying set is the unit disc and in which the distance between two points on the same radius is their usual distance, and the distance between two points on different radii is the sum of the norms of the points. The usual topology of the unit disc is coarser than this topology and is separable and metrizable. □

Problem T.12. *Suppose that all subspaces of cardinality at most $\aleph_1$ of a topological space are second-countable. Prove that the whole space is second-countable.*

Solution. We argue by contradiction. Suppose that the topological space X is not second-countable but that all subspaces of cardinality $\leq \aleph_1$ are. We construct a certain subspace Y of cardinality $\leq \aleph_1$, and second-countability of Y will lead to a contradiction.

The subspace Y is the union of subspaces Y_α ($\alpha < \omega_1$) to be defined by transfinite recursion. Together with the Y_α, we simultaneously define a family $\mathcal{G}_\alpha$ of open sets in X so that $\mathcal{G}_\alpha | Y_\alpha$ is a basis for Y_α. (If $\mathcal{H}$ is a family of subsets in X, and $Z \subset X$, then $\mathcal{H}|Z = \{H \cap Z : H \in \mathcal{H}\}$ is the trace of $\mathcal{H}$ in Z.) The recursion is as follows: $Y_0 = \mathcal{G}_0 = \emptyset$. If the countable Y_ξ and $\mathcal{G}_\xi$ have already been defined for $\xi < \alpha < \omega_1$, then define Y_α and $\mathcal{G}_\alpha$ the following way. The family $\cup_{\xi<\alpha}\mathcal{G}_\xi$, being countable, is not a basis for X, therefore there exists a point $p \in X$ and a neighborhood V of p such that, for any element G of $\cup_{\xi<\alpha}\mathcal{G}_\xi$ containing p, we have $G \not\subset V$. Choose a point $p_G \in G \setminus V$ for each such G, and put

$$Y_\alpha = \bigcup_{\xi<\alpha} Y_\xi \cup \{p\} \cup \left\{ p_G : p \in G \in \bigcup_{\xi<\alpha} \mathcal{G}_\xi \right\}.$$

It is clear that Y_α is countable; therefore, by the assumptions, it is second-countable. Let $\mathcal{G}_\alpha$ be a countable family of open sets in X such that $\mathcal{G}_\alpha | Y_\alpha$ is a basis in Y_α.

Consider the subspace $Y = \cup_{\alpha<\omega_1} Y_\alpha$. It obviously has cardinality $\leq \aleph_1$, so it is second-countable. The crucial observation now is that $\cup_{\alpha<\omega_1} \mathcal{G}_\alpha | Y$ is a basis for Y. Let us postpone the proof of this statement and show first how to complete the solution, assuming that this observation is valid.

Since Y is second-countable, the basis $\cup_{\alpha<\omega_1} \mathcal{G}_\alpha | Y$ contains a countable basis. Therefore, for a sufficiently large α, the family $\cup_{\xi<\alpha} \mathcal{G}_\xi | Y$ is a basis for Y. Then, with this α, $\cup_{\xi<\alpha} \mathcal{G}_\xi | Y_\alpha$ is a basis for Y_α. But this is a contradiction since $\cup_{\xi<\alpha} \mathcal{G}_\xi | Y_\alpha$ does not contain a basis of neighborhoods for the point p used in the definition of Y_α.

It only remains to show that $\cup_{\alpha<\omega_1} \mathcal{G}_\alpha | Y$ is a basis for Y. This statement is immediate from the following lemma.

Let Y be a second-countable topological space, and let $Y = \cup_{\alpha<\omega_1} Y_\alpha$, where $Y_\alpha \subset Y_\beta$ for $\alpha < \beta$. For $\alpha < \omega_1$, let $\mathcal{G}_\alpha$ be a family of open sets in Y such that $\mathcal{G}_\alpha | Y_\alpha$ is a basis for Y_α. Then $\cup_{\alpha<\omega_1} \mathcal{G}_\alpha$ is a basis for Y.

The proof is again by contradiction. Suppose that $\cup_{\alpha<\omega_1} \mathcal{G}_\alpha$ is not a basis for Y. We construct a sequence $\{V_\alpha : \alpha < \omega_1\}$ of open sets with the property $V_\beta \setminus \cup_{\alpha>\beta} V_\alpha \neq \emptyset$ $(\beta < \omega_1)$, but such a sequence cannot exist in a second-countable space.

The sets V_α, together with a sequence of points $r_\alpha \in V_\alpha$, are defined by transfinite recursion as follows. Since $\cup_{\alpha<\omega_1} \mathcal{G}_\alpha$ is not a basis for Y, there exists a point $q \in Y$ and a neighborhood U of q such that $G \not\subset U$ whenever $q \in G \in \cup_{\alpha<\omega_1} \mathcal{G}_\alpha$. Put $r_0 = q$ and $V_0 = Y$. If Y_ξ and r_ξ have already been defined for $\xi < \alpha < \omega_1$, then first choose a $\rho < \omega_1$ with $Y_\rho \supset \{r_\xi : \xi < \alpha\}$. Since $\mathcal{G}_\rho | Y_\rho$ is a basis for Y_ρ, there exists a set $V_\alpha \in \mathcal{G}_\rho$ such that $q \in Y_\rho \cap V_\alpha \subset U$. On the other hand, $V_\alpha \not\subset U$; therefore we can choose a point $r_\alpha \in V_\alpha \setminus U$. Thus, the sequences $\{V_\alpha\}$, $\{r_\alpha\}$ are defined.

Notice that if $\beta < \alpha$, then $r_\beta \neq V_\alpha$. Indeed, $r_\beta \in Y_\rho \setminus U$ while $V_\alpha \cap Y_\rho \subset U$. Thus, $r_\beta \in V_\beta \setminus \cup_{\alpha>\beta} V_\alpha$ $(\beta < \omega_1)$. But this contradicts that Y admits a countable basis $\{B_1, B_2, \dots\}$. Indeed, for each α choose a B_{n_α} with $r_\alpha \in B_{n_\alpha} \subset V$. Since there are only countably many B_n, there exists $\alpha > \beta$ with $n_\alpha = n_\beta = n$. But then $r_\beta \in B_n$ and $r_\beta \neq V_\alpha \supset B_n$, which is impossible. $\square$

Remark. Unfortunately, the phrase "at most" was missing from the text of the problem when it was posed for the competition. The statement is not true in that version; a counterexample is provided by the set of natural numbers endowed with the topology in which the nonempty, open sets are the sequences of density 1. This is not a second-countable space, but the hypothesis is vacuously true. The contestants usually noticed this defect and considered the correctly modified version of the problem.

Seven solutions were submitted. Best results were obtained by Emil Kiss. He proved the statement for T_3-spaces among other arguments that contained all the ideas necessary to prove the statement in full. Nándor Simányi proved the statement assuming the continuum hypothesis and regularity of the space. Vilmos Totik used the hypothesis that every point has a basis of neighborhoods of cardinality $\aleph_1$, and Zoltán Szabó assumed first-countability of the space.

Problem T.13. *Suppose that the T_3-space X has no isolated points and that in X any family of pairwise disjoint, nonempty, open sets is countable. Prove that X can be covered by at most continuum many nowhere-dense sets.*

Solution 1. By transfinite recursion on $\alpha < \omega_1$, we define a family of open sets in X indexed by sequences of length α consisting of natural numbers, that is, by functions $f : \alpha \to \omega$. For $\alpha = 0$, the only such function is the empty set; put $G_\emptyset = X$. If α is a limit ordinal and $f : \alpha \to \omega$, then put $G_f = \operatorname{int}(\cap_g G_g)$, where g ranges over restrictions of f to all smaller ordinals. If α has the form $\alpha = \beta + 1$ and $G_f = \emptyset$ for some $f : \beta \to \omega$, then we define $G_g = \emptyset$ for all extensions $g : \beta \to \omega$ of f. If $G_f \neq \emptyset$, then we can define a sequence $\{H_1, H_2, \dots\}$ of open sets such that $\overline{H_n} \subset G_f$, $H_n \cap H_m = \emptyset$, $H_n \neq \emptyset$ for all $n \neq m$, and $H_1, H_2, \dots$ is a maximal system of nonempty, open sets in G_f. (Here we used regularity of X and the property formulated in the problem. It is easy to avoid that this system is finite.) Let the sets $H_1, H_2, \dots$ be indexed with the extensions $g_1, g_2, \dots$ of f over $\beta+1$; the cardinality of these is also countably infinite: $\{H_1, H_2, \dots\} = \{G_{g_1}, G_{g_2}, \dots\}$. So the sets G_f are defined for all functions $f : \alpha \to \omega$ $(\alpha < \omega_1)$. It is obvious that $G_f \cap G_g = \emptyset$ if $f, g : \alpha \to \omega$ and $f \neq g$, and that $G_f \supseteq G_g$ if $f \subset g$.

First, we prove that there is no $p \in X$ that is contained in G_f with some $f : \alpha \to \omega$ for all $\alpha < \omega_1$. Indeed, let us suppose the contrary. Then, by our preceding remarks, for all $\alpha < \omega_1$ there exists precisely one function $f_\alpha : \alpha \to \omega$ with $p \in G_{f_\alpha}$, and these functions are restrictions of a single function $F : \omega_1 \to \omega$. Then none of the sets G_{f_α} is empty, and if we define the functions $g_\alpha : \alpha + 1 :\to \omega$ by $g_\alpha(\xi) = f_\alpha(\xi)$ $(\xi < \alpha)$ and $g_\alpha(\alpha) = f_{\alpha+1}(\alpha) + 1$, then the sets G_{g_α} are pairwise disjoint, nonempty, open sets, which contradicts the assumptions.

Thus, each point p "dies out" at a certain level below ω_1. Therefore, the space X can be covered as follows:

$$X = \bigcup_{\alpha<\omega_1} \bigcup_{f:\alpha\to\omega} \left(G_f - \bigcup_{\substack{g:\alpha+1\to\omega \\ g\supset f}} G_g \right) \cup \bigcup_{\substack{\alpha<\omega_1 \\ \text{limit}}} \bigcup_{f:\alpha\to\omega} \left(\bigcap_{g\subsetneq f} G_g - G_f \right).$$

It is immediate that this is a covering by $2^{\aleph_0}$ sets. It remains to show that the summands are nowhere-dense sets. The set

$$G_f - \bigcup_{\substack{g:\alpha+1\to\omega \\ g\supset f}} G_g$$

is nowhere dense because otherwise it would have a nonempty interior, and the family of the sets G_g would not be maximal. For the other type of summands, denoting the restriction of f to β by g_β for $\beta < \alpha$, we have

$$\overline{\bigcap_{\beta<\alpha} G_{g_\beta}} \subseteq \bigcap_{\beta+1<\alpha} \overline{G_{g_{\beta+1}}} \subseteq \bigcap_{\beta<\alpha} G_{g_\beta},$$

that is, the summand is a closed set minus its interior, which, being the boundary of a closed set, is nowhere dense. □

Solution 2. Let $\{G_\alpha : \alpha < \lambda\}$ be a well-ordering of the set of nonempty, open sets of X. Since there are no isolated points, for each $\alpha < \lambda$ there exists a pair $G_\alpha^{(0)}$, $G_\alpha^{(1)}$ of nonempty, open sets with $G_\alpha^{(0)} \cap G_\alpha^{(1)} = \emptyset$ and $G_\alpha^{(0)} \cup G_\alpha^{(1)} \subseteq G_\alpha$. Let $p \in X$ be arbitrary, and define the set $H(\alpha, p)$ by transfinite recursion on α as follows. Suppose that the sets $H(\beta, p)$ are already defined for $\beta < \alpha$. If $(\cup_{\beta<\alpha} H(\beta, p)) \cap G_\alpha \neq \emptyset$, put $H(\alpha, p) = \emptyset$. If $(\cup_{\beta<\alpha} H(\beta, p)) \cap G_\alpha = \emptyset$ and $p \in G_\alpha^{(0)}$, put $H(\alpha, p) = G_\alpha^{(1)}$; otherwise put $H(\alpha, p) = G_\alpha^{(0)}$. Finally, let $F_p = X - \cup_{\alpha<\lambda} H(\alpha, p)$. It is obvious that F_p is a closed set containing p; moreover, it is nowhere dense, because if it contained a nonempty, open set, then it would contain another nonempty, open set, say, G_α, which would not contain p, but then, by the construction, $G_\alpha^{(0)}$ would not be a subset of F_p.

It remains to prove that the number of different F_p's cannot be greater than continuum.

For each p there can only be countably many α with $H(\alpha, p) \neq \emptyset$, since the sets $H(\alpha, p)$ are pairwise disjoint. Let the set of these be $\{\alpha_\xi(p) : \xi < \varphi\}$, where $\varphi = \varphi(p) < \omega_1$. We claim that if p and q are points with $\varphi(p) = \varphi(q)$ such that, for every $\xi < \varphi(p)$, $H(\alpha_\xi(p), p) = G^{(i)}_{\alpha_\xi(p)}$ implies $H(\alpha_\xi(q), q) = G^{(i)}_{\alpha_\xi(q)}$ (that is, for each $\xi < \varphi(p)$, the points p and q "ramify in the same direction"), then $F_p = F_q$. Taking this statement for granted, we can complete the proof by the observation that the cardinality of 0–1 sequences indexed by countable ordinals is continuum; therefore the number of different F_p's is at most continuum. It remains to prove our claim. It will suffice to show, by transfinite induction, that $\alpha_\xi(p) = \alpha_\xi(q)$. If $\alpha_\zeta(p) = \alpha_\zeta(q)$ holds for all $\zeta < \xi$, then obviously $H(\alpha_\zeta(p), p) = H(\alpha_\zeta(q), q)$, and $\alpha_\xi(p)$ is the first ordinal $\alpha \geq \sup\{\alpha_\zeta : \zeta < \xi\}$ with $G_\alpha \cap (\cup_{\zeta<\xi} H(\alpha_\zeta, p)) = \emptyset$; similarly, $\alpha_\xi(q)$ is the first α with $G_\alpha \cap (\cup_{\zeta<\xi} H(\alpha_\zeta, q)) = \emptyset$, and therefore $\alpha_\xi(p) = \alpha_\xi(q)$. □

Remark. János Kollár generalized the statement of the problem; by a modified version of Solution 1, he proved the statement for T_2-spaces. In fact, Solution 2 does not use regularity, but this was not noticed by the author.

Problem T.14. *Construct an uncountable Hausdorff space in which the complement of the closure of any nonempty, open set is countable.*

Solution 1. Let X be an arbitrary uncountable set. It is easy to see that there exists a topology on X with the required properties if and only if there exists a family $\mathcal{A}$ of subsets of X that satisfies the following conditions:

(1) $|A| = \omega$ or $A = \emptyset$ for all $A \in \mathcal{A}$;
(2) $\mathcal{A}$ is closed under finite intersections;
(3) for every pair $a, b \in \mathcal{A}$, $a \neq b$, there exist $A, B \in \mathcal{A}$ such that $a \in A$, $b \in B$ and $A \cap B \neq \emptyset$;
(4) if $\mathcal{A}' \subset \mathcal{A}$ and $A \in \mathcal{A} \setminus \{\emptyset\}$, then $(\cup\mathcal{A}') \cap A = \emptyset$ implies $|\cup \mathcal{A}'| \leq \omega$.

Indeed, if $\mathcal{A}$ has properties (1)–(4), then X endowed with the topology generated by the basis $\mathcal{A}$ satisfies the requirements of the problem. Conversely, if τ is a topology required by the problem, then the family $\mathcal{A}$ of all countable τ-open sets satisfies the conditions (1)–(4).

Now choose X to be the set ω_1 of countable ordinals. Let $\{\langle b_\alpha, c_\alpha\rangle : \alpha \in \omega_1\}$ be a sequence of all pairs of elements of ω_1 with $b_\alpha \neq c_\alpha$. By transfinite recursion, we construct a sequence of families $\mathcal{A}_\alpha : \alpha \in \omega_1$ of subsets of ω_1 so that for all $\alpha \in \omega_1$, we have

(1_α) $|A| = \omega$ or $A = \emptyset$ for all $A \in \mathcal{A}_\alpha$,
(2_α) $\cup_{\beta<\alpha}\mathcal{A}_\beta \subset \mathcal{A}_\alpha$,
(3_α) $\mathcal{A}_\alpha$ is closed under finite intersections,
(4_α) $|\mathcal{A}_\alpha| \leq \omega$,
(5_α) there exist $B, C \in \mathcal{A}_\alpha$ with $b_\alpha \in B$, $c_\alpha \in C$, and $B \cap C = \emptyset$,
(6_α) if $A \in \mathcal{A}_\alpha \setminus \cup_{\beta<\alpha}\mathcal{A}_\beta$ and $A' \in \cup_{\beta<\alpha}\mathcal{A}_\beta$, then $|A \cap A'| = \omega$.

Assuming that we have constructed such a sequence, we show that $\mathcal{A} = \cup_{\alpha<\omega_1}\mathcal{A}_\alpha$ satisfies conditions (1)–(4). (1), (2), and (3) are obvious. In order to check (4), let $\mathcal{A}' \subset \mathcal{A}$ and $A \in \mathcal{A} \setminus \{\emptyset\}$, with $(\cup\mathcal{A}') \cap A = \emptyset$. Let $\alpha \in \omega_1$ be the smallest ordinal such that $A \in \mathcal{A}_\alpha$. We shall prove that $\cup\mathcal{A}' \subset \cup\mathcal{A}_\alpha$. To this end, it is sufficient to see that for each $A' \in \mathcal{A}'$, there exists $B \in \mathcal{A}_\alpha$ with $A' \subset B$. Let $A' \in \mathcal{A}'$ be arbitrary, and let $\beta \in \omega_1$ denote the smallest ordinal $B \in \mathcal{A}_\beta$ such that $B \supset A'$ and $B \cap A = \emptyset$. (The definition of β is correct since there exists $B \in \mathcal{A}$ with $B \supset A'$ and $B \cap A = \emptyset$; for example, $B = A'$ is such.) Then $\beta > \alpha$ would contradict (6_α), so $\beta \leq \alpha$, that is, $B \in \mathcal{A}_\beta \subset \mathcal{A}_\alpha$.

In order to construct the sequence $\{\mathcal{A}_\alpha : \alpha \in \omega_1\}$, we need the following simple lemma.

Lemma. If $|E| = \omega$ and ε is a nonempty, countable family of subsets of E, then there exist subsets B and C of E such that $B \cap C = \emptyset$ and $|B \cap A| = |C \cap A| = \omega$ for all $A \in \varepsilon$.

Proof. Indeed, put $\varepsilon = \{A_n : n \in \omega\}$ (possibly with multiple appearances), and for $n \in \omega$, let $A_n = \cup_{k\in\omega}A_{nk}$ be a partition of A_n into pairwise disjoint infinite sets. Arrange the sets A_{nk} $(n, k \in \omega)$ into a sequence $\{A'_n : n \in \omega\}$. Since each A'_n is infinite, by recursion we can define the elements $b_n, c_n \in \omega$ such that $b_n, c_n \in A'_n \setminus (\{b_k : k < n\} \cup \{c_k : k < n\})$ for all $n \in \omega$. Then $B = \{b_n : n \in \omega\}$ and $C = \{c_n : n \in \omega\}$ are as required.

Turning now to the construction of the sequence $\{\mathcal{A}_\alpha : \alpha < \omega_1\}$, let $B_0, C_0 \subset \omega_1$ be two arbitrary countably infinite sets such that $b_0 \in B_0$,

$c_0 \in C_0$ and $B_0 \cap C_0 = \emptyset$. Then $\mathcal{A}_0 = \{B_0, C_0, \emptyset\}$ satisfies conditions (1_0)–(6_0). Suppose that for some ordinal $\alpha \in \omega_1$, $\alpha > 0$, we already have defined the sequence $\{\mathcal{A}_\beta : \beta < \alpha\}$ so that conditions (1_β)–(6_β) are fulfilled for all $\beta < \alpha$. Then, by applying the lemma to $E = \cup_{\beta<\alpha}(\cup\mathcal{A}_\beta)$ and $\varepsilon = \cup_{\beta<\alpha}\mathcal{A}_\beta$, there exist countably infinite sets $B, C \subset \cup_{\beta<\alpha}(\cup\mathcal{A}_\beta)$ such that $|B \cap A| = |C \cap A| = \omega$ for all $A \in \cup_{\beta<\alpha}\mathcal{A}_\beta$. If $\mathcal{A}_\alpha$ is defined to consist of finite intersections from the family $\cup_{\beta<\alpha}\mathcal{A}_\beta \cup \{B \cup \{b_\alpha\}, C \cup \{c_\alpha\}\}$, then it is routine to check that conditions (1_α)–(6_α) are fulfilled, which completes the proof. □

Solution 2. Let D be a countable, dense subspace of the product space 2^{ω_1}, which exists by the Marczewski–Pondiczery theorem. Let $F = \{f_\alpha : \alpha \in \omega_1\} \subset 2^{\omega_1} \setminus D$ such that

$$f_\alpha|(\omega_1 \setminus \gamma) \neq f_\beta|(\omega_1 \setminus \gamma) \quad \text{and} \quad f|(\omega_1 \setminus \gamma) \neq f_\alpha|(\omega_1 \setminus \gamma)$$

for all $\alpha, \beta, \gamma \in \omega_1$, $\alpha \neq \beta$, and $f \in D$. Choose $D \cup F$ to be the underlying set of the space X to be defined; let D be open in X, and let its subspace topology coincide with the topology inherited from 2^{ω_1}.

It remains to define basis neighborhoods for f_α $(\alpha \in \omega_1)$. To this end, let $\mathcal{B} = \{B_\alpha : \alpha \in \omega_1\}$ be the family of elementary open sets in the product space 2^{ω_1}, let the finite sets T_α $(\alpha \in \omega_1)$ be the supports of the B_α, and put $\lambda_\alpha = \sup(\cup_{\beta<\alpha}T_\beta) + 1$. Then let the neighborhoods of f_α be the sets of the form $\{f_\alpha\} \cup (B \cap D)$, where B ranges over the elements of $\mathcal{B}$ that contain f_α and are supported in $\omega_1 \setminus \lambda$.

It is easy to check that we have defined a Hausdorff topology. By the construction, for all $B_\alpha \in \mathcal{B}$, we have

$$\mathrm{cl}_X(B_\alpha \cap D) \supset \{f_\beta : \beta \in \omega_1 \setminus (\alpha + 1)\},$$

and thus the closure of any nonempty, open set in X has a countable complement. □

Remark. It can be shown that the maximum cardinality of spaces satisfying the hypotheses of the problem is 2^ω.

Problem T.15. *Let W be a dense, open subset of the real line $\mathbb{R}$. Show that the following two statements are equivalent:*

(1) Every function $f : \mathbb{R} \to \mathbb{R}$ continuous at all points of $\mathbb{R} \setminus W$ and nondecreasing on every open interval contained in W is nondecreasing on the whole $\mathbb{R}$.

(2) $\mathbb{R} \setminus W$ is countable.

Solution. Suppose first that $F = \mathbb{R} \setminus W$ is uncountable. Then we construct a continuous function $f : \mathbb{R} \to \mathbb{R}$ that is constant on every subinterval of W and that is nonconstant and decreasing on the whole $\mathbb{R}$.

We may assume that F has no isolated points (otherwise we pass to the set of its points of condensation). Let $W = \cup\mathcal{V}$, where $\mathcal{V}$ is a countable family of pairwise disjoint, open intervals. Consider the natural ordering of $\mathcal{V}$ inherited from $\mathbb{R}$. Since F is nowhere dense and perfect, this ordering of $\mathcal{V}$ is dense (that is, between any two elements of $\mathcal{V}$, there exists a further element of $\mathcal{V}$); therefore, there exists an order-preserving map $q \mapsto V_q$ from the set Q of rational points of the interval $[0,1]$ into $\mathcal{V}$. For $x \in \mathbb{R}$, put

$$f(x) = 1 - \sup\{q \in Q : (-\infty, x] \cap V_q \neq \emptyset\}.$$

The function f is obviously constant on elements of $\mathcal{V}$, $f(x) = 1$ for $x \in V_0$, $f(x) = 0$ for $x \in V_1$, and f is monotone nonincreasing. Moreover, f takes all values $t \in [0,1]$, since $f(x) = t$ for $x = \sup \cup\{V_q : q \in Q \cap [0, 1-t]\}$. Therefore, f has no jumps and f is continuous.

Suppose now that $\mathbb{R} \setminus W = \{a_n : n \in \mathbb{N}\}$ is countable. Let $\varepsilon > 0$ be arbitrary; we show that for any pair $x, y \in \mathbb{R}$, $x \neq y$, we have $f(x) < f(y) + \varepsilon$.

Since f is continuous at the points a_n $(n \in N)$, for each $n \in N$ there exists an open neighborhood U_n of a_n such that for all $u, v \in U_n$ the inequality $f(u) - f(v) < \varepsilon/2^{n+1}$ holds. Let $W = \cup\mathcal{V}$, where $\mathcal{V}$ is a countable family of open intervals such that the restriction of f to any element of $\mathcal{V}$ is nondecreasing. The family $\mathcal{U} = \mathcal{V} \cup \{U_n : n \in N\}$ is an open cover of the real line $\mathbb{R}$; consider a finite subfamily $\mathcal{U}' = \mathcal{V}' \cup \{U_n : n \leq n_0\}$ of $\mathcal{U}$ such that $\mathcal{V}' \subset \mathcal{V}$ and $\mathcal{U}'$ covers the interval $[x, y]$. Let L be the set of points y' in $[x, y]$ for which there exists a finite sequence $x = z_0 < z_1 < \cdots < z_k = y'$ such that $[z_{i-1}, z_i] \subset U^{(i)}$ with some $U^{(i)} \in \mathcal{U}'$ for all $1 \leq i \leq k$. It is easy to see that L is an open-and-closed subset of $[x, y]$ containing x, and so $L = [x, y]$. Therefore, there exists such a sequence $z_0, \ldots, z_k$ with $z_k = y$. Then

$$f(x) - f(y) = \sum_{i=1}^{k} (f(z_{i-1}) - f(z_i)) < \sum_{n=0}^{n_0} \frac{\varepsilon}{2^{n+1}} < \varepsilon. \qquad \square$$

Remark. Statement (2)$\Rightarrow$(1) of the problem can also be proved by transfinite induction on isolated points of $\mathbb{R} \setminus W$. In connection with a few incorrect solutions, it seems worth noting that such an induction does not necessarily terminate at the first infinite ordinal.

Problem T.16. *Let $n \geq 2$ be an integer, and let X be a connected Hausdorff space such that every point of X has a neighborhood homeomorphic to the Euclidean space $\mathbb{R}^n$. Suppose that any discrete (not necessarily closed) subspace D of X can be covered by a family of pairwise disjoint, open sets of X so that each of these open sets contains precisely one element of D. Prove that X is a union of at most $\aleph_1$ compact subspaces.*

Solution. Let X be a space that satisfies the hypotheses of the problem. First, we prove the following.

Statement $(*)$. If F is an arbitrary subspace of X, then there exists a family $\mathcal{G}$ of open subsets of X homeomorphic to $\mathbb{R}^n$ such that $(\cup\mathcal{G}) \cap F$ is dense in the subspace F, and the family $\mathcal{G}$ is σ-disjoint, that is, $\mathcal{G}$ is a union of countably many families consisting of pairwise disjoint sets.

If $F = \emptyset$, then this statement is obvious. If $F \neq \emptyset$, then consider a maximal family $\mathcal{U}$ of pairwise disjoint, nonempty, separable, relative open subsets of the subspace F. (Such a family exists by Zorn's lemma.) For each $U \in \mathcal{U}$, let $S(U) = \{x_n(U) : n = 0, 1, \dots\}$ be a countable, dense subset of U. For each n, the set $S_n = \{x_n(U) : U \in \mathcal{U}\}$ is a discrete subspace of F and, thus, of X. By the hypothesis of the problem, there exists a family $\mathcal{G}_n = \{G_n(U) : U \in \mathcal{U}\}$ of pairwise disjoint, open subsets of X, such that $x_n(U) \in G_n(U)$ for all $U \in \mathcal{U}$. Since X is locally Euclidean, we may assume that $\mathcal{G}_n$ consists of sets homeomorphic to $\mathbb{R}^n$. Then the family $\mathcal{G} = \cup_{n=0}^{\infty}\mathcal{G}_n$ satisfies the requirements in $(*)$, since the closure of $(\cup\mathcal{G}) \cap F \supset \cup_{n=0}^{\infty} S_n$ in the subspace F contains the closure of $\cup\mathcal{U}$ in F, and, by maximality of $\mathcal{U}$, the set $\cup\mathcal{U}$ is dense in F.

We define a sequence $\{\langle F_\alpha, \mathcal{G}_\alpha\rangle : \alpha \in \omega_1\}$ by transfinite recursion as follows. (As usual, ω_1 denotes the set of all countable ordinals.) Let $F_0 = X$ and $\mathcal{G}_0$ be a maximal family of pairwise disjoint, open subsets of X homeomorphic to $\mathbb{R}^n$. If $\alpha \in \omega_1$, $\alpha > 0$ and $\{\langle F_\beta, \mathcal{G}_\beta\rangle : \beta < \alpha\}$ has already been defined, then put $F_\alpha = X \setminus \cup_{\beta<\alpha}(\cup\mathcal{G}_\beta)$, and let $\mathcal{G}_\alpha$ be a σ-disjoint family of open sets of X homeomorphic to $\mathbb{R}^n$ such that $(\cup\mathcal{G}_\alpha) \cap F_\alpha$ is dense in F_α. (The existence of such a family $\mathcal{G}_\alpha$ is guaranteed by $(*)$.)

We show that $X = \cup_{\alpha\in\omega_1}(\cup\mathcal{G}_\alpha)$. Arguing by contradiction, assume that there exists a point $x \in X \setminus \cup_{\alpha\in\omega_1}(\cup\mathcal{G}_\alpha)$, and consider a neighborhood V of x that is homeomorphic to $\mathbb{R}^n$. Since $(\cup\mathcal{G}_\alpha) \cap F_\alpha$ is dense in $F_\alpha = X \setminus \cup_{\beta<\alpha}(\cup\mathcal{G}_\beta)$, it follows that the sets $V_\alpha = (\cup_{\beta<\alpha}(\cup\mathcal{G}_\beta)) \cap V$ $(\alpha \in \omega_1)$ form a strictly monotone increasing sequence of order type ω_1 of open sets in V; this is impossible in a space homeomorphic to $\mathbb{R}^n$.

Therefore, the family $\mathcal{G}^* = \cup_{\alpha\in\omega_1}\mathcal{G}_\alpha$ is a covering of X by open sets homeomorphic to $\mathbb{R}^n$, and $\mathcal{G}^*$ is a union of at most $\aleph_1$ disjoint families. Since a subspace homeomorphic to $\mathbb{R}^n$ can only meet countably many disjoint open sets, each $G \in \mathcal{G}^*$ meets at most $\aleph_1$ members of $\mathcal{G}^*$.

Finally, consider the equivalence relation on $\mathcal{G}^*$ in which $G, G' \in \mathcal{G}^*$ are equivalent if and only if there exists a finite sequence $G_0, \dots, G_k \in \mathcal{G}^*$ such that $G_0 = G$, $G_k = G'$, and $G_i \cap G_{i+1} \neq \emptyset$ $(i = 0, 1, \dots, k-1)$. Using the result of the last paragraph, induction on k shows that every equivalence class contains at most $\aleph_1$ sets from $\mathcal{G}^*$. On the other hand, if $\mathcal{G} \subset \mathcal{G}^*$ is an equivalence class, then $\cup\mathcal{G}$ is open and closed in X, and then $\mathcal{G} = \mathcal{G}^*$ by connectedness of X. Therefore, the cardinality of $\mathcal{G}^*$ is at most $\aleph_1$.

Thus, X is a union of at most $\aleph_1$ subsets homeomorphic to $\mathbb{R}^n$, and therefore it is a union of at most $\aleph_1$ compact sets. □

Remarks.

1. The most well-known nonmetrizable topological manifold, the long line, is an example of a space that satisfies the hypotheses of the problem but is not a union of countably many compact subspaces.
2. Several contestants proved the statement of the problem using the assumption that all separable subspaces of X are second-countable. Gábor Moussong showed that there exists a model of ZFC where this assumption holds. Note, however, that by a construction of M. E. Rudin and P. H. Zenor in some other models of ZFC there exist separable and not second-countable topological manifolds that satisfy the hypotheses of the problem.
3. Gábor Moussong noticed that the statement of the problem follows from the continuum hypothesis, since the cardinality of a connected topological manifold of dimension ≥ 1 is continuum.

Problem T.17. *A map $F : P(X) \to P(X)$, where $P(X)$ denotes the set of all subsets of X, is called a closure operation on X if for arbitrary $A, B \subset X$, the following conditions hold:*
(i) $A \subset F(A)$;
(ii) $A \subset B \Rightarrow F(A) \subset F(B)$;
(iii) $F(F(A)) = F(A)$.
The cardinal number $\min\{|A| : A \subset X,\ F(A) = X\}$ is called the density of F and is denoted by $d(F)$. A set $H \subset X$ is called discrete with respect to F if $u \notin F(H - \{u\})$ holds for all $u \in H$. Prove that if the density of the closure operation F is a singular cardinal number, then for any nonnegative integer n, there exists a set of size n that is discrete with respect to F. Show that the statement is not true when the existence of an infinite discrete subset is required, even if F is the closure operation of a topological space satisfying the T_1 separation axiom.

Solution.

(a) We prove the statement by induction on n. For $n = 0$, the statement is obvious; assume that it is true for n. Let the density of the closure operation F on X be λ with $\mathrm{cf}(\lambda) = \kappa < \lambda$. Let $|A| = \lambda$, with $F(A) = X$, and well-order A in order type λ. By recursion, define a sequence $B = \{x_\xi : \xi < \lambda\}$ by $x_\xi = \min\left(A - F\left(\{x_\eta : \eta < \xi\}\right)\right)$. Then $F(B) \supset F(A) = X$. Let $C \subset B$ such that $|C| = \kappa$ and C is cofinal in B, and for $Y \subset B$ put $F'(Y) = F(Y \cup C) \cap B$.

It is easy to see that F' is a closure operation on B and that $d(F') = \lambda$. By the induction hypothesis, there exists a set $\{x_{\xi_i} : i < n\}$ that is discrete with respect to F'. Let $\xi_n > \xi_i$ $(i < n)$ be such that $x_{\xi_n} \in C$. Then the set $\{x_{\xi_i} : i \leq n\}$ is discrete since $x_{\xi_n} \notin F\left(\{x_{\xi_i} : i < n\}\right) \subset F\left(\{x_\eta : \eta < \xi_n\}\right)$, and $x_{\xi_i} \notin F\left(\{x_{\xi_j} : i \neq j,\ j \leq n\}\right) \subset F'\left(\{x_{\xi_j} : i \neq j,\ j < n\}\right)$ for $i < n$.

(b) Let λ be a singular, strong-limit cardinal number, and let $\mathcal{H} \subset P(\lambda)$ be a maximal, almost disjoint family of countable sets. (Such a family exists by Zorn's lemma.) Call $F \subset \lambda$ closed if, for each $H \in \mathcal{H}$, $|F \cap H| = \omega$ implies $F \supset H$. It is easy to see that by this a T_1 topological space is defined in which there are no infinite discrete subsets. It remains to prove that the density is λ. Let $A_0 \subset \lambda$, with $|A_0| < \lambda$. For $\xi \leq \omega_1$, we define a sequence A_ξ by the following recursion: if $\xi = \eta + 1$, then put $A_\xi = \cup\{H \in \mathcal{H} : |H \cap A_\eta| = \omega\}$, and if ξ is a limit ordinal, then put $A_\xi = \cup\{A_\eta : \eta < \xi\}$. Since $|A_{\eta+1}| \leq |A_\eta|^\omega$, it is easy to see that $|A_\xi| \leq |A_0|^\omega < \lambda$ for $\xi \leq \omega_1$. On the other hand, A_{ω_1} is obviously closed, and therefore A_0 is not dense. □

Problem T.18. *Suppose that K is a compact Hausdorff space and $K = \cup_{n=0}^\infty A_n$, where A_n is metrizable and $A_n \subset A_m$ for $n < m$. Prove that K is metrizable.*

Solution. First, we prove the following lemma.

Lemma. If a subspace H of a metric space A is not separable, then there exists an uncountable, discrete, closed subspace $Z \subset H$.

Proof. Let Z_n be a maximal subset of A such that $d(x, y) \geq 1/n$ for all $x, y \in Z_n$, $x \neq y$. If $a \in A$, then $d(a, x) < 1/2n$ can hold for at most one point of Z_n, so Z_n is discrete. On the other hand, by maximality of Z_n, for each $h \in H$ there exists $x \in Z_n$ such that $d(h, x) < 1/n$, so $\cup_{n=1}^\infty Z_n$ is dense. If H is not separable, then this implies that Z_n is uncountable for some n.

Now we prove that A_n is separable for all n. Indeed, assuming that it is not separable, there is an uncountable discrete closed subset Z_n in A_n; this set is dicrete in A_{n+1} and is not separable, so there exists an uncountable, discrete, closed subset Z_{n+1} in Z_n. Thus, we obtain a sequence Z_m such that $Z_{m_1} \supset Z_{m_2}$ for $m_1 < m_2$, and Z_m is an uncountable, discrete, closed subset of A_m. Since K is compact, the set Z'_m of accumulation points of Z_m is nonempty; therefore $\cap_{m=n}^\infty Z'_m \neq \emptyset$. On the other hand, $Z'_m \cap A_m = \emptyset$ since Z_m is discrete and closed in A_m, and therefore $\cap_{m=n}^\infty Z'_m \cap \cup_{m=n}^\infty A_m = \emptyset$, and which is a contradiction.

Fix k and consider the set $A_n \cap \overline{A_k}$ $(k \leq n)$, which is a separable metric space and therefore has a countable basis $\{B_{n,m} : m \in \omega\}$. Let $G_{n,m}$ be an open set in $\overline{A_k}$ such that $B_{n,m} = G_{n,m} \cap A_n$. The family $\{G_{n,m} : n > k,\ m \in \omega\}$ is a subbasis for a Hausdorff topology on $\overline{A_k}$, which is coarser than the compact topology of $\overline{A_k}$; therefore, these two topologies are equal. This means that $\overline{A_k}$ is a second-countable, compact topological space. Let $\{f_{n,k} : n \in \omega\}$ be a family of continuous functions on $\overline{A_k}$ that separates the points. Let $F_{n,k}$ be a continuous function on K for which $F_{n,k}|\overline{A_k} = f_{n,k}$. (Such functions $F_{n,k}$ exist by the Tietze theorem.) The functions $F_{n,k}$ separate the points of K; therefore the metric defined by these functions induces the topology of K. □

Problem T.19. *Let U denote the set $\{f \in C[0,1] : |f(x)| \le 1$ for all $x \in [0,1]\}$. Prove that there is no topology on $C[0,1]$ that, together with the linear structure of $C[0,1]$, makes $C[0,1]$ into a topological vector space in which the set U is compact.*

Solution. We assume that topological vector spaces are Hausdorff spaces.

Arguing by contradiction, assume that $C[0,1]$ is a topological vector space in which U is compact. Then U is closed since the topology is Hausdorff. For each $f \in C[0,1]$ and $\delta > 0$, the set

$$U_{f,\delta} = \{g \in C[0,1] : |g(x) - f(x)| \le \delta \text{ for all } x \in [0,1]\}$$

is also compact and closed.

Define the functions $f_n, g_n \in C[0,1]$ for each natural number $n \ge 2$ as

$$f_n(x) = \begin{cases} 0 & \text{if } 0 \le x \le \frac{1}{2} - \frac{1}{n}, \\ nx + 1 - \frac{n}{2} & \text{if } \frac{1}{2} - \frac{1}{n} < x < \frac{1}{2}, \\ 1 & \text{if } \frac{1}{2} \le x \le 1, \end{cases}$$

$$g_n(x) = \begin{cases} 0 & \text{if } 0 \le x \le \frac{1}{2}, \\ nx - \frac{n}{2} & \text{if } \frac{1}{2} < x < \frac{1}{2} + \frac{1}{n}, \\ 1 & \text{if } \frac{1}{2} + \frac{1}{n} \le x \le 1. \end{cases}$$

Put $K_n = \{g \in C[0,1] : g_n(x) \le g(x) \le f_n(x)$ for all $x \in C[0,1]\}$ $(n \ge 2)$. Then $K_2 \supset K_3 \supset \cdots$, and K_n is compact and closed since $K_n = U_{f_n - 1, 1} \cap U_{g_n - 1, 1}$. So $\bigcap_{n=2}^{\infty} K_n \ne \emptyset$. Let $f \in \bigcap_{n=2}^{\infty} K_n$. Then $f(x) = 0$ for $x < 1/2$, and $f(x) = 1$ for $x > 1/2$, which is impossible, since f is continuous. This contradiction shows that U cannot be compact. □

Problem T.20. *Let Φ be a family of real functions defined on a set X such that $k \circ h \in \Phi$ whenever $f_i \in \Phi$ $(i \in I)$ and $h : X \to \mathbb{R}^I$ is defined by the formula $h(x)_i = f_i(x)$, and*

(1) $k\colon h(X) \to \mathbb{R}$ is continuous with respect to the topology inherited from the product topology of $\mathbb{R}^I$. Show that $f = \sup\{g_j : j \in J, g_j \in \Phi\} = \inf\{h_m : m \in M, h_m \in \Phi\}$ implies $f \in \Phi$. Does this statement remain true if (1) is replaced with the following condition?

(2) $k\colon \overline{h(X)} \to \mathbb{R}$ is continuous on the closure of $h(X)$ in the product topology.

Solution. We may obviously assume that J and M are disjoint.

Define $p : X \to \mathbb{R}^{J \cup M}$ by $p(x)_j = g_j(x)$, for $j \in J$, and $p(x)_m = h_m(x)$, for $m \in M$. Then, for all $y \in p(X)$, we have $\sup\{y_j : j \in J\} = \inf\{y_m : m \in M\}$. Define $k : p(X) \to \mathbb{R}$ by $k(y) = \sup\{y_j : j \in J\} = \inf\{y_m : m \in M\}$. Then $f = k \circ p$, so, in view of (1), it suffices to show that k is continuous on $p(X)$.

Let $y \in p(X)$ and $\varepsilon > 0$. We shall define a neighborhood S of y in $p(X)$ such that $|k(z) - k(y)| < \varepsilon$ for all $z \in S$. Choose $j_0 \in J$ with

$g_{j_0}(y) > k(y) - (\varepsilon/2) = \sup\{g_j(y) : j \in J\} - (\varepsilon/2)$, and choose $m_0 \in M$ with $h_{m_0}(y) < k(y) + (\varepsilon/2) = \inf\{h_m(y) : m \in M\} + (\varepsilon/2)$. Such j_0 and m_0 exist by the definition of supremum and infimum. Put

$$S = \left\{ z \in p(X) : |z_{j_0} - y_{j_0}| < \frac{\varepsilon}{2} \text{ and } |z_{m_0} - y_{m_0}| < \frac{\varepsilon}{2} \right\}.$$

Then, for $z \in S$, we have

$$k(z) = \sup\{z_j : j \in J\} \geq z_{j_0} > y_{j_0} - \frac{\varepsilon}{2} > k(y) - \varepsilon$$

and

$$k(z) = \inf\{z_m : m \in M\} \leq z_{m_0} < y_{m_0} + \frac{\varepsilon}{2} < k(y) + \varepsilon,$$

that is,

$$|k(z) - k(y)| < \varepsilon.$$

Therefore, for every $y \in p(X)$ and $\varepsilon > 0$, there exists a neighborhood S of y such that $|k(z) - k(y)| < \varepsilon$ for all $z \in S$, that is, k is continuous on $p(X)$. This solves the first part of the problem.

If condition (1) is replaced with (2), then the statement is false.

Indeed, let $X = \mathbb{N}$ be the set of natural numbers, and let Φ consist of all functions $f : \mathbb{N} \to \mathbb{R}$ such that the sequence $\{f(n)\}_{n=1}^{\infty}$ is convergent. We claim that this is a counterexample.

Suppose that $f_i \in \Phi$ $(i \in I)$, that $h : \mathbb{N} \to \mathbb{R}$ is defined by $h(n)_i = f_i(n)$ $(i \in I)$, and that $k : \overline{h(X)} \to \mathbb{R}$ is continuous. Let b be the element of $\mathbb{R}^I$ such that $b_i = \lim f_i$. Then, by the definition of h, $h(n)_i \to b_i$ for all $i \in I$; therefore, $h(n) \to b$ in the product topology. This implies that b is contained in the closure of $h(X)$ since it is the limit of a sequence in $h(X)$. By continuity of k on $\overline{h(X)}$, we have that $k(h(n)) \to k(\lim h(n)) = k(b)$, that is, the sequence $k(h(n))$ is convergent, and therefore $k \circ h \in \Phi$. This shows that the hypotheses of the problem are satisfied.

Let $J = M = \mathbb{N}$; let $g_j(n) = 1$, for $n < j$ and n even, $g_j(n) = 0$ otherwise; and let $h_m(n) = 0$, for $n < m$ and n odd, $h_m(n) = 1$ otherwise. These functions are indeed in Φ, since $g_j(n) \to 0$ and $h_m(n) \to 1$ for all $j \in J$ and $m \in M$. On the other hand, $\sup\{g_j(n) : j \in J\} = \inf\{h_m(n) : m \in M\}$ equals 1 if n is even and equals 0 if n is odd. This sequence is not convergent; therefore $f = \sup\{g_j(n) : j \in J\} = \inf\{h_m(n) : m \in M\}$ does not belong to Φ. □

Problem T.21. *Characterize the sets $A \subset \mathbb{R}$ for which*

$$A + B = \{a + b : a \in A,\ b \in B\}$$

is nowhere-dense whenever $B \subset \mathbb{R}$ is a nowhere dense set.

Solution. We prove that the sets with this property are the bounded sets with countable closure. More precisely, we show that for a set $A \subset \mathbb{R}$ the following are equivalent:

(i) If $B \subset \mathbb{R}$ is nowhere dense, then so is $A + B$;

(ii) For any sequence a_n $(n = 1, 2, \dots)$ of positive numbers, there exists a finite family of intervals I_j $(j = 1, \dots, k)$ that covers A and such that the length of I_j is $|I_j| = a_j$ $(j = 1, 2, \dots, k)$;

(iii) A is bounded and its closure is countable.

(i) $\Rightarrow$ (ii): Suppose that (ii) does not hold. Then there exists a sequence consisting of positive numbers a_n $(n = 1, 2, \dots)$ such that for any finite family of intervals I_j, $|I_j| = a_j$ $(j = 1, 2, \dots, k)$ we have $A \not\subset \cup_{j=1}^{k} I_j$. Using this, we shall construct a nowhere-dense set B such that $A + B$ contains all rational numbers, which contradicts (i).

Arrange the rational numbers into a sequence r_n $(n = 1, 2, \dots)$. Now we define the numbers x_n and y_n by induction on n. Let $x_1 \in A$ be arbitrary, and $y_1 = r_1 - x_1$. Suppose that $k > 1$ and the numbers $y_1, y_2, \dots, y_{k-1}$ have already been defined. Then, by our assumption, the intervals

$$I_j = \left(r_k - y_j - \frac{a_j}{2} \, , \, r_k - y_j + \frac{a_j}{2} \right), \quad j = 1, 2, \dots, k-1$$

do not cover A. Therefore, we can choose an element $x_k \in A \setminus \cup_{j=1}^{k-1} I_j$.

Finally, put $y_k = r_k - x_k$. Then, with $B = \{y_k : k \in \mathbb{N}\}$, the set $A + B$ contains all rational numbers since $r_k = x_k + y_k \in A + B$. We show that every point of B is isolated. Indeed, for $j < k$, we have $x_j \notin (r_k - y_j - a_j/2, r_k - y_j + a_j/2)$. Therefore, $y_k = r_k - x_k \notin (y_j - a_j/2, y_j + a_j/2)$. This means that the set B contains at most j points in the neighborhood of radius $a_j/2$ of the point y_j, and thus y_j is an isolated point of B. This implies that B is nowhere dense.

(ii) $\Rightarrow$ (iii): If (ii) is true, then A is obviously bounded. Arguing by contradiction, suppose that (iii) does not hold. Then the closure of A is uncountable and, by the Cantor–Bendixson theorem, contains a nonempty, bounded, and perfect set P. Let us call a sequence $(a_1, a_2, \dots)$ of positive numbers insufficient if for any family of closed intervals I_j, $|I_j| = a_j$ $(j = 1, 2, \dots, k)$, we have $P \not\subset \cup_{j=1}^{k} I_j$. In order to arrive at a contradiction, by induction we shall define an infinite sequence of positive numbers $(a_1, a_2, \dots)$ whose all finite initial segments are insufficient. Let a_1 be a positive number that is smaller than the diameter of P; then (a_1) is obviously insufficient. Suppose that $n \geq 1$ and we have already defined an insufficient sequence $(a_1, a_2, \dots, a_n)$. We show that there exists a number a_{n+1} such that $(a_1, a_2, \dots, a_{n+1})$ is insufficient. Suppose this is not true; then for every $k \in \mathbb{N}$ and $a_{n+1} = 1/k$ there exists a sequence of intervals $I_{j,k}$, $|I_{j,k}| = a_j$ $(j = 1, 2, \dots, n+1)$ such that $P \subset \cup_{j=1}^{n+1} I_{j,k}$ holds. We may assume that the intervals $I_{j,k}$ all meet the set P, and thus they are all contained in a bounded set. By passing to a suitable subsequence, we may also assume that for each fixed $j = 1, 2, \dots, n+1$, the sequence of intervals $I_{j,k}$ converges to an interval I_j when $k \to \infty$. Then $|I_j| = a_j$, for $j = 1, 2, \dots, n$, and I_{n+1} consists of a single point. Since $(a_1, a_2, \dots, a_n)$ is insufficient, the set P is not contained in $\cup_{j=1}^{n} I_j$. Moreover, being perfect, P has infinitely many points in the complement of $\cup_{j=1}^{n} I_j$. So we can choose a point

$x \in P \setminus \cup_{j=1}^{n+1} I_j$. But then, for sufficiently large k, $x \notin \cup_{j=1}^{n+1} I_{j,k}$, which contradicts the choice of the intervals $I_{j,k}$. This proves that the sequence $(a_1, a_2, \ldots, 1/k)$ is insufficient, completing the induction.

(iii) $\Rightarrow$ (i): Suppose that, to the contrary of (i), with a nowhere-dense set B, the set $A+B$ is dense in some interval I. Let A_0 denote the closure of A; then $A_0 + B$ is also dense in I. Define the sets A_α for all countable ordinals as follows: If $\beta > 0$ is an ordinal such that A_α has already been defined for all $\alpha < \beta$, then let $A_\beta = \cap_{\alpha<\beta} A_\alpha$ if β is a limit ordinal, and let A_β be the set of accumulation points of A_α if $\beta = \alpha + 1$. Since A_0 is countable, there exists a countable ordinal γ such that $A_\gamma = \emptyset$. Then obviously $A_\gamma + B = \emptyset$ and $A_\gamma + B$ is not dense in I. Let β denote the smallest ordinal for which $A_\beta + B$ is not dense in I; then $\beta > 0$. Let J be a subinterval of I with $(A_\beta + B) \cap J = \emptyset$, and let $0 < \varepsilon < |J|/3$. Let $U(H, \delta)$ denote the neighborhood of radius δ of the set H. It is easy to see that $U(H, \delta) + B \subset U(H + B, \delta)$ holds for every set H and $\delta > 0$. Thus, $U(A_\beta, \varepsilon) + B \subset U(A_\beta + B, \varepsilon)$, and therefore $U(A_\beta, \varepsilon) + B$ does not meet the middle third K of J.

Suppose that β is a limit ordinal. Since $\{A_\alpha : \alpha < \beta\}$ is a nested sequence of compact sets whose intersection is A_β, there exists $\alpha < \beta$ with $A_\alpha \subset U(A_\beta, \varepsilon)$. Then $A_\alpha + B$ does not meet K. But this is impossible, since $\alpha < \beta$ and $A_\alpha + B$ is dense in I.

Finally, suppose that $\beta = \alpha + 1$. Then $V = A_\alpha \setminus U(A_\beta, \varepsilon)$ is a finite set, since all accumulation points of A_α are in A_β. Then the set $V + B$ is nowhere dense. Therefore, from $(A_\alpha + B) \cap K = (V + B) \cap K$, it follows that $(A_\alpha + B) \cap K$ is nowhere dense. But again this contradicts that $\alpha < \beta$ and that $A_\alpha + B$ is dense in I. This completes the proof. □

Problem T.22. *Prove that if all subspaces of a Hausdorff space X are σ-compact, then X is countable.*

Solution. Since X itself is a union of countably many compact sets, it is sufficient to prove the statement for compact spaces X.

Suppose that X has no subset that is dense in itself. Then there is a well-ordering on X in which all initial segments are open. Indeed, suppose that the points $p_\beta \in X$ have already been defined for all $\beta < \alpha$ and that $X_\alpha = \{p_\beta : \beta < \alpha\} \neq X$. Then there exists a point $p_\alpha \in X \setminus X_\alpha$ that is not an accumulation point of $X \setminus X_\alpha$, and so $p_\alpha \in G_\alpha \subseteq X_\alpha \cup \{p_\alpha\}$ with a suitable open neighborhood G_α of p_α. This recursion defines a well-ordering of X, and $\cup_{\beta<\alpha} G_\beta \subseteq X_\alpha = \cup_{\beta<\alpha}\{p_\beta\} \subseteq \cup_{\beta<\alpha} G_\beta$ shows that $X_\alpha = \cup_{\beta<\alpha} G_\beta$ is an open set for each α. An initial segment of order type ω_1 in X cannot be σ-compact since a compact subset in X must have a greatest element. Therefore, X is countable.

Thus, it suffices to prove that X does not contain a subset dense in itself. Arguing by contradiction, assume that there exists such a subset Y. Then $\overline{Y}$ is dense in itself and closed. We define a continuous function $f : \overline{Y} \to [0, 1]$ as follows. Since $\overline{Y}$ is a compact Hausdorff space, for any two distinct points p and q there exist open sets G_p, G_q and closed sets F_p,

F_q such that $p \in G_p \subseteq F_p$, $q \in G_q \subseteq F_q$, and $F_p \cap F_q = \emptyset$. So, for all finite 0–1 sequences $\{\varepsilon_i\}$, we can define closed sets $A_{\varepsilon_1,\dots,\varepsilon_n} \subseteq \overline{Y}$ so that

$$A_{\varepsilon_1,\dots,\varepsilon_n,0} \cap A_{\varepsilon_1,\dots,\varepsilon_n,1} = \emptyset \quad \text{and} \quad A_{\varepsilon_1,\dots,\varepsilon_n,0} \cup A_{\varepsilon_1,\dots,\varepsilon_n,1} \subseteq A_{\varepsilon_1,\dots,\varepsilon_n}$$

hold for all natural numbers n and $\varepsilon_1, \dots, \varepsilon_n \in \{0,1\}$. For any infinite 0–1 sequence $\varepsilon_1, \varepsilon_2, \dots$, the set

$$A(\varepsilon_1, \varepsilon_2, \dots) = \bigcap_{n=1}^{\infty} A_{\varepsilon_1,\varepsilon_2,\dots,\varepsilon_n}$$

is nonempty and closed. If $x \in A(\varepsilon_1, \varepsilon_2, \dots)$, then put

$$f(x) = \sum_{i=1}^{\infty} \varepsilon_i \frac{1}{2^i}.$$

The function f maps $A(\varepsilon_1, \varepsilon_2, \dots)$, which is a closed set since it is an intersection of closed sets, continuously onto the interval $[0,1]$. Then f extends continuously over $\overline{Y}$ with $\operatorname{Im} f = [0,1]$. The interval $[0,1]$ contains $2^{\aleph_0}$ σ-compact sets (that is, F_σ-sets), since the number of closed sets in $[0,1]$ is $2^{\aleph_0}$. Therefore, there exists a set $Z \subset [0,1]$ that is not σ-compact; then its inverse image $f^{-1}(Z) \subset \overline{Y}$ cannot be σ-compact, which contradicts the hypothesis of the problem. This proves that X contains no subset that is dense in itself. $\square$

3.11 SET THEORY

Problem ℵ.1. *Does there exist a function $f(x,y)$ of two real variables that takes natural numbers as its values and for which $f(x,y) = f(y,z)$ implies $x = y = z$?*

Solution 1. First, we show that it is sufficient to construct for each real number c, two sets A_c and B_c of natural numbers that satisfy the following conditions:
(1) $1 \notin A_c$, $1 \notin B_c$ for all c,
(2) $A_c \cap B_c = \emptyset$ for all c,
(3) $A_c \cap B_d \neq \emptyset$ for $c \neq d$.

Indeed, put $f(c,c) = 1$, and for $c \neq d$, let $f(c,d)$ be an arbitrary element of the nonempty set $A_c \cap B_d$. Now, if

$$f(x,y) = f(y,z) = m \neq 1,$$

then

$$m \in A_x \cap B_y \qquad \text{and} \qquad m \in A_y \cap B_z,$$

that is,

$$m \in A_x \cap B_y \cap A_y \cap B_z \subset A_y \cap B_y = \emptyset,$$

which is a contradiction. But $m = 1$ is only possible when $x = y = z$.

In order to construct A_c and B_c, map the set $\mathbb{R}$ of rational numbers bijectively onto the set $N = \{2, 3, \dots\}$; let $n(r)$ denote the image of $r \in \mathbb{R}$. Assign to each real number c a sequence $\{r_k(c)\}_{k=1}^{\infty}$ that converges to c and consists of rational numbers different from c. Let

$$A_c = \{n(r_k(c))\colon k = 1, 2, \dots\} \subset N,$$
$$B_c = N - A_c.$$

It is easy to see that these sets A_c and B_c satisfy conditions (1), (2), and (3). Thus, we have constructed a function with the required property. □

Solution 2. Arrange the set of rational numbers into a sequence r_1, r_2, ... Define the function $f(x,y)$ the following way:

If $x = y$, put $f(x,y) = 1$;
If $x < y$, put $f(x,y) = 2i$, where i is smallest possible with $x < r_i < y$;
If $x > y$, put $f(x,y) = 2i+1$, where i is smallest possible with $x > r_i > y$.

We show that $f(x,y)$ is as required. It is obviously defined on the reals, and its range is the set of natural numbers.

Suppose that

$$f(x,y) = f(y,z) = m.$$

If $m = 1$, then $x = y = z$ is the only possibility. If $m > 1$ and $m = 2i$, then

$$x < r_i < y \qquad \text{and} \qquad y < r_i < z,$$

which is a contradiction. Similarly, if $m > 1$ and $m = 2i + 1$, then

$$x > r_i > y \qquad \text{and} \qquad y > r_i > z,$$

which is again a contradiction. Therefore, $x = y = z$ follows. □

Solution 3. We will decompose the plane into a countable family of pairwise disjoint subsets with the following property: if one of the lines parallel to the coordinate axes through the point (y, y) intersects one of the subsets, then the other does not, except when the subset is the line $x = y$, which will be one of the subsets. It suffices to construct such a partition of the open half-plane below the line $x = y$, because by reflecting in the line $x = y$ and adding the line $x = y$, we obtain the required partition of the whole plane.

Put

$$\begin{aligned}
A_0 &= \{(x,y) : x > 0, \quad y < 0\},\\
A_k^0 &= \{(x,y) : x > k, \quad k-1 < y \le k\},\\
A_{-k}^0 &= \{(x,y) : -k < x \le -k+1, \quad y \le -k\},\\
A_n^k &= \left\{(x,y) : \frac{2k-1}{2^n} < x \le \frac{2k}{2^n}, \quad \frac{2k-2}{2^n} < y \le \frac{2k-1}{2^n}\right\},\\
A_{-k}^n &= \left\{(x,y) : \frac{-2k+1}{2^n} < x \le \frac{-2k+2}{2^n}, \quad \frac{-2k}{2^n} < y \le \frac{-2k+1}{2^n}\right\},
\end{aligned}$$

where $k, n = 1, 2, \ldots$. It is obvious that the union of these sets is the open half-plane below the line $x = y$.

This defines a partition of the whole plane, as we said above. Label the sets of this partition by the natural numbers, and assign to the point (x, y) the label of the set containing (x, y) as the value of the function $f(x, y)$.

If x, y, z are real numbers with $f(x, y) = f(y, z)$, then the points (x, y) and (y, z) are in the same subsets, but then (x, y) (and (y, z)) is on the line through the point (y, y) parallel to the x-axis (y-axis, respectively), and any of the subsets can only meet one of these, except for the subset $\{(x,y) : x = y\}$, but then $x = y = z$. □

Remarks.

1. Obviously, the existence of such a function is a question of cardinality. In general, it makes sense to ask whether there exists a function $f(x, y)$ defined on a set A of cardinality $\mathfrak{m}$ that takes its values in a set B of cardinality $\mathfrak{n}$ (both $\mathfrak{m}$ and $\mathfrak{n}$ are infinite) and for which $f(x, y) = f(y, z)$ implies $x = y = z$. J. Gerlits, L. Lovász, L. Pósa, and M. Simonovits proved that such a function exists if and only if $\mathfrak{m} \le 2^{\mathfrak{n}}$. Here we present the proof by Pósa.

Let $\mathfrak{m} \leq 2^{\mathfrak{n}}$. It suffices to give the construction for a set A of cardinality $2^{\mathfrak{n}}$. Let α be the smallest ordinal of cardinality $\mathfrak{n}$. We may assume that A is the set of sequences of the numbers 0 and 1 of order type α. Define the function $f(x, y)$ as follows:

If $x = y$, put $f(x, y) = 0$;

If $x \neq y$, then let β be the smallest ordinal where the elements in x and y are different, let j be the βth element of x ($j = 0$ or 1), and put $f(x, y) = (\beta, j)$. The range of f is the set of the symbol 0 and the ordered pairs (β, j), where $\beta < \alpha$ and $j = 0$ or 1. Therefore, it has cardinality $\mathfrak{n}$ and can be mapped bijectively onto B.

$f(x, y) = f(y, z) = (\beta, j)$ means that all elements before β agree both in x and y, and in y and z, and the βth element is j in x and y. But then the βth elements of x and y are equal, which is a contradiction. Therefore, $f(x, y) = f(y, z) = 0$, which implies that $x = y = z$. (This part of the statement is also proved by Bollobás and Juhász.)

Suppose that $\mathfrak{m} > 2^{\mathfrak{n}}$, and there exists a function required in the problem. Consider the range of $f(x, y)$ when x is kept fixed. This is a subset of B, and since the cardinality of all possible x's is greater than the cardinality of the set of all subsets of B, for some $x_1 \neq x_2$ the range of $f(x_1, y)$ equals that of $f(x_2, y)$. So, the value $f(x_1, x_2)$ is contained in the range of $f(x_2, y)$, that is, for some y_0, $f(x_1, x_2) = f(x_2, y_0)$, which is a contradiction.

2. Béla Bollobás and Miklós Simonovits notice that there exists a function $F(x_1, \ldots, x_n)$ defined on the real numbers that takes its values in the natural numbers, and for which

$$F(a, x_2, \ldots, x_n) = F(y_1, a, y_3, \ldots, y_n) = \cdots = F(u_1, \ldots, u_{n-1}, a)$$

implies $a = x_i = y_i = \cdots = u_i$ $(i = 1, 2, \ldots, n)$. For example,

$$F(x_1, \ldots, x_n) = \prod_{i,j=1}^{n} p_{ij}^{f(x_i, x_j)}$$

is such a function, where the p_{ij} $(i, j = 1, \ldots, n)$ are different primes, and $f(x, y)$ is the function defined in either solution of the problem.

Problem ℵ.2. *Prove that there exists an ordered set in which every uncountable subset contains an uncountable, well-ordered subset and that cannot be represented as a union of a countable family of well-ordered subsets.*

Solution. Let R be the set of all limit ordinals less than ω_1, and for $\alpha \in R$, let $f_\alpha(n)$ be a monotone increasing sequence (of order type ω) of ordinals converging to α. Order R as follows: for $\alpha, \beta \in R$, $\alpha \prec \beta$ if $f_\alpha(n) < f_\beta(n)$ holds for the first n with $f_\alpha(n) \neq f_\beta(n)$.

Any uncountable subset X of R contains a well-ordered subset of order type ω_1 with respect to the ordering $\prec$. Indeed, let

$$X|n = \{(\alpha_0, \alpha_1, \ldots, \alpha_n) : \text{ there exists } \alpha \in X, \text{ such that } f_\alpha(m) = \alpha(m) \text{ for all } m \le n\}.$$

For n sufficiently large, $X|n$ is uncountable, since for large enough n the ordinal

$$\gamma_n = \sup\{f_\alpha(n) : \alpha \in X\}$$

equals ω_1, since, by $f_\alpha(n) \to \alpha$,

$$\sup\{\gamma_n : n < \omega\} = \sup\{\alpha : \alpha \in X\} = \omega_1.$$

Order $X|n$ lexicographically, that is,

$$(\alpha_0, \ldots, \alpha_n) \prec' (\beta_0, \ldots, \beta_n)$$

if $\alpha_m < \beta_m$ for the smallest m with $\alpha_m \ne \beta_m$. Then $X|n$ is well-ordered and, if n is large enough for $X|n$ to be uncountable, it contains a well-ordered subset of order type ω_1. Let X' be a subset of X such that for any

$$(\alpha_0, \ldots, \alpha_n) \in X|n,$$

there exists a unique $\alpha \in X'$ with

$$f_\alpha(0) = \alpha_0, \ldots, f_\alpha(n) = \alpha_n.$$

Then the ordered set $(X', \prec)$ is isomorphic to $(X|n, \prec')$. Since we have seen that the latter contains a well-ordered subset of order type ω_1, the same is true of the former. This is what we wanted to prove.

Now we show that R is not a union of a countable family of well-ordered subsets. First we make a digression.

A function that maps a set of ordinals into the ordinals is called regressive if $f(\xi) < \xi$ holds for all ordinals $\xi \ne 0$ in the domain of f. If the domain of f is a subset of ω_1 and the set

$$\{\xi : f(\xi) < \mu\}$$

is not cofinal with ω_1 for any $\mu < \omega_1$, then f is called divergent. $X \subseteq \omega_1$ is called thin if there exists a regressive divergent function defined on X; otherwise, it is called stationary.

The set ω_1 is stationary. Indeed, assuming that f is a regressive function on ω_1, the sequence

$$\xi, f(\xi), f(f(\xi)), \ldots ,$$

being a descending sequence of ordinals, can contain only a finite number of elements different from zero. Therefore,

$$\omega_1 = \bigcup_{n<\omega} X_n$$

where
$$X_0 = \{0\},$$
and
$$X_{n+1} = \{\xi : f(\xi) \in X_n\}.$$
Assuming further that f is divergent, induction on n shows that $\sup X_n < \omega_1$ for all n, which obviously is a contradiction.

From this it easily follows that the set R of limit ordinals less than ω_1 is also stationary.

It quickly follows from the definition that the union of a countable family of thin sets is also thin. Further, if X is stationary and f is a regressive function on X, then for some $\mu < \omega_1$ the set
$$X_\mu = \{\xi : f(\xi) = \mu\}$$
is stationary (this is a special case of Fodor's theorem). Indeed, assume that all X_μ are thin, and let f_μ be a regressive divergent function defined on X_μ. Then the function g defined on X by
$$g(\xi) = \max(\mu, f_\mu(\xi)) \qquad (\xi \in X_\mu,\ \mu < \omega_1)$$
is regressive. It is divergent, too, because for all $\nu < \omega_1$ the set
$$\{\xi : g(\xi) < \nu\} \subseteq \bigcup_{\mu<\nu} \{\xi : f_\mu(\xi) < \nu\}$$
is not cofinal with ω_1, since neither one of the summands in the right-hand side is, because the functions f_μ are divergent. This contradicts the assumption that X is stationary, which proves our statement.

Let $X \subseteq R$ be stationary. We show that X is not well ordered with respect to the ordering $\prec$. Put $Y_{-1} = X$ and
$$Y_n = \{\alpha \in Y_{n-1} : f_\alpha(n) = \delta_n\},$$
where δ_n is the smallest ordinal for which Y_n is stationary (such δ_n exists by the Fodor theorem proved above).

Put
$$X' = X - \bigcup_{n<\omega} Y'_n,$$
where
$$Y'_n = \{\alpha \in Y_{n-1} : f_\alpha(n) < \delta_n\}.$$
By the choice of δ_n, Y'_n is not stationary, and neither is $X - X'$.

Put
$$\delta = \sup\{\delta_n : n < \omega\}.$$
It is obvious that $\delta < \omega_1$. Let $\alpha \in X'$ be greater than δ. (X' is stationary and so it is cofinal with ω_1.) Since $f_\alpha(n) \to \alpha$, for some $n < \omega$ we have
$$f_\alpha(n) > \delta \geq \delta_n,$$

so $f_\alpha(m) = \delta_m$ cannot hold for all m. It follows from $\alpha \in X'$ that $f_\alpha(m) > \delta_m$ for the smallest m with $f_\alpha(m) \neq \delta_m$. Thus, for this m, α is greater than all elements of Y_m with respect to the ordering $\prec$.

Put $X_0 = X$, $\alpha_0 = \alpha$, and $X_1 = Y_m$. Since Y_m is stationary, we can repeat the above argument with X_1 instead of X_0. We obtain α_1 and X_2, then α_2 and X_3, and so on. Since $\alpha_0 \succ \alpha_1 \succ \alpha_2 \succ \dots$, X is not well-ordered with respect to the ordering $\prec$, and this is what we wanted to prove.

Now, if $R = \bigcup_{n<\omega} R_n$, then some R_n is stationary, and this R_n is not well-ordered with respect to $\prec$. This proves the second statement of the problem. □

Remark. László Babai showed that for each cardinal number $\kappa > \omega$, there exists an ordered set B of cardinality κ such that all subsets of cardinality κ contain a well-ordered subset of cardinality κ and B is not a union of a family of cardinality less than κ of well-ordered subsets.

Problem ℵ.3. *Let $\leq$ be a reflexive, antisymmetric relation on a finite set A. Show that this relation can be extended to an appropriate finite superset B of A such that $\leq$ on B remains reflexive, antisymmetric, and any two elements of B have a least upper bound as well as a greatest lower bound. (The relation $\leq$ is extended to B if for $x, y \in A$, $x \leq y$ holds in A if and only if it holds in B.)*

Solution. B will consist of all the elements and subsets of A without identifying the elements with the corresponding singletons. In what follows, lowercase letters denote elements, while capitals mean subsets.

Define $\leq$ as follows:

$$\begin{aligned} a &\leq b, && \text{as given}\,,\\ a &\leq P, && \text{if } a \in P\,,\\ P &\leq a, && \text{if } a \notin P\,,\\ P &\leq Q, && \text{if } P \subseteq Q\,. \end{aligned}$$

This is obviously a reflexive and antisymmetric relation. To show the other property, it suffices to treat incomparable elements, that is, two incomparable elements in A, or two subsets of A, none containing the other, as a subset.

If $a, b \in A$ are incomparable, $\{a, b\}$, $A \setminus \{a, b\}$ are the least upper (resp. greatest lower) bounds. If $P, Q \subseteq A$, then $P \cup Q$ and $P \cap Q$ will serve as the least upper and greatest lower bounds. □

Problem ℵ.4. *Let $\mathcal{F}$ be a family of subsets of a ground set X such that $\cup_{F\in\mathcal{F}} F = X$, and*

(a) if $A, B \in \mathcal{F}$, then $A \cup B \subseteq C$ for some $C \in \mathcal{F}$;

(b) if $A_n \in \mathcal{F}$ $(n = 0, 1, \dots)$, $B \in \mathcal{F}$, and $A_0 \subset A_1 \subset \dots$, then, for some $k \geq 0$, $A_n \cap B = A_k \cap B$ for all $n \geq k$.

Show that there exist pairwise disjoint sets X_γ ($\gamma \in \Gamma$), with $X = \bigcup\{X_\gamma : \gamma \in \Gamma\}$, such that every X_γ is contained in some member of $\mathcal{F}$, and every element of $\mathcal{F}$ is contained in the union of finitely many X_γ's.

Solution. Let $<$ be a well-ordering of F. Let Γ be the set of all finite subsets of F. Well order Γ as follows:

$$\gamma_1, \gamma_2 \in \Gamma,\ \gamma_1 \neq \gamma_2,\ \gamma_1 = \{A_1, \dots, A_n\},\ \gamma_2 = \{B_1, \dots, B_m\},$$
$$A_n < \dots < A_1,\ B_m < \dots B_1,\ m \leq n\,.$$

Put $\gamma_1 > \gamma_2$ if $A_i = B_i$ for $1 \leq i \leq m$. If, on the other hand, $i \leq m$ is the least number with $A_i \neq B_i$ then put $\gamma_1 < \gamma_2$ or $\gamma_1 > \gamma_2$ if $A_i < B_i$ or $A_i > B_i$. A straightforward checking shows that this well orders Γ. We notice that $\gamma_1 \subseteq \gamma_2$ implies $\gamma_1 < \gamma_2$.

For $\gamma \in \Gamma$, we define $Y_\gamma \in F$ as follows. If $\gamma = \{A\}$, put $Y_\gamma = A$. Assume that Y_δ is defined for $\delta < \gamma$. Then let Y_γ be a member of F, covering every Y_β ($\beta < \gamma$). This is possible by (a).

Now put $X_\gamma = Y_\gamma \setminus \bigcup_{\delta<\gamma} Y_\delta$. The only nontrivial thing to prove is that every $A \in F$ is covered by finitely many X_β. We prove by transfinite induction on γ that Y_γ is covered by finitely many X_δ's. If this first fails for γ, there exists $\beta_1 < \beta_2 < \dots < \gamma$ such that $Y_\gamma \cap X_{\beta_n} \neq \emptyset$. Let $Y_\gamma = \{A_1, \dots, A_k\}$, $A_1 > \dots > A_k$. For every n there is a $1 \leq j \leq k$ such that

$$A_1, \dots, A_{j-1} \in \beta_n, \quad A_j \notin \beta_n\,.$$

We can assume, by shrinking, that j is the same for every n. Put $\eta_n = \max(\beta_1, \dots, \beta_n) < \gamma$. As γ is minimal, Y_{β_n} meets finitely many X_β. As $Y_{\eta_1} \subset Y_{\eta_2} \subset \dots$, by (b) there is an m such that $Y_{\eta_m} \cap Y_\gamma = Y_{\eta_{m+1}} \cap Y_\gamma = \dots$, so $Y_{\eta_m} \supseteq Y_{\eta_{m+i}} \cap Y_\gamma \supseteq X_{\eta_{m+i}} \cap Y_\gamma \neq \emptyset$ $(i = 0, 1, \dots)$ that is, Y_{η_m} meets $X_{\eta_m}, X_{\eta_{m+1}}, \dots$, a contradiction. □

Problem ℵ.5. *Show that there exists a tournament $(T, \to)$ of cardinality $\aleph_1$ containing no transitive subtournament of size $\aleph_1$. (A structure $(T, \to)$ is a tournament if $\to$ is a binary, irreflexive, asymmetric, and trichotomic relation. The tournament $(T, \to)$ is transitive if $\to$ is transitive, that is, if it orders T.)*

Solution. We use the existence of a Specker type, that is, an ordered set $(A, <)$ of size $\aleph_1$, not containing subsets similar to ω_1, ω_1^*, any uncountable subset of the reals. See (*P. Erdős, A. Hajnal, A. Máté, R. Rado, Combinatorial set theory: Partition Relations for Cardinals, Akadémiai Kiadó, Budapest, 1984, p. 326.*) Enumerate A as $\{a(\alpha) \colon \alpha < \omega_1\}$, and let $\{x(\alpha) \colon \alpha < \omega_1\}$ be different real numbers in $[0, 1]$. If $\alpha < \beta$, put $\alpha \leftarrow \beta$ iff either $x(\alpha) < x(\beta)$ and $a(\alpha) < a(\beta)$ or $x(\alpha) > x(\beta)$ and $a(\alpha) > a(\beta)$.

Let $B \subseteq A$ be uncountable. Let $X = \{\alpha < \omega_1 : a(\alpha) \in B\}$.

Claim 1. There is an $\alpha \in X$ such that

$$\{\beta \in X : a(\alpha) < a(\beta) \quad \text{and} \quad x(\alpha) < x(\beta)\}$$

is uncountable.

Proof. For $a \in B$, let $f(a)$ be the least $t \in [0,1]$ such that $x(\beta) < t$ for all but countably many $\beta \in X$, $a(\beta) > a$. Since f is a non-increasing real-valued function on $(A, <)$, it can only have countably many different values; otherwise, there would be an uncountable subset of A, similar to a set of reals. Hence f is constant on an uncountable $B_0 \subseteq B$. Define X_0 analogously to X. We can choose $\alpha_0 \in X_0$ such that $\{\beta \in X_0 : x(\beta) > x(\alpha_0)\}$ is uncountable; otherwise $(A, <)$ would contain a subset of type ω_1^*. Now we can choose $\alpha \in X_0$ such that $a(\alpha) > a(\alpha_0)$ and $x(\alpha) < x(\alpha_0)$. Since $g(a(\alpha)) = g(a(\alpha_0)) > x(\alpha)$, there are uncountably many $\beta \in X$ such that $a(\beta) > a(\alpha)$ and $x(\beta) > x(\alpha)$.

Claim 2. There are uncountable $X_0, X_1 \subseteq X$ such that if $\alpha \in X_0$, $\beta \in X_1$, then $a(\alpha) < a(\beta)$ and $x(\alpha) < x(\beta)$.

Proof. Let U be the set of those $\alpha \in X$ such that

$$\{\beta \in X : a(\alpha) < a(\beta) \quad \text{and} \quad x(\alpha) < x(\beta)\}$$

is countable, and let L be the set of those $\alpha \in X$ such that

$$\{\beta \in X : a(\beta) < a(\alpha) \quad \text{and} \quad x(\beta) < x(\alpha)\}$$

is countable. If U or L is uncountable, we get a contradiction by Claim 1, so we can select $\alpha \in X \setminus U \setminus L$, and put

$$X_0 = \{\beta \in X : a(\beta) < a(\alpha) \quad \text{and} \quad x(\beta) < x(\alpha)\},$$
$$X_1 = \{\beta \in X : a(\alpha) < a(\beta) \quad \text{and} \quad x(\alpha) < x(\beta)\}.$$

By an application of Claim 2 to X_0 and the Specker type $(A, >)$, one can find $Y_0, Y_1 \subseteq X_0$ such that $\alpha \in Y_0$, $\beta \in Y_1$ imply $a(\alpha) < a(\beta)$, and $x(\alpha) > x(\beta)$. Now select $\alpha \in Y_0$, $\beta \in X_1$, $\gamma \in Y_1$ with $\alpha < \beta < \gamma$. Then $\alpha \leftarrow \beta \leftarrow \gamma$ and $\gamma \leftarrow \alpha$, so our tournament is not transitive on B. □

Remark. The result in the problem was first proved by R. Laver.

Problem ℵ.6. *Assume that R, a recursive, binary relation on $\mathbb{N}$ (the set of natural numbers), orders $\mathbb{N}$ into type ω. Show that if $f(n)$ is the nth element of this order, then f is not necessarily recursive.*

Solution. We use the following well-known facts. There is a set $A \subseteq \mathbb{N}$ that is recursively enumerable, that is, the range of a recursive function, but is not recursive, namely, its characteristic function is not recursive. For every infinite r.e. set A there is a recursive function enumerating A's elements in a one-to-one manner. An infinite set possesses an increasing such enumeration if and only if it is recursive.

Take a (necessarily infinite) r.e. but not recursive set A, and a function g enumerating A without repetition. Put aRb iff $g(a) < g(b)$. Clearly, R is recursive and orders $\mathbb{N}$ into type ω. If $f(n)$=the nth element by this order, and f is recursive, then the recursive $h(n) = g(f(n))$ would enumerate A's elements in increasing order, which is impossible. □

Problem ℵ.7. *Let $\mathcal{H}$ be the class of all graphs with at most $2^{\aleph_0}$ vertices not containing a complete subgraph of size $\aleph_1$. Show that there is no graph $H \in \mathcal{H}$ such that every graph in $\mathcal{H}$ is a subgraph of H.*

Solution. We have to show that if (V, G), a graph with $|V| = 2^{\aleph_0}$, does not contain a complete graph on $\aleph_1$ vertices, then there is a graph (W, H) with these properties such that (W, H) cannot be embedded into (V, G).

Let W be the set of those functions injecting a countable ordinal (possibly $O \neq \emptyset$) into a complete subgraph of G. To define H, join two such functions if one extends the other. Clearly,

$$|W| = \sum_{\alpha<\omega_1} |\alpha \to V| \leq \aleph_1 \cdot \left(2^{\aleph_0}\right)^{\aleph_0} = 2^{\aleph_0} .$$

First, we show that (W, H) does not contain a complete $\aleph_1$-gon. Assume that $\{f_\alpha : \alpha < \omega_1\}$ are pairwise joined, and that the ordinals $\mathrm{Dom}\,(f_\alpha)$ are in increasing order. By the construction of H, for $\beta < \alpha$, f_α extends f_β, so the union of the f_α's gives a function $f : \omega_1 \to V$ onto a complete $\aleph_1$-gon in (V, G), a contradiction.

Next we prove that (W, H) may not be embedded into (V, G). Assume that $h : W \to V$ is such an embedding. By transfinite recursion on $\alpha < \omega_1$, we define $f_\alpha : \alpha \to V$, $f_\alpha \in W$, and $x_\alpha = h(f_\alpha) \in V$ in such a way that f_α extends f_β for $\beta < \alpha$. Put $f_0 = \emptyset$, $f_0 \in W$. If f_β $(\beta < \alpha)$ are defined, then

$$\{x_\beta : \beta < \alpha\} = h\{f_\beta : \beta < \alpha\}$$

is a complete subgraph, so we can define $f_\alpha(\beta) = x_\beta$. This clearly satisfies the requirements. But then $\{x_\alpha : \alpha < \omega_1\}$ is a complete subgraph in G, a contradiction. □

Remark. This is an unpublished result of R. Laver.

Problem ℵ.8. *For which cardinalities κ do antimetric spaces of cardinality κ exist?*
(X, ϱ) is called an antimetric space if X is a nonempty set, $\varrho : X^2 \to [0, \infty)$ is a symmetric map, $\varrho(x, y) = 0$ holds iff $x = y$, and for any three-element subset $\{a_1, a_2, a_3\}$ of X

$$\varrho(a_{1f}, a_{2f}) + \varrho(a_{2f}, a_{3f}) < \varrho(a_{1f}, a_{3f})$$

holds for some permutation f of $\{1, 2, 3\}$.

Solution. For $0 < \kappa \le 2^{\aleph_0}$. First, if $X \subseteq \mathbb{R}$ is nonempty, then (X, ϱ) is antimetric, where $\varrho(x, y) = (x - y)^2$ and clearly $|X|$ can take any value in the given interval.

For the other direction, let (X, ϱ) be antimetric and $|X| > 2^{\aleph_0}$. Color the complete graph on X by the integers as follows. Let $\{x, y\}$ get color k iff $2^k \le \varrho(x, y) < 2^{k+1}$. By the Erdős–Rado theorem (see *P. Erdős, A. Hajnal, A. Máté, R. Rado, Combinatorial Set Theory: Partition Relations for Cardinals, Akadémiai Kiadó, Budapest, 1984, p. 98*), there exist x, y, z such that $\{x, y\}$, $\{x, z\}$, $\{y, z\}$ get the same color. But then, x, y, z do not meet the requirement on antimetricity. □

Problem ℵ.9. *If $(A, <)$ is a partially ordered set, its dimension, $\dim(A, <)$, is the least cardinal κ such that there exist κ total orderings $\{<_\alpha : \alpha < \kappa\}$ on A with $< = \bigcap_{\alpha<\kappa} <_\alpha$. Show that if $\dim(A, <) > \aleph_0$, then there exist disjoint $A_0, A_1 \subseteq A$ with $\dim(A_0, <)$, $\dim(A_1, <) > \aleph_0$.*

Solution. Assume indirectly that $(A, <)$ is a counterexample of minimal cardinality λ. Without loss of generality, $A = \lambda$. Put

$$I = \{B \subseteq \lambda : \dim(B, <) \le \aleph_0\}.$$

Clearly, if B has $|B| < \lambda$, then $B \in I$.

Claim 1. If $B_n \in I$ $(n = 0, 1, \dots)$ and B is such that every pair of B is covered by some B_n, then $B \in I$.

Proof. Straightforward from the definition.

Claim 2. There are $B_0, B_1 \in I$ such that $B_0 \cap B_1 = \emptyset$ and $B_0 \cup B_1 \notin I$.

Proof. If not, then I is a σ-complete primideal containing every subset of cardinality $< \lambda$. For $\alpha < \lambda$, let $<_{\alpha,n}$ be a total order on α establishing $\dim(\alpha, <) \le \aleph_0$. If, for $\beta < \gamma < \lambda$ we put $\beta <_n \gamma$ if $\{\alpha < \lambda : \beta <_{\alpha,n} \gamma\} \notin I$, then $<_n$ establish that $\dim(\lambda, <) \le \aleph_0$, a contradiction.

We can assume that B_0, B_1 as above are selected with $\kappa = |B_1|$ minimal. Put

$$I_1 = \{C \subseteq B_1 : B_0 \cup C \in I\}.$$

Then $B_1 \notin I_1$ and every subset of B_1 of size $< \kappa$ is in I_1. Repeating the argument of Claim 2, we get that there exist $C_0, C_1 \in I_1$ such that $C_0 \cap C_1 = \emptyset$ and $B_0 \cup C_0$, $B_0 \cup C_1 \notin I$. Repeating the above argument for C_0, B_0 this time, one gets $D_0, D_1 \subseteq B_0$, $D_0 \cap D_1 = \emptyset$ such that $D_0 \cup C_0$, $D_1 \cup C_0 \in I$, $D_0 \cup D_1 \cup C_0 \notin I$. As every pair of $D_0 \cup D_1 \cup C_1$ is contained either in B_0 or in $D_0 \cup C_1$ or in $D_1 \cup C_1$, by Claim 1, either $D_0 \cup C_1$ or $D_1 \cup C_1$ is not in I. In the former case, $D_0 \cup C_1$ and $D_1 \cup C_0$ are two disjoint sets not in I, and we conclude similarly in the other case, too. □

Problem ℵ.10. *A binary relation $\prec$ is called a quasi-order if it is reflexive and transitive. The infimum of the quasi-order $(Q, \prec)$ is the greatest subset $J \subseteq Q$ such that*
(i) for every $B \in Q$ there is an $A \in J$ with $A \prec B$, and
(ii) $A \prec B$, $A, B \in J$ imply $B \prec A$.
Let X be a finite, nonempty alphabet, let X^* be the set of all finite words from X, and let $\mathcal{P}$ be the set of infinite subsets of X^*. For $A, B \in \mathcal{P}$, let $A \prec B$ if every element of A is a (connected) subword of some element of B. Show that $(\mathcal{P}, \prec)$ has an infimum, and characterize its elements.

Solution. As follows, subword will always mean connected subword. Let $a \le b$ denote that a is a subword of b. Call $A \in \mathcal{P}$ *minimal*, if for all $w \in A$, all but finitely many subwords of the elements of A contain w, as subword.

Claim 1. If A is minimal, $B \prec A$, then $A \prec B$.

Proof. Assume that $w \in A$ and k is so large that every subword in A longer than k, as $B \prec A$, is a subword, so $w \le b$.

Claim 2. If $A \in \mathcal{P}$ is not minimal, then there exists a $B \in \mathcal{P}$ such that $B \prec A$ but $A \not\prec B$.

Proof. Let w be a subword of A such that

$$H = \{h\colon \text{ there is an } a \in A,\ h \le a,\ w \not\le h\}$$

is infinite. Then $H \prec A$, but $A \not\prec H$ is established by w.

Claim 3. For every $A \in \mathcal{P}$ there is a minimal $B \prec A$.

Proof. Such a B is given by the following algorithm. Step 1: Let $X = \{a_1, \dots, a_n\}$. If, among the subwords of A, infinitely many do not contain a_1, let H_1 be their set. $H_1 \in \mathcal{P}$, $H_1 \prec A$, and it does not contain a_1. Then get successively $H_2, H_3, \dots$ using $a_2, a_3, \dots$. If H_{n-1} is defined, it may only contain a_n, so it is minimal. We can therefore assume the existence of H_{i-1} such that all but finitely many subwords of H_{i-1} contain a_i.

Put $b_1 = a_i$, $B_1 = H_{i-1} \prec A$, and there follows Step 2. Here Step k: If b_{k-1}, B_{k-1} are given, and $b_1 < \cdots < b_{k-1}$, $B_{k-1} \prec \cdots \prec B_1 \prec A$ are given, and all but finitely many subwords of B_t contain b_t, consider the word $b_{k-1}a_1$. If infinitely many subwords of B_{k-1} omit it, let H_1^{k-1} be this set. $b_{k-1}a_1$ therefore is not a subword of H_1^{k-1}. Repeating this to $b_{k-1}a_2$, we get H_2^{k-1}, etc. If we can reach H_{n-1}^{k-1}, then there exists a natural number m, such that in every subword of length $2m$ a b_{k-1} can be found in the first m positions, but it must be followed by a_n. So we can select H_i^{k-1} as B_k and $b_{k-1}a_i$ as b_k. By this process, we get $b_1 < b_2 < \dots$, $B_1 \succ B_2 \succ \dots$. Put $L = \{b_1, b_2, \dots\}$. Then $L \prec A$ and L is minimal. In fact, for $b_i \in L$, $b_{i+1}, b_{i+2}, \dots \in B_i$, so only finitely many subwords of them omit b_i.

We get that the minimal members constitute the infimum. Claim 3 gives (i), Claim 1 (ii), and the maximality holds as if A is nonminimal, $B \prec A$

is such that $A \not\prec B$, C is minimal, $C \prec B$, then $C \prec A$, $A \not\prec C$, so the addition of C would violate (ii). $\square$

Problem $\aleph$.11. *Define a partial order on all functions $f : \mathbb{R} \to \mathbb{R}$ by the relation $f \prec g$ if $f(x) \leq g(x)$ for all $x \in \mathbb{R}$. Show that this partially ordered set contains a totally ordered subset of size greater than $2^{\aleph_0}$ but that the latter subset cannot be well-ordered.*

Solution. For the first part of the problem, it suffices to find a family of more than $2^{\aleph_0}$ subsets of $\mathbb{R}$ totally ordered by $\subset$, as then the characteristic function will do the job. In order to show that if κ is an infinite cardinal, then there are κ^+ subsets of a set of size κ, ordered by $\subset$, let λ be the least cardinal such that $2^\lambda > \kappa$. By Cantor's theorem, λ exists and is $\leq \kappa$. If we find a totally ordered set of size κ^+, with a dense subset of cardinality $\leq \kappa$, we are done by the Dedekind cuts. For such a set, take the $\lambda \to \{0,1\}$ functions, ordered lexicographically. A dense subset is formed by those functions that take 0 from a place onward. Their number is

$$\sum_{\alpha<\lambda} 2^{|\alpha|} \leq \lambda \cdot \kappa = \kappa\,.$$

For the second part, assume that $f_\alpha < f_\beta$ for $\alpha < \beta < \left(2^{\aleph_0}\right)^+$. Let $x_\alpha \in \mathbb{R}$ satisfy $f_\alpha(x_\alpha) < f_{\alpha+1}(x_\alpha)$, for $\left(2^{\aleph_0}\right)^+$ many α, $x_\alpha = x$, and so the $f_\alpha(x)$ are $\left(2^{\aleph_0}\right)^+$ different elements of $\mathbb{R}$, which is impossible. $\square$

Index of Names

Z

Problem Books in Mathematics *(continued)*

Demography Through Problems
by *Nathan Keyfitz and John A. Beekman*

Theorems and Problems in Functional Analysis
by *A.A. Kirillov and A.D. Gvishiani*

Exercises in Classical Ring Theory
by *T.Y. Lam*

Problem-Solving Through Problems
by *Loren C. Larson*

A Problem Seminar
by *Donald J. Newman*

Exercises in Number Theory
by *D.P. Parent*

Contests in Higher Mathematics: Miklós Schweitzer Competitions 1962-1991
by *Gábor J. Székely (editor)*